Student Solutions Manual
for
Single Variable Calculus
by Laura Taalman and Peter Kohn

Roger Lipsett

W. H. Freeman and Company
New York

©2014 by W. H. Freeman and Company

ISBN-13: 978-1-4641-2538-6
ISBN-10: 1-4641-2538-4

Printed in the United States of America

First printing

W. H. Freeman and Company
41 Madison Avenue
New York, NY 10010
Houndmills, Basingstoke RG21 6XS, England

www.whfreeman.com

Contents

Part I

Differential Calculus

Chapter 0

Functions and Precalculus

0.1 Functions and Graphs

Thinking Back

Interval Notation

- ▷ $(-\infty, -3) \cup (-3, 3) \cup (3, \infty)$

- ▷ $(-\infty, 0) \cup [10, \infty)$

- ▷ $(-2, 5) \cup (5, \infty)$

- ▷ $[3, 4) \cup (4, \infty)$

Solving Equations and Inequalities

- ▷ $\dfrac{x}{x-2} = 0$ if and only if $x = 0$ but $x - 2 \neq 0$, so the solution set is $\{0\}$.

- ▷ The only x-values at which $\dfrac{x}{x-2}$ can change sign are at $x = 0$ and $x = 2$, since those are the values that make either the numerator or the denominator equal to zero. To determine the intervals on which $\dfrac{x}{x-2}$ is positive, therefore, it suffices to check its value at one point in each of the three resulting intervals:

interval	$(-\infty, 0)$	$(0, 2)$	$(2, \infty)$
x	-1	1	3
$\frac{x}{x-2}$	1	-1	3
sign	$+$	$-$	$+$

  Thus $\dfrac{x}{x-2} > 0$ for x in the range $(-\infty, 0) \cup (2, \infty)$.

- ▷ $x^3 - 5x^2 + 6x = x(x^2 - 5x + 6) = x(x - 3)(x - 2)$. This formula can change signs only at the points where it is zero, i.e., at $x = 0$, 2, or 3. To determine the intervals on which it is negative, therefore, it suffices to check its value at one point in each of the four resulting intervals:

interval	$(-\infty, 0)$	$(0, 2)$	$(2, 3)$	$(3, \infty)$
x	-1	1	$\frac{5}{2}$	4
$x^3 - 5x^2 + 6x$	-12	2	$-\frac{5}{8}$	8
sign	$-$	$+$	$-$	$+$

Thus $x^3 - 5x^2 + 6x < 0$ when x is in the range $(-\infty, 0) \cup (2, 3)$.

▷ $x^3 - 5x^2 + 6x = x(x^2 - 5x + 6) = x(x - 3)(x - 2)$, so it is zero when x takes on one of these three values, so when x is in the range $\{0\} \cup \{2\} \cup \{3\}$.

Concepts

1. (a) False. A function is a rule that transforms values into other values, while an equation expresses equality between two objects.

    (b) False. For example, a function could assign to each word in English the corresponding word in Spanish.

    (c) False. Since, for example, $f(2) = f(3) = 1$, f is definitely not one-to-one.

    (d) True. A local maximum occurs when $f(c) \geq f(x)$ for all x near $x = c$. If c is a global maximum, then $f(c) \geq f(x)$ for all x in the domain of f and thus certainly $f(c) \geq f(x)$ for x near $x = c$. Thus c is also a local maximum.

    (e) False. Consider the function $f(x) = x^3 - 3x^2$. This has a local minimum at $x = 2$, where $f(x) = -4$. But for large negative values of x, $f(x)$ gets arbitrarily large negatively, so that $x = 2$ is not a global minimum.

    (f) False. The graph at the bottom of page 8 of a function with three asymptotes is an example; its graph crosses the asymptote $y = 1$.

    (g) True. The average rate of change over an interval is the slope of the line connecting the graph of the function at the two endpoints of the interval.

    (h) True. For example, $f(x) = x^2$ has average rate of change $\dfrac{f(1) - f(-1)}{1 - (-1)} = 0$ over $[-1, 1]$ but average rate of change $\dfrac{f(2) - f(1)}{2 - 1} = 3$ over $[1, 2]$.

3. The definition of a function is in Definition 0.1. A function is a rule that associates exactly one member of a set (the range) with each member of another set (the domain). For example, associating to each person his or her eye color is a function, since every person has a single eye color. However, associating to a color the set of people with that eye color is not a function, since many people have eyes colored, say, brown.

5. The domain of a function is defined in Definition 0.2. For $f(x) = \sqrt{x}$, we have $\text{Domain}(f) = \{x \in \mathbb{R} \mid x \geq 0\}$.

7. (a) Since $\sqrt{3 + 1} = \sqrt{4} = 2$, $f(3) = 2$ so that $(3, 2)$ lies on the graph of f.

    (b) Since $\sqrt{1 + 1} = \sqrt{2} \neq 1$, $f(1) \neq 1$ so that $(1, 1)$ does not lie on the graph of f.

    (c) Since $-5 + 1$ is negative, -5 is not in the domain of f. Thus $(-5, 2)$ cannot lie on the graph of f.

9. (a) 5 is in the range of f since $f(2) = 2^2 + 1 = 5$.

    (b) 0 is not in the range of f, since if $f(x) = 0$, then $x^2 + 1 = 0$ so that $x^2 = -1$, which is impossible for $x \in \mathbb{R}$.

    (c) If $a < 1$, then $x^2 + 1 = a$ means that $x^2 = a - 1 < 0$, which is impossible. Thus the range of f must be contained in $[1, \infty)$. On the other hand, if $a \geq 1$, then $x^2 = a - 1 \geq 0$, so that $f(\sqrt{a - 1}) = a$ and then a is in the range of f. Thus the range of f is $[1, \infty)$.

11. Many answers are possible. For example, the function could consist of the pairs $(x, f(x))$ given by $\{(2,1),\ (4,2),\ (6,3),\ (8,4),\ (10,4)\}$, which in tabular form is

x	2	4	6	8	10
$f(x)$	1	2	3	4	4

and as a diagram is

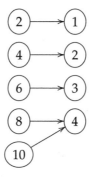

13. Yes, it could. Since there are two points on the graph with $y = 1$, it fails the horizontal line test, so is not one-to-one, but such a graph need not fail the vertical line test. For example, the function with domain $\mathbb{R}$ and rule $f(x) = 1$ is such a function.

15. The identity functions in parts (a) and (c) are well-defined since they conform to the definition. The function in part (b) is not — the domain and range are not the same in part (b).

17. If a graph fails the Vertical Line Test, then there is some number x such that there are two points on the graph above x. Thus x is not mapped to a well-defined, unique y-value, so that the graph does not define a function. Conversely, if a graph passes the Vertical Line Test, then any number x can lie under at most one point of the graph. The y-coordinate of this point is the value of the function at x. If x lies under no points of the graph, then x is not in the domain of the function.

19. For any $x \neq 0$, we have $f(-x) = (-x)^2 + 1 = x^2 + 1 = f(x)$; since $x \neq 0$, we have two distinct values of x on which f takes the same value. Thus $f(x)$ is not one-to-one.

21. (a) f is negative on I if $f(x) < 0$ for all $x \in I$; that is, if the graph of f is below the x-axis for $x \in I$.

  (b) f is decreasing on I if $f(b) < f(a)$ for all $b > a \in I$; that is, if the graph of f moves down as we look left to right across I.

23. They are the same. The average rate of change of f on $[a, b]$ is given by $\dfrac{f(b) - f(a)}{b - a}$; the numerator is the "rise", while the denominator is the "run", so that this is also the slope of the line from $(a, f(a))$ to $(b, f(b))$.

25. For the local maximum at $x = -1$, $f(x) = 19$, δ can be arbitrarily large, since $x = -1$ is in fact a global maximum. For the local maximum at $(2, -8)$, $\delta \approx 1.5$ works, since $f(2 - 1.5) = f(0.5)$, which appears to be less than $f(2)$. For the local minimum at $(1, -13)$, $\delta \approx 1.25$ works, since $f(2.25)$ appears to be greater than $f(1)$.

Skills

27. The domain of $f(x)$ is $\{x \mid x - 1 \geq 0\} = [1, \infty)$, and the range is $[0, \infty)$. The graph is (with the domain and range indicated with heavy lines on the axes)

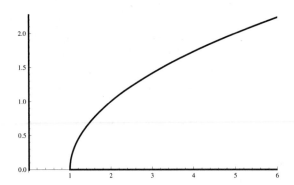

29. The domain of $f(x)$ is $\{x \mid x \neq -2\} = (-\infty, 2) \cup (2, \infty)$, and the range is $(-\infty, 0) \cup (0, \infty)$. The graph is (with the domain and range indicated with heavy lines on the axes)

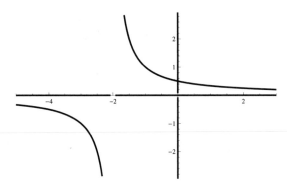

31. The domain of $f(x)$ is $\mathbb{R}$, and the range is $(0, 1]$. The graph is (with the domain and range indicated with heavy lines on the axes)

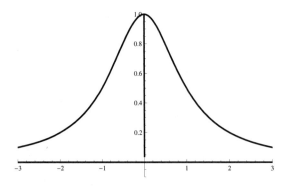

33. $f(x)$ is defined when $x(x-2) \geq 0$. Now, $x(x-2) = 0$ for $x = 0$ or $x = 2$, so $x(x-2)$ can change sign only at those points. Thus to determine the intervals on which it is positive, it suffices to consider one point in each of the resulting three intervals:

interval	$(-\infty, 0)$	$(0, 2)$	$(2, \infty)$
x	-1	1	3
$x(x-2)$	3	-1	3
sign	$+$	$-$	$+$

Thus $x(x-2) \geq 0$, so that $f(x)$ is defined, for $x \in (-\infty, 0] \cup [2, \infty)$.

35. $f(x)$ is defined when $(x-1)(x+3) \geq 0$. Now, $(x-1)(x+3) = 0$ for $x = -3$ or $x = 1$, so it can change sign only at those points. Thus to determine the intervals on which it is positive, it suffices to consider one point in each of the resulting three intervals:

interval	$(-\infty, -3)$	$(-3, 1)$	$(1, \infty)$
x	-4	0	2
$(x-1)(x+3)$	5	-3	5
sign	$+$	$-$	$+$

Thus $(x-1)(x+3) \geq 0$, so that $f(x)$ is defined, for $x \in (-\infty, -3] \cup [1, \infty)$.

37. $f(x)$ is defined when the denominator is defined and nonzero, i.e., when $(x-1)(x+3) > 0$. From Exercise 35, the domain is thus $(-\infty, -3) \cup (1, \infty)$. (Note that the ends of the interval — the points $x = -1$ and $x = 3$ — are not in the domain, in contrast to exercise 35).

39. $f(x)$ is defined when the numerator is defined and the denominator is nonzero. The numerator is defined for $x \leq -1$ and for $x \geq 1$ since then $x^2 > 1$ so that $x^2 - 1 > 0$. The denominator is nonzero for every x except for $x = \pm 3$. Thus the domain is $(-\infty, -3) \cup (-3, -1] \cup [1, 3) \cup (3, \infty)$.

41. $f(x)$ is defined when $\sqrt{x}$ is defined (i.e., $x \geq 0$), $\sqrt{x-1}$ is defined and nonzero (i.e., $x > 1$), and $x - 2 \neq 0$. The domain is given by the intersection of these three conditions: $x \geq 0$ and $x > 1$ and $x \neq 2$, so it is thus $(1, 2) \cup (2, \infty)$.

43. (a) $f(-4) = (-4)^2 + 1 = 17$.

   (b) $f(a^3) = (a^3)^2 + 1 = a^6 + 1$.

   (c) $f(f(x)) = f(x)^2 + 1 = (x^2 + 1)^2 + 1 = x^4 + 2x^2 + 2$.

45. (a) $l(5, 3, 2) = \sqrt{5^2 + 3^2 + 2^2} = \sqrt{38}$.

   (b) $l(3, 0, 4) = \sqrt{3^2 + 0^2 + 4^2} = 5$.

   (c) $l(x, y, z) = \sqrt{x^2 + y^2 + z^2}$.

47. (a) $F(2, 3, 5) = (3 \cdot 2 + 3, 2 - 5, 3 + 2 \cdot 5) = (9, -3, 13)$.

   (b) $F(5, 2, 3) = (3 \cdot 5 + 2, 5 - 3, 2 + 2 \cdot 3) = (17, 2, 8)$.

   (c) $F(a, b, 0) = (3 \cdot a + b, a - 0, b + 2 \cdot 0) = (3a + b, a, b)$.

49. (a) $f(-1) = 3(-1) + 1 = -2$, $f(0) = 3(0) + 1 = 1$, $f(1) = 4$, $f(2) = 2^3 = 8$.

   (b)

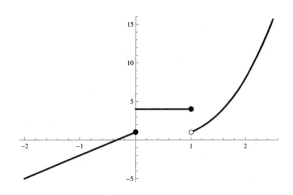

51.

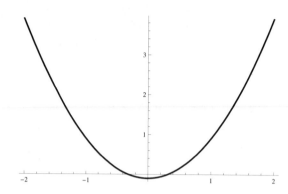

53.

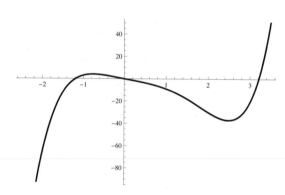

55.

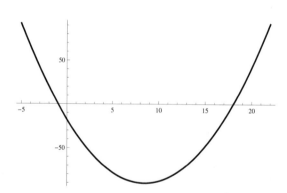

57. The domain is $\mathbb{R}$ and the range is $[-1, \infty)$. There are roots at $x = 0$ and $x = 2$, and a y-intercept at $(0, 0)$. The global minimum is at $(1, -1)$; there are no other global or local maxima or minima, and no inflection points. The graph is concave up everywhere, is decreasing on $(-\infty, 1)$ and increasing on $(1, \infty)$. There are no asymptotes.

59. The domain is $\mathbb{R}$ and the range is $[-4.75, \infty)$. There are roots at $x = 0$ and $x \approx -1.6$, and a y-intercept at $(0, 0)$. There is no global maximum since the graph increases without bound as x gets large. There is a global minimum at $(-1, -4.75)$ and a local minimum at $(2, 2)$; there is also a local maximum at $(1, 3.25)$. There are two inflection points, one near $x = 0$ and one near $x = 1.5$. The function is increasing on $(-1, 1)$ and on $(2, \infty)$ and decreasing on $(-\infty, -1)$ and on $(1, 2)$. It is concave up on $(-\infty, 0)$ and also on $(1.5, \infty)$ and concave down on $(0, 1.5)$. There are no asymptotes.

61. The domain is $\mathbb{R}$, and the range is $(1, 2]$. There are no roots, but there is a y-intercept at $(0, 2)$. The global maximum is at $(0, 2)$. There is no global minimum, since the graph approaches the line

$y = 1$ from above but never reaches it. There are no other local maxima or minima. There are two inflection points, at $\approx (\pm 1, 1.5)$. The function is increasing on $(-\infty, 0)$ and decreasing on $(0, \infty)$, is concave up on $(-\infty, -1)$ and on $(1, \infty)$, and concave down on $(-1, 1)$. There is a horizontal asymptote at $y = 1$.

63.

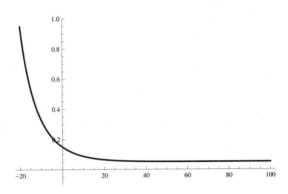

65. Not possible.

67.

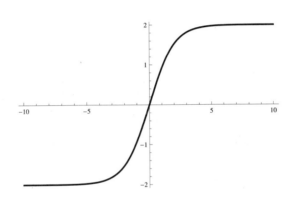

69.

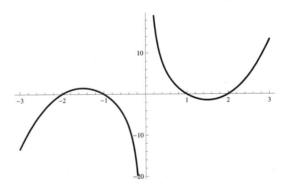

71. Not possible with the geometric definition of concavity given in the text, since an inflection point is defined to be a point at which the graph changes concavity. When an inflection point is defined mathematically later in the text, we will however see that a function such as this is indeed possible.

73. The average rate of change is $\dfrac{f(b) - f(a)}{b - a} = \dfrac{-0.5 + 4.2 \cdot 3.5 - (-0.5 + 4.2 \cdot 1)}{3.5 - 1} = 4.2$. Note that since $f(x)$ is a linear function, its average rate of change over any interval will be the slope of the line, or 4.2.

75. The average rate of change is $\dfrac{f(b)-f(a)}{b-a} = \dfrac{\sqrt{9+1}-\sqrt{1+1}}{9-1} = \dfrac{\sqrt{10}-\sqrt{2}}{8}$.

77. The average rate of change is $\dfrac{f(b)-f(a)}{b-a} = \dfrac{\frac{1}{1.1}-\frac{1}{0.9}}{1.1-0.9} = -\dfrac{.2}{.198} = -\dfrac{1}{.99} \approx -1.0101$.

79. Assuming the initial situation is at time $t = 0$, a graph could look like the following:

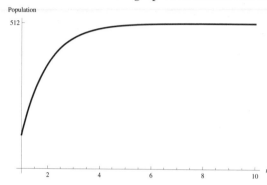

81. The graph of fraction of caffeine remaining versus time is:

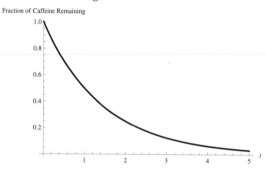

83. (a) a and b are independent variables, and H is the dependent variable; their relationship is $H(a,b) = \sqrt{a^2 + b^2}$.

(b) x, y, and z are the independent variables, and V is the dependent variable; their relationship is $V(x,y,z) = xyz$.

85. Since 5 years have passed, and the average rate of change is 4 groundhogs per year, there are $4 \cdot 5 = 20$ more groundhogs now than there were in 1996. Thus the number of initially abandoned groundhogs was 356. Poor groundhogs.

87. (a) The pretax income function is

$$i(t) = \begin{cases} 36000 & 0 \le t \le 4 \\ 38500 & 4 \le t \le 6 \\ 49000 & t > 6. \end{cases}$$

(b) Your total earnings are

$$e(t) = \begin{cases} 36000t & 0 \le t \le 4 \\ 36000 \cdot 4 + 38500(t-4) & 4 \le t \le 6 \\ 36000 \cdot 4 + 38500 \cdot 2 + 49000(t-6) & t > 6. \end{cases}$$

(c) Solve the equation $36000 \cdot 4 + 38500 \cdot 2 + 49000(t-6) = 1000000$ to get $t \approx 21.898$, so in your 22nd year after graduating you will have earned one million pretax dollars.

Proofs

89. Definition 0.2 states that the range of a function is the set of all possible outputs that it can attain. Choose $r \in \mathbb{R}$. Then for $x = \dfrac{r+1}{3}$, $f(x) = r$. Thus each $r \in \mathbb{R}$ is in the range, so that the range is $\mathbb{R}$.

91. If f is one-to-one, choose any $c \in C$ and consider the horizontal line through $y = c$. That horizontal line intersects the graph of f at each point (x, c) such that $c = f(x)$. Suppose (a, c) and (b, c) are two such points. Since both points lie on the graph, it follows that $f(a) = c$ and $f(b) = c$. But by Definition 0.5, since f is one-to-one we must have $a = b$. Thus the two points (a, c) and (b, c) are in fact the same point, so the graph of f passes the horizontal line test. Conversely, suppose that f is a function whose graph passes the horizontal line test, and suppose that $f(a) = f(b)$. Then the points (a, c) and (b, c) both lie on the graph of f, and on the horizontal line $y = c$. Since the graph passes the horizontal line test, we must have $a = b$, so that f is one-to-one.

93. Suppose that $b > a$. Then $-3b < -3a$, so that $f(b) = 1 - 3b < 1 - 3a = f(a)$. It follows that f is decreasing.

95. The average rate of change between a and b is

$$\frac{f(b) - f(a)}{b - a} = \frac{(-2b + 4) - (-2a + 4)}{b - a} = \frac{-2(b - a)}{b - a} = -2.$$

Thinking Forward

Evaluations for slopes and derivatives

▷ For $f(x) = 4 - x^2$,

- $\dfrac{f(1 + 0.1) - f(1)}{0.1} = \dfrac{(4 - 1.1^2) - (4 - 1)}{0.1} = -\dfrac{0.21}{0.1} = -2.1.$

- $\dfrac{f(1 + 0.001) - f(1)}{0.001} = \dfrac{(4 - 1.001^2) - (4 - 1)}{0.001} = -\dfrac{0.002001}{0.001} = -2.001.$

▷ For $f(x) = \sqrt{x}$,

- $\dfrac{f(1 + h) - f(1)}{h} = \dfrac{\sqrt{1 + h} - 1}{h}.$

- $\dfrac{f(x) - f(1)}{x - 1} = \dfrac{\sqrt{x} - 1}{x - 1}.$

▷ For $q(x, h) = \dfrac{(x + h)^2 - x^2}{h}$,

- $q(3, 0.5) = \dfrac{(3 + 0.5)^2 - 3^2}{0.5} = \dfrac{3.25}{0.5} = 6.5.$

- $q(3, h) = \dfrac{(3 + h)^2 - 3^2}{h} = \dfrac{h^2 + 6h}{h} = 6 + h.$

- $q(x, 0.5) = \dfrac{(x + 0.5)^2 - x^2}{0.5} = \dfrac{x + 0.25}{0.5} = 2x + 0.5.$

Evaluations for series

▷ For $S(x,n) = x - \frac{x^2}{2} + \frac{x^3}{3} + \cdots + (-1)^{n+1}\frac{x^n}{n}$,

- $S\left(\frac{1}{2},5\right) = \frac{1}{2} - \frac{1/4}{2} + \frac{1/8}{3} - \frac{1/16}{4} + \frac{1/32}{5} = \frac{391}{960}$.

- $S\left(\frac{1}{2},n\right) = \frac{1}{2} - \frac{1}{4\cdot 2} + \frac{1}{8\cdot 3} - \frac{1}{16\cdot 4} + \cdots + (-1)^{n+1}\frac{1}{2^n\cdot n}$.

- $S(x,5) = x - \frac{x^2}{2} + \frac{x^3}{3} - \frac{x^4}{4} + \frac{x^5}{5}$.

▷ For $c(x,n) = 1 - \frac{x^2}{2!} + \frac{x^4}{4!} - \frac{x^6}{6!} + \cdots + (-1)^n\frac{x^{2n}}{(2n)!}$,

- $c(\pi,3) = 1 - \frac{\pi^2}{2} + \frac{\pi^4}{24} - \frac{\pi^6}{720}$.

- $c(\pi,n) = 1 - \frac{\pi^2}{2!} + \frac{\pi^4}{4!} - \frac{\pi^6}{6!} + \cdots + (-1)^n\frac{\pi^{2n}}{(2n)!}$.

- $c(x,3) = 1 - \frac{x^2}{2!} + \frac{x^4}{4!} - \frac{x^6}{6!}$.

Tangent lines

▷ The tangent line is $f(x)$ itself, since the graph is itself a line, and the "direction" the graph is moving is given by that line.

▷ The graph is flat-topped at $x = 0$ — it is changing from increasing to decreasing there. Thus the tangent line is flat. Since $f(0) = 4$, the tangent line is $y = 4$.

▷ From the given graph, the tangent line appears to intercept the x axis at $x = 1$, and its value at $x = 2$ is 4. The slope of the line is therefore $\approx \frac{4-0}{2-1} = 4$. Since the line passes through $(1,0)$, the equation of the tangent line is $y = 4(x-1)$.

0.2 Operations, Transformations, and Inverses

Thinking Back

Evaluating Functions

With $f(x) = \frac{x}{x-1}$,

▷ $f(1.5) = \frac{1.5}{1.5-1} = 3$.

▷ $f(x^2) = \frac{x^2}{x^2-1}$.

▷ $f(x^2+1) = \frac{x^2+1}{x^2+1-1} = \frac{x^2+1}{x^2}$.

▷ $f\left(\frac{1}{x}\right) = \frac{1/x}{1/x-1} = \frac{1}{1-x}$.

▷ $f(f(x)) = \dfrac{f(x)}{f(x) - 1} = \dfrac{x/(x-1)}{\frac{x}{x-1} - 1} = \dfrac{x}{x - (x-1)} = x.$

▷ $f(f(x^2)) = \dfrac{f(x^2)}{f(x^2) - 1} = \dfrac{x^2/(x^2 - 1)}{x^2/(x^2 - 1) - 1} = \dfrac{x^2}{x^2 - (x^2 - 1)} = x^2.$

Solving Equations

▷ $s = \dfrac{r - 2}{3} \Rightarrow 3s = r - 2 \Rightarrow r = 3s + 2.$

▷ $s = \dfrac{3}{r - 2} \Rightarrow rs - 2s = 3 \Rightarrow rs = 2s + 3 \Rightarrow r = \dfrac{2s + 3}{s}.$

▷ $s = \dfrac{r + 1}{r} \Rightarrow rs = r + 1 \Rightarrow r(s - 1) = 1 \Rightarrow r = \dfrac{1}{s - 1}.$

▷ $s = \dfrac{r + 2}{r + 3} \Rightarrow rs + 3s = r + 2 \Rightarrow r(s - 1) = 2 - 3s \Rightarrow r = \dfrac{2 - 3s}{s - 1}.$

Concepts

1. (a) False. The domain of $f + g$ is the intersection of the domains of f and g. There is no reason to expect the domain of either function to be "smaller" than the other. For example, $f(x) = \dfrac{1}{x}$ and $g(x) = \dfrac{1}{x - 1}$.

   (b) True. See Definition 0.7.

   (c) True. $f(x) + C$ is the graph of $f(x)$ shifted up by C units, so if the graph of $y = f(x)$ contains (a, b), then the graph of $y = f(x) + C$ contains (a, b) shifted up by C units, which is $(a, b + C)$.

   (d) False. It contains the point $(a - C, b)$ since the graph of $f(x + C)$ is the graph of $f(x)$ shifted *left* by C units. See the table on page 24.

   (e) False. Let $g(x) = x^{-5}$; then $f(g(x)) = f(x^{-5}) = (x^{-5})^5 = x^{-25} \neq x.$

   (f) False. See the previous item for an example.

   (g) False. For example, $\cos x$ is even.

   (h) False. These points cannot both be on the graph of any function (vertical line rule) unless $b = 0$. What is true is that if f is even and $(b, 0)$ is on the graph of $y = f(x)$, then $(-b, 0)$ is also a point on the graph.

3. The *difference* of f and g is the function $f - g$ defined by $(f - g)(x) = f(x) - g(x)$ for all x in the domains of both f and g. This is the same as the sum of f and $(-1)g$.

5. (a) $(1, 2)$, since $f(4 - 3) = f(1) = 2.$

   (b) $f\left(\frac{1}{2}x\right)$, since $f\left(\frac{1}{2} \cdot 6\right) = f(3) = -2.$

   (c) $(-1, 4)$, since for an even function, $f(x) = f(-x)$ for all x. Take $x = -1.$

   (d) $(2, -5)$, since for an odd function, $f(-x) = -f(x)$ for all x. Take $x = -2.$

7. (a) $g(x)$ is defined for $x \in [-10, \infty)$, so $g(f(x))$ is defined for $f(x) \in [-10, \infty)$. But the range of f is contained in $[-10, \infty)$, so that the domain of $g \circ f$ is just the domain of f.

   (b) The domain of f is $[2, \infty)$. Since the range of g, which is $[0, \infty)$, is not a subset of the domain of f, it is impossible to tell which points in the domain of g actually map into points in $[2, \infty)$. It is those points that actually lie in the domain of $f \circ g$.

(c) For $f \circ f$, no, for the same reason as in part (b). For $g \circ g$, yes, since the range of g is a subset of its domain and thus $g \circ g$ is defined everywhere g is.

9. The transformation (call it $g(x)$) is a horizontally squeezed version of $f(x)$. Since $g(1) = f(2)$, we must have $g(x) = f(2x)$.

11. No if $f(0) = 2$, since $f(0) = -f(-0) = -f(0)$ and thus we must have $f(0) = 0$. If $f(0)$ is undefined, then 0 is not in the domain of f, and f can be an odd function (see Definition 0.9).

13. (a) Assuming that we want f to be an even function, we must have $f(-2) = f(2) = 1$ and $f(-1) = f(1) = -2$. However, $f(0)$ can be arbitrary, since all we know is that $f(0) = f(-0) = f(0)$:

x	-3	-2	-1	0	1	2	3
$f(x)$	4	1	-2	a	-2	1	4

where a is arbitrary.

(b) Assuming that we want f to be an odd function, we must have $f(-2) = -f(2) = -1$ and $f(-1) = -f(1) = 2$. If f is defined at $x = 0$, its value must be zero, since $f(0) = -f(-0) = -f(0)$ from which it follows that $f(0) = 0$.

x	-3	-2	-1	0	1	2	3
$f(x)$	4	-1	2	0	-2	1	-4

assuming f is defined at $x = 0$.

15. $(f^{-1})^{-1} = f$. To see this, note that we must have $f^{-1} \circ (f^{-1})^{-1} = x$ and $(f^{-1})^{-1} \circ f^{-1} = x$. But since f^{-1} is the inverse of f, f satisfies the equations $f^{-1} \circ f = x$ and $f \circ f^{-1} = x$. Since f^{-1} has a unique inverse, we must have $f = (f^{-1})^{-1}$.

17. The domain of f^{-1} is the range of f, which is $(-\infty, 3]$; the range of f^{-1} is the domain of f, which is $[-1, 1)$. The graph of the inverse function is simply the graph of the original function reflected through the line $y = x$.

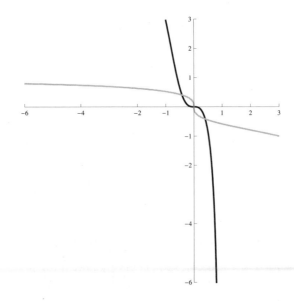

19. Since $(f + 2g - h)(0) = 4$, $f(0) = 1$, and $g(0) = 0$, it follows that $h(0) = -3$. We also have $(fg)(0) = f(0)g(0) = 0$ and $\left(\frac{g}{h}\right)(0) = \frac{g(0)}{h(0)} = 0$. Also, since $(-2f)(x) = -2f(x)$, we have

$(-2f)(0) = -2f(0) = -2$. This completes the first column. For the second column, note first that using $(-2f)(1) = 4$ we get $f(1) = -2$. Since $\left(\frac{g}{h}\right)(1) = 3$ and $h(1) = 1$, it follows that $g(1) = 3$, and then that $(fg)(1) = -2 \cdot 3 = -6$. Finally, $(f + 2g - h)(1) = f(1) + 2g(1) - h(1) = 3$. This completes the second column. For the third column, from $(fg)(2) = 2$ and $g(2) = -2$ we get $f(2) = -1$; from $\left(\frac{g}{h}\right)(2) = -1$ and $g(2) = -2$ we get $h(2) = 2$. Then $(-2f)(2) = -2f(2) = 2$ and $(f + 2g - h)(2) = f(2) + 2g(2) - h(2) = -7$. This completes the third column.

x	0	1	2
$f(x)$	1	-2	-1
$g(x)$	0	3	-2
$h(x)$	-3	1	2
$(-2f)(x)$	-2	4	2
$(f + 2g - h)(x)$	4	3	-7
$(fg)(x)$	0	-6	2
$\left(\frac{g}{h}\right)(x)$	0	3	-1

Skills

For problems 21 through 29, note that the domain of f is $\mathbb{R}$, the domain of g is $\mathbb{R} - \{2\}$, and the domain of h is $[0, \infty)$.

21. The domain of $f + g$ is the intersection of the domain of f and the domain of g, so it is $\mathbb{R} - \{2\}$. We have $(f + g)(x) = x^2 + 1 + \dfrac{1}{x - 2}$, so that $(f + g)(1) = 1^2 + 1 + \dfrac{1}{1 - 2} = 1$.

23. The domain of $\frac{g}{h}$ is the intersection of the domain of g and the places where h is defined and nonzero, which is $(0, \infty)$. Thus the domain of $\frac{g}{h}$ is $(0, 2) \cup (2, \infty)$. We have $\dfrac{g}{h}(x) = \dfrac{1}{(x - 2)\sqrt{x}}$, and $\dfrac{g}{h}(1) = \dfrac{1}{(1 - 2)\sqrt{1}} = -1$.

25. The domain of $g \circ g$ is the set of x in the domain of g such that $g(x)$ is also in the domain of g. But $g(x)$ is in the domain of g unless $g(x) = 2$, which happens when $x = \dfrac{5}{2}$. Thus the domain of $g \circ g$ is $\mathbb{R} - \left\{2, \dfrac{5}{2}\right\}$. We have $(g \circ g)(x) = \dfrac{1}{\frac{1}{x-2} - 2} = \dfrac{x - 2}{1 - 2(x - 2)} = \dfrac{x - 2}{5 - 2x}$, and $(g \circ g)(1) = -\dfrac{1}{3}$.

27. The domain of $g(x - 5)$ is the domain of g shifted right by 5 units, so it is $\mathbb{R} - \{7\}$. We have $g(x - 5) = \dfrac{1}{(x - 5) - 2} = \dfrac{1}{x - 7}$, and $g(1 - 5) = g(-4) = -\dfrac{1}{6}$.

29. The domain of $h(3x + 1)$ is the domain of h shifted left by 1 and then compressed by a factor of 3, so it is $\left[-\dfrac{1}{3}, \infty\right)$. We have $h(3x + 1) = \sqrt{3x + 1}$, and $h(3 \cdot 1 + 1) = h(4) = \sqrt{4} = 2$.

31.

x	0	1	2	3	4	5	6
$2f(x)+3$	3	5	9	7	9	3	7

33.

x	0	1	2	3	4	5	6
$(h \circ g)(x)$	2	3	2	2	3	2	3

These are arrived at by composing the functions for each value of x. For example, $(h \circ g)(3) = h(g(3)) = h(1) = 2$.

35.

x	0	1	2	3	4	5	6
$(f \circ f \circ f)(x)$	0	1	3	2	3	0	2

These are arrived at by composing the functions for each value of x. For example, $(f \circ f \circ f)(2) = f(f(f(2))) = f(f(3)) = f(2) = 3$.

37. Note here that we are computing an expression involving $h(x-1)$; to do this, $x-1$ must be in the domain of h, so that $x-1$ is an integer between 0 and 6. Thus x must be an integer between 1 and 7.

x	1	2	3	4	5	6	7
$g(x-1)$	1	0	1	1	0	1	0

39.

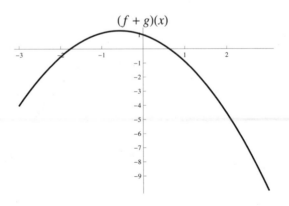

$(f + g)(x)$

41.

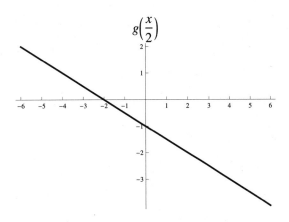

43.

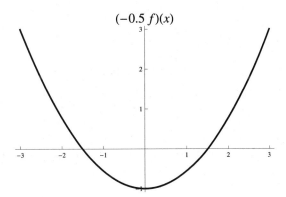

45.

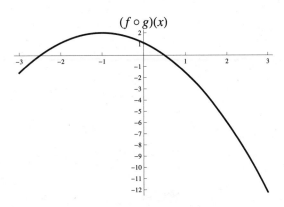

47.

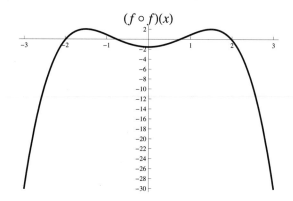

49.

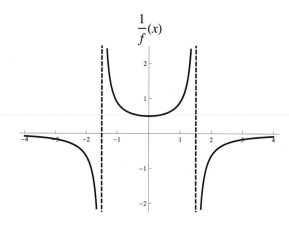

51. (a) $(f+g)(-1) = 0$, $(f+g)(0) = 1$, $(f+g)(1) = 0$, $(f+g)(2) = 9$, $(f+g)(3) = 14$.

(b)

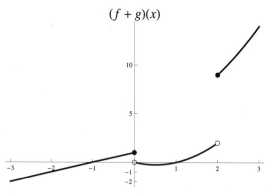

(c)

$$(f+g)(x) = \begin{cases} x+1, & x \leq 0, \\ x^2 - x, & 0 < x < 2, \\ x^2 + 5, & x \geq 2. \end{cases}$$

53. (a) $(g \circ f)(-1) = 1, \quad (g \circ f)(0) = -1, \quad (g \circ f)(1) = -1, \quad (g \circ f)(2) = 5, \quad (g \circ f)(3) = 5.$

(b)

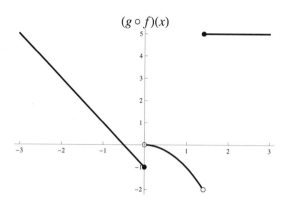

(c)

$$(g \circ f)(x) = \begin{cases} -2x - 1, & x \leq 0, \\ -x^2, & 0 < x < \sqrt{2}, \\ 5 & x \geq \sqrt{2}. \end{cases}$$

55. (a) Write $h(x) = f(x-1)$: $h(-1) = -3, \quad h(0) = -1, \quad h(1) = 1, \quad h(2) = 1, \quad h(3) = 4.$

(b)

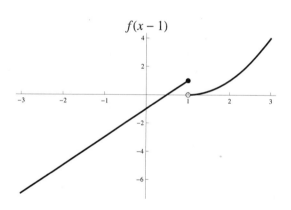

(c)

$$h(x) = \begin{cases} 2x - 1, & x \leq 1, \\ (x-1)^2, & x > 1. \end{cases}$$

57. For example, let $g(x) = 3x + 5$ and $h(x) = x^2$, or let $g(x) = x + 10$ and $h(x) = 3x^2 - 5$.

59. For example, let $g(x) = \dfrac{6}{x}$ and $h(x) = x + 1$, or let $g(x) = 6x$ and $h(x) = \dfrac{1}{x+1}$.

61. Even, since $f(-x) = (-x)^4 + 1 = x^4 + 1 = f(x)$. The graph of $f(x)$ is symmetric around the y-axis.

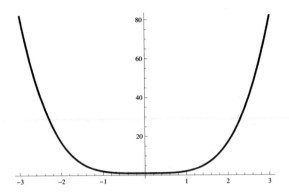

63. Neither, since for example $f(1) = 2$ while $f(-1) = 0$, so that $f(1) \neq f(-1)$ and $f(1) \neq -f(-1)$. The graph is neither rotationally symmetric around the origin nor symmetric around the y-axis.

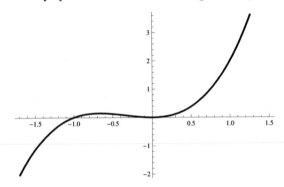

65. Odd, since $f(-x) = \dfrac{(-x)^3}{(-x)^2 + 1} = -\dfrac{x^3}{x^2 + 1} = -f(x)$. The graph is rotationally symmetric around the origin.

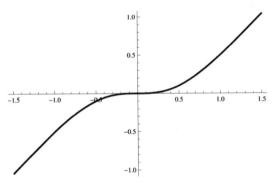

67.
$$g(f(x)) = g(2 - 3x) = -\frac{1}{3}(2 - 3x) + \frac{2}{3} = -\frac{2}{3} + x + \frac{2}{3} = x,$$
$$f(g(x)) = f\left(-\frac{1}{3}x + \frac{2}{3}\right) = 2 - 3\left(-\frac{1}{3}x + \frac{2}{3}\right) = 2 + x - 2 = x.$$

69.
$$g(f(x) = g\left(\frac{x}{1 - x}\right) = \frac{\frac{x}{1-x}}{1 + \frac{x}{1-x}} = \frac{x}{(1 - x) + x} = x,$$
$$f(g(x)) = f\left(\frac{x}{1 + x}\right) = \frac{\frac{x}{1+x}}{1 - \frac{x}{1+x}} = \frac{x}{(1 + x) - x} = x.$$

71. f is invertible since it is one-to-one on its domain. The graph of its inverse is the reflection of its graph around $y = x$.

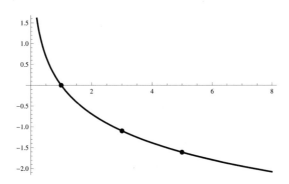

73. f is not invertible on its entire domain since it is not one-to-one. Restrict the domain to $[0, \infty)$, where f is one-to-one; then its inverse is the reflection of that portion of its graph through $y = x$.

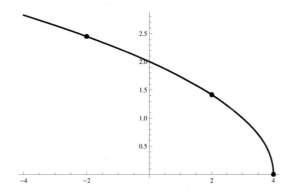

75. Solving for x gives $f^{-1}(x) = \dfrac{1 - 2x}{5}$. The graph of f and f^{-1} are below, with f in black and f^{-1} in gray:

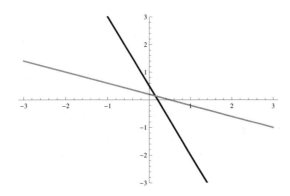

77. Solving for x gives $f^{-1}(x) = \dfrac{1}{x - 1}$. The graph of f and f^{-1} are below, with f in black and f^{-1} in gray:

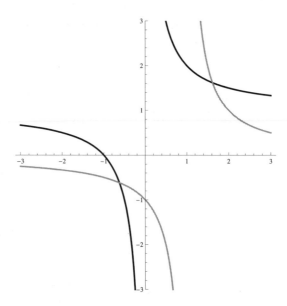

79. Solving for x gives $f^{-1}(x) = \dfrac{1+x}{1-x}$. The graph of f and f^{-1} are below, with f in black and f^{-1} in gray:

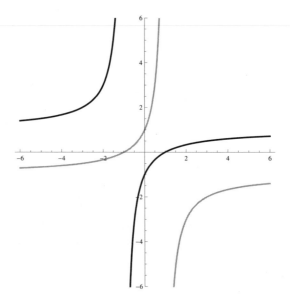

Applications

81. (a) Since the area of a square is the square of its side length, we have $S(x) = x^2$.

(b) $C(S) = 200 + 4.25S$.

(c) $C(x) = 200 + 4.25x^2$. This is the composition $C \circ S(x)$.

83. (a) $C(n) = 20 + 12n$.

(b) The graph of $C(n)$ is a line that is not horizontal, so it passes the horizontal line test and is one-to-one. Thus $C(n)$ is invertible.

(c) Solving for n gives $n(C) = \dfrac{C - 20}{12}$. This represents the number of shirts that can be purchased for $\$C$.

(d) $n(150) = \dfrac{150 - 20}{12} = \dfrac{130}{12} \approx 10.83$, so you can make 10 shirts and have some material left over.

Proofs

85. If $f(x) = x^k$, then $f(2x) = (2x)^k = 2^k x^k = 2^k f(x)$, so the graphs of these two functions are the same.

87. (a) Since $f(x) = y$, also $kf(x) = ky$, so that (x, ky) is on the graph of $kf(x)$.

    (b) Write $g(x) = f(kx)$. Then $g\left(\dfrac{1}{k}x\right) = f\left(\dfrac{1}{k} \cdot kx\right) = f(x) = y$, so that $\left(\dfrac{1}{k}x, y\right)$ is on the graph of $g(x)$, which is the graph of $f(kx)$.

89. If f is odd and defined at 0, then by the definition of an odd function, we have $f(0) = -f(-0) = -f(0)$. Solving gives $f(0) = 0$.

91. Since f^{-1} is the inverse of f, we have by definition 0.10 that $f^{-1}(f(x)) = x$ for all x in the domain of f. Thus f^{-1} is defined for all $f(x)$; that is, the domain of f^{-1} contains the range of f. Now suppose y is in the domain of f^{-1} but not in the range of f. Then clearly $y = f(f^{-1}(y))$ cannot be satisfied, so that f^{-1} cannot be the inverse of f. Thus the domain of f^{-1} is equal to the range of f.

Similarly, since $f(f^{-1}(x)) = x$, f is defined for all $f^{-1}(x)$; that is, the domain of f contains the range of f^{-1}. Now suppose x is in the domain of f but not in the range of f^{-1}. Then clearly the equation $x = f^{-1}(f(x))$ cannot be satisfied, and f^{-1} cannot be the inverse of f. Thus the range of f^{-1} is equal to the domain of f.

93. Since $x = f^{-1} \circ f(x)$, we see, setting $x = a$, that $a = f^{-1}(f(a)) = f^{-1}(b)$.

Thinking Forward

Transformations of trigonometric graphs

▷

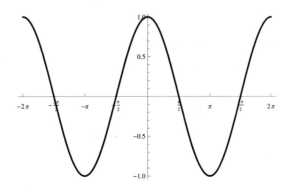

▷

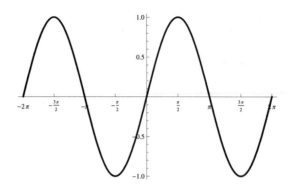

Since the graph is identical to the graph of $f(x)$ we see that $\sin(x + 2\pi) = \sin(x)$.

▷

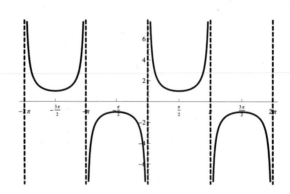

▷ In the graph below, $f(x) = \sin x$ is in black while $f^{-1}(x) = \sin^{-1} x$ is in gray:

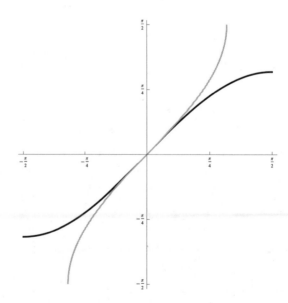

0.3 Algebraic Functions

Thinking Back

Basic Algebra Review

▷ $\left(\dfrac{8}{27}\right)^{-\frac{2}{3}} = \left(\dfrac{27}{8}\right)^{\frac{2}{3}} = \left(\dfrac{3}{2}\right)^2 = \dfrac{9}{4}.$

▷ Since $x^{-3} - x^{-2} = (x^{-1} - 1)x^{-2}$, it follows that $\dfrac{x^{-3} - x^{-2}}{x^{-1} - 1} = x^{-2}$, so $A = 1$ and $k = -2$.

▷ $16x^6 - 81x^2 = x^2(4x^2 - 9)(4x^2 + 9) = x^2(2x - 3)(2x + 3)(4x^2 + 9).$

▷ The equation is satisfied when the numerator vanishes but the denominator does not. The numerator factors as $x(x^2 + x - 2) = x(x - 1)(x + 2)$, so it vanishes at $x = 0$, $x = 1$, and $x = -2$. The denominator factors as $(x - 3)(x - 1)$, so it vanishes at $x = 1$ and $x = 3$. Thus the equation is satisfied only for $x = 0$ and for $x = -2$.

▷ Simplifying gives $1 = (2x - 1)(3x^2 - 2x - 1) = 6x^3 - 7x^2 + 1$, or $6x^3 - 7x^2 = 0$. This equation holds for $x = 0$ and for $x = \dfrac{7}{6}$, so these are the solutions to the original equation.

Factoring after Root-Guessing

▷ Examination shows $x = 1$ to be a root; -1 is also a root since the equation is an even function. Dividing by $(x - 1)(x + 1) = x^2 - 1$ gives $2x^2 + 8$, which has no real roots. The factorization is therefore $2(x - 1)(x + 1)(x^2 + 4)$.

▷ Examination shows $x = 1$ to be a root. Dividing by $x - 1$ gives $x^2 + 5x - 6$, which factors as $(x - 1)(x + 6)$. Thus the original function factors as $(x - 1)^2(x + 6)$.

▷ $x = 1$ is a root of the numerator; the result of dividing by $x - 1$ is $x^2 + 7x + 10 = (x + 5)(x + 2)$, so the numerator factors as $(x - 1)(x + 2)(x + 5)$. $x = 1$ is also a root of the denominator; the result of dividing by $x - 1$ is $x^4 + 4x^3 + 4x^2 + 4x + 3$, which has -1 as a recognizable root. Again dividing gives $x^3 + 3x^2 + x + 3$, which has $x = -3$ as a root. The quotient by $x + 3$ is $x^2 + 1$, which has no real roots. Thus the original rational function factors as

$$\frac{x^3 + 6x^2 + 3x - 10}{x^5 + 3x^4 - x - 3} = \frac{(x - 1)(x + 2)(x + 5)}{(x - 1)(x + 1)(x + 3)(x^2 + 1)}.$$

Concepts

1. (a) True, since if both f and g can be expressed using the algebraic operations, so can $f + g$.

   (b) False. A polynomial function uses only integral powers of x, but Ax^k may have k rational but not an integer.

   (c) True, since a polynomial function $p(x)$ can be written as the quotient of the two polynomial functions $p(x)$ and 1, so is a rational function.

   (d) False. For example, if $a = b = 0$, then $f(x) = x^4 + 7$, which clearly has no real roots.

   (e) False. By Theorem 0.1.15(a), it must be at *least* k.

   (f) Assuming that by a "turning point" is meant local extrema, this is true. See Theorem 0.15(b) for the equivalent contrapositive statement.

   (g) True. See Definition 0.16.

(h) False. See Theorem 0.18(d). For example, any nonconstant polynomial, such as $f(x) = x$, has no horizontal asymptote.

3.

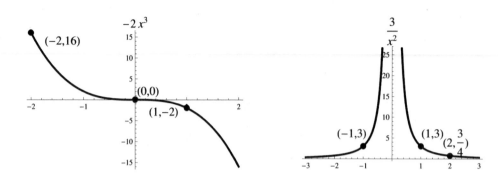

5. x^2 is in black, x^4 in light gray, and x^6 in dark gray.

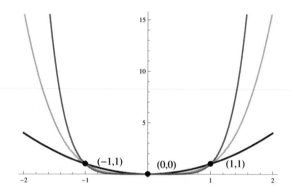

7. Assuming $A \neq 0$, yes, it is. $\dfrac{1}{f(x)} = \dfrac{1}{Ax^k} = A^{-1}x^{-k}$, so it is a power function Cx^r where $C = A^{-1}$ and $r = -k$.

9. The term with the highest power of x appearing in $f(x)$ is $x(3x)(x^2) = 3x^4$, so that the leading coefficient is 3, the leading term is $3x^4$, and the polynomial has degree 4. Since the polynomial is a multiple of x, the constant term is zero. The coefficient a_1 of x will be the constant term of $(3x + 1)(x - 2)^2$, which is 4. Finally, the coefficient a_3 of x^3 is the coefficient of x^2 in $(3x + 1)(x - 2)^2 = (3x + 1)(x^2 - 4x + 4)$, which is -11.

11. The quadratic formula tells us that $f(x) = ax^2 + bx + c$ has the two roots $x = \dfrac{-b \pm \sqrt{b^2 - 4ac}}{2a}$. If $b^2 - 4ac$ is positive, then both roots are real numbers, and we can write

$$f(x) = \left(x - \frac{-b + \sqrt{b^2 - 4ac}}{2a} \right) \left(x - \frac{-b - \sqrt{b^2 - 4ac}}{2a} \right)$$

Conversely, if $b^2 - 4ac < 0$, then neither root is real, so the quadratic cannot be factored (since the factors are determined by the roots). For $3x^2 + 2x + 6$, note that the discriminant is $b^2 - 4ac = 4 - 72 = -68 < 0$.

13. Since the multiplicity of the root -2 is higher in the denominator than it is in the numerator, $f(x)$ does not have a hole at $x = -2$, by Theorem 0.17(c), but instead has a vertical asymptote there, by Theorem 0.18(a).

15. Using Theorems 0.17 and 0.18, one possibility is $\dfrac{3x^2+1}{x^2-4}$. It has a horizontal asymptote at $y=3$ by 0.3.18(c) and vertical asymptotes at $x=\pm 2$ by 0.3.18(a). It has no roots since $3x^2+1$ has no roots, and no holes by 0.3.17(c).

17. (a)

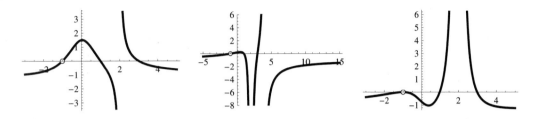

(b) The three functions are

$$f(x) = -\frac{(x-1)(x-3)(x+1)^2}{(x+1)(x-2)(x^2+1)},$$

$$f(x) = -\frac{(x-1)(x-3)(x+1)^2}{(x+1)(x-2)^2(x-4)},$$

$$f(x) = -\frac{(x-1)(x-3)(x+1)^3}{(x+1)(x-2)^2(x^2+1)}.$$

19. This is the same as the given graph, except that all points below the x axis are reflected through the x axis.

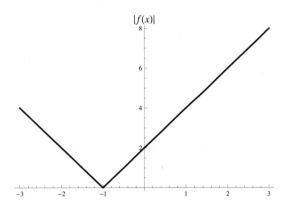

21. This is the same as the given graph, except that all points below the x axis are reflected through the x axis.

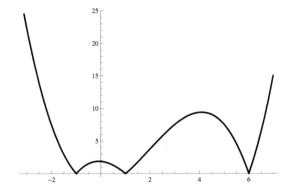

23. Different answers are possible. One explanation is that since the square root function always takes the positive square root, it follows that $\sqrt{x^2}$ is always the positive number whose square is x^2. If x is positive, this number is x itself; if x is negative, this number is $-x$. In other words, that positive number is $|x|$.

Skills

25. The only zero is $x = 0$. Since $x^{2/3}$ is defined for all x, the domain is $\mathbb{R}$.

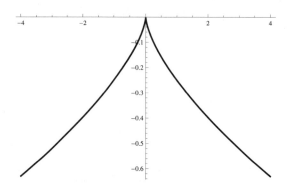

27. We can rewrite f as $f(x) = \dfrac{2}{x^{7/4}}$. Since the numerator never vanishes, there are no zeros. f is defined for $x > 0$, but not for $x \leq 0$ (at $x = 0$, the denominator is zero, while for $x < 0$, you cannot take an even root of a negative number). Thus the domain is $(0, \infty)$.

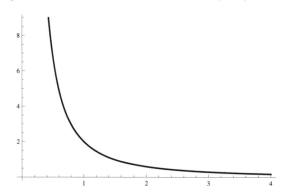

29. We have $f(x) = \dfrac{1}{(x^2 - 1)^{1/4}}$. Since the numerator never vanishes, f has no zeroes. f is defined except at ± 1, where the denominator vanishes, and on $(-1, 1)$, where $x^2 - 1$ is negative and thus does not have a fourth root. The domain is therefore $(-\infty, -1) \cup (1, \infty)$.

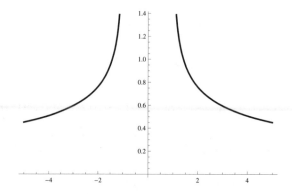

31. The numerator vanishes at $x = 3$, while the denominator vanishes at $x = -1$ and at $x = 4$. The numerator is defined only for $x \leq 3$. The denominator is, however, defined for all x, since all numbers have cube roots. Thus $f(x)$ has a single zero, at $x = 3$, and is defined on $(-\infty, -1) \cup (-1, 3]$.

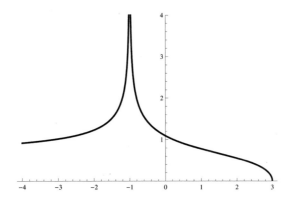

33. The domain of f consists of all points where $g(x) = \dfrac{x^2 - 1}{x^3 - 7x + 6}$ is defined and is nonnegative (since otherwise you cannot take its square root). g can change sign only at the points where the numerator or denominator is zero. Since $g(x) = \dfrac{(x-1)(x+1)}{(x-1)(x-2)(x+3)}$, those points are $x = 1$, $x = -1$, $x = 2$, and $x = -3$. We can thus determine where g is nonnegative by checking its value at one point of each of the resulting five intervals:

interval	$(-\infty, -3)$	$(-3, -1)$	$(-1, 1)$	$(1, 2)$	$(2, \infty)$
x	-4	-2	0	$\frac{3}{2}$	3
$g(x)$	$-\frac{1}{2}$	$\frac{1}{4}$	$-\frac{1}{6}$	$-\frac{10}{9}$	$\frac{2}{3}$
sign	$-$	$+$	$-$	$-$	$+$

Thus $g(x) > 0$ for $x \in (-3, -1) \cup (2, \infty)$. Since g is undefined at $x = -3$ and at $x = 2$, but is zero at $x = -1$, we have $g(x) \geq 0$ for $x \in (-3, 1] \cup (2, \infty)$, which is therefore the domain of f.

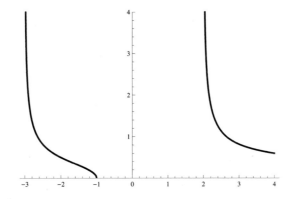

35. Use the point-slope form to get $y - (-2) = -1(x - 3)$ and simplify to get $y = -x + 1$.

37. Since the line is parallel to $y = 2x + 1$, it must have slope 2; since it passes through $(-1, 4)$, its equation must be $y - 4 = 2(x - (-1)) = 2x + 2$, or $y = 2x + 6$.

39. For example, $y = x^3$.

41. For example, $y = x(x-2)(x-4)$.

43. $f(x)$ has no holes or asymptotes. It has a root at $x = -1$.

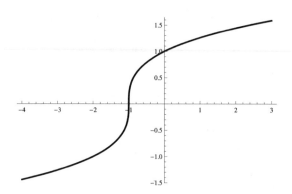

45. $f(x)$ has no zeroes or holes. Since the denominator vanishes at $x = \pm 2$, it has vertical asymptotes there. It also has a horizontal asymptote at $y = 0$, since the denominator goes to infinity while the numerator is constant.

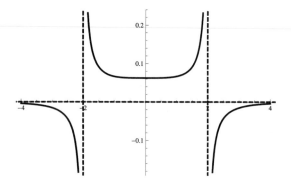

47. $f(x) = 2x^4 - x^3 - x^2 = x^2(2x^2 - x - 1) = x^2(x-1)(2x+1)$, so $f(x)$ has no holes or asymptotes, but has zeroes at $x = 0$, $x = 1$, and $x = -\dfrac{1}{2}$.

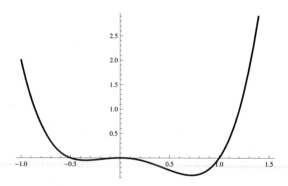

49. $f(x) = x^3 - 2x^2 - 4x + 8 = (x-2)^2(x+2)$, so $f(x)$ has no holes or asymptotes, but has zeroes at $x = \pm 2$.

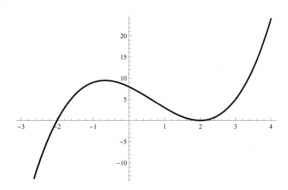

51. $f(x) = \dfrac{2x^3 + 4x^2 - 6x}{x^2 - 4} = \dfrac{2x(x+3)(x-1)}{(x-2)(x+2)}$. Thus $f(x)$ has no holes, but has zeroes at $x = 0$, $x = 1$, and $x = -3$, and vertical asymptotes at $x = \pm 2$. $f(x)$ has no horizontal asymptotes since the degree of the numerator is larger than the degree of the denominator.

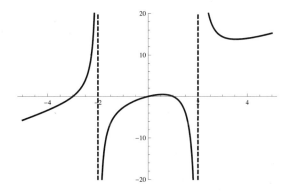

53. $f(x)$ has holes at $x = 3$ and $x = -2$, and a zero at $x = -1$.

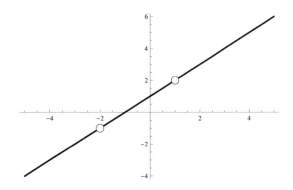

55. $f(x) = \dfrac{2x^3 + 3x^2 - 2x - 3}{x^2 - 2x - 3} = \dfrac{(x+1)(x-1)(2x+3)}{(x+1)(x-3)}$. Thus $f(x)$ has a hole at $x = -1$, roots at $x = 1$ and $x = -\dfrac{3}{2}$, and a vertical asymptote at $x = 3$.

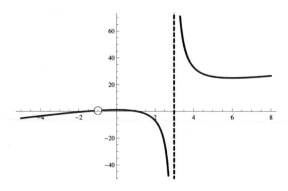

57. This is the same as the graph of f, but with all points below the x axis reflected through the x axis.

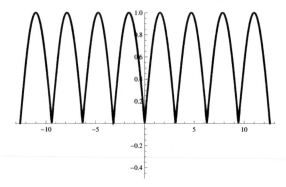

59. This is the graph from Exercise 57 shifted down by 2 units.

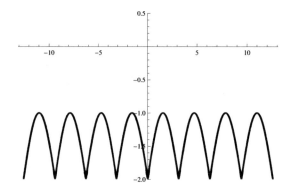

61. For example, $y = \dfrac{2}{x}$.

63. $f(x)$ has zeroes at $x = -2$ and $x = 1$; the zero at $x = 1$ looks like a multiple zero. The entire equation looks like an even power of x, since both ends of the graph point down. So try $f(x) = -(x + 2)(x - 1)^3$.

65. This looks like a parabola with constant term 6. The vertex of the parabola $ax^2 + bx + c$ is at $x = -\dfrac{b}{2a}$, so that $4a = -b$. Thus the parabola is $ax^2 - 4ax + 6$. Since it passes through $(2, 2)$, we must have $a = 1$, so the equation is $y = x^2 - 4x + 6$.

67. This has no roots, but has vertical asymptotes at $x = \pm 2$, so we first try $\dfrac{1}{x^2 - 4}$. This has the right general shape, but is upside-down. Then $\dfrac{1}{4 - x^2}$ is closer, but the y-intercept and the value

at $x = 4$ are wrong. So assume the function is of the form $\frac{a}{4-x^2} + b$, and use the fact that $f(0) = 1$ and $f(4) = -\frac{17}{3}$ to get $a = 20$ and $b = -4$. Finally, we need to insert a hole at $x = 4$. Thus

$$f(x) = \left(\frac{20}{4-x^2} - 4\right)\frac{x-4}{x-4} = \frac{4(x^2+1)(x-4)}{(4-x^2)(x-4)}.$$

69. There are roots at $x = \pm 1$, a y-intercept at $y = 4$, and a horizontal asymptote of 0. This suggests that the numerator is a multiple of $x^2 - 1$, and the denominator is of degree higher than 2 with no zeroes. This suggests $\frac{x^2-1}{x^4+1}$, which satisfies everything except for the y-intercept. Multiply by -4 to correct that and we have $f(x) = \frac{-4(x^2-1)}{x^4+1}$.

71. This looks like the absolute value of a quadratic function. The quadratic must be of the form $f(x) = a(x+1)(x-3)$ given the roots; then $8 = f(1) = a(2)(-2) = -4a$ so that $a = -2$. The function is $g(x) = |-2(x+1)(x-3)| = |2(x+1)(x-3)|$.

73. Written as a piecewise function, we have

$$f(x) = \begin{cases} 5 - 3x, & x \le \frac{5}{3}, \\ 3x - 5, & x > \frac{5}{3}. \end{cases}$$

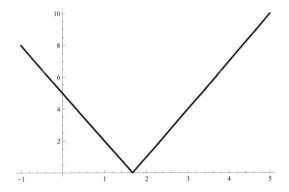

75. Written as a piecewise function, we have

$$f(x) = x^2 + 1.$$

Note that $x^2 + 1$ is always positive, so it is its own absolute value and the "piecewise" function has only one piece.

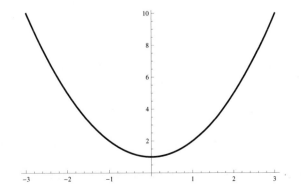

77. Written as a piecewise function, we have

$$f(x) = \begin{cases} x^2 - 9, & x \le -3, \\ 9 - x^2, & -3 < x < 3, \\ x^2 - 9, & x \ge 3. \end{cases}$$

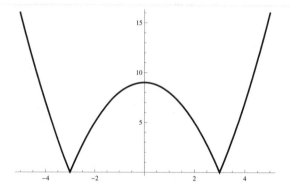

79. Written as a piecewise function, we have

$$f(x) = \begin{cases} x^2 - 3x - 4, & x \le -1, \\ -x^2 + 3x + 4, & -1 < x < 4, \\ x^2 - 3x - 4, & x \ge 4. \end{cases}$$

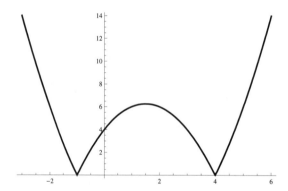

Applications

81. The points, together with a curve that fits the points, are:

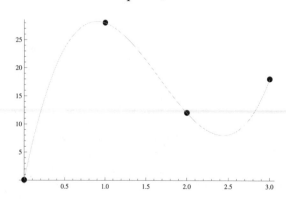

Since the curve that seems to fit these points has a maximum at around $x = 1$ and a minimum at around $x = 2$, it must be at least of degree 3.

83. If the concentration will approach a constant value, then a polynomial is a poor choice of function, since polynomials grow without bound as t grows.

Proofs

85. Suppose $g(x) = Ax^k$ and $f(x) = Bx^l$ are two power functions, where $A, B \in \mathbb{R}$ and k, l are rational numbers. Then

$$(g \circ f)(x) = A(f(x))^k = A(Bx^l)^k = (AB^k)x^{kl}$$

which is Cx^r for $C = AB^k$ and $r = kl$, so the composition is also a power function.

87. Suppose $g(x) = Ax + B$ and $f(x) = Cx + D$ are two linear functions. Then

$$(g \circ f)(x) = Af(x) + B = A(Cx + D) + B = ACx + (AD + B)$$

which is $Ex + F$ for $E = AC$ and $F = AD + B$. Thus $g \circ f$ is also linear.

89. Suppose that $f(x) = a_3x^3 + a_2x^2 + a_1x + a_0$ and $g(x) = b_3x^3 + b_2x^2 + b_1x + b_0$, where $a_3b_3 \neq 0$. Then fg is a sum of terms each of which is a product of one term from f and one term from g, so the highest-degree term in fg is the product of the highest-degree terms in f and g, so it is $a_3b_3x^6$. Since $a_3b_3 \neq 0$, fg is a sixth-degree polynomial.

91. (a) A constant function is simply a linear function with $m = 0$.

    (b) A linear function is $f(x) = mx + b$. If $m \neq 0$, then f is a polynomial of degree 1, while if $m = 0$, then f is a polynomial of degree zero.

93. At any point a, $f(a)$ is the quotient of the two numbers $p(a)$ and $q(a)$. A quotient of two numbers is zero if and only if the numerator is zero and the denominator is not. Thus $f(a) = 0$ precisely when $p(a) = 0$ but $q(a) \neq 0$; that is, the roots of $f(x)$ are the points that are roots of $p(x)$ but not roots of $q(x)$.

Thinking Forward

Algebra for Derivatives

$\triangleright$ $\dfrac{(x+h)^3 - x^3}{h} = \dfrac{x^3 + 3x^2h + 3xh^2 + h^3 - x^3}{h} = \dfrac{3x^2h + 3xh^2 + h^3}{h} = 3x^2 + 3xh + h^2.$

$\triangleright$

$$\begin{aligned} \frac{(x+h)^{1/2} - x^{1/2}}{h} &= \frac{((x+h)^{1/2} - x^{1/2})((x+h)^{1/2} + x^{1/2})}{h((x+h)^{1/2} + x^{1/2})} \\ &= \frac{x + h - x}{h((x+h)^{1/2} + x^{1/2})} \\ &= \frac{h}{h((x+h)^{1/2} + x^{1/2})} \\ &= \frac{1}{(x+h)^{1/2} + x^{1/2}}. \end{aligned}$$

▷

$$\frac{(x+h)^{-2} - x^{-2}}{h} = \frac{\frac{1}{(x+h)^2} - \frac{1}{x^2}}{h}$$

$$= \frac{x^2 - (x+h)^2}{hx^2(x+h)^2}$$

$$= \frac{-2hx - h^2}{hx^2(x+h)^2}$$

$$= -\frac{2x + h}{x^2(x+h)^2}.$$

▷

$$\frac{(x+h)^{-1/2} - x^{-1/2}}{h} = \frac{((x+h)^{-1/2} - x^{-1/2})((x+h)^{-1/2} + x^{-1/2})}{h((x+h)^{-1/2} + x^{-1/2})}$$

$$= \frac{(x+h)^{-1} - x^{-1}}{h((x+h)^{-1/2} + x^{-1/2})}$$

$$= \frac{\frac{1}{x+h} - \frac{1}{x}}{h((x+h)^{-1/2} + x^{-1/2})}$$

$$= \frac{x - (x+h)}{h((x+h)^{-1/2} + x^{-1/2})(x+h)x}$$

$$= -\frac{h}{hx(x+h)((x+h)^{-1/2} + x^{-1/2})}$$

$$= -\frac{1}{x(x+h)((x+h)^{-1/2} + x^{-1/2})}.$$

Alternative algebra for derivatives

▷ $\dfrac{t^3 - x^3}{t - x} = \dfrac{(t-x)(t^2 + tx + x^2)}{t - x} = t^2 + tx + x^2.$

▷ $\dfrac{t^{1/2} - x^{1/2}}{t - x} = \dfrac{(t^{1/2} - x^{1/2})(t^{1/2} + x^{1/2})}{(t - x)(t^{1/2} + x^{1/2})} = \dfrac{t - x}{(t - x)(t^{1/2} + x^{1/2})} = \dfrac{1}{t^{1/2} + x^{1/2}}.$

▷ $\dfrac{t^{-2} - x^{-2}}{t - x} = \dfrac{\frac{1}{t^2} - \frac{1}{x^2}}{t - x} = \dfrac{x^2 - t^2}{(t - x)t^2 x^2} = \dfrac{(x - t)(x + t)}{(t - x)t^2 x^2} = -\dfrac{x + t}{t^2 x^2}.$

▷

$$\frac{t^{-1/2} - x^{-1/2}}{t - x} = \frac{(t^{-1/2} - x^{-1/2})(t^{-1/2} + x^{-1/2})}{(t - x)(t^{-1/2} + x^{-1/2})}$$

$$= \frac{\frac{1}{t} - \frac{1}{x}}{(t - x)(t^{-1/2} + x^{-1/2})}$$

$$= \frac{x - t}{(t - x)(t^{-1/2} + x^{-1/2})tx}$$

$$= -\frac{1}{(t - x)(t^{-1/2} + x^{-1/2})tx}.$$

0.4   Exponential and Trigonometric Functions

Thinking Back

Algebra with Exponents

▷ $3^{2x+1} = 3^1 \cdot 3^{2x} = 3 \cdot 9^x$.

▷ $5^x 2^{3-x} = 5^x 2^3 2^{-x} = 8 \left(\dfrac{5}{2}\right)^x$.

▷ $\left(2^{3x-5}\right)^4 = (2^4)^{3x-5} = 2^{-20} \cdot 2^{12x} = 2^{-20}(2^{12})^x$.

▷ $\dfrac{1}{2(3^{x-4})} = 2^{-1}3^{4-x} = \dfrac{81}{2}3^{-x} = \dfrac{81}{2} \cdot \left(\dfrac{1}{3}\right)^x$.

▷ $\dfrac{4(3^x)^2}{2^x} = 4\dfrac{3^{2x}}{2^x} = 4\left(\dfrac{9}{2}\right)^x$.

▷ $\dfrac{\left(\frac{1}{8}\right)^x}{3(2^{3x+1})} = \dfrac{2^{-3x}}{3 \cdot 2 \cdot 2^{3x}} = \dfrac{1}{6}\dfrac{1}{2^{6x}} = \dfrac{1}{6} \cdot \left(\dfrac{1}{64}\right)^x$.

Inverse functions

▷ If f and g are inverses of each other, then $f(g(x)) = x$ for all x in the domain of g, and $g(f(x)) = x$ for all x in the domain of f. See Definition 0.10.

▷ The graph of $g(x) = f^{-1}(x)$ is the graph of $f(x)$ reflected in the line $y = x$, so that if $f(x)$ has a horizontal asymptote at $y = 0$, then $g(x)$ has a vertical asymptote at $x = 0$.

▷ The graph of $g(x) = f^{-1}(x)$ is the graph of $f(x)$ reflected in the line $y = x$, so that if $f(x)$ has a y-intercept at $y = 1$, then $g(x)$ has an x-intercept at $x = 1$.

Famous triangles, degrees, and radians

▷ Since $\sin 30° = \dfrac{\sqrt{3}}{2}$ and $\cos 30° = \frac{1}{2}$, the legs of the triangle have length $\frac{\sqrt{3}}{2}$ and $\frac{1}{2}$.

▷ The triangle is isosceles, so the other two sides are equal, and the sum of their squares is 1. Thus they must both be equal to $\sqrt{\dfrac{1}{2}} = \dfrac{\sqrt{2}}{2}$.

▷ A radian is $\dfrac{1}{2\pi} \approx \dfrac{1}{6.3}$ of a circle. Since a degree is $\dfrac{1}{360}$ of a circle, a radian is much larger than a degree.

▷

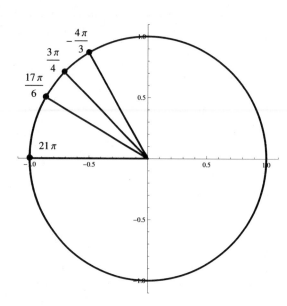

Concepts

1. (a) False. An exponential function is of the form Ab^x; there is no additive constant term.

    (b) True. If $k > 0$, then e^{kx} approaches zero as x gets large negatively, while if $k < 0$, then e^{kx} approaches zero as x gets large positively.

    (c) True. See Theorem 0.24(d).

    (d) True. By Theorem 0.24(h), they are both equal to $\log_3 x$.

    (e) True. The angle $-\dfrac{7\pi}{3}$ is the same as the angle $-\dfrac{\pi}{3} = -60°$, which is in the fourth quadrant.

    (f) False. Its sine is $\pm$ the sine of the reference angle. The reference angle is the angle with the same size, but positioned in the first quadrant.

    (g) True. For any y, $\cos^2 y + \sin^2 y = 1$; substitute $y = 5x^3$ and rearrange terms.

    (h) False. It is true that $\sec x = \dfrac{1}{\cos x}$ whenever $\cos x \neq 0$, but this is not true for the inverse functions. In fact, for $\cos^{-1} x$ to even be defined, we must have $|x| \leq 1$; for $\sec^{-1} x$ to be defined, we must have $|x| \geq 1$.

3. An exponential function is a function of the form Ab^x where A and b are real numbers with $A \neq 0$ and $b > 0, b \neq 1$. A power function is a function of the form Ax^b where A is a nonzero real number and b is rational. Aside from the restrictions on A and b, the two differ in that in an exponential function, the variable is in the exponent, while in a power function, the variable is in the base. x^x is neither a power function nor an exponential function, since neither the base nor the exponent is a constant.

5. Using the sequence of rational approximations for $\sqrt{3}$, we get

r	1.7	1.73	1.732	1.7321	1.73205
2^r	3.24901	3.31728	3.32188	3.32211	3.322

It appears that $2^{\sqrt{3}} \approx 3.322$.

7. We have $3e^{-2x} = 3(e^{-2})^x$; here $A = 3$ and $b = e^{-2}$. For the second part, note that $3 = e^{\ln 3}$, so that $-2(3^x) = -2e^{(\ln 3)x}$.

9. We must have $b > 0$ and $b \neq 1$, since $\log_b x$ is the inverse of the exponential function b^x, and those are precisely the restrictions required on b to make b^x an exponential function.

11. These are the graphs of $y = 2^x$ and $y = 4^x$ reflected in the line $y = x$. Since 4^x grows faster than 2^x, its reflection will grow more slowly. Thus the graph whose value is approximately 1 for $x = 4$ is the graph of $\log_4 x$, and the other is the graph of $\log_2 x$.

13. Note that $\ln(x^2) = 2 \ln x$ and $\ln \dfrac{1}{x} = -\ln x$ by Theorem 0.24 parts (d) and (f).

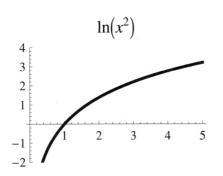

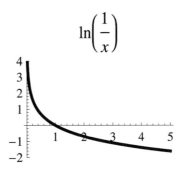

15. Given an angle θ, let θ' be the angle in standard position corresponding to θ, and let (x, y) be the point where the terminal edge of θ' intersects the unit circle. Then $\sin \theta = \sin \theta' = y$.

17. Since $\frac{\pi}{4}$, $\frac{9\pi}{4}$, and $-\frac{7\pi}{4}$ all have the same standard angle, namely $\frac{\pi}{4}$, they all have the same sine.

19. If $y = -\frac{1}{3}$, since $x^2 + y^2 = 1$, we must have $x = \pm\sqrt{\frac{8}{9}} = \frac{2\sqrt{2}}{3}$, so that $\cos \theta = \pm\frac{2\sqrt{2}}{3}$. If the terminal edge of θ is in the third quadrant, then $x < 0$, so that $\cos \theta = -\frac{2\sqrt{2}}{3}$. If the terminal edge of θ is in the fourth quadrant, then $x > 0$, so that $\cos \theta = \frac{2\sqrt{2}}{3}$. The terminal edge of θ cannot be in the first or second quadrants, since $y > 0$ there and we know that $y = -\frac{1}{3}$.

21. $\sin x$ is invertible on $\left[-\dfrac{\pi}{2}, \dfrac{\pi}{2}\right]$, $\tan x$ is invertible on $\left(-\dfrac{\pi}{2}, \dfrac{\pi}{2}\right)$, and $\sec x$ is invertible on $\left[0, \dfrac{\pi}{2}\right) \cup \left(\dfrac{\pi}{2}, \pi\right]$. The domain for arcsin is clearly the range of sin, which is $[-1, 1]$, and the range of arcsin is then the same as the restricted domain for sin. The domain for arctan is $(-\infty, \infty)$, and its range is the same as the restricted range for tan. Finally, the domain of arcsec is $(-\infty, -1] \cup [1, \infty)$, and its range is the restricted domain for sec.

23. Only (a) and (c) are defined. The domain of $\sin^{-1}$ is $[-1, 1]$, and the domain of $\tan^{-1}$ is $(-\infty, \infty)$, so both (a) and (c) are defined. (b) is not defined again since the domain of $\sin^{-1}$ is $[-1, 1]$. Finally, (d) is not defined since the domain of $\sec^{-1}$ is $(-\infty, -1] \cup [1, \infty)$, and $0 < \frac{\pi}{4} < 1$.

25. (a) $\sin \theta$ is negative on its restricted domain for $\theta \in \left[-\frac{\pi}{2}, 0\right)$, so that the value of this expression is negative.

   (b) For the same reason as in part (a), the value of this expression is negative.

   (c) $\tan \theta$ is positive on its restricted domain for $\theta \in \left(0, \frac{\pi}{2}\right)$, so the value of this expression is positive.

   (d) Since the restricted domain for secant is $\left[0, \dfrac{\pi}{2}\right) \cup \left(\dfrac{\pi}{2}, \pi\right]$, the value of this expression is positive.

Skills

27. $f(x)$ is defined where the numerator or denominator are both defined and the denominator is nonzero. The numerator is defined for $x + 1 > 0$, i.e., for $x > -1$. The denominator is defined for $x - 2 > 0$, i.e. for $x > 2$, and is nonzero there except at $x = 3$. Thus the domain of $f(x)$ is the intersection of these two ranges, which is $(2, 3) \cup (3, \infty)$.

29. $f(x)$ is defined where the denominator is defined and nonzero, i.e. when $\ln(x - 1) > 0$. But $\ln(x - 1) > 0$ for $x > 2$. Thus the domain of $f(x)$ is $(2, \infty)$.

31. $f(\theta)$ is defined where $\sec \theta \geq 0$, i.e. where $\cos \theta > 0$. But $\cos \theta > 0$ for θ in the first or fourth quadrants, i.e. for $\theta \in \left(-\dfrac{\pi}{2} + 2n\pi, \dfrac{\pi}{2} + 2n\pi \right)$ where n is an integer.

33. This is the power to which we must raise e to get e^{-2}, which is -2.

35. We have $4 \log_2 6 - 2 \log_2 9 = \log_2 6^4 - \log_2 9^2 = \log_2 \frac{36^2}{9^2} = \log_2 4^2 = 2 \log_2 4 = 4$.

37. Since $\cos \left(-\frac{\pi}{4} \right) = \frac{\sqrt{2}}{2}$ and $\sin \left(-\frac{\pi}{4} \right) = -\frac{\sqrt{2}}{2}$, we have $\tan \left(-\frac{\pi}{4} \right) = -1$.

39. $\csc \left(-\dfrac{5\pi}{4} \right) = \csc \left(\dfrac{3\pi}{4} \right) = \dfrac{1}{\sin \left(\frac{3\pi}{4} \right)} = \dfrac{2}{\sqrt{2}} = \sqrt{2}$.

41. $\cos^{-1}(-1)$ is the angle in $[0, \pi]$ whose cosine is -1; that angle is π.

43. $\operatorname{arcsec} \left(-\dfrac{2}{\sqrt{2}} \right) = \arccos \left(-\dfrac{\sqrt{2}}{2} \right) = \dfrac{3\pi}{4}$.

45. Take natural logs to get $x \ln 2 = (x - 1) \ln 3$; collect terms and simplify to get $x = -\frac{\ln 3}{\ln 2 - \ln 3} \approx 2.70951$.

47. Raise 2 to both sides of the equation to get $\dfrac{x - 1}{x + 1} = 16$. Clearing fractions gives $x - 1 = 16x + 16$, so that $x = -\dfrac{17}{15}$.

49. $\cos 2x = 1$ means that $2x$ is an integral multiple of 2π, so that x is an integral multiple of π. The solutions are $x = n\pi$ for n an integer.

51. $\sin \theta = \sqrt{1 - \cos^2 \theta} = \sqrt{1 - \dfrac{1}{36}} = \dfrac{\sqrt{35}}{6}$.

53. $\cos(2\theta) = \cos^2 \theta - \sin^2 \theta = \dfrac{1}{36} - \dfrac{35}{36} = -\dfrac{17}{18}$, using the results of Exercise 51.

55. Start with $\cos(\theta + \phi) = \cos \theta \cos \phi - \sin \theta \sin \phi$. Now, $\cos \theta > 0$ and $\cos \phi < 0$, so the first term is negative. Also, $\sin \theta > 0$ and $\sin \phi > 0$, so the second term is positive. Thus their difference is negative, so the sign of $\cos(\theta + \phi)$ is $-$.

57. If $\theta = \cos^{-1} x$, then $x = \cos \theta$; since $\sin^2 \theta + \cos^2 \theta = 1$, it follows that $\sin(\cos^{-1} x) = \sin \theta = \pm\sqrt{1 - x^2}$. Since the range of $\cos^{-1}$ is $[0, \pi]$, we see that $\sin(\cos^{-1} x)$ is always positive, so we choose the positive square root to get $\sqrt{1 - x^2}$.

59. In the triangle below, $\tan \theta = x$, so $\theta = \tan^{-1} x$. Clearly $\sec \theta = \sqrt{x^2 + 1}$, so $\sec^2(\tan^{-1} x) = x^2 + 1$.

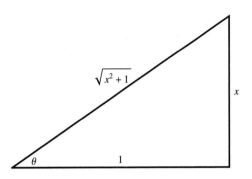

61. In the triangle below, $\sec \theta = \dfrac{x}{3}$, so that $\theta = \sec^{-1} \dfrac{x}{3}$. Then it is easy to see that $\sin\left(\sec^{-1} \dfrac{x}{3}\right) = \sin \theta = \dfrac{\sqrt{x^2 - 9}}{x}$.

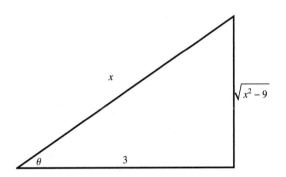

63. We have $\cos(2 \sin^{-1}(5x)) = 1 - 2 \sin^2(\sin^{-1}(5x)) = 1 - 2 \cdot (5x)^2 = 1 - 50x^2$.

65. We have $f(x) = -2^{-x} + 10$, so this is the graph of 2^x reflected in the y axis and in the x axis, and then shifted up by 10 units.

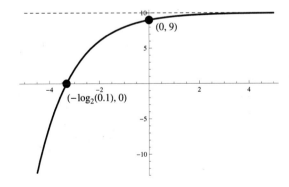

67. This is the graph of e^x compressed horizontally by a factor of 2, stretched vertically by a factor of 5, reflected in the x axis and finally shifted up by 20 units.

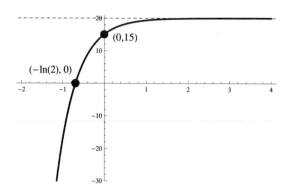

69. This is the graph of $\log_2 x$ shifted right by 3 units and reflected in the x axis.

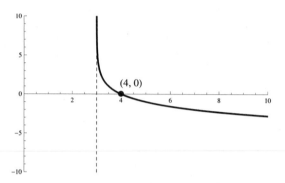

71. This is the graph of $\cos x$ shifted right by $\dfrac{\pi}{4}$ and stretched vertically by a factor of 2.

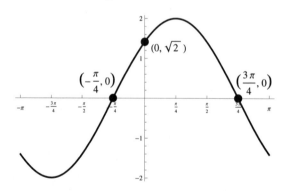

73. This is exponential decay. Moving the graph up by three units to get the asymptote at $y = 0$ gives a y-intercept at 2, so this could be $2e^{-x} - 3$.

75. This looks like the negation of an exponential decay curve, and the y-intercept is 5 units below the asymptote, so this could be $-5e^{-x} + 10$.

77. If negated, this looks like a compressed cosine curve, that reaches its minimum, -1, at $\pi/2$ instead of π, so the negation is $\cos(2x)$ and the original curve is $-\cos(2x)$.

Applications

79. (a) From the given data, we know that $I(0) = Ae^{k \cdot 0} = A = 10000$, so that $I(t) = 10000e^{kt}$. Also $I(10) = 10000e^{k \cdot 10} = 22609.80$ and thus $e^{k \cdot 10} = 0.226098$. Take logs and simplify to get $10k = \ln 2.26098 \approx 0.815798$, so that $k \approx 0.08158$. Thus $I(t) = 10000e^{0.08158t}$.

(b) Since $e^{0.08158} \approx 1.085$, we have $I(t) = 10000 \cdot 1.085^t$.

81. (a) From $S(0) = 250$ we get $A = 250$, so that $S(t) = 250e^{kt}$. Now, $S(29) = 125 = 250e^{29k}$, so that $e^{29k} = 0.5$ and $k = \frac{\ln 0.5}{29} \approx -0.0239$, so that $S(t) = 250e^{-0.0239t}$.

    (b) $S(1) = e^{-0.0239} \approx 0.976$, so after one year, 97.6% of the original amount will remain. Thus 2.4% decays each year.

    (c) Solve $6 = 250e^{-0.0239t}$ to get $t = \frac{\ln(6/250)}{-0.0239} \approx 156$, so after 156 years.

83. Alina, the kite, and the point on the ground directly under the kite form a right triangle with hypotenuse 400. The ratio between the height of the kite and 400 is $\sin 32° \approx 0.53$, so the height of the kite is $400 \cdot 0.53 \approx 212$ feet.

Proofs

85. Suppose $f(c) = Ab^c = 0$ for some c. Since $A \neq 0$ we must have $b^c = 0$. Since $b \neq 0$, we cannot have $c = 0$, else $b^c = b^0 = 1 \neq 0$. But then $0 = 0^{\frac{1}{c}} = (b^c)^{\frac{1}{c}} = b^1 = b$ and again $b = 0$. This is a contradiction, so $f(x) = Ab^x$ is never zero.

87. The base conversion formula tells us that $\log_3 x = \frac{\log_2 x}{\log_2 3}$ so that $\log_2 x = \log_2 3 \cdot \log_3 x$. Assume that $\log_2 x = \log_3 x$ and that their common value is nonzero. Then dividing through by that value gives $1 = \log_2 3$. But $\log_2$ is one-to-one since it is the inverse of an exponential function. Since $\log_2 2 = 1$, we get the contradiction that $2 = 3$. Thus $\log_2 x = \log_3 x$ only if both are zero, which occurs only when $x = 1$.

89. $\log_b x$ is the power to which you must raise b to get x. Thus $\log_b 1$ is the power to which you must raise b to get 1. Clearly 0 is that power, so that $\log_b 1 = 0$.

91. We have by Exercise 90(c) and (b) that $\log_b(x^a) = \log_b\left((b^{\log_b x})^a\right) = \log_b\left(b^{a \log_b x}\right) = a \log_b x$.

93. (a) By Exercise 92, $\log_b\left(\frac{1}{x}\right) + \log_b x = \log_b\left(x \cdot \frac{1}{x}\right) = 0$, so that $\log_b\left(\frac{1}{x}\right) = -\log_b x$.

    (b) We have, using Exercise 92 and part (a) of this exercise,

    $$\log_b\left(\frac{x}{y}\right) = \log_b\left(x \cdot \frac{1}{y}\right) = \log_b x + \log_b\left(\frac{1}{y}\right) = \log_b x - \log_b y.$$

95. If θ is an angle with (x, y) the point where the terminal edge of θ intersects the unit circle, then $x = \cos\theta$ and $y = \sin\theta$. But since (x, y) lies on the unit circle, which has equation $x^2 + y^2 = 1$, we have $1 = x^2 + y^2 = \cos^2\theta + \sin^2\theta$.

97. Suppose θ is an angle with (x, y) the point where the terminal edge of θ intersects the unit circle. $-\theta$ is the same angle measured in the other direction (clockwise); thinking about the geometry shows that the terminal edge of $-\theta$ intersects the unit circle at $(-x, y)$. Then $\sin(-\theta) = -x = -\sin(\theta)$, $\cos(-\theta) = y = \cos(\theta)$, and $\tan(-\theta) = \frac{y}{-x} = -\frac{y}{x} = -\tan(\theta)$. For the shift identities, note that if (x, y) lies on the unit circle, then rotating the point by $\frac{\pi}{2}$ gives the point $(-y, x)$. Now, let θ be any angle, and (x, y) the point where its terminal edge intersects the unit circle. Then $\cos\theta = x$, and since $\theta + \frac{\pi}{2}$ intersects the unit circle at $(-y, x)$, it follows that $\sin\left(\theta + \frac{\pi}{2}\right) = x$ as well. This proves the second shift identity. For the first shift identity, note that (using the second shift identity)

    $$\cos\left(\theta - \frac{\pi}{2}\right) = \sin\left(\left(\theta - \frac{\pi}{2}\right) + \frac{\pi}{2}\right) = \sin\theta$$

    The final two shift identities follow since the terminal edge of $\theta + 2\pi$ coincides with the terminal edge of θ.

99. We have

$$\sin(2\theta) = \sin(\theta + \theta) = \sin\theta\cos\theta + \sin\theta\cos\theta = 2\sin\theta\cos\theta$$
$$\cos(2\theta) = \cos\theta\cos\theta - \sin\theta\sin\theta = \cos^2\theta - \sin^2\theta$$

Thinking Forward

▷ For $h = 0.1$, $h = 0.01$, and $h = 0.001$, the values of $\dfrac{e^h - 1}{h}$ are approximately 1.05171, 1.00502, and 1.0005. They appear to be approaching 1.

▷

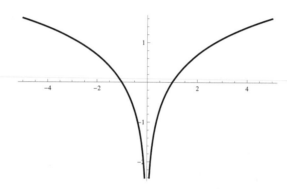

The domain of the function is $\mathbb{R} - \{0\}$; it is an even function since $f(-x) = \ln|-x| = \ln|x| = f(x)$ where f is defined.

▷ Start with

$$\sin^4 x\cos^2 x = \sin^2 x(\sin^2 x\cos^2 x)$$
$$= \frac{1 - \cos 2x}{2}\left(\frac{1}{2}\sin 2x\right)^2$$
$$= \frac{1}{16}(1 - \cos 2x)(1 - \cos 4x)$$
$$= \frac{1}{16}(1 - \cos 2x - \cos 4x - \cos 2x\cos 4x)$$

Now use the formulas

$$\cos 6x = \cos 2x\cos 4x - \sin 2x\sin 4x \quad \text{and} \quad \cos 2x = \cos(4x - 2x) = \cos 2x\cos 4x + \sin 2x\sin 4x:$$

$$\sin^4 x \cos^2 x = \frac{1}{16}(1 - \cos 2x - \cos 4x - \cos 2x \cos 4x)$$

$$= \frac{1}{16}\left(1 - \frac{1}{2}\cos 2x - \frac{1}{2}\cos 2x - \cos 4x - \cos 2x \cos 4x\right)$$

$$= \frac{1}{16}\left(1 - \frac{1}{2}\cos 2x - \frac{1}{2}\cos(4x - 2x) - \cos 4x - \cos 2x \cos 4x\right)$$

$$= \frac{1}{16}\left(1 - \frac{1}{2}\cos 2x - \frac{1}{2}(\cos 2x \cos 4x + \sin 2x \sin 4x) - \cos 4x - \cos 2x \cos 4x\right)$$

$$= \frac{1}{16}\left(1 - \frac{1}{2}\cos 2x + \frac{1}{2}(\cos 2x \cos 4x - \sin 2x \sin 4x) - \cos 4x\right)$$

$$= \frac{1}{16}\left(1 - \frac{1}{2}\cos 2x + \frac{1}{2}\cos 6x - \cos 4x\right)$$

$$= \frac{1}{16} - \frac{1}{32}\cos 2x - \frac{1}{16}\cos 4x + \frac{1}{32}\cos 6x$$

0.5   Logic and Mathematical Thinking

Thinking Back

Basic definitions

▷ A *nonnegative* number x is a number such that $x \geq 0$.

▷ An *integer* is a member of the set $\{\ldots, -2, -1, 0, 1, 2, \ldots\}$.

▷ A *rational number* is a number that can be written as $\frac{a}{b}$ where a and b are integers with $b \neq 0$.

▷ An *irrational number* is a real number that is not rational.

▷ $|x|$ is the absolute value, or magnitude, of x. It is equal to x for $x \geq 0$, and to $-x$ for $x < 0$.

▷ dist(a, b) is the distance from a to b along the real line; it is equal to $|b - a|$.

Types of numbers:

All integers are rational since any integer n may be written as $\frac{n}{1}$. But not all rational numbers are integers; for example, $\frac{1}{2}$. Yes, 0 is a rational number, since it is an integer.

Inequality opposites:

If $x > 9$ is false, then $x \leq 9$. This does not necessarily mean that $x < 9$, since x could equal 9.

Concepts

1. (a) False. You must show P holds for *every* possible value of x.

    (b) True. Since the statement says that P holds for every value of x, exhibiting just one value of x where it does not is sufficient to disprove it.

    (c) True. That is what the statement says.

    (d) False. The statement only says that P needs to hold for some value of x. If you exhibit some value of x for which it does not hold, there may yet be another value of x for which it does.

    (e) True. The converse of $A \Rightarrow B$ is $B \Rightarrow A$.

(f) True, since the implication says that B is true if A is.

(g) False. The implication only says that *if* A is true, then B is. It says nothing about what must happen if A is false.

(h) False. See previous point.

3. If C is true, then D must be true, since "$C \Rightarrow D$" says precisely that. If C is false, we have no information about D.

5. The statement can be written "For all positive real numbers x, we have $x > -2$," or as "If $x > 0$ is a real number, then $x > -2$."

7. Yes, the statement is true, since a square has four sides and four right angles. The converse, "Every rectangle is a square", is false — any rectangle in which the sides are unequal is a counterexample. The contrapositive is "Every shape that is not a rectangle is not a square" is true — if a shape is not a rectangle, then either it does not have four sides, or it does not have four right angles. In either case, it cannot be a square.

9. The contrapositive of $P \Rightarrow Q$ is $(\text{Not } Q) \Rightarrow (\text{Not } P)$. These statements are equivalent. For, suppose $P \Rightarrow Q$ is true and Not Q holds. Then Q is false, so since $P \Rightarrow Q$, we must have P false as well. That is, Not P is true. Thus the second implication holds. The reverse argument is similar.

11. If $0 < |x - 3| < 0.1$, then $0 < |x - 3| < 0.1 < 0.2$, so that $|f(x) + 5| < 0.2$. However, there is no implication then that it is less than 0.1. Perhaps $|x - 3|$ is always 0.15, for example.

13. If $0 < |x - 3| < 0.05$, then $0 < |x - 3| < 0.05 < 0.1$, so that $|f(x) + 5| < 0.2$, so the statement holds.

15. If $0 < |x - 3| < 0.05 < 0.1$, then $|f(x) + 5| < 0.2 < 0.4$, so the statement is true.

Skills

17. True; for example, 2.5 is such a number.

19. True, since every rational number can be written as the quotient of two integers and no irrational number can be so written.

21. True. Simply square x.

23. True, since the integers are $\{\dots, -2, -1, 0, 1, 2, 3, \dots\}$.

25. False. The concept of even and odd applies only to integers.

27. False. Since $2 > 1$ any number that is ≤ 1 must be < 2.

29. False. For example, -2.

31. True. Any number not in the interval $(1, 2)$ is either at most 1 or at least 2.

33. True, since multiplying an inequality by a positive number preserves the inequality, as does adding something to both sides. Thus $a < b \Longrightarrow 3a < 3b \Longrightarrow 3a + 1 < 3b + 1$.

35. True. $|x| \geq 0$ by definition, and $x^2 \geq 0$ since the product of two positive numbers, two negative numbers, or $0 \cdot 0$, is always nonnegative.

37. False, since if $x, y < 0$ their product is positive.

39. True, since multiplying an inequality by a positive number preserves the inequality, as does adding something to both sides. Thus $x < y \Longrightarrow 2x < 2y \Longrightarrow 2x - 1 < 2y - 1$.

41. True. Since all absolute values are nonnegative, any negative number works for x.

43. True. For example, choose $y = x + 1$.

45. False. If $y = 0$, then $\dfrac{x}{y}$ is not even defined. If we assume $y \neq 0$, then the statement is true, since we can multiply both sides of $\dfrac{x}{y} = 0$ by y to get $x = 0$, and divide by y for the reverse implication.

47. The converse is $B \Rightarrow (\text{Not } A)$. The contrapositive is $(\text{Not } B) \Rightarrow (\text{Not Not } A)$, which is the same as $(\text{Not } B) \Rightarrow A$.

49. The converse is $(\text{Not } A) \Rightarrow (\text{Not } B)$. The contrapositive is $(\text{Not Not } A) \Rightarrow (\text{Not Not } B)$, which is the same as $A \Rightarrow B$.

51. The converse is $C \Rightarrow (A \text{ and } B)$. The contrapositive is $(\text{Not } C) \Rightarrow (\text{Not } (A \text{ and } B))$, which is the same as $(\text{Not } C) \Rightarrow ((\text{Not } A) \text{ or } (\text{Not } B))$.

53. The converse is $(B \text{ and } C) \Rightarrow A$. The contrapositive is $(\text{Not } (B \text{ and } C) \Rightarrow (\text{Not } A)$, which is the same as $((\text{Not } B) \text{ or } (\text{Not } C)) \Rightarrow (\text{Not } A)$.

55. The converse is "If x is a rational number, then x is real." The contrapositive is "If x is an irrational number, then x is not real.". The converse is true, but the original statement and the contrapositive are false. $x = \sqrt{2}$ provides one counterexample.

57. The converse is "If $x \geq 3$, then $x > 2$." The contrapositive is "If $x < 3$, then $x \leq 2$." The converse is true, but the original statement and the contrapositive are false. $x = 2.5$ provides one counterexample.

59. The converse is "If $\sqrt{x}$ is not a real number, then x is negative." The contrapositive is "If $\sqrt{x}$ is a real number, then x is nonnegative." All statements are true.

61. The converse is "If $|x| = -x$, then $x \leq 0$," and the contrapositive is "If $|x| \neq -x$, then $x > 0$." All statements are true.

63. The converse is "If $x^2 > x$, then $x \neq 0$," and the contrapositive is "If $x^2 \leq x$, then $x = 0$." The converse is true, but the original statement and the contrapositive are both false. $x = \frac{1}{2}$ provides one counterexample.

65. The converse is "If there is some integer n such that $x = 2n + 1$, then x is odd." The contrapositive is "If there is no integer n such that $x = 2n + 1$, then x is not odd." All statements are true.

67. Number the statements (1) through (5). From statements (1) and (4), the only person who can wear the red hat is Linda. From statement (5), Stuart does not wear the blue hat; from statements (2) and (5), he does not wear the green hat. Thus he wears the yellow hat. Finally, from statements (2) and (3), Alina cannot wear the green hat, so she wears the blue hat and Phil, the oldest, wears the green hat.

69. Rein cannot be telling the truth, since if he were, then Liz would be lying and so would Zubin, so in fact only one person would be telling the truth. Thus Rein is lying. Since at least one person (Rein) is lying, Liz is clearly lying as well. Finally, that means that Zubin is telling the truth. Hence Zubin is the only truth-teller.

71. If Hyun is a truth-teller, then by his statement, Kate is as well, which means that Hyun is lying, a contradiction. Thus Hyun is lying, so that Kate must be a liar as well. But Kate said that both Hyun and Jaan were liars; since she is lying, at least one of them must be a truth-teller. It is not Hyun, whom we know to be a liar, so Jaan must be the only truth-teller.

73. Since r is rational, write $r = \dfrac{a}{b}$ where a and b are integers with b nonzero. Now suppose $x - r$ is rational, say $x - r = \dfrac{c}{d}$ with c and d integers and $d \neq 0$. Then

$$x = r + \frac{c}{d} = \frac{a}{b} + \frac{c}{d} = \frac{ad + bc}{bd}$$

Now, $ad + bc$ and bd are integers, and $bd \neq 0$ since neither b nor d is zero. Thus we conclude that x is rational, a contradiction. Thus $x - r$ is irrational.

75. Suppose $x = 2n$ is even and $y = 2m + 1$ is odd, where m and n are integers. Then $x + y = 2n + 2m + 1 = 2(n + m) + 1$, so that $x + y$ is odd.

77. Suppose $x = 2n$ is even and $y = 2m + 1$ is odd, where m and n are integers. Then $xy = 2n(2m + 1) = 4mn + 2n = 2(2mn + n)$, so that xy is even.

79. The square of the distance from the midpoint to P is, by Exercise 78,

$$\left(\frac{y_1 + y_2}{2} - y_1\right)^2 + \left(\frac{x_1 + x_2}{2} - x_1\right)^2 = \left(\frac{y_1 + y_2 - 2y_1}{2}\right)^2 + \left(\frac{x_1 + x_2 - 2x_1}{2}\right)^2$$
$$= \left(\frac{y_2 - y_1}{2}\right)^2 + \left(\frac{x_2 - x_1}{2}\right)^2$$
$$= \frac{1}{4}\left((y_2 - y_1)^2 + (x_2 - x_1)^2\right).$$

Taking square roots shows that the distance from the midpoint to P is $\dfrac{1}{2}\sqrt{(y_2 - y_1)^2 + (x_2 - x_1)^2}$, which is half the distance from P to Q.

81. $\dfrac{\left(\frac{a}{b}\right)}{c} = \dfrac{\left(\frac{a}{b}\right)}{\frac{c}{1}} = \dfrac{a \cdot 1}{bc} = \dfrac{a}{bc}.$

83. For any real number x, $\sqrt{x^2}$ is the positive square root of x^2. If x is positive, this is x itself, while if x is negative, this is $-x$. Thus $\sqrt{x^2} = |x|$. Now,

$$(a + b)^2 = a^2 + 2ab + b^2 \leq |a|^2 + 2|a||b| + |b|^2 = (|a| + |b|)^2$$

since $a \leq |a|$ and $b \leq |b|$. Now, $\sqrt{(a + b)^2} = |a + b|$ by the above, so taking square roots of both sides of the above equation gives $|a + b| \leq |a| + |b|$.

85. Write $x = x - y + y = (x - y) + y$; then

$$|x| = |(x - y) + y| \leq |x - y| + |y|.$$

Rearrange the terms to get $|x - y| \geq |x| - |y|$.

87. $|x - c| > \delta \iff -\delta > x - c$ or $\delta < x - c \iff x < c - \delta$ or $x > c + \delta \iff x \in (-\infty, c - \delta) \cup (c + \delta, \infty)$.

Thinking Forward

Quantified statements about distance

▷ We want $6 < 3x + 1 < 8$, so $5 < 3x < 7$, so $\dfrac{5}{3} < x < \dfrac{7}{3}$. So if $\delta = \dfrac{1}{3}$, then if $x \in (2 - \delta, 2 + \delta) = \left(\dfrac{5}{3}, \dfrac{7}{3}\right)$, the required condition is satisfied.

▷ We want $-0.3 < 3x - 6 < 0.3$, so $5.7 < 3x < 6.3$, so $1.9 < x < 2.1$. Choose $\delta = 0.1$.

▷ We want $6 - \epsilon < 2x < 6 + \epsilon$, so $3 - \dfrac{\epsilon}{2} < x < 3 + \dfrac{\epsilon}{2}$. Choose $\delta = \dfrac{\epsilon}{2}$.

Negating quantified statements about distance

▷ There exists some $M > 0$ such that for all $N > 0$ there exists an $x > N$ with $x^2 \leq M$.

▷ There exists some $\epsilon > 0$ such that for all $\delta > 0$ there exists an x with $0 < |x - 2| < \delta$ but $|x^2 - 4| \geq \epsilon$.

▷ There exists some $\epsilon > 0$ such that for all $\delta > 0$ there exists an x with $0 < |x - 4| < \delta$ but $|\sqrt{x} - 2| \geq \epsilon$.

Chapter Review and Self-Test

Skill Certification — Algebra and Functions

1. $\dfrac{x^3 - 2^3}{x - 2} = \dfrac{(x-2)(x^2 + x + 2^2)}{x - 2} = x^2 + x + 4$, assuming $x \neq 2$.

3. Note that 2 is a root of the denominator; assuming $x \neq 2$ and then dividing by $x - 2$ gives $x^2 + x - 2 = (x+2)(x-1)$. Thus $\dfrac{x - 2}{x^3 - x^2 - 4x + 4} = \dfrac{1}{(x+2)(x-1)}$.

5. Since $x^2 + 1 \geq 0$, we see that $-2(x^2 + 1) \leq 0$, so that $\left|-2(x^2 + 1)\right| = 2(x^2 + 1)$.

7. $\dfrac{x^{1/4} + x^{1/3}}{x^2} = (x^{1/4} + x^{1/3})x^{-2} = x^{-7/4} + x^{-5/3}$.

9. $e^{2\ln x} = (e^{\ln x})^2 = x^2$.

11. $\tan\left(\dfrac{\pi}{3}\right) + \tan\left(\dfrac{\pi}{4}\right) = \sqrt{3} + 1$.

13. $2x^2 - 7x + 3 = (x - 3)(2x - 1) > 0$, so it can change sign only at the points where it is zero, which is at $x = 3$ and $x = \frac{1}{2}$. So to determine where it is positive, it suffices to check its value at one point in each of the resulting three intervals:

interval	$\left(-\infty, \frac{1}{2}\right)$	$\left(\frac{1}{2}, 3\right)$	$(3, \infty)$
x	0	1	4
$2x^2 - 7x + 3$	3	-2	7
sign	$+$	$-$	$+$

Thus the solution set is $\left(-\infty, \dfrac{1}{2}\right) \cup (3, \infty)$.

15. $\dfrac{3}{x - 2} < 1$ means $\dfrac{3}{x - 2} - 1 = \dfrac{5 - x}{x - 2} < 0$. To determine the places where this inequality holds, note that $\dfrac{5 - x}{x - 2}$ can change sign only at those places where either the numerator or denominator

is zero, i.e., at $x = 2$ and $x = 5$. So to determine where it is negative, it suffices to check its value at one point in each of the resulting three intervals:

interval	$(-\infty, 2)$	$(2, 5)$	$(5, \infty)$
x	0	3	6
$\frac{5-x}{x-2}$	$-\frac{5}{2}$	2	$-\frac{1}{4}$
sign	$-$	$+$	$-$

Thus the fraction is negative for $x \in (-\infty, 2) \cup (5, \infty)$, which is the solution set of the original inequality.

17. The solution set is the set of points in $\mathbb{R}$ not satisfying $|5x - 2| \leq 1$, which is the set $-1 \leq 5x - 2 \leq 1$, or $1 \leq 5x \leq 3$, or $\frac{1}{5} \leq x \leq \frac{3}{5}$. Thus the solution set to the original inequality is $\left(-\infty, \frac{1}{5}\right) \cup \left(\frac{3}{5}, \infty\right)$.

19. Factoring, $f(x) = \dfrac{(x-1)(2x-3)}{x}$. $f(x)$ is undefined at $x = 0$, since the denominator vanishes, and it is zero for $x = 1$ and $x = \frac{3}{2}$, since the numerator is zero but the denominator is not.

21. $f(x)$ is defined everywhere. It is zero when $|x - 2| = 5$, so when $x - 2 = 5$ or $2 - x = 5$. Thus the zeroes are at $x = 7$ and at $x = -3$.

23. $f(x)$ is undefined at $x = 1$, where the denominator vanishes. In addition, it has zeroes where $2(x-1)^2 - 4 = 0$, i.e., when $(x-1)^2 = 2$, or $x = 1 \pm \sqrt{2}$.

25. $f(x)$ is defined everywhere, since exponentials are. Since e^x is never zero, $f(x) = 0$ only when $1 - 2e^x = 0$, which occurs when $e^x = \frac{1}{2}$, so that $x = \ln\frac{1}{2} = -\ln 2$.

27. The denominator vanishes at $x = 0$, so $f(x)$ is undefined there. Note that $x = 0$ is also a root of the numerator. Zeros of $f(x)$ occur when the numerator vanishes but the denominator does not. The numerator vanishes for x any integer, so the zeros of $f(x)$ are all integers except 0.

29. The numerator never vanishes, so the function has no zeroes. $f(x)$ is undefined when $\sin^{-1} x = 0$. Since the domain of $\sin^{-1}$ is $\left(-\frac{\pi}{2}, \frac{\pi}{2}\right)$, the only possible value of x is $x = 0$. Thus $f(x)$ is undefined at $x = 0$.

31. $f(x)$ is defined when $\sqrt{x+2}$ is defined and where the denominator does not vanish. $\sqrt{x+2}$ is defined for $x \geq -2$, and the denominator is zero only at $x = 0$. Thus the domain of $f(x)$ is $[-2, 0) \cup (0, \infty)$.

33. Factoring gives $f(x) = \dfrac{x+2}{(x-3)(x+2)}$. $f(x)$ is defined everywhere except at $x = 3$ and $x = -2$; since $x = -2$ is a root of the numerator as well, its graph has a hole there.

35. $f(x)$ is defined when both $\sqrt{x}$ and $\sqrt{2-x}$ are defined, i.e., when $x \geq 0$ and $2 - x \geq 0$. The second inequality is equivalent to $x \leq 2$. Thus the domain of $f(x)$ is $[0, 2]$.

37. Rewriting gives $f(x) = e^{\frac{1}{2}x}$. Since exponentials are everywhere defined, so is $f(x)$.

39. The function is defined everywhere except when $\sin x = \frac{1}{2}$; this happens for $x = \dfrac{\pi}{6} + n\pi$ and for $x = \dfrac{5\pi}{6} + n\pi$, where n is an integer. The domain is thus $\mathbb{R} - \left\{ \frac{\pi}{6} + n\pi \mid n \text{ an integer} \right\} - \left\{ \frac{5\pi}{6} + n\pi \mid n \text{ an integer} \right\}$.

41. $f(x)$ is a linear function with y intercept $(0,3)$ and slope -2. $f(x)$ has a zero at $x = \frac{3}{2}$.

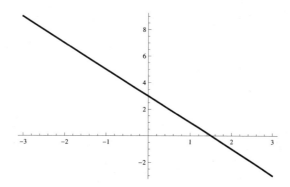

43. This is the graph of $\sqrt{x}$ shifted right by 3 units. $f(x)$ is defined for $x \geq 3$, and has a zero at $x = 3$.

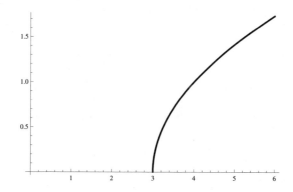

45. This is the graph of $y = x^3$ stretched vertically by a factor of 2 and shifted down by 1 unit. $f(x)$ is defined everywhere, and has a zero at $x = \dfrac{1}{\sqrt[3]{2}}$.

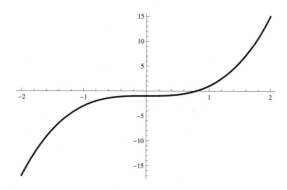

47. $f(x)$ is defined except at $x = 0$, where it has a vertical asymptote, and has a zero where $\dfrac{1}{x} = 2$, which is for $x = \frac{1}{2}$. Finally, since $f(x) = \dfrac{1}{x} - 2 = \dfrac{1 - 2x}{x}$, it has a horizontal asymptote at $y = -2$.

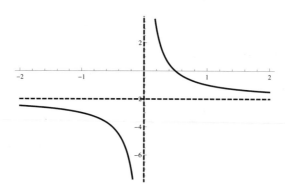

49. Since $|4x - 3| = 4\left|x - \dfrac{3}{4}\right|$, we see that this is the graph of $|x|$ shifted right by $\frac{3}{4}$ and stretched vertically by a factor of 4. $f(x)$ is defined everywhere, and has a zero where $4x - 3 = 0$, which is for $x = \dfrac{3}{4}$.

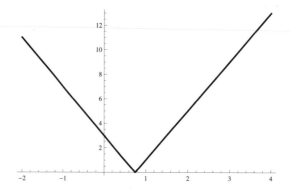

51. This is the graph of $x^{1/3}$ stretched vertically by a factor of 2 and then reflected through the x axis. $f(x)$ is defined everywhere, since $x^{1/3}$ is; it has a zero at $x = 0$.

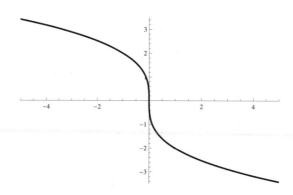

53. $f(x)$ is defined everywhere, and has zeroes at $x = 0$ and at $x = -1$.

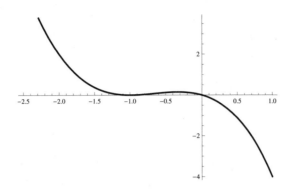

55. $f(x)$ is defined everywhere except at $x = 1$. Since 1 is a root of the denominator with multiplicity 2, but only a root of the numerator with multiplicity 1, $f(x)$ has a vertical asymptote at $x = 1$. $f(x)$ also has a horizontal asymptote at $y = 1$, and a zero at $x = -2$.

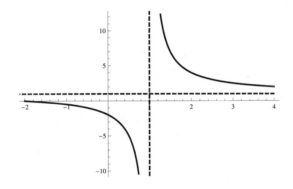

57. $f(x)$ has no zeroes. It has a horizontal asymptote at $y = 0$.

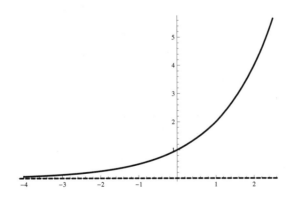

59. This is the graph of $y = e^x$ compressed horizontally by a factor of 3. $f(x)$ has no zeroes. It has a horizontal asymptote at $y = 0$.

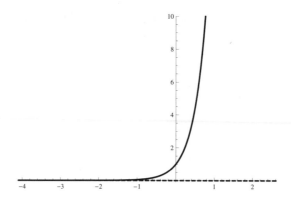

61. This is the graph of $y = e^x$ stretched vertically by a factor of 5, reflected in the x-axis, and shifted up by one unit. $f(x)$ has a zero when $e^x = \frac{1}{5}$, which is for $x = -\ln 5$. It has a horizontal asymptote at $y = 1$.

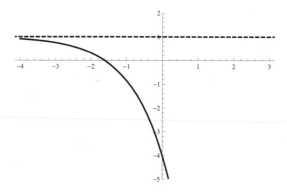

63. $f(x)$ is defined for $x > 0$, and has a vertical asymptote at $x = 0$ and a zero at $x = 1$.

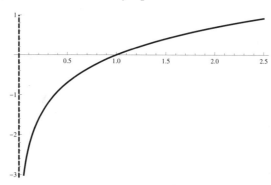

65. $f(x)$ is defined for $x > 0$, and has a vertical asymptote at $x = 0$ and a zero at $x = 1$.

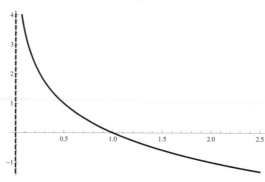

67. $f(x)$ is defined for all x and has zeroes at every integral multiple of π. It is periodic with period 2π.

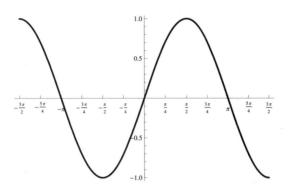

69. $f(x)$ is defined except where $\cos x = 0$, which is at odd multiples of $\dfrac{\pi}{2}$, where it has vertical asymptotes. It is periodic with period 2π.

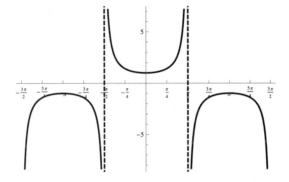

71. $f(x)$ is defined on $[-1, 1]$ and is zero at $x = 0$.

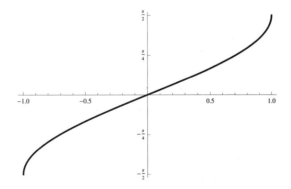

Chapter 1

Limits

1.1 An Intuitive Introduction to Limits

Thinking Back

Finding the pattern in a sequence

▷ The next two terms are 26 and 30; a formula for the general term is $2 + 4(k-1)$ for $k \geq 1$.

▷ The next two terms are 343 and 512; a formula for the general term is k^3.

▷ The next two terms are $\dfrac{3}{49}$ and $\dfrac{3}{64}$; a formula for the general term is $\dfrac{3}{k^2}$.

▷ The next two terms are $\dfrac{1}{729}$ and $\dfrac{1}{2187}$; a formula for the general term is $\dfrac{1}{3^{k-1}}$.

▷ The next two terms are $\dfrac{9}{17}$ and $\dfrac{10}{19}$; a formula for the general term is $\dfrac{k+2}{2k+3}$.

▷ The next two terms are $\dfrac{15}{50}$ and $\dfrac{17}{65}$; a formula for the general term is $\dfrac{2k+1}{k^2+1}$.

Distance, rate, and time

▷ The watermelon hits the ground when $s(t) = 50 - 16t^2 = 0$. Solving for t gives $t_0 = \dfrac{5\sqrt{2}}{4}$. The average rate of change over the entire fall is thus $-\dfrac{50}{5\sqrt{2}/4} = -20\sqrt{2}$ feet per second. Considering the first half of its fall, we let $t_h = \dfrac{5\sqrt{2}}{8}$; then $s(t_h) = 50 - 16\left(\frac{5\sqrt{2}}{8}\right)^2 = \frac{75}{2}$, so that it falls a distance of $50 - \dfrac{75}{2} = \dfrac{25}{2}$ and its average rate of change over the first half of the fall is $-\dfrac{25/2}{5\sqrt{2}/8} = -10\sqrt{2}$ feet per second. Considering the second half of its fall, the watermelon falls from $s(t_h) = \dfrac{75}{2}$ to height 0, so its average rate of change over the second half of the fall is $-\dfrac{75/2}{5\sqrt{2}/8} = -30\sqrt{2}$ feet per second.

▷ Let t_1, $t_{1/2}$, and $t_{1/4}$ be the times one second, half a second, and a quarter of a second before the watermelon hits the ground. Then the average rates of change are given by the following table:

t	$s(t)$	$t_0 - t$	$-\dfrac{s(t)}{t_0 - t}$
$t_1 = \frac{5\sqrt{2}}{4} - 1$	$-16 + 40\sqrt{2}$	1	$16 - 40\sqrt{2} \approx -40.5685$
$t_{1/2} = \frac{5\sqrt{2}}{4} - \frac{1}{2}$	$-4 + 20\sqrt{2}$	$\frac{1}{2}$	$8 - 40\sqrt{2} \approx -48.5685$
$t_{1/4} = \frac{5\sqrt{2}}{4} - \frac{1}{4}$	$-1 + 10\sqrt{2}$	$\frac{1}{4}$	$4 - 40\sqrt{2} \approx -52.5685$

Concepts

1. (a) True. See Definition 1.1 and the paragraph after it.

   (b) False. For example, the function $f(x) = 1$ for $x \neq 2$ and $f(2) = 3$ has $\lim\limits_{x \to 2} f(x) = 1 \neq f(2)$.

   (c) True. For example, the function $f(x) = \dfrac{x-2}{x-2}$ has a limit of 1 as $x \to 2$, but $f(x)$ has a "hole" at 2.

   (d) False. It exists only if the left and right limits exist *and are equal*.

   (e) False. It could also be $-\infty$, or not exist.

   (f) True. See Definition 1.3.

   (g) False. It has a *vertical* asymptote at $x = 2$. See Definitions 1.3 and 1.4.

   (h) True. See Definition 1.4.

3. By Definition 1.2 and the material following, since the one-sided limits exist and are equal, we see that $\lim\limits_{x \to 1} f(x) = 5$ as well. However, there is no information about $f(1)$ — it may not exist at all, or it may be equal to something else.

5. If $\lim\limits_{x \to 2^-} f(x)$ were equal to 8, then the two-sided limit would exist. Since it does not, either $\lim\limits_{x \to 2^-} f(x)$ does not exist, or it exists but is unequal to 8.

7. By Definition 1.4, f has a horizontal asymptote at $y = 3$, and by Definition 1.3, it has a vertical asymptote at $x = 1$.

9. (a) The terms approach 0 as a limit: $\lim\limits_{k \to \infty} \dfrac{1}{3^k} = 0$.

   (b) For $k = 8$, $\dfrac{1}{3^k} \approx 0.000152416$, while for $k = 9$, $\dfrac{1}{3^k} \approx 0.0000508053$.

11. (a) The sequence of sums increases without bound; $\lim\limits_{k \to \infty} (1 + 2 + \cdots + k) = \infty$. See part (b) for some partial sums supporting this conclusion.

   (b) For $k = 44$, the sum is 990, while for $k = 45$, it is 1035.

13. (a)

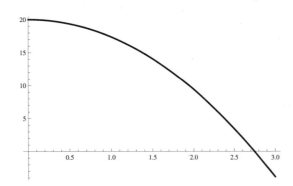

The orange hits the moon when $s(t) = 0$. Solving $20 - 2.65t^2 = 0$ for t gives $t \approx 2.74721$ seconds.

(b) The average rates of change are given by the following table:

t	$-\dfrac{s(t)}{2.74721-t}$
$2.74721 - 1$	-11.9102
$2.74721 - \frac{1}{2}$	-13.2353
$2.74721 - \frac{1}{4}$	-13.8978
$2.74721 - \frac{1}{8}$	-14.2291

(c) The instantaneous velocity is in fact ≈ -14.5602.

15. (a) Using four rectangles, note that the base of each is $\dfrac{1}{2}$, so the total area is

$$\frac{1}{2}\left((4-0^2) + \left(4 - \left(\frac{1}{2}\right)^2\right) + (4 - 1^2) + \left(4 - \left(\frac{3}{2}\right)^2\right)\right) = \frac{25}{4} = 6.25$$

Using eight rectangles the base of each is $\dfrac{1}{4}$, and the total of the areas is

$$\frac{1}{4}\left((4-0^2) + \left(4 - \left(\frac{1}{4}\right)^2\right) + \left(4 - \left(\frac{1}{2}\right)^2\right) + \left(4 - \left(\frac{3}{4}\right)^2\right) + (4 - 1^2)\right.$$

$$\left. + \left(4 - \left(\frac{5}{4}\right)^2\right) + \left(4 - \left(\frac{3}{2}\right)^2\right) + \left(4 - \left(\frac{7}{4}\right)^2\right)\right) = \frac{93}{16} = 5.8125.$$

(b) The rectangles get closer and closer to "matching" the curve, so the sum of the areas should get closer to the area under the curve. The resulting limit is $\dfrac{16}{3} \approx 5.3333$.

17.

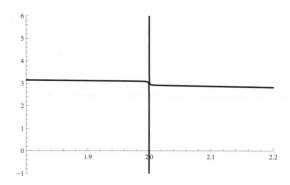

Many other graphs are possible.

19. (a) $f(x)$ is not defined at $x = 1$ since 1 is a zero of the denominator. Since the multiplicity of the zero at $x = 1$ in the numerator is higher than that in the denominator, $x = 1$ is a "hole" in the graph. The calculator graph does not show that hole.

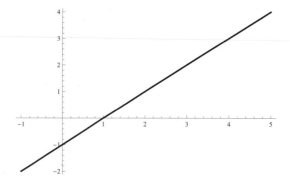

(b) Looking at the graph, the curve approaches 0 as $x \to 1$ from both the left and the right. Thus $\lim\limits_{x \to 1} g(x) = 0$.

Skills

21.

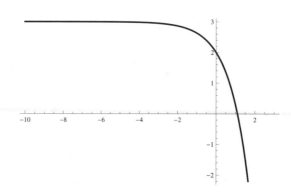

23.

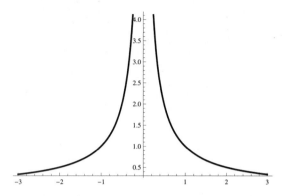

25.

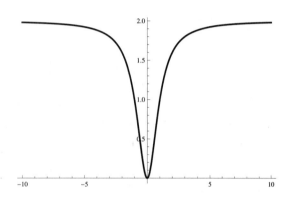

27.

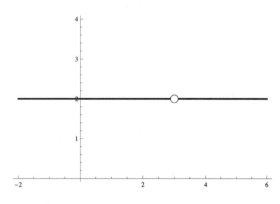

29.

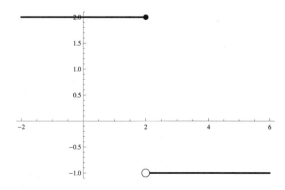

31. From the graph, $\lim\limits_{x \to -2^-} f(x) = \lim\limits_{x \to -2^+} f(x) = -2$ and thus $\lim\limits_{x \to -2} f(x) = -2$. Also from the graph, $f(-2) = -2$.

33. From the graph, $\lim\limits_{x \to 2^-} f(x) = -2$ and $\lim\limits_{x \to 2^+} f(x) = 2$, so that $\lim\limits_{x \to 2} f(x)$ is undefined. From the graph, $f(2) = 2$.

35. From the graph, $\lim\limits_{x \to -1^-} g(x) = \infty$ and $\lim\limits_{x \to -1^+} g(x) = -\infty$, so that $\lim_{x \to -1} g(x)$ does not exist. The graph shows that $g(1)$ is undefined.

37. From the graph, $\lim\limits_{x \to 2^-} g(x) = \lim\limits_{x \to 2^+} g(x) = -1$, so that $\lim\limits_{x \to 2} g(x) = -1$, but from the graph $g(2) = -2$.

39. Many choices of values for a table are possible. The limit is 7.

41. Many choices of values for a table are possible. The limit does not exist (the function approaches $-\infty$ from the left and ∞ from the right.)

43. Many choices of values for a table are possible. The limit is $\dfrac{1}{7}$.

45. Many choices of values for a table are possible. The limit does not exist (the function approaches ∞ from the left and $-\infty$ from the right.)

47. Many choices of values for a table are possible. The limit is -3.

49. Many choices of values for a table are possible. The limit is 0.

51. Many choices of values for a table are possible. The limit does not exist (the function continues to oscillate between -1 and 1.)

53. From the graph, the limit does not exist. The limit from the left is $-\infty$ and from the right is ∞.

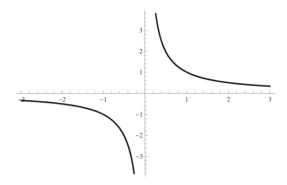

55. From the graph, the limit exists and appears to be 2.

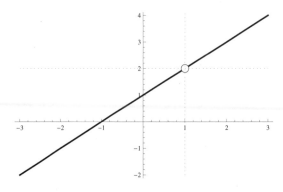

57. From the graph, the limit at 1 exists and is equal to $\dfrac{1}{2}$.

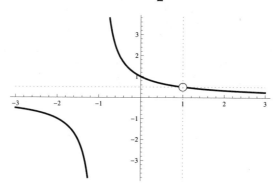

59. From the graph, the limit exists and is equal to 1.

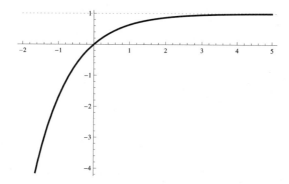

61. From the graph, the limit does not exist. The limit from the left is ∞ and from the right is $-\infty$.

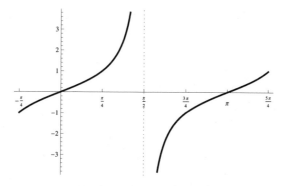

63. From the graph, the limit does not exist. The limit from the left is 4 and from the right is -5.

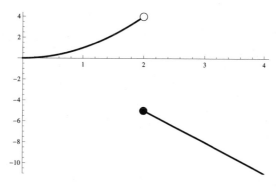

65. From the graph, the limit exists and is equal to 2. The fact that $f(1) = 3$ is irrelvant.

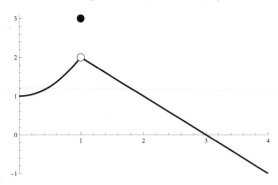

67. From the graph, the limit is about -100. The exact value is -93.

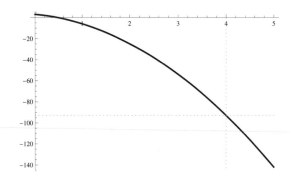

69. From the graph, the limit does not exist. The limit from the left is $-\infty$ and from the right is ∞.

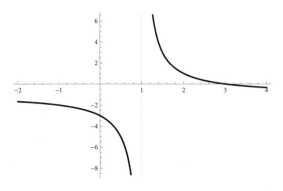

71. From the graph, the limit is zero. Note that the function grows *very* large before coming back down and approaching zero, so it is necessary to graph it for large values of x to see the actual behavior.

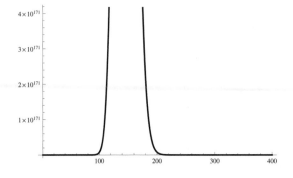

73. From the graph, the limit exists and is 1.

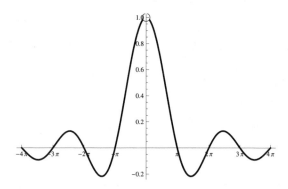

Applications

75. (a) $S(0) = \dfrac{12 + 5.5 \cdot 0}{3 + 0.25 \cdot 0} = \dfrac{12}{3} = 4.$

    (b) The number of squirrels is

t	$S(t)$
30	$16.86 \approx 17$
60	19
365	$21.43 \approx 21$

    (c) Evaluating $S(t)$ for large values of t gives the following:

t	$S(t)$
1000	21.7866
5000	21.9569
10000	21.9784

It seems as though the number of squirrels approaches 22, so there are 22 squirrels in the attic in the long run.

    (d)

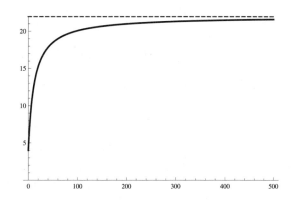

77. (a)

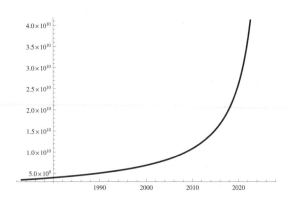

(b) The limit appears to be infinite.

(c) Presumably because the model predicts that human population will increase without bound as we get closer to the year 2027, which is the doomsday year.

(d) The graph to the right of $t = 2027$ is meaningless. The model is valid at most up to the year 2027.

Proofs

79. Since the sum of the first 100 integers is $\dfrac{100(101)}{2} = 5050$, it follows that for $k > 100$, the sum of the first k integers is greater than 5050 and thus greater than 5000. Since the sequence of sums increases without bound, the limit of the sequence is infinite.

81. If x is within 0.01 of $x = 1$, then $|x - 1| < 0.01$, so that, squaring both sides, $|x - 1|^2 < 0.0001$. But the left-hand side is just $(x - 1)^2$, so that $(x - 1)^2 < 0.0001$. Taking reciprocals gives $\dfrac{1}{(x - 1)^2} > 10000$. For values closer to 1 than 0.01, we get even larger lower bounds for $\dfrac{1}{(x - 1)^2}$, so that
$$\lim_{x \to 1} \frac{1}{(x - 1)^2} = \infty.$$

Thinking Forward

Convergence and divergence of sequences:

▷ $\lim\limits_{k \to \infty} \left(\dfrac{1}{4}\right)^k = 0$. Writing out the value of the kth term for large k will convince you of this; alternatively, note that raising any number less than one to higher and higher powers gives a number closer and closer to zero.

▷ $\lim\limits_{k \to \infty} \left(\dfrac{5}{4}\right)^k = \infty$. Writing out the value of the kth term for large k will convince you of this; alternatively, note that raising any number greater than one to higher and higher powers gives larger and larger results.

▷ $\lim\limits_{k \to \infty} \dfrac{k}{k + 2} = 1$. For example, $\frac{98}{100} = 0.98$ and $\frac{998}{1000} = 0.998$.

▷ Since $\dfrac{k + 1}{k} = 1 + \dfrac{1}{k}$, and $\lim\limits_{k \to \infty} \dfrac{1}{k} = 0$, we have $\lim\limits_{k \to \infty} \dfrac{k + 1}{k} = 1$.

Convergence and divergence of series:

- ▷ For example, the sum up to $k = 10$ is ≈ 1.54977, for $k = 100$ it is 1.63498, and for $k = 1000$ it is 1.64393. This is not conclusive, but it does appear to be converging to something between 1.64 and 1.65. (The actual limit is $\dfrac{\pi^2}{6} \approx 1.64493$).

- ▷ For example, the sum up to $k = 1$ is 1, up to $k = 4$ is ≈ 2.08, up to $k = 11$ is ≈ 3.02, and up to $k = 31$ is ≈ 4.03. There is no real evidence that this series converges. (In fact it diverges).

1.2 Formal Definition of Limit

Thinking Back

Logical Quantifiers

- ▷ True. This says that every number has a square.

- ▷ False. This says that every real number has a real square root (negative numbers don't).

- ▷ True.

- ▷ True. For example, take $b = a - 1$.

- ▷ False. It cannot be that every $x > a$ is equal to a single real number b.

Solving function equations

- ▷ $f(x) = 7.5$ when $x = \sqrt[3]{7.5} \approx 1.957$, and $f(x) = 8.5$ when $x = \sqrt[3]{8.5} \approx 2.04$. A graph of the situation is

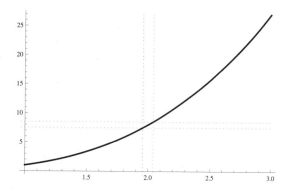

- ▷ $f(x) = 1.8$ when $x = 1 + 1.8^2 = 4.24$. $f(x) = 2.2$ when $x = 1 + 2.2^2 = 5.84$. A graph of the situation is

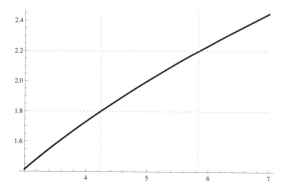

▷ $f(x) = 90$ when $x \approx 37.61$ and when $x \approx 72.39$. A graph of the situation is

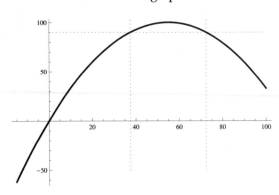

▷ $f(x) = -7.01$ when $x^2 = 9.01$, so when $x \approx 3.00167$. $f(x) = -6.99$ when $x^2 = 8.99$, so when $x \approx 2.99833$. A graph of the situation (for $x > 0$ as in the problem statement) is

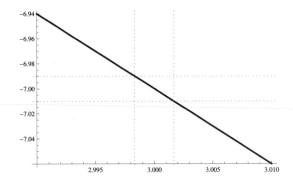

▷ $f(x) = -2$ when $x^2 - 2x - 3 = -2(x - 1)$. Simplifying gives $x^2 - 5 = 0$, so $x = \pm\sqrt{5}$. $f(x) = 0$ when the numerator vanishes but the denominator does not. The numerator factors as $(x + 1)(x - 3)$, so this happens for $x = -1$ and for $x = 3$. Since we are restricting attention to $x < 0$, we need look only at $x = -1$ and $x = -\sqrt{5}$. A graph of the situation is

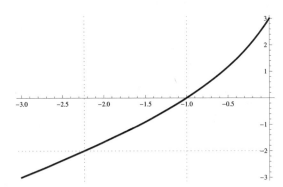

Concepts

1. (a) False. The existence and value of the limit do not depend on the function's behavior at $x = c$. See Definition 1.7 and note that x is in an *open* interval not containing c.

   (b) False. See part (a), or consider the function $f(x)$ which is 2 everywhere except at $x = 0$, where it is 3 and look at $\lim_{x \to 0} f(x)$.

   (c) False. Again see parts (a) and (b).

(d) False. See part (a).

(e) False. The left and right limits might exist and be different, but the function does not approach a limit unless these two are equal. See Theorem 1.8.

(f) False. We can make $f(x)$ as close to 10 as we like by choosing x sufficiently close to 4.

(g) True. See Definition 1.7 and Theorem 1.8.

(h) True. Since the limit is 100, there is some N such that for $x > N$, we have $|f(x) - 100| < 0.1$, i.e. that $f(x) \in (99.9, 100.1)$.

3. A punctured interval is an interval minus some interior point. Punctured intervals are important when discussing limits because they give us a way to talk about values of a function at points close to, but not equal to, some specific value (which corresponds to the missing interior point).

5. $f(x)$ is undefined exactly where the denominator vanishes, which occurs for $x = 0$ and for $x = 1$. So any punctured interval not containing these points will solve the problem. For example (other intervals are possible):

(a) $(1, 1.5) \cup (1.5, 2)$.

(b) $(0, 0.25) \cup (0.25, 0.5)$.

(c) $(0, 1) \cup (1, 2)$.

7. If $\lim_{x \to 2} f(x) = 5$ then for all $\epsilon > 0$, there is some $\delta > 0$ such that if $x \in (2 - \delta, 2) \cup (2, 2 + \delta)$ then $f(x) \in (5 - \epsilon, 5 + \epsilon)$.

9. If $\lim_{x \to \infty} f(x) = 2$ then for all $\epsilon > 0$, there is some $N > 0$ such that if $x \in (N, \infty)$, then $f(x) \in (2 - \epsilon, 2 + \epsilon)$.

11. If $\lim_{x \to 1^+} f(x) = \infty$ then for all $M > 0$, there is some $\delta > 0$ such that if $x \in (1, 1 + \delta)$, then $f(x) \in (M, \infty)$.

13. Let $f(x) = x^2$ and $c = 2$, and consider the diagram

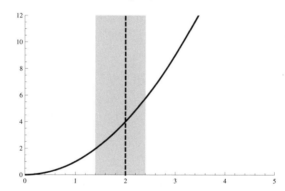

The delta bar to the left of 2 has a width of 0.6, while the delta bar to the right has a width of 0.4. There is a punctured interval around 2 of radius 0.4 contained in the union of these two delta bars.

15. Since $1.5^2 = 2.25$ and $2.5^2 = 6.25$, we have $4 - 1.75 = 1.5^2$ and $4 + 2.25 = 2.5^2$. Thus the smallest possible ϵ is $\epsilon = 2.25$. If x is between 1.5 and 2, its square will be closer than 1.75 to 4, so certainly closer than 2.25, while if x is between 2 and 2.5, then its square will be closer than 2.25 to 4.

17. If the limit statement is true, then for all $M > 0$ there exists an $N > 0$ such that if $x \in (N, \infty)$, then $\dfrac{1000}{x} \in (M, \infty)$. But choose any M, say $M = 1$, and suppose N satisfies the limit statement. Choose $x = \max(N, 2000) \in (N, \infty)$. Then $\dfrac{1000}{x} < \dfrac{1}{2} \notin (1, \infty) = (M, \infty)$.

Skills

19.

For all $\epsilon > 0$ there exists $\delta > 0$ such that if $x \in (-3 - \delta, -3) \cup (-3, -3 + \delta)$ then $\sqrt{x + 7} \in (2 - \epsilon, 2 + \epsilon)$. The graph shows the situation for $\epsilon = 0.5$, when $\delta \approx 1.75$.

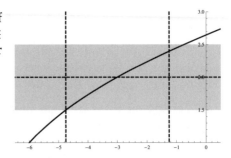

21. For all $\epsilon > 0$ there exists $\delta > 0$ such that if $x \in (-1 - \delta, -1) \cup (-1, -1 + \delta)$ then $x^3 - 2 \in (-3 - \epsilon, -3 + \epsilon)$. The graph shows the situation for $\epsilon = 0.5$; here $\delta \approx 0.145$.

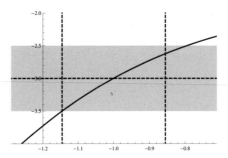

23. For all $\epsilon > 0$ there exists $\delta > 0$ such that if $x \in (2 - \delta, 2) \cup (2, 2 + \delta)$ then $\dfrac{x^2 - 4}{x - 2} \in (4 - \epsilon, 4 + \epsilon)$. The graph shows the situation for $\epsilon = 0.5$; here $\delta = \epsilon$.

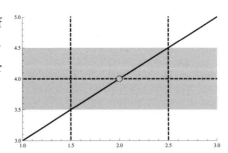

25. For all $\epsilon > 0$ there exists $\delta > 0$ such that if $x \in (0, \delta)$ then $\sqrt{x} \in (-\epsilon, \epsilon)$. The graph shows the situation for $\epsilon = 0.5$; here $\delta = \epsilon^2 = 0.25$.

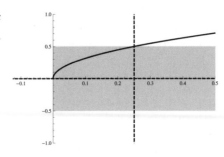

27. For all $M > 0$ there exists $\delta > 0$ such that if $x \in (-2, -2 + \delta)$ then $\dfrac{1}{x+2} \in (M, \infty)$. The graph shows the situation for $M = 3$; here $\delta \approx 0.33$.

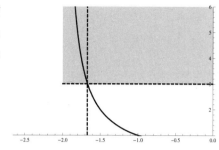

29. For all $M > 0$ there exists $\delta > 0$ such that if $x \in (2, 2 + \delta)$ then $\dfrac{1}{2^x - 4} \in (M, \infty)$. The graph shows the situation for $M = 3$; here $\delta \approx 0.117$.

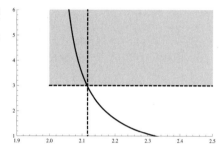

31. For all $\epsilon > 0$ there exists $N > 0$ such that if $x \in (N, \infty)$ then $\dfrac{x}{1 - 2x} \in (-.5 - \epsilon, -.5 + \epsilon)$. The graph shows the situation for $\epsilon = 0.01$; here $N = 26$.

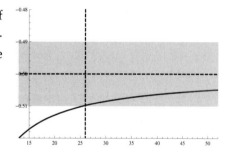

33. For all $M > 0$ there exists $N > 0$ such that if $x \in (N, \infty)$ then $x^3 + x + 1 \in (M, \infty)$. The graph shows the situation for $M = 3$; here $N = 1$.

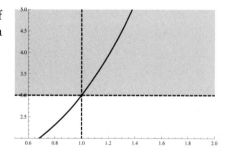

35. For all $\epsilon > 0$ there exists $\delta > 0$ such that if $x \in (-\delta, 0) \cup (0, \delta)$ then $\dfrac{(2 + h)^2 - 4}{h} \in (4 - \epsilon, 4 + \epsilon)$. The graph shows the situation for $\epsilon = 0.5$; here $\delta = \epsilon = 0.5$.

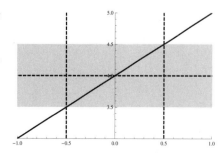

37. For all $M > 0$ there exists $\delta > 0$ such that if $x \in (c, c + \delta)$ then $f(x) \in (-\infty, -M)$. As an example, let $f(x) = \dfrac{1}{e - e^x}$ and $c = 1$; the graph shows the situation for $M = 3$; here $\delta \approx 0.115$.

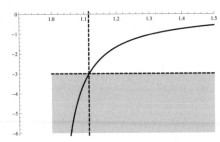

39. For all $\epsilon > 0$ there exists $N > 0$ such that if $x \in (-\infty, -N)$ then $f(x) \in (L - \epsilon, L + \epsilon)$. As an example, let $f(x) = e^x$ and $L = 0$; the graph shows the situation for $\epsilon = 0.5$; here $N \approx 0.7$.

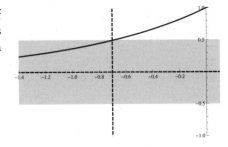

41. For all $M > 0$ there exists $N > 0$ such that if $x \in (-\infty, -N)$ then $f(x) \in (M, \infty)$. As an example, let $f(x) = -x$. The graph shows the situation for $M = 3$, where $N = M = 3$.

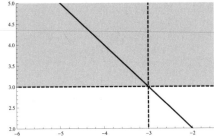

43. A graph of the situation is

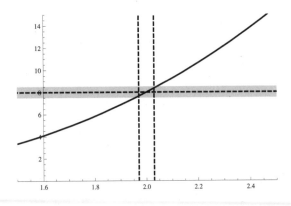

It appears that $\delta \approx 0.03$. Algebraically, we want $\left| x^3 - 8 \right| < 0.5$, so that $-0.5 < x^3 - 8 < 0.5$. Simplifying gives $7.5 < x^3 < 8.5$, so that approximately $1.957 < x < 2.041$. Thus in fact $\delta = 0.04$ will suffice.

45. A graph of the situation is

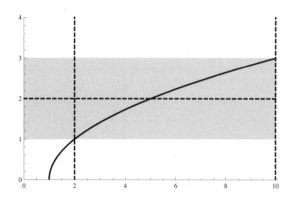

It appears that $\delta \approx 3$. Algebraically, we want $\left|\sqrt{x-1}-2\right| < 1$, so that $-1 < \sqrt{x-1}-2 < 1$. Simplifying gives $1 < \sqrt{x-1} < 3$, so that $1 < x-1 < 9$ or $2 < x < 10$. Thus $\delta = 3$ is sufficient.

47. A graph of the situation is

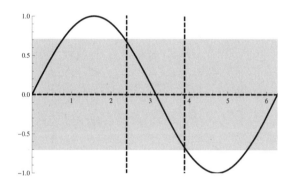

It appears that $x \approx 2.4$ and $x \approx 3.9$ are the boundaries for δ; since $\pi \approx 3.1416$, we get $\delta \approx 0.7$. Algebraically, we want $|\sin x| < \dfrac{\sqrt{2}}{2}$ so that $-\dfrac{\sqrt{2}}{2} < \sin x < \dfrac{\sqrt{2}}{2}$ and $\dfrac{3\pi}{4} < x < \dfrac{5\pi}{4}$. Thus $\delta = \frac{\pi}{4} \approx 0.785$ works.

49. A graph of the situation is

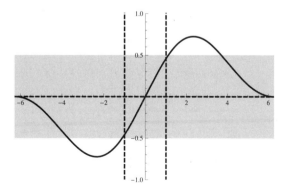

It appears that $\delta \approx 1$. Algebraically, we want $\left|\dfrac{1-\cos x}{x}\right| < \dfrac{1}{2}$ so that $-\dfrac{1}{2} < \dfrac{1-\cos x}{x} < \dfrac{1}{2}$. Simplifying gives $-x < 2 - 2\cos x < x$, so that $-2 - x < -2\cos x < x - 2$ and thus $1 - \frac{x}{2} < \cos x < 1 + \frac{x}{2}$. Solving this equation numerically gives $x \approx \pm 1.109$, so that $\delta = 1.1$ works.

51. A graph of the situation is (note the magnification)

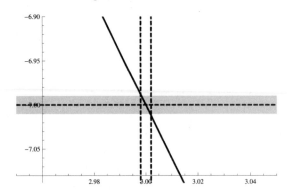

It appears that $\delta \approx 0.02$. Algebraically, we want $\left|2 - x^2 + 7\right| < 0.01$ so that $-0.01 < 9 - x^2 < 0.01$. Simplifying gives $8.99 < x^2 < 9.01$, or $2.99833 < x < 3.00167$. Thus $\delta \approx 0.0016$ is sufficient.

53. A graph of the situation is

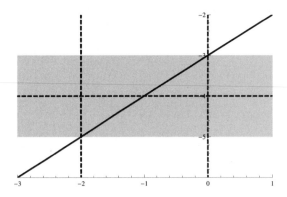

It appears that $\delta = 1$. Algebraically, note that $x^2 - 2x - 3 = (x + 1)(x - 2)$. Since we are considering only $x \neq -1$, we may write $\dfrac{x^2 - 2x - 3}{x + 1} = x - 2$. Then we want $|x - 2 - (-4)| < 1$, so that $-1 < x + 2 < 1$, Simplifying gives $-3 < x < -1$, so indeed $\delta = 1$ is sufficient.

55. A plot of the situation is

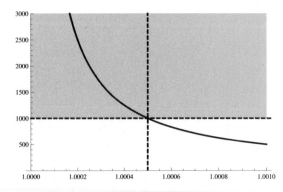

Algebraically, we want $\left|\dfrac{1}{x^2 - 1}\right| > 1000$; since x approaches 1 from the right, the expression is positive, so we want $\dfrac{1}{x^2 - 1} > 1000$ so that $0 < x^2 - 1 < 0.001$. This means that $1 < x^2 < 1.001$, or $1 < x < 1.0005$ to four decimal places. Thus $\delta = 0.0005$ works.

57. A plot of the situation is

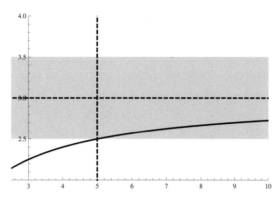

Algebraically, we want $\left| \dfrac{3x}{x+1} - 3 \right| < 0.5$, so that $-0.5 < \dfrac{3x}{x+1} - 3 < 0.5$, or $2.5 < \dfrac{3x}{x+1} < 3.5$. Since $x \to \infty$, we assume x is positive; clearing fractions then gives $2.5x + 2.5 < 3x < 3.5x + 3.5$. The second inequality is obviously always satisfied for $x > 0$, so we are left with $2.5x + 2.5 < 3x$, or $x > 5$. Thus $N = 5$ is the smallest N.

59. A plot of the situation is

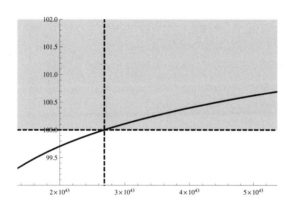

Algebraically, we want $N > 0$ such that for $x > N$, $\ln x > 100$. Clearly $N = e^{100}$ is sufficient.

61. A plot of the situation is

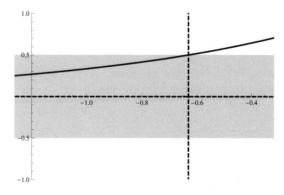

Algebraically, we want $|3^x| < \dfrac{1}{2}$. Since $3^x > 0$, this means that $3^x < \dfrac{1}{2}$. Taking logs gives $x \ln 3 < \ln 0.5$, so that $x \approx -0.631$. Thus $N = -0.631$ works.

63. A plot of the situation is

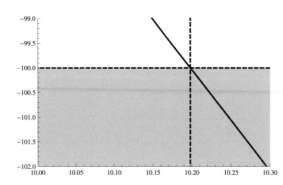

Algebraically, we want $4 - x^2 < -100$, so that $x^2 > 104$ and $x > \sqrt{104} \approx 10.198$. Thus $N = 10.198$ works.

Applications

65. (a) Solving $F(t) = 7465$ for t gives $50t = 7465$, or $t = 149.3$. Thus it will take 150 months, at which time Jack will in fact have \$7500.

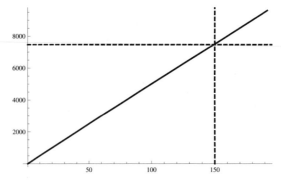

(b) There will always be such a value. Since $\lim\limits_{t \to \infty} F(t) = \infty$, for any M there is an N such that for all $t > N$, $f(t) > M$. Thus eventually $f(t)$ will exceed M and Jack will have saved enough. On the graph above, replace 150 by N and 7465 by M to see the picture.

67. (a) We want $N(p) = 6000$. Solving this equation algebraically gives $p = \$18.68$ and $p = \$60.12$. Since the price is limited to \$50, we must have $p = \$18.68$ in order to sell 6000 paintings.

(b) Solving $N(p) = 5000$ gives $p = \$21.49$; solving $N(p) = 7000$ gives $p = \$16.21$. Thus any price between \$16.21 and \$21.49 will result in between 5000 and 7000 sales.

(c) Here c is the price needed to sell 6000 paintings, which is $c = 18.68$; clearly $L = N(c) = 6000$. ϵ represents the allowable error in the range, which is $\epsilon = 1000$. To determine δ, note that the range of prices is $(16.21, 21.49) = (18.68 - 2.47, 18.68 + 2.81)$, so that $\delta = 2.47$. A graph would be similar to the graph preceding Definition 1.5.

Proofs

69. (a) We must show that for any $\epsilon > 0$ there is a $\delta > 0$ such that if $x \in (2 - \delta, 2) \cup (2, 2 + \delta)$ then $7 - x \in (5 - \epsilon, 5 + \epsilon)$.

(b) $x \in (2 - \delta, 2) \cup (2, 2 + \delta)$ means $x \in (2 - \delta, 2 + \delta)$ and $x \neq 2$. But $x \in (2 - \delta, 2 + \delta)$ is equivalent to $2 - \delta < x < 2 + \delta$; subtracting 2 gives $-\delta < x - 2 < \delta$. The definition of absolute value shows that this is the same as $|x - 2| < \delta$, while $x \neq 2$ means $0 < |x - 2|$. Thus $0 < |x - 2| < \delta$.

(c) $7 - x \in (5 - \epsilon, 5 + \epsilon)$ means $5 - \epsilon < 7 - x < 5 + \epsilon$. Simplifying gives $-\epsilon < 2 - x < \epsilon$. The definition of absolute value shows that this is the same as $|2 - x| < \epsilon$. But $|2 - x| = |x - 2|$, so this is the same as $|x - 2| < \epsilon$.

(d) We may choose $\delta = \epsilon$, since if $0 < |x - 2| < \delta = \epsilon$, then obviously $|x - 2| < \epsilon$.

(e) Given $\epsilon > 0$, choose $\delta = \epsilon$. Then if $x \in (2 - \delta, 2) \cup (2, 2 + \delta)$, part (d) shows that $7 - x \in (5 - \epsilon, 5 + \epsilon)$.

71. (a) We must show that for any $\epsilon > 0$ there is an $N > 0$ such that if $x \in (N, \infty)$ then $\dfrac{1}{x} \in (0 - \epsilon, 0 + \epsilon)$.

(b) $x \in (N, \infty)$ means that $x > N$, by the definition of an open interval.

(c) $\dfrac{1}{x} \in (0 - \epsilon, 0 + \epsilon)$ means that $-\epsilon < \dfrac{1}{x} < \epsilon$. Since we are considering $x \to \infty$, we may clearly assume that $x > 0$, so that in fact $0 < \dfrac{1}{x} < \epsilon$.

(d) Given ϵ, if we choose $N = \dfrac{1}{\epsilon}$, then $x > N$ means $x > \frac{1}{\epsilon}$, which means that $0 < x < \epsilon$ as desired.

(e) Given ϵ, choose $N = \dfrac{1}{\epsilon}$. Then if $x \in (N, \infty)$, then $x > N = \frac{1}{\epsilon}$, so that $\frac{1}{x} < \epsilon$ and $\dfrac{1}{x} \in (0, \epsilon)$.

Thinking Forward

Continuity

▷ If $f(c) = \lim\limits_{x \to c} f(x)$, then as your pencil gets closer to c, you can continue drawing right through the point at $x = c$. If the limit and the value were not equal, you'd have to move your pencil to a different point on the page when you got to $x = c$.

▷ $f(x)$ is *continuous* at $x = c$ if for any $\epsilon > 0$ there is some $\delta > 0$ such that whenever $0 < |x - c| < \delta$, then $|f(x) - f(c)| < \epsilon$.

▷ Since $f(x)$ is continuous, we can evaluate these limits by evaluating the function at the limit point. Thus the values of the limits are 2^2, 5^2, and $(-4)^2$, or 4, 25, and 16.

Limits of Sequences

▷ $|a_k - L| < \epsilon$ means that $-\epsilon < a_k - L < \epsilon$, so that $L - \epsilon < a_k < L + \epsilon$. Thus $a_k \in (L - \epsilon, L + \epsilon)$.

▷ The sequence a_k converges to a limit L if for any $\epsilon > 0$ there is some $N > 0$ such that whenever $k > N$ is an integer, then $a_k \in (L - \epsilon, L + \epsilon)$.

1.3 Delta-Epsilon Proofs

Thinking Back

Inequalities

▷ $-0.5 < x - 2 < 0.5$, so $1.5 < x < 2.5$.

▷ $-0.1 < x + 5 < 0.1$, so $-5.1 < x < -4.9$.

▷ $-0.5 < x^2 - 4 < 0.5$, so $3.5 < x^2 < 4.5$. The solution set is then $(-\sqrt{4.5}, -\sqrt{3.5}) \cup (\sqrt{3.5}, \sqrt{4.5})$.

  ▷ $-1 < (3x + 1) - 10 < 1$, so $8 < 3x < 10$. The solution set is $\left(\dfrac{8}{3}, \dfrac{10}{3}\right)$.

  ▷ Since $\dfrac{1}{x^2} > 0$, the given inequality means that $x^2 > 100$, so that $x > 10$ or $x < -10$. Thus the solution set is $(-\infty, -10) \cup (10, \infty)$.

  ▷ $1000 < \dfrac{1}{x^2}$ means that $x^2 < \frac{1}{1000}$, so that $|x| < \frac{1}{10\sqrt{10}}$. The solution set is $\left(-\frac{1}{10\sqrt{10}}, \frac{1}{10\sqrt{10}}\right)$.

Logical implications

  ▷ B must be true. "If A, then B" means that whenever A is true, B must be true as well.

  ▷ Nothing. B may be either true or false.

  ▷ Nothing. A may be either true or false.

Concepts

1. (a) True. $|a| \geq 0$ for any a, and is zero only if $a = 0$. Thus $|x - c| > 0$ if $x \neq c$.

   (b) True. The same reasoning as in part (a) shows this.

   (c) False. x is a solution if and only if $c - \delta < x < c + \delta$ *and* $x \neq c$.

   (d) True. See Theorem 1.11.

   (e) True. See Theorem 1.11.

   (f) False. Suppose that in fact $f(x) = L$.

   (g) False. This demonstrates the $\epsilon - \delta$ argument for a particular choice of ϵ, namely 0.5. To show that the limit is 5 requires that we execute the argument for every value of $\epsilon > 0$.

   (h) False. It means that for all $\epsilon > 0$ there is some $\delta > 0$ such that ...

3. For every $\epsilon > 0$ there is a $\delta > 0$ such that whenever $0 < |x - (-2)| < \delta$, then $\left|\dfrac{3}{x+1} - (-3)\right| < \epsilon$.

5. $0 < |x - 5| < 0.01$. For the second equation, $|f(x) - (-2)| < 0.5$.

7. $0 < |x - 2| < 0.1$ means $x \neq 2$ and $-0.1 < x - 2 < 0.1$, so that $x \neq 2$ and $1.9 < x < 2.1$. Thus $x \in (1.9, 2) \cup (2, 2.1)$.

9. $\left|(x^2 - 1) + 3\right| < 0.5$ means $-0.5 < x^2 + 2 < 0.5$, so that $-2.5 < x^2 < -1.5$. The solution set is empty since $x^2 \geq 0$ for all x.

11. $|f(x) - L| < \epsilon$ means $-\epsilon < f(x) - L < \epsilon$, so that $L - \epsilon < f(x) < L + \epsilon$. The solution set is $f(x) \in (L - \epsilon, L + \epsilon)$.

13. False. For example, let $x = 2.5$; then $x^2 = 6.25$ and $x^2 - 4 = 2.25 > 0.5$.

15. False. $0.74^2 = 0.5476 > 0.5$.

17. True. See the graph. The vertical lines at a distance of 0.075 from $x = 2$ intersect the curve $y = x^2$ within the shaded region, which is at a distance of 0.4 from $y = 4$.

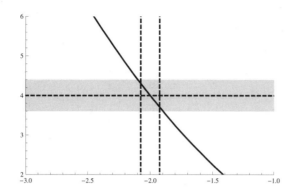

19. Example 4 showed that choosing $\delta = \min\left(1, \dfrac{\epsilon}{325}\right)$ was sufficient. So for $\epsilon = 5$, $\delta = \dfrac{5}{325} = \dfrac{1}{65}$; for $\epsilon = 0.01$, $\delta = \dfrac{0.01}{325} = \dfrac{1}{32500}$, and for $\epsilon = 350$, $\delta = \min\left(1, \frac{350}{325}\right) = 1$. A graph illustrating the solution for $\epsilon = 5$ is

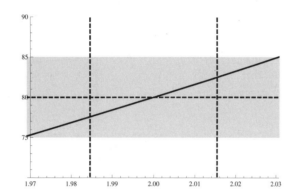

21. $|f(x) - L| < \epsilon$ means that $-\epsilon < f(x) - L < \epsilon$, so that $L - \epsilon < f(x) < L + \epsilon$. In interval notation, this is $f(x) \in (L - \epsilon, L + \epsilon)$.

Skills

23. If $|(3x - 1) - 5| < 0.25$, then $-0.25 < 3x - 6 < 0.25$, so that $5.75 < 3x < 6.25$, or $1.9167 < x < 2.0833$. Thus $\delta = 0.0833$ works.

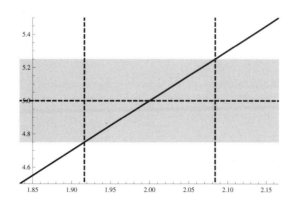

25. If $\left|\sqrt{x - 1} - 0\right| < 0.5$, then $-0.5 < \sqrt{x - 1} < 0.5$. But $\sqrt{x - 1} \geq 0$, so this is equivalent to $0 \leq \sqrt{x - 1} < 0.5$, or $0 \leq x - 1 < 0.25$. Thus $1 \leq x < 1.25$. Thus $\delta = 0.25$ works.

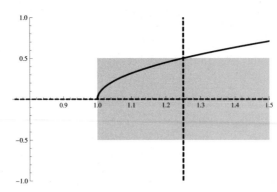

27. If $\left|\dfrac{1}{x^2} - 0\right| < 0.001$, then $-0.001 < \dfrac{1}{x^2} < 0.001$. However, $\dfrac{1}{x^2} > 0$, so this is the same as $0 < \dfrac{1}{x^2} <$ 0.001. Inverting gives $1000 < x^2$, so that $x > 10\sqrt{10} \approx 31.62$. Thus $N = 31.62$ is sufficient.

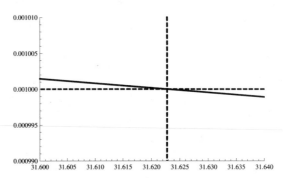

29. $|x + 5 - 8| < \epsilon$ means that $-\epsilon < x - 3 < \epsilon$, $|x - 3| < \epsilon$. Thus $\delta = \epsilon$ works.

31. $|3 - 4x - 3| < \epsilon$ gives $4|x| < \epsilon$, so that $|x| < \dfrac{\epsilon}{4}$. Let $\delta = \dfrac{\epsilon}{4}$; then if $0 < |x| < \delta$, we have $|(3 - 4x) - 3| = |-4x| = 4|x| < 4\delta = \epsilon$.

33. $|5x^2 - 1 - (-1)| < \epsilon$ gives $|5x^2| < \epsilon$, so that $0 < x^2 < \dfrac{\epsilon}{5}$. Let $\delta = \sqrt{\dfrac{\epsilon}{5}}$. Then if $0 < |x| < \delta$, then $x^2 < \dfrac{\epsilon}{5}$, so that $|(5x^2 - 1) - (-1)| = 5x^2 < 5\delta^2 = \epsilon$.

35. $|x^2 - 4x + 6 - 2| < \epsilon$ gives $|x^2 - 4x + 4| = |(x - 2)^2| < \epsilon$; since $(x - 2)^2 \geq 0$, this means $0 \leq x - 2 < \sqrt{\epsilon}$. Let $\delta = \sqrt{\epsilon}$; then if $|x - 2| < \delta$, we get $|(x^2 - 4x + 6) - 2| = |x^2 - 4x + 4| = |(x - 2)^2| < \delta^2 = \epsilon$.

37. $\left|\dfrac{1}{x} - \dfrac{1}{2}\right| < \epsilon$ means that $\dfrac{|x - 2|}{2|x|} < \epsilon$. Assuming $\delta \leq 1$, then $0 < |x - 2| < \delta$ means that $|x - 2| < 1$, so that $1 < x < 3$ and therefore $\dfrac{1}{x} < 1$. Now, choose $\delta = \min(1, 2\epsilon)$. Then if $0 < |x - 2| < \delta$, we have
$$\left|\dfrac{1}{x} - \dfrac{1}{2}\right| = \dfrac{|x - 2|}{2|x|} < \dfrac{1}{2}|x - 2| < \dfrac{1}{2}\delta \leq \epsilon.$$

39. $|x^2 - 2x - 3| < \epsilon$ means that $|(x - 3)(x + 1)| < \epsilon$. If we assume $\delta \leq 1$, then $0 < |x - 3| < \delta$ means that $|x - 3| < 1$, so that x is between 2 and 4. Thus $x + 1$ is bounded by 5. Choose $\delta = \min\left(1, \dfrac{\epsilon}{5}\right)$. Then if $0 < |x - 3| < \delta$, we get
$$|x^2 - 2x - 3| = |(x - 3)(x + 1)| = |x - 3| \cdot |x + 1| \leq 5|x - 3| < 5\delta \leq 5 \cdot \dfrac{\epsilon}{5} = \epsilon.$$

41. If $\left|1 + \sqrt{x+2} - 1\right| < \epsilon$, then $0 \le \sqrt{x+2} < \epsilon$ so that $0 \le x + 2 < \epsilon^2$. Letting $\delta = \epsilon^2$ is sufficient, for then if $0 < |x - 2| < \delta$, we get

$$\left|1 + \sqrt{x+2} - 1\right| = \left|\sqrt{x+2}\right| < \left|\sqrt{\delta}\right| = \epsilon.$$

43. If $\dfrac{1}{1-x} > M$, then $1 - x < \dfrac{1}{M}$, so that $x > 1 - \dfrac{1}{M}$. Choose $\delta = \dfrac{1}{M}$.

Applications

45. The curved sides form a cylinder of height 5 and radius r, so the surface area of the sides is $2\pi r \cdot 5 = 10\pi r$, so the cost is $10\pi r \cdot 0.25 = 2.5\pi r$ cents. The top and bottom are each disks of radius r, so the area of each is πr^2. Thus the cost of the top and bottom is $2\pi r^2 \cdot 0.5 = \pi r^2$ cents. The top and bottom welds are each circles with radius r, so the cost of these welds is $2 \cdot 2\pi r \cdot 0.10 = 0.4\pi r$. Finally, the weld along the side is a line of length 5, with cost $5 \cdot 0.10 = 0.5$. Thus the total cost is

$$C(r) = 2.5\pi r + \pi r^2 + 0.4\pi r + 0.5 = \pi r^2 + 2.9\pi r + 0.5$$

(a) Solving $C(r) = 30$ gives $\pi r^2 + 2.9\pi r + 0.5 - 30 = 0$, so that $r \approx -4.84$ or $r \approx 1.94$. Discarding the negative root gives a radius of 1.94 inches.

(b) $|C(r) - 30| \le 10$; this is an interval of radius 10 around 30, so is $20 < C(r) < 40$.

(c) $C(r) = 20$ gives $r \approx 1.43$, while $C(r) = 40$ gives $r \approx 2.38$. An absolute value inequality contained in this range is $|r - 1.94| < 0.44$ (there are many other possibilities).

Proofs

47. Choose $\epsilon > 0$, and let $\delta = \dfrac{\epsilon}{2}$. Then if $0 < |x - 1| < \delta$, we have

$$|(2x + 4) - 6| = |2x - 2| = 2|x - 1| < 2\delta = \epsilon.$$

49. Choose $\epsilon > 0$, and let $\delta = \epsilon$. Then if $0 < |x - (-6)| < \delta$, we have

$$|(x + 2) - (-4)| = |x + 6| < \delta = \epsilon.$$

51. Choose $\epsilon > 0$, and let $\delta = \dfrac{\epsilon}{6}$. Then if $0 < |x - 4| < \delta$, we have

$$|(6x - 1) - 23| = |6x - 24| = 6|x - 4| < 6\delta = \epsilon.$$

53. Choose $\epsilon > 0$, and let $\delta = \sqrt{\dfrac{\epsilon}{3}}$. Then if $0 < |x| < \delta$, we have

$$\left|(3x^2 + 1) - 1\right| = \left|3x^2\right| = 3\left|x^2\right| < 3\delta^2 = \epsilon.$$

55. Choose $\epsilon > 0$, and let $\delta = \sqrt{\dfrac{\epsilon}{2}}$. Then if $0 < |x - 1| < \delta$, we have

$$\left|(2x^2 - 4x + 3) - 1\right| = \left|2x^2 - 4x + 2\right| = 2\left|(x - 1)^2\right| < 2\delta^2 = \epsilon.$$

57. Choose $\epsilon > 0$, and let $\delta = \epsilon$. Then if $0 < |x - 1| < \delta$, we have (writing $\dfrac{x^2 - 1}{x - 1} = x + 1$ since $x \ne 1$)

$$|(x + 1) - 2| = |x - 1| < \delta = \epsilon.$$

59. Choose $\epsilon > 0$, and let $\delta = \epsilon^2$. Then if $0 < x - 5 < \delta$, we have

$$\left| \sqrt{x-5} - 0 \right| = \left| \sqrt{x-5} \right| < \sqrt{\delta} = \epsilon.$$

61. Choose $M > 0$, and let $\delta = \dfrac{1}{M}$. Then if $0 < x - (-2) < \delta$, we have

$$\frac{1}{x+2} > \frac{1}{\delta} = M.$$

63. Choose $\epsilon > 0$ and let $N = \dfrac{1}{\epsilon}$. Then if $x > N$, we have (assuming $x > 0$ since we are examining limits as $x \to \infty$)

$$\left| \frac{2x-1}{x} - 2 \right| = \left| 2 - \frac{1}{x} - 2 \right| = \frac{1}{x} < \frac{1}{N} = \epsilon.$$

65. Choose $M > 0$ and let $N = \dfrac{M+5}{3}$, so that $M = 3N - 5$. Then if $x > N$, we have

$$3x - 5 > 3N - 5 = M.$$

67. Choose $\epsilon > 0$, and let $\delta = \min\left(1, \frac{\epsilon}{5}\right)$. Now, $x^2 - 2x - 3 = (x-3)(x+1)$, so that if $0 < |x - 3| < \delta \le 1$, then $-1 < x - 3 < 1$ so that $2 < x < 4$ and thus $x + 1 < 5$. Then

$$\left| (x^2 - 2x - 3) - 0 \right| = |x-3| \cdot |x+1| \le 5|x-3| < 5\delta \le 5\frac{\epsilon}{5} = \epsilon.$$

69. Choose $\epsilon > 0$, and let $\delta = \min\left(1, \frac{\epsilon}{5}\right)$. Now, if $0 < |x-5| < \delta \le 1$, then $-1 < x - 5 < 1$ so that $4 < x < 6$, and thus $x - 1$ is bounded above by 5. Then

$$\left| (x^2 - 6x + 7) - 2 \right| = \left| x^2 - 6x + 5 \right| = |(x-5)(x-1)| \le 5|x-5| < 5\delta \le 5\frac{\epsilon}{5} = \epsilon.$$

71. Choose $\epsilon > 0$, and let $\delta = \min\left(1, \dfrac{\epsilon}{5}\right)$. Then if $0 < |x-2| < \delta$, we have $|x-2| < 1$ so that $1 < x < 3$ and thus $\dfrac{1}{x} < 1$. Thus, if $0 < |x-2| < \delta$, we get

$$\left| \frac{4}{x^2} - 1 \right| = \left| \frac{4 - x^2}{x^2} \right| = \frac{|x+2|}{|x|^2} |x-2| < 5|x-2| < 5\delta \le 5 \cdot \frac{\epsilon}{5} = \epsilon.$$

Thinking Forward

Calculating Limits

▷ This statement says that if x is approaching c, then it approaches c, which is intuitively true. Given $\epsilon > 0$, choose $\delta = \epsilon$. Then if $0 < |x - c| < \delta$, we have $|x - c| < \delta = \epsilon$.

▷ This statement says that if x approaches c, then x^2 approaches c^2, which is intuitively reasonable. Given $\epsilon > 0$, choose $\delta = \min\left(1, \frac{\epsilon}{|2c|+1}\right)$. Then if $0 < |x-c| < \delta \le 1$, we have $-1 < x - c < 1$, so that $c - 1 < x < c + 1$. Thus $2c - 1 < x + c < 2c + 1$, so that $|x + c| < |2c| + 1$. Now

$$\left| x^2 - c^2 \right| = |x - c| \cdot |x + c| < (|2c| + 1)|x - c| < (|2c| + 1)\delta \le (|2c| + 1)\frac{\epsilon}{|2c| + 1} = \epsilon.$$

▷   ▷ $\lim\limits_{x \to -1} x = -1$.

▷ $\lim\limits_{x \to 0} x^2 = 0^2 = 0$.

▷ $\lim\limits_{x \to 0} (2x - 3) = 2 \cdot 0 - 3 = -3$.

▷ $\lim\limits_{x \to 4} x = 4$.

▷ $\lim\limits_{x \to 5} x^2 = 5^2 = 25$.

▷ $\lim\limits_{x \to 1} (1 - x) = 1 - 1 = 0$.

▷ $\lim\limits_{x \to \pi} x = \pi$.

▷ $\lim\limits_{x \to -2} x^2 = (-2)^2 = 4$.

▷ $\lim\limits_{x \to 3} (3x + 1) = 3 \cdot 3 + 1 = 10$.

▷ No, it will not. This worked only because for these particular cases, $\lim\limits_{x \to c} f(x) = f(c)$.

1.4   Continuity and its Consequences

Thinking Back

Finding roots of piecewise functions

▷ For $x < 0$, $f(x) = 0$ when $4 - x^2 = 0$, i.e. when $x = -2$. For $x \geq 0$, $x + 1 > 0$ so that there are no zeros in the range $x \geq 0$. The only zero is $x = -2$.

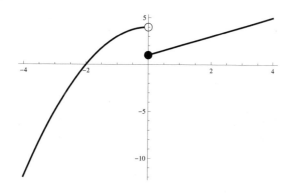

▷ For $x < 0$, $f(x) = 0$ when $x + 1 = 0$, i.e. when $x = -1$. For $x \geq 0$, $4 - x^2 = 0$ when $x = 2$. The zeros are thus $x = -1$ and $x = 2$.

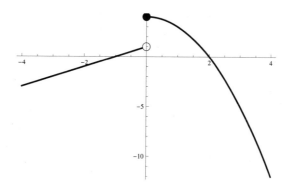

▷ For $x \leq 1$, $f(x) = 0$ when $2x - 1 = 0$, i.e. when $x = \dfrac{1}{2}$. We can factor $2x^2 + x - 3 = (x - 1)(2x + 3)$, so it has roots $x = 1, \dfrac{-3}{2}$. Neither of these is in the range $x > 1$. Hence the only root is at $x = \dfrac{1}{2}$.

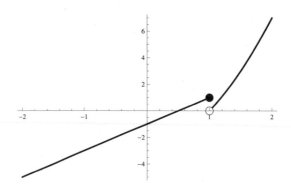

Logical existence statements

▷ False. The statement does not hold if x is negative.

▷ True. Define $y = -x$.

▷ False. If $x = 2$, then $x^2 = 4 \notin (0,4)$ (the open interval).

▷ True. Define $y = \sqrt{x}$ or $y = -\sqrt{x}$.

Concepts

1. (a) True. Since $\lim_{x \to c^-} f(x) = f(c) = \lim_{x \to c^-} f(x)$, it follows that $\lim_{x \to c} f(x) = f(c)$.

   (b) True. See Definition 1.14.

   (c) False. See Definition 1.14; f need be only left-continuous at 5 and only right-continuous at 0 (along with being continuous on the interior). For example, let $f(x) = 3$ for $x < 0$ and $f(x) = 4$ for $x \geq 0$; then f is continuous on $[0,5]$ but is only right-continuous at $x = 0$.

   (d) False. $(2,4)$ is not a closed interval, so the Extreme Value Theorem does not apply. For example, let $f(x) = x$; then f has neither a minimum nor a maximum on $(2,4)$.

   (e) False, unless f is continuous on $[3,9]$. For example, let $f(x) = -5$ for $x \in (-\infty, 6]$ and $f(x) = -2$ for $x \in (6, \infty)$.

   (f) False. Since $f(-2)$ and $f(1)$ have the same sign, there is no reason why f has to change sign somewhere in that interval. For example, let $f(x) = -\frac{1}{3}x + \frac{7}{3}$.

   (g) True. See Theorem 1.18.

   (h) False. f could have another root in $(0,6)$, and the function could be negative somewhere in the interval. For example, $x(x-3)(x-6)$ has the required properties, yet $f(4) < 0$.

3. Since f is continuous, $\lim_{x \to 1} f(x) = f(1)$. See Definition 1.12.

5. For a constant function $f(x) = c$, $|f(x) - c| < \epsilon$ is always true, since $f(x) - c = 0$. Thus it does not matter which x we choose, which is to say there are no restrictions on $|x - c|$, so that any $\delta > 0$ will work. For example, consider $f(x) = 8$ at the point $x = 2$. The graph below shows that $\delta = 1$ works, since the value of the function is always in the shaded bar. Clearly the same is true for any value of δ.

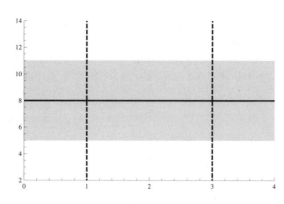

7. Since $f(x) = \dfrac{x^2 - 2x + 1}{x^2 - 6x + 5} = \dfrac{(x-1)^2}{(x-5)(x-1)}$, f has a hole at $x = 1$. Away from 1, $f(x) = \dfrac{x-1}{x-5}$, so defining $f(1) = \lim\limits_{x \to 1} \dfrac{x-1}{x-5} = \dfrac{1-1}{1-5} = 0$ will remove the discontinuity.

9. This is a removable discontinuity since the two one-sided limits are equal, but not equal to the value of f at -1.

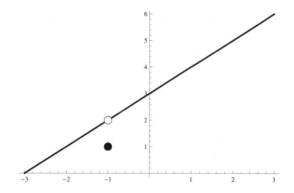

11. This is a jump discontinuity since the two one-sided limits are unequal. Since neither is equal to $f(0)$, f is neither right-continuous nor left-continuous at $x = 0$.

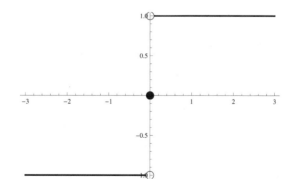

13. It means that for any $\epsilon > 0$, there is some $\delta > 0$ such that whenever $0 < |x - c| < \delta$ then $|f(x) - f(c)| < \epsilon$, since this sentence exactly expresses the requirement that $\lim\limits_{x \to c} f(x) = f(c)$.

15. It means that for any $\epsilon > 0$, there is some $\delta > 0$ such that whenever $0 < x - c < \delta$ then $|f(x) - f(c)| < \epsilon$, since this sentence exactly expresses the requirement that $\lim\limits_{x \to c^+} f(x) = f(c)$.

17. The graph below shows a function continuous on $[-1, 3]$. Since $f(-1) = -3$ and $f(3) = 1$, the Intermediate Value Theorem predicts that there will be at least one point c in $(-1, 3)$ with $f(c) = 0$ (since $-3 < 0 < 1$). In fact there are three such marked points. Note that the Intermediate Value Theorem also predicts that for *any* value y between -3 and 1 there will be at least one point $c \in (-1, 3)$ with $f(c) = y$.

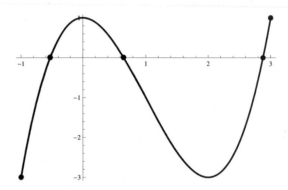

19. See, for example, the first and third graphs following Theorem 1.18.

21.

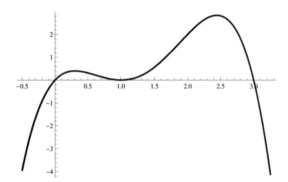

Skills

23. f is continuous on $(0, 2)$ and on $(2, \infty)$. It has an infinite discontinuity at $x = 0$ since $\lim_{x \to 0^+} f(x) = \infty$, and a removable discontinuity at $x = 2$ since $\lim_{x \to 2^-} f(x) = \lim_{x \to 2^+} f(x) = 1$ while $f(2) = 3$.

25. f is continuous on $(-\infty, -1)$, on $(-1, 1]$, and on $(1, \infty)$. (It is left-continuous at $x = 1$. It has a removable discontinuity at $x = -1$ since $\lim_{x \to -1^-} f(x) = \lim_{x \to -1^+} f(x) = 2$ but $f(-1) = -1$. It has an infinite discontinuity at $x = 1$ since $\lim_{x \to 1^+} f(x) = -\infty$.

27. This is not possible. If it is both left-continuous and right-continuous at $x = 1$, then it must be continuous there.

29.

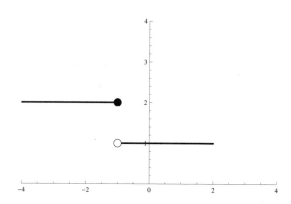

31. This is not possible. Since f is right continuous at $x = -2$, we must have $f(-2) = \lim\limits_{x \to 2^-} f(x)$. But since f has a removable discontinuity at $x = -2$, we also have $\lim\limits_{x \to c} f(x)$ exists, so it must be $f(-2)$. Thus f must be continuous, which is a contradiction.

33. By Theorem 1.16(a), $\lim\limits_{x \to -1} 6 = 6$.

35. By Theorem 1.16(a), $\lim\limits_{x \to -5} (3x - 2) = 3(-5) - 2 = -17$.

37. Since x^{-3} is not defined at $x = 0$, Theorem 1.16 does not apply.

39. By Theorem 1.16, we have $\lim\limits_{x \to 3^-} f(x) = 3 - 3 = 0$ and $\lim\limits_{x \to 3^+} f(x) = -(3 - 3) = 0$, so that the one-sided limits exist and are equal. Since they are equal to $f(3)$, we see that f is continuous at 3.

41. By Theorem 1.16, $\lim\limits_{x \to 2^-} f(x) = 2^2 = 4$ while $\lim\limits_{x \to 2^+} f(x) = 2 \cdot 2 + 1 = 5$. Since the one-sided limits exist but are unequal, this is a jump discontinuity. f is left-continuous at $x = 2$.

43. By Theorem 1.16, $\lim\limits_{x \to 1^-} f(x) = 1 + 1 = 2$ and $\lim\limits_{x \to 1^+} f(x) = 3 \cdot 1 - 1 = 2$. Thus the one-sided limits exist and are equal, and they are equal to $f(1)$. Thus f is continuous at $x = 1$. At $x = 2$, $\lim\limits_{x \to 2^-} f(x) = 3 \cdot 2 - 1 = 5$ while $\lim\limits_{x \to 2^+} 2 + 2 = 4$, so the limits exist but are unequal. Thus f has a jump discontinuity at $x = 2$. Also, f is right-continuous at $x = 2$.

45. Not continuous, since $f(2) = 0$, but there are irrational numbers close to 2; their squares are close to 4. Thus $\lim\limits_{x \to 2} f(x)$ does not even exist, so that f cannot be continuous there.

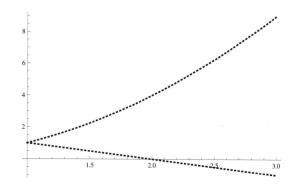

47. $f(1) = 2 - 1 = 1$. For x close to 1, either x is rational, in which case $f(x) = 2 - x$ is also close to 1, or x is irrational, in which case $f(x) = x^2$ is close to 1. In either case, if x is close to 1, so is $f(x)$. So f is continuous at 1.

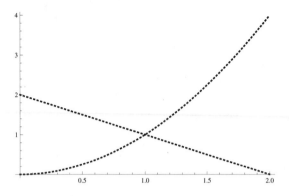

49. Since f is continuous on $[-2,2]$, the Extreme Value Theorem shows that f has both a maximum and a minimum on $[-2,2]$. The maximum occurs at $x = \pm 2$, and the minimum occurs at $x \approx \pm 1.2247$.

51. Since f is continuous on $[-1,1]$, the Extreme Value Theorem shows that f has both a maximum and a minimum on $[-1,1]$. The maximum occurs at $x = 0$, and the minimum occurs at $x = \pm 1$.

53. Since f is continuous on $[0,2]$, the Extreme Value Theorem shows that f has both a maximum and a minimum on $[0,2]$. The maximum occurs at $x = 0$ and $x = 2$, and the minimum occurs at $x \approx 1.333$.

55. Since $f(0) = 5 > 0$ and $f(2) = -11 < 0$, there is some $c \in (0,2)$ such that $f(c) = 0$. There is in fact one such c, namely $c = \sqrt[4]{5} \approx 1.495$.

57. Since $f(-2) = (-2)^3 - 3(-2)^2 - 2 = -22 < -4$ and $f(4) = 4^3 - 3 \cdot 4^2 - 2 = 14 > -4$, it follows that there is some $c \in (-2,4)$ such that $f(c) = -4$. In fact there are three such values: $x = 1$, $x = 1 - \sqrt{3} \approx -0.732$, and $x = 1 + \sqrt{3} \approx 2.732$.

59. Since $f(2) = 2^3 - 3 \cdot 2^2 - 2 = -6 < -4$ and $f(4) = 14 > -4$, it follows that there is some $c \in (2,4)$ such that $f(c) = -4$. In fact there is one such value, namely $c = 1 + \sqrt{3} \approx 2.732$.

61. Use the interval $[-3,0]$. Since $f(-3) = (-3)^3 + 2 = -25 < -15$ and $f(0) = 2 > -15$, it follows that there is some $c \in (-3,0)$ with $f(c) = -15$. In fact there is one such value, namely $c \approx -2.57128$.

63. Use the interval $\left[0, \dfrac{\pi}{2}\right]$. Since $f(0) = 0 < \dfrac{1}{2}$ and $f\left(\dfrac{\pi}{2}\right) = 1 > \dfrac{1}{2}$, it follows that there is some $c \in \left(0, \dfrac{\pi}{2}\right)$ with $f(c) = \dfrac{1}{2}$. In fact there is one such c, namely $c = \dfrac{\pi}{6}$.

65. Use the interval $\left[-\dfrac{1}{3}, 1\right]$. Since $f\left(-\dfrac{1}{3}\right) = 0 < 1$ and $f(1) = 4 > 1$, it follows that there is some $c \in \left(-\dfrac{1}{3}, 1\right)$ with $f(c) = 1$. In fact there is exactly one such c, namely $c = 0$.

67. Since $f(x) = 2 + 5x + 2x^2 = (2x + 1)(x + 2)$, f is zero at $x = -2$ and at $x = -\dfrac{1}{2}$. Since it is continuous, it can change sign only at those points, so to see where it is positive or negative, it suffices to check one point in each of the three resulting intervals.

Interval	$(-\infty, -2)$	$\left(-2, -\tfrac{1}{2}\right)$	$\left(-\tfrac{1}{2}, \infty\right)$
x	-3	-1	0
$f(x)$	5	-1	2
Sign of f	$+$	$-$	$+$

69. We have $f(x) = \dfrac{(x-2)(x+2)}{(x-1)(x+1)}$. By Theorem 1.19, f can change sign only at its zeroes, which are ± 2, and its points of discontinuity, which are ± 1. We can determine the sign of f by testing one point in each of the five resulting intervals:

Interval	$(-\infty, -2)$	$(-2, -1)$	$(-1, 1)$	$(1, 2)$	$(2, \infty)$
x	-3	$-\frac{3}{2}$	0	$\frac{3}{2}$	3
$f(x)$	$\frac{5}{8}$	$-\frac{7}{5}$	4	$-\frac{7}{5}$	$\frac{5}{8}$
Sign of f	$+$	$-$	$+$	$-$	$+$

71. Note that f is continuous at $x = 1$ since both one-sided limits are equal to -3 and $f(1) = -3$. Thus f is continuous everywhere, since each of the pieces is continuous on its domain. Thus f can change sign only at its zeroes. But $x - 4 < 0$ when $x \le 1$, so it has no zeroes, and the zeroes of $x^2 - 4$ are $x = \pm 2$, so its only zero for $x > 1$ is at $x = 2$. Since $f(1) = -3 < 0$ and $f(3) = 9 - 4 = 5 > 0$, we conclude that f is negative on $(-\infty, 2)$ and positive on $(2, \infty)$.

73. Note that f is discontinuous at $x = 2$ since the left-hand limit is 8 while the right-hand limit is 0. Now, x^3 has a zero at $x = 0$, but the zeroes of $4x - x^3$ are 0 and ± 2, none of which is in the range $x > 2$. Thus the only zero of f is at $x = 0$. By Theorem 1.19, we can determine the sign of f by testing one point in each of the resulting three intervals:

Interval	$(-\infty, 0)$	$(0, 2]$	$(2, \infty)$
x	-1	1	3
$f(x)$	-1	1	-15
Sign of f	$-$	$+$	$-$

Applications

75. The Extreme Value Theorem tells us that at some time during those six years, Alina's hair was the longest, and at some time it was the shortest. Realistically, since hair grows longer in the absence of cutting, the minimum was 2 inches and the maximum is 42 inches. The Intermediate Value Theorem tells us that for any hair length between 2 and 42 inches, at some time Alina's hair was that length.

77. The Extreme Value Theorem tells us that at some time during the past year the amount of gasoline in the tank was at a minimum, and at some time it was at a maximum. The minimum is clearly zero, which happened today. The maximum is at least 19 gallons, but may be more if the tank capacity was larger and Phil filled up at some point. The Intermediate Value Theorem tells us that for any given amount of gasoline between 0 and 19 gallons, there was a time at which the tank held that amount of gasoline.

79. (a) Such a function is

$$M(v) = \begin{cases} 8500, & 0 \le v < 30, \\ 10000, & 30 \le v < 60, \\ 11500, & 60 \le v < 90, \\ 13000, & v = 90. \end{cases}$$

(b) We have

$$M(0) = 8500, \quad M(30) = 10000, \quad M(59) = 10000, \quad M(61) = 11500, \quad M(90) = 13000.$$

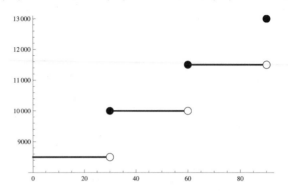

(c) $M(v)$ is discontinuous at $v = 30$, $v = 60$, and $v = 90$. At $v = 30$, the left-hand limit is 8500 while the right-hand limit is 10000. At $v = 60$, the left-hand limit is 10000 while the right-hand limit is 11500. At $v = 90$, the left-hand limit is 11500 while the right-hand limit is 13000.

Proofs

81. Choose $\epsilon > 0$, and let $\delta = \dfrac{\epsilon}{3}$. Then if $0 < |x - 2| < \delta$, we have

$$|(3x - 5) - 1| = |3x - 6| = 3\,|x - 2| < 3\delta = \epsilon,$$

so that $\lim\limits_{x \to 2}(3x - 5) = 1$. Since $f(2) = 3 \cdot 2 - 5 = 1$, we see that $\lim\limits_{x \to 2} f(x) = f(2)$, so that f is continuous at $x = 2$.

83. Choose any point $c \in \mathbb{R}$, and choose $\epsilon > 0$, and let $\delta = \epsilon$. Then if $0 < |x - c| < \delta$, we have

$$\big||x| - |c|\big| \le |x - c| < \delta = \epsilon,$$

so that $\lim\limits_{x \to c} f(x) = |c|$. Since $f(c) = |c|$, we see that $\lim\limits_{x \to c} f(x) = f(c)$, so that f is continuous at $x = c$. Since c was arbitrary, we have proved that f is continuous everywhere.

85. Since $f(2) = 2 - 2 = 0$ while for irrational numbers near 2, $f(x) = x^2$ is close to 4, we do not expect $f(x)$ to be continuous at $x = 2$. If $f(x)$ were continuous there, then $\lim\limits_{x \to 2} f(x) = f(2) = 0$. Choose $\epsilon = 0.5$, and let $\delta > 0$. Choose an irrational number $x \in (2, 2 + \delta)$. Then

$$|f(x) - 0| = \left|x^2\right| > \left|2^2\right| = 4 > \epsilon,$$

so that $\lim\limits_{x \to 2} f(x) \ne f(c)$, and f is not continuous at $x = 2$.

87. Since the equation is a cubic, we may assume that $A \ne 0$. Define $g(x) = x^3 + \dfrac{B}{A}x^2 + \dfrac{C}{A}x + \dfrac{D}{A}$; clearly $f(x)$ and $g(x)$ have the same roots. Thus we are reduced to the case of a cubic $f(x) = x^3 + Ex^2 + Fx + G$. If $E = F = G = 0$, then $f(0) = 0$ and f has a real root. So assume that at least one of E, F, and G is nonzero. Choose any $c > \max(1, |E| + |F| + |G|) > 0$. We claim that $f(c) > 0$. Well, look at the trailing three terms of $f(c)$:

$$\left|Ec^2 + Fc + G\right| \le \left|Ec^2\right| + |Fc| + |G| < \left|Ec^2\right| + \left|Fc^2\right| + \left|Gc^2\right| = c^2(|E| + |F| + |G|) < c^3$$

where the strict inequality follows since $c > 1$. Thus $-c^3 < Ec^2 + Fc + G < c^3$, and then $f(c) = c^3 + (Ec^2 + Fc + G) > 0$. We also claim that $f(-c) < 0$. Again look at the three trailing terms:

$$\left| E(-c)^2 + F(-c) + G \right| \leq \left| Ec^2 \right| + |-Fc| + |G| < \left| Ec^2 \right| + |-Fc(-c)| + \left| Gc^2 \right|$$
$$= c^2 (|E| + |F| + |G|) < c^3$$

so that $-c^3 < E(-c)^2 + F(-c) + G < c^3$. But then $f(-c) = (-c)^3 + (E(-c)^2 + F(-c) + G) = -c^3 +$ something smaller than c^3 in absolute value < 0. Since $f(-c) < 0 < f(c)$, it follows from the Intermediate Value Theorem that f has a real zero between $-c$ and c.

89. Choose $c \in \mathbb{R}$ and choose $\epsilon > 0$. Let $\delta = \min\left(1, \frac{\epsilon}{2|c|+1}\right)$ (note that $2|c| + 1 > 0$, so that this is well-defined). Then if $0 < |x - c| < \delta$, we have also $|x - c| < 1$ so that $-1 < x - c < 1$ and $c - 1 < x < c + 1$. It follows that $2c - 1 < x + c < 2c + 1$, so that $x + c$ is bounded in absolute value by $2|c| + 1$. Now, if $0 < |x - c| < \delta$, then

$$|f(x) - f(c)| = \left| x^2 - c^2 \right| = |x - c| \cdot |x + c| \leq (2|c| + 1) |x - c| < (2|c| + 1)\delta \leq \epsilon,$$

so that $\lim_{x \to c} f(x) = f(c)$ and f is continuous at c. Since $c \in \mathbb{R}$ was arbitrary, f is continuous everywhere.

91. Choose $c \in (0, \infty)$ and choose $\epsilon > 0$. Let $\delta = \min\left(1, \frac{c}{2}, \frac{4(2|c|+1)}{c^4}\right)$. Then if $0 < |x - c| < \delta \leq 1$, we have $-1 < x - c < 1$, so that $2c - 1 < x + c < 2c + 1$. It follows that $x + c$ is bounded in magnitude by $2|c| + 1$. Similarly, if $0 < |x - c| < \delta < \frac{c}{2}$, then $-\frac{c}{2} < x - c < \frac{c}{2}$, so that $x > \frac{c}{2}$ and $\frac{1}{x^2} < \frac{4}{c^2}$. Then

$$|f(x) - f(c)| = \left| \frac{1}{x^2} - \frac{1}{c^2} \right| = \left| \frac{(c-x)(c+x)}{c^2 x^2} \right| = \left| \frac{1}{c^2 x^2} \right| \cdot |c + x| \cdot |c - x| < \frac{4}{c^4} \cdot (2|c| + 1) \cdot \delta \leq \epsilon.$$

It follows that $\lim_{x \to c} f(x) = f(c)$, so that f is continuous at c. Since c was arbitrary, f is continuous everywhere on $(0, \infty)$.

93. Choose $c \in (0, \infty)$ and choose $\epsilon > 0$. Let $\delta = \min(c, \epsilon\sqrt{c})$. Then if $0 < |x - c| < \delta \leq c$, we have $-c < x - c < c$ so that $0 < x < 2c$ and $\sqrt{x} < \sqrt{2c}$. Thus $\sqrt{c} < \sqrt{x} + \sqrt{c} < \sqrt{c} + \sqrt{2c}$, so that $\frac{1}{\sqrt{x} + \sqrt{c}} < \frac{1}{\sqrt{c}}$. Now,

$$|f(x) - f(c)| = \left| \sqrt{x} - \sqrt{c} \right| = \left| \frac{x - c}{\sqrt{x} + \sqrt{c}} \right| = \left| \frac{1}{\sqrt{x} + \sqrt{c}} \right| \cdot |x - c| < \frac{1}{\sqrt{c}} |x - c| < \frac{\delta}{\sqrt{c}} \leq \epsilon.$$

Thus $\lim_{x \to c} f(x) = f(c)$, so that $f(x)$ is continuous at c. Since c was arbitrary, f is continuous on $(0, \infty)$. It remains to show that f is right-continuous at 0. Choose $\epsilon > 0$, and let $\delta = \epsilon^2$. Then if $0 < x < \delta$, we have

$$|f(x) - f(0)| = \left| \sqrt{x} \right| < \sqrt{\delta} = \epsilon,$$

so that $\lim_{x \to 0^+} f(x) = f(0)$, and f is right-continuous at 0. Thus f is continuous everywhere on its domain.

Thinking Forward

Interesting Trigonometric limits

    ▷ Yes, f is continuous at $x = 0$. A graph of f near $x = 0$ is

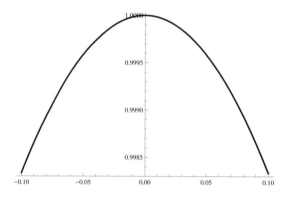

    ▷ No, f oscillates between -1 and $+1$ as $x \to 0$. A graph of f near $x = 0$ is

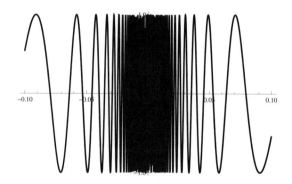

    ▷ Yes, f is continuous at $x = 0$. A graph of f near $x = 0$ is

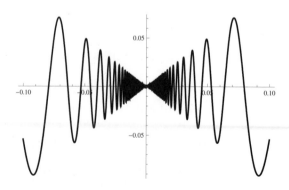

▷ Yes, f is continuous at $x = 0$. A graph of f near $x = 0$ is

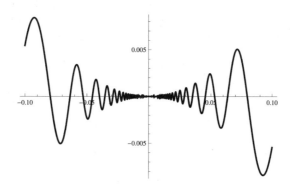

1.5   Limit Rules and Calculating Basic Limits

Thinking Back

Values of transcendental functions

▷ First, $f(\pi) = \dfrac{1}{\sin \pi}$, which is undefined. Second, $f\left(\dfrac{\pi}{2}\right) = \dfrac{1}{\sin \frac{\pi}{2}} = 1$.

▷ First, $f(\pi) = \tan^2(\pi) = 0$. Second, $f\left(\frac{\pi}{2}\right) = \tan^2\left(\frac{\pi}{2}\right)$, which is undefined.

▷ First, $\sin^{-1}(-1) = -\dfrac{\pi}{2}$. Second, $\sin^{-1}\left(\dfrac{1}{2}\right) = \dfrac{\pi}{6}$.

▷ First, $\tan^{-1}(\sqrt{1}) = \tan^{-1}(1) = \dfrac{\pi}{4}$. Second, $\tan^{-1}(\sqrt{3}) = \dfrac{\pi}{3}$.

The $\delta - \epsilon$ definition of limit

▷ For any $\epsilon > 0$, there exists $\delta > 0$ such that whenever $0 < |x - 2| < \delta$, then $|(3x - 2) - 4| < \epsilon$. The δ corresponding to $\epsilon = 0.5$ is $\delta = \dfrac{0.5}{3}$.

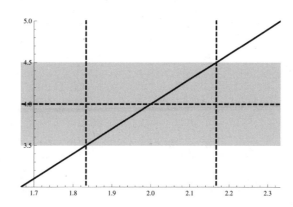

▷ For any $\epsilon > 0$, there exists $\delta > 0$ such that whenever $0 < |x - 1| < \delta$, then $|x^3 - 1| < \epsilon$. The δ corresponding to $\epsilon = 0.5$ is $\delta \approx 0.14$.

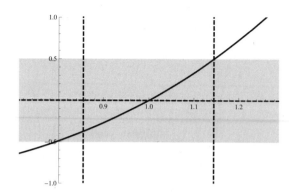

▷ For any $\epsilon > 0$, there exists $\delta > 0$ such that whenever $0 < |x - 1| < \delta$, then $\left|\dfrac{x^2 - 1}{x + 3}\right| < \epsilon$. The δ corresponding to $\epsilon = 0.5$ is $\delta \approx 0.8$.

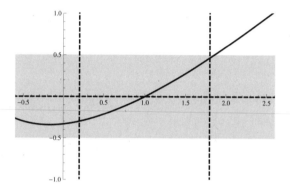

▷ For any $\epsilon > 0$, there exists $\delta > 0$ such that whenever $0 < x < \delta$, then $\left|\sqrt{x + 4} - 2\right| < \epsilon$. The δ corresponding to $\epsilon = 0.5$ is $\delta \approx 1.7$.

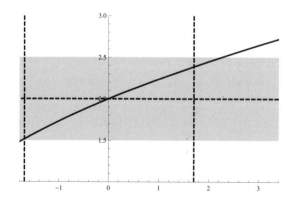

Concepts

1. (a) True. See Theorem 1.20.

    (b) True. The given conditions mean $|f(x) - 7| < 0.25$ and $|g(x) - 2| < 0.25$, so that $-0.25 < f(x) - 7 < 0.25$ and $-0.25 < g(x) - 2 < 0.25$. Adding these two inequalities gives the required result.

    (c) False. For example, if $f(x) = 7$ and $g(x) = 2$, then $f(x)g(x) = 14$, which is not within 0.5 units of 9.

(d) False. It is continuous only at numbers in its domain. For example, $\dfrac{1}{x-1}$ is algebraic, but is not continuous at $x = 1$.

(e) True since 2 is in the domain of $f(x) = Ax^k$. See Theorem 1.27.

(f) False. Since $\cos\dfrac{\pi}{2} = 0$, $\sec x$ is not defined at $\dfrac{\pi}{2}$.

(g) False. The given function is undefined at $x = c$.

(h) True. In computing $\displaystyle\lim_{x\to c} \dfrac{(x-c)f(x)}{(x-c)g(x)}$, we assume that $x \neq c$, so that we may cancel the $x - c$ in the numerator and denominator.

3. See Theorem 1.20.

5. For example, f might be undefined at $x = c$, or it might have a removable discontinuity there. As one example, consider $f(x) = \dfrac{1}{x}$ at $x = 0$. As a second example, see the graph below, with $c = 2$: we have $f(2) = 3$ while $\displaystyle\lim_{x\to 2} f(x) = 2$.

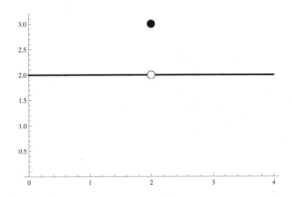

7. Let $f(x) = \dfrac{1}{x}$ and $g(x) = x$. Then $\displaystyle\lim_{x\to 0} f(x)$ does not exist, and $\displaystyle\lim_{x\to 0} g(x) = 0$, but $\displaystyle\lim_{x\to 0}(f(x)g(x)) = \displaystyle\lim_{x\to 0} 1 = 1$. This does not contradict the product rule for limits since the conditions of Theorem 1.20 are not satisfied ($\displaystyle\lim_{x\to 0} f(x)$ does not exist).

9. For any $\epsilon > 0$ there exists a $\delta > 0$ such that whenever $0 < |x - c| < \delta$, then

$$\left|(f(x) - g(x)) - \left(\lim_{x\to c} f(x) - \lim_{x\to c} g(x)\right)\right| < \epsilon.$$

(Note that both $\displaystyle\lim_{x\to c} f(x)$ and $\displaystyle\lim_{x\to c} g(x)$ exist by hypothesis.)

11. Let $c(x) = x^3$ and $s(x) = \sqrt{x}$ be the power functions for $k = 3$ and $k = \frac{1}{2}$, and let $i(x) = 1$ be the constant function. Then $f(x) = c(s(x) + i(x))$, so that $f(x)$ is a combination of these functions. Since $c(x)$, $s(x)$, and $i(x)$ are all continuous on their domain, it follows from Theorem 1.20 that $f(x)$ is as well.

13. Using the constant multiple rule and the difference rule, we have

$$\lim_{x\to 3}(2f(x) - 3g(x)) = \lim_{x\to 3} 2f(x) - \lim_{x\to 3} 3g(x) = 2\lim_{x\to 3} f(x) - 3\lim_{x\to 3} g(x) = 2\cdot 5 - 3\cdot 4 = -2.$$

15. We have no information about the limit of $f(x)$ at $x = 7$, so this is not possible to compute.

17. Using the difference rule and the quotient rule, we have

$$\lim_{x\to 3}\frac{f(x)-3}{g(x)}=\frac{\lim_{x\to 3}(f(x)-3)}{\lim_{x\to 3}g(x)}=\frac{\lim_{x\to 3}f(x)-\lim_{x\to 3}3}{\lim_{x\to 3}g(x)}=\frac{5-3}{4}=\frac{1}{2}.$$

19. The graphs of the two functions are:

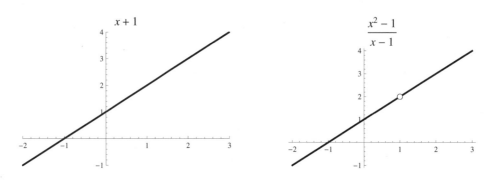

The graphs are identical except at $x=1$, where $f(x)=2$ but $g(x)$ is undefined. Note that away from $x=1$, we can write $g(x)=\dfrac{x^2-1}{x-1}=\dfrac{(x-1)(x+1)}{x-1}=x+1=f(x)$ so that they have the same limit as $x\to 1$.

21. Because finding limits as $x\to c$ does not depend on the value of any of the functions at the point $x=c$, only in the vicinity of $x=c$.

23. Take the figure from the text showing that $\lim_{\theta\to 0^+}\cos\theta=1$ and reflect it through the x-axis. This is a diagram for the situation $\lim_{\theta\to 0^-}\cos\theta$. But the argument is almost the same: it is clear that $0<\cos\theta<|\theta|$; then apply the squeeze theorem.

Skills

25.

$$\lim_{x\to 1}15(3-2x)=15\lim_{x\to 1}(3-2x)\quad\text{constant multiple rule}$$
$$=15(3-2\cdot 1)\qquad\text{continuity of linear functions}$$
$$=15$$

27.

$$\lim_{x\to 3}(3x+x^2(2x+1))=\lim_{x\to 3}3x+\lim_{x\to 3}(x^2(2x+1))\qquad\text{sum rule}$$
$$=\lim_{x\to 3}3x+\left(\lim_{x\to 3}x^2\right)\left(\lim_{x\to 3}(2x+1)\right)\quad\text{product rule}$$
$$=3\cdot 3+3^2(2\cdot 3+1)\qquad\text{continuity of linear and power functions}$$
$$=72\qquad\text{algebra}$$

29. $\lim_{x\to 0}(x^2-1)=0^2-1=-1$ by continuity of algebraic functions.

31. $\lim_{x\to -1}\left(x^2-\dfrac{3x}{x+2}\right)=(-1)^2-\dfrac{3(-1)}{-1+2}=4$ by continuity of algebraic functions.

33. $\lim\limits_{x \to 2} \dfrac{4 + 2x}{x + 2} = \dfrac{4 + 2 \cdot 2}{2 + 2} = 2$ by continuity of algebraic functions.

35. Since $\dfrac{x^2 - 1}{x - 1} = \dfrac{(x + 1)(x - 1)}{x - 1}$, and for $x \neq 1$ this is equal to $x + 1$, the Cancellation Theorem

   shows that $\lim\limits_{x \to 1} \dfrac{x^2 - 1}{x - 1} = \lim\limits_{x \to 1}(x + 1) = 2$ by continuity of algebraic functions.

37. Since $\dfrac{x + 3}{3x^2 + 8x - 3} = \dfrac{x + 3}{(x + 3)(3x - 1)}$, and for $x \neq -3$ this is equal to $\dfrac{1}{3x - 1}$, the Cancellation

   Theorem shows that
   $$\lim_{x \to -3} \frac{x + 3}{3x^2 + 8x - 3} = \lim_{x \to -3} \frac{1}{3x - 1} = -\frac{1}{10}$$
   by continuity of algebraic functions.

39. Since $\dfrac{4x - 2}{6x^2 + x - 2} = \dfrac{4\left(x - \frac{1}{2}\right)}{\left(x - \frac{1}{2}\right)(6x + 4)}$, and for $x \neq \frac{1}{2}$ this is equal to $\dfrac{4}{6x + 4}$, the Cancellation

   Theorem shows that
   $$\lim_{x \to \frac{1}{2}} \frac{4x - 2}{6x^2 + x - 2} = \lim_{x \to \frac{1}{2}} \frac{4}{6x + 4} = \frac{4}{7}.$$

41. Since $\dfrac{x - 1}{\sqrt{x - 1}} = \dfrac{\left(\sqrt{x - 1}\right)^2}{\sqrt{x - 1}}$, and for $x \neq 1$ this is equal to $\sqrt{x - 1}$, the Cancellation Theorem shows

   that
   $$\lim_{x \to 1^+} \frac{x - 1}{\sqrt{x - 1}} = \lim_{x \to 1^+} \sqrt{x - 1} = 0.$$

43. Since $\dfrac{x + x^2 - 2x^3}{x - x^2} = \dfrac{x(2x + 1)(1 - x)}{x(1 - x)}$, and for $x \neq 1$ this is equal to $\dfrac{x(2x + 1)}{x}$, the Cancellation

   Theorem shows that
   $$\lim_{x \to 1} \frac{x + x^2 - 2x^3}{x - x^2} = \lim_{x \to 1} \frac{x(2x + 1)}{x} = 3.$$

45. Since $\dfrac{(1 + h)^2 - 1}{h} = \dfrac{2h + h^2}{h} = \dfrac{h(2 + h)}{h}$, and for $h \neq 0$ this is equal to $2 + h$, the Cancellation

   Theorem shows that
   $$\lim_{h \to 0} \frac{(1 + h)^2 - 1}{h} = \lim_{h \to 0}(2 + h) = 2.$$

47. $\lim\limits_{x \to 0} \dfrac{2^x - 3^x}{4^x} = \dfrac{2^0 - 3^0}{4^0} = 0.$

49. $\lim\limits_{x \to 4}\left(3e^{1.7x} + 1\right) = 3e^{1.7 \cdot 4} + 1 \approx 2694.54.$

51. We have $\dfrac{e^x - 1}{e^{2x} + 2e^x - 3} = \dfrac{e^x - 1}{(e^x - 1)(e^x + 3)}$, and for $e^x \neq 1$ this is equal to $\dfrac{1}{e^x + 3}$. Note that $e^x \neq 1$

   is the same as $x \neq 0$. Then the Cancellation Theorem shows that
   $$\lim_{x \to 0} \frac{e^x - 1}{e^{2x} + 2e^x - 3} = \lim_{x \to 0} \frac{1}{e^x + 3} = \frac{1}{e^0 + 3} = \frac{1}{4}.$$

53. Since by Theorem 1.26, $\lim\limits_{x \to 0} \dfrac{e^x - 1}{x} = 1$, we also have $\lim\limits_{x \to 0} \dfrac{x}{e^x - 1} = 1$ (to see this, note that the prod-

   uct of the two functions is 1, which has limit 1, and apply the quotient rule). Thus $\lim\limits_{x \to 0} \dfrac{2x}{e^x - 1} = 2$

   by the constant multiple rule.

55. For x near π but unequal to π, $\sin(x - \pi) \neq 0$, so that $\dfrac{1}{\csc(x - \pi)} = \sin(x - \pi)$. Then

$$\lim_{x \to \pi} \frac{1}{\csc(x - \pi)} = \lim_{x \to \pi} \sin(x - \pi) = \sin 0 = 0.$$

57. Note that $\dfrac{\sin x}{\tan x} = \dfrac{\sin x}{\sin x / \cos x}$, and for $\sin x \neq 0$ this is equal to $\dfrac{1}{1/\cos x} = \cos x$. Note that if $x \neq 0$ (but x is near zero), then $\sin x \neq 0$. Thus the Cancellation Theorem tells us that

$$\lim_{x \to 0} \frac{\sin x}{\tan x} = \lim_{x \to 0} \cos x = 1.$$

59. Since both x and $\sin^{-1} \dfrac{x}{2}$ are continuous at $x = 1$, we have $\lim\limits_{x \to 1} x \sin^{-1} \dfrac{x}{2} = 1 \cdot \sin^{-1} \dfrac{1}{2} = \dfrac{\pi}{6}$.

61. Since $\sin^{-1} x$ is left-continuous at $x = 1$, we have $\lim\limits_{x \to 1^-} \dfrac{1}{\sin^{-1} x} = \dfrac{1}{\sin^{-1} 1} = \dfrac{2}{\pi}$.

63. $\lim\limits_{h \to 0} \dfrac{(3 + h)^2 - 3^2}{h} = \lim\limits_{h \to 0} \dfrac{9 + 6h + h^2 - 9}{h} = \lim\limits_{h \to 0} \dfrac{6h + h^2}{h} = \lim\limits_{h \to 0} (6 + h) = 6.$

65. $\lim\limits_{h \to 0} \dfrac{(1 + h)^3 - 1^3}{h} = \lim\limits_{h \to 0} \dfrac{1 + 3h + 3h^2 + h^3 - 1}{h} = \lim\limits_{h \to 0} \dfrac{3h + 3h^2 + h^3}{h} = \lim\limits_{h \to 0} (3 + 3h + h^2) = 3.$

67. $\lim\limits_{h \to 0} \dfrac{\frac{1}{1+h} - 1}{h} = \lim\limits_{h \to 0} \dfrac{\frac{1-1-h}{1+h}}{h} = \lim\limits_{h \to 0} \dfrac{-h}{h(1 + h)} = \lim\limits_{h \to 0} \dfrac{-1}{h + 1} = -1.$

69. $\lim\limits_{h \to 0} \dfrac{\frac{4}{(2+h)^2} - 1}{h} = \lim\limits_{h \to 0} \dfrac{\frac{4-4-4h-h^2}{(2+h)^2}}{h} = \lim\limits_{h \to 0} \dfrac{-4h - h^2}{h(2 + h)^2} = \lim\limits_{h \to 0} \dfrac{-4 - h}{(2 + h)^2} = -1.$

71. Since both parts of f are continuous on their domains, the only possible point of discontinuity is at $x = 0$. The left-hand limit is $\lim\limits_{x \to 0^-} f(x) = \lim\limits_{x \to 0^-} (x^2 + 1) = 0^2 + 1 = 1$, while the right-hand limit is $\lim\limits_{x \to 0^+} f(x) = \lim\limits_{x \to 0^+} (1 - x) = 1 - 0 = 1$. Since the left-hand and right-hand limits are equal, and are both equal to $f(1) = 1$, f is continuous everywhere.

73. Since both parts of f are continuous on their domains, the only possible point of discontinuity is at $x = -2$. The left-hand limit is $\lim\limits_{x \to -2^-} (x^2 - 3x - 1) = (-2)^2 - 3(-2) - 1 = 9$, and the right-hand limit is similarly 9. Since $f(-2) = 3 \neq 9$, we see that f is discontinuous at $x = -2$ with a removable discontinuity there. Since neither one-sided limit is equal to $f(-2)$, the function is neither left- nor right-continuous at $x = -2$.

75. The only possible point of discontinuity is at $x = 1$, since the pieces of f are all defined elsewhere. The left-hand limit at $x = 1$ is $\lim\limits_{x \to 1^-} \dfrac{x^2 - 1}{x - 1} = \lim\limits_{x \to 1^-} (x + 1) = 2$, while the right-hand limit is $\lim\limits_{x \to 1^+} (3x - 1) = 3 \cdot 1 - 1 = 2$. However, $f(1) = 0$. Thus f has a removable discontinuity at $x = 1$, and is neither left- nor right-continuous there.

77. The pieces of f are continuous everywhere, so $x = \pi$ is the only possible point of discontinuity. The left-hand limit is $\lim\limits_{x \to \pi^-} f(x) = \lim\limits_{x \to \pi^-} \sin x = \sin \pi = 0$, while the right-hand limit is $\lim\limits_{x \to \pi^+} f(x) = \lim\limits_{x \to \pi^+} \cos x = \cos \pi = -1$. Further, $f(\pi) = \cos \pi = -1$. The two one-sided limits exist but are unequal, and the right-hand limit is equal to $f(\pi)$. Thus f has a jump discontinuity at $x = \pi$, and is right-continuous there.

79. For any nonzero value of x, we know that $-1 < \sin \frac{1}{x} < 1$ since $[-1, 1]$ is the range of sin. Thus, for $x > 0$, we have $-x \le x \sin \frac{1}{x} \le x$. Then by the Squeeze Theorem,

$$0 = \lim_{x \to 0^+} -x \le \lim_{x \to 0^+} x \sin \frac{1}{x} \le \lim_{x \to 0^+} x = 0.$$

Similarly, for $x < 0$, we have $x \le x \sin \frac{1}{x} \le -x$, and a similar argument shows that $\lim_{x \to 0^-} x \sin \frac{1}{x} = 0$. Since the one-sided limits are both equal to zero, the limit exists and is also zero.

81. Since the range of $\sin \frac{1}{x}$ is $[-1, 1]$, and it is defined for $x \ne 0$, we have for $x > 0$ that $-(e^x - 1) \le (e^x - 1) \sin \frac{1}{x} \le e^x - 1$, so by the Squeeze Theorem,

$$0 = \lim_{x \to 0^+} -(e^x - 1) \le \lim_{x \to 0^+} (e^x - 1) \sin \frac{1}{x} \le \lim_{x \to 0^+} e^x - 1 = 0,$$

so that the right-hand limit is zero. Similarly, for $x < 0$, where $e^x - 1 < 0$, we have $e^x - 1 \le (e^x - 1) \sin \frac{1}{x} \le -(e^x - 1)$, and again by the Squeeze Theorem,

$$0 = \lim_{x \to 0^-} e^x - 1 \le \lim_{x \to 0^-} (e^x - 1) \sin \frac{1}{x} \le \lim_{x \to 0^-} -(e^x - 1) = 0,$$

so that the left-hand limit is also zero. Thus the limit exists and is zero.

83. The range of $\cos \frac{1}{x-1}$ is $[-1, 1]$, and it is defined for $x \ne 1$. Then for $x > 1$, we have $-(x - 1) \le (x - 1) \cos \frac{1}{x-1} \le x - 1$, so that by the Squeeze Theorem,

$$0 = \lim_{x \to 1^+} -(x - 1) \le \lim_{x \to 1^+} (x - 1) \cos \frac{1}{x - 1} \le \lim_{x \to 1^+} x - 1 = 0,$$

so that the right-hand limit is zero. An argument similar to that used in the previous several exercises shows that the left-hand limit is zero as well, so that the limit exists and is zero.

85. For $x \ne 0$, $\tan^{-1} \frac{1}{x}$ is defined, and the range of $\tan^{-1}$ is $\left(-\frac{\pi}{2}, \frac{\pi}{2}\right)$. Thus for $x > 0$, we have $-x \frac{\pi}{2} \le x \tan^{-1} \frac{1}{x} \le x \frac{\pi}{2}$, so by the Squeeze Theorem,

$$0 = \lim_{x \to 0^+} -x \frac{\pi}{2} \le \lim_{x \to 0^+} x \tan^{-1} \frac{1}{x} \le \lim_{x \to 0^+} x \frac{\pi}{2} = 0,$$

so that the right-hand limit exists and is zero. Similarly to the preceding exercises, the left-hand limit also exists and is zero, so the limit exists and is zero as well.

Applications

87. (a) The second line of the table applies, so that $T(63550) = 3937 + 0.28(63550 - 26250) = 14381$. The left-hand limit $\lim_{m \to 63550^-} T(m)$ is also 14381 by continuity. The right-hand limit is $\lim_{m \to 63550^+} T(m) = \lim_{m \to 63550^+} (14381 + 0.31(m - 63550)) = 14381$, again by continuity.

   (b) Since the left and right limits are equal, and are equal to $T(63550)$, it follows that T is continuous at $m = 63550$. In practice, this means that the tax code is reasonable in the sense that if you earn \$63551 rather than \$63449, your taxes will not change by a large amount.

Proofs

89. Let $f(x) = a_n x^n + a_{n-1} x^{n-1} + \cdots + a_1 x + a_0$, and choose $c \in \mathbb{R}$. Then

$$\lim_{x \to c} f(x) = \lim_{x \to c} (a_n x^n + a_{n-1} x^{n-1} + \cdots + a_1 x + a_0)$$

$$= \lim_{x \to c} a_n x^n + \lim_{x \to c} a_{n-1} x^{n-1} + \cdots + \lim_{x \to c} a_1 x + \lim_{x \to c} a_0 \quad \text{(sum rule)}$$

$$= a_n c^n + a_{n-1} c^{n-1} + \cdots + a_1 c + a_0 \quad\quad\quad \text{(continuity of power functions)}$$

$$= f(c)$$

Since $\lim_{x \to c} f(x) = f(c)$ and c was arbitrary, it follows that f is everywhere continuous.

91. Choose $\epsilon > 0$. Since $\lim_{x \to c} f(x) = L$, there is some δ such that if $0 < |x - c| < \delta$, then $|f(x) - L| < \dfrac{\epsilon}{|k|}$. Then if $0 < |x - c| < \delta$, we have

$$|kf(x) - kL| = |k|\,|f(x) - L| < |k| \cdot \frac{\epsilon}{|k|} = \epsilon.$$

This shows that $\lim_{x \to c} kf(x) = kL$.

93. Suppose that $\lim_{x \to c} f(x)$ and $\lim_{x \to c} g(x)$ exist and that $\lim_{x \to c} g(x) \neq 0$. Then

$$\lim_{x \to c} \frac{f(x)}{g(x)} = \lim_{x \to c} \left(f(x) \cdot \frac{1}{g(x)} \right)$$

$$= \lim_{x \to c} f(x) \cdot \lim_{x \to c} \frac{1}{g(x)} \quad \text{(product rule)}$$

$$= \lim_{x \to c} f(x) \cdot \frac{1}{\lim_{x \to c} g(x)} \quad \text{(given reciprocal rule)}$$

$$= \frac{\lim_{x \to c} f(x)}{\lim_{x \to c} g(x)}$$

95. Assuming $b > 0$ so that the definition of f makes sense, we have $f(x) = Ab^x = Ae^{x \ln b}$. This is continuous by Exercise 94.

97.

$$\lim_{x \to c} \cos x = \lim_{h \to 0} \cos(c + h) \quad\quad\quad \text{(change variables)}$$

$$= \lim_{h \to 0} (\cos c \cos h - \sin c \sin h) \quad\quad \text{(sum identity for cos)}$$

$$= \cos c \lim_{h \to 0} \cos h - \sin c \lim_{h \to 0} \sin h \quad \text{(sum, constant multiple rules)}$$

$$= (\cos c)(1) - (\sin c)(0) = \cos c.$$

99. Choose c in the domain of sec. Then using the continuity of cos and the quotient rule, we have

$$\lim_{x \to c} \sec x = \lim_{x \to c} \frac{1}{\cos x} = \frac{\lim_{x \to c} 1}{\lim_{x \to c} \cos x} = \frac{1}{\cos c} = \sec c.$$

Thinking Forward

Limits for Derivatives

▷ $\lim\limits_{h \to 0} \dfrac{f(0+h) - f(0)}{h} = \lim\limits_{h \to 0} \dfrac{h^2}{h} = \lim\limits_{h \to 0} h = 0.$

▷ $\lim\limits_{h \to 0} \dfrac{f(2+h) - f(2)}{h} = \lim\limits_{h \to 0} \dfrac{4 + 4h + h^2 - 4}{h} = \lim\limits_{h \to 0} \dfrac{4h + h^2}{h} = \lim\limits_{h \to 0} (4 + h) = 4.$

▷ $\lim\limits_{h \to 0} \dfrac{f(4+h) - f(4)}{h} = \lim\limits_{h \to 0} \dfrac{16 + 8h + h^2 - 16}{h} = \lim\limits_{h \to 0} \dfrac{8h + h^2}{h} = \lim\limits_{h \to 0} (8 + h) = 8.$

▷

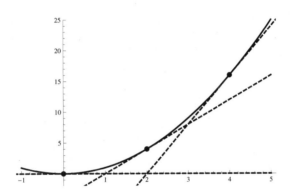

The slopes of these three lines are equal to the values determined in the first three parts.

▷

$$\begin{aligned}
\lim_{h \to 0} \frac{f(4+h) - f(4)}{h} &= \lim_{h \to 0} \frac{\sqrt{4+h} - 2}{h} \\
&= \lim_{h \to 0} \frac{(\sqrt{4+h} - 2)(\sqrt{4+h} + 2)}{h(\sqrt{4+h} + 2)} \\
&= \lim_{h \to 0} \frac{4 + h - 4}{h(\sqrt{4+h} + 2)} \\
&= \lim_{h \to 0} \frac{1}{\sqrt{4+h} + 2} = \frac{1}{4}.
\end{aligned}$$

▷

$$\lim_{h \to 0} \frac{f(4+h) - f(4)}{h} = \lim_{h \to 0} \frac{\frac{1}{\sqrt{4+h}} - \frac{1}{2}}{h}$$

$$= \lim_{h \to 0} \frac{\frac{2 - \sqrt{4+h}}{2\sqrt{4+h}}}{h}$$

$$= \lim_{h \to 0} \frac{2 - \sqrt{4+h}}{2h\sqrt{4+h}}$$

$$= \lim_{h \to 0} \frac{(2 - \sqrt{4+h})(2 + \sqrt{4+h})}{2h\sqrt{4+h}(2 + \sqrt{4+h})}$$

$$= \lim_{h \to 0} \frac{4 - 4 - h}{2h\sqrt{4+h}(2 + \sqrt{4+h})}$$

$$= \lim_{h \to 0} \frac{-1}{2\sqrt{4+h}(2 + \sqrt{4+h})}$$

$$= \frac{-1}{2\sqrt{4}(2 + \sqrt{4})} = -\frac{1}{16}.$$

▷ Using Theorem 1.26, we have

$$\lim_{h \to 0} \frac{f(0+h) - f(0)}{h} = \lim_{h \to 0} \frac{e^h - e^0}{h} = \lim_{h \to 0} \frac{e^h - 1}{h} = 1.$$

1.6   Infinite Limits and Indeterminate Forms

Thinking Back

Behavior of transcendental functions

▷ Since $\csc x = \dfrac{1}{\sin x}$, it becomes infinite as $x \to 0$ and as $x \to \pi$.

▷ We have $\tan^2 x = \dfrac{\sin^2 x}{\cos^2 x} \to \dfrac{0^2}{(\pm 1)^2} = 0$ as $x \to 0$ and as $x \to \pi$.

▷ $\sin^{-1} 0 = 0$ and $\sin^{-1} 1 = \dfrac{\pi}{4}$; since $\sin^{-1}$ is continuous, these are the limits as well.

▷ $\tan^{-1} 0 = 0$ and $\tan^{-1} \sqrt{3} = \dfrac{\pi}{3}$. Since $\tan^{-1}$ is continuous, these are the limits as well.

The definition of infinite limits and limits at infinity

▷ For any $\epsilon > 0$ there is some $N > 0$ such that whenever $x > N$, then $\left| \dfrac{2x}{x-1} - 2 \right| < \epsilon$. For $\epsilon = 0.5$, we want $\left| \dfrac{2x}{x-1} - 2 \right| = \left| \dfrac{2}{x-1} \right| < 0.5$. Since we are taking limits as $x \to \infty$, we may assume $x > 0$, so that $0 < \frac{2}{x-1} < 0.5$. Thus $2 < 0.5(x-1)$, so that $x > 5$. It follows that $N = 5$ is the smallest such N.

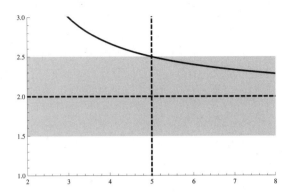

▷ For any $M > 0$ there is some $\delta > 0$ such that whenever $0 < x - 2 < \delta$ then $\dfrac{1}{x^2 - 4} > M$. For $M = 100$, we want $x^2 - 4 < \dfrac{1}{100}$, so that $x^2 < 4.01$ and $x < \sqrt{4.01} \approx 2.0025$. Thus $\delta \approx 0.0025$.

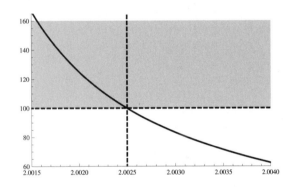

▷ For any $M > 0$ there is some $N > 0$ such that whenever $x > N$ then $\sqrt{x - 1} > M$. For $M = 100$, we want $\sqrt{x - 1} > 100$, so that $x > 10001$. Thus $N = 10001$ is the smallest such N.

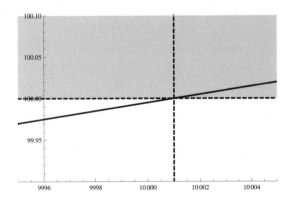

▷ For any $\epsilon > 0$ there is some $N > 0$ such that whenever $x < -N$, then $\left|\dfrac{1}{x}\right| < \epsilon$. For $\epsilon = 0.5$, we want $-0.5 < \dfrac{1}{x} < 0.5$, so that $x > 2$ or $x < -2$. Thus $N = 2$ is the smallest such N.

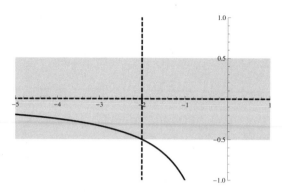

Concepts

1. (a) True. See Theorem 1.29 (a).

   (b) Mostly true. See Theorem 1.30(a). However, ∞^{+} does not really make sense.

   (c) False. See for example Theorem 1.35.

   (d) True. See Definition 1.1 and the paragraph following.

   (e) True. See Theorem 1.34 (b).

   (f) True. $\lim\limits_{x \to \infty} 2^x = \infty$.

   (g) False. For example, $\lim\limits_{x \to \infty} (2x - x) \neq 0$.

   (h) False, if f is discontinuous at $x = c$.

3. Since $M \neq 0$, the Quotient Rule implies that $\lim\limits_{x \to c} \dfrac{f(x)}{g(x)} = \dfrac{L}{M}$.

5. By Theorem 1.29, $\lim\limits_{x \to c} \dfrac{f(x)}{g(x)} = \pm\infty$.

7. Only $\infty - \infty$ is indeterminate (see Theorem 1.33). The other forms must all approach ∞.

9. The forms $\dfrac{0}{0}$ and $\dfrac{\infty}{\infty}$ are indeterminate, by Theorem 1.33. The forms $\dfrac{0}{\infty}, \dfrac{0}{1}$, and $\dfrac{1}{\infty}$ all approach zero, while the forms $\dfrac{\infty}{0}, \dfrac{1}{0}$, and $\dfrac{\infty}{1}$ all approach ∞.

11. For $\dfrac{1}{\infty}$, since the denominator gets arbitrarily large while the numerator remains close to 1, the quotient gets arbitrarily small, so approaches zero. For $\dfrac{0}{\infty}$, the same argument holds: the denominator gets arbitrarily large while the denominator remains close to zero, so that the quotient gets arbitrarily small, so approaches zero. Finally for $\dfrac{0}{1}$, since the numerator remains close to zero while the denominator is close to 1, the quotient is close to $\dfrac{0}{1} = 0$.

13. These arguments are exactly the arguments in Exercise 11 with numerator and denominator reversed.

15. For example, (many other solutions are possible)

(a) $\lim\limits_{x \to 0} \dfrac{x^2}{x} = 0.$
(b) $\lim\limits_{x \to 0} \dfrac{2x}{x} = 2.$
(c) $\lim\limits_{x \to 0} \dfrac{x}{x^2} = \infty.$

17. For example, (many other solutions are possible)

(a) $\lim\limits_{x \to \infty} (x^{-1} \cdot \ln x) = 0.$
(b) $\lim\limits_{x \to \infty} (x^{-1} \cdot x) = 1.$
(c) $\lim\limits_{x \to \infty} \left((\ln x)^{-1} \cdot x \right) = \infty.$

19. For example, (many other solutions are possible)

(a) $\lim\limits_{x \to 0} x^x = 1.$
(b) $\lim\limits_{x \to \infty} \left(x^{-x} \right)^{1/x} = 0.$
(c) $\lim\limits_{x \to \infty} \left(x^{-x} \right)^{-1/x} = \infty.$

21. For example, (many other solutions are possible)

(a) $\lim\limits_{x \to \infty} x^{1/x} = 1.$
(b) $\lim\limits_{x \to \infty} (2^x)^{1/x} = 2.$
(c) $\lim\limits_{x \to \infty} (x^x)^{1/x} = \infty.$

Skills

23. Since $x^2 - 2x - 3 = (x-3)(x+1)$, we conclude that $f(x)$ has a root at $x = -1$ and a discontinuity at $x = 3$. At $x = 3$, $\lim\limits_{x \to 3} f(x) = \lim\limits_{x \to 3} (x+1) = 4$, so the limit exists and the discontinuity is removable. Since the degree of the numerator exceeds that of the denominator, there are no horizontal asymptotes.

25. $f(x)$ has a root at $x = -1$. At $x = -2$, the limit form is $\dfrac{4}{-4 \cdot 0^-} = \dfrac{1}{0^+}$ from the left, and $\dfrac{4}{-4 \cdot 0^+} = \dfrac{1}{0^-}$ from the right, so that $x = -2$ is a vertical asymptote with $\lim\limits_{x \to -2^-} f(x) = \infty$ and $\lim\limits_{x \to -2^+} f(x) = -\infty$. At $x = 2$, $\lim\limits_{x \to 2} f(x) = \lim\limits_{x \to 2} \dfrac{x+1}{x+2} = \dfrac{3}{4}$, so the limit exists and f has a removable discontinuity at $x = 2$. Finally, since the degree of the numerator and denominator are equal, there is a two-sided horizontal asymptote, at $y = 1$:

$$\lim_{x \to \pm\infty} \frac{(x+1)(x-2)}{(x-2)(x+2)} = \lim_{x \to \pm\infty} \frac{x+1}{x+2} = \lim_{x \to \pm\infty} \frac{1+1/x}{1+2/x} = 1.$$

27. Since the multiplicity of $x - 2$ in the denominator is higher than that in the numerator, f has a vertical asymptote at $x = 2$, as it does at $x = -2$. f does have a zero at $x = -1$. At $x = 2$, the limit from the left side is of the form (after canceling one pair of $x - 2$'s) $\dfrac{3}{4 \cdot 0^-}$, and from the right is $\dfrac{3}{4 \cdot 0^+}$. Thus $\lim\limits_{x \to 2^-} f(x) = -\infty$ and $\lim\limits_{x \to 2^+} f(x) = \infty$. At $x = -2$, the left-hand limit form is $\dfrac{4}{16 \cdot 0^-} = \dfrac{1}{4 \cdot 0^-}$, while the right-hand form is $\dfrac{1}{4 \cdot 0^+}$, so that $\lim\limits_{x \to -2^-} f(x) = -\infty$ and $\lim\limits_{x \to -2^+} f(x) = \infty$. Finally, since the degree of the denominator exceeds that of the numerator, $y = 0$ is a two-sided horizontal asymptote:

$$\lim_{x \to \pm\infty} f(x) = \lim_{x \to \pm\infty} \frac{x+1}{x^2-4} = \lim_{x \to \pm\infty} \frac{1+1/x}{x-4/x} = \frac{1}{\infty} = 0.$$

29. $f(x)$ has no roots since the numerator never vanishes, and no vertical asymptotes since the denominator never vanishes. However, f has a one-sided horizontal asymptote as $x \to \infty$ at $y = \dfrac{1}{2}$:

$$\lim_{x \to \infty} f(x) = \frac{\displaystyle\lim_{x \to \infty} 2}{\displaystyle\lim_{x \to \infty} (4 + e^{-2x})} = \frac{2}{4 + 0} = \frac{1}{2}.$$

It also has a one-sided horizontal asymptote as $x \to -\infty$ at $y = 0$:

$$\lim_{x \to -\infty} f(x) = \frac{\displaystyle\lim_{x \to -\infty} 2}{\displaystyle\lim_{x \to -\infty} (4 + e^{-2x})} = \frac{2}{4 + \infty} = 0.$$

31. f has a root at $x = 0$. Since the denominator never vanishes, it has no vertical asymptotes. It also has no horizontal asymptotes, since

$$\lim_{x \to \infty} f(x) = \lim_{x \to \infty} \left(\left(\frac{2}{3}\right)^x - \left(\frac{4}{3}\right)^x \right) = 0 - \infty = -\infty, \text{ and}$$

$$\lim_{x \to -\infty} f(x) = \lim_{x \to -\infty} \left(\left(\frac{2}{3}\right)^x - \left(\frac{4}{3}\right)^x \right) = \infty - 0 = \infty.$$

33. f has a root where $\tan^{-1}(3x) = -1$, which occurs when $3x = \tan(-1)$, so that $x = -\dfrac{\tan 1}{3} \approx$ -0.519. f has a horizontal asymptote at $1 + \dfrac{\pi}{2}$ as $x \to \infty$ and one at $1 - \dfrac{\pi}{2}$ as $x \to -\infty$, since $\displaystyle\lim_{x \to \infty} \tan^{-1}(3x) = \dfrac{\pi}{2}$ and $\displaystyle\lim_{x \to -\infty} \tan^{-1}(3x) = -\dfrac{\pi}{2}$.

35.

$$\lim_{x \to 0^-} -4x^{-3} = \lim_{x \to 0^-} \frac{-4}{x^3} = \frac{-4}{0^-} = \infty$$

$$\lim_{x \to 0^+} -4x^{-3} = \lim_{x \to 0^+} \frac{-4}{x^3} = \frac{-4}{0^+} = -\infty$$

Since the two one-sided limits differ, the limit as $x \to 0$ does not exist.

37. $\displaystyle\lim_{x \to \infty} 2x^{-4/3} = \lim_{x \to \infty} \frac{2}{x^{4/3}} = \frac{2}{\infty} = 0.$

39. $\displaystyle\lim_{x \to \infty} (\sqrt{x} - x) = \lim_{x \to \infty} \sqrt{x}(1 - \sqrt{x}) = \infty \cdot (-\infty) = -\infty.$

41. $\displaystyle\lim_{x \to \infty} (-3x^5 + 4x + 11) = \lim_{x \to \infty} x^5(-3 + 4/x^4 + 11/x^5) = -\infty.$

43. $\displaystyle\lim_{x \to -4} \frac{x^2 + 8x + 16}{(x+4)^2(x+1)} = \lim_{x \to -4} \frac{(x+4)^2}{(x+4)^2(x+1)} = \lim_{x \to -4} \frac{1}{x+1} = -\frac{1}{3}.$

45. $\dfrac{x^2 + 1}{x(x-1)} \to \dfrac{1}{0^-(-1)} = \dfrac{1}{0^+}$ from the left, and $\dfrac{1}{0^-(1)} = \dfrac{1}{0^-}$ from the right. Thus $\displaystyle\lim_{x \to 0^-} \dfrac{x^2 + 1}{x(x-1)} = \infty$ and $\displaystyle\lim_{x \to 0^+} \dfrac{x^2 + 1}{x(x-1)} = -\infty$. Since the two one-sided limits are unequal, the limit as $x \to 0$ does not exist.

47. $\displaystyle\lim_{x \to 0} \frac{x}{x^2 - x} = \lim_{x \to 0} \frac{1}{x - 1} = -1.$

49.

$$\lim_{x \to \infty} \frac{(3x+1)^2(x-1)}{1-x^3} = \lim_{x \to \infty} \frac{(3x+1)^2(x-1)}{(1-x)(1+x+x^2)}$$

$$= -\lim_{x \to \infty} \frac{9x^2+6x+1}{x^2+x+1}$$

$$= -\lim_{x \to \infty} \frac{9+6/x+1/x^2}{1+1/x+1/x^2} = -9$$

51. $\lim_{x \to 0^+} \left(x^{-1/3} - x^{-1/2} \right) = \lim_{x \to 0^+} x^{-1/2}(x^{1/6} - 1) = \infty \cdot (-1) = -\infty.$

53. $\lim_{x \to \infty} \dfrac{x^{-3}}{x^2 - x^{-1}} = \lim_{x \to \infty} \dfrac{1}{x^5 - x^4} = \lim_{x \to \infty} \dfrac{1/x^5}{1 - 1/x} = 0.$

55. $\lim_{x \to 0^+} \dfrac{x^{7/2} - x^{8/3}}{x^2} = \lim_{x \to 0^+} (x^{3/2} - x^{2/3}) = 0 - 0 = 0.$

57. $\lim_{x \to \infty} \dfrac{4^x - 3^x}{5^x} = \lim_{x \to \infty} \left(\left(\dfrac{4}{5}\right)^x - \left(\dfrac{3}{5}\right)^x \right) = 0 - 0 = 0.$

59. $\lim_{x \to -\infty} \dfrac{3^x - 5^x}{4^x} = \lim_{x \to -\infty} \left(\left(\dfrac{3}{4}\right)^x - \left(\dfrac{5}{4}\right)^x \right) = \infty - 0 = \infty.$

61. $\lim_{x \to \infty} \dfrac{2e^{1.5x}}{3e^{2x} + e^{1.5x}} = \lim_{x \to \infty} \dfrac{2}{3e^{0.5x} + 1} = \dfrac{2}{3\infty + 1} = 0.$

63. $\lim_{x \to 3^+} \ln(x^2 - 9) = \lim_{x \to 3^+} \ln(x-3)\ln(x+3) = \ln 6 \cdot \lim_{x \to 3^+} \ln(x-3) = -\infty.$

65. $\lim_{x \to \infty} (\ln x^2 - \ln(2x+1)) = \lim_{x \to \infty} \ln \dfrac{x^2}{2x+1} = \ln \infty = \infty.$

67. $\lim_{x \to 0} \dfrac{1 - \cos 2x}{7x} = \dfrac{2}{7} \lim_{x \to 0} \dfrac{1 - \cos 2x}{2x} = \dfrac{2}{7} \cdot 0 = 0.$

69.

$$\lim_{x \to 0^-} \frac{x}{1 - \cos x} = \lim_{x \to 0^-} \frac{1}{\frac{1 - \cos x}{x}} = \frac{1}{0^-} = -\infty$$

$$\lim_{x \to 0^+} \frac{x}{1 - \cos x} = \lim_{x \to 0^+} \frac{1}{\frac{1 - \cos x}{x}} = \frac{1}{0^+} = \infty$$

Since the two one-sided limits differ, the two sided limit at $x = 0$ does not exist.

71. $\lim_{x \to 0} \dfrac{\sin^2 3x}{x^3 - x} = \lim_{x \to 0} \left(\dfrac{\sin 3x}{x} \right) \left(\dfrac{\sin 3x}{x} \right) \left(\dfrac{1}{x - 1/x} \right) = 3 \cdot 3 \cdot 0 = 0.$

73. $\lim_{x \to 0^+} \dfrac{x^2 \csc 3x}{1 - \cos 2x} = \lim_{x \to 0^+} \dfrac{x^2}{\sin 3x(1 - \cos 2x)} = \dfrac{1}{2} \lim_{x \to 0^+} \left(\dfrac{x}{\sin 3x} \right) \left(\dfrac{2x}{1 - \cos 2x} \right) = \dfrac{1}{2} \cdot \dfrac{1}{3} \cdot \infty = \infty.$

75. $\lim_{x \to 0} \dfrac{\sec x \tan x}{x} = \lim_{x \to 0} \dfrac{\sin x}{x \cos^2 x} = \lim_{x \to 0} \left(\dfrac{\sin x}{x} \right) \left(\dfrac{1}{\cos^2 x} \right) = 1.$

77. $\lim_{x \to 0} (1 + x)^{2/x} = \lim_{x \to 0} ((1 + x)^{1/x})^2 = e^2.$

79. Write $y = 1/x$; then we want to evaluate $\lim_{1/y \to \infty} (1 + y)^{3/y} = \lim_{y \to 0} \left((1 + y)^{1/y} \right)^3 = e^3.$

Applications

81. $\lim_{t \to 2027^-} P(t) = \lim_{t \to 2027^-} \dfrac{179 \times 10^9}{(2027 - t)^{0.99}} = \dfrac{179 \times 10^9}{0} = \infty.$ What the model predicts is that the population will grow without bound as the year gets close to 2027. What it most likely means is that the model is invalid in that range.

83. (a) $\lim_{t \to \infty} s(t)$ does not exist, since $s(t) = B$ when $t = 2\pi n \sqrt{\dfrac{m}{k}}$ where n is an integer, and $s(t) = -B$ when $t = (2n+1)\pi \sqrt{\dfrac{m}{k}}$ where n is an integer. (Should B be zero, choose values of t where sin is either 1 or -1). Thus the mass oscillates forever.

    (b) Air resistance would act to damp, or retard, the motion of the mass, so that it would eventually come to rest. Thus this model does not accurately predict the real world behavior.

    (c) The graph below, of position versus time, shows that the mass oscillates with a constant amplitude, so it is not being damped and will not come to rest.

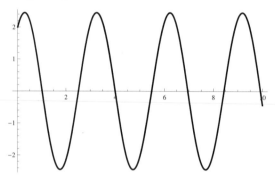

Proofs

85. Suppose $f(x)$ is as in the problem statement. Then

$$f(x) = a_n x^n \left(1 + \frac{a_{n-1}}{a^n} x^{-1} + \frac{a^{n-2}}{a_n} x^{-2} + \cdots + \frac{a_1}{a_n} x^{1-n} + \frac{a_1}{a_n} x^{-n} \right).$$

Now take the limit as $x \to \pm\infty$. The terms in parentheses all go to zero as $x \to \pm\infty$ except for the first term, which is 1. Thus $\lim_{x \to \pm\infty} f(x) = 1 \cdot \lim_{x \to \pm\infty} a_n x^n$, which is what was to be proved.

87. Choose $\epsilon_1 > 0$. Since $\lim_{x \to c} f(x) = 1$, there exists a $\delta_1 > 0$ such that if $0 < |x - c| < \delta_1$, then $|f(x) - 1| < \epsilon_1$. In particular, if $-\delta_1 < x - c < 0$, then $|f(x) - 1| < \epsilon_1$. Now, $0 < x - c < \delta_1$ means $c - \delta_1 < x < c$, and $|f(x) - 1| < \epsilon_1$ means that $-\epsilon_1 < f(x) - 1 < \epsilon_1$, so that $1 - \epsilon_1 < f(x) < 1 + \epsilon_1$. Similarly, since $\lim_{x \to c^-} g(x) = 0$, we know that for any $\epsilon_2 > 0$ there is a $\delta_2 > 0$ such that if $c - \delta_2 < x < c$, then $-\epsilon_2 < g(x) < 0$. Choose $M > 0$, and let δ_1 correspond to $\epsilon_1 = \dfrac{1}{2}$ and δ_2 correspond to $\epsilon_2 = \dfrac{1}{2M}$. Define $\delta = \min(\delta_1, \delta_2)$. Then if $c - \delta < x < c$, we have

$$\frac{f(x)}{g(x)} < -\frac{1 - \epsilon_1}{\epsilon_2} = -\frac{1/2}{1/(2M)} = -M,$$

so that $\lim_{x \to c^-} \dfrac{f(x)}{g(x)} = -\infty.$

89. Suppose $\lim_{x \to \infty} f(x) = L$ and $\lim_{x \to \infty} g(x) = M$. Choose $\epsilon > 0$, and choose $N_1 > 0$ such that if $x > N_1$ then $|f(x) - L| < \frac{\epsilon}{2}$. Similarly, choose $N_2 > 0$ such that if $x > N_2$ then $|g(x) - M| < \frac{\epsilon}{2}$. Let $N = \max(N_1, N_2)$. Then if $x > N$, we have

$$|(f(x) + g(x) - L - M| \leq |f(x) - L| + |g(x) - M| < \frac{\epsilon}{2} + \frac{\epsilon}{2} = \epsilon,$$

so that $\lim_{x \to \infty} (f(x) + g(x)) = L + M$.

91. Use $f(x) = 1$ and $g(x) = x^k$ for $k > 0$ in Theorem 1.30(a). Then

$$\lim_{x \to \infty} x^{-k} = \lim_{x \to \infty} \frac{1}{x^k} = 0,$$

since $x^k \to \infty$ by the previous exercise.

93. Use $f(x) = 1$ and $g(x) = e^x$ in Theorem 1.30(a). Then

$$\lim_{x \to \infty} e^{-x} = \lim_{x \to \infty} \frac{1}{e^x} = 0,$$

since $e^x \to \infty$ by the previous exercise.

Thinking Forward

A limit representing an instantaneous rate of change

▷ See the table below:

h	Average rate of change
0.5	-104.00
0.25	-100.00
0.10	-97.60
0.01	-96.16

▷

$$\frac{h(3 + h) - h(3)}{h} = \frac{350 - 16(3 + h)^2 - (350 - 16 \cdot 3^2)}{h}$$

$$= \frac{-16 \cdot 3^2 - 16 \cdot 6h - 16h^2 + 16 \cdot 3^2}{h}$$

$$= \frac{-96h - 16h^2}{h} = -96 - 16h.$$

▷ $\lim_{h \to 0}(-96 - 16h) = -96 - 16 \cdot 0 = -96$. This says that the instantaneous rate of change in the height of the bowling ball is -96 after three seconds. This is the instantaneous velocity of the ball (downwards) after three seconds.

Taylor Series

▷ Let $n = 1/h$. Then as $h \to 0$, also $n \to \infty$, so that

$$\lim_{h \to 0}(1 + h)^{1/h} = \lim_{1/n \to 0}\left(1 + \frac{1}{n}\right)^n = \lim_{n \to \infty}\left(1 + \frac{1}{n}\right)^n.$$

▷ Applying the Binomial Theorem to $\left(1 + \frac{1}{n}\right)^n$ gives

$$\left(1 + \frac{1}{n}\right)^n = 1 + \binom{n}{1}\frac{1}{n} + \cdots + \binom{n}{k}\frac{1}{n^k} + \cdots + \binom{n}{n}\frac{1}{n^n}.$$

Now examine the general term in this expansion, $\binom{n}{k}\frac{1}{n^k}$. Since $\binom{n}{k} = \frac{n!}{k!(n-k)!} = \frac{n(n-1)(\ldots)(n-(k-1))}{k!}$, it is the quotient of a kth degree polynomial in n by $k!$. Thus the entire expression $\binom{n}{k}\frac{1}{n^k}$ is the quotient of two k^{th} degree polynomials in n, each of which has the coefficient of n^k equal to 1, multiplied by $\frac{1}{k!}$. The quotient of these two polynomials approaches 1 as $n \to \infty$, by Theorem 1.32(b), so that

$$\lim_{n\to\infty} \binom{n}{k}\frac{1}{n^k} = \frac{1}{k!}.$$

As n grows, there will be more and more terms in the expansion, but the kth term will always be converging to $\frac{1}{k!}$. Thus the entire expansion approaches

$$1 + \frac{1}{1!} + \frac{1}{2!} + \frac{1}{3!} + \frac{1}{4!} + \frac{1}{5!} + \ldots$$

▷ The sum of the six terms above is 2.71667, and $e \approx 2.18281828459045$.

Chapter Review and Self-Test

Skill Certification - Basic Limits

1. $\lim_{x\to 0} 3x^{-4} = \lim_{x\to 0} \frac{3}{x^4} = \frac{3}{0^+} = \infty.$

3. $\lim_{x\to 2^+} \frac{1}{2 - x} = \frac{1}{0^-} = -\infty,$ while $\lim_{x\to 2^-} \frac{1}{2 - x} = \frac{1}{0^+} = \infty.$ Since the two one-sided limits are unequal, the limit as $x \to 2$ does not exist.

5. $\lim_{x\to 1} \frac{2x^3 - x^2 - 2x + 1}{x^2 - 2x + 1} = \lim_{x\to 1} \frac{(x - 1)(x + 1)(2x + 1)}{(x - 1)^2} = \lim_{x\to 1} \frac{(x + 1)(2x + 1)}{x - 1}.$ Thus we have different one-sided limits: $\lim_{x\to 1^-} \frac{(x + 1)(2x + 1)}{x - 1} = \frac{2 \cdot 3}{0^-} = -\infty$ while $\lim_{x\to 1^+} \frac{(x + 1)(2x + 1)}{x - 1} = \frac{2 \cdot 3}{0^+} = \infty.$ Since the two one-sided limits are unequal, the limit as $x \to 2$ does not exist.

7. $\lim_{x\to 0} \frac{3^x - 4^x}{3^x} = \lim_{x\to 0} \frac{1 - 1}{1} = 0.$

9. $\lim_{x\to 0^+} \frac{\ln x}{x} = \frac{-\infty}{0^+} = -\infty.$

11. $\lim_{x\to\infty} \frac{1 - e^x}{e^{2x}} = \lim_{x\to\infty} \frac{e^{-x} - 1}{e^x} = \frac{-1}{\infty} = 0.$

13. $\lim_{x\to\infty} (\sqrt{x} - x) = \lim_{x\to\infty} \sqrt{x}(1 - \sqrt{x}) = \infty \cdot (-\infty) = -\infty.$

15. $\lim_{x\to 3} \frac{\frac{1}{x-3} - \frac{1}{x}}{x - 3} = \lim_{x\to 3} \frac{\frac{x-(x-3)}{x(x-3)}}{x - 3} = \lim_{x\to 3} \frac{3}{x(x - 3)^2} = \frac{3}{3 \cdot 0^+} = \infty.$

17. $\lim_{x\to\infty} (-2x^3 + x^2 - 10) = \lim_{x\to\infty} \left(x^3(-2 + 1/x - 10/x^3)\right) = -\infty.$

19. $\lim\limits_{x\to\infty} \dfrac{(2x-1)(x^2+1)}{x^2-4} = \lim\limits_{x\to\infty} \dfrac{2x^3 - x^2 + 2x - 1}{x^2 - 4} = \lim\limits_{x\to\infty} \dfrac{2 - 1/x + 2/x^2 - 1/x^3}{1/x - 4/x^2} = \dfrac{2}{0} = \infty.$

21. $\lim\limits_{x\to\infty} \dfrac{\sqrt{x}}{1 - \sqrt{x}} = \lim\limits_{x\to\infty} \dfrac{1}{x^{-1/2} - 1} = \dfrac{1}{0 - 1} = -1.$

23. $\lim\limits_{x\to 0} \dfrac{3x}{\sin 2x} = \dfrac{3}{2} \lim\limits_{x\to 0} \dfrac{2x}{\sin 2x} = \dfrac{3}{2}.$

25. $\lim\limits_{x\to 0} \dfrac{1 - \cos x}{\sin x} = \lim\limits_{x\to 0} \left(\dfrac{1 - \cos x}{x} \cdot \dfrac{x}{\sin x} \right) = 0 \cdot 1 = 0.$

27. Let $y = 1/x$. Then as $x \to \infty$, also $y \to 0$, so that

$$\lim\limits_{x\to\infty} \left(1 + \dfrac{1}{x}\right)^x = \lim\limits_{1/x\to 0} \left(1 + \dfrac{1}{x}\right)^x = \lim\limits_{y\to 0}(1 + y)^{1/y} = e.$$

29. $\lim\limits_{x\to\infty} \sin x$ does not exist, since $\sin x$ oscillates between -1 and $+1$ forever.

31. Use the Squeeze Theorem: since $-1 \le \sin x \le 1$, we have $-\dfrac{1}{x} \le \dfrac{1}{x}\sin x \le \dfrac{1}{x}$; taking limits as $x \to \infty$ gives $\lim\limits_{x\to\infty} \dfrac{1}{x}\sin x = 0.$

Chapter 2

Derivatives

2.1 An Intuitive Introduction to Derivatives

Thinking Back

Slope and linear functions

    ▷ Since f is linear, when x increases by 2, $f(x)$ will decrease by $3 \cdot 2 = 6$, so that $f(4) = -5$.

    ▷ Since f is linear, when x increases by 5, $f(x)$ will decrease by $3 \cdot 5 = 18$, so that $f(4) = -17$.

    ▷ Since f is linear, when x decreases by 4, $f(x)$ will increase by $3 \cdot 4 = 12$, so that $f(4) = 13$.

Approximating limits

    ▷

x	1.9	1.99	1.999	2.001	2.01	2.1
$\frac{4-x^2}{2-x}$	3.9	3.99	3.999	4.001	4.01	4.1

It appears that the limit is 4.

    ▷

z	2.9	2.99	2.999	3.001	3.01	3.1
$\frac{z^3-27}{z-3}$	26.11	26.9101	26.991	27.009	27.901	27.91

It appears that the limit is 27.

Identifying increasing and decreasing behavior

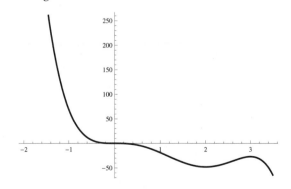

From the graph, it appears that f is increasing on $(2,3)$ and decreasing on approximately $(-\infty, 0) \cup (0,2) \cup (3,\infty)$; in fact these are the exact answers.

Interpreting distance graphically

▷

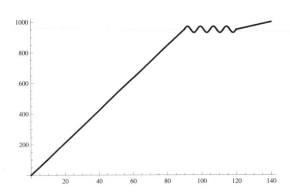

The long linear portion represents the flight until she is 50 miles from ORD. The squiggly part shows how her distance from DIA changes while she is circling Chicago. Finally, she lands. The final value of 1000 is approximately the distance from Denver to Chicago.

▷

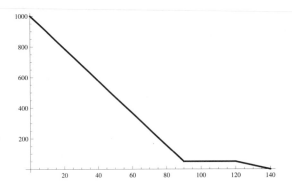

In this graph, the long linear portion represents the flight until she is 50 miles from ORD. The graph is then flat, since the plane is circling and thus maintains a constant distance from ORD. Finally, the plane lands; this is the final segment. The initial value of 1000 is approximately the distance from Denver to Chicago.

Concepts

1. Justifications for most of these questions may be found in the opening subsection of this section.

    (a) True.

    (b) True.

    (c) False. It is the slope of a *tangent* line.

    (d) False. When f is *increasing*, the slope function f' is *positive*.

    (e) True.

    (f) True.

    (g) True.

(h) True. See the discussion of Position and Velocity.

3. If the equation of the line is known, then yes, it is a simple matter to calculate its slope. However, the difficulty arises in finding which line is in fact tangent to the curve at a point, and an equation for that line, since we know only one point on the line.

5. Since $s(b) - s(a)$ measures the change in position and $b - a$ measures the total time taken, then $\dfrac{s(b) - s(a)}{b - a} = \dfrac{d}{t}$ measures the average velocity over the interval $[a, b]$. Note that the average rate of change is also the slope of the line l, since this formula is exactly the "rise over the run" for l.

7. They are the same concept and have the same value. See the introductory subsection in this section for a discussion.

9.

z	2.0001	2.001	2.01
$-\frac{1}{2}z^2 + 3z$	4.0001	4.001	4.01
$\frac{f(z) - 4}{z - 2}$	0.99995	0.9995	0.995

11.

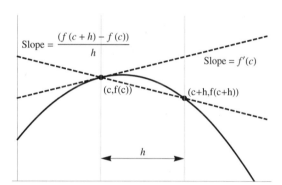

13.

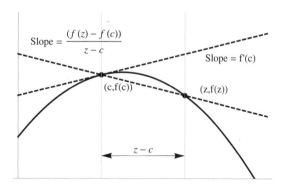

15. It appears visually that the tangent line at $x = 1$, which is the instantaneous rate of change there, falls below the line connecting $(-1, f(-1))$ and $(1, f(1))$, so it has a larger slope and $a < b$. Next, $f'(-1)$, the slope of the tangent at $x = -1$, is a fairly large negative number, and $\dfrac{f(2) - f(-1)}{2 - (-1)}$, the average rate of change on $[-1, 2]$, is close to zero. Thus, in increasing order, we have $c < d < a < b$.

17. The instantaneous rate of change is the greatest when the slope is the greatest, which looks to be at about $x = 0.5$. The smallest instantaneous rate of change is where the slope is the least; this appears to happen at about $x = 2$ and at $x = -1.5$. The instantaneous rate of change is zero where the slope is zero, which is at $x \approx -\dfrac{3}{4}$ and $\dfrac{5}{4}$.

19. In the graph below, the tangent line at $x = 1$ is the black dashed line; secant lines through $x = 1$ and $x = 2$, 1.5, and 1.2 are the gray dashed lines. The secant lines appear to be approaching the tangent line.

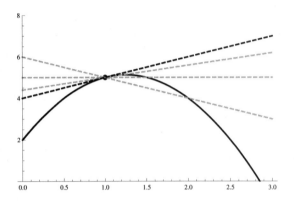

Skills

21.

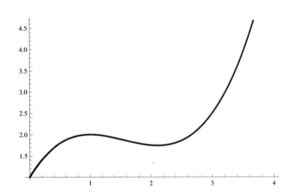

23.

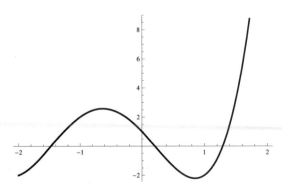

25. The slope of f is constantly increasing.

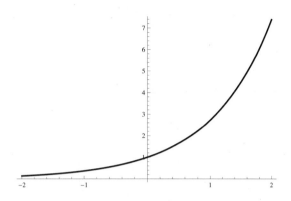

27. The slope of f is positive and decreasing from $x = -\infty$ to about $x = -1.5$, then negative and decreasing until about $x = -1$, then negative and increasing to $x = 0$, then positive and increasing to about $x = 1$, then positive and decreasing to about $x = 1.5$, then negative and decreasing.

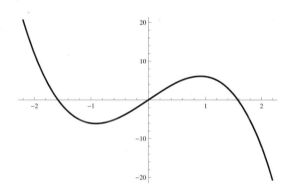

29. The slope of f is positive and increasing from $x = -\infty$ to about $x = -0.4$, then positive and decreasing to $x = 0$, then negative and decreasing to about $x = 0.4$, then negative and increasing to $x = \infty$. Since the curve has a horizontal asymptote at $y = 0$, the slope approaches zero as $x \to \infty$, so that the slope curve also has a horizontal asymptote at $y = 0$.

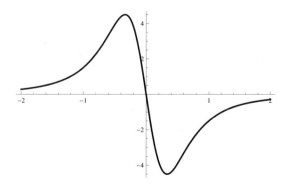

31. f is decreasing on $(-\infty, -2) \cup (0, \infty)$ since the slope is negative there, and is increasing on $(-2, 0)$. Further, f is concave upwards on $(-\infty, -1)$ since the slope is increasing there, and is concave downwards on $(-1, \infty)$.

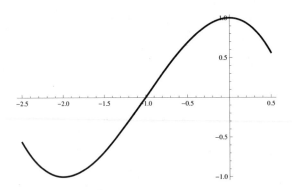

33. f is increasing on $(-1.4, 0)$ and on $(1.4, \infty)$ approximately since the slope is positive there; it is decreasing on $(-\infty, -1.4) \cup (0, 1.4)$. Also, f is concave upwards on $(-\infty, -0.8) \cup (0.8, \infty)$ since the slope is increasing there, and concave downwards on $(-0.8, 0.8)$.

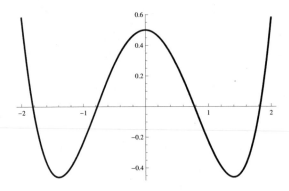

35. A table of approximations for various values of h is:

h	1	0.5	0.1	0.01
$\frac{f(c+h)-f(c)}{h}$	-3	-2.5	-2.1	-2.01

and a plot of the tangent together with various secant approximations is:

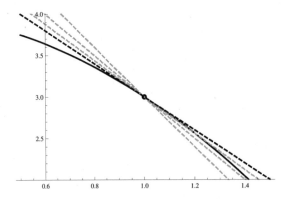

It appears that $f'(1) = -2$.

37. A table of approximations for various values of h is:

h	1	0.5	0.1	0.01
$\frac{f(c+h)-f(c)}{h}$	2	1.25	1.01	1.0001

and a plot of the tangent together with various secant approximations is:

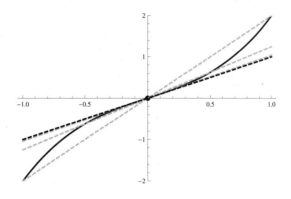

It appears that $f'(0) = 1$.

39. A table of approximations for various values of h is:

h	1	0.5	0.1	0.01
$\frac{f(c+h)-f(c)}{h}$	0.6931	0.4463	0.0995	0.0100

and a plot of the tangent together with various secant approximations is:

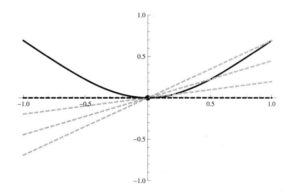

It appears that $f'(0) = 0$.

41. A table of approximations for various values of h is:

h	1	0.5	0.1	0.01
$\frac{f(c+h)-f(c)}{h}$	-0.4597	-0.2448	-0.0500	-0.005

and a plot of the tangent together with various secant approximations is:

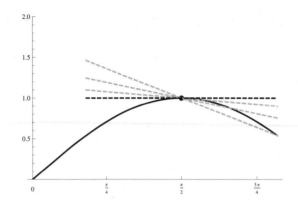

It appears that $f'\left(\frac{\pi}{2}\right) = 0$.

43. A table of approximations for various values of h is:

h	1	0.5	0.1	0.01
$\frac{f(c+h)-f(c)}{h}$	1.0	1.0	1.0	1.0

and a plot of the tangent together with various secant approximations is:

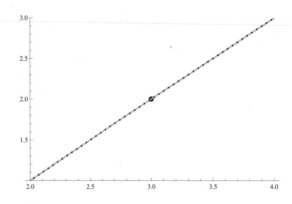

It appears that $f'(3) = 1$.

45. A table of approximations for various values of h at $t = 0$, with $s(t) = 400 - 16t^2$, is:

h	0.5	0.25	0.1	0.01
$\frac{f(0+h)-f(0)}{h}$	−8	−4	−1.6	−0.16

It appears that $v(0) = 0$.

47. A table of approximations for various values of h at $t = 2$, with $s(t) = 400 - 16t^2$, is:

h	0.1	0.01	−0.1	−0.01
$\frac{f(2+h)-f(2)}{h}$	−65.6	−64.16	−62.4	−63.84

It appears that $v(2) = -64$.

49. Answers will vary.

51. (a) Since she began and ended at the oak tree, her average rate of change of distance was $\dfrac{0-0}{30} =$ 0 feet per minute.

    (b) The change in the height of the graph between 10 and 20 minutes was greatest; since the three time intervals are equal, this is the interval with the largest average change.

    (c) At $t = 5$ she was ≈ 300 feet from the tree; at $t = 10$ she was ≈ 100 feet, so her average change of distance from the oak tree over that time was $\dfrac{100-300}{10-5} = -40$ feet per minute. The fact that the sign is negative means that she was on average moving towards the oak tree.

    (d) Her velocity was zero where the slope of the curve is zero, so at about 5, 12, and 24 minutes. These are the times at which she changed direction.

    (e) Her velocity was negative where the slope of the curve is negative, so on about $(5, 12)$ and $(24, 30)$. These are the times during which she was moving towards the tree.

53. (a) $h(12)$ represents the height in feet of an average 12-year old, while $h'(12)$ represents the growth rate of an average 12-year old, measured in feet per year.

    (b) $h(12)$ is positive, since people have positive height. Also, $h'(12)$ is positive since 12-year olds are, on average, growing.

    (c) Probably $h(t)$ has a maximum at $t = 25$ or $t = 30$ when most people are full-grown. After that time, they may begin slowly shrinking. $h'(t)$ is probably a maximum for $t = 15$ or $t = 16$, when most teenagers are in their biggest growth spurt. It could also be near $t = 0$.

55. Since Katie walks at 3 mph for 20 minutes, she walks one mile. She then runs at 10 mph for 8 minutes, so runs $10 \cdot \dfrac{8}{60} = \dfrac{4}{3}$ miles. She therefore goes a total of $\frac{7}{3}$ miles in 28 minutes. Dave's speed must therefore be $\dfrac{\frac{7}{3}}{\frac{28}{60}} = 5$ mph. In the graph below, Katie's path is the solid line and Dave's the dashed line.

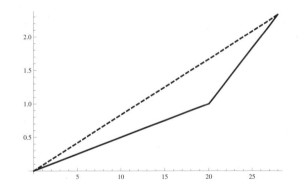

57. (a) Total yearly expenditures in 1995 are given by $E(5) \approx 139.163$ billion dollars.

    (b) The average rate of change between 1990 and 2000 is given by $\dfrac{E(10)-E(0)}{10-0} \approx 3.445$ billion dollars per year.

    (c) The average rate of change between 1995 and 1996 is given by $\dfrac{E(6)-E(5)}{6-5} \approx 3.479$ billion dollars per year.

    (d) The rate at which expenditures were increasing is the ratio between expenditures in 1996 and those in 1995, which is simply 1.025, so 2.5%. In dollars, this is about the same as the answer to part (c).

Proofs

59. Suppose $f(x) = mx + c$ with $m > 0$. Then the average rate of change on $[a, b]$ is

$$\frac{f(b) - f(a)}{b - a} = \frac{(mb + c) - (ma + c)}{b - a} = \frac{m(b - a)}{b - a} = m > 0.$$

Since the average rate of change on any interval is positive, it follows that for $a < b$ we have $\frac{f(b) - f(a)}{b - a} > 0$ so that $f(b) - f(a) > b - a > 0$ and f is increasing.

Thinking Forward

Taking the limit

▷ $f'(4) = \lim\limits_{h \to 0} \dfrac{(4 + h)^2 - 4^2}{h} = \lim\limits_{h \to 0} \dfrac{16 + 8h + h^2 - 16}{h} = \lim\limits_{h \to 0}(8 + h) = 8$, so that $f'(4) = 8$.

▷ $\lim\limits_{z \to c} \dfrac{f(z) - f(c)}{z - c}$.

▷ $f'(4) = \lim\limits_{x \to 4} \dfrac{f(x) - f(4)}{x - 4} = \lim\limits_{x \to 4} \dfrac{x^2 - 4^2}{x - 4} = \lim\limits_{x \to 4} \dfrac{(x - 4)(x + 4)}{x - 4} = \lim\limits_{x \to 4}(x + 4) = 8$.

2.2 Formal Definition of the Derivative

Thinking Back

Simplifying quotients

▷

$$\frac{(x + h)^4 - x^4}{h} = \frac{x^4 + 4x^3h + 6x^2h^2 + 4xh^3 + h^4 - x^4}{h} = \frac{4x^3h + 6x^2h^2 + 4xh^3 + h^4}{h}$$
$$= 4x^3 + 6x^2h + 4xh^2 + h^3.$$

▷

$$\frac{(x + h)^{-2} - x^{-2}}{h} = \frac{\frac{1}{(x+h)^2} - \frac{1}{x^2}}{h} = \frac{x^2 - (x + h)^2}{hx^2(x + h)^2} = \frac{x^2 - x^2 - 2hx - h^2}{hx^2(x + h)^2} = -\frac{2hx + h^2}{hx^2(x + h)^2}$$
$$= -\frac{2x + h}{x^2(x + h)^2}$$

▷ $\dfrac{z^4 - x^4}{z - x} = \dfrac{(z - x)(z + x)(z^2 + x^2)}{z - x} = (z + x)(z^2 + x^2).$

▷ $\dfrac{z^{-2} - x^{-2}}{z - x} = \dfrac{\frac{1}{z^2} - \frac{1}{x^2}}{z - x} = \dfrac{x^2 - z^2}{x^2z^2(z - x)} = \dfrac{(x - z)(x + z)}{x^2z^2(z - x)} = -\dfrac{x + z}{x^2z^2}.$

Limit calculations

▷ $\lim\limits_{h \to 0} \dfrac{(1-h) - 1}{h} = \lim\limits_{h \to 0} \dfrac{-h}{h} = \lim\limits_{h \to 0}(-1) = -1.$

▷ $\lim\limits_{h \to 0} \dfrac{(3(-1+h)^2 + 1) - 4}{h} = \lim\limits_{h \to 0} \dfrac{3 - 6h + 3h^2 + 1 - 4}{h} = \lim\limits_{h \to 0} \dfrac{-6h + 3h^2}{h} = \lim\limits_{h \to 0}(-6 + 3h) = -6.$

▷ $\lim\limits_{h \to 0} \dfrac{\frac{1}{2+h} - \frac{1}{2}}{h} = \lim\limits_{h \to 0} \dfrac{2 - (2+h)}{2h(2+h)} = \lim\limits_{h \to 0} \dfrac{-h}{2h(2+h)} = -\lim\limits_{h \to 0} \dfrac{1}{2(2+h)} = -\dfrac{1}{4}.$

▷ $\lim\limits_{z \to 2} \dfrac{z^2 - 4}{z - 2} = \lim\limits_{z \to 2} \dfrac{(z-2)(z+2)}{z-2} = \lim\limits_{z \to 2}(z+2) = 4.$

▷ $\lim\limits_{z \to 4} \dfrac{(1 - 3z) + 11}{z - 4} = \lim\limits_{z \to 4} \dfrac{12 - 3z}{z - 4} = 3\lim\limits_{z \to 4}(-1) = -3.$

▷ $\lim\limits_{z \to 1} \dfrac{\frac{1}{z} - 1}{z - 1} = \lim\limits_{z \to 1} \dfrac{1 - z}{z(z - 1)} = -\lim\limits_{z \to 1} \dfrac{1}{z} = -1.$

Concepts

1. (a) False. $f'(x)$ is the limit of this expression as $h \to 0$.

   (b) False. $f'(x)$ is the limit of this expression as $h \to 0$, not as $x \to 0$.

   (c) False. $f'(x)$ is the limit of this expression as $z \to x$, not as $z \to 0$.

   (d) False. $f(x + h) = (x + h)^3 = x^3 + 3x^2 h + 3x h^2 + h^3$.

   (e) False. $f'(x) = \lim\limits_{h \to 0} \dfrac{f(x + h) - f(x)}{h}$.

   (f) False. The left and right derivatives must exist *and be equal*.

   (g) False. This is the converse of Theorem 2.5. $|x|$ at $x = 0$ is a counterexample.

   (h) True. This is the contrapositive of Theorem 2.5.

3. (a) $\lim\limits_{h \to 0} \dfrac{f(5 + h) - f(5)}{h}$ and $\lim\limits_{z \to 5} \dfrac{f(z) - f(5)}{z - 5}$.

   (b) $\lim\limits_{h \to 0} \dfrac{f(x + h) - f(x)}{h}$ and $\lim\limits_{z \to x} \dfrac{f(z) - f(x)}{z - x}$.

   (c) $\lim\limits_{h \to 0^+} \dfrac{f(-2 + h) - f(-2)}{h}$ and $\lim\limits_{z \to -2^+} \dfrac{f(z) - f(-2)}{z + 2}$.

5. Simply substitute $x + h$ for z. Note that $z \to x$ becomes $x + h \to x$, which is the same as $h \to 0$.

7. Continuity means that $\lim\limits_{x \to 2}(4x^3 - 5x + 1) = f(2) = 23$, while differentiability means that the limit $\lim\limits_{h \to 0} \dfrac{f(2+h) - f(2)}{h}$ exists.

9. This limit defines $f'(c)$, so if it exists, f is differentiable at $x = c$. By Theorem 2.5, this implies continuity of f at $x = c$ as well.

11. (a) Since the left and right limits of $f(x)$ as $x \to 1$ are equal and both equal $f(1)$, it follows that f is continuous at $x = 1$. Since both one-sided limits of $\dfrac{f(1+h) - f(1)}{h}$ exist, the function is left- and right-differentiable at $x = 1$. Since they are unequal, the function is not differentiable there.

(b)

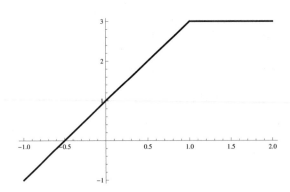

13. The graph with secant lines is

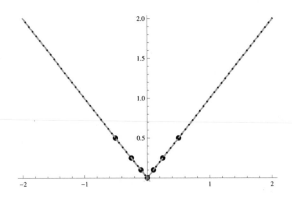

The secant lines from the left are all $y = -x$, and from the right are all $y = x$. Thus the left derivative at 0 is -1 and the right derivative at 0 is 1. Since the two are unequal, $|x|$ is not differentiable there.

15. (a) $f'(-4) = -24$.   (b) $\dfrac{df}{dx}\Big|_{-4} = -24$.   (c) $\dfrac{d}{dx}(f(x))\Big|_{-4} = -24$.

17. The first three are all different ways of expressing the first derivative of $f(x)$, so are all equal to $3x^2 - 2$. The last is the derivative of $f(x)$, which is $\dfrac{dy}{dx}$, multiplied by $f(x)$, so is $(3x^2 - 2)(x^3 - 2x + 1)$.

19. For every $\epsilon > 0$ there is some $\delta > 0$ such that if $0 < |h| < \delta$, then $\left| \dfrac{(3+h)^2 - 9}{h} - f'(3) \right| < \epsilon$.

21. Since $f'(4) = \dfrac{1}{2\sqrt{4}} = \dfrac{1}{4}$, the tangent line to $f(x)$ at $x = 4$ is

$$y = f(4) + \frac{1}{4}(x - 4) = \frac{1}{4}x + 1.$$

Then $\sqrt{4.1} \approx \frac{1}{4}(4.1) + 1 = 2.025$. A more precise answer is $\sqrt{4.1} \approx 2.02485$, so the linear approximation is not bad.

Skills

23. (a) $\lim\limits_{h\to 0} \dfrac{(-3+h)^2 - (-3)^2}{h} = \lim\limits_{h\to 0} \dfrac{9 - 6h + h^2 - 9}{h} = \lim\limits_{h\to 0} \dfrac{-6h + h^2}{h} = \lim\limits_{h\to 0}(-6 + h) = -6.$

(b) $\lim\limits_{z\to -3} \dfrac{z^2 - (-3)^2}{z - (-3)} = \lim\limits_{z\to -3} \dfrac{(z-3)(z+3)}{z+3} = \lim\limits_{z\to -3}(z-3) = -6.$

25. (a) $\lim\limits_{h\to 0} \dfrac{\frac{1}{-1+h} - \frac{1}{-1}}{h} = \lim\limits_{h\to 0} \dfrac{-h}{h(1-h)} = \lim\limits_{h\to 0} \dfrac{1}{h-1} = -1.$

(b) $\lim\limits_{z\to -1} \dfrac{\frac{1}{z} - \frac{1}{-1}}{z - (-1)} = \lim\limits_{z\to -1} \dfrac{-1-z}{-z(z+1)} = \lim\limits_{z\to -1} \dfrac{1}{z} = -1.$

27. (a) $\lim\limits_{h\to 0} \dfrac{(1-(-1+h)^3) - (1-(-1)^3)}{h} = \lim\limits_{h\to 0} \dfrac{-3h + 3h^2 - h^3}{h} = \lim\limits_{h\to 0}(-3 + 3h - h^3) = -3.$

(b)

$$\lim\limits_{z\to -1} \dfrac{(1-z^3) - (1-(-1)^3)}{z - (-1)} = \lim\limits_{z\to -1} \dfrac{-1-z^3}{z+1} = -\lim\limits_{z\to -1} \dfrac{(z+1)(z^2+z+1)}{z+1}$$

$$= -\lim\limits_{z\to -1}(z^2 + z + 1) = -3.$$

29. (a)

$$\lim\limits_{h\to 0} \dfrac{\sqrt{9+h} - 3}{h} = \lim\limits_{h\to 0} \dfrac{(\sqrt{9+h} - 3)(\sqrt{9+h} + 3)}{h(\sqrt{9+h} + 3)} = \lim\limits_{h\to 0} \dfrac{9 + h - 9}{h(\sqrt{9+h} + 3)}$$

$$= \lim\limits_{h\to 0} \dfrac{1}{\sqrt{9+h} + 3} = \dfrac{1}{6}.$$

(b) $\lim\limits_{z\to 9} \dfrac{\sqrt{z} - \sqrt{9}}{z - 9} = \lim\limits_{z\to 9} \dfrac{(\sqrt{z} - 3)(\sqrt{z} + 3)}{(z-9)(\sqrt{z} + 3)} = \lim\limits_{z\to 9} \dfrac{z - 9}{(z-9)(\sqrt{z} + 3)} = \lim\limits_{z\to 9} \dfrac{1}{\sqrt{z} + 3} = \dfrac{1}{6}.$

31. (a)

$$\lim\limits_{h\to 0} \dfrac{\frac{(2+h)-1}{(2+h)+3} - \frac{2-1}{2+3}}{h} = \lim\limits_{h\to 0} \dfrac{\frac{h+1}{h+5} - \frac{1}{5}}{h}$$

$$= \lim\limits_{h\to 0} \dfrac{5(h+1) - (h+5)}{5h(h+5)}$$

$$= \lim\limits_{h\to 0} \dfrac{4h}{5h(h+5)}$$

$$= \lim\limits_{h\to 0} \dfrac{4}{5(h+5)} = \dfrac{4}{25}.$$

(b) $\lim\limits_{z\to 2} \dfrac{\frac{z-1}{z+3} - \frac{1}{5}}{z - 2} = \lim\limits_{z\to 2} \dfrac{5(z-1) - (z+3)}{5(z-2)(z+3)} = \lim\limits_{z\to 2} \dfrac{4(z-2)}{5(z-2)(z+3)} = \lim\limits_{z\to 2} \dfrac{4}{5(z+3)} = \dfrac{4}{25}.$

33. (a) $\lim\limits_{h\to 0} \dfrac{e^{0+h} - e^0}{h} = \lim\limits_{h\to 0} \dfrac{e^h - 1}{h} = 1$ by Theorem 1.26.

(b) $\lim\limits_{z\to 0} \dfrac{e^z - e^0}{z - 0} = \lim\limits_{z\to 0} \dfrac{e^z - 1}{z} = 1$ by Theorem 1.26.

35. (a) $\displaystyle\lim_{h\to 0}\frac{\sin(0+h)-\sin 0}{h}=\lim_{h\to 0}\frac{\sin h}{h}=1$ by Theorem 1.35.

    (b) $\displaystyle\lim_{z\to 0}\frac{\sin z-\sin 0}{z-0}=\lim_{z\to 0}\frac{\sin z}{z}=1$ by Theorem 1.35.

37. (a) $\displaystyle\lim_{h\to 0}\frac{\tan(0+h)-\tan 0}{h}=\lim_{h\to 0}\frac{\frac{\sin h}{\cos h}}{h}=\lim_{h\to 0}\left(\frac{\sin h}{h}\right)\cos h=1\cdot 1=1.$

    (b) $\displaystyle\lim_{z\to 0}\frac{\tan z-\tan 0}{z-0}=\lim_{z\to 0}\frac{\frac{\sin z}{\cos z}}{z}=\lim_{z\to 0}\left(\frac{\sin z}{z}\right)\cos z=1\cdot 1=1.$

39. $\displaystyle\lim_{h\to 0}\frac{-2(x+h)^2-2x^2}{h}=\lim_{h\to 0}\frac{-4xh-2h^2}{h}=\lim_{h\to 0}(-4x-2h)=-4x.$

41. $\displaystyle\lim_{h\to 0}\frac{(x+h)^3+2-(x^3+2)}{h}=\lim_{h\to 0}\frac{3x^2h+3xh^2+h^3}{h}=\lim_{h\to 0}(3x^2+3xh+h^2)=3x^2.$

43.
$$\lim_{h\to 0}\frac{\frac{2}{x+h+1}-\frac{2}{x+1}}{h}=\lim_{h\to 0}\frac{2x+2-2x-2h-2}{h(x+1)(x+h+1)}=\lim_{h\to 0}\frac{-2h}{h(x+1)(x+h+1)}$$
$$=-\lim_{h\to 0}\frac{2}{(x+1)(x+h+1)}=-\frac{2}{(x+1)^2}.$$

45. $\displaystyle\lim_{z\to x}\frac{\frac{1}{z^2}-\frac{1}{x^2}}{z-x}=\lim_{z\to x}\frac{x^2-z^2}{(z-x)z^2x^2}=\lim_{z\to x}\frac{(x-z)(x+z)}{(z-x)z^2x^2}=-\lim_{z\to x}\frac{z+x}{z^2x^2}=-\frac{2}{x^3}.$

47.
$$\lim_{h\to 0}\frac{3\sqrt{x+h}-3\sqrt{x}}{h}=3\lim_{h\to 0}\frac{(\sqrt{x+h}-\sqrt{x})(\sqrt{x+h}+\sqrt{x})}{h(\sqrt{x+h}+\sqrt{x})}$$
$$=3\lim_{h\to 0}\frac{h}{h(\sqrt{x+h}+\sqrt{x})}=3\lim_{h\to 0}\frac{1}{\sqrt{x+h}+\sqrt{x}}=\frac{3}{2\sqrt{x}}.$$

49.
$$\lim_{h\to 0}\frac{\sqrt{2(x+h)+1}-\sqrt{2x+1}}{h}=\lim_{h\to 0}\frac{(\sqrt{2(x+h)+1}-\sqrt{2x+1})(\sqrt{2(x+h)+1}+\sqrt{2x+1})}{h(\sqrt{2(x+h)+1}+\sqrt{2x+1})}$$
$$=\lim_{h\to 0}\frac{2(x+h)+1-2x-1}{h(\sqrt{2(x+h)+1}+\sqrt{2x+1})}=\lim_{h\to 0}\frac{2}{\sqrt{2(x+h)+1}+\sqrt{2x+1}}=\frac{1}{\sqrt{2x+1}}.$$

51.
$$\lim_{h\to 0}\frac{\frac{(x+h)-1}{(x+h)+3}-\frac{x-1}{x+3}}{h}=\lim_{h\to 0}\frac{(x+3)(x+h-1)-(x+h+3)(x-1)}{h(x+3)(x+h+3)}$$
$$=\lim_{h\to 0}\frac{x^2+xh-x+3x+3h-3-(x^2+xh+3x-x-h-3)}{h(x+3)(x+h+3)}$$
$$=\lim_{h\to 0}\frac{4h}{h(x+3)(x+h+3)}$$
$$=\lim_{h\to 0}\frac{4}{(x+3)(x+h+3)}=\frac{4}{(x+3)^2}$$

53.

$$\lim_{h \to 0} \frac{\frac{(x+h)^3}{x+h+1} - \frac{x^3}{x+1}}{h} = \lim_{h \to 0} \frac{(x+1)(x+h)^3 - (x+h+1)x^3}{h(x+1)(x+h+1)}$$

$$= \lim_{h \to 0} \frac{x^4 + 3hx^3 + x^3 + 3h^2x^2 + 3hx^2 + h^3x + 3h^2x + h^3 - x^4 - hx^3 - x^3}{h(x+1)(x+h+1)}$$

$$= \lim_{h \to 0} \frac{2hx^3 + 3h^2x^2 + 3hx^2 + h^3x + 3h^2x + h^3}{h(x+1)(x+h+1)}$$

$$= \lim_{h \to 0} \frac{2x^3 + 3hx^2 + 3x^2 + h^2x + 3hx + h^2}{(x+1)(x+h+1)} = \frac{2x^3 + 3x^2}{(x+1)^2}$$

55.

$$\frac{d}{dx}(2x^3) = \lim_{h \to 0} \frac{2(x+h)^3 - 2x^2}{h} = 2 \lim_{h \to 0} \frac{3x^2h + 3xh^2 + h^3}{h} = 2 \lim_{h \to 0}(3x^2 + 3xh + h^2) = 6x^2$$

$$\frac{d^2}{dx^2}(2x^3) = \frac{d}{dx}(6x^2) = \lim_{h \to 0} \frac{6(x+h)^2 - 6x^2}{h} = 6 \lim_{h \to 0} \frac{2xh + h^2}{h} = 6 \lim_{h \to 0}(2x + h) = 12x$$

$$\frac{d^3}{dx^2}(2x^3) = \frac{d}{dx}(12x) = \lim_{h \to 0} \frac{12(x+h) - 12x}{h} = \lim_{h \to 0} \frac{12h}{h} = 12.$$

57.

$$\frac{d^4 f}{dx^4} = \frac{d}{dx}(3x^2 + 1) = \lim_{h \to 0} \frac{3(x+h)^2 + 1 - 3x^2 - 1}{h} = \lim_{h \to 0} \frac{6xh + 3h^2}{h} = \lim_{h \to 0}(6x + 3h) = 6x,$$

$$\frac{d^4 f}{dx^4}\bigg|_2 = (6x)|_2 = 12.$$

59. We have $\frac{d}{dx}x^2 = \lim_{h \to 0} \frac{(x+h)^2 - x^2}{h} = \lim_{h \to 0} \frac{2xh + h^2}{h} = \lim_{h \to 0}(2x + h) = 2x$, so that the slope of $f(x) = x^2$ at $x = -3$ is $2 \cdot (-3) = -6$. Thus the tangent line at $x = -3$, where $f(x) = 9$, is $y - 9 = -6(x - (-3))$, or $y = -6x - 9$.

61. We first compute the derivative of $f(x) = 1 - x - x^2$:

$$\lim_{h \to 0} \frac{(1 - (x+h) - (x+h)^2) - (1 - x - x^2)}{h}$$

$$= \lim_{h \to 0} \frac{-h - 2xh - h^2}{h} = -\lim_{h \to 0}(1 + 2x + h) = -2x - 1.$$

Thus the slope of $f(x)$ at $x = 1$ is $-2 \cdot 1 - 1 = -3$, and the tangent line there is $y - (-1) = -3(x - 1)$, or $y = -3x + 2$.

63. The derivative of $f(x) = \frac{1}{x}$ at $x = -1$ is -1 by Exercise 25, so we are looking for a line of slope -1 passing through $(3, 2)$. This line is $y - 2 = -(x - 3)$, or $y = -x + 5$.

65. $f(x)$ fails to be differentiable at $x = 2$, since it is not continuous there, and at $x = -1$. At $x = 2$, secant lines have slopes approaching $\pm\infty$. At $x = -1$, secants from the left have slope 1, while secants from the right have negative slope, so they cannot have the same limit. However, f both left-and-right differentiable there since both limits exist.

67. $f(x)$ fails to be differentiable only at $x = -1$, since it is discontinuous there. Secants from the left have positive slope for x close to -1, while secants from the right have negative slope. f is left-differentiable at $x = -1$. Finally, f is not differentiable at $x = 1$ since the tangent line is vertical.

69. $f(x)$ is not defined at $x = 0$, so it cannot be differentiable there, and cannot be either left- or right-continuous, or left- or right-differentiable either.

71. $f(x)$ is continuous at $x = 2$ but is not differentiable there, since (noting that from the right, $|x^2 - 4| = x^2 - 4$ while from the left $|x^2 - 4| = 4 - x^2$),

$$\lim_{h \to 0^+} \frac{|(2+h)^2 - 4| - |2^2 - 4|}{h} = \lim_{h \to 0^+} \frac{4h + h^2}{h} = \lim_{h \to 0^+} (4 + h) = 4$$

$$\lim_{h \to 0^-} \frac{|(2+h)^2 - 4| - |2^2 - 4|}{h} = \lim_{h \to 0^-} \frac{-4h - h^2}{h} = \lim_{h \to 0^-} (-4 - h) = -4$$

The two one-sided derivatives exist but are unequal.

73. $f(x)$ is continuous at $x = 2$ since the two one-sided limits are both equal to $f(2) = 6$. However, $f(x)$ is not differentiable at $x = 2$, since the one-sided derivatives exist but are unequal:

$$\lim_{h \to 0^-} f(x) = \lim_{h \to 0^-} \frac{(x + h + 4) - (x + 4)}{h} = \lim_{h \to 0^-} \frac{h}{h} = 1$$

$$\lim_{h \to 0^+} f(x) = \lim_{h \to 0^+} \frac{3(x + h) - 3x}{h} = \lim_{h \to 0^+} \frac{3h}{h} = 3$$

75. f is continuous at $x = 1$ since the two one-sided limits are each equal to $f(1) = 1$. It is also differentiable at $x = 1$, since the two one-sided derivatives exist and are equal:

$$\lim_{h \to 0^+} \frac{f(1 + h) - f(1)}{h} = \lim_{h \to 0^+} \frac{2(1 + h) - 1 - 1^2}{h} = \lim_{h \to 0^+} \frac{2h}{h} = 2$$

$$\lim_{h \to 0^-} \frac{f(1 + h) - f(1)}{h} = \lim_{h \to 0^-} \frac{(1 + h)^2 - 1^2}{h} = \lim_{h \to 0^-} \frac{2h + h^2}{h} = 2$$

77. A graph of $f(x)$ is

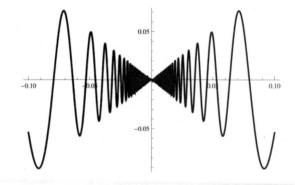

It is apparent that $f(x)$ is continuous at $x = 0$, but is not differentiable there - the secant slopes oscillate between a slope of 1, achieved whenever $\sin \frac{1}{x} = 1$, and -1, achieved whenever $\sin \frac{1}{x} = -1$.

79. A graph of $f(x)$ is

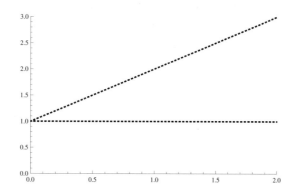

Clearly $f(x)$ is not continuous at $x = 1$ (and in fact this is obvious since there are irrational numbers arbitrarily close to $x = 1$; at those numbers, $f(x)$ is close to 2, while $f(1) = 1$). Thus $f(x)$ cannot be differentiable there, and from the graph, it seems clear that neither one-sided derivative exists either.

81. Note that f is continuous on $[1, 3]$, so that the Intermediate Value Theorem applies on this interval. Since $f(1) = 1^2 - 5 = -4 < 0$ and $f(3) = 3^2 - 5 = 4 > 0$, it follows that f has a root between 1 and 3. Apply Newton's method with an initial guess of 2 and use four iterations:

i	0	1	2	3	4
x_i	2	2.25	2.23611	2.23607	2.23607

The true root is $\sqrt{5} \approx 2.23607$.

83. Note that f is continuous on $[0, 1]$, so that the Intermediate Value Theorem applies on this interval. Since $f(0) = 1 > 0$ and $f(1) = 1^3 - 3 \cdot 1 + 1 = -1 < 0$, there is a root between 0 and 1. Apply Newton's method with an initial guess of 0.5 and use four iterations:

i	0	1	2	3	4
x_i	0.5	0.3333	0.347222	0.347296	0.347296

The true root is ≈ 0.347296.

85. Note that f is continuous on $[-2, 1]$, so that the Intermediate Value Theorem applies on this interval. Since $f(-2) = (-2)^3 + 1 = -7 < 0$ and $f(1) = 1^3 + 1 = 2 > 0$, there is a root between -2 and 1. Apply Newton's method with an initial guess of -0.5 (halfway between -2 and -1 and use five iterations:

i	0	1	2	3	4	5
x_i	-0.5	-1.66667	-1.23111	-1.04067	-1.00157	-1

The true root is -1.

Applications

87. We have

$$v(1) = \lim_{h \to 0} \frac{(-16(1 + h)^2 + 100) - (-16 \cdot 1^2 + 100)}{h} = \lim_{h \to 0} \frac{-32h - 16h^2}{h} = \lim_{h \to 0}(-32 - 16h)$$

$$= -32,$$

so that $v(1) = -32$ feet per second. The approximation from Section 2.1 was -33 feet per second.

89. (a) A plot of $v(t)$ is:

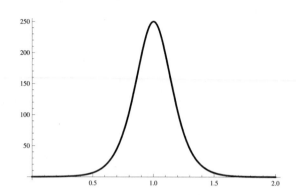

The velocity starts close to zero, builds to a maximum at about $t = 1$, and then decreases rapidly towards near zero again.

(b) A plot of $a(t)$ is

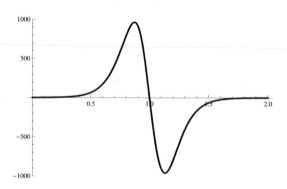

While the velocity is increasing rapidly, the acceleration is very positive; as the velocity starts to level off, the acceleration decreases to zero. Then the reverse happens as the velocity rapidly decreases.

91. (a) After three weeks you have been paid $80 per week for $240, so $S(3) = 240 + 200 = 440$. After six weeks you have been paid $6 \cdot \$80 = \$480$, so that $S(6) = 480 + 200 = 680$. After eight weeks you have been paid $480 for the first six weeks, plus two weeks at $11.50 per hour, or $115 per week, so that $S(8) = 230 + 480 + 200 = 910$. For the derivatives, $S'(3)$ is the weekly rate of change of your savings account after 3 weeks; that is the amount of money you make in one week, so it is $80. $S'(6)$ is not defined, since $S(t)$ is not continuous there. Finally, $S'(8) = \$115$.

(b) We have

$$S(t) = \begin{cases} 200 + 80t, & 0 \le t \le 6, \\ 680 + 115(t - 6), & 6 < t. \end{cases}$$

(c) A graph is

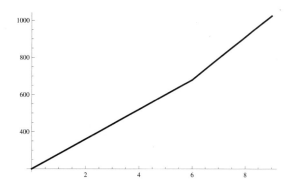

From the graph, $S(t)$ is continuous and differentiable except at $t = 6$, where it is continuous but not differentiable.

(d) For continuity, note that $S(t)$ is obviously continuous everywhere except possibly $t = 6$, while at $t = 6$,

$$\lim_{t \to 6^-} S(t) = \lim_{t \to 6^-} 200 + 80t = 200 + 80 \cdot 6 = 680$$

$$\lim_{t \to 6^+} S(t) = \lim_{t \to 6^+} 680 + 115(t - 6) = 680 + 115(6 - 6) = 680$$

$$S(6) = 280 + 80 \cdot 6 = 680.$$

For differentiability, we have

$$\lim_{h \to 0^-} \frac{S(6 + h) - S(6)}{h} = \lim_{h \to 0^-} \frac{200 + 80(6 + h) - 680}{h} = \lim_{h \to 0^-} \frac{80h}{h} = 80$$

$$\lim_{h \to 0^+} \frac{S(6 + h) - S(6)}{h} = \lim_{h \to 0^+} \frac{680 + 115((6 + h) - 6) - 680}{h} = \lim_{h \to 0^+} \frac{115h}{h} = 115$$

Since the left and right derivatives are unequal, $S(t)$ is not differentiable at $t = 6$.

Proofs

93. Let $f(x) = mx + b$. Then

$$f'(x) = \lim_{h \to 0} \frac{m(x + h) + b - (mx + b)}{h} = \lim_{h \to 0} \frac{mh}{h} = \lim_{h \to 0} m = m.$$

95. The slope of the tangent line to $f(x)$ at $x = c$ is given by $f'(c)$, by definition. Since the line goes through the point $(c, f(c))$, the equation of this line is $y - f(c) = f'(c)(x - c)$, so that $y = f'(c)(x - c) + f(c)$.

97. If $f'(c)$ exists, then $\lim_{h \to 0} \frac{f(c + h) - f(c)}{h}$ exists. By Theorem 1.8, this means that both of the one-sided limits $\lim_{h \to 0^+} \frac{f(c+h)-f(c)}{h}$ and $\lim_{h \to 0^-} \frac{f(c+h)-f(c)}{h}$ exist, and they are equal. Conversely, if both of the one-sided limits exist and are equal, then by Theorem 1.8, $\lim_{h \to 0} \frac{f(c + h) - f(c)}{h}$ exists so that $f'(c)$ exists.

Thinking Forward

Derivatives of power functions

▷ $\lim_{z \to x} \dfrac{z^4 - x^4}{z - x} = \lim_{z \to x} \dfrac{(z - x)(z + x)(z^2 + x^2)}{z - x} = \lim_{z \to x}((z + x)(z^2 + x^2)) = 4z^3.$

▷ $\lim\limits_{z \to x} \dfrac{z^8 - x^8}{z - x} = \lim\limits_{z \to x} \dfrac{(z-x)(z^7 + z^6 x + \cdots + zx^6 + x^7)}{z - x} = \lim\limits_{z \to x}(z^7 + z^6 x + \cdots + zx^6 + x^7) = 8x^7.$

▷ $\dfrac{d}{dx}(x^n) = nx^{n-1}.$

Derivatives of combinations of functions

▷ By Exercise 93, $\dfrac{d}{dx}(3x) = 3$. By the material following Definition 2.2, $\dfrac{d}{dx}(x^2) = 2x$. Finally,

$$\frac{d}{dx}(3x + x^2) = \lim_{h \to 0} \frac{3(x+h) + (x+h)^2 - (3x + x^2)}{h}$$

$$= \lim_{h \to 0} \frac{3h + 2xh + h^2}{h} = \lim_{h \to 0}(3 + 2x + h) = 2x + 3.$$

At least in this case, $\dfrac{d}{dx}(f(x) + g(x)) = \dfrac{d}{dx}f(x) + \dfrac{d}{dx}g(x).$

▷ By Exercise 93, $\dfrac{d}{dx}(x - 3) = 1$ and $\dfrac{d}{dx}(2x + 1) = 2$. However, $(x-3)(2x+1) = 2x^2 - 5x - 3$, and

$$\frac{d}{dx}(2x^2 - 5x - 3) = \lim_{h \to 0} \frac{2(x+h)^2 - 5(x+h) - 3 - (2x^2 - 5x - 3)}{h}$$

$$= \lim_{h \to 0} \frac{4xh + 2h^2 - 5h}{h} = \lim_{h \to 0}(4x - 5 + 2h) = 4x - 5,$$

which is not equal to $1 \cdot 2$. Thus $\dfrac{d}{dx}(f(x)g(x)) \neq \dfrac{df}{dx} \cdot \dfrac{dg}{dx}.$

2.3   Rules for Calculating Basic Derivatives

Thinking Back

Factoring formulas

▷ $z^2 - 100 = (z - 10)(z + 10).$

▷ $z^3 - 27 = (z - 3)(z^2 + 3z + 9).$

▷ $z^6 - 64 = (z - 2)(z^5 + 2z^4 + 4z^3 + 8z^2 + 16z + 32).$

Definition-of-derivative calculations

▷ $f'(x) = \lim\limits_{z \to x} \dfrac{z^4 - x^4}{z - x} = \lim\limits_{z \to x} \dfrac{(z-x)(z+x)(z^2 + x^2)}{z - x} = \lim\limits_{z \to x}((z+x)(z^2 + x^2)) = 4x^3.$

▷ See Exercise 45 in Section 2.2.

▷ Since $x^2(x + 1) = x^3 + x^2$,

$$f'(x) = \lim_{h \to 0} \frac{(x+h)^3 + (x+h)^2 - x^3 - x^2}{h} = \lim_{h \to 0} \frac{3x^2 h + 3xh^2 + h^3 + 2xh + h^2}{h}$$

$$= \lim_{h \to 0}(3x^2 + 2x + 3xh + h^2 + h) = 3x^2 + 2x.$$

▷

$$f'(x) = \lim_{h \to 0} \frac{\frac{(x+h)^2}{x+h+1} - \frac{x^2}{x+1}}{h} = \lim_{h \to 0} \frac{(x+1)(x+h)^2 - (x+h+1)x^2}{h(x+1)(x+h+1)}$$

$$= \lim_{h \to 0} \frac{x^3 + x^2 + 2x^2h + 2xh + xh^2 + h^2 - x^3 - x^2h - x^2}{h(x+1)(x+h+1)}$$

$$= \lim_{h \to 0} \frac{x^2h + 2xh + xh^2 + h^2}{h(x+1)(x+h+1)}$$

$$= \lim_{h \to 0} \frac{x^2 + 2x + xh + h}{(x+1)(x+h+1)}$$

$$= \frac{x^2 + 2x}{(x+1)^2}$$

Associated slope functions

▷ The slope function is positive and increasing on approximately $(-\infty, -1)$, positive and decreasing on $(-1, 0)$, negative and decreasing on $(0, 1)$, and negative and increasing on $(1, \infty)$. Since f has a horizontal asymptote at $y = 2$, its slope approaches zero as $x \to \pm\infty$, so that the slope curve has a horizontal asymptote at $y = 0$.

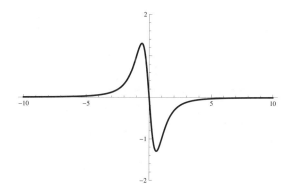

▷ The slope function is positive and increasing on approximately $(-\infty, -0.8) \cup (2, \infty)$, is positive and decreasing on $(-0.8, 0.8)$, is negative and decreasing on $(0.8, 1.4)$, and is negative and increasing on $(1.4, 2)$. Since f has a horizontal asymptote at $y = 0$, its slope approaches 0 as $x \to -\infty$, so that the slope curve also has a horizontal asymptote at $y = 0$.

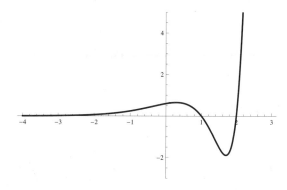

Concepts

1. (a) True. The derivative of any constant function is zero, by Theorem 2.8(a).

   (b) False. See Theorem 2.8(c); here r plays the role of x, $m = 1$, and $b = ks$, so the derivative is 1.

   (c) True. Again see Theorem 2.8(c), but here s plays the role of x.

   (d) False. For example, let $k = 2$. Then

   $$\frac{d}{dx}(3x+1)^2 = \frac{d}{dx}(9x^2 + 6x + 1) = 18x + 6 = 3(3x+1)^1.$$

   (e) False. $\frac{d}{dx}\frac{1}{x^3} = \frac{d}{dx}x^{-3} = -3x^{-4}$ by Theorem 2.9.

   (f) True. See Theorem 2.11, the Product Rule.

   (g) True. See Theorem 2.11, the Quotient Rule.

   (h) True. See the proof.

3. In Leibniz notation, they are

$$\frac{d(kf)}{dx} = k\frac{df}{dx}, \quad \frac{d(f+g)}{dx} = \frac{df}{dx} + \frac{dg}{dx}, \quad \frac{d(f-g)}{dx} = \frac{df}{dx} - \frac{dg}{dx}.$$

In operator notation, they are

$$\frac{d}{dx}(kf) = k\frac{d}{dx}(f), \quad \frac{d}{dx}(f+g) = \frac{d}{dx}(f) + \frac{d}{dx}(g), \quad \frac{d}{dx}(f-g) = \frac{d}{dx}(f) - \frac{d}{dx}(g).$$

5. The derivative of a function represents the slope of its graph at each point in its domain. Constant functions are straight lines parallel to the x-axis, so have zero slope everywhere. The identity function clearly has an instantaneous rate of change equal to 1, since x and $f(x)$ change at the same rate, so it makes sense that its derivative is 1. Similarly, for a general linear function $f(x) = mx + b$, any change in x gives a change of mx in $f(x)$, so that the ratio of the change in $f(x)$ to the change in x is constant at m; this is the instantaneous rate of change and thus the derivative.

7. The power rule applies when the variable in the derivative is the base of the power, not the exponent.

9.

$$(f(x)g(x)h(x))' = (f(x)g(x))h'(x) + (f(x)g(x))'h(x)$$
$$= f(x)g(x)h'(x) + (f(x)g'(x) + f'(x)g(x))h(x)$$
$$= f(x)g(x)h'(x) + f(x)g'(x)h(x) + f'(x)g(x)h(x)$$

Then

$$y' = (2x+1)(x^2+x+1)(-12x^3) + (2x-1)(2x+1)(1-3x^4) + 2(x^2+x+1)(1-3x^4)$$
$$= -42x^6 - 54x^5 - 45x^4 - 12x^3 + 6x^2 + 6x + 3$$

To check this answer, expand y and differentiate:

$$y = (2x+1)(x^2+x+1)(1-3x^4) = -6x^7 - 9x^6 - 9x^5 - 3x^4 + 2x^3 + 3x^2 + 3x + 1$$

and differentiating this polynomial using the sum and power rules gives the same answer.

11. $(3f + 4h)'(2) = 3f'(2) + 4h'(2) = 3 \cdot 3 + 4 \cdot (-1) = 5.$

13. $(fh)'(2) = f(2)h'(2) + f'(2)h(2) = 1 \cdot (-1) + 3 \cdot 3 = 8.$

15. $f(x)$ is decreasing and then increasing, so that $f'(x)$ should be negative and then positive. Also, $f(x)$ has a horizontal tangent at 0, so that $f'(0) = 0$. This behavior is reflected in $f'(x)$. Similar analyses apply for the remaining curves.

17.

$$\frac{(2x)(x^{-3}\sqrt{x}) - (x^2)(-3x^{-4}\sqrt{x} + x^{-3}(\frac{1}{2}x^{-1/2}))}{(x^{-3}\sqrt{x})^2} \qquad \leftarrow \text{first formula}$$

$$= \frac{2x^{-3/2} + 3x^{-3/2} - \frac{1}{2}x^{-3/2}}{x^{-5}} \qquad \leftarrow \text{multiply powers of } x$$

$$= \frac{9}{2} \cdot \frac{x^{-3/2}}{x^{-5}} \qquad \leftarrow \text{combine like terms}$$

$$= \frac{9}{2}x^{7/2} \qquad \leftarrow \text{multiply powers of } x$$

19. No, not if the right derivative of $h(x)$ at $x = 2$ differs from the left derivative of $g(x)$ at $x = 2$. For example, let $g(x) = 2x$ and $h(x) = x + 2$.

21. In order for f to be continuous at $x = 1$, we must have $3 \cdot 1 + a = 1$, so that $a = -2$. Now, the left derivative of $f(x)$ at $x = 1$ is the derivative of $3x - 2$, which is 3, so that the right derivative of $f(x)$ at $x = 1$ must also be 3. But this right derivative is the derivative of $x^{b/2}$ at $x = 1$, which is $\frac{b}{2}1^{b/2-1} = \frac{b}{2}$. Thus $\frac{b}{2} = 3$ so that $b = 6$. Hence setting $a = -2$ and $b = 6$ produces a function that is continuous and differentiable at $x = 1$. Since the pieces of f are continuous and differentiable for all other values of x, the resulting f is continuous and differentiable everywhere.

Skills

23. $f'(3) = (3j(x) - 2h(x))'(3) = 3j'(3) - 2h'(3) = 3 \cdot 2 - 2 \cdot 1 = 4$ by the Difference and Constant Multiple rules.

25. $f'(2) = g(2)(h(x) + j(x))'(2) + g'(2)(h(2) + j(2)) = g(2)(h'(2) + j'(2)) + g'(2)(h(2) + j(2)) = -2 \cdot (0 + 2) + (-1) \cdot (-2 + -3) = 1$ by the Product and Sum rules.

27. $f'(0) = \frac{(g(0)+j(0))(3h(x))'(0) - 3h(0)(g(x)+j(x))'(0)}{(g(0)+j(0))^2} = \frac{(2)(3h'(0)) - 9(g'(0)+j'(0))}{4} = \frac{18 - 9(-4)}{4} = \frac{27}{2}$ by the Quotient, Sum, and Constant Multiplier rules.

29.

$$f'(x) = x^2(x+1)' + (x^2)'(x+1) = x^2(1) + 2x(x+1) = 3x^2 + 2x$$
$$f'(x) = (x^3 + x^2)' = 3x^2 + 2x$$

31.

$$f'(x) = x^{7/2}(2 - 5x^3)' + \left(x^{7/2}\right)'(2 - 5x^3) = x^{7/2}(-15x^2) + \frac{7}{2}x^{5/2}(2 - 5x^3)$$

$$= -15x^{11/2} + 7x^{5/2} - \frac{35}{2}x^{11/2} = 7x^{5/2} - \frac{65}{2}x^{11/2}$$

$$f'(x) = (2x^{7/2} - 5x^{13/2})' = 7x^{5/2} - \frac{65}{2}x^{11/2}$$

33.

$$f'(x) = \frac{x^{1/2}(x^2 - x^3)' - (x^2 - x^3)(x^{1/2})'}{x} = \frac{x^{1/2}(2x - 3x^2) - (x^2 - x^3)\left(\frac{1}{2}x^{-1/2}\right)}{x}$$

$$= 2x^{1/2} - 3x^{3/2} - \frac{1}{2}x^{1/2} + \frac{1}{2}x^{3/2} = \frac{3}{2}x^{1/2} - \frac{5}{2}x^{3/2}$$

$$f'(x) = (x^{3/2} - x^{5/2})' = \frac{3}{2}x^{1/2} - \frac{5}{2}x^{3/2}$$

In Exercises 35-64, the Constant Multiple, Sum, Difference, Product, and Quotient Rules are used without mention.

35. $f'(x) = 0 - 21x^6 = -21x^6$.

37. $f(x) = 2 + 6x^2$, so that $f'(x) = 0 + 12x = 12x$.

39. $f(x) = 6x + 2 - 4x^5$, so that $f'(x) = 6 + 0 - 20x^4 = -20x^4 + 6$.

41. $f(x) = x^2 + x - 2$, so that $f'(x) = 2x + 1 - 0 = 2x + 1$.

43. $f(x) = 27x^3 + 54x^2 + 36x + 8$, so that $f'(x) = 81x^2 + 108x + 36$.

45. $f(x) = \frac{1}{3} - 2x^3$, so that $f'(x) = -6x^2$.

47. $f(x) = \frac{x^2 - 1}{x + 1} = \frac{(x-1)(x+1)}{x+1} = x - 1$, so that $f'(x) = 1$ for $x \neq -1$.

49. Since $f(x)$ is a constant, $f'(x) = 0$.

51. For $x \neq 0$, we can simplify to get $f(x) = x^2 \cdot x^{-2/5} = x^{8/5}$, so that $f'(x) = \frac{8}{5}x^{3/5}$.

53. For $x \neq 0$, this can be simplified to $f(x) = x^{7/5}x^{-3} - 2x^4 x^{-3} = x^{-8/5} - 2x$, so that $f'(x) = -\frac{8}{5}x^{-13/5} - 2$.

55.

$$f'(x) = \frac{(1 - 3x^4)(x^7 - 3x^5 + 4)' - (x^7 - 3x^5 + 4)(1 - 3x^4)'}{(1 - 3x^4)^2}$$

$$= \frac{(1 - 3x^4)(7x^6 - 15x^4) - (x^7 - 3x^5 + 4)(-12x^3)}{(1 - 3x^4)^2}$$

$$= \frac{7x^6 - 15x^4 - 21x^{10} + 45x^8 + 12x^{10} - 36x^8 + 48x^3}{(1 - 3x^4)^2}$$

$$= \frac{-9x^{10} + 9x^8 + 7x^6 - 15x^4 + 48x^3}{(1 - 3x^4)^2}.$$

This is valid only where the denominator does not vanish.

57. $f(x) = \sqrt{x^{-3}} + \left(x^{-1/2}\right)^3 = 2x^{-3/2}$, so that $f'(x) = -3x^{-5/2}$. This is only valid for $x > 0$, since $f(x)$ is undefined for $x < 0$.

59. $f'(x) = \frac{(5x + 4)(2) - (2x - 3)(5)}{(5x + 4)^2} = \frac{23}{(5x + 4)^2}$. This is valid only where the denominator does not vanish.

61. $f(x) = \dfrac{1}{x^2 - 5x + 6}$, so that $f'(x) = \dfrac{(x^2 - 5x + 6)(1)' - (x^2 - 5x + 6)'}{(x^2 - 5x + 6)^2} = \dfrac{5 - 2x}{(x^2 - 5x + 6)^2}$. This is valid only where the denominator does not vanish.

63. For $x \neq 0$, $f(x) = \dfrac{x^2}{x^3 + 5x^2 - 3x} = \dfrac{x}{x^2 + 5x - 3}$, so

$$f'(x) = \frac{(x^2 + 5x - 3)(x)' - x(x^2 + 5x - 3)'}{(x^2 + 5x - 3)^2}$$

$$= \frac{x^2 + 5x - 3 - 2x^2 - 5x}{(x^2 + 5x - 3)^2}$$

$$= \frac{-x^2 - 3}{(x^2 + 5x - 3)^2}$$

This is valid only where the denominator does not vanish.

65. $f(x)$ is not differentiable at zero. For $x < 0$, we have $f(x) = -x$ so that $f'(x) = -1$. For $x > 0$, we have $f(x) = x$, so that $f'(x) = 1$. Hence

$$f'(x) = \begin{cases} -1, & x < 0, \\ \text{DNE}, & x = 0, \\ 1, & x > 0. \end{cases}$$

67. We have

$$f(x) = \begin{cases} 1 - 2x, & x < \frac{1}{2}, \\ 0, & x = \frac{1}{2}, \\ 2x - 1, & x > \frac{1}{2}. \end{cases}$$

For the first case, $f'(x) = -2$, while for the third case, $f'(x) = 2$. Since the limits of these derivatives as $x \to \dfrac{1}{2}$ are unequal, f is not differentiable there. Hence

$$f'(x) = \begin{cases} -2, & x < \frac{1}{2}, \\ \text{DNE}, & x = \frac{1}{2}, \\ 2, & x > \frac{1}{2}. \end{cases}$$

69. The left derivative of $f(x)$ at $x = 1$ is $3x^2|_{x=1} = 3$, and the right derivative is 1, so f is not differentiable at $x = 1$. Thus

$$f'(x) = \begin{cases} 3x^2, & x < 1, \\ \text{DNE}, & x = 1, \\ 1, & x > 1. \end{cases}$$

71. The left derivative of $f(x)$ at $x = 0$ is $(-2x)|_{x=0} = 0$, and the right derivative is $(2x)|_{x=0} = 0$, so that $f(x)$ is differentiable at 0 and has derivative zero. Thus

$$f'(x) = \begin{cases} -2x, & x \leq 0, \\ 2x, & x > 0. \end{cases}$$

73. Since $(x^6)' = 6x^5$, $(x^3)' = 3x^2$, and $x' = 1$, we start with $\dfrac{1}{2}x^6 - \dfrac{2}{3}x^3 + 4x$, which has the right derivative. However, its value at 0 is 0, so we must add one to get $f(x) = \dfrac{1}{2}x^6 - \dfrac{2}{3}x^3 + 4x + 1$.

75. Since $x' = 1$ and $(x^7)' = 7x^6$, we start with $x - \frac{4}{7}x^7$, which has the right derivative. However, its value at 1 is $\frac{3}{7}$; since we want it to be 3, we add $3 - \frac{3}{7} = \frac{18}{7}$ to get $f(x) = -\frac{4}{7}x^7 + x + \frac{18}{7}$.

77. We have $f'(x) = -3x^9 + 24x^5 + x^4 - 8$, so we start with $-\frac{3}{10}x^{10} + 4x^6 + \frac{1}{5}x^5 - 8x$. This has the right derivative, but its value at $x = 0$ is 0, so we add 2 to get $f(x) = -\frac{3}{10}x^{10} + 4x^6 + \frac{1}{5}x^5 - 8x + 2$.

Applications

79. (a) The initial position, s_0, is the surface of Venus, and the initial velocity is $v(0) = 400$ thousand miles per hour. Since $v(t) = 0.012t^2 + 400$, it follows that $a(t) = v'(t) = 0.024t$. Finally, we find $s(t)$ by noting that it must be an antiderivative of $v(t)$. Such an antiderivative is $0.004t^3 + 400t$. Since we want $s(0) = s_0$, we must have $s(t) = 0.004t^3 + 400t + s_0$.

  (b) $v(t)$ is always positive, since $t^2 \geq 0$. Thus the spaceship is always moving towards the sun (until it gets there!).

  (c) The acceleration is not constant; it is $0.024t$. But $a(t) > 0$ when $t > 0$, so the spaceship is always speeding up.

  (d) To find out when the spaceship gets there, we want $s(t) = s_0 + 67000$. Simplifying gives $0.004t^3 + 400t - 67000 = 0$. This has the solution $t \approx 140.04$, so the spaceship reaches the sun after about 140 hours. Its velocity at that time is $v(140) = 0.012 \cdot 140^2 + 400 \approx 635.2$ thousand miles per hour.

81. Since acceleration is given by $a(t) = -32$, then $v(t) = -32t$ except for a possible added constant. Since we start with $v(0) = v_0$, we must have $v(t) = -32t + v_0$. An antiderivative of $v(t)$ is then $s(t) = -16t^2 + v_0t$. Since we know that $s(0) = s_0$, we must add the constant s_0 in to get $s(t) = -16t^2 + v_0t + s_0$.

83. (a) Since $s(t)$ is cubic, let $s(t) = at^3 + bt^2 + ct + d$. Then $s'(t) = 3at^2 + 2bt + c$, and $s''(t) = 6at + 2b$. Using the given data,

$$
\begin{aligned}
s(5) &= 100 = 125a + 25b + 5c + d \\
s'(0) &= 0 = c \\
s'(5) &= -200 = 75a + 10b + c \\
s''(5) &= -46 = 30a + 2b
\end{aligned}
$$

Solving these equations gives $a = -\frac{2}{5}$, $b = -17$, $c = 0$, and $d = 575$, so that $s(t) = -\frac{2}{5}t^3 - 17t^2 + 575$.

  (b) Noting that $s'(t) = -\frac{6}{5}t^2 - 34t$ and $s''(t) = -\frac{12}{5}t - 34$,

$$s(5) = -\frac{2}{5} \cdot 5^3 - 17 \cdot 5^2 + 575 = 100,$$

$$s'(0) = -\frac{6}{5} \cdot 0^2 - 34 \cdot 0 = 0,$$

$$s'(5) = -\frac{6}{5} \cdot 5^2 - 34 \cdot 5 = -200,$$

$$s''(5) = \frac{12}{5} \cdot 5 - 34 = -46.$$

  (c) The height of the tower is the initial height of the kiwi, which is $s(0) = 575$.

  (d) $a(t) = s''(t) = -\frac{12}{5}t - 34$. This acceleration is most definitely non-constant.

Proofs

85. Suppose $f(x) = x^k$ for k a positive integer. Then (using the Binomial Theorem to expand $(x+h)^k$)

$$
\begin{aligned}
f'(x) &= \lim_{h \to 0} \frac{(x+h)^k - x^k}{h} \\
&= \lim_{h \to 0} \frac{x^k + \binom{k}{1}x^{k-1}h + \binom{k}{2}x^{k-2}h^2 + \cdots + \binom{k}{k-1}xh^{k-1} + h^k - x^k}{h} \\
&= \lim_{h \to 0} \frac{kx^{k-1}h + \binom{k}{2}x^{k-2}h^2 + \cdots + \binom{k}{k-1}xh^{k-1} + h^k}{h} \\
&= \lim_{h \to 0} \left(kx^{k-1} + \binom{k}{2}x^{k-2}h + \cdots + \binom{k}{k-1}xh^{k-2} + h^{k-1} \right) \\
&= kx^{k-1}.
\end{aligned}
$$

87. (a)

$$
(f-g)'(x) = \lim_{h \to 0} \frac{(f-g)(x+h) - (f-g)(x)}{h} = \lim_{h \to 0} \left(\frac{f(x+h) - f(x)}{h} - \frac{g(x+h) - g(x)}{h} \right)
$$

Since both f and g are differentiable by hypothesis, this limit is equal to the difference of the two limits, so that, continuing,

$$
(f-g)'(x) = \lim_{h \to 0} \frac{f(x+h) - f(x)}{h} - \frac{g(x+h) - g(x)}{h} = f'(x) - g'(x).
$$

(b)

$$
(f-g)'(x) = (f + (-1)g)'(x) = f'(x) + ((-1)g)'(x) = f'(x) - g'(x),
$$

where the second equality follows from the sum rule and the third from the constant multiple rule.

89.

$$
\begin{aligned}
\frac{d}{dx}\left(\frac{f(x)}{g(x)} \right) &= \lim_{h \to 0} \frac{\frac{f(x+h)}{g(x+h)} - \frac{f(x)}{g(x)}}{h} \\
&= \lim_{h \to 0} \frac{f(x+h)g(x) - f(x)g(x+h)}{g(x)g(x+h)h} \\
&= \lim_{h \to 0} \frac{(f(x+h)g(x) - f(x)g(x)) - (f(x)g(x+h) - f(x)g(x))}{hg(x)g(x+h)} \\
&= \lim_{h \to 0} \left(\frac{1}{g(x)g(x+h)} \cdot \left(g(x)\frac{f(x+h) - f(x)}{h} - f(x)\frac{g(x+h) - g(x)}{h} \right) \right) \\
&= \lim_{h \to 0} \left(\frac{g(x)}{g(x)g(x+h)} \frac{f(x+h) - f(x)}{h} \right) - \lim_{h \to 0} \left(\frac{f(x)}{g(x)g(x+h)} \frac{g(x+h) - g(x)}{h} \right) \\
&= \frac{g(x)}{g(x)^2} \lim_{h \to 0} \frac{f(x+h) - f(x)}{h} - \frac{f(x)}{g(x)^2} \lim_{h \to 0} \frac{g(x+h) - g(x)}{h} \\
&= \frac{g(x)f'(x) - f(x)g'(x)}{g(x)^2},
\end{aligned}
$$

where again splitting limits into two limits is justified by the fact that each individual limit exists.

91. (a) Let $f(x) = \dfrac{1}{k+1}x^{k+1}$; assume $k \neq -1$ so that the denominator does not vanish and this definition makes sense. Then clearly $f'(x) = x^k$ by the Power Rule, so that $f(x)$ is an antiderivative of x^k. Since all antiderivatives differ by a constant, then any antiderivative of x^k with $k \neq -1$ is of the form $f(x) = \dfrac{1}{k+1}x^{k+1} + C$ for some constant C.

(b) $f(x)$ clearly cannot be defined in this way when $k = -1$, as the coefficient of x^{k+1} is $\dfrac{1}{0}$.

Thinking Forward

▷ We have $f'(x) = 3x^2 - 2$, so that $f'(x) = 0$ when $x = \pm\sqrt{\frac{2}{3}}$. $f'(x)$ is defined everywhere.

▷ We have $f'(x) = \dfrac{1}{2}x^{-1/2} - 1$. $f'(x)$ is undefined when $x = 0$, and is equal to zero when $x^{-1/2} = 2$, so when $x = \dfrac{1}{4}$.

▷ We have

$$f'(x) = \frac{(1+\sqrt{x})(1)' - 1 \cdot (1+\sqrt{x})'}{(1+\sqrt{x})^2}$$

$$= -\frac{\frac{1}{2}x^{-1/2}}{(1+\sqrt{x})^2}$$

$$= -\frac{1}{2\sqrt{x}(1+\sqrt{x})^2}.$$

Clearly $f'(x)$ is never zero, but it is undefined at $x = 0$, where the denominator vanishes.

▷ Since $f(x) = \dfrac{x^3 - x^2}{x^2 - 4x + 4}$, we have

$$f'(x) = \frac{(x^2 - 4x + 4)(3x^2 - 2x) - (x^3 - x^2)(2x - 4)}{(x-2)^4}$$

$$= \frac{3x^4 - 2x^3 - 12x^3 + 8x^2 + 12x^2 - 8x - 2x^4 + 4x^3 + 2x^3 - 4x^2}{(x-2)^4}$$

$$= \frac{x^4 - 8x^3 + 16x^2 - 8x}{(x-2)^4}$$

$$= \frac{x(x^2 - 6x + 4)}{(x-2)^3}.$$

Then $f'(x) = 0$ at $x = 0$ and at the roots of $x^2 - 6x + 4$, which are $x = 3 \pm \sqrt{5}$, and $f'(x)$ is undefined at $x = 2$. However, $x = 2$ is not a critical point, since f is not defined there either.

Taylor polynomials

▷ If $f(x) = ax^2 + bx + c$, then by the power rule, $f'(x) = 2ax + b$ and $f''(x) = 2a$. From the last formula, it follows that $a = \dfrac{f''(x)}{2} = \dfrac{f''(0)}{2}$. The second formula implies (setting $x = 0$) that $f'(0) = b$. Finally, clearly $f(0) = c$.

▷ A similar analysis gives

$$f(x) = ax^3 + bx^2 + cx + d, \quad f'(x) = 3ax^2 + 2bx + c, \quad f''(x) = 6ax + 2b, \quad f'''(x) = 6a$$

Substituting $x = 0$ throughout gives

$$f(0) = d, \quad f'(0) = c, \quad f''(0) = 2b, \quad f'''(0) = 6a,$$

so that

$$a = \frac{f'''(0)}{6}, \quad b = \frac{f''(0)}{2}, \quad c = f'(0), \quad d = f(0).$$

▷ Using the equation from part (b), we have

$$f(2) = 8a + 4b + 2c + d$$
$$f'(2) = 12a + 4b + c$$
$$f''(2) = 12a + 2b$$
$$f'''(2) = 6a$$

This linear system in the variables $a, b, c,$ and d clearly has a solution, by back-substitution starting with the last equation. Thus each of $a, b, c,$ and d can be expressed in terms of $f(2), f'(2), f''(2),$ and $f'''(2)$.

▷ Consider $f^{[k]}(x)$. The terms of $f(x)$ that have powers of x smaller than k will vanish when the kth derivative is taken, since each derivative reduces the exponent by one. The powers of x larger than k will still have a nonzero power of x. Finally, the term $a_k x^k$ will turn into $a_k \cdot k \cdot (k-1) \ldots 2 \cdot 1 = k! a_k$, since each derivative pulls down the exponent. Thus $f^{[k]}(x)$ consists of some terms that have powers of x in them, plus a constant term of $k! a_k$. Thus $f^{[k]}(0) = k! a_k$, so that $a_k = \dfrac{f^{[k]}(0)}{k!}$.

2.4 The Chain Rule and Implicit Differentiation

Thinking Back

Differentiation review

▷ Since $f(x) = 81x^4 + 108x^3 + 54x^2 + 12x + 1$, we have $f'(x) = 324x^3 + 324x^2 + 108x + 12$.

▷ Since $f(x) = \dfrac{1}{x^2} + \dfrac{2}{x^3} + \dfrac{1}{x^4} = x^{-2} + 2x^{-3} + x^{-4}$, we have $f'(x) = -2x^{-3} - 6x^{-4} - 4x^{-5}$.

▷ We have $f(x) = (x^2 + 2x + 1)x^{1/2} = x^{5/2} + 2x^{3/2} + x^{1/2}$; then $f'(x) = \dfrac{5}{2}x^{3/2} + 3x^{1/2} + \dfrac{1}{2}x^{-1/2}$.

▷ We have $f(x) = \dfrac{x^2 + 2x + 1}{x^{1/2}} = x^{3/2} + 2x^{1/2} + x^{-1/2}$, so that $f'(x) = \dfrac{3}{2}x^{1/2} + x^{-1/2} - \dfrac{1}{2}x^{-3/2}$.

Decomposing functions

▷ Let $h(x) = x^2$, $g(x) = x + 1$, and $f(x) = x^3$. Then $f(g(h(x))) = f(g(x^2)) = f(x^2 + 1) = (x^2 + 1)^3$.

▷ Let $h(x) = x - 2$, $g(x) = 1 + x^2$, and $f(x) = \sqrt{x}$. Then $f(g(h(x))) = f(g(x - 2)) = f(1 + (x - 2)^2) = \sqrt{1 + (x - 2)^2}$.

▷ Let $h(x) = 3x + 1$, $g(x) = \dfrac{1}{x}$, and $f(x) = \sqrt{x}$. Then $f(g(h(x))) = f(g(3x+1)) = f\left(\dfrac{1}{3x+1}\right) = \sqrt{\dfrac{1}{3x+1}}$.

▷ Let $h(x) = 3x + 1$, $g(x) = \sqrt{x}$, and $f(x) = \dfrac{1}{x}$. Then $f(g(h(x))) = f(g(3x+1)) = f(\sqrt{3x+1}) = \dfrac{1}{\sqrt{3x+1}}$.

Concepts

1. (a) True. See Theorem 2.12.

   (b) False. For example, let $f(x) = x^2$ and $g(x) = x + 1$. Then $(f \circ g)(x) = (x+1)^2 = x^2 + 2x + 1$, which has derivative $2x + 2$, while $(g \circ f)(x) = x^2 + 1$, which has derivative $2x$.

   (c) False. It is $f'(g(x))g'(x)$. See Theorem 2.12.

   (d) False. It is $u'(v(x))v'(x)$. See Theorem 2.12.

   (e) True. See Theorem 2.12.

   (f) True. Consider $x^2 + y^2 = 1$ and the x-value of 0.

   (g) True. Consider $x^2 + y^2 = 1$ and the x-value of 1.

   (h) Generally false. While the graph of $y = f(x)$ has a horizontal tangent line at $x = 2$, the point on the graph need not be $(2, 0)$.

3. (a) $(g \circ h)'(x) = g'(h(x))h'(x)$.

   (b) $\dfrac{dg}{dx} = \dfrac{dg}{dh} \cdot \dfrac{dh}{dx}$.

5. (a) $(f \circ u \circ v \circ w)'(x) = f'(u(v(w(x))))u'(v(w(x)))v'(w(x))w'(x)$.

   (b) $\dfrac{df}{dx} = \dfrac{df}{du} \cdot \dfrac{du}{dv} \cdot \dfrac{dv}{dw} \cdot \dfrac{dw}{dx}$.

7. (a) Let $g(x) = x^2$ and $h(x) = 3x + \sqrt{x}$. Then $f(x) = g(h(x))$, so that

$$f'(x) = g'(h(x))h'(x) = 2h(x)\left(3 + \frac{1}{2}x^{-1/2}\right) = 2(3x + x^{1/2})\left(3 + \frac{1}{2}x^{-1/2}\right).$$

   (b) With $h(x)$ as above, $f(x) = h(x)h(x)$, so that

$$f'(x) = h(x)h'(x) + h'(x)h(x) = 2h(x)h'(x) = 2(3x + x^{1/2})\left(3 + \frac{1}{2}x^{-1/2}\right).$$

   (c) We have $f(x) = (3x + x^{1/2})(3x + x^{1/2}) = 9x^2 + 6x^{3/2} + x$, so that $f'(x) = 18x + 9x^{1/2} + 1$. To see that these are the same, multiply out the second formula:

$$2(3x + x^{1/2})\left(3 + \frac{1}{2}x^{-1/2}\right) = 18x + 3x^{1/2} + 6x^{1/2} + 1 = 18x + 9x^{1/2} + 1.$$

9. $f'(3) = g'(h(3))h'(3) = g'(0) \cdot 1 = -2$.

11. $f'(-2) = 3g(-2)^2 g'(-2) = 3 \cdot 1^2 \cdot 2 = 6$.

13. $f'(1) = h'(g(j(1)))g'(j(1))j'(1) = h'(1)g'(-2)j'(1) = (-2)(2)(-1) = 4$.

15. $f'(0) = h'(g(0)j(0))(g(x)j(x))'(0) = h'(0)(g(0)j'(0) + j(0)g'(0)) = 3(2 \cdot (-2) + 0 \cdot (-2)) = -12$.

17. If $y = y(x)$, let $f(y) = y^3$. Then $\frac{dy^3}{dx} = \frac{df}{dx} = \frac{df}{dy} \cdot \frac{dy}{dx} = 3y^2 \frac{dy}{dx}$ by the chain rule.

19. The graph on the left matches $xy^2 + y = 1$, since when $x = 0$, we get $y = 1$ as the only solution. Setting $x = 0$ in $(x + 1)(y^2 + y - 1) = 1$ gives $y^2 + y - 2 = 0$, which has the solutions $y = 1$, $y = -2$, matching the right graph.

Skills

21. Rewrite $f(x) = (x^3 + 1)^{-1}$; then $f'(x) = -1(x^3 + 1)^{-2}3x^2 = -\dfrac{3x^2}{(x^3 + 1)^2}$.

23. Let $g(x) = x$ and $h(x) = (3x^2 + 1)^9$. Then

$$f'(x) = g(x)h'(x) + g'(x)h(x)$$
$$= x \cdot 9(3x^2 + 1)^8(3x^2 + 1)' + 1 \cdot (3x^2 + 1)^9$$
$$= 54x^2(3x^2 + 1)^8 + (3x^2 + 1)^9 = (57x^2 + 1)(3x^2 + 1)^8.$$

25. $f(x) = (x^2 + 1)^{-1/2}$, so $f'(x) = -\frac{1}{2}(x^2 + 1)^{-3/2}(x^2 + 1)' = -x(x^2 + 1)^{-3/2}$.

27. Rewrite $f(x) = (x\sqrt{x+1})^{-2} = \dfrac{1}{x^2(x+1)} = (x^3 + x^2)^{-1}$. Then

$$f'(x) = -(x^3 + x^2)^{-2}(x^3 + x^2)' = -\dfrac{3x^2 + 2x}{(x^3 + x^2)^2}.$$

29. Multiply top and bottom by x to get $f(x) = \dfrac{1 - 3x^3}{x^6 - \sqrt{x}}$, then use the quotient rule:

$$f'(x) = \frac{(x^6 - \sqrt{x})(-9x^2) - (1 - 3x^3)\left(6x^5 - \frac{1}{2}x^{-1/2}\right)}{(x^6 - \sqrt{x})^2}$$
$$= \frac{-9x^8 + 9x^{5/2} - 6x^5 + \frac{1}{2}x^{-1/2} + 18x^8 - \frac{3}{2}x^{5/2}}{(x^6 - \sqrt{x})^2}$$
$$= \frac{9x^8 - 6x^5 + \frac{15}{2}x^{5/2} + \frac{1}{2}x^{-1/2}}{(x^6 - x^{1/2})^2}$$
$$= \frac{18x^8 - 12x^5 + 15x^{5/2} + x^{-1/2}}{2(x^6 - x^{1/2})^2}$$

31. $f'(x) = -1(x^{1/3} - 2x)^{-2}(x^{1/3} - 2x)' = (x^{1/3} - 2x)^{-2}\left(2 - \frac{1}{3}x^{-2/3}\right)$.

33.

$$f'(x) = -\frac{1}{2}x^{-3/2}(x^2 - 1)^3 + x^{-1/2}(3(x^2 - 1)^2(x^2 - 1)')$$
$$= -\frac{1}{2}x^{-3/2}(x^2 - 1)^3 + 6x^{1/2}(x^2 - 1)^2$$

Alternatively, one could expand $(x^2 - 1)^3$ and avoid the chain and product rules altogether.

35. $f(x) = (3x - 4(2x + 1)^6)^{1/2}$, so

$$f'(x) = \frac{1}{2}(3x - 4(2x + 1)^6)^{-1/2}(3x - 4(2x + 1)^6)'$$
$$= \frac{1}{2}(3x - 4(2x + 1)^6)^{-1/2}(3 - 24(2x + 1)^5(2))$$
$$= \frac{1}{2}(3x - 4(2x + 1)^6)^{-1/2}(3 - 48(2x + 1)^5)$$

37.

$$f'(x) = 100(5(3x^4 - 1)^3 + 3x - 1)^{99}(5(3x^4 - 1)^3 + 3x - 1)'$$
$$= 100(5(3x^4 - 1)^3 + 3x - 1)^{99}(5 \cdot 3(3x^4 - 1)^2(3x^4 - 1)' + 3)$$
$$= 100(5(3x^4 - 1)^3 + 3x - 1)^{99}(15(3x^4 - 1)^2(12x^3) + 3)$$
$$= 100(180x^3(3x^4 - 1)^2 + 3)(5(3x^4 - 1)^3 + 3x - 1)^{99}$$

39.

$$f'(x) = -2((x^2 + 1)^8 - 7x)^{-5/3}((x^2 + 1)^8 - 7x)'$$
$$= -2((x^2 + 1)^8 - 7x)^{-5/3}(8(x^2 + 1)^7(2x) - 7)$$
$$= -2(16x(x^2 + 1)^7 - 7)((x^2 + 1)^8 - 7x)^{-5/3}$$

41.

$$f'(x) = (5x^4 - 3x^2)^7(2x^3 + 1)' + ((5x^4 - 3x^2)^7)'(2x^3 + 1)$$
$$= (5x^4 - 3x^2)^7(6x^2) + (7(5x^4 - 3x^2)^6(5x^4 - 3x^2)')(2x^3 + 1)$$
$$= 6x^2(5x^4 - 3x^2)^7 + 7(2x^3 + 1)(20x^3 - 6x)(5x^4 - 3x^2)^6$$

43.

$$f'(x) = -9((2x + 1)^{-5} - 1)^{-10}((2x + 1)^{-5} - 1)'$$
$$= -9((2x + 1)^{-5} - 1)^{-10}\left(-5(2x + 1)^{-6}(2)\right)$$
$$= 90(2x + 1)^{-6}((2x + 1)^{-5} - 1)^{-10}$$

45.

$$f'(x) = 8(x^4 - \sqrt{3 - 4x})^7(x^4 - \sqrt{3 - 4x})' + 5$$
$$= 8(x^4 - \sqrt{3 - 4x})^7\left(4x^3 - \frac{1}{2}(3 - 4x)^{-1/2}(-4)\right) + 5$$
$$= 5 + 8(x^4 - \sqrt{3 - 4x})^7\left(4x^3 + 2(3 - 4x)^{-1/2}\right)$$

47. Expand to get $(x\sqrt{x+1})^{-2} = \dfrac{1}{x^2(x+1)} = \dfrac{1}{x^2+x^3}$. Then

$$\frac{d}{dx}\frac{1}{x^2+x^3} = -\frac{2x+3x^2}{(x^2+x^3)^2}$$

$$\frac{d^2}{dx^2}\frac{1}{x^2+x^3} = -\frac{d}{dx}\frac{2x+3x^2}{(x^2+x^3)^2}$$

$$= -\frac{(x^2+x^3)^2(2+6x) - (2x+3x^2)(2(x^2+x^3)(2x+3x^2))}{(x^2+x^3)^4}$$

$$= -\frac{(x^2+x^3)(2+6x) - 2(2x+3x^2)(2x+3x^2)}{(x^2+x^3)^3}$$

$$= -\frac{2x^2+6x^3+2x^3+6x^4-8x^2-24x^3-18x^4}{(x^2+x^3)^3}$$

$$= -\frac{-12x^4-16x^3-6x^2}{(x^2+x^3)^3}$$

$$= \frac{12x^4+16x^3+6x^2}{(x^2+x^3)^3}$$

49. Simplifying, $\dfrac{3}{x^{-3/2}\sqrt{x}} = \dfrac{3}{x^{-1}} = 3x$, so that $\dfrac{d^2}{dx^2}(3x) = \dfrac{d}{dx}(3) = 0$. Thus its value at $x = 2$ is also zero.

51. Write $g(x) = (3x^2+1)^9$. Then

$$\frac{d^3}{dx^3}(x(3x^2+1)^9) = \frac{d^3}{dx^3}(xg(x))$$

$$= \frac{d^2}{dx^2}\left(g(x) + xg'(x)\right)$$

$$= \frac{d}{dx}\left(g'(x) + g'(x) + xg''(x)\right) = \frac{d}{dx}\left(2g'(x) + xg''(x)\right)$$

$$= 2g''(x) + g''(x) + xg'''(x) = 3g''(x) + xg'''(x).$$

Thus

$$\frac{d^3}{dx^3}\left(x(3x^2+1)^9\right)\Big|_{x=0} = 3g''(0) + 0\cdot g'''(0) = 3g''(0).$$

So we need to find $g''(x)$:

$$\frac{d^2}{dx^2}\left((3x^2+1)^9\right) = \frac{d}{dx}\left(9(3x^2+1)^8(6x)\right) = \frac{d}{dx}\left(54x(3x^2+1)^8\right)$$

$$= 54(3x^2+1)^8 + 54x\cdot 8(3x^2+1)^7(6x)$$

Evaluating this expression at $x = 0$, the second term vanishes, and the first term becomes $54(3\cdot 0^1+1)^8 = 54$. The answer is $3g''(0) = 3\cdot 54 = 162$.

53. $\dfrac{d}{dr}(s^3) = 3s^2\dfrac{ds}{dr}$.

55. $\dfrac{d}{dr}(q^3) = 0$ since q is a constant.

57. $\dfrac{d}{dr}(rs^2) = r\dfrac{d}{dr}(s^2) + s^2\dfrac{d}{dr}(r) = s^2 + 2rs\dfrac{ds}{dr}$.

59. Solving for y gives $y^2 = 9 - \frac{1}{4}x^2$, so that $y = \pm\sqrt{9 - \frac{1}{4}x^2}$.

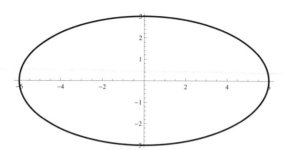

61. Solving for y gives $3y^2 = x^2 - 16$, or $y = \pm\sqrt{\dfrac{x^2 - 16}{3}}$.

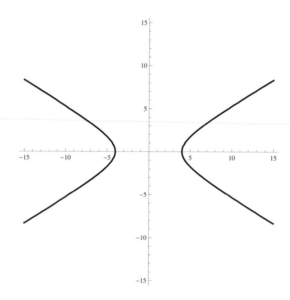

63. Since $2x$ is the derivative of $x^2 + 1$, try $f(x) = (x^2 + 1)^5$. This has the correct derivative, and also $f(0) = 1$.

65. The derivative of $x^2 + 1$ is $2x$; we have an x, so the 2 must have been canceled by a factor of $\frac{1}{2}$. Try $f(x) = \frac{1}{2}(x^2 + 1)^5$. This does not quite have the correct derivative — its derivative is $\frac{1}{2} \cdot 5(x^2 + 1)^4 \cdot 2x = 5x(x^2 + 1)^4$. We must deal with the 5 by dividing the original function by 5. So try $f(x) = \frac{1}{10}(x^2 + 1)^5$. This has the correct derivative, but $f(0) = \frac{1}{10}$. Thus the function we want is $f(x) = \frac{1}{10}(x^2 + 1)^5 - \frac{1}{10}$.

67. We have $f'(x) = (3x + 1)^{1/2}$, so the natural thing to try is $f(x) = (3x + 1)^{3/2}$. The derivative of this function is, however, $\frac{3}{2}(3x + 1)^{1/2}(3) = \frac{9}{2}(3x + 1)^{1/2}$. To correct for the fraction, let $f(x) = \frac{2}{9}(3x + 1)^{3/2}$. This function has the correct derivative. However, $f(0) = \frac{2}{9}$, so the function we want is $f(x) = \frac{2}{9}(3x + 1)^{3/2} + \frac{7}{9}$.

69. Differentiating gives $8x - 2y\dfrac{dy}{dx} = 0$, so that $\dfrac{dy}{dx} = \dfrac{8x}{2y} = \dfrac{4x}{y}$.

71. Differentiating gives

$$(xy^2)' + (3x^2)' = 0 \Rightarrow y^2 + 2xyy' + 6x = 0 \Rightarrow 2xyy' = -y^2 - 6x \Rightarrow y' = -\frac{y^2 + 6x}{2xy}$$

73. Differentiating using the product rule gives $3(y^2 - y + 6) + (3x + 1)(2yy' - y') = 0$. Thus

$$(2y - 1)y' = -\frac{3(y^2 - y + 6)}{3x + 1} \implies y' = -\frac{3(y^2 - y + 6)}{(3x + 1)(2y - 1)}.$$

75. Differentiating gives $\dfrac{1}{2}(3y - 1)^{-1/2}(3y') = 5xy' + 5y$. Collecting terms involving y' gives

$$y'\left(\frac{3}{2}(3y - 1)^{-1/2} - 5x\right) = 5y \implies y' = \frac{5y}{\frac{3}{2}(3y - 1)^{-1/2} - 5x} = \frac{10y\sqrt{3y - 1}}{3 - 10x\sqrt{3y - 1}}.$$

77. Multiply through to get $y^2 + 1 = x(3y - 1)$. Differentiating gives $2yy' = (3y - 1) + x(3y') = 3xy' + 3y - 1$; simplify to get $y' = \dfrac{3y - 1}{2y - 3x}$.

79. First multiply through by $xy(y + 1)$ to get $x(y + 1) - (y^2 + y) = x^3y$. Now differentiate both sides to get $(y + 1) + xy' - 2yy' - y' = 3x^2y + x^3y'$, and collect terms to get $(x^3 - x + 2y + 1)y' = y - 3x^2y + 1$. Thus $y' = \dfrac{y - 3x^2y + 1}{x^3 - x + 2y + 1}$.

81. (a) The points with $x = \dfrac{1}{2}$ are solutions to $\dfrac{1}{4} + y^2 = 1$, or $y = \pm\dfrac{\sqrt{3}}{2}$. Differentiating the equation gives $2x + 2yy' = 0$, so that $y' = -\dfrac{x}{y}$. Thus at $\left(\dfrac{1}{2}, \dfrac{\sqrt{3}}{2}\right)$, the slope is $-\dfrac{1}{\sqrt{3}}$, while at $\left(\dfrac{1}{2}, -\dfrac{\sqrt{3}}{2}\right)$, the slope is $\dfrac{1}{\sqrt{3}}$.

   (b) The points with $y = \dfrac{\sqrt{2}}{2}$ are solutions to $x^2 + \dfrac{1}{2} = 1$, or $x = \pm\dfrac{\sqrt{2}}{2}$. Using y' as computed in part (a), we see that at $\left(\dfrac{\sqrt{2}}{2}, \dfrac{\sqrt{2}}{2}\right)$, the slope is -1, while at $\left(-\dfrac{\sqrt{2}}{2}, \dfrac{\sqrt{2}}{2}\right)$, the slope is 1.

   (c) The tangent line is vertical where y' does not exist, which happens when $y = 0$. These points are solutions to $x^2 = 1$, so are $(1, 0)$ and $(-1, 0)$.

   (d) The tangent line has a slope of -1 when $-\dfrac{x}{y} = -1$, so that $x = y$. If $x = y$, then $2x^2 = 1$, so that $x = \pm\dfrac{\sqrt{2}}{2}$. Thus the two points are $\left(\dfrac{\sqrt{2}}{2}, \dfrac{\sqrt{2}}{2}\right)$ and $\left(-\dfrac{\sqrt{2}}{2}, -\dfrac{\sqrt{2}}{2}\right)$.

83. Note that differentiating gives $3y^2y' + y + xy' = 0$, so that $y' = -\dfrac{y}{3y^2 + x}$.

   (a) When $x = 1$, the corresponding y-coordinates are solutions of $y^3 + 1 \cdot y + 2 = 0$, or $y^3 + y + 2 = 0$. The only real solution is $y = -1$, so the only such point is $(1, -1)$. The slope of the tangent line at $(1, -1)$ is $-\dfrac{y}{3y^2 + x} = -\dfrac{-1}{3 + 1} = \dfrac{1}{4}$.

(b) When $y = 1$, the corresponding x-coordinates are solutions of $1 + x + 2 = 0$, so that $x = -3$. The only such point is $(-3, 1)$. The slope of the tangent line at $(-3, 1)$ is $-\frac{y}{3y^2+x} = -\frac{1}{3-3}$, which is infinite. The tangent line is vertical at $(-3, 1)$.

(c) The tangent line is horizontal when $y' = 0$, so when the numerator vanishes; this is when $y = 0$. Corresponding x coordinates are solutions of $2 = 0$, so there are none, and the tangent line is never horizontal.

(d) The tangent line is vertical when y' does not exist, which happens when $3y^2 + x = 0$, so that $x = -3y^2$. Substituting this value of x into the original equation gives $y^3 - 3y^3 + 2 = 0$, so that $2y^3 = 2$ and $y = 1$. Since $x = -3y^2$, we have $x = -3$, and the only point with a vertical tangent line is $(-3, 1)$.

Applications

85. Let $s(t) = 12t$ be the number of subscriptions Linda has sold after t weeks, and let $p(s) = 4s$ be the amount she has made from selling s subscriptions. Then the amount Linda has made after t weeks is $p(s(t))$, so the incremental amount she makes each week is $(p \circ s)'(t)$, which, by the chain rule, is $(p \circ s)'(t) = p'(s(t))s'(t) = 12 \cdot 4 = 48$.

87. (a) $\frac{dA}{dr} = 2\pi r$. This is the rate at which area is expanding at a given radius.

(b) No, $\frac{dA}{dr} = 2\pi r$ does not depend on the rate at which the radius is increasing (this rate is $\frac{dr}{dt}$). It does depend on the radius of the circle, however, from the formula.

(c) $\frac{dA}{dt} = 2\pi r r'$. This is the rate at which the area of the circle is expanding after a given period of time.

(d) $2\pi r r'$ depends both on the radius of the circle and the rate at which the radius is increasing, since it is a function of both r and $r' = \frac{dr}{dt}$.

(e) Since $\frac{dA}{dt} = 2\pi r \frac{dr}{dt}$ and $r' = 2$, then when the radius is 24, $\frac{dA}{dt} = 2\pi \cdot 24 \cdot 2 = 96\pi$ square inches per second.

Proofs

89.

$$
\begin{aligned}
\left(\frac{f(x)}{g(x)}\right)' &= \left(f(x)g(x)^{-1}\right)' && \leftarrow \text{algebra} \\
&= f(x)(g(x)^{-1})' + f'(x)g(x)^{-1} && \leftarrow \text{product rule} \\
&= f(x)(-1 \cdot g(x)^{-2} \cdot g'(x)) + f'(x)g(x)^{-1} && \leftarrow \text{power, chain rules} \\
&= f'(x)g(x) \cdot g(x)^{-2} - f(x)g'(x) \cdot g(x)^{-2} && \leftarrow \text{algebra} \\
&= \frac{g(x)f'(x) - f(x)g'(x)}{g(x)^2} && \leftarrow \text{algebra}
\end{aligned}
$$

91. Let $y = x^{3/5}$; then $y^5 = x^3$. Differentiating gives $5y^4 y' = 3x^2$, so that

$$
y' = \frac{d}{dx}\left(x^{3/5}\right) = \frac{3x^2}{5y^4} = \frac{3x^2}{5(x^{3/5})^4} = \frac{3}{5}x^{2-12/5} = \frac{3}{5}x^{-2/5}.
$$

93. Suppose $y = x^{-k}$ where $k > 0$. Then $yx^k = 1$; differentiating implicitly gives $y'x^k + kx^{k-1}y = 0$. Simplifying gives

$$y' = \frac{d}{dx}\left(x^{-k}\right) = \frac{-kx^{k-1}y}{x^k} = -kyx^{-1} = -kx^{-k-1}.$$

Thinking Forward

Finding critical points

▷ We have $f'(x) = 3x^2\sqrt{3x+1} + \frac{3}{2}x^3(3x+1)^{-1/2}$. Thus $f'(x)$ is undefined for $x = -\frac{1}{3}$ (it is also undefined for x smaller than that, but f is undefined there as well). $f'(x) = 0$ means that $3x^2\sqrt{3x+1} + \frac{3}{2}x^3(3x+1)^{-1/2} = 0$; multiply through by $2(3x+1)^{-1/2}$ to get $6x^2(3x+1) + 3x^3 = 3x^2(7x+2) = 0$. Thus $f'(x) = 0$ for $x = 0$ as well as for $x = -\frac{2}{7}$.

▷ Since $f'(x) = 7(1-4x)^6(-4) = -28(1-4x)^6$, $f'(x)$ is defined everywhere, and $f'(x) = 0$ for $x = \frac{1}{4}$.

▷ We have

$$f'(x) = 2x(x-2)^{3/2} + \frac{3}{2}(x^2+3)(x-2)^{1/2} = (x-2)^{1/2}\left(2x(x-2) + \frac{3}{2}(x^2+3)\right)$$

$$= (x-2)^{1/2}\left(\frac{7}{2}x^2 - 4x + \frac{9}{2}\right).$$

Then $f'(x)$ is defined for $x \geq 2$, and is zero where $x = 2$, or at the roots of $\frac{7}{2}x^2 - 4x + \frac{9}{2}$. But this quadratic has no real roots, so the only critical point of f is at $x = 2$.

▷ Multiply through to get $3x^2 + 3$, so that $f'(x) = 6x$ and the only candidate for a critical point is at $x = 0$. However, $f(x)$ is undefined there, so it has no critical points.

▷ We have $f(x) = \frac{\sqrt{x}}{x\sqrt{x-1}} = \frac{1}{\sqrt{x(x-1)}} = (x^2-x)^{-1/2}$. Thus

$$f'(x) = -\frac{1}{2}(x^2-x)^{-3/2}(2x-1) = \frac{1}{2}(1-2x)(x^2-x)^{-3/2}.$$

The domain of $f(x)$ is $x > 1$. Since the only zero of the numerator of f' is $\frac{1}{2}$, which is not in the domain of f, we see that f has no critical points.

▷ Expanding $f(x)$ gives $f(x) = x^{-2}(x+1)^{-1}$, so that

$$f'(x) = -2x^{-3}(x+1)^{-1} + (-1)x^{-2}(x+1)^{-2} = x^{-3}(x+1)^{-2}(-2(x+1) - x) = -\frac{3x+2}{x^3(1+x)^2}.$$

Thus $f'(x)$ is undefined at $x = 0$ and at $x = -1$, but those are not in the domain of f. $f'(x) = 0$ at $x = -\frac{2}{3}$, which is the only critical point of f.

Finding antiderivatives by undoing the chain rule

▷ Since $(1+x^2)' = 2x$, try first $(1+x^2)^{3/2}$; this has derivative $\frac{3}{2}(2x)(1+x^2)^{1/2} = 3x\sqrt{1+x^2}$. Thus $f(x) = \frac{1}{3}(1+x^2)^{3/2}$.

▷ Since $(1 + x^3)' = 3x^3$, try first $(1 + x^3)^{3/2}$; this has derivative $\frac{3}{2}(3x^2)(1 + x^3)^{1/2} = \frac{9}{2}x^2(1 + x^3)^{1/2}$. Thus $f(x) = \frac{2}{9}(1 + x^3)^{3/2}$.

▷ Use $(2 - 5x)^{-2}$ as a first try; this gives as derivative $-2(2 - 5x)^{-3}(-5) = 10(2 - 5x)^{-3}$. Thus $f(x) = \dfrac{1}{10(2 - 5x)^{-2}}$.

▷ Use $(1 + 3x)^{1/2}$ as a first try; this gives as derivative $\dfrac{1}{2}(1 + 3x)^{-1/2}(3) = \dfrac{3}{2}(1 + 3x)^{-1/2}$. Thus $f(x) = \dfrac{2}{3}\sqrt{1 + 3x}$.

2.5   Derivatives of Exponential and Logarithmic Functions

Thinking Back

Solving exponential and logarithmic equations

▷ Taking logs of both sides gives $x \ln 1.2 + \ln 3 = \ln 500$, so that $x = \dfrac{\ln 500 - \ln 3}{\ln 1.2} = \dfrac{\ln \frac{500}{3}}{\ln 1.2}$.

▷ Exponentiating both sides gives $x^2 + x - 5 = 1$, so that $x^2 + x - 6 = (x + 3)(x - 2) = 0$, so that $x = 2$ or $x = -3$. But substituting $x = 2$ in the original equation gives -1 as the argument of ln, so we reject that root. $x = -3$ is the only solution.

▷ This equation holds whenever the numerator vanishes, which happens when $x = 0$. But at $x = 0$, the denominator is undefined. Thus this equation has no solutions.

▷ This equation holds whenever the numerator vanishes. But $2^x \geq 0$, so the numerator is always positive. This equation has no solutions.

Compositions

▷ For example, $h(x) = x^2 + 5$, $g(x) = \sqrt{x}$, and $f(x) = \ln x$. Then $f(g(h(x))) = \ln\left(\sqrt{x^2 + 5}\right)$.

▷ For example, $h(x) = \sqrt{x + 1}$, $g(x) = \dfrac{1}{x}$, and $f(x) = e^x$. Then $f(g(h(x))) = e^{\frac{1}{\sqrt{x+1}}}$.

▷ For example, $h(x) = 3x$, $g(x) = \ln x$, and $f(x) = x^2$. Then $f(g(h(x))) = (\ln 3x)^2$.

▷ For example, $h(x) = 5^x$, $g(x) = x + 2$, and $f(x) = 3 \ln x$. Then $f(g(h(x))) = 3 \ln(5^x + 2)$.

Concepts

1. (a) True. e^π is a constant.

    (b) True. See Theorem 2.13(a).

    (c) False. $\dfrac{d}{dx}\dfrac{1}{x} = \dfrac{d}{dx}x^{-1} = -x^{-2}$. What *is* true is that $\dfrac{d}{dx}\ln x = \dfrac{1}{x}$.

    (d) False. By Theorem 2.15(c), $\dfrac{d}{dx}\ln |x| = \dfrac{1}{x}$.

    (e) True. See Theorem 2.14.

    (f) True. See Theorem 2.14 (that theorem is an if and only if theorem).

    (g) False. It is not required, but it may make things easier.

(h) True.

3. The exponential rule does not apply since the base is not a constant. The power rule does not apply since the exponent is not a constant.

5. Differentiating e^x is the special case of differentiating e^{kx} when $k = 1$. It is a special case of differentiating b^x when $b = e$.

7. Differentiating $\ln x$ is the special case of differentiating $\log_b x$ when $b = e$.

9. Only $2(2^x)$ has the value 2 at $x = 0$, so the blue curve is $2(2^x)$. 4^x increases faster than 2^x for $x > 0$, so the red curve is 4^x and the gray curve is 2^x.

11. We have $f(3) = 2^{(3^2)} = 2^9$, while $g(3) = (2^3)^2 = 8^2 = 2^6$, so the two are unequal. Note that $g(x) = (2^x)^2 = 2^{2x}$, so that $f(x) = g(x)$ means that $x^2 = 2x$, so that $x = 0$ or $x = 2$. These are the only two values for which $f(x) = g(x)$.

13. Since exponential functions are one-to-one, and logarithmic functions are their inverses, it follows that logarithmic functions are one-to-one as well. By definition, this means that $A = B$ if and only if $\log_b A = \log_b B$.

15. Given $y = f(x)$, the process of logarithmic differentiation consists of first taking logs of both sides to get $\ln y = \ln f(x)$ and then differentiating both sides: $\dfrac{y'}{y} = (\ln f(x))'$. Then $y' = y (\ln f(x))' = f(x) (\ln f(x))'$. This is useful if the log of $f(x)$ has a simpler form than $f(x)$; for example, when $f(x)$ is a complicated rational function, or when $f(x)$ has x's in both base and exponent.

Skills

17. $f(x) = (2 - e^{5x})^{-1}$, so that $f'(x) = -(2 - e^{5x})^{-2}(2 - e^{5x})' = 5e^{5x}(2 - e^{5x})^{-2}$.

19. $f'(x) = 3 \cdot 2xe^{-4x} + 3x^2 \cdot (-4e^{-4x}) = e^{-4x} \left(6x - 12x^2 \right)$.

21. $f(x) = e^{-x}(1 - x)$, so $f'(x) = -e^{-x}(1 - x) + e^{-x}(-1) = (x - 2)e^{-x}$.

23. $f'(x) = (e^x)'(x^2 + 3x - 1) + e^x(2x + 3) = e^x(x^2 + 5x + 2)$.

25. $f(x) = e^{3 \ln x} = \left(e^{\ln x} \right)^3 = x^3$, so that $f'(x) = 3x^2$.

27. $f'(x) = e^{(e^x)}(e^x)' = e^x e^{(e^x)}$.

29. $f(x) = e^{ex}$, so $f'(x) = e \cdot e^{ex} = e^{ex+1}$.

31. Simplifying, we get $f(x) = \dfrac{5 \ln x}{4 \ln x} = \dfrac{5}{4}$, so that $f'(x) = 0$.

33. $f'(x) = -3x^{-4}e^{2x} + x^{-3}(2e^{2x}) = e^{2x}(-3x^{-4} + 2x^{-3})$.

35. $f(x) = x^2(\log_2 x + \log_2(2^x)) = x^2(\log_2 x + x) = x^3 + x^2 \log_2 x$, so that

$$f'(x) = 3x^2 + 2x \log_2 x + x^2 \frac{1}{x \ln 2} = 3x^2 + 2x \log_2 x + \frac{x}{\ln 2}.$$

37.

$$f'(x) = \frac{1}{x^2 + e^{\sqrt{x}}}\left(x^2 + e^{\sqrt{x}}\right)'$$

$$= \frac{1}{x^2 + e^{\sqrt{x}}}\left(2x + e^{\sqrt{x}}(\sqrt{x})'\right)$$

$$= \frac{1}{x^2 + e^{\sqrt{x}}}\left(2x + e^{\sqrt{x}}\frac{1}{2}x^{-1/2}\right)$$

$$= \frac{2x + \frac{e^{\sqrt{x}}}{2\sqrt{x}}}{x^2 + e^{\sqrt{x}}}$$

39.

$$f'(x) = \frac{1}{2}\left(\log_2(3^x - 5)\right)^{-1/2}\left(\log_2(3^x - 5)\right)'$$

$$= \frac{1}{2}\left(\log_2(3^x - 5)\right)^{-1/2}\frac{1}{(3^x - 5)\ln 2}(3^x - 5)'$$

$$= \frac{1}{2}\left(\log_2(3^x - 5)\right)^{-1/2}\frac{1}{(3^x - 5)\ln 2}(3^x \ln 3)$$

$$= \frac{\ln 3}{\ln 4}\left(\log_2(3^x - 5)\right)^{-1/2}\frac{3^x}{3^x - 5}.$$

41.

$$f'(x) = 2x\ln(\ln x) + x^2\frac{1}{\ln x}(\ln x)'$$

$$= 2x\ln(\ln x) + x^2\frac{1}{\ln x}\cdot\frac{1}{x}$$

$$= 2x\ln(\ln x) + \frac{x}{\ln x}.$$

43. $f(x) = (2^x)^x = 2^{(x^2)}$. Thus $f'(x) = 2^{(x^2)}(x^2)'\ln 2 = 2x\ln(2)2^{(x^2)}$.

45. The derivative when $x < -2$ is $\frac{d}{dx}(2^x) = 2^x\ln 2$, and the derivative when $x > -2$ is $\frac{d}{dx}(x^{-2}) = -2x^{-3}$. At $x = -2$, the function is continuous, since $2^{-2} = \frac{1}{(-2)^2}$, but it is not differentiable, since $2^{-2}\ln 2 \neq -2(-2)^{-3}$. Thus

$$f'(x) = \begin{cases} 2^x\ln 2, & x < -2, \\ \text{undefined}, & x = -2, \\ -2x^{-3}, & x > -2. \end{cases}$$

47. The derivative when $x < 1$ is $2x$; when $x \geq 1$, the derivative is $\frac{1}{x}$. The function is continuous, since both one-sided limits at $x = 1$ are equal to 1. However, it is not differentiable, since the left derivative is $2\cdot 1 = 2$ while the right derivative is 1. Thus

$$f'(x) = \begin{cases} 2x, & x < 1, \\ \text{undefined}, & x = 1, \\ \frac{1}{x}, & x > 1. \end{cases}$$

49. With $f(x) = \sqrt{x \ln |2^x + 1|} = (x \ln |2^x + 1|)^{1/2}$, taking logs (noting that $2^x + 1 > 0$) gives

$$\ln f(x) = \frac{1}{2} \ln(x \ln |2^x + 1|) = \frac{1}{2} \left(\ln x + \ln \ln(2^x + 1) \right).$$

The derivative of the left-hand side is $\frac{f'(x)}{f(x)}$, while the derivative of the right-hand side is

$$\frac{1}{2} \left(\frac{1}{x} + \frac{1}{\ln(2^x + 1)} (\ln(2^x + 1))' \right) = \frac{1}{2x} + \frac{1}{2(2^x + 1) \ln(2^x + 1)} (2^x + 1)'$$

$$= \frac{1}{2x} + \frac{2^{x-1} \ln 2}{(2^x + 1) \ln(2^x + 1)}.$$

Thus

$$f'(x) = f(x) \left(\frac{1}{2x} + \frac{2^{x-1} \ln 2}{(2^x + 1) \ln(2^x + 1)} \right) = \sqrt{x \ln(2^x + 1)} \left(\frac{1}{2x} + \frac{2^{x-1} \ln 2}{(2^x + 1) \ln(2^x + 1)} \right).$$

51. Taking logs gives $f(x) = x \ln 2 + \frac{1}{2} \ln(x^3 - 1) - \frac{1}{2} \ln x - \ln(2x - 1)$. Thus

$$\frac{f'(x)}{f(x)} = \ln 2 + \frac{3x^2}{2(x^3 - 1)} - \frac{1}{2x} - \frac{2}{2x - 1},$$

so that

$$f'(x) = f(x) \left(\ln 2 + \frac{3x^2}{2(x^3 - 1)} - \frac{1}{2x} - \frac{2}{2x - 1} \right)$$

$$= \frac{2^x \sqrt{x^3 - 1}}{\sqrt{x}(2x - 1)} \left(\ln 2 + \frac{3x^2}{2(x^3 - 1)} - \frac{1}{2x} - \frac{2}{2x - 1} \right).$$

53. Taking logs gives $\ln f(x) = \ln x \cdot \ln x = (\ln x)^2$, so that

$$f'(x) = f(x) \left((\ln x)^2 \right)' = x^{\ln x} \left(2 \ln x \cdot \frac{1}{x} \right) = 2x^{-1 + \ln x} \ln x.$$

55. Taking logs gives $\ln f(x) = x \ln \left(\frac{x}{x-1} \right) = x (\ln x - \ln(x - 1))$. Thus

$$f'(x) = f(x) (x (\ln x - \ln(x - 1)))'$$

$$= \left(\frac{x}{x-1} \right)^x \left((\ln x - \ln(x - 1)) + x \left(\frac{1}{x} - \frac{1}{x-1} \right) \right)$$

$$= \left(\frac{x}{x-1} \right)^x \left(1 + \ln x - \ln(x - 1) - \frac{x}{x-1} \right).$$

57. Taking logs gives $\ln f(x) = \ln x \cdot \ln(\ln x)$. Differentiating the right-hand side gives

$$(\ln x \cdot \ln(\ln x))' = \frac{1}{x} \ln(\ln x) + \ln x \cdot \frac{1}{\ln x} (\ln x)'$$

$$= \frac{\ln(\ln x)}{x} + \frac{1}{x} = \frac{1 + \ln(\ln x)}{x}.$$

Thus

$$f'(x) = f(x) \left(\frac{1 + \ln(\ln x)}{x} \right) = (\ln x)^{\ln x} \left(\frac{1 + \ln(\ln x)}{x} \right).$$

59. e^{4x} is a factor of $f'(x)$, so presumably it is also a factor of $f(x)$. This looks like a quotient rule application with $3x^5 - 1$ in the denominator. Try $f(x) = \dfrac{e^{4x}}{3x^5 - 1}$; then

$$f'(x) = \frac{(3x^5 - 1)(4e^{4x}) - e^{4x}(15x^4)}{(3x^5 - 1)^2},$$

which is exactly what we are looking for.

61. The derivative of $x^2 + 3$ is $2x$, and the derivative of $\ln(x^2 + 3)$ has an $x^2 + 3$ in the denominator. Try $\ln(x^2 + 3)$ for the function. Its derivative is $\dfrac{1}{x^2 + 3}(x^2 + 3)' = \dfrac{2x}{x^2 + 3}$, which is too large by a factor of 2. So let $f(x) = \dfrac{1}{2}\ln(x^2 + 3)$.

63. $\ln(1 + e^x)$ has a derivative with $1 + e^x$ in the denominator, so try $f(x) = \ln(1 + e^x)$. Since $(e^x)' = e^x$, this has the desired derivative.

Applications

65. (a) Since the total amount of money increases at a rate proportional to the amount, the function $A(t)$ must be an exponential function, $A(t) = Ce^{kt}$. Since $A(0) = Ce^0 = C = 1000$, we have $C = 1000$. We also know that $A(3) = 1260 = 1000e^{3k}$, so that $1.26 = e^{3k}$. Taking logs gives $k = \frac{\ln 1.26}{3} \approx 0.077$. Thus $A(t) = 1000e^{0.077t}$.

(b) After 30 years, the amount is $A(30) = 1000e^{0.077 \cdot 30} \approx \$10074.40$.

(c) Her investment will quadruple when $A(t) = 4000 = 1000e^{0.077t}$, or when $e^{0.077t} = 4$. Taking logs gives $t = \frac{\ln 4}{0.077} \approx 18.004$ years, so about 18 years.

67. (a) The number of people who knew about him before the advertising started is

$$P(0) = \frac{45000}{1 + 35e^{-0.12 \cdot 0}} = \frac{45000}{36} = 1250.$$

(b) As $t \to \infty$, the value of $35e^{-0.12t} \to 0$, so that $\displaystyle\lim_{t \to \infty} P(t) = \frac{45000}{1 + 0} = 45000$. The population of Hamtown is 45000.

(c) We want to solve $P(t) = 44999$, so that

$$44999 = \frac{45000}{1 + 35e^{-0.12t}} \quad \Rightarrow \quad 1 + 35e^{-0.12t} = \frac{45000}{44999} \quad \Rightarrow \quad e^{-0.12t} = \frac{1}{44999 \cdot 35}.$$

Taking logs gives $-0.12t = -\ln 44999 - \ln 35$, so that $t = \frac{\ln 44999 + \ln 35}{.12} \approx 118.9$, so about 119 days.

(d) We have $P(t) = 45000(1 + 35e^{-0.12t})^{-1}$, so that

$$P'(t) = 45000(-1)(1 + 35e^{-0.12t})^{-2}(1 + 35e^{-0.12t})'$$

$$= -45000(1 + 35e^{-0.12t})^{-2}]\left(-35 \cdot 0.12e^{-0.12t}\right)$$

$$= 189000 \frac{e^{-0.12t}}{(1 + 35e^{-0.12t})^2}.$$

A plot of $P'(t)$ over time is

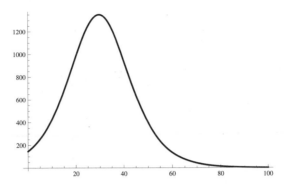

So clearly the derivative increases to a maximum at around $t = 30$, then decreases again with an asymptote at $y = 0$. This corresponds to $P(t)$ increasing more and more quickly until about $t = 30$, then continuing to increase, but more and more slowly. To see more precisely where the rate of increase is the fastest, we must find the location of the local maximum of $P'(t)$. To do this, compute $P''(t)$:

$$P''(t) = 189000 \frac{(1 + 35e^{-0.12t})^2(-0.12e^{-0.12t}) - e^{-0.12t}(2(1 + 35e^{-0.12t})(-35 \cdot 0.12e^{-0.12t})}{(1 + 35e^{-0.12t})^4}$$

$$= 189000 \frac{e^{-0.12t}(-0.12 - 4.2e^{-0.12t} + 8.4e^{-0.12t})}{(1 + 35e^{-0.12t})^3}$$

$$= 189000 \frac{e^{-0.12t}(-0.12 + 4.2e^{-0.12t})}{(1 + 35e^{-0.12t})^3}.$$

Then $P''(t) = 0$ when $e^{-0.12t} = \frac{0.12}{4.2}$, so when $t = -\frac{1}{0.12} \ln \frac{0.12}{4.2} \approx 29.628$. So in fact $t \approx 30$ is the time when the number of people who have heard of the candidate is increasing most rapidly.

Proofs

69. Let $f(x) = e^x$ and $g(x) = x^k$. Then $e^{kx} = (e^x)^k = (g \circ f)(x)$, so that

$$\frac{d}{dx}\left(e^{kx}\right) = (g \circ f)'(x) = g'(f(x))f'(x) = kf(x)^{k-1}e^x = k(e^x)^{k-1}e^x = ke^{kx}.$$

71. Let $f(x) = Ab^x$. Then since $\frac{d}{dx}b^x = (\ln b)b^x$, we have

$$f'(x) = A(\ln b)b^x = (\ln b)\left(Ab^x\right) = (\ln b)f(x).$$

73. Let $g(x) = |x|$ and $h(x) = \ln x$. Then

$$\frac{d}{dx}(\ln|x|) = (h \circ g)'(x) = h'(g(x))g'(x) = \frac{1}{g(x)}g'(x) = \frac{1}{|x|} \cdot |x|'.$$

Now, from Exercise 65 in Section 3, we know that

$$|x|' = \begin{cases} 1, & x > 0, \\ \text{undefined}, & x = 0, \\ -1, & x < 0. \end{cases}$$

Thus we get

$$\frac{1}{|x|} \cdot |x|' = \begin{cases} \frac{1}{x} \cdot 1 = \frac{1}{x}, & x > 0, \\ \text{undefined}, & x = 0, \\ \frac{1}{-x} \cdot (-1) = \frac{1}{x}, & x < 0. \end{cases}$$

Thus $\frac{d}{dx} \ln|x| = \frac{1}{x}$ for $x \neq 0$.

Thinking Forward

L'Hôpital's rule

▷ Since $\lim\limits_{x \to 0} x^3 = \lim\limits_{x \to 0} (1 - 2^x) = 0$, this limit is of the form $\frac{0}{0}$, so that

$$\lim_{x \to 0} \frac{x^3}{1 - 2^x} \overset{\text{L'H}}{=} \lim_{x \to 0} \frac{3x^2}{(\ln 2)2^x} = \frac{0}{\ln 2} = 0.$$

▷ Since $\lim\limits_{x \to 1} (3^x - 3) = \lim\limits_{x \to 1} (1 - x^2) = 0$, this limit is of the form $\frac{0}{0}$, so that

$$\lim_{x \to 1} \frac{3^x - 3}{1 - x^2} \overset{\text{L'H}}{=} \lim_{x \to 1} \frac{(\ln 3)3^x}{-2x} = -\frac{3 \ln 3}{2}.$$

▷ Since $\lim\limits_{x \to 1} \ln x = \lim\limits_{x \to 1} (x - 1) = 0$, this limit is of the form $\frac{0}{0}$, so that

$$\lim_{x \to 1} \frac{\ln x}{x - 1} \overset{\text{L'H}}{=} \lim_{x \to 1} \frac{1/x}{1} = 1.$$

▷ Since $\lim\limits_{x \to 3} (x - 3)^2 = (3 - 3)^2 = 0$ and $\lim\limits_{x \to 3} (1 - e^{x-3}) = 1 - e^{3-3} = 0$, this limit is of the form $\frac{0}{0}$, so that

$$\lim_{x \to 3} \frac{(x - 3)^2}{1 - e^{x-3}} \overset{\text{L'H}}{=} \lim_{x \to 3} \frac{2(x - 3)}{-e^{x-3}} = \frac{2(3 - 3)}{-e^{3-3}} = 0.$$

Differential equations

▷ Note that $y' = (4e^{3x})' = 4 \cdot 3e^{3x} = 3y$.

▷ This is true since $y' = 1.7 \cdot (-2.1)e^{-2.1x} = -2.1 \left(1.7e^{-2.1x}\right) = -2.1y$.

▷ This holds since $y' = 3 \left(\ln 2 \cdot 2^x\right) = (\ln 2)(3(2^x)) = (\ln 2)y$.

▷ Because the derivative is proportional to the function, by Theorem 2.14, all solutions are of the form Ae^{3x}.

▷ Because the derivative is proportional to the function, by Theorem 2.14, the general solution is $y = Ae^{3x}$. Since $y(0) = 2 = Ae^{3 \cdot 0} = A$, we have $A = 2$ so that $y = 2e^{3x}$.

2.6   Derivatives of Trigonometric and Hyperbolic Functions

Thinking Back

Trigonometric and inverse trigonometric values

▷ $-\frac{\sqrt{3}}{2}$.

▷ -1.

▷ $\dfrac{1}{\cos \frac{5\pi}{6}} = -\dfrac{2}{\sqrt{3}}$.

▷ $\frac{\pi}{2}$.

▷ $\dfrac{\pi}{3}$.

▷ $\cos^{-1}\left(\frac{-1}{2}\right) = \frac{2\pi}{3}$.

Compositions

▷ For example, let $h(x) = x^3$, $g(x) = \sin x$, and $f(x) = \dfrac{1}{x}$.

▷ For example, let $h(x) = 3x + 1$, $g(x) = \tan x$, and $f(x) = x^2$.

▷ For example, let $h(x) = \cos x$, $g(x) = x^2$, and $f(x) = \sin^{-1}(x)$.

▷ Since $k(x) = \dfrac{\sin(x^3)}{\cos^2(x^3)}$, let $h(x) = x^3$, $g(x) = \dfrac{\sqrt{\sin x}}{\cos x}$, and $f(x) = x^2$.

Writing trigonometric compositions algebraically

▷ $\sin^{-1}(x)$ is an angle of a right triangle with hypotenuse 1 and opposite leg x, so that the adjacent leg is $\sqrt{1 - x^2}$. Thus $\cos(\sin^{-1} x) = \frac{\sqrt{1-x^2}}{1} = \sqrt{1 - x^2}$.

▷ $\tan^{-1} x$ is an angle of a right triangle with nearest leg 1 and the other leg x, so that its hypotenuse is $\sqrt{1 + x^2}$. Thus $\sec(\tan^{-1} x) = \sqrt{1 + x^2}$, so that $\sec^2(\tan^{-1} x) = 1 + x^2$.

▷ $\cos^{-1}(x)$ is the other acute angle of the same right triangle as in the first item, so that $\sin(\cos^{-1} x) = \sqrt{1 - x^2}$.

▷ $\sec^{-1} x$ is an angle of a right triangle whose hypotenuse is x and whose nearest leg is 1, so the other leg is $\sqrt{x^2 - 1}$. Thus $\tan x = \dfrac{\sqrt{x^2 - 1}}{1} = |x| \sqrt{1 - \dfrac{1}{x^2}}$.

Concepts

1. (a) True. See the proof of Theorem 2.17(a).

   (b) False. We can use the quotient rule.

   (c) False. One must use the quotient rule, so the derivative is $\dfrac{4x^3 \sin x - x^4 \cos x}{\sin^2 x}$.

   (d) True.

   (e) False. For example, $\ln x$ or $\sin^{-1} x$ have derivatives that are algebraic functions.

   (f) True (it can of course be a product of trigonometric functions).

   (g) False. See Theorem 2.18.

   (h) True (it can be a product of hyperbolic functions). See Theorem 2.20.

3. The sum identity for sin is used as well as basic limit rules and the facts that $\lim\limits_{h \to 0} \frac{\cos h - 1}{h} = 0$ and $\lim\limits_{h \to 0} \frac{\sin h}{h} = 1$.

5. The graph of $\sin x$ where x is in degrees is

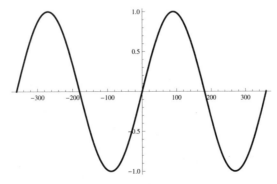

The slope of this graph near $x = 0$ is clearly not 1 — using the nearby maximum and minimum as an approximation, the slope is $\approx \frac{2}{200} = 0.01$.

7. (a) That $f(x) = \sin(x) \cdot 3x^2$.

(b) That you can differentiate this function composition by just differentiating each part.

9. No. We would then have the equation $\sin^{-1}(\sin x) = x$; proceeding as before gives

$$\sin^{-1}(\sin x) = x \qquad \leftarrow \sin x \text{ is the inverse of } \sin^{-1} x$$

$$\frac{d}{dx}\left(\sin^{-1}(\sin x)\right) = \frac{d}{dx}(x) \qquad \leftarrow \text{differentiate both sides}$$

$$(\sin^{-1})'(\sin x) \cdot \cos x = 1 \qquad \leftarrow \text{apply the chain rule}$$

and we do not get a formula for $\frac{d}{dx} \sin^{-1} x$.

11. By the identities in the *Writing trigonometric compositions algebraically* Thinking Back questions, $\cos(\sin^{-1} x) = \sqrt{1 - x^2}$, so that the two expressions are equal. The algebraically written expression, $\frac{1}{\sqrt{1-x^2}}$, is probably easier to manipulate, depending on the context.

13. (a) Since $\sinh x$ is the lower curve and $\cosh x$ is the upper curve, this inequality holds graphically. Algebraically, this says that

$$\frac{e^x - e^{-x}}{2} \leq \frac{e^x}{2} \leq \frac{e^x + e^{-x}}{2},$$

which is clearly true since $e^{-x} \geq 0$.

(b) Again from looking at the graphs, it appears that all three graphs converge to a single curve as $x \to \infty$, so their ratio approaches 1. Algebraically, we have

$$\lim_{x \to \infty} \frac{\sinh x}{\frac{1}{2}e^x} = \lim_{x \to \infty} \frac{e^x - e^{-x}}{2\frac{1}{2}e^x} = \lim_{x \to \infty}\left(1 - e^{-2x}\right) = 1$$

$$\lim_{x \to \infty} \frac{\cosh x}{\frac{1}{2}e^x} = \lim_{x \to \infty} \frac{e^x + e^{-x}}{2\frac{1}{2}e^x} = \lim_{x \to \infty}\left(1 + e^{-2x}\right) = 1$$

since $\lim_{x \to \infty} e^{-2x} = 0$.

15. (a) $\sinh x$ is the solid curve, $\frac{1}{2}e^x$ is the upper dotted curve, and $-\frac{1}{2}e^{-x}$ is the lower dotted curve. From looking at the graph, it is plausible that the identity holds, since, for example, as $x \to \infty$, the lower curve approaches zero while the two other curves seem to converge. Algebraically,

$$\sinh x = \frac{e^x - e^{-x}}{2} = \frac{1}{2}e^x - \frac{1}{2}e^{-x}$$

(b) As $x \to -\infty$, the solid curve and the lower dotted curve appear to be converging, so this identity is plausible. Algebraically,

$$\lim_{x \to -\infty} \frac{\sinh x}{-\frac{1}{2}e^{-x}} = \lim_{x \to -\infty} \frac{e^x - e^{-x}}{-2 \cdot \frac{1}{2}e^{-x}} = \lim_{x \to -\infty}\left(-e^{2x} + 1\right) = 1.$$

Skills

17. $f'(x) = \dfrac{(\cos x)(2x) - (x^2 + 1)(-\sin x)}{\cos^2 x} = \dfrac{2x\cos x + (x^2 + 1)\sin x}{\cos^2 x}.$

19. $f'(x) = -\csc^2 x - \csc x \cot x = -\dfrac{1}{\sin^2 x} - \dfrac{\cos x}{\sin^2 x} = -\dfrac{\cos x + 1}{\sin^2 x}.$

21. Since $f(x) = 4(\sin^2 x + \cos^2 x) = 4$, it follows that $f'(x) = 0$.

23. $f'(x) = 3(\sec x(\sec^2 x) + (\sec x \tan x)\tan x) = 3\sec^3 x + 3\sec x \tan^2 x.$

25.

$$\begin{aligned} f'(x) &= \cos(\cos(\sec x))(\cos(\sec x))' \\ &= \cos(\cos(\sec x))(-\sin(\sec x))(\sec x)' \\ &= -\cos(\cos(\sec x))(\sin(\sec x))(\sec x \tan x). \end{aligned}$$

27. $f'(x) = e^{\csc^2 x}\left(\csc^2 x\right)' = e^{\csc^2 x}\left(2\csc x(\csc x)'\right) = -2\csc^2 x \cot x\, e^{\csc^2 x}.$

29. $f'(x) = \dfrac{5x\sin x(-2^x \ln 2) - (-2^x)(5\sin x + 5x\cos x)}{(5x\sin x)^2} = \dfrac{2^x\left(\sin x - x\ln(2)\sin x + x\cos x\right)}{5x^2\sin^2 x}.$

31.

$$\begin{aligned} f'(x) &= 1\cdot\sqrt{\sin x\cos x} + x\cdot\frac{1}{2}\left(\sin x\cos x\right)^{-1/2}\left(\sin x\cos x\right)' \\ &= (\sin x\cos x)^{1/2} + \frac{1}{2}x(\sin x\cos x)^{-1/2}(\cos^2 x - \sin^2 x) \end{aligned}$$

33.

$$\begin{aligned} f'(x) &= \frac{\tan x\left(6x\ln x + 3x^2\cdot\frac{1}{x}\right) - 3x^2\ln x(\sec^2 x)}{\tan^2 x} \\ &= \frac{3x\tan x(1 + 2\ln x) - 3x^2\sec^2 x\ln x}{\tan^2 x} \\ &= 3x\cot x + 6x\cot x\ln x - 3x^2\csc^2 x\ln x. \end{aligned}$$

35. $f'(x) = \cos(\ln x)(\ln x)' = \dfrac{\cos(\ln x)}{x}.$

37. $f'(x) = \dfrac{1}{\sqrt{1 - (3x^2)^2}}\cdot 6x = \dfrac{6x}{\sqrt{1 - 9x^4}}.$

39. $f'(x) = 2x\arctan(x^2) + x^2\dfrac{1}{1 + (x^2)^2}(x^2)' = 2x\arctan(x^2) + \dfrac{2x^3}{1 + x^4}.$

41. $f'(x) = \dfrac{1}{|x^2|\sqrt{(x^2)^2 - 1}}\cdot(x^2)' = \dfrac{2}{x}\sqrt{x^4 - 1}.$

43. $f'(x) = \dfrac{1}{\sqrt{1 - (\sec^2 x)^2}}(\sec^2 x)' = \dfrac{2\sec x(\sec x)'}{\sqrt{1 - \sec^4 x}} = \dfrac{2\sec^2 x\tan x}{\sqrt{1 - \sec^4 x}}.$

45. $f'(x) = \dfrac{\tan^{-1}x\left(\frac{1}{\sqrt{1-x^2}}\right) - \sin^{-1}x\left(\frac{1}{1+x^2}\right)}{\left(\tan^{-1}x\right)^2} = \dfrac{1}{\tan^{-1}x\sqrt{1 - x^2}} - \dfrac{\sin^{-1}x}{\left(\tan^{-1}x\right)^2(1 + x^2)}.$

47.

$$f'(x) = \frac{1}{\text{arcsec}(\sin^2 x)}\left(\text{arcsec}(\sin^2 x)\right)'$$

$$= \frac{1}{\text{arcsec}(\sin^2 x)}\frac{1}{|\sin^2 x|\sqrt{\sin^4 x - 1}}\left(\sin^2 x\right)'$$

$$= \frac{2\cos x}{\sin x\,\text{arcsec}(\sin^2 x)\sqrt{\sin^4 x - 1}}.$$

49. $f'(x) = \sec(1 + \tan^{-1}x)\tan(1 + \tan^{-1}x)(1 + \tan^{-1}x)' = \dfrac{\sec(1 + \tan^{-1}x)\tan(1 + \tan^{-1}x)}{1 + x^2}.$

51. $f'(x) = \sinh(x^3) + x\cosh(x^3)\cdot 3x^2 = \sinh(x^3) + 3x^3\cosh(x^3).$

53. Since $\cosh x = \dfrac{e^x + e^{-x}}{2}$, we have

$$f(x) = \cosh(\ln(x^2 + 1)) = \frac{e^{\ln(x^2+1)} + e^{-\ln(x^2+1)}}{2} = \frac{x^2 + 1}{2} + \frac{1}{2(x^2 + 1)}.$$

Thus

$$f'(x) = x + \frac{1}{2}\left((x^2 + 1)^{-1}\right)' = x - \frac{1}{2(x^2 + 1)^2}(2x) = x - \frac{x}{(x^2 + 1)^2}.$$

55. $f'(x) = \dfrac{1}{2}\left(\cosh^2 x + 1\right)^{-1/2}\left(\cosh^2 x + 1\right)' = \dfrac{1}{2\sqrt{\cosh^2 x + 1}}(2\cosh x \sinh x) = \dfrac{\sinh x \cosh x}{\sqrt{\cosh^2 x + 1}}.$

57. $f'(x) = \dfrac{1}{\sqrt{(x^3)^2 + 1}}(x^3)' = \dfrac{3x^2}{\sqrt{x^6 + 1}}.$

59.

$$f'(x) = \frac{\cosh^{-1}x\left(\frac{1}{\sqrt{x^2+1}}\right) - \sinh^{-1}x\left(\frac{1}{\sqrt{x^2-1}}\right)}{\left(\cosh^{-1}x\right)^2}$$

$$= \frac{\sqrt{x^2 - 1}\cosh^{-1}x - \sqrt{x^2 + 1}\sinh^{-1}x}{\sqrt{x^4 - 1}\left(\cosh^{-1}x\right)^2}.$$

61.

$$f'(x) = \cos\left(e^{\sinh^{-1}x}\right)\left(e^{\sinh^{-1}x}\right)'$$

$$= \cos\left(e^{\sinh^{-1}x}\right)e^{\sinh^{-1}x}\left(\sinh^{-1}x\right)'$$

$$= \frac{\cos\left(e^{\sinh^{-1}x}\right)e^{\sinh^{-1}x}}{\sqrt{x^2 + 1}}.$$

63. With $f(x) = (\sin x)^x$, taking logs gives $\ln f(x) = x\ln\sin x$, so that

$$f'(x) = f(x)(x\ln\sin x)' = f(x)\left(\ln\sin x + x\frac{(\sin x)'}{\sin x}\right) = (\sin x)^x(\ln\sin x + x\cot x)$$

65. With $f(x) = (\sin x)^{\cos x}$, taking logs gives $\ln f(x) = \cos x \ln \sin x$, so that

$$f'(x) = f(x) (\cos x \ln \sin x)'$$

$$= f(x) \left(-\sin x \ln \sin x + \cos x \frac{(\sin x)'}{\sin x} \right)$$

$$= (\sin x)^{\cos x} \left(\frac{\cos^2 x}{\sin x} - \sin x \ln \sin x \right).$$

67. In contrast to the previous exercise, the derivative of $4x^2$ is not a linear multiple of the numerator. This derivative fits the pattern for a derivative of arcsin, so try $f(x) = \arcsin(2x)$. This has derivative $\frac{1}{\sqrt{1-4x^2}} \cdot (2x)' = \frac{2}{\sqrt{1-4x^2}}$. This is correct.

69. Here the derivative of the denominator is a linear multiple of the numerator. Since the denominator appears to the first power, its antiderivative is most likely a log. Try $\ln(1+9x^2)$. This has derivative $\frac{1}{1+9x^2}(9x^2)' = \frac{18x}{1+9x^2}$, so let $f(x) = \frac{1}{6} \ln(1+9x^2)$.

71. Dividing numerator and denominator by 2 gives $f'(x) = \frac{3/2}{\sqrt{1-\left(\frac{3}{2}x\right)^2}}$, so this fits the pattern for a derivative of $\sin^{-1} x$. Try $f(x) = \sin^{-1}\left(\frac{3}{2}x\right)$. This has derivative $\frac{1}{\sqrt{1-\left(\frac{3}{2}x\right)^2}}\left(\frac{3}{2}x\right)' = \frac{3/2}{\sqrt{1-\left(\frac{3}{2}x\right)^2}}$, which is exactly right.

73. Here the derivative of $1 + 4x^2$ is not a linear multiple of the numerator. The derivative fits the pattern for a derivative of $\sinh^{-1}$; we try $f(x) = \sinh^{-1}(2x)$. This has derivative $\frac{1}{\sqrt{1+4x^2}}(2x)' = \frac{2}{\sqrt{1+4x^2}}$, which is what we want.

75. The derivative of the denominator, $1 - 9x^2$, is a linear multiple of the numerator, and the denominator appears to the first power, so try $\ln(1 - 9x^2)$. This has derivative $\frac{1}{1-9x^2}(1 - 9x^2)' = \frac{-18x}{1-9x^2}$, which is off by a factor of -6. So let $f(x) = -\frac{1}{6} \ln(1 - 9x^2)$.

77. Dividing numerator and denominator by 2 gives $f'(x) = \frac{3/2}{\sqrt{1+\left(\frac{3}{2}x\right)^2}}$. This fits the pattern for a derivative of $\sinh^{-1}$. Try $f(x) = \sinh^{-1}\frac{3}{2}x$, which has derivative $\frac{3/2}{\sqrt{1+\left(\frac{3}{2}x\right)^2}}$, which is exactly right.

Applications

79. (a) Write $C = \frac{\sqrt{4km - f^2}}{2m}$. Then $s(t) = e^{-\frac{f}{2m}t}(A\sin(Ct) + B\cos(Ct))$, and

$$s'(t) = e^{-\frac{f}{2m}t}(AC\cos(Ct) - BC\sin(Ct)) - \frac{f}{2m}e^{-\frac{f}{2m}t}(A\sin(Ct) + B\cos(Ct))$$

$$= e^{-\frac{f}{2m}t}\left(\left(AC - B\frac{f}{2m}\right)\cos(Ct) - \left(BC + A\frac{f}{2m}\right)\sin(Ct)\right).$$

Continuing,

$$s''(t) = e^{-\frac{f}{2m}t}\left(-C\left(AC - B\frac{f}{2m}\right)\sin(Ct) - C\left(BC + A\frac{f}{2m}\right)\cos(Ct)\right)$$

$$- \frac{f}{2m}e^{-\frac{f}{2m}t}\left(\left(AC - B\frac{f}{2m}\right)\cos(Ct) - \left(BC + A\frac{f}{2m}\right)\sin(Ct)\right)$$

$$= e^{-\frac{f}{2m}t}\left(\left(BC\frac{f}{2m} - AC^2\right)\sin(Ct) + \left(-BC^2 - AC\frac{f}{2m}\right)\cos(Ct)\right)$$

$$+ e^{-\frac{f}{2m}t}\left(\left(A\frac{f^2}{4m^2} + BC\frac{f}{2m}\right)\sin(Ct) + \left(-AC\frac{f}{2m} + B\frac{f^2}{4m^2}\right)\cos(Ct)\right)$$

Now, $C^2 = \frac{4km}{4m^2} - \frac{f^2}{4m^2} = \frac{k}{m} - \frac{f^2}{4m^2}$, so continuing to simplify, this is

$$s''(t) = e^{-\frac{f}{2m}t}\left(\left(BC\frac{f}{m} - A\frac{k}{m} + A\frac{f^2}{2m^2}\right)\sin(Ct) + \left(-AC\frac{f}{m} - B\frac{k}{m} + B\frac{f^2}{2m^2}\right)\cos(Ct)\right).$$

We now get

$$s''(t) = e^{-\frac{f}{2m}t}\left(\left(BC\frac{f}{m} - A\frac{k}{m} + A\frac{f^2}{2m^2}\right)\sin(Ct) + \left(-AC\frac{f}{m} - B\frac{k}{m} + B\frac{f^2}{2m^2}\right)\cos(Ct)\right)$$

$$\frac{f}{m}s'(t) = e^{-\frac{f}{2m}t}\left(-\left(BC\frac{f}{m} + A\frac{f^2}{2m^2}\right)\sin(Ct) + \left(AC\frac{f}{m} - B\frac{f^2}{2m^2}\right)\cos(Ct)\right)$$

$$\frac{k}{m}s(t) = e^{-\frac{f}{2m}t}\left(A\frac{k}{m}\sin(Ct) + B\frac{k}{m}\cos(Ct)\right)$$

Adding these three gives zero.

(b) Since $s(0) = s_0$ and $s'(0) = v(0) = v_0$, substitute into $s(t)$ and $s'(t)$ to get

$$s(0) = s_0 = e^{-\frac{f}{2m}\cdot 0}\left(A\sin(C \cdot 0) + B\cos(C \cdot 0)\right) = B$$

$$s'(0) = v_0 = e^{-\frac{f}{2m}\cdot 0}\left(\left(AC - B\frac{f}{2m}\right)\cos(C \cdot 0) - \left(BC + A\frac{f}{2m}\right)\sin(C \cdot 0)\right) = AC - B\frac{f}{2m}$$

Rearranging gives

$$A = \frac{v_0 + B\frac{f}{2m}}{C} == \frac{2mv_0 + Bf}{2m\frac{\sqrt{4km-f^2}}{2m}} = \frac{2mv_0 + fs_0}{\sqrt{4km - f^2}}.$$

81. (a) The top of the arch is the point where the slope of the curve describing the arch is zero, i.e. when $A'(x) = 0$. But

$$A'(x) = -68.8\sinh\left(\frac{x - 299.22}{99.7}\right)\left(\frac{1}{99.7}\right) \approx -0.69\sinh\left(\frac{x - 299.22}{99.7}\right).$$

Thus $A'(x)$ vanishes when $x = 299.22$, since $\sinh z = 0$ only for $z = 0$. The arch is at its highest point when $x = 299.22$; at that point,

$$A(x) = 693.8 - 68.8\cosh(0) = 693.8 - 68.8 = 625 \text{ feet.}$$

so that he height of the arch is 625 feet.

(b) The ground is at $x = 0$; at that point, $A'(x) = -.69\sinh\left(\frac{-299.22}{99.7}\right) = 6.9207$. Thus the slope of the curve at the ground is 6.9207, so the angle the curve makes with the ground at that point is $\tan^{-1}(6.9207) \approx 1.4273$ radians $\approx 81.778°$.

(c) The center of the arch is at 299.22 feet, so 33 feet from there is at $x = 266.22$. At that point, $A'(x) = -.69\sinh\left(\frac{-33}{99.7}\right) \approx 0.2326$, so the angle the tangent to the curve makes with the horizontal at that point is $\tan^{-1}(0.2326) \approx 0.2285$ radians $\approx 13.09°$.

Proofs

83.

$$\frac{d}{dx}(\cos x) = \lim_{h \to 0} \frac{\cos(x+h) - \cos x}{h}$$

$$= \lim_{h \to 0} \frac{\cos x \cos h - \sin x \sin h - \cos x}{h}$$

$$= \cos x \left(\lim_{h \to 0} \frac{\cos h - 1}{h} \right) - \sin x \lim_{h \to 0} \frac{\sin h}{h}$$

$$= -\sin x,$$

since $\lim_{h \to 0} \frac{\cos h - 1}{h} = 0$ and $\lim_{h \to 0} \frac{\sin h}{h} = 1$.

85. $\frac{d}{dx}(\csc x) = \frac{d}{dx}\left(\frac{1}{\sin x} \right) = \frac{\sin x \cdot (1)' - \cos x}{\sin^2 x} = -\csc x \cot x.$

87. Let $y = \tan^{-1} x$; then $\tan y = \tan(\tan^{-1} x) = x$. Differentiate to get $y' \sec^2 y = 1$, so that $y' = \cos^2 y = \cos^2(\tan^{-1} x)$. Now, $\tan^{-1} x$ is the angle of a right triangle whose opposite leg is x and whose adjacent leg is 1, so its hypotenuse is $\sqrt{x^2 + 1}$. Thus the cosine of that angle is $\frac{1}{\sqrt{x^2 + 1}}$, so that $\tan^{-1} x = y' = \left(\frac{1}{\sqrt{x^2+1}} \right)^2 = \frac{1}{x^2+1}$.

89. Since

$$\cosh t = \frac{e^t + e^{-t}}{2} \quad \text{and} \quad \sinh t = \frac{e^t - e^{-t}}{2},$$

we also have

$$\cosh^2 t = \frac{e^{2t} + 2 + e^{-2t}}{4} \quad \text{and} \quad \sinh^2 t = \frac{e^{2t} - 2 + e^{-2t}}{4}.$$

Thus the point $(\cosh t, \sinh t)$ satisfies

$$x^2 - y^2 = \cosh^2 t - \sinh^2 t = \frac{4}{4} = 1.$$

The point always lies on the right-hand branch of the hyperbola (where $x > 0$) since $\cosh t > 0$ always (both e^t and e^{-t} are always positive).

91.

$$\sinh x \cosh y = \frac{e^x - e^{-x}}{2} \cdot \frac{e^y + e^{-y}}{2} = \frac{e^x e^y + e^x e^{-y} - e^{-x} e^y - e^{-x} e^{-y}}{4}$$

$$\cosh x \sinh y = \frac{e^x + e^{-x}}{2} \cdot \frac{e^y - e^{-y}}{2} = \frac{e^x e^y - e^x e^{-y} + e^{-x} e^y - e^{-x} e^{-y}}{4},$$

so that

$$\sinh x \cosh y + \cosh x \sinh y = \frac{2e^x e^y - 2e^{-x} e^{-y}}{4} = \frac{e^{x+y} - e^{-x-y}}{2} = \sinh(x + y).$$

93. $\frac{d}{dx}(\cosh x) = \frac{d}{dx}\frac{e^x + e^{-x}}{2} = \frac{e^x - e^{-x}}{2} = \sinh x.$

95. Using the fact that $\sinh(\sinh^{-1} x) = x$, we differentiate:

$$\cosh(\cosh^{-1} x) = x \qquad\qquad \leftarrow \cosh^{-1} x \text{ is the inverse of } \cosh x$$

$$\frac{d}{dx}(\cosh(\cosh^{-1} x)) = \frac{d}{dx}(x) \qquad\qquad \leftarrow \text{differentiate both sides}$$

$$\sinh(\cosh^{-1} x) \cdot \frac{d}{dx}(\cosh^{-1} x) = 1 \qquad\qquad \leftarrow \text{chain rule, derivative of } \cosh x$$

$$\frac{d}{dx}(\cosh^{-1} x) = \frac{1}{\sinh(\cosh^{-1} x)} \qquad\qquad \leftarrow \text{algebra}$$

$$\frac{d}{dx}(\cosh^{-1} x) = \frac{1}{\sqrt{\cosh^2(\cosh^{-1} x) - 1}} \qquad\qquad \leftarrow \text{since } \cosh^2 x - \sinh^2 x = 1$$

$$\frac{d}{dx}(\cosh^{-1} x) = \frac{1}{\sqrt{x^2 - 1}} \qquad\qquad \leftarrow \cosh x \text{ is the inverse of } \cosh^{-1} x$$

97. If $y = \sinh^{-1} x$, then $\sinh y = x$, so that $\dfrac{e^y - e^{-y}}{2} = x$. Simplifying gives $e^y - e^{-y} = 2x$. Multiply through by e^y and collect terms to get $e^{2y} - 2xe^y - 1 = 0$. Solving as a quadratic in e^y gives $e^y = \dfrac{2x \pm \sqrt{4x^2 + 4}}{2} = x \pm \sqrt{x^2 + 1}$. Since $e^y > 0$ for any y, the correct root is the positive sign, so that $e^y = x + \sqrt{x^2 + 1}$ and $y = \ln\left(x + \sqrt{x^2 + 1}\right)$.

99. If $y = \tanh^{-1} x$, then $\tanh y = x$, so that $\dfrac{e^y - e^{-y}}{e^y + e^{-y}} = x$. Simplifying gives $e^y - e^{-y} = xe^y + xe^{-y}$. Multiply through by e^y and collect terms to get $(1 - x)e^{2y} - (1 + x) = 0$, so that $e^{2y} = \dfrac{1 + x}{1 - x}$. Take logs and simplify to get $y = \dfrac{1}{2}\ln\left(\dfrac{1 + x}{1 - x}\right)$.

Thinking Forward

Local extrema and inflection points

  ▷ If f has a maximum or a minimum value at $x = c$, then the function can be neither increasing nor decreasing there. Thus, if $f'(c)$ exists, it must be zero.

  ▷ For $f(x) = \sin x$, we have $f'(x) = \cos x$. Now, $\cos x$ changes sign at each of its zeroes, so these zeroes are the locations of the local extrema of $f(x) = \sin x$. These are the points $\dfrac{\pi}{2} + n\pi$ for n an integer. The corresponding graph with marked points is

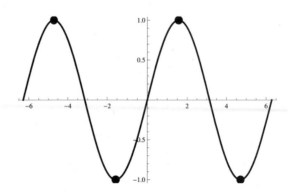

▷ For $f(x) = \sin x$, we have $f'(x) = \cos x$ and $f''(x) = -\sin x$. Now, $-\sin x$ changes sign at each of its zeroes, so those zeroes are the locations of the inflection points of $f(x) = \sin x$. These points are $n\pi$ for n an integer. The corresponding graph with marked points is

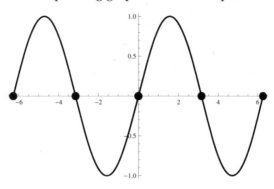

Chapter Review and Self-Test

Skill Certification: Basic Derivatives

1. (a) $\displaystyle\lim_{h\to 0}\frac{(x+h)^2 - x^2}{h} = \lim_{h\to 0}\frac{x^2 + 2hx + h^2 - x^2}{h} = \lim_{h\to 0}\frac{2hx + h^2}{h} = \lim_{h\to 0}(2x+h) = 2x.$

    (b) $\displaystyle\lim_{z\to x}\frac{z^2 - x^2}{z - x} = \lim_{z\to x}\frac{(z-x)(z+x)}{z-x} = \lim_{z\to x}(z+x) = 2x.$

3. (a) $\displaystyle\lim_{h\to 0}\frac{\frac{1}{x+h} - \frac{1}{x}}{h} = \lim_{h\to 0}\frac{-h}{hx(x+h)} = -\lim_{h\to 0}\frac{1}{x(x+h)} = -\frac{1}{x^2}.$

    (b) $\displaystyle\lim_{z\to x}\frac{\frac{1}{z} - \frac{1}{x}}{z - x} = \lim_{z\to x}\frac{x-z}{xz(z-x)} = -\lim_{z\to x}\frac{1}{xz} = -\frac{1}{x^2}.$

5. (a)

    $$\lim_{h\to 0}\frac{\sqrt{x+h} - \sqrt{x}}{h} = \lim_{h\to 0}\frac{(\sqrt{x+h} - \sqrt{x})(\sqrt{x+h} + \sqrt{x})}{h(\sqrt{x+h} + \sqrt{x})}$$

    $$= \lim_{h\to 0}\frac{h}{h(\sqrt{x+h} + \sqrt{x})} = \lim_{h\to 0}\frac{1}{\sqrt{x+h} + \sqrt{x}} = \frac{1}{2\sqrt{x}}.$$

    (b)

    $$\lim_{z\to x}\frac{\sqrt{z} - \sqrt{x}}{z - x} = \lim_{z\to x}\frac{(\sqrt{z} - \sqrt{x})(\sqrt{z} + \sqrt{x})}{(z-x)(\sqrt{z} + \sqrt{x})}$$

    $$= \lim_{z\to x}\frac{z - x}{(z-x)(\sqrt{z} + \sqrt{x})} = \lim_{z\to x}\frac{1}{\sqrt{z} + \sqrt{x}} = \frac{1}{2\sqrt{x}}.$$

7. Since $f(x) = (x^4 - 5x^3 + 2)^{-1}$, we have

    $$f'(x) = -(x^4 - 5x^3 + 2)^{-2}(4x^3 - 15x^2) = (15x^2 - 4x^3)(x^4 - 5x^3 + 2)^{-2}.$$

9.

$$f'(x) = -3(x\sqrt{2x-1})^{-4}\left(x\sqrt{2x-1}\right)' = -3(x\sqrt{2x-1})^{-4}\left(\sqrt{2x-1} + x\cdot\frac{1}{2}(2x-1)^{-1/2}\cdot 2\right)$$

$$= -3(x\sqrt{2x-1})^{-4}\left(\sqrt{2x-1} + \frac{x}{\sqrt{2x-1}}\right).$$

11. Simplify to get $f(x) = x^{-1/2} - x^{-3/2}$; then $f'(x) = -\frac{1}{2}x^{-3/2} + \frac{3}{2}x^{-5/2}$.

13. $f(x)$ is not differentiable at zero. For $x < 0$, we have $f(x) = -x$ so that $f'(x) = -1$. For $x > 0$, we have $f(x) = x$, so that $f'(x) = 1$. Hence

$$f'(x) = \begin{cases} -1, & x < 0, \\ \text{DNE}, & x = 0, \\ 1, & x > 0. \end{cases}$$

15. $f'(x) = e^x \cos x + e^x \sin x = e^x (\sin x + \cos x)$.

17. $f'(x) = \cos(e^x)(e^x)' = e^x \cos(e^x)$.

19. $f'(x) = \dfrac{1}{\tan^2 x} \left(\tan^2 x\right)' = \dfrac{2\tan x(\tan x)'}{\tan^2 x} = \dfrac{2\sec^2 x}{\tan x} = \dfrac{2}{\sin x \cos x}$.

21. $f'(x) = \dfrac{1}{2}(\sin^{-1} x^2)^{-1/2}\left(\sin^{-1} x^2\right)' = \dfrac{1}{2}(\sin^{-1} x^2)^{-1/2}\dfrac{1}{\sqrt{1-x^4}}(2x) = \dfrac{x}{\sqrt{\sin^{-1} x^2}\sqrt{1-x^4}}$.

23. $f'(x) = 2^{3x+1} + x(\ln 2)2^{3x+1} \cdot 3 = 2^{3x+1}(1 + 3x\ln 2)$.

25.

$$\begin{aligned} f'(x) &= \frac{1}{2}\left(\tanh^3(x^5)\right)^{-1/2}\left(\tanh^3(x^5)\right)' \\ &= \frac{1}{2}\left(\tanh^3(x^5)\right)^{-1/2}\left(3\tanh^2(x^5)\left(\tanh(x^5)\right)'\right) \\ &= \frac{3}{2}\tanh^2(x^5)\left(\tanh^3(x^5)\right)^{-1/2}\operatorname{sech}^2(x^5) \cdot 5x^4 \\ &= \frac{15x^4}{2}\tanh^2(x^5)\operatorname{sech}^2(x^5)\left(\tanh^3(x^5)\right)^{-1/2}. \end{aligned}$$

27. Since $f'(x) = -32$, we know that $f(x) = -32x + C$; since $f(0) = 4$, we get $C = 4$ so that $f(x) = -32x + 4$.

29. Simplifying, $f'(x) = 3x^2 + x$. x^3 is an antiderivative of $3x^2$, and $\frac{1}{2}x^2$ is an antiderivative of x. Thus $f(x) = x^3 + \frac{1}{2}x^2 + C$; since $f(2) = 2^3 + 2 + C = 4$, we get $C = -6$ so that $f(x) = x^3 + \frac{1}{2}x^2 - 6$.

31. Since the derivative of e^{4x} is $4e^{4x}$, we get $f(x) = 2e^{4x} + x + C$; using $f(0) = 3$ gives $3 = 2e^0 + 0 + C$ so that $C = 1$ and $f(x) = 2e^{4x} + x + 1$.

33. We recognize this as the derivative of a log, since the denominator is a first power. Try $\ln(1 + 4x)$; this has derivative $\dfrac{4}{1+4x}$, which is too big by a factor of 4. Thus let $f(x) = \dfrac{1}{4}\ln(1 + 4x) + C$. Since $f(0) = 1$, we get $C = 1$ and $f(x) = \dfrac{1}{4}\ln(1 + 4x) + 1$.

35. The numerator is a linear multiple of the derivative of the denominator, which is a first power, so the antiderivative is probably a log. Try $\ln(1 + 4x^2)$. This has derivative $\dfrac{8x}{1+4x^2}$, which is too large by a factor of 8. Thus we let $f(x) = \dfrac{1}{8}\ln(1 + 4x^2) + C$. Since $f(0) = 1$, we get $C = 1$, so that $f(x) = \dfrac{1}{8}\ln(1 + 4x^2) + 1$.

37. Using implicit differentiation, we get $0 = 6vv' + (xv)' = 6vv' + v + xv'$, so that $v' = -\dfrac{v}{x+6v}$.

39. $\dfrac{dA}{dr} = 2\pi r \dfrac{dr}{dr} = 2\pi r$.

41. This is a polynomial in t; its derivative is $y' = 3x^2 - kt^{k-1}$.

Chapter 3

Applications of the Derivative

3.1 The Mean Value Theorem

Thinking Back

Review of definitions and theorems

See the appropriate definition or theorem in the text.

Concepts

1. (a) True. Rolle's Theorem is simply the Mean Value Theorem applied to the case where $f(a) = f(b) = 0$.

   (b) True. See the discussion following the statement of the Mean Value Theorem.

   (c) True. By Theorem 3.3, $x = -2$ is a critical point, so either $f'(-2) = 0$ or $f'(-2)$ is undefined. But since f is differentiable on $\mathbb{R}$, we must have $f'(-2) = 0$.

   (d) False. Consider $f(x) = x^3$. Then $f'(x) = 3x^2$, and $f'(0) = 0$, so that 0 is a critical point of f. However, if you look at the graph of f, it is clear that f has neither a local minimum nor a local maximum at 0.

   (e) False. For example, it is false for

   $$f(x) = \begin{cases} x - 2, & 0 < x \leq 4, \\ 4 - \frac{1}{2}x, & 4 < x \leq 8. \end{cases}$$

   Here $f(x)$ is not differentiable on (a, b), so does not satisfy the conditions of Rolle's Theorem.

   (f) True. By the Mean Value Theorem, whose conditions apply in this case, there is some $c \in (-2, 2)$ with $f'(c) = \dfrac{f(b) - f(a)}{b - a} = \dfrac{f(2) - f(-2)}{2 - (-2)} = -\dfrac{4}{4} = -1$.

   (g) False. Consider $f(x) = (x - 5)^3$. Then $f'(x) = 3(x - 5)^2$ so that $f'(5) = 0$. However, if you look at the graph of f, it is clear that f has neither a local minimum nor a local maximum at $x = 5$.

   (h) False. Consider $f(x) = (x - 5)^2 + 1$. Then $f'(5) = 0$, but $f(x) \geq 1$ for all x, so it cannot be zero on $(0, 10)$.

3. By Theorem 3.3, if $x = 1$ is the location of a local maximum, then either $f'(1) = 0$ or $f'(1)$ does not exist. If f is differentiable at $x = 1$, then we must have $f'(1) = 0$.

5. The critical points of f, by Definition 3.2, are $x = -2$, $x = 0$, $x = 4$, and $x = 5$. A possible graph of f is

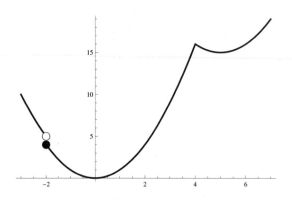

7. Since f is continuous and differentiable, Rolle's Theorem applies on both $[-4, 1]$ and $[1, 2]$, so that we know that there exist $c \in (-4, 1)$ and $d \in (1, 2)$ such that $f'(c) = f'(d) = 0$.

9. Since f is continuous and differentiable, the Mean Value Theorem applies on $[-2, 4]$, so that there exists $c \in (-2, 4)$ with $f'(c) = \dfrac{f(4) - f(-2)}{4 - (-2)} = \dfrac{-2}{6} = -\dfrac{1}{3}$.

11. The statement that $f'(c) = 0$ is equivalent to the statement that the tangent line to f at c is horizontal. Thus a restatement of Rolle's Theorem is: If f is continuous on $[a, b]$ and differentiable on (a, b), and if $f(a) = f(b) = 0$, then there exists at least one value $c \in (a, b)$ such that the tangent line to f at $x = c$ is horizontal.

13. In the plot below, f is continuous on $[2, 6]$ and differentiable on $(2, 6)$. Further, $f(2) = f(6) = 0$, and the average change of f between 2 and 6 is zero as shown by the black dotted line connecting the two points. The gray dotted line shows a point on the curve where the slope of the tangent line is zero, so that $f' = 0$.

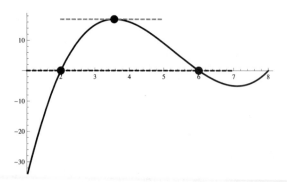

15. The graph shown below has $f(-2) = f(2) = 0$, and f is clearly continuous on $[-2, 2]$ and differentiable on $(-2, 2)$, so that Rolle's Theorem applies. The three points in $(-2, 2)$ with $f'(c) = 0$ are shown, with their tangent lines in gray.

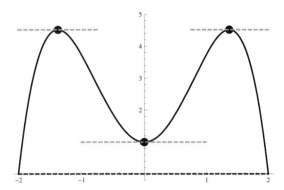

17. The graph shown below satisfies the conditions of the problem. Since it is not differentiable at −1 ∈ (−2, 2), Rolle's Theorem does not apply, and clearly there is no point on the curve with horizontal tangent line.

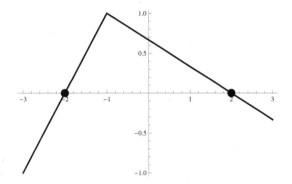

19. The graph shown below satisfies the conditions of the problem. Since it is not continuous at −1 ∈ [−3, −1], Rolle's Theorem does not apply, and clearly there is no point on the curve with $x \in (-3, -1)$ with horizontal tangent line.

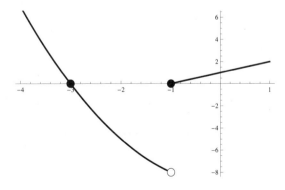

21. The graph shown below satisfies the conditions of the problem; the secant from $x = -3$ to $x = 3$ is drawn as a dashed line. Since it is not continuous at $1 \in [-3, 3]$, the Mean Value Theorem does not apply, and clearly there is no point on the curve with $x \in (-3, 3)$ such that the slope of the tangent line at that point is equal to the slope of the secant.

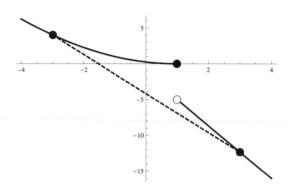

Skills

23. The derivative of f is zero where the tangent line is horizontal. This appears to happen only at $x \approx \frac{3}{2}$; this is a local maximum. The derivative does not exist at $x \approx 3$; this is a local minimum.

25. The derivative of f exists everywhere. The derivative of f is zero where the tangent line is horizontal. This appears to happen at $x \approx \frac{1}{2}$, which is a local minimum, at $x \approx 2$, which is neither a local minimum nor a local maximum, and at $x \approx \frac{7}{2}$, which is a local maximum.

27. Since f is a polynomial, its derivative exists everywhere. By the product rule, we have $f'(x) = (x - 1.7) + (x + 3) = 2x + 1.3$, so that $f'(x) = 0$ only at $x = -0.65$. From the graph, f has a local minimum there.

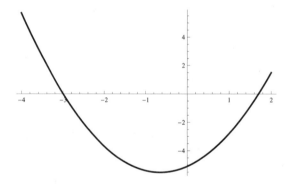

29. Since f is a polynomial, its derivative exists everywhere. We have $f'(x) = 12x^3 + 24x^2 - 36x = 12x(x + 3)(x - 1)$, so that $f'(x) = 0$ for $x = -3$, $x = 0$, and $x = 1$. From the graph, $x = 0$ is a local maximum while the other two critical points are local minima.

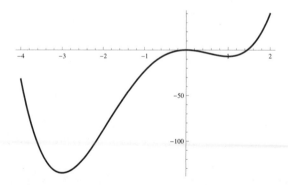

31. Since both $3x$ and $2e^x$ are differentiable everywhere, so is f. We have $f'(x) = 3 - 2e^x$, so that $f'(x) = 0$ for $x = \ln \frac{3}{2} \approx 0.4055$. From the graph, this is a local maximum.

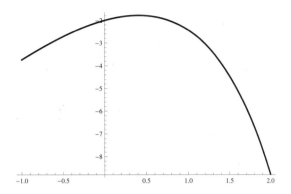

33. $f(x)$ is defined when the numerator is defined (which is for $x > 0$) and when the denominator is defined and nonzero (which is for $x \neq 0$). Thus the domain of f is $x > 0$. There, we have

$$f'(x) = \frac{x \cdot \frac{1}{2x} \cdot 2 - \ln 2x}{x^2} = \frac{1 - \ln 2x}{x^2}.$$

Thus $f'(x) = 0$ for $\ln 2x = 1$, which is for $x = \frac{e}{2} \approx 1.3591$. From the graph, this is a local maximum.

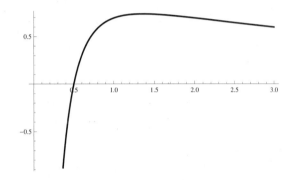

35. $f(x) = \cos x$ is differentiable everywhere, and $f'(x) = -\sin x$. Thus the critical points of $\cos x$ are the points where $\sin x = 0$, so at $\pm n\pi$. From the graph below, even multiples of π are local maxima while odd multiples are local minima:

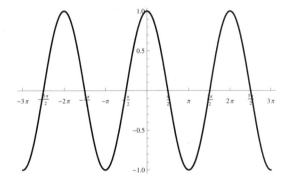

37. Since f is continuous on $[-3, 1]$ and differentiable on $(-3, 1)$, and since $f(-3) = f(1) = 0$, it follows that f satisfies the hypotheses of Rolle's Theorem on $[-3, 1]$. The values of x at which $f'(x) = 0$ appear to be $x \approx -2.25$, $x \approx -1$, and $x \approx 0.3$.

39. Since f is continuous on $[0, 4]$ and differentiable on $(0, 4)$, and since $f(0) = f(4) = 0$, it follows that f satisfies the hypotheses of Rolle's Theorem on $[0, 4]$. The values of x at which $f'(x) = 0$ appear to be $x \approx 0.5$, $x = 2$, and $x \approx 3.5$.

41. Since f is a polynomial, it is continuous and differentiable everywhere. Since $f(0) = 0$ and $f(3) = 3^3 - 4 \cdot 3^2 + 9 = 0$, we see that f satisfies the conditions of Rolle's Theorem on $[0, 3]$. Now, $f'(x) = 3x^2 - 8x + 3$, which has roots $x = \dfrac{8 \pm \sqrt{28}}{6} = \dfrac{4 \pm \sqrt{7}}{3} \approx 0.451, 2.215$. Thus these are the points at which $f'(x) = 0$ by Rolle's Theorem.

43. Since f is a polynomial, it is continuous and differentiable everywhere. Also, $f(2) = 2^4 - 3.24 \cdot 2^2 - 3.04 = 16 - 12.96 - 3.04 = 0$; since f is an even function, $f(-2) = 0$ as well. Thus f satisfies the conditions of Rolle's Theorem on $[-2, 2]$. We have $f'(x) = 4x^3 - 6.48x = 4x(x^2 - 1.62)$, so that $f'(x) = 0$ for $x = 0$ and for $x = \pm\sqrt{1.62} \approx \pm 1.273$. Since all of these points lie in $(-2, 2)$, they are all points where the tangent line to f is horizontal.

45. $f(x) = \cos x$ is continuous and differentiable everywhere, and $\cos\left(-\frac{\pi}{2}\right) = \cos\left(\frac{3\pi}{2}\right) = 0$. Thus f satisfies the conditions of Rolle's Theorem on $\left[-\frac{\pi}{2}, \frac{3\pi}{2}\right]$. Now, $f'(x) = -\sin x$, and $f'(x) = 0$ in $\left(-\frac{\pi}{2}, \frac{3\pi}{2}\right)$ for $x = 0$ and $x = \pi$. These are the two points in the given range where the tangent line to f is horizontal.

47. Since $f(x) = e^x(x^2 - 2x)$ is the product of two functions that are continuous and differentiable everywhere, so is f. Also, $f(0) = f(2) = 0$, so that f satisfies the conditions of Rolle's Theorem on $[0, 2]$. We have $f'(x) = e^x(2x - 2) + e^x(x^2 - 2x) = e^x(x^2 - 2)$, so that $f'(x) = 0$ for $x = \pm\sqrt{2}$. Of these, only $x = \sqrt{2} \in (0, 2)$, so this is the only point with a horizontal tangent line.

49. Since f is continuous and differentiable on $[0, 2]$, it satisfies the hypotheses of the Mean Value Theorem there. It looks as though $x \approx \frac{3}{2}$ is the point where the tangent line to $f(x)$ has the same slope as the secant line through $(0, f(0))$ and $(2, f(2))$.

51. Since f is continuous and differentiable on $[-3, 0]$, it satisfies the hypotheses of the Mean Value Theorem there. There are two points at which the slope of the tangent line is the same as the slope of the secant through $(-3, f(-3))$ and $(0, f(0))$; they are $x \approx -2.7$ and $x \approx -0.6$.

53. Since f is not defined at $x = 0 \in [-3, 2]$, we see that f does not satisfy the conditions of the Mean Value Theorem.

55. Since f is a polynomial, it is continuous and differentiable everywhere, so that it satisfies the conditions of the Mean Value Theorem on $[-2, 3]$. Since $f(-2) = 13$ and $f(3) = -7$, we want to find $x \in (-2, 3)$ such that $f'(x) = -3x^2 + 6x = \dfrac{f(3) - f(-2)}{5} = -4$. So we want to solve $3x^2 - 6x - 4 = 0$. This quadratic has roots $x = \dfrac{6 \pm \sqrt{84}}{6} = \dfrac{3 \pm \sqrt{21}}{3} \approx -0.528, 2.528$. Both of these roots are in $(-2, 3)$, so they are both points satisfying the conclusion of the Mean Value Theorem there.

57. $f(x)$ is defined and is continuous on $[0, 1]$. We have $f'(x) = \frac{2x}{x^2+1}$, so that f' is defined on $(0, 1)$. Thus f is differentiable on $(0, 1)$, so that f satisfies the conditions of the Mean Value Theorem on $[0, 1]$. Since $f(0) = 0$ and $f(1) = \ln 2$, we want to find x such that $f'(x) = \frac{2x}{x^2+1} = \ln 2$. This becomes the quadratic $(\ln 2)x^2 - 2x + \ln 2 = 0$, which has solutions

$$x = \frac{1 \pm \sqrt{1 - (\ln 2)^2}}{\ln 2} \approx 0.403, 2.843.$$

Of these, only the first is in the interval $(0, 1)$, so it is the only value of x satisfying the conclusions of the Mean Value Theorem.

59. Since $f(x) = \sin x$ is continuous and differentiable everywhere, it satisfies the conditions of the Mean Value Theorem on $\left[0, \frac{\pi}{2}\right]$. Now, $f\left(\frac{\pi}{2}\right) = 1$ and $f(0) = 0$, so we want to find x such that $f'(x) = \cos x = \frac{1}{\frac{\pi}{2} - 0} = \frac{2}{\pi}$. Thus $x = \cos^{-1}\frac{2}{\pi} \approx 0.881$. This is the only value of x in $\left(0, \frac{\pi}{2}\right)$ satisfying the conclusions of the Mean Value Theorem.

Application

61. Since $f'(h) = 2h - 7.4$, we see that $f'(4) = 0.6 \neq 0$, so that $h = 4$ cannot be a critical point of f, and thus by Theorem 3.3, $h = 4$ cannot be a local extremum.

63. (a) Let $s(t)$ be Alina's distance from home at time t. Then $s(0) = 0$ and $s(0.5) = 20$. Since Alina's motion was continuous and differentiable, the Mean Value Theorem implies that there is some point t_0 such that
$$v(t_0) = s'(t_0) = \frac{s(0.5) - s(0)}{0.5 - 0} = 40.$$
That is, at some time her velocity was 40 miles per hour.

    (b) This makes sense - since her average velocity was 40 miles per hour, she could not have exceeded that velocity for the entire time nor been below it for the entire time. Since velocity changes smoothly, it must have been exactly 40 miles per hour at some time.

Proofs

65. This proof is given in the text, q.v.

67. Since f is continuous and differentiable everywhere, it satisfies the hypotheses of Rolle's Theorem on any interval. Let the roots be x_1, x_2, and x_3, and assume that $x_1 < x_2 < x_3$. Then since f satisfies the conditions of Rolle's Theorem on $[x_1, x_2]$, there is some $a_1 \in (x_1, x_2)$ with $f'(a_1) = 0$; since it satisfies the conditions on $[x_2, x_3]$, there is some $a_2 \in (x_2, x_3)$ with $f'(a_2) = 0$. Note that $a_1 \neq a_2$ since they lie in disjoint intervals.

69. Consider the function $g(x) = f(x) - f(a)$. Then g is also continuous on $[a, b]$ and differentiable on (a, b) since f is. Further $g(a) = f(a) - f(a) = 0$, and $g(b) = f(b) - f(a) = 0$ since $f(a) = f(b)$. Thus g satisfies the hypotheses of Rolle's Theorem on $[a, b]$, so there is some point $c \in (a, b)$ with $g'(c) = 0$. But $g'(x) = f'(x)$, so that $f'(c) = 0$ as well.

Thinking Forward

Sign analyses for derivatives

▷ For $f(x) = \frac{x}{x^2+1}$, we have $f'(x) = \frac{(x^2+1)-x(2x)}{(x^2+1)^2} = \frac{1-x^2}{(1+x^2)}$. Then $f'(x) > 0$ for $|x| < 1$, and $f'(x) < 0$ for $|x| = 1$. A sign chart is:

▷ For $f(x) = x^2 3^x$, we have $f'(x) = 2x3^x + x^2(\ln 3)3^x = x3^x(2 + x\ln 3)$. Then $f(x) = 0$ for $x = 0$ and for $x = -\frac{2}{\ln 3} \approx -1.82$, so we can determine the sign of f' by testing one point in each of the resulting three intervals: $f'(-2) = \frac{4}{9}(-1 + \ln 3) \approx 0.044 > 0$, $f'(-1) = -\frac{1}{3}(2 - \ln 3) \approx -0.3 < 0$, and $f'(1) = 3(2 + \ln 3) \approx 9.396 > 0$. A sign chart is:

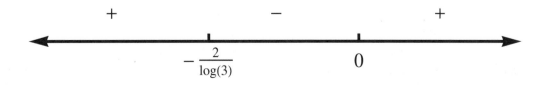

▷ For $f(x) = \frac{\sin x}{e^x}$, we have $f'(x) = \frac{e^x \cos x - e^x \sin x}{e^{2x}} = \frac{\cos x - \sin x}{e^x}$. Then $f'(x) = 0$ where $\sin x = \cos x$, i.e., for $x = \frac{\pi}{4} + n\pi$. Since $e^x > 0$, it follows that $f(x) > 0$ where $\cos x > \sin x$, which is on $\left(0, \frac{\pi}{4}\right) \cup \left(\frac{5\pi}{4}, 2\pi\right)$. A sign chart is:

▷ For $f(x) = \ln \ln x$, we have $f'(x) = \frac{1}{\ln x} \cdot \frac{1}{x} = \frac{1}{x \ln x}$. Then $f'(x)$ is never zero, so it suffices to check its value at any point to see what its sign is (note that the domain of $f(x)$ is $x > 1$ and of $f'(x)$ is $x > 0$). For example, at $x = 2$, we get $f'(2) = \frac{1}{2 \ln 2} > 0$, so that $f'(x)$ is always positive. A sign chart is:

Sign analyses for second derivatives

▷ From the preceding problem, we get

$$f''(x) = \frac{(1+x^2)^2(-2x) - (1-x^2)(2(1+x^2)(2x))}{(1+x^2)^4} = \frac{2x^3 - 6x}{(1+x^2)^3}.$$

Then $f''(x) = 0$ for $x = 0$ and for $x = \pm\sqrt{3}$, so it suffices to check its value at any point in each of the resulting intervals to determine its sign chart. Note that f'' is an odd function, so it suffices to check its value at $x = 1$ and $x = 2$. $f''(1) = -\frac{1}{2} < 0$ and $f''(2) = \frac{4}{125} > 0$, so that $f''(-1) > 0$ and $f''(-2) < 0$. The sign chart is:

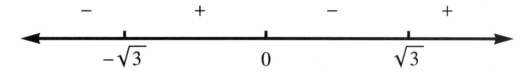

▷ From the preceding problem, we get

$$f''(x) = 3^x(2 + x \ln 3) + x(\ln 3)3^x(2 + x \ln 3) + x3^x(\ln 3) = 3^x(2 + 4x \ln 3 + x^2(\ln 3)^2).$$

The roots of $f''(x)$ are thus the roots of the quadratic in x, which are $\frac{-2 \pm \sqrt{2}}{\log 3}$. Doing interval analysis as before, we get for a sign chart:

▷ From the preceding problem, we get

$$f''(x) = \frac{e^x(-\sin x - \cos x) - (\cos x - \sin x)e^x}{e^{2x}} = \frac{-2\cos x}{e^x}.$$

Then $f''(x) = 0$ where $\cos x = 0$, i.e., for $x = \frac{\pi}{2} + n\pi$. Further, since e^x is positive, $f'(x) > 0$ when $\cos x < 0$, i.e., when x lies in the second or third quadrants, and negative when x lies in the first or fourth quadrants. A sign chart on $[0, 2\pi]$ is:

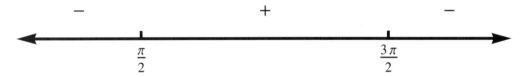

▷ From the preceding problem, since $\left(\frac{1}{f(x)}\right)' = -\frac{1}{f(x)^2}f'(x)$, we get

$$f''(x) = -\frac{1}{(x\ln x)^2} \cdot \left(x \cdot \frac{1}{x} + \ln x\right) = -\frac{1+\ln x}{(x\ln x)^2}.$$

The only zero of $f''(x)$ is thus $x = e^{-1}$, where $\ln x = -1$. However, the domain of $f(x)$ is $x > 1$, so that the second derivative is defined only for $x > 1$; in that range, it has no zeros. To test its sign, it suffices to evaluate it at one point, say $x = 2$; there it is obviously negative. So a sign chart is:

3.2 The First Derivative and Curve Sketching

Thinking Back

Differentiation

▷ Using the product rule,

$$f'(x) = 3(2x-1)^2(2)(3x+1)^2 + (2x-1)^3(2(3x+1)(3))$$
$$= (2x-1)^2(3x+1)(6(3x+1) + 6(2x-1))$$
$$= 30x(2x-1)^2(3x+1).$$

▷ Using the quotient rule,

$$f'(x) = \frac{(3x+1)^2(3(2x-1)^2(2)) - (2x-1)^3(2(3x+1)(3))}{(3x+1)^4}$$
$$= \frac{(2x-1)^2(6(3x+1) - 6(2x-1))}{(3x+1)^3}$$
$$= \frac{6(x+2)(2x-1)^2}{(3x+1)^3}.$$

▷ Using the product rule,

$$f'(x) = 3(2x)e^{-4x} + 3x^2(-4e^{-4x}) = xe^{-4x}(6 - 12x).$$

▷ Using the chain rule,

$$f'(x) = \cos(\ln x) \cdot \frac{1}{x} = \frac{\cos(\ln x)}{x}.$$

Solving equations

▷ Multiply numerator and denominator of $g(x)$ by $2\sqrt{x}$ to get

$$g(x) = \frac{(1 + 5x) - 2x(5)}{2(1 + 5x)^2\sqrt{x}} = \frac{1 - 5x}{2(1 + 5x)^2\sqrt{x}}.$$

Now, $g(x)$ is undefined when the denominator vanishes, i.e., for $x = 0$ and $x = -\frac{1}{5}$. Also, $g(x) = 0$ when the numerator vanishes but the denominator does not; this happens only for $x = \frac{1}{5}$.

▷ Factor out an $(x - 1)^{-1/2}$ to get

$$g(x) = (x - 1)^{-1/2}\left(2x(x - 1) + x^2\left(\frac{1}{2}\right)\right) = \frac{5x^2 - 4x}{2\sqrt{x - 1}}.$$

Now, $g(x)$ is undefined when the denominator vanishes or is undefined, i.e. for $x \le 1$. Also, $g(x) = 0$ when the numerator vanishes but the denominator is defined and nonzero. The numerator vanishes at $x = 0$ and at $x = \frac{4}{5}$; since neither of these values is in the range $x > 1$, the denominator is undefined at both points, so that g has no zeros.

▷ Multiply out the numerator and simplify to get

$$g(x) = \frac{e^x}{(1 - e^x)^2}.$$

Now, $g(x)$ is undefined when the denominator vanishes or is undefined, i.e. for $x = 0$. Since the numerator is never zero, g has no zeros.

Sign analyses

▷ Since $g(x) = 6x(x - 3)$, it can change sign only where $g(x) = 0$, i.e., at $x = 0$ and at $x = 3$. Checking one point in each of the resulting three intervals, for example $x = -1$, $x = 1$ and $x = 4$, gives the following sign chart:

▷ By inspection, $x = 1$ is a root, and $g(x)$ factors as $g(x) = (x - 1)(x^2 - x - 2) = (x - 1)(x + 1)(x - 2)$. Checking the sign of g at $x = -2$, $x = 0$, $x = \frac{3}{2}$, and $x = 3$ gives the following sign chart:

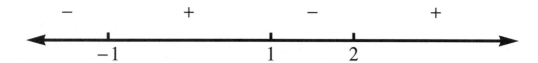

▷ Since the denominator is always positive, the sign of $g(x)$ depends solely on the sign of the numerator. $(x+2)(x-1)^4 = 0$ for $x = 1$ and for $x = -2$. Further, $(x-1)^4 \geq 0$, so that when $g(x) \neq 0$, its sign depends only on $x+2$. Thus $g(x) < 0$ for $x < -2$ and $g(x) \geq 0$ for $g > -2$. The sign chart is:

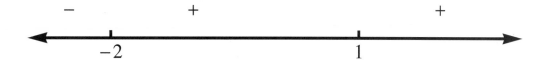

▷ $g(x)$ is undefined when the denominator vanishes, i.e., at $x = n\pi$ for n any integer. Since the denominator is always positive on the domain of g, the sign of g depends solely on the sign of the numerator. $(3x^2 - 5x - 2) = (x-2)(3x+1)$, so that $g(x) = 0$ only at $x = 2$ and at $x = -\frac{1}{3}$. On $(-\pi, \pi)$, then, there are four intervals on which we must check the sign of g; they are $\left(-\pi, -\frac{1}{3}\right)$, $\left(-\frac{1}{3}, 0\right)$, $(0, 2)$, and $(2, \pi)$. Checking one point in each of those intervals gives the following sign chart:

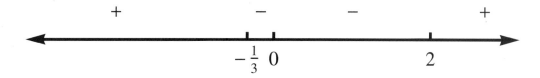

Concepts

1. (a) True. See Theorem 3.6 and the discussion preceding it.

   (b) False. f need not be differentiable or even continuous on $(-2, 2)$; for example, $f(x) = x$ for $-2 < x \leq 0$ and $f(x) = x+1$ for $0 < x \leq 2$. However, if f is differentiable on $(-2, 2)$, then this is a true statement.

   (c) True. Since x^2 is an antiderivative of $2x$, Theorem 3.7 assures us that any antiderivative has the form $x^2 + C$.

   (d) False. f need not be differentiable on $(1, 8)$. But even if it is, the statement need not hold; for example, look at $y = -(x-2)^2$.

   (e) False. f' need not be defined at $x = 3$.

   (f) False. Unless f' changes sign at $x = -2$, there need not be a local extremum there.

   (g) False. Consider $f(x) = (x-1)^3$. Then $x = 1$ is a critical point, but both $f'(0)$ and $f'(2)$ are positive.

   (h) False. Assuming f' is continuous, $f'(c) = 0$ for some $c \in (1, 3)$, but that c need not be 2. If f' is not continuous, there need not be any point in $c \in (1, 3)$ with $f'(c) = 0$.

3. See the first paragraph of this section.

5. Yes, a point can be both a critical point and a local extremum; for example, $f(x) = x^2$ and $c = 0$ (which is a critical point since $f'(0) = 0$). Yes, a point can be both an inflection point and a critical point; for example, $f(x) = x^3$ and $c = 0$ (which is a critical point since $f'(0) = 0$). No, a point cannot be both an inflection point and a local extremum. If a point is a local extremum, then the curve is either concave up or concave down in a neighborhood of that point, so the concavity does not change there and hence it cannot be an inflection point. Graphs of the first two functions are:

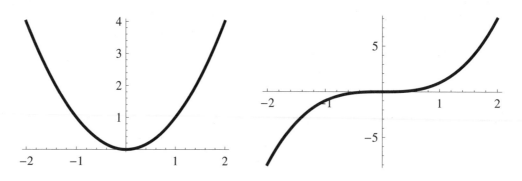

7. Suppose $x = a$ and $x = b$ with $a < b$ are critical points, with no critical points in the interval (a, b). Then $f'(c) \neq 0$ for all $c \in (a, b)$. Since f' is assumed continuous, it can change sign only at points where it is zero. Thus f' does not change sign in (a, b), so must be always positive or always negative. Hence testing its value at one point determines its sign on the entire interval (a, b).

9. For example,

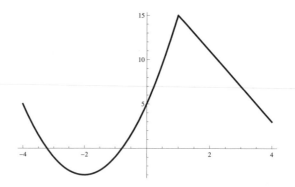

11. Assume first that $0 < a < b$. Then $f(b) = b^4 > a^4 = f(a)$, so that $b > a$ implies $f(b) > f(a)$. Thus f is increasing on $[0, \infty)$. Next assume $a < b < 0$. Then $f(b) = b^4 < |b|^4 < |a|^4 = a^4 = f(a)$, so that $b > a$ implies $f(b) < f(a)$. Thus f is decreasing on $(-\infty, 0]$. Using derivatives, we apply Theorem 3.6 and note that $f'(x) = 4x^3$. With $I = [0, \infty)$, clearly $f'(x) > 0$ on the interior of I (which is simply $(0, \infty)$), so that by Theorem 3.6(a), f is increasing on $[0, \infty)$. Similarly, with $I = (-\infty, 0]$, we have $f'(x) < 0$ on the interior of I (which is $(-\infty, 0)$), so that by Theorem 3.6(b), f is decreasing on $(-\infty, 0]$.

13. The critical points of f are those points where f' is either undefined or zero. Now, f' is defined everywhere since $1 + x^2 \geq 0$, and $f'(x) = 0$ when $\sqrt{1 + x^2} = 4$, so that $1 + x^2 = 16$. Thus $f'(x) = 0$ for $x = \pm\sqrt{15}$, and those are the critical points of f.

15. By Theorem 3.7, since $g(x)$ and $h(x)$ have the same derivative, they differ by a constant, so that $g(x) - h(x) = C$ for some $C \in \mathbb{R}$.

17. Graph I is the derivative, f'. To see this, note that the critical points of Graph II are at $x = -1$, $x = 1$, and $x = 2.5$, where the tangent line is horizontal. Those are precisely the zeros of Graph I, so Graph I gives the slope of the tangent lines to Graph II, i.e., its derivative.

Skills

19. Since $f(x)$ has critical points at $x \approx 0.5$ and $x \approx 2.25$, it follows that f' will be zero at those points. Further, $f'(x) > 0$ for $x \in (0, 0.5) \cup (2.25, 3)$ and $f'(x) < 0$ for $x \in (0.5, 2.25)$. A possible graph of f' is

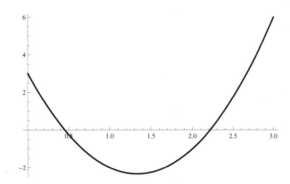

21. Since $f(x)$ has critical points at $x \approx 0$ and $x \approx 2$, it follows that f' will be zero at those points. Further, $f'(x) < 0$ only on $(-\infty, 0)$, and $f'(x) > 0$ on $(0, 2) \cup (2, \infty)$. A possible graph of f' is

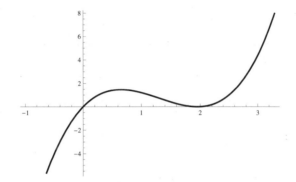

23. Since $f'(-2) = f'(0) = f'(2) = 0$, it follows that these are critical points of f. Using Theorem 3.6, we see that $x = -2$ and $x = 2$ are local minima while $x = 1$ is a local maximum. A possible graph of f is

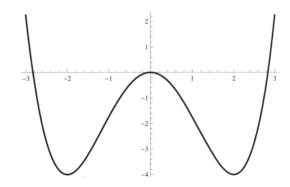

25. Since $f'(1) = f'(3) = 0$, it follows that these are critical points of f; from Theorem 3.6 we see that $x = 1$ is not a local extremum, but the slope of f remains negative near $x = 1$. On the other hand, $x = 3$ is a local minimum. A possible graph of f is

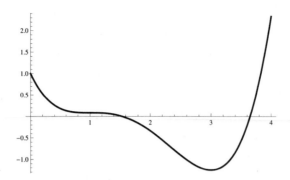

27. We have $f'(x) = 6x^2 - 18x = 6x(x-3)$, which vanishes at $x = 0$ and $x = 3$; evaluating f' at one point in each of the resulting three intervals (say, $f'(-1) = 24$, $f'(1) = -12$, and $f'(4) = 24$) gives the following sign chart:

From the sign chart, $f(x)$ is increasing on $(-\infty, 0] \cup [3, \infty)$ and decreasing on $[0, 3]$. A graph of $f(x)$ verifying this behavior is

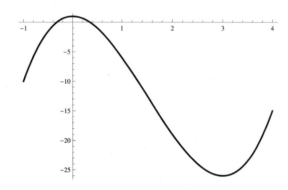

29. We have $f'(x) = \frac{x^2 + 4 - x(2x)}{(x^2+4)^2} = \frac{4 - x^2}{(4+x^2)^2}$. Then $f'(x)$ is defined everywhere, and is zero at $x = \pm 2$. Evaluating f' at one point in each of the resulting three intervals (say, $f'(-3) = f'(3) = \frac{-5}{169}$ and $f'(0) = \frac{1}{4}$ gives the following sign chart:

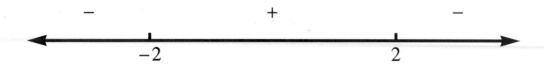

From the sign chart, $f(x)$ is increasing on $[-2, 2]$ and decreasing elsewhere. A graph of $f(x)$ verifying this behavior is

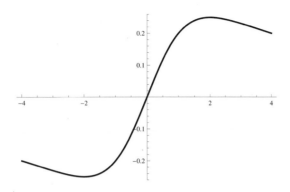

31. We have $f'(x) = e^x(x-2) + e^x = e^x(x-1)$. Then $f'(x) = 0$ only at $x = 1$; evaluating $f'(x)$ at one point in each of the resulting two intervals (say, $f'(0) = -1 < 0$ and $f'(2) = e^2 > 0$) gives the following sign chart:

Thus $f(x)$ is increasing on $[1, \infty)$ and decreasing on $(-\infty, 1]$. A graph of $f(x)$ verifying this behavior is

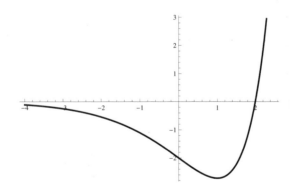

33. We have $f'(x) = \frac{1}{x^2+1}(2x) = \frac{2x}{x^2+1}$. Now, $f'(x)$ is defined everywhere since the denominator is always positive, and is zero only at $x = 0$. Obviously $f'(x) > 0$ for $x > 0$ and $f'(x) < 0$ for $x < 0$. A sign chart for f' is thus

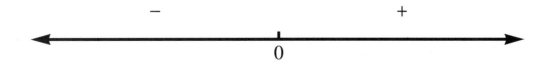

It follows that f is increasing on $(0, \infty)$ and decreasing on $(-\infty, 0)$. A graph of $f(x)$ verifying this behavior is

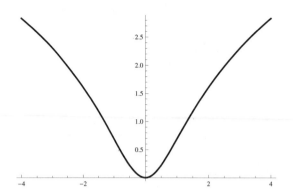

35. We have $f'(x) = \frac{\pi}{2}\cos\left(\frac{\pi}{2}x\right)$. Restricting our attention to $(-4, 4)$, which is two full periods for both f and f', we see that $f'(x) = 0$ for $x = -3, -1, 1$, and 3, so that these are the critical points of f. Evaluating $f'(x)$ at points in the resulting intervals (say, $f'(-3.5) = f(3.5) = \frac{\pi}{2}\cos\left(\frac{7\pi}{4}\right) > 0$, $f(-2) = f(2) = \frac{\pi}{2}\cos(-\pi) < 0$ and $f(0) = \frac{\pi}{2} > 0$, since $f'(x)$ is an even function) gives the following sign chart:

Thus $f(x)$ is increasing on $[-4, -3] \cup [-1, 1] \cup [3, 4]$ and decreasing on $[-3, -1] \cup [1, 3]$. A graph of $f(x)$ verifying this behavior is

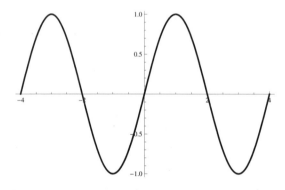

37. We have $f'(x) = \cos^2 x - \sin^2 x$. Restricting our attention to $(0, 2\pi)$, which is two full periods for both f and f', we see that $f'(x) = 0$ whenever $\cos x = \pm \sin x$, which is at $x = \frac{\pi}{4}, x = \frac{3\pi}{4}, x = \frac{5\pi}{4}$, and $x = \frac{7\pi}{4}$, so that these are the critical points of f. Evaluating $f'(x)$ at points in the resulting intervals gives the following sign chart:

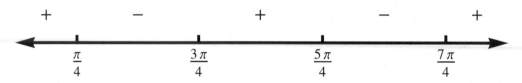

Thus $f(x)$ is increasing on $\left[0, \frac{\pi}{4}\right] \cup \left[\frac{3\pi}{4}, \frac{5\pi}{4}\right] \cup \left[\frac{7\pi}{4}, 2\pi\right]$ and decreasing elsewhere. A graph of $f(x)$ verifying this behavior is

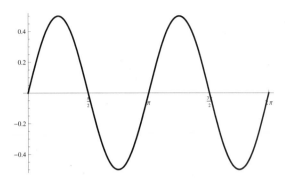

39. $f'(x) = 2(x-2)(1+x) + (x-2)^2 = 3x^2 - 6x = 3x(x-2)$, so that $f'(x) = 0$ at $x = 0$ and $x = 2$. Testing points, we see that $f'(-1) = 9 > 0$, $f'(1) = -3 < 0$, and $f'(3) = 9 > 0$. Thus by Theorem 3.8, $x = 0$ is a local maximum and $x = 2$ is a local minimum. A graph of f is

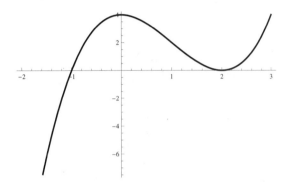

41. $f'(x) = \frac{(x^2+x-2)(2x+1)-(x^2+x+1)(2x+1)}{(x^2+x-2)^2} = -\frac{3(2x+1)}{(x^2+x-2)^2}$. Thus $f'(x) = 0$ at $x = -\frac{1}{2}$, and $f'(x)$ is undefined at the roots of $x^2 + x - 2$, which are $x = -2$ and $x = 1$. Since f is discontinuous at both $x = -2$ and $x = 1$, we need only test a point on either side of $x = -\frac{1}{2}$ not containing either of the other two critical points. Thus for example $f'(-1) = \frac{3}{4}$ and $f'(0) = -\frac{3}{4}$. Thus by Theorem 3.8, $x = -\frac{1}{2}$ is a local maximum. A graph of f is

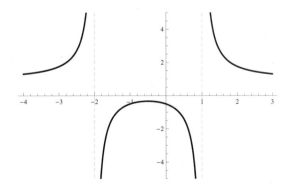

43. $f'(x) = \frac{2e^x}{(3-2e^x)^2}$. Thus $f(x)$ is everywhere nonzero, and is undefined at $x = \ln\frac{3}{2}$, so there are no critical points and f has no local extrema. A graph of f is

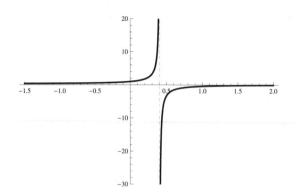

45. $f'(x) = -\pi \sin(\pi(x+1))$, so that $f'(x)$ is zero where $\sin(\pi(x+1)) = 0$, i.e., for x any integer. Considering $(-2,2)$, we get $f'(x) = 0$ for $x = -1, 0, 1$; checking points gives $f'\left(-\frac{3}{2}\right) = \pi > 0$, $f'\left(-\frac{1}{2}\right) = -\pi < 0$, $f'\left(\frac{1}{2}\right) = \pi > 0$, and $f'\left(\frac{3}{2}\right) = -\pi < 0$. Thus by Theorem 3.8, $x = -1$ and $x = 1$ are local maxima, while $x = 0$ is a local minimum. A graph of f is

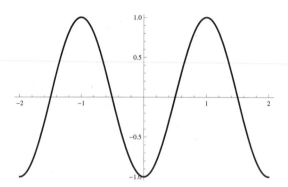

47. $f'(x) = \frac{1}{1+x^2}$, so that $f'(x)$ is defined everywhere and is nonzero everywhere. Thus $f(x)$ has no local extrema. A graph of f is

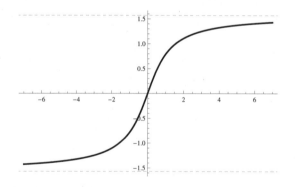

49. $f'(x) = \cos(\cos^{-1} x) \cdot \left(-\frac{1}{\sqrt{1-x^2}}\right) = -\frac{x}{\sqrt{1-x^2}}$. Then $f'(x) = 0$ only at $x = 0$, and $f'(x)$ is undefined for $|x| \geq 1$. Checking points we find $f'\left(-\frac{1}{2}\right) = \frac{1/2}{3/4} > 0$ and $f'\left(\frac{1}{2}\right) = -\frac{1/2}{3/4} < 0$, so that $x = 0$ is a local maximum. A graph of f is

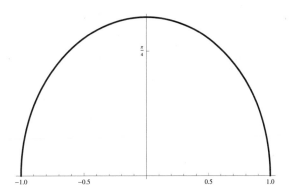

51. Since $f'(x) > 0$ for $x < 0$ and for $x > 5$, and $f'(x) < 0$ for $0 < x < 5$, we see that f is increasing on $(-\infty, 0)$ and on $(5, \infty)$ and decreasing on $(0, 5)$. It has a local maximum at $x = 0$ and a local minimum at $x = 5$. Possible graphs for f' and f are

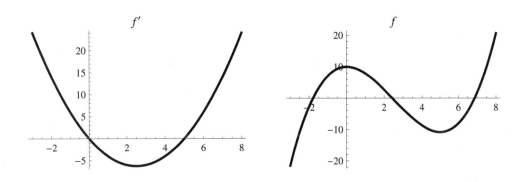

53. Since $f'(x) > 0$ for $x > 3$, and $f'(x) < 0$ on $(-\infty, -3) \cup (-3, 3)$, we see that f is decreasing on $(-\infty, 3)$ with an inflection point at $x = -3$, and increasing on $(3, \infty)$. It has a local minimum at $x = 3$. Possible graphs for f' and f are

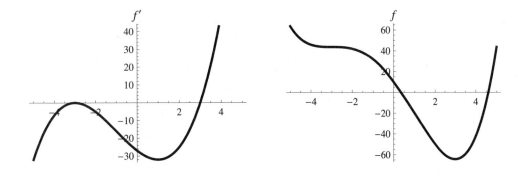

55. Since $f'(x)$ changes sign from positive to negative at $x = 1$, that is a local maximum of f; since the sign of f' does not change at $x = 8$, that is an inflection point of f. Finally, f is not differentiable at $x = 3$. Possible graphs of f' and f are

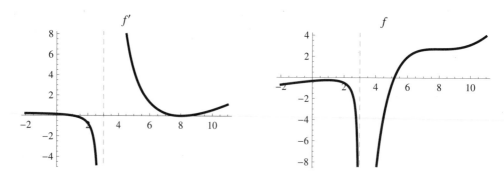

57. f has zeros at $x = 2$ and $x = -\frac{1}{3}$; testing points for f gives $f(-1) = 6 > 0$, $f(0) = -2 < 0$, and $f(3) = 10 > 0$, so we have the following sign chart:

$f'(x) = (3x + 1) + (x - 2)(3) = 6x - 5$, so that f' has a zero at $x = \frac{5}{6}$. Testing points gives $f'(0) = -5 < 0$ and $f'(1) = 1 > 0$, and we get the following sign chart:

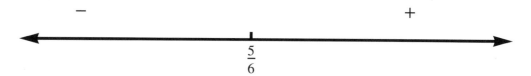

We see that f has a local minimum at $x = \frac{5}{6}$. Since f is an even-degree polynomial with positive leading term, it goes to ∞ as $x \to \pm\infty$. Graphs for f and f' are

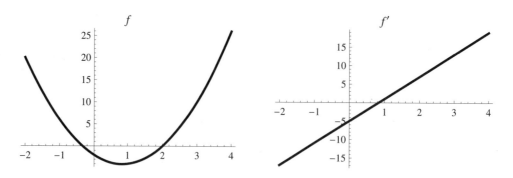

59. $f(x) = (x - 1)^2(x + 1)$ has zeros at $x = \pm 1$. Checking points we get $f(-2) = -9 < 0$, $f(0) = 1 > 0$, and $f(2) = 3 > 0$, giving the sign chart

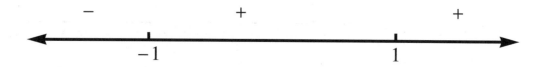

$f'(x) = 3x^2 - 2x - 1 = (x - 1)(3x + 1)$, so that f' has zeros at $x = 1$ and $x = -\frac{1}{3}$. Testing points gives $f'(-1) = 4 > 0$, $f'(0) = -1 < 0$, and $f'(2) = 7 > 0$, giving the sign chart

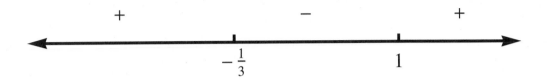

We see that f has a local minimum at $x = 1$ and a local maximum at $x = -\frac{1}{3}$. Since f is a polynomial, it has no horizontal or vertical asymptotes. Since f is an odd-degree polynomial with positive leading term, it goes to $-\infty$ as $x \to -\infty$ and to ∞ as $x \to \infty$. Graphs of f and f' are

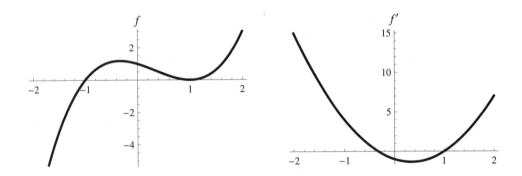

61. $f(x) = x(x^2 - 6x + 12)$ has a zero at $x = 0$. Since the discriminant of the quadratic factor is negative, there are no other roots. Checking points we get $f(-1) = -19 < 0$ and $f(1) = 7 > 0$, giving the sign chart

$f'(x) = 3x^2 - 12x + 12 = 3(x - 2)^2$, so that f' has a root only at $x = 2$. Clearly $f' > 0$ elsewhere, so that its sign chart is

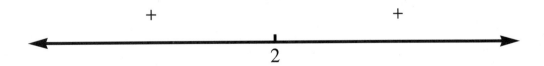

Since f' does not change sign at $x = 2$, we conclude that f has neither a minimum nor a maximum there. Since f is an odd-degree polynomial with positive leading term, it goes to $-\infty$ as $x \to -\infty$ and to ∞ as $x \to \infty$. Graphs of f and f' are

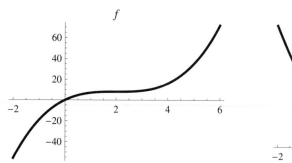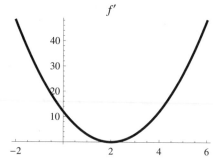

63. $f(x)$ has its only zero at $x = -\frac{11}{2}$ since $x^2 + 10 > 0$. Checking points we get $f(-6) = -46 < 0$ and $f(0) = 110 > 0$. Thus the sign chart for f is

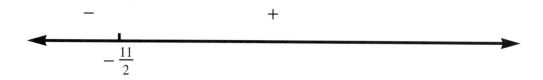

$f'(x) = 6x^2 + 22x + 20 = 2(x+2)(3x+5)$, so that f' has zeros at $x = -2$ and $x = -\frac{5}{3}$. Checking points we get $f'(-3) = 8 > 0$, $f'\left(-\frac{11}{6}\right) = -\frac{1}{6} < 0$, and $f'(0) = 20 > 0$, so its sign chart is

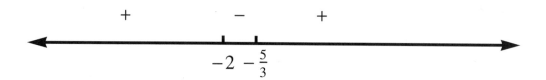

We see that f has a local minimum at $x = -\frac{5}{3}$ and a local maximum at $x = -2$. Since f is an odd-degree polynomial with positive leading term, it goes to $-\infty$ as $x \to -\infty$ and to ∞ as $x \to \infty$. Graphs of f and f' are

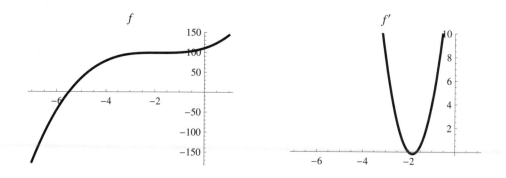

65. $f(x) = (1 - x^4)^7$ has zeros at $x = \pm 1$. Checking points gives $f(-2) = f(2) = (-15)^7 < 0$ and $f(0) = 1 > 0$. Thus the sign chart for f is

$f'(x) = -28x^3(1-x^4)^6$, which has zeros at $x = 0$ and at $x = \pm 1$. Checking signs gives $f'(-2) \approx 2 \times 10^6 > 0$ and $f'\left(-\frac{1}{2}\right) \approx 2.4 > 0$. Since f' is an odd function, we also have $f'(2) < 0$ and $f'\left(\frac{1}{2}\right) < 0$. Thus its sign chart is

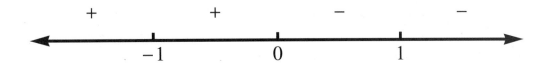

Since f' changes sign only at $x = 0$, that is the only local extremum of f, and it is a local maximum. f does not have local extrema at $x = \pm 1$. Since f is an even-degree polynomial with negative leading term, it goes to $-\infty$ as $x \to \pm\infty$. Graphs of f and f' are

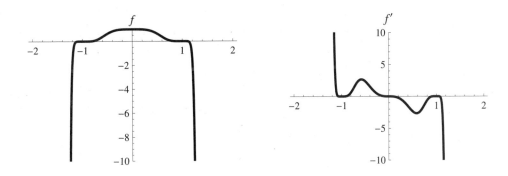

67. $f(x)$ has domain $x \neq 1$ and has as its only root $x = -1$. Checking points gives $f(-2) = \frac{1}{3} > 0$, $f(0) = -1 < 0$, and $f(2) = 3 > 0$. Thus the sign chart for f is

$f'(x) = \frac{(x-1)-(x+1)}{(x-1)^2} = \frac{-2}{(x-1)^2}$. Clearly $f'(x)$ has no zeros, is defined wherever f is, and is everywhere negative. Thus its sign chart looks like

Since f' is always negative, f is decreasing everywhere and has no local extrema. f has a two-sided horizontal asymptote at $y = 1$. Also, it has a vertical asymptote at $x = 1$, with $\lim\limits_{x \to 1^-} f(x) = -\infty$ and $\lim\limits_{x \to 1^+} f(x) = \infty$. Graphs of f and f' are

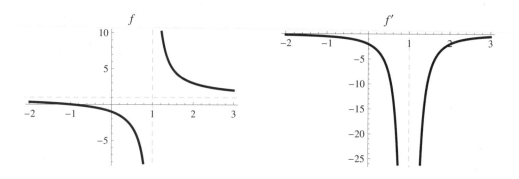

69. $f(x)$ is defined everywhere and has no zeros, since $x^2 + 1 > 0$. It is always positive. Thus the sign chart for f is

$f'(x) = \frac{1}{2}(x^2 + 1)^{-1/2}(2x) = \frac{x}{\sqrt{x^2+1}}$. Clearly f' is defined everywhere, and its only zero is at $x = 0$. Also obviously $f'(x) < 0$ for $x < 0$ and $f'(x) > 0$ for $x > 0$. Thus its sign chart looks like

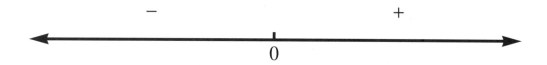

Since f' changes from negative to positive at $x = 0$, it follows that f has a local minimum there. It has no other local extrema. $\lim\limits_{x \to \pm\infty} f(x) = \infty$, and since f is defined everywhere, it has no vertical asymptotes. Graphs of f and f' are

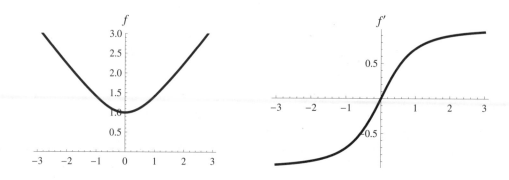

71. $f(x) = \frac{(x-1)^2}{(x+3)(x-2)}$, so that the domain of f is $\mathbb{R} - \{2, -3\}$, and $f(x) = 0$ for $x = 1$. Checking points gives $f(-4) = \frac{25}{6} > 0$, $f(0) = -\frac{1}{6} < 0$, $f\left(\frac{3}{2}\right) = -\frac{1}{9} < 0$, and $f(3) = \frac{2}{3} > 0$. Thus the sign chart for f is

$$f'(x) = \frac{(x^2 + x - 6)(2(x-1)) - (x-1)^2(2x+1)}{(x^2 + x - 6)^2} = \frac{3x^2 - 14x + 11}{(x^2 + x - 6)^2} = \frac{(3x - 11)(x - 1)}{(x^2 + x - 6)^2}.$$

Clearly f' is defined on the same domain as is f, and has zeros at $x = 1$ and $x = \frac{11}{3}$. The denominator of f' is positive where f' is defined, so that the sign of f' depends on the sign of its numerator. Checking points, we get $f'(0) > 0$, $f'(2) < 0$, and $f'(4) > 0$, so that the sign chart for f' looks like

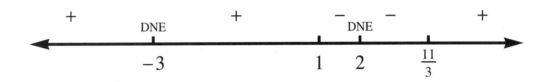

f has a two-sided horizontal asymptote at $y = 1$. Further, it has vertical asymptotes at $x = -3$ and at $x = 2$. At $x = -3$ the left-hand limit is ∞ while the right-hand limit is $-\infty$; at $x = 2$ the left-hand limit is $-\infty$ while the right-hand limit is ∞. Graphs of f and f' are

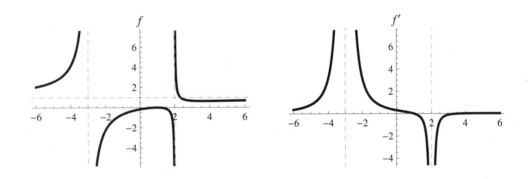

73. $f(x)$ has zeros at $x = 0$ and $x = 1$, and is undefined at $x = 2$. Checking points gives $f(-1) = -\frac{2}{9} < 0$, $f\left(\frac{1}{2}\right) = -\frac{1}{18} < 0$, $f\left(\frac{3}{2}\right) = \frac{9}{2} > 0$, and $f(3) = 18 > 0$. Thus the sign chart for f is

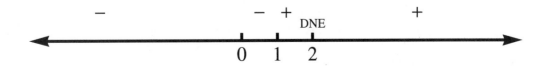

Now,

$$f'(x) = \frac{(x-2)^2(3x^2 - 2x) - (x^3 - x^2)(2(x-2))}{(x-2)^4} = \frac{x(x^2 - 6x + 4)}{(x-2)^3}.$$

Then f' has zeros at $x = 0$ and at $x = 3 \pm \sqrt{5} \approx 0.76, 5.24$ and is undefined at $x = 2$. Checking points gives $f'(-1) = \frac{11}{27} > 0$, $f'\left(\frac{1}{2}\right) = -\frac{5}{27} < 0$, $f'(1) = 1 > 0$, $f'(3) = -15 < 0$, and $f'(6) = \frac{3}{8} > 0$, so that the sign chart for f' is

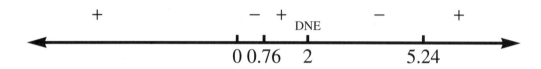

From the sign chart for f', we see that f has a local maximum at $x = 0$ and local minima at $3 \pm \sqrt{5}$. It has no horizontal asymptotes since the degree of the numerator exceeds that of the denominator. Finally, it has a vertical asymptote at $x = 2$, where both one-sided limits are ∞. Graphs of f and f' are

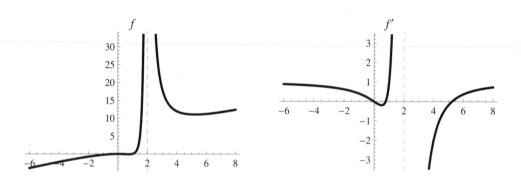

75. $f(x)$ is defined for $x > 0$ and has its only zero at $x = 1$. Checking signs we see that $f(x) < 0$ for $x < 1$ and $f(x) > 0$ for $x > 1$. Thus the sign chart for f is

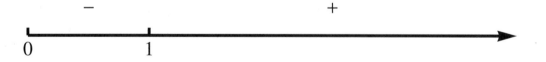

$f'(x) = x \cdot \frac{1}{x} + \ln x = 1 + \ln x$, so that the domain of f' is the same as the domain of f, and f' has its only zero at $e^{-1} \approx 0.37$ (since $\ln e^{-1} = -1$). Checking points, we find $f'(0.25) \approx -0.39 < 0$ and $f'(e) = 2 > 0$, so that the sign chart for f' is

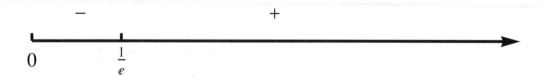

From the sign chart for f', we see that f has a local minimum at $x = \frac{1}{e}$ and no other local extrema. It has no horizontal asymptotes and no vertical asymptotes, since $\lim\limits_{x \to 0^-} x \ln x = \lim\limits_{x \to 0^+} \frac{x}{1/\ln x} = 0$.

Graphs of f and f' are

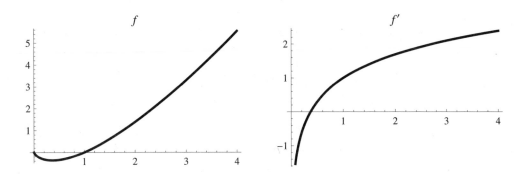

77. $f(x)$ is defined for all x and is zero only for $x = 0$. Clearly $f(x) \geq 0$ for all x. Thus the sign chart for f is

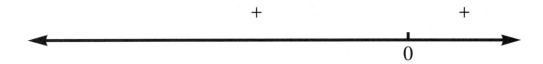

Now, $f'(x) = 2x3^x + x^2(\ln 3)3^x = x3^x(2 + x \ln 3)$. Thus $f'(x) = 0$ at $x = 0$ and at $x = -\frac{2}{\ln 3} \approx -1.82$. Checking points, we have $f'(-2) \approx 0.044 > 0$, $f'(-1) \approx -0.30 < 0$, and $f'(1) = \approx 9.30 > 0$. The sign chart for f' is

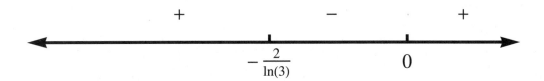

From the sign chart for f', we see that f has a local maximum at $x = -\frac{2}{\ln 3}$ and a local minimum at $x = 0$. Since $\lim\limits_{x \to -\infty} x^2 3^x = 0$, we have a horizontal asymptote at $y = 0$. There are no vertical asymptotes. Graphs of f and f' are

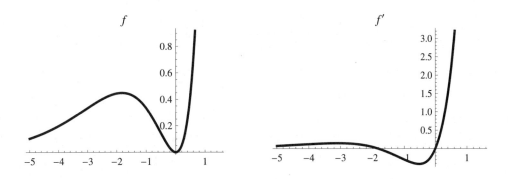

79. $f(x)$ is defined for all $x > 0$ and is zero only for $x = 1$. Clearly $f(x) > 0$ for $x > 1$ and $f(x) < 0$ for $x < 1$ since the sign of f depends on the sign of $\ln x$. Thus the sign chart for f is

Now, $f'(x) = \frac{x \cdot \frac{1}{x} - \ln x}{x^2} = \frac{1 - \ln x}{x^2}$. Thus f' is defined where f is, and $f'(x) = 0$ at $x = e$. Checking points, we have $f'(1) = 1 > 0$ and $f'(3) = \frac{1 - \ln 3}{9} < 0$. Thus the sign chart for f' is

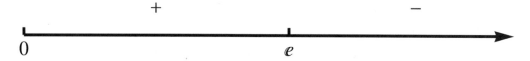

From the sign chart for f', we see that f is always increasing and has a local maximum at $x = e$. Since $\lim\limits_{x \to \infty} f(x) = 0$, f has a horizontal asymptote at $y = 0$ as $x \to \infty$. Also, f has a vertical asymptote at $x = 0$, and the one-sided limit is $-\infty$. Graphs of f and f' are

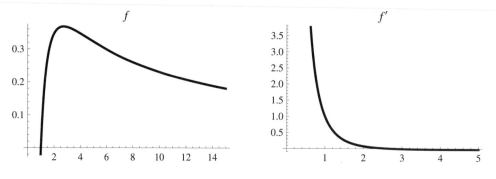

81. $f(x)$ is defined and positive everywhere; it is never zero. This is because $e^z > 0$ everywhere and the exponent is defined everywhere. Thus the sign chart for f is

Now, $f'(x) = (3x^2 - 6x + 2)e^{x^3 - 3x^2 + 2x}$. f' is defined everywhere as well, and is zero only when $3x^2 - 6x + 2 = 0$, i.e., at $x = \frac{3 \pm \sqrt{3}}{3} \approx 0.42, 1.58$. Checking points, $f'(0) = 2e^0 > 0$, $f'(1) = -e^0 = -1 < 0$, and $f'(3) = 11e^6 > 0$. Thus the sign chart for f' is

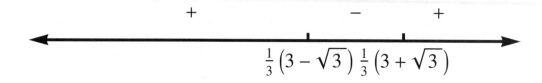

From the sign chart for f', we see that f has a local maximum at $x = \frac{3-\sqrt{3}}{3}$ and a local minimum at $x = \frac{3+\sqrt{3}}{3}$. Since the exponent goes to ∞ as $x \to \infty$ and to $-\infty$ as $x \to -\infty$, we see that f has a one-sided horizontal asymptote at $y = 0$ as $x \to -\infty$. There are no vertical asymptotes. Graphs of f and f' are

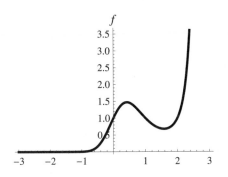
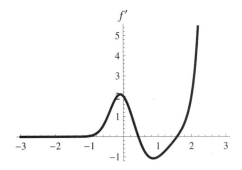

Applications

83. (a) Bubbles is moving towards the right-hand side when its velocity is positive, which is on $(0, 2.5)$.

   (b) Bubbles is furthest away from the left-hand side of the tunnel when it stops moving towards the right. This happens when its velocity is no longer positive, so at $t = 2.5$. No, Bubbles never returns to the left-hand side of the tunnel. The rat is moving towards the right-hand side for a longer time, and at higher velocities, than it is moving back (from $t = 2.5$ to $t = 4$), so it never gets all the way back.

   (c) The acceleration is positive when the first derivative of the velocity curve is positive, i.e., when its slope is positive. This is on $(0, 1) \cup (3.3, 4)$.

   (d) Since velocity is negative on $(2.5, 4)$, velocity is negative but acceleration positive on $(2.5, 4) \cap ((0, 1) \cup (3.3, 4)) = (3.3, 4)$. During this time, Bubbles is moving back towards the left-hand side of the tunnel at an decreasing rate (velocity is getting closer to zero).

85. (a) $c(t) = 0$ when $\cos(0.51t + 2.04) = 0$, so when $0.51t + 2.04 = (2n + 1)\frac{\pi}{2}$ for n an integer. Solving for t gives

$$t = \frac{(2n + 1)\frac{\pi}{2} - 2.04}{0.51}$$

   The first such time, for $n = 0$, is $t = \frac{\frac{\pi}{2} - 2.04}{0.51} \approx -0.92$ hours. The first time where $t > 0$ corresponds to $n = 1$, and is ≈ 5.24 hours. Each such time is a time when the tide is going neither in nor out, so it corresponds to either a low tide or a high tide.

   (b) High tides occur for those solutions to $c(t) = 0$ where $c(t)$ changes from positive (coming in) to negative (going out). Since cos changes from positive to negative at $(4n + 1)\frac{\pi}{2}$ and from negative to positive at $(4n + 3)\frac{\pi}{2}$, we see that the values of t corresponding to $n = 1, 5, 9...$ are high tides, while those corresponding to $n = 3, 7, 11...$ are low tides.

   (c) She would like to head towards the point just before low tide, as the current is moving south; these would be at times t just before solutions of $c(t) = 0$ corresponding to $(4n + 3)\frac{\pi}{2}$. Then she will reach the point at low tide, and will turn north around the point as the tide starts back in.

Proofs

87. Let $f(x) = mx + b$ be a linear function with $m \neq 0$ (so that it is nonconstant). Then $f'(x) = m$, so that either $f'(x) > 0$ for all x or $f'(x) < 0$ for all x depending on the sign of m. Then by Theorem 3.6, since f is differentiable on all of $\mathbb{R}$, we see that f is either increasing everywhere on $\mathbb{R}$ or decreasing everywhere on $\mathbb{R}$.

89. Suppose f is differentiable on an interval I and that f' is negative on the interior of I. Choose $a, b \in I$ with $a < b$. By the definition of decreasing, we must show that $f(a) > f(b)$. Since f is differentiable and thus continuous on I, and since $[a, b] \subset I$, we see that f satisfies the conditions of the Mean Value Theorem on $[a, b]$. Thus there exists some point $c \in (a, b)$ such that

$$f'(c) = \frac{f(b) - f(a)}{b - a}.$$

Rearranging gives $f(b) - f(a) = f'(c)(b - a)$. Now, $b - a > 0$ since $a < b$, and $f'(c) < 0$ since f' is negative on (a, b) and $c \in (a, b)$. Thus $f(b) - f(a) < 0$ as well, so that $f(a) > f(b)$.

91. Let f, a, b, c, and I be as in the statement of the theorem. Suppose $f'(x) < 0$ for $x \in (a, c)$ and $f'(x) > 0$ for $x \in (c, b)$. By Theorem 3.6, this means that f is decreasing on $(a, c]$ and increasing on $[c, b)$. We will show that $f(c) \leq f(x)$ for all $c \in (a, b)$, which will prove that f has a local minimum at $x = c$. First, if $x = c$, then clearly $f(c) = f(x) \leq f(x)$. Second, if $a < x < c$, then since f is decreasing on $(a, c]$, we have $f(x) \geq f(c)$, i.e., $f(c) \leq f(x)$. Third, if $c < x < b$, then since f is increasing on $[c, b)$, we have $f(c) \leq f(x)$. Since in each case $f(c) \leq f(x)$, we conclude that f has a local minimum at $x = c$.

Thinking Forward

Second-derivative graphs

Graph I is the graph of f, Graph II of f', and Graph III of f''. Note that the slope of the curve in graph I is negative for $x < 0$, zero at $x = 0$, and positive for $x > 0$. This matches the values of the curve in graph II. Similarly, the slope of graph II is at a maximum at $x = 0$, is positive between -1 and 1, and is negative otherwise; this matches the values of the curve in graph III.

More second-derivative graphs

Again Graph I is the graph of f, Graph II of f', and Graph III of f''. Note that the slope of the curve in graph I is negative for $x < 1$ and positive for $x > 1$; this matches the curve in graph II. Similarly, the slope of the curve in graph II is negative for $x < 0$ and positive for $x > 0$, which matches the curve in graph III.

3.3   The Second Derivative and Curve Sketching

Thinking Back

Finding the second derivative

▷ We have

$$f'(x) = \frac{(3x^2 - 4x + 2)(1) - (x-1)(6x-4)}{(3x^2 - 4x + 2)^2} = \frac{-3x^2 + 6x - 2}{(3x^2 - 4x + 2)^2}$$

$$f''(x) = \frac{(3x^2 - 4x + 2)(-6x + 6) - (-3x^2 + 6x - 2)(2(3x^2 - 4x + 2)(6x-4)}{(3x^2 - 4x + 2)^4}$$

$$= \frac{18x^3 - 54x^2 + 36x - 4}{(3x^2 - 4x + 2)^3}.$$

▷ We have

$$f'(x) = \frac{(x^2 + 1)^{1/2}(1) - x\left(\frac{1}{2}(x^2 + 1)^{-1/2}(2x)\right)}{x^2 + 1} = \frac{(x^2 + 1)^{1/2} - x^2(x^2 + 1)^{-1/2}}{x^2 + 1}$$

$$= \frac{1}{(x^2 + 1)^{3/2}}$$

$$f''(x) = \left((x^2 + 1)^{-3/2}\right)' = -\frac{3}{2}(x^2 + 1)^{-5/2}(2x) = -\frac{3x}{(x^2 + 1)^{5/2}}.$$

▷ We have

$$f'(x) = \frac{e^x(-1) - (1-x)e^x}{e^{2x}} = \frac{x-2}{e^x}$$

$$f''(x) = \frac{e^x(1) - (x-2)e^x}{e^{2x}} = \frac{3-x}{e^x}.$$

▷ We have

$$f'(x) = 3e^{3x}\ln(x^2 + 1) + e^{3x}\frac{1}{x^2 + 1}\cdot 2x = e^{3x}\left(3\ln(x^2 + 1) + \frac{2x}{x^2 + 1}\right)$$

$$f''(x) = 3e^{3x}\left(3\ln(x^2 + 1) + \frac{2x}{x^2 + 1}\right) + e^{3x}\left(3\frac{1}{x^2 + 1}\cdot 2x + \frac{(x^2 + 1)(2) - 2x(2x)}{(x^2 + 1)^2}\right)$$

$$= e^{3x}\left(9\ln(x^2 + 1) + \frac{12x}{x^2 + 1} + \frac{2 - 2x^2}{(x^2 + 1)^2}\right).$$

Solving for zeros and non-domain points

▷ Factoring gives $g(x) = \dfrac{(3x + 2)(x - 1)}{(x^2 + 3)(x + 1)(x - 1)}$. Then $g(x)$ has a zero at $x = -\frac{2}{3}$. It has a hole at $x = 1$ since $x - 1$ divides both numerator and denominator to degree 1. Finally, it is undefined at $x = -1$. The factor of $x^2 + 3$ in the denominator has no roots.

▷ Simplifying,

$$g(x) = \frac{1}{x - 2} - \frac{3x + 1}{x + 2} = \frac{x + 2 - (3x + 1)(x - 2)}{(x + 2)(x - 2)} = \frac{-3x^2 + 6x + 4}{(x + 2)(x - 2)}.$$

The roots of the numerator, which are the roots of g, are $\dfrac{3 \pm \sqrt{21}}{3}$. Also, g is undefined at $x = \pm 2$ since those are roots of the denominator.

> $g(x)$ has roots where $\sin x = 0$, i.e., at $x = n\pi$ for n any integer. It is undefined where $\cos x = 0$, i.e., at $x = (2n+1)\frac{\pi}{2}$ for n any integer.

> $g(x)$ is undefined where $\ln x$ is zero or undefined, since the numerator is defined everywhere. Thus $g(x)$ is undefined for $x \leq 0$ and for $x = 1$. Since the numerator vanishes only at $x = 1$, we see that $g(x)$ has no roots.

Concepts

1. (a) False. $x = 2$ is an inflection point only if the sign of f'' changes at 2. For example, $f(x) = x$ has no inflection points, yet $f''(x) = 0$ so that in particular $f''(2) = 0$.

   (b) True. Since f' has a maximum at $x = -1$, its slope (which is f'') changes from positive to negative at $x = -1$, so that f has an inflection point there.

   (c) False. By Definition 3.9, if f is concave up on I, then f' is increasing on I, so that f'' is nonnegative on I. However, it could be zero.

   (d) False. $x = 2$ might not be in the domain of f'; for example, if $f(x) = |x - 2|$.

   (e) True. Since f'' is zero at $x = 3$, we see that f' has a local extremum there.

   (f) False. It has a local *maximum* there. See Theorem 3.11 (b).

   (g) False. It involves checking the sign of the second derivative *at* each critical point. See Theorem 3.11

   (h) False. For example, the first-derivative test will tell you the location of the local extrema of the function, which the second-derivative test will not. Also, even if the second derivative is zero, that does not imply a local extremum; that information can be gleaned from the first derivative.

3. Consider the graph

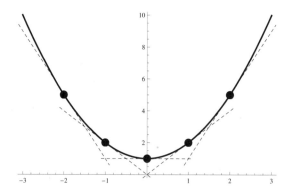

   Since the slopes of the tangent lines start out negative and increase monotonically until at the end they are very positive, we see that f' is increasing everywhere.

5. The converse is: Suppose f and f' are both differentiable on an interval I. If f is concave up on I, then f'' is positive on I. The converse is false. Suppose f is concave up on I. Then by Definition 3.9, f' is increasing on I. Since f'' is the derivative of f', it follows that f'' is nonnegative on I. However, it could be zero. Consider the curve $f(x) = x^4$ at $x = 0$.

7. Consider the graph (here $c = 0$)

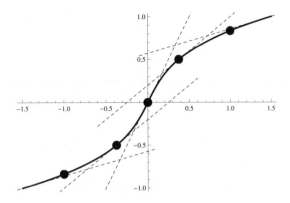

Clearly the concavity of the graph changes at $x = 0$. The slopes of the tangent lines increase up until $x = 0$, and then start to decrease. Thus f' as a local maximum at $x = 0$.

9. For $f(x) = x^6$, we have $f'(x) = 6x^5$ and $f''(x) = 30x^4$. Then clearly $f''(0) = 0$. However, since $f''(x) \geq 0$ everywhere, clearly it does not change sign at $x = 0$, so that 0 is not an inflection point of f.

11. If f'' is zero on I, then by Theorem 3.6(c), f' is constant on I, say $f'(x) = m$. Then $f(x) = mx + C$. Thus if f'' is zero on I, then f is linear on I.

13. If f has four critical points, then there are five intervals determined by those critical points. We need test the value of f' at one point in each of those five intervals, so that five calculations are required. The second derivative test, by contrast, requires calculations *at* the critical points, so that four calculations are required.

15. For example,

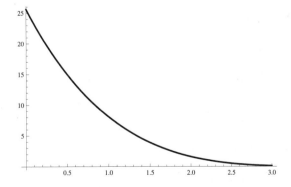

f' is negative since the slope is negative. However, the slope becomes less negative as $x \to 3$, so that f'' is positive.

17. For example,

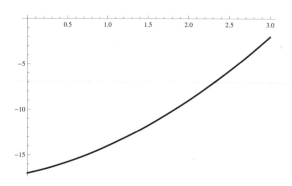

f' is positive since the slope is positive. Since the slope becomes increasingly positive, $f'' > 0$ as well.

19. For example,

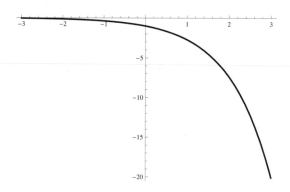

f' is negative since the slope is negative. Since the slope becomes increasingly negative as $x \to \infty$, we see that $f'' > 0$ as well.

Skills

21. Graphs of f, f', and f'' are:

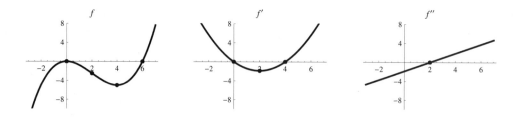

On the graph of f'', the marked point at $x = 2$ is a root. Since f'' changes sign from negative to positive there, the marked point at $x = 2$ on f' is a local minimum. The marked points on f' at $x = 0$ and $x = 4$ are roots. Since f' changes sign at each of those, they are also local extrema of f — $x = 0$ is a local maximum while $x = 4$ is a local minimum. Since f'' changes sign at $x = 2$, that point is an inflection point of f. Finally, $x = 0$ and $x = 6$ are roots of f.

23. Graphs of f, f', and f'' are:

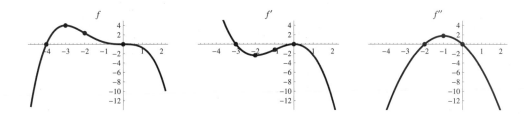

f'' has roots at $x = -2$ and $x = 0$; since it changes sign there, those are local extrema of f' and inflection points of f. f'' also has a local maximum at $x = -1$; this means that f' has an inflection point at $x = -1$. The point $x = -3$ on f' is a root; since f' changes sign from positive to negative there, $x = -3$ is a local maximum of f. Since f' does not change sign at $x = 0$, that point is not a local extremum of f. Finally, the points $x = -4$ and $x = 0$ on f are roots.

25. Graphs of f, f', and f'' are:

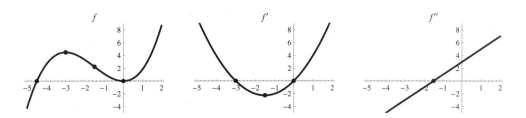

f'' has one root, at -1.5; it changes sign from negative to positive there, so that f' has a local minimum at $x = -1.5$ and f has an inflection point there. The other two marked points on f' are roots, at $x = -3$ and $x = 0$; those correspond to a local maximum of f at $x = 3$ and a local minimum at $x = 0$. There are also marked roots on f at $x = -4.5$ and $x = 0$. Note that any vertical shift of f will also work.

27. Graphs of f, f', and f'' are:

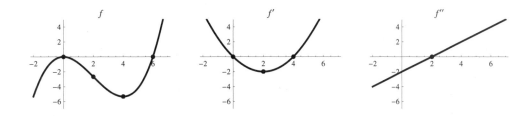

f'' has one root, at $x = 2$; it changes sign from negative to positive there, so that f' has a local minimum at $x = 2$ and f has an inflection point there. The other two marked points on f' are roots, at $x = 0$ and $x = 4$; those correspond to a local maximum of f at $x = 0$ and a local minimum at $x = 4$. There are also roots of f, at $x = 0$ and $x = 6$. Note that any vertical shift for f' will also work; this will produce different function for f. Those functions as well may be shifted vertically.

29. From Exercise 39 in Section 3.2, we have $f'(x) = 3x^2 - 6x$ with critical points at $x = 0$ and at $x = 2$. Then $f''(x) = 6x - 6$. Since $f''(0) = -6 < 0$ and $f''(2) = 12 - 6 = 6 > 0$, we conclude from the second-derivative test that f has a local maximum at $x = 0$ and a local minimum at $x = 2$. For a graph, see Exercise 39 in Section 3.2.

31. From Exercise 41 in Section 3.2, we have $f'(x) = -\frac{3(2x+1)}{(x^2+x-2)^2}$, with critical points at $x = -\frac{1}{2}$, where $f' = 0$. Also, both f and f' are undefined at $x = -2$ and $x = 1$. Then $f''(x) = \frac{18(x^2+x+1)}{(x^2+x-2)^3}$, and $f''\left(-\frac{1}{2}\right) = -\frac{32}{27} < 0$, so that f has a local maximum at $x = -\frac{1}{2}$. For a graph, see Exercise 41 in Section 3.2.

33. From Exercise 43 in Section 3.2, we have $f'(x) = -\frac{2e^x}{(3-2e^x)^2}$. Since $f'(x)$ is never zero, and is undefined only where $f(x)$ is undefined, the function has no critical points. Thus the second-derivative test does not give us any local extrema. For a graph, see Exercise 43 in Section 3.2.

35. From Exercise 45 in Section 3.2, we have $f'(x) = -\pi\sin(\pi(x+1))$, with critical points where x is any integer; then $f'(x) = 0$. On the interval $(-2, 2)$, then, we get $f'(x) = 0$ for $x = -1, 0, 1$. Now, $f''(x) = -\pi^2\cos(\pi(x+1))$. Since $f''(-1) = -\pi^2\cos 0 < 0$, $f''(0) = -\pi^2\cos \pi > 0$, and $f''(1) = -\pi^2\cos(2\pi) < 0$, the second-derivative test tells us that there are local maxima at $x = \pm1$ and a local minimum at $x = 0$. For a graph, see Exercise 45 in Section 3.2.

37. From Exercise 47 in Section 3.2, we have $f'(x) = \frac{1}{1+x^2}$, so that there are no critical points. The second-derivative test therefore gives us no local extrema. For a graph, see Exercise 47 in Section 3.2.

39. From Exercise 49 in Section 3.2, we have $f'(x) = -\frac{x}{\sqrt{1-x^2}}$, with the only critical point at $x = 0$. Then $f''(x) = -\frac{1}{(1-x^2)^{3/2}}$. Since $f''(0) = -1 < 0$, the second-derivative test tells us that f has a local maximum at $x = 0$. For a graph, see Exercise 49 in Section 3.2.

41. Since $f''(x) = 12(x-2)^2$, a sign chart for f'' is

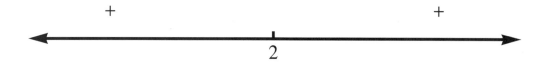

This shows that f is concave up everywhere, with no inflection points. A graph of f demonstrating this is

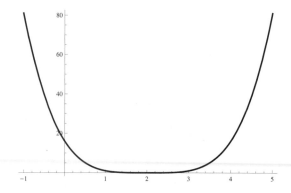

43. Differentiating gives $f''(x) = 12(x^2 - x) = 12x(x - 1)$, with zeros at $x = 0$ and $x = 1$. Testing points in each of the resulting three intervals gives the following sign chart for f'':

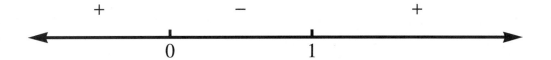

This shows that f is concave up on $(-\infty, 0) \cup (1, \infty)$ and concave down on $(0, 1)$, with inflection points at $x = 0$ and at $x = 1$. A graph of f demonstrating this is

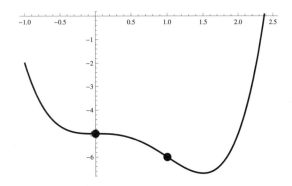

45. Differentiating gives $f''(x) = \frac{6x(x+1)}{(x^2+x+1)^3}$, with zeros at $x = 0$ and $x = -1$. Testing points in each of the resulting three intervals gives the following sign chart for f'':

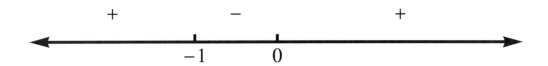

This shows that f is concave up on $(-\infty, -1) \cup (0, \infty)$ and concave down on $(0, 1)$, with inflection points at $x = 0$ and $x = 1$. A graph of f demonstrating this is

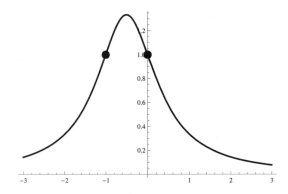

47. Differentiating gives $f''(x) = e^{3x}(9 - 16e^x)$, so that the point where f'' can change sign is where $e^x = \frac{9}{16}$, which is when $x = \ln \frac{9}{16} \approx -0.575$. Testing points in each of the resulting two intervals gives the following sign chart for f'':

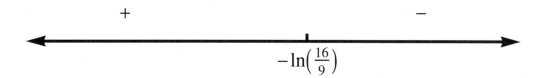

This shows that f is concave up on $\left(-\infty, \ln\frac{9}{16}\right)$ and concave down on $\left(\ln\frac{9}{16}, \infty\right)$, with an inflection point at $x = \ln\frac{9}{16}$. A graph of f demonstrating this is

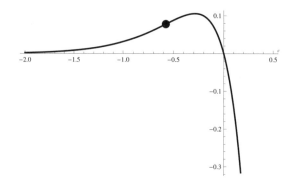

49. Differentiating gives $f'(x) = (2 - 2x)e^{1+2x-x^2}$ and then $f''(x) = 2(2x^2 - 4x + 1)e^{1+2x-x^2}$, so that f'' can change sign at the roots of the quadratic factor, which are $x = \frac{1}{2}(2 \pm \sqrt{2})$. Testing points in each of the resulting three intervals gives the following sign chart for f'':

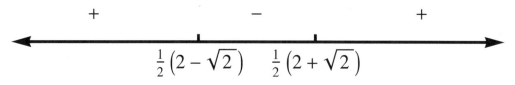

This shows that f is concave down on $\left(\frac{1}{2}(2 - \sqrt{2}), \frac{1}{2}(2 + \sqrt{2})\right)$, and that it is concave up on $\left(-\infty, \frac{1}{2}(2 - \sqrt{2})\right) \cup \left(\frac{1}{2}(2 + \sqrt{2}), \infty\right)$, with inflection points at $\frac{1}{2}(2 \pm \sqrt{2})$. A graph of f demonstrating this is

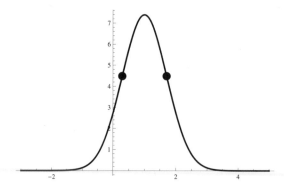

51. Differentiating gives $f''(x) = -\sin\left(x - \frac{\pi}{4}\right)$, so that $f''(x)$ is defined everywhere, and is zero for $x = \frac{\pi}{4} + n\pi$ for n any integer. Considering $(-2\pi, 2\pi)$, $f''(x) = 0$ at $-\frac{7\pi}{4}$, $-\frac{3\pi}{4}$, $\frac{\pi}{4}$, and $\frac{5\pi}{4}$. Testing points in each of the resulting five intervals gives the following sign chart for f'':

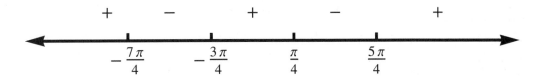

This shows that f is concave down on $\left(-\frac{7\pi}{4}, -\frac{3\pi}{4}\right) \cup \left(\frac{\pi}{4}, \frac{5\pi}{4}\right)$ and concave up on $\left(-2\pi, -\frac{7\pi}{4}\right) \cup \left(-\frac{3\pi}{4}, \frac{\pi}{4}\right) \cup \left(\frac{5\pi}{4}, 2\pi\right)$. A graph of f demonstrating this is

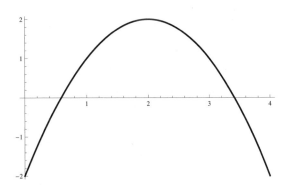

53. From the sign chart for f', we see that f is increasing for $x < 2$, has a local maximum at $x = 2$, and is decreasing for $x > 2$. Since f'' is always negative, f is concave down everywhere.

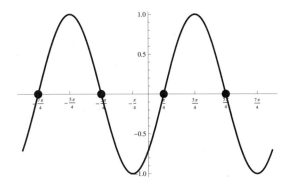

55. From the sign chart for f', we see that f is increasing everywhere but is flat at $x = -2$. From the sign chart for f'', f is concave down for $x < -2$, has an inflection point at $x = -2$, and is concave up after that. A possible graph of f is

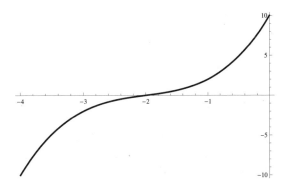

57. From the sign chart for f'', we see that f is concave down for $x < 1$ and concave up for $x > 1$. From the sign chart for f', we see that it has a local maximum at $x = 0$ and a local minimum at $x = 2$. A possible graph of f is

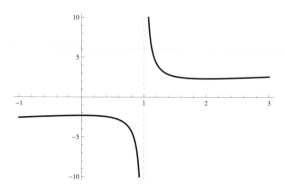

59. From the sign chart for f', we see that f decreases to a local minimum at $x = -2$, then increases after that, with a flat spot at $x = 0$. The sign chart for f'' tells us that f is concave up to an inflection point at $x = -1$, then concave down to another inflection point at $x = 0$, then concave up again. Finally, the sign chart for f tells us that f has zeros at $x = -3$ and $x = -1$, and is negative only on $(-3, -1)$. A possible graph of f is

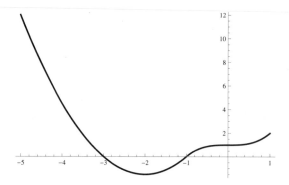

61. The graph is decreasing on $(-2, 3)$ and increasing elsewhere, with inflection points at -1, 1, and 2. It starts out concave down, and changes concavity at each inflection point. Adjusting for the required signs of f itself, a possible graph for f is

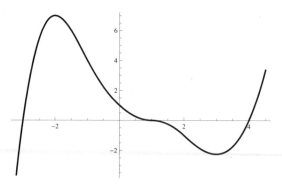

63. Since $f(x) = x(x + 3)$, it has roots at $x = 0$ and $x = -3$. Testing points in the resulting three intervals gives the following sign chart for f:

Differentiating gives $f'(x) = 2x + 3$, which has the root $x = -\frac{3}{2}$. Testing points in the resulting two intervals gives the following sign chart for f':

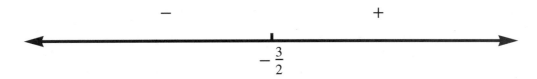

Thus f has a local minimum at $x = -\frac{3}{2}$. Differentiating again gives $f''(x) = 2$, which has the obvious sign chart

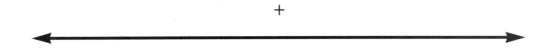

Since f is a nonconstant polynomial, it is defined everywhere and has no asymptotes. A graph of f is

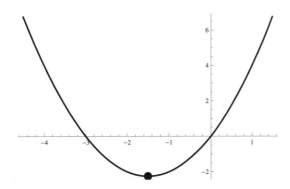

65. $f(x) = x^2(x + 3)$ has roots at $x = 0$ and $x = -3$. Testing points in the resulting three intervals gives the following sign chart for f:

Differentiating gives $f'(x) = 3x^2 + 6x = 3x(x + 2)$, which has the roots $x = 0$ and $x = -2$. Testing points in the resulting two intervals gives the following sign chart for f':

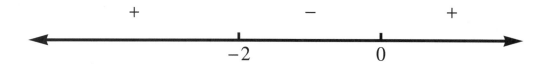

Thus f has a local maximum at $x = -2$ and a local minimum at $x = 0$. Differentiating again gives $f''(x) = 6x + 6$, which has the single root $x = -1$ and the sign chart

Thus f has an inflection point at $x = -1$. Since f is a nonconstant polynomial, it is defined everywhere and has no asymptotes. A graph of f is

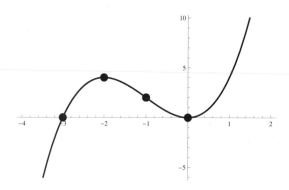

67. $f(x) = x^3 - 2x^2 + x = x(x-1)^2$ has roots at $x = 0$ and at $x = 1$; since it is a polynomial, it is defined everywhere. Checking points in the resulting three intervals gives the sign chart

Differentiating gives $f'(x) = 3x^2 - 4x + 1 = (x-1)(3x-1)$, which has the roots $x = 1$ and $x = \frac{1}{3}$. Testing points in the resulting three intervals gives the following sign chart for f':

Thus f has a local maximum at $x = \frac{1}{3}$ and a local minimum at $x = 1$. Differentiating again gives $f''(x) = 6x - 4$, which has the root $x = \frac{2}{3}$ and the sign chart

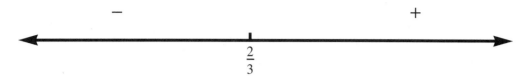

Thus f has an inflection point at $x = \frac{2}{3}$ and is concave down for $x < \frac{2}{3}$ and concave up for $x > \frac{2}{3}$. f is defined everywhere and has no horizontal or vertical asymptotes. A graph of f is

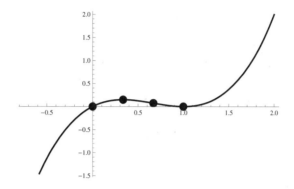

69. $f(x) = x^3(x+2) = x^4 + 2x^3$ has roots at $x = 0$ and $x = -2$; since it is a polynomial, it is defined everywhere. Checking points in the resulting three intervals gives the sign chart

Differentiating gives $f'(x) = 4x^3 + 6x^2 = 2x^2(2x + 3)$, which has the roots $x = 0$ and $x = -\frac{3}{2}$. Testing points gives the following sign chart for f':

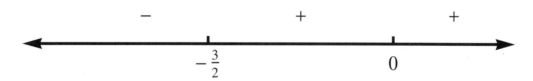

Thus f has a local minimum at $x = -\frac{3}{2}$. Although $f'(0) = 0$, it does not change sign at $x = 0$, so that point is not a local extremum. Differentiating again gives $f''(x) = 12x^2 + 12x = 12x(x + 1)$, which has the roots $x = -1$ and $x = 0$ and the sign chart

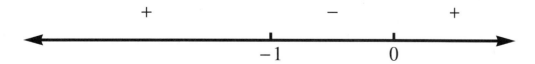

Thus f is concave up on $(-\infty, -1)$ and on $(0, \infty)$ and concave down on $(-1, 0)$, with inflection points at $x = 0$ and $x = -1$. Since f is a polynomial, it has no asymptotes. A graph of f is

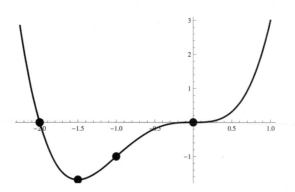

71. $f(x) = \sqrt{x}(4-x) = 4x^{1/2} - x^{3/2}$ has domain $x \geq 0$. It has roots at $x = 0$ and $x = 4$. Checking points in the resulting two intervals $(0, 4)$ and $(4, \infty)$ gives the sign chart

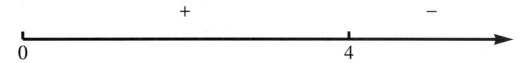

Differentiating gives $f'(x) = 2x^{-1/2} - \frac{3}{2}x^{1/2} = \frac{4-3x}{2\sqrt{x}}$, which has the root $x = \frac{4}{3}$. Testing points gives the following sign chart for f':

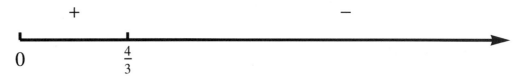

Thus f has a local maximum at $x = \frac{4}{3}$. Differentiating again gives $f''(x) = -x^{-3/2} - \frac{3}{4}x^{-1/2} = -\frac{3x+4}{4x^{3/2}}$, which has no roots for $x \geq 0$. Since $f''(1) < 0$, we see that f'' is everywhere negative and has the sign chart

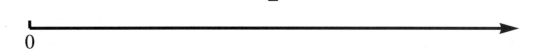

Thus f is concave down everywhere. f has no horizontal or vertical asymptotes. A graph of f is

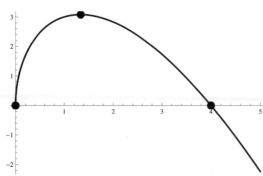

73. $f(x) = \frac{x^2-1}{x^2-5x+4} = \frac{(x-1)(x+1)}{(x-1)(x-4)}$ is defined except at $x = 1$ and $x = 4$. It has a hole at $x = 1$ and a vertical asymptote at $x = 4$. It also has a root at $x = -1$. Checking points in the resulting four intervals gives the sign chart

Going forward, we assume $x \neq 1$, since f is undefined there. Then $f(x)$ simplifies to $\frac{x+1}{x-4}$, and $f'(x) = -\frac{5}{(x-4)^2}$. This has no roots, but is undefined at $x = 4$ (and also at $x = 1$). Testing points in the resulting three intervals gives the following sign chart for f':

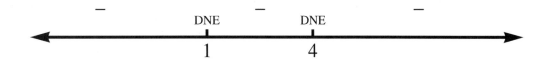

Thus f has no local extrema and is decreasing everywhere. Differentiating again gives $f''(x) = \frac{10}{(x-4)^3}$, which has no roots, but is undefined at $x = 4$ (and $x = 1$). Testing points in the three resulting intervals gives the following sign chart for f'':

Thus f is concave down on $(-\infty, 1) \cup (1, 4)$ and concave up on $(4, \infty)$, with no inflection points. Checking the vertical asymptote, we have $\lim\limits_{x \to 4^-} f(x) = -\infty$ and $\lim\limits_{x \to 4^+} f(x) = \infty$. Finally, f has a two-sided horizontal asymptote at $y = 1$. A graph of f is

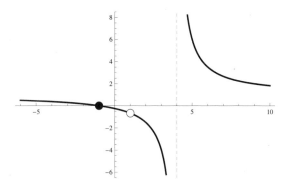

75. $f(x) = \frac{1-x}{e^x} = e^{-x}(1-x)$ is defined everywhere and has a root at $x = 1$. Checking points in the resulting two intervals gives the sign chart

Differentiating gives $f'(x) = e^{-x}(x-2)$, which has a root at $x = 2$. Checking points gives the sign chart

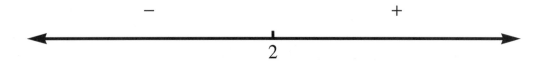

Thus f has a local minimum at $x = 2$. Differentiating again gives $f''(x) = e^{-x}(3-x)$, which has a root at $x = 3$. Checking points gives the sign chart

Thus f is concave up on $(-\infty, 3)$ and concave down on $(3, \infty)$, with an inflection point at $x = 3$. Finally, f has a horizontal asymptote at $y = 0$ as $x \to \infty$. A graph of f is

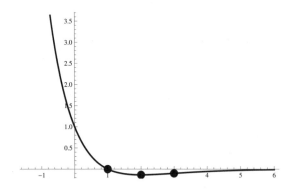

77. $f(x) = e^x \frac{1}{x}$ is defined except at $x = 0$, where it has a vertical asymptote. It has no roots since e^x never vanishes. Checking points on either side of 0 gives the following sign chart for f:

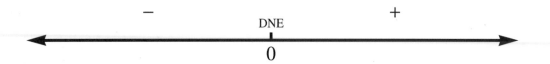

Differentiating gives $f'(x) = e^x \frac{x-1}{x^2}$, which is defined except at $x = 0$ and has a root at $x = 1$. Checking points in the resulting three intervals gives the following sign chart for f':

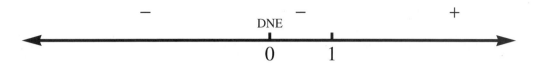

Thus f has a local minimum at $x = 1$. Differentiating again gives $f''(x) = e^x \frac{x^2 - 2x + 2}{x^3}$, which is defined except at $x = 0$ and has no roots since $x^2 - 2x + 2$ has negative discriminant. Checking a point on either side of zero gives the following sign chart for f'':

Thus f is concave up on $(0, \infty)$ and concave down on $(-\infty, 0)$, with no inflection points. Checking the vertical asymptote, we find $\lim\limits_{x \to 0^-} f(x) = -\infty$ and $\lim\limits_{x \to 0^+} f(x) = \infty$. Finally, $\lim\limits_{x \to -\infty} f(x) = 0$, so that f has a one-sided asymptote at $y = 0$. A graph of f is

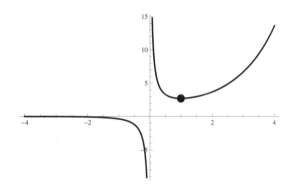

79. $f(x) = x^{2/3} - x^{1/3}$ is defined everywhere, and has roots at $x = 0$ and $x = 1$. Checking points in the resulting three intervals gives the following sign chart for f:

Differentiating gives $f'(x) = \frac{2}{3}x^{-1/3} - \frac{1}{3}x^{-2/3} = \frac{2x^{1/3} - 1}{3x^{2/3}}$, which is defined except at $x = 0$ and has a root where $2x^{1/3} = 1$, i.e., at $x = \frac{1}{8}$. Checking points in the resulting three intervals gives the following sign chart for f':

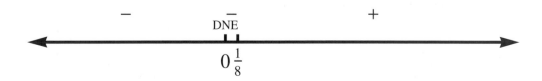

Thus f has a local minimum at $x = \frac{1}{8}$. Differentiating again gives $f''(x) = \frac{2(1-x^{1/3})}{9x^{5/3}}$, which is defined except at $x = 0$ and has a root at $x = 1$. Checking points in the resulting three intervals gives the following sign chart for f'':

Thus f is concave up on $(0, 1)$ and concave down elsewhere, with inflection points at $x = 0$ and $x = 1$. Since f is defined everywhere, it has no vertical asymptotes. It also has no horizontal asymptotes. A graph of f is

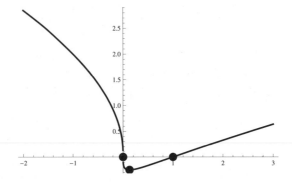

81. $f(x) = (\ln x)^2 + 1$ is defined for $x > 0$, and has no roots since $(\ln x)^2 \geq 0$. Since $f(1) = 1 > 0$, we see that f is positive everywhere, so that we get the following sign chart for f:

Differentiating gives $f'(x) = \frac{2\ln x}{x}$, which is also defined for $x > 0$. It has a root at $x = 1$. Checking points in the resulting two intervals gives the following sign chart for f':

Thus f has a local minimum at $x = 1$. Differentiating again gives $f''(x) = \frac{2 - 2\ln x}{x^2}$, which is also defined for $x > 0$, and has a root at $x = e$ (where $\ln x = 1$). Checking points in the resulting two intervals gives the following sign chart for f'':

Thus f is concave up on $(0, e)$ and concave down on (e, ∞), with an inflection point at $x = e$. Testing the vertical asymptote, we get $\lim\limits_{x \to 0^+} f(x) = \infty$. Finally, f has no horizontal asymptotes. A graph of f is

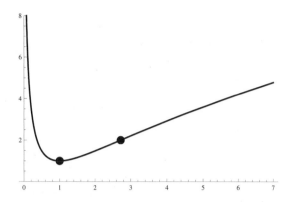

83. Since $f'(x) = x^3 - 3x^2 + 3x = x(x^2 - 3x + 3)$, the only root of f' is at $x = 0$. Testing points, we get for a sign chart for f'

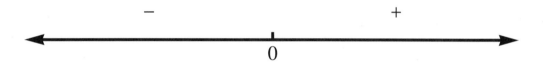

Thus f must have a local minimum at $x = 0$. Differentiating gives $f''(x) = 3x^2 - 6x + 3 = 3(x-1)^2$, so $f''(x) = 0$ at $x = 1$. However, testing points, we get for a sign chart for f''

so that f'' does not change sign at $x = 1$, and f must be concave up everywhere. Summarizing, f has a local minimum at $x = 0$ and is concave up everywhere. At $x = 1$, the slope of the curve is not changing, so the curve must be "flat" there, although with a nonzero slope. A possible graph of f is

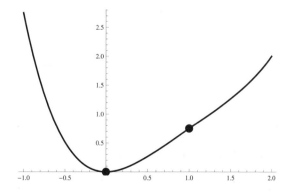

85. Since $f'(x) = \frac{1}{x}$, it is defined for $x \neq 0$ but has no roots. Testing points on either side of 0, we get for a sign chart for f'

Thus f has no local extrema. Differentiating gives $f''(x) = -\frac{1}{x^2}$, so f'' is also defined for $x \neq 0$ and has no roots. Testing points, we get for a sign chart for f''

so that f is concave downwards everywhere. Summarizing, f has no local extrema and is concave downwards everywhere. We may as well assume it is not defined at $x = 0$ since neither f' nor f'' is. A possible graph of f is

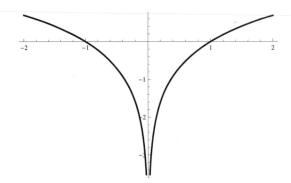

Applications

87. (a) From about $t = 0$ to $t = 15$, Jason's velocity is positive (the slope of $s(t)$ is positive, but decreasing since the slope is decreasing. During this time, Jason is moving north away from Main and High, at a decreasing rate.

   (b) From about $t = 60$ to $t = 80$ minutes, Jason is moving north. Also, his velocity is increasing, since the slope of the curve is increasing. During this time, Jason is south of Main and High and is moving towards it at an increasing rate.

   (c) From about $t = 15$ to $t = 40$, Jason's velocity is negative since the slope of $s(t)$ is negative. Since the slope is getting increasingly negative, his acceleration, which is the change in velocity, is also negative. During this time, Jason is walking south through the intersection at an increasing rate.

   (d) Jason's velocity is at a minimum at $t = 40$. At that point he is walking south. He has been speeding up, but has just started slowing down, so that his velocity is maximally negative.

89. Since $h(x) = 0.00046x^4 - 0.0072x^3 + 0.037x^2 + 0.0095x + 1.2$, we get

$$h'(x) = 0.00184x^3 - 0.0216x^2 + 0.074x + 0.095 \text{ and } h''(x) = 0.0552x^2 - 0.0432x + 0.074.$$

Solving $h''(x) = 0$ gives $x \approx 2.532$ and $x \approx 5.294$. Testing points, we get $h''(0) = 0.074 > 0$, $h''(4) = -0.010 < 0$, and $h''(6) = 0.01352 > 0$, so that both of these points are inflection points. The glacier is concave down where $h''(x) < 0$, which is in the interval $(2.532, 5.294)$.

Proofs

91. Suppose that f and f' are differentiable on I, and that $f''(x) < 0$ for all $x \in I$. Then since the derivative of f' is f'', it follows from Theorem 3.6 that f' is decreasing on I. By the definition of concavity, this means that f is concave down on the interval I.

93. Suppose $f(x) = ax^2 + bx + c$ is a quadratic function, with $a \neq 0$. Then $f'(x) = 2ax + b$ and $f''(x) = 2a$. Note that both f and f' are differentiable on all of $\mathbb{R}$. Then if $a > 0$, we see that $f''(x) > 0$ for all x, so that by Theorem 3.10 (a), f is concave up on $\mathbb{R}$. Alternatively, if $a < 0$, then $f''(x) < 0$ for all x, so that by Theorem 3.10(b), f is concave down on $\mathbb{R}$.

95. Since $f''(x) = 0$, then $f'(x) = c$ is an antiderivative of f'', so that by Theorem 3.7, all antiderivatives are of the form $f'(x) = c$ for some constant c. But then $f(x) = cx$ is an antiderivative of f', so that again by Theorem 3.7, all antiderivatives are of the form $f(x) = cx + d$ for some constant d. Thus $f(x)$ is linear.

Thinking Forward

Global extrema on an interval

Yes, $x = -\frac{3}{2}$ is a global minimum. We have $f\left(-\frac{3}{2}\right) = \frac{9}{4} - 3\frac{3}{2} = -\frac{9}{4}$. Now suppose $x^2 + 3x < -\frac{9}{4}$; then $x^2 + 3x + \frac{9}{4} < 0$, so that (clearing fractions) $4x^2 + 12x + 9 < 0$. But $4x^2 + 12x + 9 = (2x + 3)^2 \geq 0$, which is a contradiction. Thus $-\frac{9}{4}$ is a global minimum. There is no global maximum, since $\lim_{x \to \infty} (x^2 + 3x) = \infty$. Now, $f'(x) = 2x + 3$, so that f is decreasing to the left of $-\frac{3}{2}$ and increasing to the right of $-\frac{3}{2}$. Thus the global maximum of f on $[-3, 3]$ must occur at either -3 or 3. Since $f(-3) = 0$ while $f(3) = 18$, we conclude that the global maximum of $f(x)$ on $[-3, 3]$ is 18 at $x = 3$. The global minimum is $-\frac{9}{4}$ at $x = -\frac{3}{2}$.

Global extrema and derivatives

Neither of these local extrema is a global extremum. Since $\lim_{x \to \infty} (x^3 - x^2 + x) = \lim_{x \to \infty} (x^3(1 - 1/x + 1/x^2)) = \infty$ and $\lim_{x \to -\infty} (x^3 - x^2 + x) = \lim_{x \to -\infty} (x^3(1 - 1/x + 1/x^2)) = -\infty$, we see that $f(x)$ has no global maximum or minimum. The first and second derivatives say nothing about whether a point is a global extremum, only about whether it is a local extremum.

3.4   Optimization

Thinking Back

Local and global extrema

  ▷ There exists $\epsilon > 0$ such that if $x \in (2 - \epsilon, 2 + \epsilon) \cap [-3, 5]$, then $f(x) \leq f(2)$.

  ▷ If $x \in [-3, 5]$, then $f(x) \leq f(2)$.

  ▷ If $x \in [-4, 4]$, then $f(x) \geq f(0)$.

  ▷ For all $N < 0$, there exists $x \in [0, 5]$ such that $f(x) < N$.

Critical points

Recall that the critical points of a function f are those points in the domain of f where $f'(x) = 0$ or where $f'(x)$ does not exist.

▷ $f'(x) = \dfrac{(x-1)(3) - (3x-2)(1)}{(x-1)^2} = \dfrac{-1}{(x-1)^2}$. Clearly $f'(x)$ is never zero, and it is undefined at $x = 1$. However, $x = 1$ is not in the domain of f, so that f has no critical points.

▷ $f'(x) = \dfrac{(\sqrt{x}+1)(0) - \frac{1}{2}x^{-1/2}(1)}{(1+\sqrt{x})^2} = -\dfrac{1}{2\sqrt{x}(1+\sqrt{x})^2}$. Clearly $f'(x)$ is never zero, and it is un-defined for $x \le 0$. Since the domain of f is $x \ge 0$, it follows that the only critical point of f is $x = 0$.

▷ $f'(x) = \dfrac{\pi}{2}\cos\left(\dfrac{\pi}{2}x\right)$. Then $f'(x)$ is defined everywhere, and it is zero for x any odd integer. Thus the critical points of f are $2n+1$ for n any integer.

▷ $f'(x) = e^x + e^x(x-2) = e^x(x-1)$. Then $f'(x)$ is defined everywhere, and has the single root $x = 1$. Thus the only critical point of f is $x = 1$.

Concepts

1. (a) False. For example, the function $f(x) = x^3 + 6x^2$ has a local maximum at $x = -4$ by either the first or second derivative test, but $\lim_{x \to \infty} f(x) = \infty$, so that f has no global maximum.

    (b) True. If $x = c$ is a global minimum, then $f(x) \ge f(c)$ for all x in the domain of f, so certainly $f(x) \ge f(c)$ for all $x \in (c - \delta, c + \delta)$ for any $\delta > 0$. Thus c is a local minimum.

    (c) True. Since f has a global maximum at $x = 2$ on $(-\infty, \infty)$, then for any $x \in (-\infty, \infty)$, we have $f(x) \le f(2)$. Since this is true for $x \in (-\infty, \infty)$, it is also true for $x \in [0, 4]$. Since $2 \in [0, 4]$, it follows that $x = 2$ is a global maximum on $[0, 4]$.

    (d) False. There are many counterexamples. One simple one is $f(x) = 2$ for $x \in [0, 4]$ and $f(x) = 3$ for $x \notin [0, 4]$.

    (e) False. This is true only if I is a *closed* interval. For example, $f(x) = x$ has no global maximum or minimum on $(0, 1)$.

    (f) False. For example, we could have $f(0) = f(10) = 0$, so that the global minimum of f on $[0, 10]$ is at most zero.

    (g) False. For example, let $f(x) = x$. Then f has no local maxima on $(-\infty, \infty)$, but the global maximum of f on $[0, 5]$ is at $x = 5$.

    (h) False. For example, let $f(x) = (x-3)^3$. Then $f'(3) = 0$, but f does not have a local extremum at $x = 3$; it has an inflection point.

3. The endpoints may not be critical points, so that they will not show up when evaluating the critical points of f. For example, look at the function in Exercise 2(c).

5. The answer is no to both parts. For the first part, note that f might not even be defined at the endpoints, since I is an open interval. But even if it is, consider the example $f(x) = x$ on $(-3, 3)$. Since f has no local extrema, certainly $x = \pm 3$ are not local extrema. For the second part, consider the same function, and choose any $c \in (-3, 3)$. Choose $\delta > 0$ such that $c - \delta > -3$ and $c + \delta < 3$, and choose any $d \in (c, c + \delta)$ and any $e \in (c - \delta, c)$. Then $f(d) = d > c = f(c)$ so that $x = c$ is not a local maximum, and $f(e) = e < c = f(c)$, so that $x = c$ is not a local minimum. Thus f has no local extrema on $(-3, 3)$.

7. For example, f could have a vertical asymptote on the interior of I. This will not be a critical point, yet clearly either the global minimum will be $-\infty$, the global maximum will be ∞, or both. However, evaluating f at the endpoints and the critical points will produce some finite value as the global extremum. For example, consider the function $f(x) = \frac{1}{x}$ on $[-1, 1]$.

9. (a) On $[0, 4]$, there is no global maximum since the function increases without bound near $x = 1$. The global minimum of 0 occurs at $x = 0$.

   (b) On $[2, 5]$, the global minimum is ≈ 6 at $x = 2$, and the global minimum of ≈ 2 occurs at $x = 5$.

   (c) On $(-2, 1)$, there is no global maximum since the function increases without bound near $x = 1$. The global minimum of 0 occurs at $x = 0$.

   (d) On $[0, \infty)$, the global minimum of 0 occurs at $x = 0$. Assuming the function continues the behavior implied by the graph, there is no global maximum since the function's values increase without bound as $x \to 1$.

   (e) On $(-\infty, 0]$, assuming the function continues the behavior implied by the graph, there is no global maximum since the function's values approach 1 from below as $x \to -\infty$. The global minimum of 0 occurs at $x = 0$.

   (f) On $(-\infty, \infty)$, assuming the function continues the behavior implied by the graph, there is no global maximum since the function's values increase without bound as $x \to 1$. The global minimum of 0 occurs at $x = 0$.

Skills

11. Since $f'(x) = 6x^2 - 6x - 12 = 6(x - 2)(x + 1)$, the critical points of f are at $x = 2$ and $x = -1$. Values of f at these points as well as the endpoints of the intervals in parts (a)-(d) are:

x	-3	-2	-1	0	1	2	3
$f(x)$	-45	-4	7	0	-13	-20	-9

   (a) On $[-3, 3]$, the global minimum occurs at $x = -3$ and the global maximum occurs at $x = -1$.

   (b) On $[0, 3]$, the global minimum occurs at $x = 2$ and the global maximum occurs at $x = 0$.

   (c) On $(-1, 2]$, the global minimum occurs at $x = 2$. There is no global maximum since the function values approach 7 from below as $x \to -1^+$.

   (d) On $(-2, 1)$, there is no global minimum since the function values approach -4 from above as $x \to -2^+$. The global maximum occurs at $x = -1$.

   A graph of f is

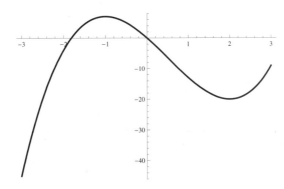

13. Since $f'(x) = -12x^3 + 12x^2 + 12x - 12 = -12(x - 1)^2(x + 1)$, the critical points of f are at $x = \pm 1$. Values of f at these points as well as the endpoints of the intervals in parts (a)-(d) are:

x	-3	-1	0	1	3
$f(x)$	-261	11	0	-5	-117

(a) On $[-1, 1]$, the global minimum occurs at $x = 1$ and the global maximum occurs at $x = -1$.

(b) On $(-1, 1)$, there is no global minimum or maximum, since the function values approach 11 from below as $x \to -1^+$, and approach -5 from above as $x \to 1^-$.

(c) On $(-3, 0]$, there is no global minimum since the function values approach -261 from above as $x \to -3^+$. The global maximum occurs at $x = -1$.

(d) On $[0, 3]$, the global minimum occurs at $x = 3$ and the global maximum occurs at $x = 0$.

A graph of f is

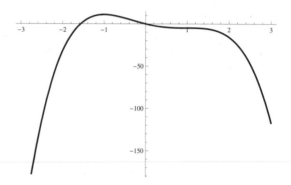

15. Since $f'(x) = -\frac{1}{2\sqrt{x}(1+\sqrt{x})^2}$, we see that $f'(x)$ never vanishes, and is undefined at $x = 0$, which is thus the only critical point of f. Note that the domain of f is $x \geq 0$. Values of f at 0 and at the endpoints of the intervals in parts (a)-(d) are:

x	0	1	2	3
$f(x)$	1	$\frac{1}{2}$	$\frac{1}{1+\sqrt{2}}$	$\frac{1}{1+\sqrt{3}}$
$f(x) \approx$	1	0.5	0.414	0.366

Note that $\lim\limits_{x \to \infty} f(x) = 0$, so that x has a horizontal asymptote of $y = 0$, approached from above, as $x \to \infty$.

(a) On $[0, 3]$, the global minimum occurs at $x = 3$ and the global maximum occurs at $x = 0$.

(b) On $(0, 3)$, there is no global minimum since the function values approach $f(3)$ from above as $x \to 3^-$. There is no global maximum since the function values approach $f(0)$ from below as $x \to 0^+$.

(c) On $[1, 2]$, the global minimum occurs at $x = 2$ and the global maximum occurs at $x = 1$.

(d) On $[0, \infty)$, there is no global minimum since the function values approach 0 from above as $x \to \infty$. The global maximum occurs at $x = 0$.

A graph of f is

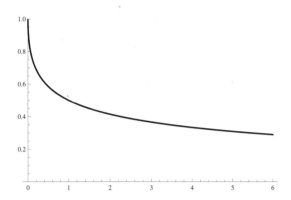

17. Since $f'(x) = \frac{15}{2}\sqrt{x}(x-1)$, we see that $f'(x)$ vanishes at $x = 0$ and at $x = 1$. The domain of f is $x \geq 0$, so that the critical points of f are $x = 0$ and $x = 1$. Values of f at these points and at the endpoints of the intervals in parts (a)-(d) are:

x	0	1	4
$f(x)$	0	-2	56

(a) On $[0,4]$, the global minimum occurs at $x = 1$ and the global maximum occurs at $x = 4$.

(b) On $[0,4)$, the global minimum occurs at $x = 1$. There is no global maximum since the function values approach 56 from below as $x \to 4^-$.

(c) On $[0,1)$, there is no global minimum since the function values approach -2 from above as $x \to 1^-$. The global maximum is at $x = 0$.

(d) On $(0,1)$, there is no global minimum since the function values approach -2 from above as $x \to 1^-$. There is no global maximum since the function values approach 0 from below as $x \to 0^+$.

A graph of f is

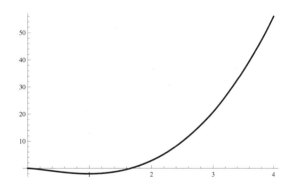

19. Note that the domain of f is all x such that $x^2 - 4x + 3 > 0$. Sign analysis reveals that this is for $x \in (-\infty, 1) \cup (3, \infty)$. Now, $f'(x) = \frac{2-x}{(x^2-4x+3)^{3/2}}$, so that $f'(x)$ vanishes at $x = 2$; however, this is not in the domain of f. Thus f has no critical points. Note that $\lim\limits_{x \to 1^-} f(x) = \lim\limits_{x \to 3^+} f(x) = \infty$. Further, $\lim\limits_{x \to \infty} f(x) = 0$ so that f has a horizontal asymptote at $y = 0$, approached from above as $x \to \infty$. The values of f at the endpoints of the intervals in parts (a)-(d) are:

x	0	3	3.5	4	10
$f(x) \approx$	0.577	undefined	0.894	0.577	0.126

(a) On $[0, 4]$, the global minimum occurs at both $x = 0$ and $x = 4$. There is no global maximum since $\lim\limits_{x \to 1^{-}} f(x) = \infty$.

(b) On $[0, 10]$, the global minimum occurs at $x = 10$. But f does not have a global maximum since $\lim\limits_{x \to 1^{-}} f(x) = \infty$.

(c) On $[0, 3.5]$, the global minimum occurs at $x = 0$. But f does not have a global maximum since $\lim\limits_{x \to 1^{-}} f(x) = \infty$.

(d) On $(3, \infty)$, there is no global minimum since the function values approach zero from above as $x \to \infty$. There is no global maximum since $\lim\limits_{x \to 3^{+}} f(x) = \infty$.

A graph of f is

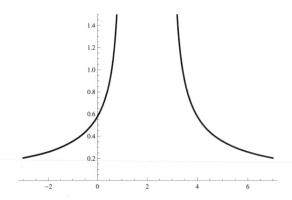

21. Let the length of the straight line segments be l and the radius of the semicircular segments be r. Then the constraint on the edge length means that $2l + 2\pi r = 120$, so that $l + \pi r = 60$. The area of the figure is composed of a rectangular segment with length l and height $2r$ together with two semicircular regions of radius r. Thus the total area is $2lr + \pi r^2$. So we want to maximize $2lr + \pi r^2$ subject to the constraint $l + \pi r = 60$. Substitute $l = 60 - \pi r$ into the area formula to get $2(60 - \pi r)r + \pi r^2 = 120r - \pi r^2$. This is the expression we wish to maximize. The smallest possible value for r is $r = 0$, since the radius must be nonnegative. The smallest possible value of l is also zero, so that $60 - \pi r \geq 0$ and thus $r \leq \frac{60}{\pi}$. So, in summary, the problem is to maximize $A(r) = 120r - \pi r^2$ on the interval $\left[0, \frac{60}{\pi}\right]$. Now, $A'(r) = 120 - 2\pi r$, so that the only critical point of A is at $r = \frac{60}{\pi}$. Since $A''(r) = -2\pi$, we see that A'' is negative everywhere so that $r = \frac{60}{\pi}$ is indeed a local maximum. Evaluating $A(r)$ at the critical point as well as at the other endpoint gives

$$A(0) = 0, \qquad A\left(\frac{60}{\pi}\right) = 120\frac{60}{\pi} - \pi\frac{60^2}{\pi^2} = \frac{60^2}{\pi}.$$

Thus the maximum area is $\frac{60^2}{\pi} = \frac{3600}{\pi}$ square inches, achieved when $r = \frac{60}{\pi}$ inches and $l = 0$ inches (Note that setting $l = 0$ does not result in a figure with zero area, since the circles still contribute. But setting $r = 0$ does give a figure with zero area).

23. Use the same notation as in the previous problem; then the total edge length is $2l + 3(2r) + 2\pi r = 2l + (6 + 2\pi)r$, so that $2l + (6 + 2\pi)r = 120$ and $l + (3 + \pi)r = 60$. The area of the figure, as before, is $2lr + \pi r^2$. So we want to maximize $2lr + \pi r^2$ subject to the constraint $l + (3 + \pi)r = 60$. Substitute $l = 60 - (3 + \pi)r$ into the area formula to get $A(r) = 2(60 - (3 + \pi)r)r + \pi r^2 = 120r - (6 + \pi)r^2$. The smallest possible value for r is 0. The smallest possible value for l is also zero, so that $60 - (3 + \pi)r \geq 0$ and thus $r \leq \frac{60}{3+\pi}$. So, in summary, the problem is to maximize $A(r) = 120r - (6 + \pi)r^2$ on the interval $\left[0, \frac{60}{3+\pi}\right]$. Now, $A'(r) = 120 - (12 + 2\pi)r$, so that the only critical point of A is at $r = \frac{60}{6+\pi}$. Since $A''(r) = -12 - 2\pi < 0$, we see that A'' is negative

everywhere, so that $r = \frac{60}{6+\pi}$ is indeed a local maximum. Evaluating $A(r)$ at the critical point as well as at the endpoints gives

$$A(0) = 0, \qquad A\left(\frac{60}{3+\pi}\right) = 120\frac{60}{3+\pi} - (6+\pi)\frac{60^2}{(3+\pi)^2} = \frac{3600\pi}{(3+\pi)^2} \approx 299.84,$$

$$A\left(\frac{60}{6+\pi}\right) = 120\frac{60}{6+\pi} - (6+\pi)\frac{60^2}{(6+\pi)^2} = \frac{60^2}{6+\pi} = \frac{3600}{6+\pi} \approx 393.80.$$

Thus the maximum area is $\frac{3600}{6+\pi}$ square inches, achieved when $r = \frac{60}{6+\pi}$ inches and $l = 60 - (3 + \pi)\frac{60}{6+\pi}$ inches.

25. We want to maximize xy subject to the constraint that $x + y = 36$. Then $y = 36 - x$; substitute for y to get $P(x) = x(36 - x) = 36x - x^2$. If $x < 0$, then we must have $y > 0$ in order for the sum to be 36, so that $xy < 0$. This will clearly not be a maximum product. Thus we must have $x \geq 0$ and $y = 36 - x \geq 0$, so that $x \leq 36$. In summary, we wish to maximize $P(x) = 36x - x^2$ on $[0, 36]$. Now, $P'(x) = 36 - 2x$, so that the only critical point is $x = 18$. Further, $P''(x) = -2 < 0$, so that $x = 18$ is indeed a local maximum. Evaluating $P(x)$ at the critical point and at the endpoints gives

$$P(0) = 0, \qquad P(36) = 0, \qquad P(18) = 36 \cdot 18 - 18^2 = 18^2 = 324.$$

Thus the maximum product is achieved when $x = y = 18$; that product is 324.

27. We wish to minimize $a^2 + b^2$ subject to the constraint $a + b = 100$ (or $b = 100 - a$). Substitute for b to get $s(a) = a^2 + (100 - a)^2 = 10000 - 200a + 2a^2$. There are no constraints on either a or b, so that we wish to minimize $s(a)$ on $(-\infty, \infty)$. Now, $s'(a) = 4a - 200$, so that the only critical point is $a = 50$. Since $s''(a) = 4 > 0$, we see that $a = 50$ is indeed a local minimum. Thus $a^2 + b^2$ is minimized when $a = b = 50$.

29. This is very similar to Exercise 28. Given any rectangle fitting inside a circle of radius 10, clearly all four vertices must touch the circle or it would be possible to enlarge the rectangle. Any rectangle with all four vertices lying on the circle can be characterized by the smallest angle θ between a side of the rectangle and a line connecting one of the vertices B of the rectangle with the center O of the circle (see the figure).

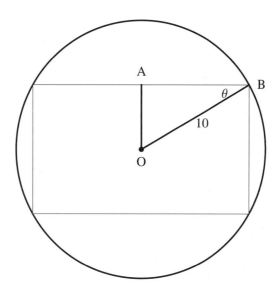

Note that $0 \leq \theta \leq \frac{\pi}{2}$. So, we wish to maximize the area of such a rectangle as θ varies. Now, the rectangle is composed of eight triangles congruent to the triangle OAB in the figure. Since this is a right triangle with $OB = 10$, we have $OA = 10\sin\theta$ and $OB = 10\cos\theta$, so that its area is $50\sin\theta\cos\theta$. Thus the area of the rectangle is $A(\theta) = 8 \cdot 50\sin\theta\cos\theta = 400\sin\theta\cos\theta$. We wish to maximize $A(\theta)$ on $\left[0, \frac{\pi}{2}\right]$.

Now, $A'(\theta) = 400(\cos^2\theta - \sin^2\theta) = 400(\cos\theta - \sin\theta)(\cos\theta + \sin\theta)$. Thus the critical points are the points where $\cos\theta = \sin\theta$ or where $\cos\theta = -\sin\theta$. Since both $\cos\theta$ and $\sin\theta$ are nonnegative on the interval we are interested in, and since they do not both vanish at the same point, the second of these conditions has no solutions. The first condition, $\cos\theta = \sin\theta$, is satisfied for $\theta = \frac{\pi}{4}$, so this is the only critical point. Now, $A''(\theta) = -1600\cos\theta\sin\theta$, and this is negative at $\theta = \frac{\pi}{4}$, so that this is in fact a local maximum. Thus to determine the global maximum on $\left[0, \frac{\pi}{2}\right]$, we evaluate $A(\theta)$ at the critical point and at the endpoints:

$$A(0) = 400\sin 0\cos 0 = 0, \quad A\left(\frac{\pi}{2}\right) = 400\sin\frac{\pi}{2}\cos\frac{\pi}{2} = 0, \quad A\left(\frac{\pi}{4}\right) = 400\sin\frac{\pi}{4}\cos\frac{\pi}{4} = 200.$$

Thus the rectangle of maximum area has a corresponding angle of $\frac{\pi}{4}$, so that $OA = OB$ and it is a square. Its area is 200, so that its side length is $10\sqrt{2}$.

31. Let $(x, 3x + 1)$ be an arbitrary point on the graph of $f(x)$. The square of the distance from this point to $(-2, 1)$ is $d(x) = (x - (-2))^2 + (3x + 1 - 1)^2 = (x + 2)^2 + 9x^2 = 10x^2 + 4x + 4$. Now, $d'(x) = 20x + 4$, so that $x = -\frac{1}{5}$ is the only critical point of d. Since $d''(x) = 20 > 0$, this is in fact a local minimum. We want to minimize $d(x)$ over all $x \in \mathbb{R}$; since $\lim_{x \to \pm\infty} d(x) = \infty$, the global minimum occurs at the point $\left(-\frac{1}{5}, \frac{2}{5}\right)$; the square of the distance is $d\left(-\frac{1}{5}\right) = \frac{18}{5}$.

33. Let $(x, x^2 - 2x + 1)$ be an arbitrary point on the graph of $f(x)$. The square of the distance from this point to $(1, 2)$ is $d(x) = (x - 1)^2 + (x^2 - 2x + 1 - 2)^2 = x^4 - 4x^3 + 3x^2 + 2x + 2$. We wish to minimize $d(x)$. Now, $d'(x) = 4x^3 - 12x^2 + 6x + 2 = 2(x - 1)(2x^2 - 4x - 1)$, with roots $x = 1$ and $x = \frac{1}{2}(2 \pm \sqrt{6})$. Since $d''(x) = 12x^2 - 24x + 6$, we have $d''(1) = -6 < 0$, $d''\left(\frac{1}{2}(2 + \sqrt{6})\right) = d''\left(\frac{1}{2}(2 - \sqrt{6})\right) = 12 > 0$, so that $\frac{1}{2}(2 \pm \sqrt{6})$ are local minima. Since $\lim_{x \to \pm\infty} d(x) = \infty$, we see that the global minimum occurs at $x = \frac{1}{2}(2 \pm \sqrt{6})$, where the minimum squared distance is $\frac{7}{4}$.

Applications

35. Let the dimensions of the pen by x by y. The perimeter is $2x + 2y$, which must be 350 by the conditions of the problem, so that $y = 175 - x$. The area of the pen is xy. So we wish to minimize $xy = x(175 - x) = A(x)$. The minimum possible value of x is 0. Also, the smallest possible value of y is 0, so that $x \leq 175$. So we want to maximize $A(x)$ on $[0, 175]$. Now, $A'(x) = 175 - 2x$, so that $A'(x) = 0$ at $x = \frac{175}{2}$. Since $A''(x) = -2 < 0$, this is in fact a local maximum. So we need to test $A(x)$ at this point as well as at the endpoints of the interval:

$$A(0) = 0, \quad A(175) = 0, \quad A\left(\frac{175}{2}\right) = \frac{175}{2}\left(175 - \frac{175}{2}\right) = \frac{175^2}{4}.$$

Thus the maximum possible area is $\frac{175^2}{4}$ square feet formed from a square pen of side length $\frac{175}{2}$.

37. Let x be the length of the horizontal sides in the diagram and y the length of the vertical sides. Then the total amount of fencing is $1000 = 2x + 4y$, so that $500 = x + 2y$ and $x = 500 - 2y$. The area of the pen is xy. Substituting for x gives $A(y) = (500 - 2y)y = 500y - 2y^2$ for the area. The minimum possible value for y is 0. Also, the minimum possible value for x is 0, so that $0 \leq 500 - 2y$ and thus $y \leq 250$. Thus we wish to maximize $A(y) = 500y - 2y^2$ on the interval $[0, 250]$. Now, $A'(y) = 500 - 4y$, so that $y = 125$ is the only critical point. Since $A''(y) = -4 < 0$, this is indeed

a local maximum. So to find the global maximum, we need to test $A(y)$ at this point as well as at the interval endpoints.

$$A(0) = 0, \qquad A(250) = 500 \cdot 250 - 2 \cdot 250^2 = 0,$$
$$A(125) = 500 \cdot 125 - 2 \cdot 125^2 = 2 \cdot 125^2 = 31250.$$

Thus the maximum possible area is 31250 square feet, achieved by making four vertical segments of length 125 and two horizontal segments of length 250.

39. Let the width of the habitat be x feet and its height y feet (all ignoring the hutch). Then the area is xy, so that $xy = 2500$. The perimeter is $2y + x + (x - 20) = 2x + 2y - 20$. Substituting $y = \frac{2500}{x}$ gives

$$P(x) = 2x + \frac{5000}{x} - 20.$$

The minimum value for x is 20, to make room for the hutch. The minimum value for y is 0, so that there is no maximum for $x = \frac{2500}{y}$. Thus we want to minimize $P(x)$ on $[20, \infty)$. Now,

$$P'(x) = 2 - \frac{5000}{x^2},$$

so that $P'(x) = 0$ for $x^2 = 2500$, or $x = 50$, which is thus the only critical point. Since $P''(x) = \frac{10000}{x^3}$, we see that $x = 50$ is a local minimum. Evaluating there as well as at the interval endpoint gives

$$P(20) = 40 + \frac{5000}{20} - 20 = 270$$
$$P(100) = 100 + \frac{5000}{50} - 20 = 180,$$

and since $\lim_{x \to \infty} P(x) = \infty$, we see that $x = 50$ is a global minimum. Thus the minimal perimeter is 180 feet, with the horizontal side 50 feet and each vertical side 50 feet.

41. Consider the redrawn and labeled figure

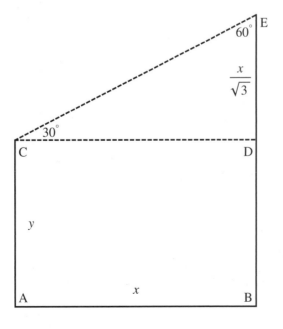

Denoting AB by x and AC by y, the fact that $\angle ECD = 30°$ means that $\tan 30° = \frac{1}{\sqrt{3}} = \frac{DE}{CD} = \frac{DE}{x}$, so that $DE = \frac{x}{\sqrt{3}}$. The area of the pen is then the sum of the areas of the rectangular and triangular regions, so is $xy + \frac{1}{2}x \cdot \frac{x}{\sqrt{3}} = xy + \frac{x^2}{2\sqrt{3}}$. The given constraint means that this must equal 2500. Solving for y gives $y = \frac{2500 - \frac{x^2}{2\sqrt{3}}}{x} = \frac{2500}{x} - \frac{x}{2\sqrt{3}}$. The amount of fencing needed is $x + 2y + \frac{x}{\sqrt{3}}$; substituting the above value for y gives for the amount of fencing

$$P(x) = x + 2\left(\frac{2500}{x} - \frac{x}{2\sqrt{3}}\right) + \frac{x}{\sqrt{3}} = x + \frac{5000}{x}.$$

The smallest possible value for x is 0; similarly, the smallest possible value for y is 0, so that $\frac{2500}{x} - \frac{x}{2\sqrt{3}} = \frac{5000\sqrt{3} - x^2}{2\sqrt{3}x} \geq 0$. This means that $x^2 \leq 5000\sqrt{3}$, so that $x \leq 50\sqrt[4]{12}$. Hence we wish to minimize $P(x)$ on the interval $(0, 50\sqrt[4]{12}]$ (note that $x = 0$ is not a valid value, since $P(x)$ does not exist). Now, $P'(x) = 1 - \frac{5000}{x^2}$, so that $P'(x) = 0$ for $x = \pm 50\sqrt{2}$. Taking only the positive root since we need $x \geq 0$, gives only $50\sqrt{2}$ as a critical point. Since $P''(x) = \frac{10000}{x^3}$, we see that $P''(x) > 0$ for $x > 0$, so that this is indeed a local minimum. Also, since $\lim_{x \to 0} P(x) = \infty$, the global minimum must occur at either the critical point or at the closed endpoint of the interval:

$$P(50\sqrt[4]{12}) = 50\sqrt[4]{12} + \frac{100}{\sqrt[4]{12}} \approx 146.789$$

$$P\left(50\sqrt{2}\right) = 50\sqrt{2} + \frac{5000}{50\sqrt{2}} = 100\sqrt{2} \approx 141.421$$

Thus the minimum amount of fencing needed is approximately 141.421 feet, which is achieved when the horizontal side is $50\sqrt{2} \approx 70.71$ feet and the vertical sides are $25\sqrt{2} - \frac{25}{3}\sqrt{6} \approx 14.943$ feet.

43. Let x be the length of a side of the base and y be the height of the box. Then there are four wood sides of area xy and one wood bottom of area x^2; there is also a metal top of area x^2. Using the costs given, we want $2000 = 4xy(5) + x^2(5) + x^2(12)$, or $2000 = 17x^2 + 20xy$. Solving for y gives $y = \frac{100}{x} - \frac{17x}{20}$. Now, the total volume of the box is x^2y; substituting for y from the above formula gives

$$V(x) = x^2y = x^2\left(\frac{100}{x} - \frac{17x}{20}\right) = -\frac{17}{20}x^3 + 100x.$$

The smallest possible value for x is zero. The smallest possible value for y is also zero, so that $0 \leq \frac{100}{x} - \frac{17x}{20}$. Solving for x gives $x \leq 20\sqrt{\frac{5}{17}}$. Thus we wish to maximize $V(x)$ on the interval $\left[0, 20\sqrt{\frac{5}{17}}\right]$. Now, $V'(x) = -\frac{51}{20}x^2 + 100$, so that $V'(x) = 0$ for $x > 0$ at $x = 20\sqrt{\frac{5}{51}}$. Since $V''(x) = -\frac{51}{10}x$, we see that $V''(x) > 0$ for $x > 0$, so that this critical point is indeed a local maximum. So to find the global maximum we must evaluate $V(x)$ at the critical point as well as at the endpoints:

$$V(0) = 0$$

$$V\left(20\sqrt{\frac{5}{17}}\right) = -\frac{17}{20} \cdot 8000\frac{5}{17}\sqrt{\frac{5}{17}} + 2000\sqrt{\frac{5}{17}} = 0$$

$$V\left(20\sqrt{\frac{5}{51}}\right) = -\frac{17}{20} \cdot 8000\frac{5}{51}\sqrt{\frac{5}{51}} + 2000\sqrt{\frac{5}{51}} = \frac{4000}{3}\sqrt{\frac{5}{51}} \approx 417.483.$$

Thus the maximum possible volume is about 417 cubic inches, achieved with a square base of side $20\sqrt{\frac{5}{51}} \approx 6.26$ inches and a height of ≈ 10.65 inches.

45. Clearly to get the largest volume or surface area, Linda wants the sum of the length and girth to be 108 inches, since otherwise she could enlarge the package. Now, let the width and height of the package be x and the length be y. Then the girth is $4x$. The constraints mean that $y + 4x = 108$, so that $y = 108 - 4x$.

The volume of such a package is $x^2 y$; substituting gives $V(x) = 108x^2 - 4x^3$. The smallest possible value for x is 0. The smallest possible value for y is also 0; this means that $0 \leq 108 - 4x$ so that $x \leq 27$. Thus we wish to maximize $V(x)$ on $[0, 27]$. Now, $V'(x) = 216x - 12x^2$, so that $V'(x) = 0$ for $x = 0$ and for $x = 18$. Since $V''(x) = 216 - 24x$, we see that $V''(18) < 0$, so that $x = 18$ is in fact a local maximum. (0 is an interval endpoint, so we will test it regardless). So to determine the global maximum, we must evaluate $V(x)$ at $x = 18$ as well as at the two endpoints:

$$V(0) = 0, \qquad V(27) = 108 \cdot 27^2 - 4 \cdot 27^3 = 0, \qquad V(18) = 108 \cdot 18^2 - 4 \cdot 18^3 = 11664.$$

Thus the maximum package volume is 11664 cubic inches, achieved with a width and height of 18 inches and a length of 36 inches.

The surface area of such a package is $2x^2 + 4xy$ (the $2x^2$ is for the two ends; the $4xy$ for the four sides). Substituting $y = 108 - 4x$ gives $A(x) = 2x^2 + 4x(108 - 4x) = 432x - 14x^2$. The smallest possible value for x is zero; as above, the largest possible value is 27. Thus we want to maximize $A(x)$ on $[0, 27]$. Now, $A'(x) = 432 - 28x$, so that $A'(x) = 0$ at $x = \frac{108}{7}$. Since $A''(x) = -28 < 0$, we see that this is in fact a local maximum. So to determine the global maximum, we must evaluate $A(x)$ at $x = \frac{108}{7}$ as well as at the two endpoints:

$$A(0) = 0, \qquad A(27) = 432 \cdot 27 - 14 \cdot 27^2 = 1458,$$

$$A\left(\frac{108}{7}\right) = 432 \cdot \frac{108}{7} - 14 \cdot \left(\frac{108}{7}\right)^2 = \frac{23328}{7} \approx 3333.$$

Thus the largest surface area is ≈ 3333 square inches, achieved with a width and height of $\frac{108}{7}$ inches and a length of $\frac{296}{7}$ inches.

47. Clearly to get the largest volume or surface area, Linda wants the sum of the length and girth to be 108 inches, since otherwise she could enlarge the package. Let the radius of the cylinder be r and the length of the cylinder be h. Then the girth is the circumference of a circle with radius r, or $2\pi r$, so that we have $108 = h + 2\pi r$. Solving for h gives $h = 108 - 2\pi r$.

The volume of such a package is $\pi r^2 h$; substituting for h gives $V(r) = \pi r^2 (108 - 2\pi r) = 108\pi r^2 - 2\pi^2 r^3$. The smallest possible value for r is zero. The smallest possible value for h is also zero, so that $0 \leq 108 - 2\pi r$ and thus $r \leq \frac{54}{\pi}$. So we want to maximize $V(r)$ on the interval $\left[0, \frac{54}{\pi}\right]$. Now, $V'(r) = 216\pi r - 6\pi^2 r^2$, so that $V'(r) = 0$ for $r = 0$ and $r = \frac{36}{\pi}$. Since $V''(r) = 216\pi - 12\pi^2 r$, we have $V''\left(\frac{36}{\pi}\right) = -216\pi$, so that $\frac{36}{\pi}$ is indeed a local maximum. Since 0 is an interval endpoint, we will test its value in any case. So to find the global maximum, we must evaluate $V(r)$ at the critical point as well as at the interval endpoints:

$$V(0) = 0, \qquad V\left(\frac{54}{\pi}\right) = 108\pi \frac{54^2}{\pi^2} - 2\pi^2 \frac{54^3}{\pi^3} = 0,$$

$$V\left(\frac{36}{\pi}\right) = 108\pi \frac{36^2}{\pi^2} - 2\pi^2 \frac{36^3}{\pi^3} = \frac{36^3}{\pi} \approx 14851.$$

Thus the maximum possible package volume is about 14851 cubic inches, achieved with a radius of $\frac{36}{\pi} \approx 11.46$ inches and a length of 36 inches.

The surface area of such a package is the sum of the areas of the two ends and the area of the sides of the cylinder. This is $2(\pi r^2) + 2\pi r h$. Substituting $h = 108 - 2\pi r$ gives $A(r) = 2\pi r^2 + 2\pi r(108 -$

$2\pi r) = (2\pi - 4\pi^2)r^2 + 216\pi r$. The interval on which we want to maximize $A(r)$ is still $\left[0, \frac{54}{\pi}\right]$. Now, $A'(r) = (4\pi - 8\pi^2)r + 216\pi$, so that $A'(r) = 0$ for $r = \frac{216}{8\pi - 4}$. Since $A''(r) = 4\pi - 8\pi^2 < 0$, we see that this is in fact a local maximum. So to determine the global maximum, we must evaluate $A(r)$ at the critical point as well as at the interval endpoints:

$$A(0) = 0, \qquad A\left(\frac{54}{\pi}\right) = (2\pi - 4\pi^2)\frac{54^2}{\pi^2} + 216\pi\frac{54}{\pi} = \frac{2 \cdot 54^2}{\pi} \approx 1856$$

$$A\left(\frac{216}{8\pi - 4}\right) = (2\pi - 4\pi^2)\frac{216^2}{(8\pi - 4)^2} + 216\pi\frac{216}{8\pi - 4} = \frac{5832\pi}{2\pi - 1} \approx 3468.$$

Thus the maximum possible surface area is ≈ 3468 square inches, achieved with a radius of $\frac{216}{8\pi - 4} \approx 10.2$ inches and a length of ≈ 43.8 inches.

49. With $s(t)$ as above, the car is moving fastest to the right when the velocity is maximized, and is moving fastest to the left when the velocity is minimized. Now, $v(t) = s'(t) = -12t^3 + 84t^2 - 168t + 96$, so that $v'(t) = -36t^2 + 168t - 168$, which has roots $t = \frac{1}{3}(7 \pm \sqrt{7})$. Evaluating $v(t)$ at these points together with the interval endpoints gives

t	0	$\frac{1}{3}(7 - \sqrt{7})$	$\frac{1}{3}(7 + \sqrt{7})$	4
$v(t) \approx$	96	-7.57	25.35	0

Thus the car is moving fastest to the right at $t = 0$, when it is moving at 96 centimeters per minute, and fastest to the left at $t = \frac{1}{3}(7 - \sqrt{7})$, when it is moving left at ≈ 7.57 centimeters per second.

51. Let the piece used to construct the square be of length $4x$, so that the remaining piece is of length $10 - 4x$. The length of one side of the square is x, so its area is x^2. Then the circle has circumference $10 - 4x$, so that $2\pi r = 10 - 4x$ and the radius of the circle is $\frac{10-4x}{2\pi} = \frac{5-2x}{\pi}$. Thus the area of the circle is $\pi r^2 = \pi\frac{(5-2x)^2}{\pi^2} = \frac{(5-2x)^2}{\pi}$. Hence the total area is

$$A(x) = x^2 + \frac{(5 - 2x)^2}{\pi} = \frac{1}{\pi}((4 + \pi)x^2 - 20x + 25)$$

Since $10 - 4x$ must be nonnegative, x is in the range $\left[0, \frac{5}{2}\right]$, so we want to find the extrema of $A(x)$ on this interval. Now, $A'(x) = \frac{1}{\pi}((8 + 2\pi)x - 20)$, so that $A'(x) = 0$ for $x = \frac{20}{8+2\pi} = \frac{10}{4+\pi}$. Evaluating $A(x)$ at this point as well as at the interval endpoints gives

$$A(0) = \frac{25}{\pi} \approx 7.96$$

$$A\left(\frac{10}{4 + \pi}\right) = \frac{1}{\pi}\left((4 + \pi)\frac{100}{(4 + \pi)^2} - 20\frac{10}{4 + \pi} + 25\right) = \frac{1}{\pi}\left(-\frac{100}{4 + \pi} + 25\right)$$

$$= \frac{1}{\pi}\left(\frac{-100 + 100 + 25\pi}{4 + \pi}\right) = \frac{25}{4 + \pi} \approx 3.5$$

$$A\left(\frac{5}{2}\right) = \frac{1}{\pi}\left((4 + \pi)\frac{25}{4} - 50 + 25\right) = \frac{25}{4} = 6.25.$$

Thus

(a) The minimum area of $\frac{25}{4+\pi} \approx 3.5$ square inches is achieved by forming a square of side length $\frac{10}{4+\pi} \approx 1.4$ inches and using the remaining wire to form a circle.

(b) The maximum area of $\frac{25}{\pi} \approx 7.96$ square inches is achieved by using the entire wire to form a circle (since $x = 0$).

53. Let the piece used to construct the triangle be of length $3x$, so that the remaining piece is of length $10 - 3x$. The length of one side of the triangle is x, so that its area is $x^2 \frac{\sqrt{3}}{4}$. The circumference of the circle is $2\pi r = 10 - 3x$, so that the radius is $\frac{10-3x}{2\pi}$. Thus the area of the circle is $\pi \frac{(10-3x)^2}{4\pi^2}$. Hence the total area is

$$A(x) = x^2 \frac{\sqrt{3}}{4} + \frac{(10 - 3x)^2}{4\pi} = \frac{1}{4\pi}\left((9 + \pi\sqrt{3})x^2 - 60x + 100\right)$$

The minimum possible value for x is 0, and the maximum is $\frac{10}{3}$, so that we want to find the extrema of $A(x)$ on $\left[0, \frac{10}{3}\right]$. Now, $A'(x) = \frac{1}{4\pi}\left((18 + 2\pi\sqrt{3})x - 60\right)$, so that $A'(x) = 0$ for $x = \frac{60}{18+2\pi\sqrt{3}} = \frac{30}{9+\pi\sqrt{3}}$. Evaluating $A(x)$ at this point as well as at the interval endpoints gives after a calculation

$$A(0) = \frac{25}{\pi} \approx 7.96$$

$$A\left(\frac{30}{9 + \pi\sqrt{3}}\right) = \frac{25\sqrt{3}}{9 + \pi\sqrt{3}} \approx 2.998$$

$$A\left(\frac{10}{3}\right) = \frac{25}{3\sqrt{3}} \approx 4.81.$$

Thus

(a) The minimum possible area is ≈ 2.998 square inches, achieved by constructing an equilateral triangle of side $\frac{30}{9+\pi\sqrt{3}} \approx 2.08$ inches and using the remaining wire to construct a circle.

(b) The maximum possible area is ≈ 7.96 square inches, achieved by using the entire 10 inch wire to construct a circle.

55. Let the radius of the cylinder be r and its height be h. Then the volume is $40 = \pi r^2 h$, so that $h = \frac{40}{\pi r^2}$. The surface area of the cylinder is $2\pi r^2 + 2\pi rh$; substituting for h gives

$$A(r) = 2\pi r^2 + 2\pi r \frac{40}{\pi r^2} = \frac{80}{r} + 2\pi r^2.$$

The possible values for r are bounded below by zero (note that $A(r)$ is undefined at $r = 0$). As $h \to 0$, however, r gets arbitrarily large in order to maintain the volume at 40. Thus we want to find the extrema of $A(r)$ on $(0, \infty)$. Note that $\lim_{r\to 0} A(r) = \lim_{r\to\infty} A(r) = \infty$, so that the answer to the second part is no, it is not possible to *maximize* the area of the barrel. Choosing r very large or very small will make the area arbitrarily large. However, note that $A'(r) = 4\pi r - \frac{80}{r^2}$, so that $A'(r) = 0$ for $r = \left(\frac{20}{\pi}\right)^{1/3}$. Since $A''(r) = 4\pi + \frac{80}{r^3}$, we have $A''(r) > 0$ for $r > 0$, so that this critical point is in fact a local minimum. Since the function increases without bound near the endpoints, it follows that the critical point is the global minimum. Thus the minimum possible surface area is

$$A\left(\left(\frac{20}{\pi}\right)^{1/3}\right) = 12(50\pi)^{1/3} \approx 64.74 \text{ cubic feet,}$$

achieved with a barrel radius of $\left(\frac{20}{\pi}\right)^{1/3} \approx 1.85$ feet and a height of ≈ 3.71 feet.

57. Let the radius of the cylinder be r and the height be h. Then since the volume is 400, we have $400 = \pi r^2 h$, so that $h = \frac{400}{\pi r^2}$. Now, the top and bottom each cost $5 \cdot \pi r^2$ cents, while the side of the can costs $2\pi rh$. Substituting for h gives for the total cost of the can

$$C(r) = 2 \cdot 5\pi r^2 + 2 \cdot 2\pi rh = 10\pi r^2 + 4\pi r \frac{400}{\pi r^2} = 10\pi r^2 + \frac{1600}{r}.$$

The possible values for r are bounded below by zero (note that $C(r)$ is undefined at $r = 0$). As $h \to 0$, however, r gets arbitrarily large in order to maintain the volume at 400. Thus we want to find the minimum of $C(r)$ on $(0, \infty)$. Now, $C'(r) = 20\pi r - \frac{1600}{r^2}$, so that $C'(r) = 0$ at $r = 2 \left(\frac{10}{\pi} \right)^{1/3}$. Further, $C''(r) = 20\pi + \frac{3200}{r^3}$, which is positive for $r > 0$, so that this point is indeed a local minimum. Note that $\lim_{r \to 0} C(r) = \lim_{r \to \infty} C(r) = \infty$, so that $r = 2 \left(\frac{10}{\pi} \right)^{1/3}$ is the global minimum on $(0, \infty)$. Thus the minimum possible cost is

$$C \left(2 \left(\frac{10}{\pi} \right)^{1/3} \right) = 40\pi \left(\frac{10}{\pi} \right)^{2/3} + 800 \left(\frac{\pi}{10} \right)^{1/3} \approx 815.8,$$

achieved with a radius of $2\frac{10}{\pi}^{1/3} \approx 2.9$ inches and a height of ≈ 14.71 inches. Since $\lim_{r \to 0} C(r) = \lim_{r \to \infty} C(r) = \infty$, there is no maximum possible cost — making either r or h very large will increase the cost without bound.

59. Assume we are laying the pipe from the southwest (lower left) to the northeast (upper right) corner. Then we can lay pipe along the ground eastwards for length a, then lay pipe under the parking lot to the other side, then lay pipe along the ground eastwards for the remainder, say b, to get to the northeast corner. This has exactly the same cost as laying pipe along the ground for length $a + b$ and then going under the parking lot to the northeast corner. So this is the situation we will consider.

Assume that the distance we lay pipe along the ground from the southwest corner is x. Then the remaining horizontal distance to be covered is $800 - x$, and the vertical distance to be covered is 500, so that the amount of pipe needed under the parking lot is just the hypotenuse of a triangle with those legs, or $\sqrt{500^2 + (800 - x)^2}$. Given the costs involved, we have

$$C(x) = 3x + 5\sqrt{500^2 + (800 - x)^2} = 3x + 5\sqrt{890000 - 1600x + x^2}.$$

The minimum value for x is 0, and the maximum is 800. So we want to minimize $C(x)$ on the interval $[0, 800]$. Now,

$$C'(x) = 3 + \frac{5(x - 800)}{\sqrt{890000 - 1600x + x^2}} = \frac{5x - 4000 + 3\sqrt{890000 - 1600x + x^2}}{\sqrt{890000 - 1600x + x^2}}.$$

Solving $C'(x) = 0$ involves solving $5x - 4000 + 3\sqrt{890000 - 1600x + x^2} = 0$. Rewrite this as $4000 - 5x = 3\sqrt{890000 - 1600x + x^2}$, and square both sides to get

$$16000000 - 40000x + 25x^2 = 8010000 - 14400x + 9x^2$$
$$16x^2 - 25600x + 7990000 = 0$$
$$16(x - 425)(x - 1175) = 0$$

Since $x = 1175$ is outside of the interval, we are left with $x = 425$ as the only critical point. Since $C''(x) = \frac{125000}{(890000 - 1600x + x^2)^{3/2}}$, we see that $C''(x) > 0$ at $x = 425$ since the denominator is positive, so that $x = 425$ represents a local minimum. To determine the global minimum, we must evaluate $C(x)$ at this point as well as at the interval endpoints:

$$C(0) = 5\sqrt{890000} = 500\sqrt{89} \approx 4717$$
$$C(425) = 3 \cdot 425 + 5\sqrt{390625} = 1275 + 3125 = 4400$$
$$C(800) = 3 \cdot 800 + 5\sqrt{250000} = 2400 + 2500 = 4900.$$

Thus the minimum cost of $4400 is achieved by laying pipe along the ground for 425 feet, then laying pipe in a straight line from there to the northeast corner, under the parking lot.

61. (a) We want to find the extrema of $N(c)$ on the given interval, $[5, 55]$. Now, $N'(c) = 1.2c - 54$, so that $N'(c) = 0$ for $c = 45$. Now evaluate $N(c)$ at $c = 45$ as well as at the interval endpoints:

$$N(5) = 0.6 \cdot 25 - 54 \cdot 5 + 1230 = 975$$
$$N(45) = 0.6 \cdot 2025 - 5445 + 1230 = 15$$
$$N(55) = 0.6 \cdot 3025 - 54 \cdot 55 + 1230 = 75.$$

Thus the largest number of paintings, 975, will be sold at \$5 per painting, and the smallest number, 15, will be sold at \$45 per painting.

(b) Since at a price of c per painting you will sell $N(c)$ paintings, your revenue at that price will be $c \times N(c)$ dollars. Thus $R(c) = c \times N(c) = 0.6c^3 - 54c^2 + 1230c$.

(c) We want to find the extrema of $R(c)$ on the given interval, $[5, 55]$. Now, $R'(c) = 1.8c^2 - 108c + 1230$, so that $R'(c) = 0$ for $c \approx 15.28$ and $c \approx 44.72$. Now evaluate $R(c)$ at these points as well as at the interval endpoints:

c	5	15.28	44.72	55
$r(c)$	4875	8327.10	672.90	4125

Thus the maximum revenue, \$8327.10, is achieved at a price of \$15.28, while the minimum revenue of \$672.90 is achieved at a price of \$44.72.

(d) In general, quantity and revenue are not the same thing. The revenue is the quantity weighted by the price. So, for example, if you were able to sell 50 widgets at \$1, but only one widget at \$100, you would clearly prefer the latter.

Proofs

63. Suppose a rectangle with sides x and y has perimeter P. Then $P = 2x + 2y$, so that $y = \frac{P-2x}{2}$. The area is xy; substituting for y gives

$$A(x) = xy = \frac{Px - 2x^2}{2}.$$

The minimum possible value for x is zero. The minimum value for y is also zero, so that $x \leq \frac{P}{2}$. Thus we want to maximize $A(x)$ for $x \in \left[0, \frac{P}{2}\right]$. Now, $A'(x) = \frac{P-4x}{2}$, so that $A'(x) = 0$ for $x = \frac{P}{4}$. Since $A''(x) = -2$, we have $A''(x) < 0$ everywhere so that this point is a local maximum. To find the global maximum, we evaluate $A(x)$ at this point as well as at the interval endpoints:

$$A(0) = 0, \qquad A\left(\frac{P}{2}\right) = \frac{\frac{P^2}{2} - 2\frac{P^2}{4}}{2} = 0, \qquad A\left(\frac{P}{4}\right) = \frac{\frac{P^2}{4} - 2\frac{P^2}{16}}{2} = \frac{P^2}{16}.$$

Thus the maximum area of $\frac{P^2}{16}$ occurs when $x = \frac{P}{2}$ and $y = \frac{P-2x}{2} = \frac{P}{2}$. Hence $x = y$ and the rectangle is a square.

Thinking Forward

▷ From the graph, $A(0) = 0$. For $A(1)$, we have $f(1) \approx 90$; since f is concave downwards, the area should be a little more than the area of the triangle with vertices $(0,0)$, $(1,0)$, and $(1,90)$, which is 45. So say $A(1) \approx 50$. For $A(2)$, the portion of the graph above $y = f(2)$ appears to be roughly the same size as the area above the graph and below $y = f(2)$. So perhaps $A(2) \approx 2 \cdot f(2) \approx 140$. We can similarly estimate $A(3)$ to be perhaps one half of the area of the rectangle with width three and height the maximum height of f, so $3 \cdot 50 = 150$. Continuing, we can estimate $A(4) \approx 130$ and $A(5) \approx 110$. (Note that the area under the graph between $x = 3$ and $x = 5$ is *negative*, since the graph lies below the x axis.

▷ Since the area increases as long as the graph is above the x axis, and decreases when it is not, clearly $x = 3$ is a local maximum for $A(x)$. Also $x = 0$ is a local minimum, since $A(x)$ increases as you move away from $x = 0$, and $x = 5$ is a local minimum since $A(x)$ increases as you move away from $x = 5$.

▷ The global maximum of $A(x)$ appears to be $A(3)$ — that is the most positive that the function gets. $A(0)$ is a global minimum, since $A(0) = 0$ and $A(x) > 0$ for all other values of x.

▷ Computing derivatives shows that $f(x) = A'(x)$.

▷ We want to compute the extrema of $A(x)$ on $[0, 5]$. Now, $A'(x) = f(x) = 12x^3 - 96x^2 + 180x = 12x(x - 3)(x - 5)$, which has zeros at $x = 0$, $x = 3$, and $x = 5$. These are the critical points of A. Evaluating A at these points gives

$$A(0) = 0, \qquad A(3) = 189, \qquad A(5) = 125,$$

so that $x = 3$ is a local and global maximum, $x = 5$ is a local minimum, and $x = 0$ is a global minimum on $[0, 5]$.

3.5   Related Rates

Thinking Back

Using the chain rule

▷ $\dfrac{d}{dt}(\pi u^2) = \dfrac{d}{dt}(\pi u(t)^2) = 2\pi u(t)\dfrac{du}{dt}$.

▷ $\dfrac{d}{dt}(3r + 2s) = \dfrac{d}{dt}(3r(t) + 2s(t)) = \dfrac{d}{dt}(3r(t)) + \dfrac{d}{dt}(2s(t)) = 3\dfrac{dr}{dt} + 2\dfrac{ds}{dt}$.

▷ $\dfrac{d}{dt}(cu + rs) = \dfrac{d}{dt}(cu(t) + r(t)s(t)) = \dfrac{d}{dt}(cu(t)) + \dfrac{d}{dt}(r(t)s(t)) = c\dfrac{du}{dt} + r(t)\dfrac{ds}{dt} + s(t)\dfrac{dr}{dt}$.

▷ $\dfrac{d}{dt}(k + cu^3) = \dfrac{d}{dt}(k + cu(t)^3) = \dfrac{d}{dt}(k) + \dfrac{d}{dt}(cu^3) = c \cdot 3u(t)^2\dfrac{du}{dt}$.

▷

$$\begin{aligned}
\frac{d}{dt}(cr^2 u) &= \frac{d}{dt}(cr(t)^2 u(t)) \\
&= c\frac{d}{dt}(r(t)^2 u(t)) \\
&= c\left(r(t)^2\frac{du}{dt} + u(t)\frac{d}{dt}(r(t)^2))\right) \\
&= c\left(r(t)^2\frac{du}{dt} + 2u(t)r(t)\frac{dr}{dt}\right).
\end{aligned}$$

▷

$$\begin{aligned}
\frac{d}{dt}\left(\frac{c+s}{k+u}\right) &= \frac{d}{dt}\left(\frac{c+s(t)}{k+u(t)}\right) \\
&= \frac{(k+u(t))\frac{d}{dt}(c+s(t)) - (c+s(t))\frac{d}{dt}(k+u(t))}{(k+u(t))^2} \\
&= \frac{(k+u(t))\frac{ds}{dt} - (c+s(t))\frac{du}{dt}}{(k+u(t))^2} \\
&= \frac{(k+u)\frac{ds}{dt} - (c+s)\frac{du}{dt}}{(k+u)^2}.
\end{aligned}$$

Evaluation in Leibniz notation

▷ Here $\frac{ds}{dt}\big|_{s=2} = 3 \cdot 2^2 - 4 = 8$.

▷ Solving for $\frac{dr}{dt}$ gives $\frac{dr}{dt} = \frac{2}{r}$, so that $\frac{dr}{dt}\big|_{r=3} = \frac{2}{3}$.

▷ Since $4 = 2u\frac{du}{dt}$, we have $\frac{du}{dt} = \frac{2}{u} = \frac{2}{2+3t}$. Thus $\frac{du}{dt}\big|_{t=4} = \frac{2}{2+3\cdot4} = \frac{1}{7}$.

Concepts

1. (a) False. If the side x of a square increases at a constant rate, then $\frac{dx}{dt} = c$. But the area is $A = x^2$, so that $\frac{dA}{dt} = 2x\frac{dx}{dt} = 2cx$; this is not a constant rate.

   (b) True. If the side x of a square increases at a constant rate, then $\frac{dx}{dt} = c$. The perimeter is $P = 4x$, so that $\frac{dP}{dt} = 4\frac{dx}{dt} = 4c$, which is also a constant.

   (c) True. If r is the radius, then $\frac{dr}{dt} = c$. The circumference is $C = 2\pi r$, so that $\frac{dC}{dt} = 2\pi\frac{dr}{dt} = 2c\pi$, which is also a constant.

   (d) False. If r is the radius then $\frac{dr}{dt} = c$. The volume is $V = \frac{4}{3}\pi r^3$, so that $\frac{dV}{dt} = \frac{4}{3}\pi \cdot 3r^2\frac{dr}{dt} = 4c\pi r^2$. This is not a constant rate.

   (e) True. See Theorem 3.12 parts (c) and (d). The volume of the cylinder is $\pi r^2 h$; the volume of the cone is $\frac{1}{3}\pi r^2 h$.

   (f) True. See Theorem 3.12 (b). The volume is $V(r) = \frac{4}{3}\pi r^3$, so that $V'(r) = \frac{4}{3}\pi \cdot 3r^2 = 4\pi r^2 = S(r)$.

   (g) True. The ends of the cylinder are circles with radius r and thus with circumference $2\pi r$.

   (h) False. We must have $a^2 + c^2 = b^2$, since b is the hypotenuse.

3. The volume is $\pi y^2 s$; the surface area is $2\pi ys + 2\pi y^2$. See Theorem 3.12(a).

5. The volume is $\pi r^2 h = \pi\left(\frac{h}{2}\right)^2 h = \frac{\pi}{4}h^3$. The surface area is $2\pi rh + 2\pi r^2 = 2\pi\frac{h}{2}h + 2\pi\left(\frac{h}{2}\right)^2 = \frac{3\pi}{2}h^2$. See Theorem 3.12(b).

7. The statement of the Pythagorean Theorem is Theorem 3.13(a). For example, a triangle with legs 7 and 24 and hypotenuse 25, since $7^2 + 24^2 = 49 + 576 = 625 = 25^2$.

9. Since $V = \frac{4}{3}\pi r^3$, we have

$$\frac{d}{dt}V(t) = \frac{d}{dt}\left(\frac{4}{3}\pi r^3\right) = \frac{4}{3}\pi\frac{d}{dt}(r^3) = \frac{4}{3}\pi \cdot 3r^2\frac{dr}{dt} = 4\pi r^2\frac{dr}{dt}.$$

11. A sphere of radius r has volume $V = \frac{4}{3}\pi r^3$. The equator of such a sphere is the circumference of a circle of radius r, so is $E = 2\pi r$. Solve for r to get $r = \frac{E}{2\pi}$ and substitute in the formula for V to get $V = \frac{4}{3}\pi\left(\frac{E}{2\pi}\right)^3 = \frac{1}{6\pi^2}E^3$. Then

$$\frac{d}{dt}V(t) = \frac{d}{dt}\left(\frac{1}{6\pi^2}E(t)^3\right) = \frac{1}{6\pi^2} \cdot 3E^2\frac{d}{dt}E(t) = \frac{E^2}{2\pi^2} \cdot \frac{dE}{dt}.$$

13. The volume of a sphere of radius r is $V = \frac{4}{3}\pi r^3$, and its surface area is $4\pi r^2$ (see Theorem 3.12 (b)). Then

   (a) $\frac{dV}{dt} = \frac{d}{dt}\left(\frac{4}{3}\pi r^3\right) = \frac{4}{3}\pi \cdot \frac{d}{dt}r(t)^3 = \frac{4}{3}\pi \cdot 3r^2\frac{dr}{dt} = 4\pi r^2\frac{dr}{dt}.$

(b) $\dfrac{dS}{dt} = \dfrac{d}{dt}(4\pi r^2) = 4\pi \dfrac{d}{dt}r(t)^2 = 4\pi \cdot 2r\dfrac{dr}{dt} = 8\pi r\dfrac{dr}{dt}.$

15. Since $V = \pi r^2 h$ and h is constant, we have

$$\frac{dV}{dt} = \frac{d}{dt}(\pi r^2 h) = \pi h \frac{d}{dt}r^2 = \pi h \cdot 2r\frac{dr}{dt} = 2\pi rh\frac{dr}{dt}.$$

17. For a cylinder, $V = \pi r^2 h$; since $h = 2r$, substitute $r = \frac{h}{2}$ to get $V = \frac{\pi}{4}h^3$. Then

$$\frac{dV}{dt} = \frac{d}{dt}\left(\frac{\pi}{4}h^3\right) = \frac{\pi}{4}\frac{d}{dt}h^3 = \frac{\pi}{4} \cdot 3h^2\frac{dh}{dt}.$$

Skills

19. Let x be the side of the square. Then $P = 4x$, so that $x = \frac{P}{4}$. Also $A = x^2 = \left(\frac{P}{4}\right)^2 = \frac{1}{16}P^2$. Thus

$$\frac{dA}{dt} = \frac{d}{dt}\left(\frac{1}{16}P^2\right) = \frac{1}{8}P\frac{dP}{dt}.$$

21. The surface area of a cylinder with radius r and height h is $2\pi rh + 2\pi r^2$. Since $r = 2$, this reduces to $S = 4\pi h + 8\pi$. Then

$$\frac{dS}{dt} = \frac{d}{dt}(4\pi h + 8\pi) = 4\pi\frac{dh}{dt}.$$

23. The surface area of a cone with radius r and height h is $S = \pi r\sqrt{r^2 + h^2} + \pi r^2$. Since $h = 5$, this becomes $S = \pi r\sqrt{r^2 + 25} + \pi r^2$. Then

$$\begin{aligned}
\frac{dS}{dt} &= \frac{d}{dt}(\pi r\sqrt{r^2 + 25}) + \frac{d}{dt}(\pi r^2) \\
&= \pi\left(r\frac{d}{dt}\sqrt{r^2 + 25} + \sqrt{r^2 + 25}\frac{dr}{dt}\right) + \pi\frac{d}{dt}r^2 \\
&= \pi\left(r \cdot \frac{1}{2}(r^2 + 25)^{-1/2} \cdot 2r\frac{dr}{dt} + \sqrt{r^2 + 25}\frac{dr}{dt}\right) + 2\pi r\frac{dr}{dt} \\
&= \left(\frac{\pi r^2}{\sqrt{r^2 + 25}} + \pi\sqrt{r^2 + 25} + 2\pi r\right)\frac{dr}{dt}.
\end{aligned}$$

25. If the hypotenuse of the isosceles right triangle is c, then the length of each leg is $\frac{c}{\sqrt{2}}$, so that the area is

$$A = \frac{1}{2}\left(\frac{c}{\sqrt{2}}\right)\left(\frac{c}{\sqrt{2}}\right) = \frac{1}{4}c^2.$$

Then

$$\frac{dA}{dt} = \frac{d}{dt}\left(\frac{1}{4}c^2\right) = \frac{1}{4} \cdot 2c\frac{dc}{dt} = \frac{1}{2}c\frac{dc}{dt}.$$

27. $\dfrac{df}{dt} = \dfrac{d}{dt}(u^2 + kv) = \dfrac{d}{dt}u^2 + \dfrac{d}{dt}(kv) = 2u\dfrac{du}{dt} + k\dfrac{dv}{dt}.$

29. $\dfrac{df}{dt} = \dfrac{d}{dt}(tv) + \dfrac{d}{dt}(kv) = \left(t\dfrac{dv}{dt} + v\dfrac{d}{dt}t\right) + k\dfrac{dv}{dt} = (k + t)\dfrac{dv}{dt} + v.$

31.

$$\frac{df}{dt} = \frac{d}{dt}(2v\sqrt{u+w}) = 2v\frac{d}{dt}\sqrt{u+w} + \sqrt{u+w}\frac{d}{dt}(2v)$$

$$= 2v\left(\frac{1}{2}(u+w)^{-1/2}\frac{du}{dt} + \frac{1}{2}(u+w)^{-1/2}\frac{dw}{dt}\right) + 2\sqrt{u+w}\frac{dv}{dt}$$

$$= v(u+w)^{-1/2}\left(\frac{du}{dt} + \frac{dw}{dt}\right) + 2(u+w)^{1/2}\frac{dv}{dt}.$$

33.

$$\frac{df}{dt} = \frac{d}{dt}\left(w(u+t)^2\right)$$

$$= w\frac{d}{dt}((u+t)^2) + (u+t)^2\frac{d}{dt}(w)$$

$$= w\left(2(u+t)\frac{du}{dt} + 2(u+t)\frac{dt}{dt}\right) + (u+t)^2\frac{dw}{dt}$$

$$= 2w(u+t)\left(\frac{du}{dt} + 1\right) + (u+t)^2\frac{dw}{dt}.$$

35.

$$\frac{df}{dt} = \frac{d}{dt}\left(\frac{ut+w}{k}\right) = \frac{1}{k}\frac{d}{dt}(ut+w)$$

$$= \frac{1}{k}\left(\frac{d}{dt}(ut) + \frac{d}{dt}(w)\right)$$

$$= \frac{1}{k}\left(u\frac{d}{dt}(t) + t\frac{d}{dt}(u) + \frac{dw}{dt}\right)$$

$$= \frac{1}{k}\left(u + t\frac{du}{dt} + \frac{dw}{dt}\right).$$

Applications

37. If the radius of the circle of ripples is r, then the area is $A = \pi r^2$. Thus $\frac{dA}{dt} = 2\pi r\frac{dr}{dt}$. We are given that the rate of expansion of the radius is 4 inches per second; this is $\frac{dr}{dt}$. Therefore $\frac{dA}{dt} = 8\pi r$. So the rate of expansion of the area is

r	12	24	100
$\frac{dA}{dt}$	96π	192π	800π

When the radius is large, a unit increase in the radius adds an annular region with a large average radius, so a large area. When the radius is small, the annular region has a smaller average radius, so a smaller area. Thus the rate of change of the area is larger for larger radii.

39. The volume of the cube is $V = r^3$ when the side length is r, so that $\frac{dV}{dt} = 3r^2\frac{dr}{dt}$. We are given that $\frac{dr}{dt} = 2$, so that $\frac{dV}{dt} = 6r^2$. If the side length is 8, then $\frac{dV}{dt} = 384$ cubic inches per second.

41. We have an expression, $\frac{dV}{dt} = 6r^2$, for the rate of change of the volume in terms of the radius, but we want an expression in terms of the volume. Since $V = r^3$, we have $r = V^{1/3}$. Substituting this into the expression for $\frac{dV}{dt}$ gives $\frac{dV}{dt} = 6V^{2/3}$. If the volume is 55 cubic inches, then the rate of change of the volume is $6 \cdot 55^{2/3} \approx 86.8$ cubic inches per second.

43. The volume of a balloon of radius r is $V = \frac{4}{3}\pi r^3$, so that

$$\frac{dV}{dt} = \frac{d}{dt}\left(\frac{4}{3}\pi r^3\right) = \frac{4}{3}\pi \frac{d}{dt}(r^3) = \frac{4}{3}\pi \cdot 3r^2 \frac{dr}{dt} = 4\pi r^2 \frac{dr}{dt}.$$

We are given that $\frac{dV}{dt} = 120$ and $r = 12$; substituting gives $120 = 576\pi \frac{dr}{dt}\Big|_{r=12}$, or $\frac{dr}{dt}\Big|_{r=12} = \frac{5}{24\pi}$ inches per second.

45. The surface area is $S = 4\pi r^2$, so that $\frac{dS}{dt} = 8\pi r \frac{dr}{dt}$. Since $r = 15$, this gives $\frac{dS}{dt} = 120\pi \frac{dr}{dt}\Big|_{r=15}$. From Exercise 43, we have $\frac{dV}{dt} = 4\pi r^2 \frac{dr}{dt}$. Now, since $\frac{dV}{dt} = 120$ and $r = 15$, this equation becomes $120 = 900\pi \frac{dr}{dt}\Big|_{r=15}$, so that $\frac{dr}{dt}\Big|_{r=15} = \frac{2}{15\pi}$. Substitute this value into the formula for $\frac{dS}{dt}$ to get

$$\frac{dS}{dt} = 120\pi \frac{dr}{dt}\Big|_{r=15} = 120\pi \cdot \frac{2}{15\pi} = 16.$$

47. The tip of Stuart's shadow is at $s + l$, so its rate of change is

$$\frac{d}{dt}(s+l) = \frac{ds}{dt} + \frac{dl}{dt} = 4 + \frac{12}{7} = \frac{40}{7}$$

using the value of $\frac{dl}{dt}$ from the previous problem. Thus the velocity of the tip of his shadow does not depend on how far Stuart is from the streetlight.

49. Let x be the distance of the base of the ladder from the house. Then the top of the ladder is $T(x) = \sqrt{144 - x^2}$ feet off the ground. So we have

$$\frac{dT}{dt} = \frac{d}{dt}(\sqrt{144 - x^2}) = \frac{1}{2}(144 - x^2)^{-1/2} \cdot (-2x)\frac{dx}{dt} = -x(144 - x^2)^{-1/2}\frac{dx}{dt}.$$

We are given that $\frac{dx}{dt} = \frac{1}{2}$. So when the base of the ladder is 4 feet from the house, $x = 4$, and substituting gives

$$\frac{dT}{dt} = -4(144 - 4^2)^{-1/2} \cdot \frac{1}{2} = -\frac{2}{\sqrt{128}} = -\frac{1}{4\sqrt{2}} = -\frac{\sqrt{2}}{8} \text{ feet per second}$$

for the rate at which the top of the ladder is moving down the side of the house.

51. The triangle formed by the ladder, the house, and the ground is a right triangle with legs x and $\sqrt{144 - x^2}$, so its area is $A = \frac{1}{2}x(144 - x^2)^{1/2}$. Then

$$\frac{dA}{dt} = \frac{d}{dt}\left(\frac{1}{2}x(144 - x^2)^{1/2}\right)$$
$$= \frac{1}{2}\left(x\frac{d}{dt}(144 - x^2)^{1/2} + (144 - x^2)^{1/2}\frac{d}{dt}(x)\right)$$
$$= \frac{1}{2}\left(x \cdot \frac{1}{2}(144 - x^2)^{-1/2} \cdot (-2x)\frac{dx}{dt} + (144 - x^2)^{1/2}\frac{dx}{dt}\right)$$
$$= \frac{1}{2}\left((144 - x^2)^{1/2} - \frac{x^2}{(144 - x^2)^{-1/2}}\right)\frac{dx}{dt}.$$

When the top of the ladder is 6 feet from the ground, we have $\sqrt{144 - x^2} = 6$, so that $x^2 = 144 - 36 = 108$. Further, we are given $\frac{dx}{dt} = \frac{1}{2}$. Substituting all of these values into the above equation gives

$$\frac{dA}{dt} = \frac{1}{2}\left((144 - x^2)^{1/2} - \frac{x^2}{(144 - x^2)^{-1/2}}\right)\frac{dx}{dt} = \frac{1}{2}\left(6 - \frac{108}{6}\right)\frac{1}{2} = \frac{1}{4}(-12) = -3.$$

So the area of the triangle is decreasing by 3 square feet per second at that time.

53. Since the height is $\frac{2}{3}$ of the radius, we see that when the height is 4, then the radius is 6. From the previous problem we have

$$\frac{dV}{dt} = \frac{2}{3}\pi r^2 \frac{dr}{dt}.$$

Substituting $\frac{dV}{dt} = 3$ and $r = 6$, we have $3 = \frac{2}{3}\pi \cdot 6^2 \frac{dr}{dt}\Big|_{h=4}$, so that

$$\frac{dr}{dt}\Big|_{h=4} = \frac{1}{8\pi} \text{ inches per second.}$$

55. We know that $h = \frac{2}{3}r$, so that $r = \frac{3}{2}h$. Since $V = \frac{1}{3}\pi r^2 h$, substituting for r gives $V = \frac{1}{3}\pi\left(\frac{3}{2}h\right)^2 h = \frac{3}{4}\pi h^3$. Differentiating, we get

$$\frac{dV}{dt} = \frac{d}{dt}\left(\frac{3}{4}\pi h^3\right) = \frac{3}{4}\pi \cdot 3h^2 \frac{dh}{dt} = \frac{9}{4}\pi h^2 \frac{dh}{dt}.$$

Now, $\frac{dV}{dt} = 3$ and $h = 4$; substituting gives $3 = \frac{9}{4}\pi \cdot 4^2 \frac{dh}{dt}$, so that

$$\frac{dh}{dt} = \frac{1}{12\pi} \text{ inches per second.}$$

57. As the height of the cone decreases, the ratio between radius and height will continue to be $\frac{2}{5}$; that is, $r = \frac{2}{5}h$. Now, the volume of the cone is $V = \frac{1}{3}\pi r^2 h = \frac{1}{3}\pi\left(\frac{2}{5}h\right)^2 h = \frac{4}{75}\pi h^3$, so that

$$\frac{dV}{dt} = \frac{d}{dt}\left(\frac{4}{75}\pi h^3\right) = \frac{4}{75}\pi \cdot 3h^2 \frac{dh}{dt} = \frac{4}{25}\pi h^2 \frac{dh}{dt}.$$

Since the ice cream drips out at a rate of one half cubic inch per minute, we have $\frac{dV}{dt} = -\frac{1}{2}$. When $h = 3$, substituting gives

$$-\frac{1}{2} = \frac{4}{25}\pi \cdot 3^2 \frac{dh}{dt}\Big|_{h=3}, \text{ so that } \frac{dh}{dt}\Big|_{h=3} = -\frac{1}{2}\left(\frac{25}{36\pi}\right) = -\frac{25}{72\pi}.$$

Thus the height is decreasing at a rate of $\frac{25}{72\pi}$ inches per second.

59. (a) The area of the triangle is $A = \frac{1}{2}ab$, so

$$\frac{dA}{dt} = \frac{d}{dt}\left(\frac{1}{2}ab\right) = \frac{1}{2}\left(a\frac{db}{dt} + b\frac{da}{dt}\right) = \frac{1}{2}(-2a + 4b) = 2b - a,$$

since a is increasing at 4 inches per second and b is decreasing at 2 inches per second.

(b) Since a is increasing from 1 inch, it makes sense on $[1, \infty)$. Since b is decreasing from 10 inches, it makes sense on $[0, 10]$. Since b decreases at 2 inches per second, it reaches zero at 5 seconds, so that the problem makes sense for $t \in [0, 5]$.

(c) The area of the triangle is increasing when $2b - a > 0$, and decreasing when $2b - a < 0$. Since $a = 1 + 4t$ and $b = 10 - 2t$, we see that $2b - a > 0$ when $20 - 4t - (1 + 4t) > 0$, so when $8t < 19$. Thus the area of the triangle is increasing until $t = \frac{19}{8}$ seconds and decreases after that.

61. (a) Let the angle between the horizontal and the line to the top of the cliff, at the tip of the shadow, be $\rho = \rho(t)$. Then clearly $\cot \rho(t) = \frac{x(t)}{13200 - 11710}$, so that $x(t) = 1490 \cot \rho(t)$. Note that $\rho(t)$ is changing at $11°$ degrees per hour, so at $\frac{11 \cdot \pi}{180 \cdot 60} = \frac{11\pi}{10800}$ radians per minute. It follows that

$$x'(t) = 1490(\cot \rho(t))' = -1490 \csc^2 \rho(t) \cdot \rho'(t) = \frac{11\pi}{10800} \cdot 1490 \csc^2 \rho(t)$$

$$= \frac{1639\pi}{1080} \csc^2 \rho(t).$$

(b) When the tip of the shadow reaches Ian, then $\theta = \rho$, and at that point $\csc \theta = \frac{\sqrt{1490^2 + 3100^2}}{1490}$, so that

$$x'(t) = \frac{1639\pi}{1080} \cdot \frac{1490^2 + 3100^2}{1490^2} \approx 25.4 \text{ feet per minute.}$$

Proofs

63. The volume of a sphere with radius r is $V(r) = \frac{4}{3}\pi r^3$. Then

$$\frac{d}{dr} V(r) = \frac{4}{3}\pi \cdot 3r^2 = 4\pi r^2 = A(r).$$

Think of enlarging a sphere by enclosing it in successively larger hollow spheres. Intuitively, each time you enlarge the sphere, you are adding a hollow sphere with surface area $A(r)$.

65. Since the length of the hypotenuse does not change, we must have $\frac{dh}{dt} = 0$. But $h = \sqrt{a^2 + b^2}$, so that

$$\frac{dh}{dt} = \frac{d}{dt}\left((a^2 + b^2)^{1/2}\right)$$

$$= \frac{1}{2}(a^2 + b^2)^{-1/2} \cdot 2a\frac{da}{dt} + \frac{1}{2}(a^2 + b^2)^{-1/2} \cdot 2b\frac{db}{dt}$$

$$= \left(a\frac{da}{dt} + b\frac{db}{dt}\right)(a^2 + b^2)^{-1/2}.$$

If $\frac{dh}{dt} = 0$, then $a\frac{da}{dt} + b\frac{db}{dt} = 0$. Rearranging gives $a\frac{da}{dt} = -b\frac{db}{dt}$, so that $\frac{\frac{da}{dt}}{\frac{db}{dt}} = -\frac{b}{a}$.

Thinking Forward

Parametric curves

▷ A table of the coordinates against z is

z	-3	-2	-1	0	1	2	3
$(3z + 1, z^2 - 4)$	$(-8, 5)$	$(-5, 0)$	$(-2, -3)$	$(1, -4)$	$(4, -3)$	$(7, 0)$	$(10, 5)$

and a plot of these points is

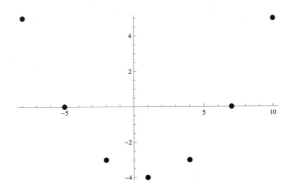

▷ Since $z = 3t$, this is the same as evaluating the coordinates $(x, y) = (3z + 1, z^2 - 4)$ at $z = 0, 3, 6, 9$, and 12. A table of these coordinates is

z	0	3	6	9	12
$(3z + 1, z^2 - 4)$	$(1, -4)$	$(10, 5)$	$(19, 32)$	$(28, 77)$	$(37, 140)$

and a plot of these points is

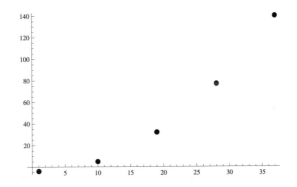

▷ The curve passes through $(7, 0)$ when $3z + 1 = 7$, so when $z = 2$. Since z moves at 5 units per second, this happens at $t = \frac{2}{5}$. The statement that z moves at 5 units per second means that $\frac{dz}{dt} = 5$. Then the rate of change of the x and y coordinates is

$$\frac{dx}{dt} = \frac{d}{dt}(3z + 1) = 3\frac{dz}{dt} = 15,$$

$$\frac{dy}{dt} = \frac{d}{dt}\left(z^2 - 4\right) = 2z\frac{dz}{dt} = 10z = 20.$$

▷ Since $x = 3z + 1$, we have $z = \frac{x}{3} - \frac{1}{3}$. Thus $\frac{dz}{dt} = \frac{d}{dt}\left(\frac{x}{3} - \frac{1}{3}\right) = \frac{1}{3}\frac{dx}{dt}$. Since $\frac{dx}{dt} = 5$, we have $\frac{dz}{dt} = \frac{5}{3}$.
Now, $y = z^2 - 4$, so that

$$\frac{dy}{dt} = \frac{d}{dt}\left(z^2 - 4\right) = 2z\frac{dz}{dt} = 2z \cdot \frac{5}{3} = \frac{10}{3}z.$$

Finally, when the curve passes through $(7, 0)$, we have $z = 2$, so that $\frac{dy}{dt} = \frac{20}{3}$.

3.6   L'Hôpital's Rule

Thinking Back

Indeterminate Forms

Theorems 1.33 and 1.34 present the indeterminate forms and the forms that are not indeterminate. Of the ones given, the following are indeterminate:

$$\frac{\infty}{\infty} \quad 0^0 \quad \frac{0}{0} \quad 1^\infty \quad \infty - \infty \quad \infty \cdot 0 \quad \infty^0$$

For the forms that are not indeterminate, we have the following behaviors:

- $\infty \times \infty$ is ∞ since both factors go to ∞.

- ∞^∞ is ∞ since both the base and the exponent go to ∞.

- $\frac{0}{\infty}$ is 0 since the numerator is small while the denominator is large.

- 0^∞ is 0 since the base is small while the exponent is large.

- $\frac{\infty}{0}$ is ∞ since the numerator is large while the denominator is small.

Simple limit calculations

▷ Since 2^x gets large as x does, $\lim\limits_{x\to\infty} 2^x = \infty$.

▷ Since x gets large as x does, so does x^5, so that $\frac{1}{x^5} = x^{-5}$ gets close to zero: $\lim\limits_{x\to\infty} x^{-5} = 0$.

▷ Since $3x \to -\infty$ as $x \to -\infty$, $e^{3x} \to e^{-\infty} = 0$, so that $\lim\limits_{x\to-\infty} e^{3x} = 0$.

▷ Since $x^2 \to 0$, $\lim\limits_{x\to 0} \frac{1}{x^2} = \infty$.

▷ Since $\ln x$ gets increasingly negative as $x \to 0^+$, we have $\lim\limits_{x\to 0^+} \ln x = -\infty$.

▷ Large powers of $\frac{1}{2}$ get closer to zero, so that $\lim\limits_{x\to\infty} \left(\frac{1}{2}\right)^x = 0$.

▷ The graph of $\tan x$ goes off to ∞ as $x \to \frac{\pi}{2}$ from the left, and to $-\infty$ as $x \to \frac{\pi}{2}$ from the right. Thus the limit does not exist.

▷ This limit does not exist since $\sin x$ oscillates between 1 and -1.

▷ $\log_{1/2} x$ is the power to which you must raise $\frac{1}{2}$ to get x. As x gets increasingly large, this power gets increasingly large negatively. Alternatively, the graph of $\log_{1/2} x$ is the graph of $\log_2 x$ reflected through the x axis. Thus $\lim\limits_{x\to\infty} \log_{1/2} x = -\infty$.

Concepts

1. (a) False. For example, $\lim\limits_{x\to\infty} \frac{x}{x}$ has the indeterminate form $\frac{\infty}{\infty}$, yet the limit is 1.

    (b) False. It can be used only when the quotient has the indeterminate form $\frac{0}{0}$ or $\frac{\infty}{\infty}$ (and when f and g satisfy some other technical conditions; see Theorem 3.14).

    (c) False. L'Hôpital's rule involves differentiating the numerator and denominator separately.

    (d) False. For example, it applies to $\lim\limits_{x\to 1} \frac{x-1}{x-1}$.

(e) True (see Theorem 3.14), although quite often other indeterminate forms can be manipulated into one of these forms.

(f) False. If $\lim\limits_{x \to 2} \ln(f(x)) = 4$, then $\lim\limits_{x \to 2} f(x) = e^4$. See Theorem 3.15(a).

(g) True. See Theorem 3.15(b).

(h) False. If $\lim\limits_{x \to 2} \ln(f(x)) = -\infty$, then $\lim\limits_{x \to 2} f(x) = 0$. See Theorem 3.15(c).

3. See the discussion at the beginning of the chapter. In summary, near $x = c$, $f(x)$ is closely approximated by its tangent line, which has slope $f'(x)$, and $g(x)$ is closely approximated by its tangent line, which has slope $g'(x)$.

5. Since $f'(x) = 2x$, the slope of the tangent line at $x = 1$ is 2, so that the equation of the tangent line is $y - f(1) = 2(x - 1)$, or $y = 2x - 2$. Since $g'(x) = \frac{1}{x}$, the slope of the tangent line at $x = 1$ is 1, so that the equation of the tangent line is $y - g(1) = 1(x - 1)$, or $y = x - 1$. Since $f(x)$ is well-approximated by its tangent line as $x \to 1$, and similarly for $g(x)$, the limit of the quotient might be well-approximated by the quotient of the tangent lines.

7. (a) $\lim\limits_{x \to \infty} (2^{-x} x) = \lim\limits_{x \to \infty} \frac{2^{-x}}{1/x}$ is in the form $\frac{0}{0}$.

(b) $\lim\limits_{x \to \infty} (2^{-x} x) = \lim\limits_{x \to \infty} \frac{x}{2^x}$ is in the form $\frac{\infty}{\infty}$.

(c) The second form is clearly simpler.

9. (a) $\lim\limits_{x \to \infty} (x^{-2} \ln x) = \lim\limits_{x \to \infty} \frac{x^{-2}}{1/\ln x}$ is in the form $\frac{0}{0}$.

(b) $\lim\limits_{x \to \infty} (x^{-2} \ln x) = \lim\limits_{x \to \infty} \frac{\ln x}{x^2}$ is in the form $\frac{\infty}{\infty}$.

(c) The second form is clearly simpler.

11. (a) $\lim\limits_{x \to 0} (x \csc x) = \lim\limits_{x \to 0} \frac{x}{1/\csc x} = \lim\limits_{x \to 0} \frac{x}{\sin x}$ is in the form $\frac{0}{0}$.

(b) $\lim\limits_{x \to 0} (x \csc x) = \lim\limits_{x \to 0} \frac{\csc x}{1/x}$ is in the form $\frac{\infty}{\infty}$.

(c) The first form is clearly simpler.

13. The error is in the second equality. L'Hôpital's rule does not apply to $\lim\limits_{x \to 0} \frac{2x-1}{(\ln 2)2^x}$ since this is not an indeterminate form. In fact, neither the numerator nor the denominator vanishes, so the limit can be easily evaluated:

$$\lim_{x \to 0} \frac{x^2 - x}{2^x - 1} \overset{\text{L'H}}{=} \lim_{x \to 0} \frac{2x - 1}{(\ln 2)2^x} = \frac{2 \cdot 0 - 1}{(\ln 2)2^0} = \frac{-1}{\ln 2} = -\frac{1}{\ln 2}.$$

Skills

15. (a) The form of the limit is $\frac{0}{0}$, so that L'Hôpital's rule applies, and

$$\lim_{x \to 1} \frac{x^2 + x - 2}{x - 1} \overset{\text{L'H}}{=} \lim_{x \to 1} \frac{2x + 1}{1} = \lim_{x \to 1} (2x + 1) = 3.$$

(b) Since $x^2 + x - 2 = (x + 2)(x - 1)$, we have

$$\lim_{x \to 1} \frac{x^2 + x - 2}{x - 1} = \lim_{x \to 1} \frac{(x + 2)(x - 1)}{x - 1} = \lim_{x \to 1} (x + 2) = 3.$$

17. (a) The form of the limit is $\frac{\infty}{-\infty}$, so that L'Hôpital's rule applies, and

$$\lim_{x\to\infty}\frac{x-1}{2-3x^2}\overset{\text{L'H}}{=}\lim_{x\to\infty}\frac{1}{-6x}=\frac{0}{-\infty}=0.$$

(b) Since the degree of the denominator exceeds that of the numerator, we have

$$\lim_{x\to\infty}\frac{x-1}{2-3x^2}=\lim_{x\to\infty}\frac{1/x-1/x^2}{2/x^2-3}=\frac{0}{0-3}=0.$$

19. (a) Since the form of the limit is $\frac{\infty}{-\infty}$, L'Hôpital's rule applies, and

$$\lim_{x\to\infty}\frac{e^{3x}}{1-e^{2x}}\overset{\text{L'H}}{=}\lim_{x\to\infty}\frac{3e^{3x}}{-2e^{2x}}=-\frac{3}{2}\lim_{x\to\infty}e^x=-\infty.$$

(b) $\lim_{x\to\infty}\dfrac{e^{3x}}{1-e^{2x}}=\lim_{x\to\infty}\dfrac{e^x}{e^{-2x}-1}=\dfrac{\infty}{-1}=-\infty.$

21. The form of the limit is $\frac{2^3-8}{3-3}=\frac{0}{0}$, so that L'Hôpital's rule applies. Then

$$\lim_{x\to3}\frac{2^x-8}{3-x}\overset{\text{L'H}}{=}\lim_{x\to3}\frac{(\ln 2)2^x}{-1}=-(\ln 2)\lim_{x\to3}2^x=-8\ln 2=-\ln 256.$$

23. The form of the limit is $\frac{e^{4-4}-1}{2^2-4}=\frac{0}{0}$, so that L'Hôpital's rule applies, and

$$\lim_{x\to2}\frac{e^{2x-4}-1}{x^2-4}\overset{\text{L'H}}{=}\lim_{x\to2}\frac{2e^{2x-4}}{2x}=\frac{e^{4-4}}{2}=\frac{1}{2}.$$

25. The numerator goes to zero, and the denominator to ∞, as $x\to\infty$, so that $\lim_{x\to\infty}\dfrac{2^{-x}}{x^2+1}=0.$

27. $\lim_{x\to0}\dfrac{x^{-2}}{x^{-3}+1}=\lim_{x\to0}\dfrac{x}{1+x^3}=0.$

29. Rewriting allows us to use L'Hôpital's rule :

$$\lim_{x\to\infty}\left(e^x-\frac{e^x}{x+1}\right)=\lim_{x\to\infty}\frac{(x+1)e^x-e^x}{x+1}=\lim_{x\to\infty}\frac{xe^x}{x+1}\overset{\text{L'H}}{=}\lim_{x\to\infty}\frac{e^x+xe^x}{1}=\infty.$$

31. This has the form $\infty\cdot0$, but rewriting allows us to use L'Hôpital's rule :

$$\lim_{x\to\infty}x\left(\frac{1}{2}\right)^x=\lim_{x\to\infty}\frac{x}{2^x}\overset{\text{L'H}}{=}\lim_{x\to\infty}\frac{1}{(\ln 2)2^x}=0.$$

33. Multiply numerator and denominator by e^{-x} and then use L'Hôpital's rule :

$$\lim_{x\to\infty}\frac{xe^x}{e^{2x}+1}=\lim_{x\to\infty}\frac{x}{e^x+e^{-x}}\overset{\text{L'H}}{=}\lim_{x\to\infty}\frac{1}{e^x-e^{-x}}=\frac{1}{\infty-0}=0.$$

35. Since this has the form $\frac{0}{\infty}$, the limit is zero.

37. This has the form $\frac{-\infty}{-\infty}$, so that L'Hôpital's rule applies. However, it is simpler to note that $\log_2 3x=\log_2 3+\log_2 x$:

$$\lim_{x\to0^+}\frac{\log_2 x}{\log_2 3x}=\lim_{x\to0^+}\frac{\log_2 x}{\log_2 3+\log_2 x}=\lim_{x\to0^+}\frac{1}{(\log_2 3)/(\log_2 x)+1}=1.$$

39. This has the form $\frac{-\infty}{-\infty}$, so that L'Hôpital's rule applies:

$$\lim_{x \to 2^+} \frac{\ln(x-2)}{\ln(x^2-4)} \overset{\text{L'H}}{=} \lim_{x \to 2^+} \frac{1/(x-2)}{(2x)/(x^2-4)}$$

$$= \lim_{x \to 2^+} \frac{x^2-4}{2x(x-2)} = \lim_{x \to 2^+} \frac{x^2-4}{2x^2-4x}$$

$$\overset{\text{L'H}}{=} \lim_{x \to 2^+} \frac{2x}{4x-4} = 1.$$

41. This has the form $\frac{0}{0}$, so that L'Hôpital's rule applies:

$$\lim_{x \to 0} \frac{\cos x - 1}{\sin x} \overset{\text{L'H}}{=} \lim_{x \to 0} \frac{-\sin x}{\cos x} = 0.$$

43. This has the form $\frac{0}{0}$, so that L'Hôpital's rule applies:

$$\lim_{x \to 0} \frac{1-\cos x}{\tan x} \overset{\text{L'H}}{=} \lim_{x \to 0} \frac{\sin x}{\sec^2 x} = \lim_{x \to 0} (\sin x \cos^2 x) = 0.$$

45. This has the form $\frac{0}{0}$, so that L'Hôpital's rule applies:

$$\lim_{x \to 0} \frac{x \cos x}{1-e^x} \overset{\text{L'H}}{=} \lim_{x \to 0} \frac{\cos x - x \sin x}{-e^x} = \frac{1-0}{-1} = -1.$$

47. This has the form $\frac{0}{0}$, so that L'Hôpital's rule applies:

$$\lim_{x \to 0} \frac{\tan^{-1} x}{\sin x} \overset{\text{L'H}}{=} \lim_{x \to 0} \frac{1/(x^2+1)}{\cos x} = \lim_{x \to 0} \frac{1}{(x^2+1)\cos x} = 1.$$

49. This has the form $0^{-\infty} = \frac{1}{0^\infty}$, which is not indeterminate. The limit is ∞.

51. This has the indeterminate form 0^0. Take logs and then compute the limit:

$$\lim_{x \to 2^+} (x^2-4)\ln(x-2) = \lim_{x \to 2^+} \frac{\ln(x-2)}{(x^2-4)^{-1}}$$

$$\overset{\text{L'H}}{=} \lim_{x \to 2^+} \frac{1/(x-2)}{-2x(x^2-4)^{-2}} = \lim_{x \to 2^+} \frac{(x^2-4)^2}{-2x(x-2)}$$

$$\overset{\text{L'H}}{=} \lim_{x \to 2^+} \frac{4x(x^2-4)}{-4x+4} = 0.$$

Since the limit of the log is zero, Theorem 3.15 tells us that the original limit is 1.

53. This has the indeterminate form 1^∞. Computing the limit of the log, we get

$$\lim_{x \to 1^+} \frac{1}{x-1} \ln x = \lim_{x \to 1^+} \frac{\ln x}{x-1} \overset{\text{L'H}}{=} \lim_{x \to 1^+} \frac{1/x}{1} \lim_{x \to 1^+} \frac{1}{x} = 1.$$

Since the limit of the log is 1, Theorem 3.15 tells us that the original limit is $e^1 = e$.

55. This has the indeterminate form ∞^0. Computing the limit of the log, we get

$$\lim_{x \to \infty} \frac{1}{x} \ln x = \lim_{x \to \infty} \frac{\ln x}{x} \overset{\text{L'H}}{=} \lim_{x \to \infty} \frac{1/x}{1} = 0.$$

Since the limit of the log is zero, Theorem 3.15 tells us that the original limit is 1.

57. This has the form 0^∞, so that the limit is zero.

59. This has the indeterminate form 0^0. Taking the limit of the log gives

$$\lim_{x \to 1^+} \ln x \cdot \ln(x - 1) = \lim_{x \to 1^+} \frac{\ln(x - 1)}{(\ln x)^{-1}}$$
$$\overset{\text{L'H}}{=} \lim_{x \to 1^+} \frac{(x - 1)^{-1}}{-(\ln x)^{-2} \cdot x^{-1}} = -\lim_{x \to 1^+} \frac{x(\ln x)^2}{x - 1}$$
$$\overset{\text{L'H}}{=} -\lim_{x \to 1^+} \frac{(\ln x)^2 + 2x \ln x \cdot x^{-1}}{1} = -\lim_{x \to 1^+} \left(2 \ln x + (\ln x)^2\right) = 0.$$

Since the limit of the log is 0, Theorem 3.15 tells us that the original limit is 1.

61. This has the indeterminate form 0^0. Taking the limit of the log gives

$$\lim_{x \to 0^+} \sin x \cdot \ln x = \lim_{x \to 0^+} \frac{\ln x}{(\sin x)^{-1}}$$
$$\overset{\text{L'H}}{=} \lim_{x \to 0^+} \frac{1}{x(-\cos x(\sin x)^{-2})} = -\lim_{x \to 0^+} \frac{\sin^2 x}{x \cos x}$$
$$\overset{\text{L'H}}{=} -\lim_{x \to 0^+} \frac{2 \sin x \cos x}{\cos x - x \sin x} = 0.$$

Since the limit of the log is zero, Theorem 3.15 tell us that the original limit is 1.

63. This has the indeterminate form 1^∞. Taking the limit of the log gives

$$\lim_{x \to 0^+} \frac{\ln \cos x}{x} \overset{\text{L'H}}{=} \lim_{x \to 0} \frac{\frac{1}{\cos x}(-\sin x)}{1} = \lim_{x \to 0}(-\tan x) = 0.$$

Since the limit of the log is zero, Theorem 3.15 tells us that the original limit is 1.

65. Since

$$\lim_{x \to \infty} \frac{u(x)}{v(x)} = \lim_{x \to \infty} \frac{x + 100}{x} = \lim_{x \to \infty} \left(1 + \frac{100}{x}\right) = 1$$
$$\lim_{x \to \infty} \frac{v(x)}{u(x)} = \lim_{x \to \infty} \frac{x}{x + 100} = \lim_{x \to \infty} \frac{1}{1 + 100/x} = 1,$$

neither function dominates the other.

67. Since

$$\lim_{x \to \infty} \frac{u(x)}{v(x)} = \lim_{x \to \infty} \frac{100x^2}{2x^{100}} = \lim_{x \to \infty} 50x^{-98} = 0$$
$$\lim_{x \to \infty} \frac{v(x)}{u(x)} = \lim_{x \to \infty} \frac{2x^{100}}{100x^2} = \lim_{x \to \infty} \frac{1}{50} x^{98} = \infty,$$

$v(x)$ dominates $u(x)$.

69. Since $e > 2$, we have

$$\lim_{x \to \infty} \frac{u(x)}{v(x)} = \lim_{x \to \infty} \left(\frac{2}{e}\right)^x = 0, \qquad \lim_{x \to \infty} \frac{v(x)}{u(x)} = \lim_{x \to \infty} \left(\frac{e}{2}\right)^x = \infty,$$

so that $v(x)$ dominates $u(x)$.

71. Since $u(x) = \log_2 x = \frac{\ln x}{\ln 2}$ and $v(x) = \log_{30} x = \frac{\ln x}{\ln 30}$, we see that $\frac{u(x)}{v(x)} = \frac{\ln 30}{\ln 2}$ and $\frac{v(x)}{u(x)} = \frac{\ln 2}{\ln 30}$.
    Since these ratios are constant real numbers, neither function dominates the other.

73. Since

$$\lim_{x \to \infty} \frac{u(x)}{v(x)} = \lim_{x \to \infty} \frac{0.001 e^{0.001x}}{100 x^{100}}$$

$$\overset{\text{L'H}}{=} 10^{-5} \lim_{x \to \infty} \frac{10^{-3} e^{0.001x}}{100 x^{99}}$$

$$\overset{\text{L'H}}{=} \ldots \overset{\text{L'H}}{=} 10^{-5} \lim_{x \to \infty} \frac{10^{-3 \cdot 100} e^{0.001x}}{100!} = \infty,$$

$$\lim_{x \to \infty} \frac{v(x)}{v(x)} = \lim_{x \to \infty} \frac{100 x^{100}}{0.001 e^{0.001x}}$$

$$\overset{\text{L'H}}{=} 10^5 \lim_{x \to \infty} \frac{100 x^{99}}{10^{-3} e^{0.001x}}$$

$$\overset{\text{L'H}}{=} \ldots \overset{\text{L'H}}{=} 10^5 \lim_{x \to \infty} \frac{100!}{10^{-3 \cdot 100} e^{0.001x}} = 0,$$

$u(x)$ dominates $v(x)$.

75. Since e^x dominates $x^{101} + 500$, we see that

$$\lim_{x \to \infty} \frac{x^{101} + 500}{e^x} = \lim_{x \to \infty} \frac{1}{\frac{e^x}{x^{101} + 500}} = \frac{1}{\infty} = 0.$$

77. Since $\sqrt{x} = x^{1/2}$ is a positive power of x, it dominates $\ln x$ and thus dominates $300 \ln x$, so that
    $\lim_{x \to \infty} \frac{\sqrt{x}}{300 \ln x} = \infty.$

79. Since exponential growth functions dominate power functions,

$$\lim_{x \to \infty} 2^x x^{-100} = \lim_{x \to \infty} \frac{2^x}{x^{100}} = \infty.$$

81. We have $f'(x) = 1 + \ln x$, so that $f'(x) = 0$ at $x = e^{-1}$. Further, $f''(x) = x^{-1}$ is positive on $(0, \infty)$,
    so that $x = e^{-1}$ is a local minimum. Now, $f(e^{-1}) = e^{-1}(-1) = -e^{-1}$, and $f(1) = 1 \ln 1 = 0$.
    Further,

$$\lim_{x \to 0} x \ln x = \lim_{x \to 0} \frac{\ln x}{x^{-1}} \overset{\text{L'H}}{=} \lim_{x \to 0} \frac{x^{-1}}{-x^{-2}} = -\lim_{x \to 0} x = 0,$$

$$\lim_{x \to \infty} x \ln x = \infty.$$

So on $I = (0, 1]$, we have a global minimum at $x = e^{-1}$, with value $f(x) = -e^{-1}$, and a global
maximum at $x = 1$ with value $f(x) = 0$. Values of $f(x)$ as $x \to 0^+$ approach, but never reach, that
global maximum value. On $J = (0, \infty)$, we still have a global minimum at $x = e^{-1}$, but there is no
global maximum since f increases without bound as $x \to \infty$.

83. We have $f'(x) = (3x^2 - x^3)e^{-x}$, so that $f'(x) = 0$ for $x = 0$ and for $x = 3$. Since $f''(x) = (x^3 - 6x^2 + 6x)e^{-x}$, we see that $f''(0) = 0$ and $f''(3) < 0$, so that $x = 3$ is a local maximum while
    $x = 0$ is an inflection point. We have $f(3) = 27e^{-3} \approx 1.344$ and $f(0) = 0$. Further, since e^x
    dominates x^3, we have

$$\lim_{x \to \infty} x^3 e^{-x} = \lim_{x \to \infty} \frac{x^3}{e^x} = 0,$$

$$\lim_{x \to -\infty} x^3 e^{-x} = \lim_{x \to \infty} (-x)^3 e^x = -\lim_{x \to \infty} \frac{e^x}{x^3} = -\infty.$$

Thus f has a horizontal asymptote at $y = 0$ as $x \to \infty$, but f grows negatively without bound as $x \to -\infty$. Finally, note that $f(x) > 0$ for $x > 0$, so that f approaches its horizontal asymptote from above. Thus on $[0, \infty)$ there is a global maximum at $x = 3$ and a global minimum at $x = 0$. On $(-\infty, \infty)$, there is a global maximum at $x = 3$, but no global minimum since f gets increasingly small as $x \to -\infty$.

85. We have $f'(x) = \frac{(1-\cos x)(\cos x) - \sin x(\sin x)}{(1-\cos x)^2} = \frac{\cos x - 1}{(1-\cos x)^2} = \frac{1}{\cos x - 1}$. Since f' never vanishes, f has no critical points, so we need only examine its behavior at the interval endpoints. We have $f(\pi) = \frac{\sin \pi}{1 - \cos \pi} = 0$, and

$$\lim_{x \to 0^+} \frac{\sin x}{1 - \cos x} \overset{\text{L'H}}{=} \lim_{x \to 0^+} \frac{\cos x}{\sin x} \overset{\text{L'H}}{=} \lim_{x \to 0^+} \cot x = \infty,$$

$$\lim_{x \to 2\pi^-} \frac{\sin x}{1 - \cos x} = \lim_{x \to 0^-} \frac{\sin x}{1 - \cos x} \overset{\text{L'H}}{=} \lim_{x \to 0^-} \frac{\cos x}{\sin x} \overset{\text{L'H}}{=} \lim_{x \to 0^-} \cot x = -\infty.$$

So on $I = (0, \pi]$, there is no global maximum since f goes to ∞ near $x = 0$. The global minimum is thus $x = \pi$, with $f(\pi) = 0$. On $J = (0, 2\pi)$, there is neither a global minimum nor a global maximum, since f goes to ∞ near 0 and to $-\infty$ near 2π. (Note that $f'(x)$ does not exist at either endpoint, 0 or 2π, but those points are not contained in the interval of interest).

Applications

87. Since $w(0) = 835(1 - e^{-0.006 \cdot 0}) = 0$, the formula $\frac{w(t)}{t}$ is the same as $\frac{w(t) - w(0)}{t - 0}$, which is clearly the average rate of change of the number of wolves from time 0 to time t. Since $t = 0$ corresponds to 1994, it follows that $\frac{w(t)}{t}$ is the average rate of change of the number of wolves since 1994.

89. We want $\lim\limits_{t \to \infty} E(t)$:

$$\lim_{t \to \infty} E(t) = \lim_{t \to \infty} \left(\frac{1}{1 + 0.2W_0} \left(E_0 + 72e^{-0.006t} \sin\left(\frac{\pi t}{4}\right) + 8te^{-0.006t} \sin\left(\frac{\pi t}{4}\right) \right) \right)$$

$$= \lim_{t \to \infty} \frac{E_0}{1 + 0.2W_0} + \lim_{t \to \infty} \frac{(72 + 8t)\sin\left(\frac{\pi t}{4}\right)}{(1 + 0.2W_0)e^{0.006t}}$$

assuming that both limits exist. The first limit is clearly just $\frac{E_0}{1 + 0.2W_0}$. Now, since $-1 \le \sin\left(\frac{\pi t}{4}\right) \le 1$, we have

$$-\frac{72 + 8t}{e^{0.006t}} \le \frac{(72 + 8t)\sin\left(\frac{\pi t}{4}\right)}{(1 + 0.2W_0)e^{0.006t}} \le \frac{72 + 8t}{e^{0.006t}}.$$

To evaluate the limit as $t \to \infty$ of the expression on either end of this inequality, note that exponential growth functions such as $e^{0.006t}$ dominate power functions, so that

$$\lim_{t \to \infty} \left(-\frac{72 + 8t}{e^{0.006t}} \right) = \lim_{t \to \infty} \frac{72 + 8t}{e^{0.006t}} = 0.$$

By the Squeeze Theorem, the limit we are interested in is zero as well. Thus $\lim\limits_{t \to \infty} E(t) = \frac{E_0}{1 + 0.2W_0}$.

Attempting to use L'Hôpital's rule will be complicated, since the derivative of the numerator will consist of two terms: one of the form $K \sin \frac{\pi t}{4}$ and the other of the form $C(72 + 8t) \cos \frac{\pi t}{4}$. Taking further derivatives of the numerator will continue to produce a term containing the factor $72 + 8t$ multiplied by either $\sin \frac{\pi t}{4}$ or $\cos \frac{\pi t}{4}$, and the denominator will be a multiple of $e^{0.006t}$. Thus without appealing to the fact that exponentials dominate power functions, L'Hôpital's rule will not produce the answer (at least not without some algebraic manipulations).

Proofs

91. Let $f(x) = Ae^{kx}$ for $k > 0$ be an exponential growth function with $A > 0$. Then

$$\lim_{x\to\infty} \frac{Ae^{kx}}{x^2} = A \lim_{x\to\infty} \frac{e^{kx}}{x^2}.$$

Since this limit has the form $\frac{\infty}{\infty}$ because $k > 0$, L'Hôpital's rule applies so we get

$$\lim_{x\to\infty} \frac{Ae^{kx}}{x^2} \stackrel{\text{L'H}}{=} A \lim_{x\to\infty} \frac{ke^{kx}}{2x} \stackrel{\text{L'H}}{=} A \lim_{x\to\infty} \frac{k^2 e^{kx}}{2} = \infty.$$

Thus the exponential function dominates x^2.

93. We want to show that $\lim_{x\to\infty} \frac{Ae^{kx}}{Bx^r} = \infty$ for $k, r, A, B > 0$. First assume that r is an integer. Then since $k, r > 0$, L'Hôpital's rule applies, so that

$$\lim_{x\to\infty} \frac{Ae^{kx}}{Bx^r} \stackrel{\text{L'H}}{=} \frac{A}{B} \lim_{x\to\infty} \frac{ke^{kx}}{rx^{r-1}} \stackrel{\text{L'H}}{=} \frac{A}{B} \lim_{x\to\infty} \frac{k^2 e^{kx}}{r(r-1)x^{r-2}} \stackrel{\text{L'H}}{=} \cdots \stackrel{\text{L'H}}{=} \frac{A}{B} \lim_{x\to\infty} \frac{k^r e^{kx}}{r!} = \infty.$$

Thus exponential functions dominate power functions with positive integral exponents. Now let r be an arbitrary positive number, and let $\lceil r \rceil$ be the smallest integer less than or equal to r. Then for $x > 1$ we have

$$\frac{Ae^{kx}}{Bx^r} > \frac{A}{B} \cdot \frac{e^{kx}}{x^{\lceil r \rceil}},$$

so that, since $\lceil r \rceil$ is an integer,

$$\lim_{x\to\infty} \frac{Ae^{kx}}{Bx^r} \geq \frac{A}{B} \cdot \lim_{x\to\infty} \frac{e^{kx}}{x^{\lceil r \rceil}} = \infty.$$

This proves the result.

95. This limit has the indeterminate form 1^∞. Taking the limit of the log gives

$$\lim_{x\to\infty} x \ln\left(1 + \frac{r}{x}\right) = \lim_{x\to\infty} \frac{\ln\left(1 + \frac{r}{x}\right)}{x^{-1}} \stackrel{\text{L'H}}{=} \lim_{x\to\infty} \frac{\frac{1}{1+\frac{r}{x}} \cdot rx^{-2}}{-x^{-2}} = \lim_{x\to\infty} \frac{r}{1+\frac{r}{x}} = r.$$

Theorem 3.15 then tells us that the original limit is e^r.

Thinking Forward

Convergence and divergence of sequences

▷ The general term is $a_k = \frac{k^2}{2^k}$. Since exponential growth functions dominate power functions, we have $\lim_{k\to\infty} a_k = 0$.

▷ The general term is $a_k = \frac{\ln k}{\ln(k+1)}$. Then $\lim_{k\to\infty} a_k$ has the form $\frac{\infty}{\infty}$, so applying L'Hôpital's rule gives

$$\lim_{k\to\infty} a_k = \lim_{k\to\infty} \frac{\ln k}{\ln(k+1)} \stackrel{\text{L'H}}{=} \lim_{k\to\infty} \frac{k^{-1}}{(k+1)^{-1}} = \lim_{k\to\infty} \frac{k+1}{k} = \lim_{k\to\infty} \frac{1+1/k}{1} = 1.$$

▷ The general term is $a_k = \frac{2^k-1}{10^k} = \frac{2^k}{10^k} - \frac{1}{10^k} = \left(\frac{1}{5}\right)^k - \left(\frac{1}{10}\right)^k$. Since increasing positive powers of numbers less than one go to zero, this limit is zero.

▷ The general term is $a_k = \frac{k^3}{300+k^2}$. Using L'Hôpital's rule is overkill, but we can:

$$\lim_{k\to\infty} \frac{k^3}{300+k^2} \stackrel{\text{L'H}}{=} \lim_{k\to\infty} \frac{3k^2}{2k} \stackrel{\text{L'H}}{=} \lim_{k\to\infty} \frac{6k}{2} = \lim_{k\to\infty} 3k = \infty.$$

Chapter Review and Self-Test

Skill Certification: Curve Sketching and L'Hôpital's Rule

Intervals of behavior

1. Since $f(x) = x^3 + 3x^2 - 9x - 27 = (x-3)(x+3)^2$, we see that f has roots at $x = \pm 3$. Testing points in the resulting three intervals shows that f is positive on $(3, \infty)$ and negative on $(-\infty, -3) \cup (-3, 3)$. Then $f'(x) = 3x^2 + 6x - 9 = 3(x+3)(x-1)$, so that $f'(x) = 0$ at $x = 1$ and $x = -3$. Testing points in the resulting three intervals shows that f' is positive on $(-\infty, -3) \cup (1, \infty)$ and negative on $(-3, 1)$, so that f is increasing on $(-\infty, -3) \cup (1, \infty)$ and decreasing on $(-3, 1)$. Finally, $f''(x) = 6x + 6$, so that $f'(x) = 0$ at $x = -1$; testing points shows that it is positive on $(-1, \infty)$ and negative on $(-\infty, -1)$, so that f is concave up on $(-1, \infty)$ and concave down on $(-\infty, -1)$.

3. Since $f(x) = \frac{x-1}{x+3}$, we see that f has its only root at $x = 1$ and is undefined at $x = -3$. Testing points in the resulting three intervals shows that f is positive on $(-\infty, -3) \cup (1, \infty)$ and negative on $(-3, 1)$. Then $f'(x) = \frac{4}{(x+3)^2}$, so that $f'(x)$ is never zero, but also is undefined at $x = -3$. Away from $x = -3$, it is positive, so that f is increasing where it is defined. Finally, $f''(x) = -\frac{8}{(x+3)^3}$, so that $f''(x)$ has no zeros and is undefined at $x = -3$. Testing points shows that f'' is positive on $(-\infty, -3)$ and negative on $(-3, \infty)$, so that f is concave up on $(-\infty, -3)$ and concave down on $(-3, \infty)$.

5. Since $f(x) = e^{1+x^2}$, we see that f has no roots, and is defined and positive everywhere. Then $f'(x) = 2xe^{1+x^2}$, so that $f'(x)$ is zero at $x = 0$ and is defined everywhere. Testing points shows that f' is positive for $x > 0$ and negative for $x < 0$, so that f is increasing on $(0, \infty)$ and decreasing on $(-\infty, 0)$. Finally, $f''(x) = (4x^2 + 2)e^{1+x^2}$, so that $f''(x)$ is never zero. Testing a point shows that f'' is always positive, so that f is concave up everywhere.

7. Since $f(x) = 2^x(2^x - 1) = 4^x - 2^x$, we see that f has a root at $x = 0$. Testing points in the resulting two intervals shows that f is positive on $(0, \infty)$ and negative on $(-\infty, 0)$. Then $f'(x) = (\ln 2)2^x(2^x - 1) + 2^x((\ln 2)2^x) = (\ln 2)2^x(2^{x+1} - 1)$, so that $f'(x)$ is zero at $x = -1$. Testing points shows that f' is positive on $(-1, \infty)$ and negative on $(-\infty, -1)$, so that f is increasing on $(-1, \infty)$ and decreasing on $(-\infty, -1)$. Finally, $f''(x) = (\ln 2)^2 2^x(2^{x+2} - 1)$, so that $f''(x)$ is zero at $x = -2$. Testing points shows that f'' is positive on $(-2, \infty)$ and negative on $(-\infty, -2)$, so that f is concave up on $(-2, \infty)$ and concave down on $(-\infty, -2)$.

Important points

9. With $f(x) = 3x^4 - 8x^3 = x^3(3x - 8)$, we see that f has roots at $x = 0$ and at $x = \frac{8}{3}$. Then $f'(x) = 12x^3 - 24x^2 = 12x^2(x - 2)$, so that the critical points are at $x = 0$ and at $x = 2$. Next, $f''(x) = 36x^2 - 48x = 12x(3x - 4)$, so that $f''(x) = 0$ at $x = 0$ and at $x = \frac{4}{3}$ and these are inflection points since f'' changes sign at each. Since $f''(0) = 0$, the second derivative test is inconclusive regarding whether $x = 0$ is a local extremum; however, since f' does not change sign at $x = 0$, it is not. Since $f''(2) > 0$, we see that $x = 2$ is a local minimum. The local minimum is also a global minimum, since $\lim_{x \to -\infty} f(x) = \lim_{x \to \infty} f(x) = \infty$.

11. With $f(x) = \frac{x}{x^2+1}$, we see that f has its only root at $x = 0$ and it is defined everywhere. Now, $f'(x) = \frac{1-x^2}{(x^2+1)^2}$, so that $f'(x) = 0$ at $x = \pm 1$, and it is defined everywhere. Next, $f''(x) = \frac{2x(x^2-3)}{(x^2+1)^3}$, so that $f''(x) = 0$ at $x = 0$ and at $x = \pm\sqrt{3}$, which are inflection points since f'' changes sign at each. Since $f''(1) < 0$ and $f''(-1) > 0$, we see that $x = -1$ is a local minimum while $x = 1$ is a local maximum. Since $\lim_{x \to \infty} f(x) = \lim_{x \to -\infty} f(x) = 0$, both of these local extrema are global extrema.

13. With $f(x) = x \ln x$, the domain of f is $(0, \infty)$, and its only root is at $x = 1$. Now, $f'(x) = 1 + \ln x$, which also has domain $(0, \infty)$, and $f'(x) = 0$ only for $x = e^{-1}$, so this is the only critical point. Next, $f''(x) = \frac{1}{x} > 0$ for $x > 0$, so that there are no inflection points, and also $x = e^{-1}$ is a local minimum. Since $\lim\limits_{x \to 0} x \ln x = \lim\limits_{x \to 0} \frac{\ln x}{x^{-1}} \overset{L'H}{=} \lim\limits_{x \to 0} \frac{1}{x(-x^{-2})} = -\lim\limits_{x \to 0} x = 0 > f(e^{-1}) = -e^{-1}$ and $\lim\limits_{x \to \infty} x \ln x = \infty$, we see that the local minimum is also a global minimum.

15. With $f(x) = \frac{e^x}{1 - e^x}$, we see that f has no zeros and is defined except at $x = 0$. Now, $f'(x) = \frac{e^x}{(1 - e^x)^2}$, so that $f'(x)$ is never zero and is also defined except at $x = 0$. Thus there are no critical points. Next, $f''(x) = \frac{e^x(e^x + 1)}{(e^x - 1)^3}$, so that $f''(x)$ is never zero and is defined except at $x = 0$. So f has no roots, no local extrema, and no inflection points.

17. We have $f(x) = x^3 - 2x^2 - 4x + 8 = (x + 2)(x - 2)^2$, so that f has roots at $x = \pm 2$. Testing points gives for a sign chart for f

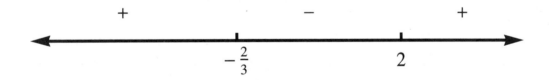

Now, $f'(x) = 3x^2 - 4x - 4 = (3x + 2)(x - 2)$, so that $f'(x) = 0$ at $x = -\frac{2}{3}$ and at $x = 2$. Testing points gives for a sign chart for f'

Since $f''(x) = 6x - 4$, we have $f''\left(-\frac{2}{3}\right) < 0$ and $f''(2) > 0$, so that $-\frac{2}{3}$ is a local maximum and 2 is a local minimum. Also, $f''\left(\frac{2}{3}\right) = 0$, and since f'' changes sign there, that is an inflection point. Testing points gives for a sign chart for f''

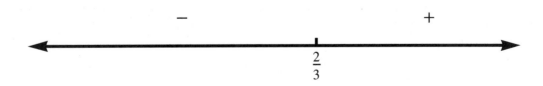

Since f is a polynomial, it is defined everywhere and has no asymptotes. Also, $\lim\limits_{x \to \pm\infty} f(x) = \pm\infty$. A graph of $f(x)$ is

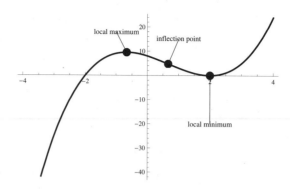

19. We have $f(x) = x\sqrt{x+1}$, so that f is defined on $[-1, \infty)$ and has roots at $x = -1$ and $x = 0$. Testing points gives for a sign chart for f

Now, $f'(x) = \frac{3x+2}{2\sqrt{x+1}}$, so that $f'(x)$ is defined on $(-1, \infty)$ and has a root at $x = -\frac{2}{3}$. Testing points gives for a sign chart for f'

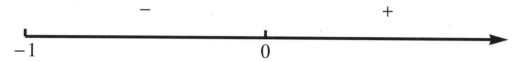

Finally, $f''(x) = \frac{3x+4}{4(x+1)^{3/2}}$ is not zero anywhere in its domain ($x = -\frac{4}{3}$ is not in its domain, since the denominator is undefined there). Testing points gives for a sign chart for f''

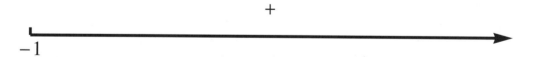

Since f'' is always positive, $x = -\frac{2}{3}$ is a local minimum. f has no horizontal or vertical asymptotes, and $\lim\limits_{x \to \infty} f(x) = \infty$. A graph of $f(x)$ is

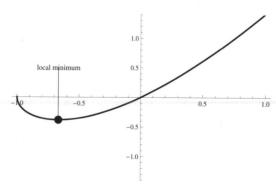

21. We have $f(x) = xe^x$, which has a root at $x = 0$. Thus a sign chart for f is

Now, $f'(x) = (x+1)e^x$, so that $f'(x)$ is defined everywhere and has a root at $x = -1$. Testing points gives for a sign chart for f'

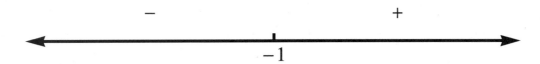

Finally, $f''(x) = (x+2)e^x$ is defined everywhere and has a root at $x = -2$. Testing points gives for a sign chart for f''

Since $f''(-1) > 0$, we see that $x = -1$ is a local minimum for f. Since f'' changes sign at $x = -2$, that is a point of inflection. Since $\lim\limits_{x \to -\infty} f(x) = 0$, f has a one-sided horizontal asymptote of $y = 0$ as $x \to -\infty$. Finally, $\lim\limits_{x \to \infty} f(x) = \infty$. f has no vertical asymptotes. A graph of $f(x)$ is

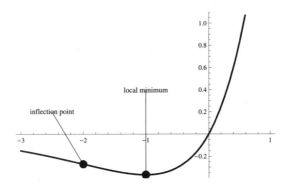

23. We have $f(x) = \cos x$, which has roots at $x = \frac{\pi}{2} + n\pi$ for n an integer. Going forward, we consider the interval $(0, 3\pi)$, which contains a full period of cos; on that interval, f has zeros at $\frac{\pi}{2}$, $\frac{3\pi}{2}$, and $\frac{5\pi}{2}$. A sign chart for f is

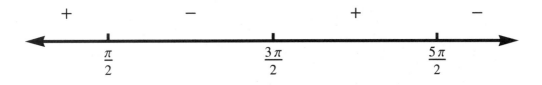

Now, $f'(x) = -\sin x$, so that $f'(x)$ has roots at $x = \pi$ and $x = 2\pi$. Testing points gives for a sign chart for f'

Finally, $f''(x) = -\cos x$ has roots at $\frac{\pi}{2}$, $\frac{3\pi}{2}$, and $\frac{5\pi}{2}$. Testing points gives for a sign chart for f''

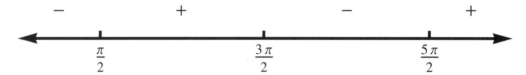

Since $f''(\pi) > 0$, we see that $x = \pi$ is a local minimum for f; since $f''(2\pi) < 0$, we see that $x = 2\pi$ is a local maximum. Since f'' changes sign at each of its roots, those are points of inflection. f has no vertical asymptotes. Since f oscillates between -1 and 1, it has no horizontal asymptotes either. A graph of $f(x)$ is

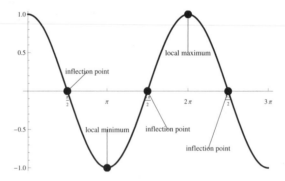

(Note that while $x = 0$ and $x = 3\pi$ are local extrema of $f(x)$, they fall outside the open interval $(0, 3\pi)$, so they are not noted in the analysis above or marked on the graph).

25. $\displaystyle\lim_{x\to\infty} \frac{x^{-2}}{x^{-3} + 1} = \lim_{x\to\infty} \frac{1}{x^{-1} + x^2} = \frac{1}{\infty} = 0.$

27. Since this limit has the form $\frac{0}{0}$, L'Hôpital's rule applies, and

$$\lim_{x\to0^+} \frac{\sin x}{\cos x - 1} \stackrel{\text{L'H}}{=} \lim_{x\to0^+} \frac{\cos x}{-\sin x} = -\infty, \qquad \lim_{x\to0^-} \frac{\sin x}{\cos x - 1} \stackrel{\text{L'H}}{=} \lim_{x\to0^-} \frac{\cos x}{-\sin x} = \infty.$$

29. Since this limit has the form $\frac{\infty}{\infty}$, L'Hôpital's rule applies, and

$$\lim_{x\to\infty} \frac{\ln x}{\ln(x + 1)} \stackrel{\text{L'H}}{=} \lim_{x\to\infty} \frac{x^{-1}}{(x + 1)^{-1}} = \lim_{x\to\infty} \frac{x + 1}{x} = 1.$$

31. Since this limit has the indeterminate form 1^∞, compute the limit of the log:

$$\lim_{x\to1^+} \frac{\ln x}{x - 1} \stackrel{\text{L'H}}{=} \lim_{x\to1^+} \frac{x^{-1}}{1} = \lim_{x\to1^+} \frac{1}{x} = 1.$$

Then Theorem 3.15 tells us that the original limit is $e^1 = e$.

Part II

Integral Calculus

Chapter 4

Definite Integrals

4.1   Addition and Accumulation

Thinking Back

Approximations and limits

The slope of a nearby secant is close to the slope of the tangent line because the two points that determine the secant are both close to the point of tangency. Since the curve is smooth at the point of tangency if it is differentiable, moving a point a short distance away only changes the slope a little bit.

In this section, sums of some sequence a_k for large values of k approximate the sum S of the sequence for k up to ∞: $\lim_{k \to \infty} \sum_{i=1}^{k} a_k = S$.

Properties of addition

If $x, y,$ and z are real numbers, then

$$x + y = y + x \qquad \text{commutative law}$$
$$(x + y) + z = x + (y + z) \qquad \text{associative law}$$
$$(x + y) \cdot z = x \cdot z + y \cdot z \qquad \text{distributive law}$$

Sum and constant-multiple rules

(a) If f and g are differentiable at $x = c$, and if k is a real number, then

$$(f + g)'(c) = f'(c) + g'(c) \qquad \text{sum rule}$$
$$(kf)'(c) = kf'(c) \qquad \text{constant-multiple rule}$$

(b) If $\lim_{x \to c} f(x)$ and $\lim_{x \to c} g(x)$ exist, and if k is a real number, then

$$\lim_{x \to c}(f(x) + g(x)) = \lim_{x \to c} f(x) + \lim_{x \to c} g(x) \qquad \text{sum rule}$$
$$\lim_{x \to c}(kf)(x) = k \lim_{x \to c} f(x) \qquad \text{constant-multiple rule}$$

Concepts

1. (a) True. For example, $\sum_{k=2}^{3} 1 = 1 + 1 = 2$; if we applied the formula, we would get 3.

   (b) True.

$$\sum_{k=0}^{n} \frac{1}{k+1} + \sum_{k=1}^{n} k^2 = \frac{1}{0+1} + \sum_{k=1}^{n} \frac{1}{k+1} + \sum_{k=1}^{n} k^2$$

$$= 1 + \sum_{k=1}^{n} \left(\frac{1}{k+1} + k^2 \right)$$

$$= 1 + \sum_{k=1}^{n} \frac{k^3 + k^2 + 1}{k+1}$$

$$= \sum_{k=0}^{n} \frac{k^3 + k^2 + 1}{k+1}.$$

   (c) False.

$$\sum_{k=1}^{n} \frac{1}{k+1} + \sum_{k=0}^{n} k^2 = 0^2 + \sum_{k=1}^{n} \frac{1}{k+1} + \sum_{k=1}^{n} k^2$$

$$= 0 + \sum_{k=1}^{n} \left(\frac{1}{k+1} + k^2 \right)$$

$$= \sum_{k=1}^{n} \frac{k^3 + k^2 + 1}{k+1}$$

$$\neq \sum_{k=0}^{n} \frac{k^3 + k^2 + 1}{k+1}.$$

   since the term in the final sum for $k = 0$ is 1, so that the two sums are not the same.

   (d) False. For example, let $n = 2$; then

$$\left(\sum_{k=1}^{n} \frac{1}{k+1} \right) \left(\sum_{k=1}^{n} k^2 \right) = \left(\frac{1}{2} + \frac{1}{3} \right) (1+4) = \frac{25}{6}$$

$$\sum_{k=1}^{n} \frac{k^2}{k+1} = \frac{1}{2} + \frac{4}{3} = \frac{11}{6}.$$

   (e) False. This is *almost* true, but the term for $k = m$ is counted twice.

   (f) False. What is true is that $\sum_{k=0}^{n} a_k = a_0 + a_n + \sum_{k=1}^{n-1} a_k$.

   (g) False. For example, let $a_k = 1$ for all k. Then

$$\left(\sum_{k=1}^{10} a_k \right)^2 = \left(\sum_{k=1}^{10} 1 \right)^2 = 10^2 = 100$$

$$\sum_{k=1}^{10} a_k^2 = \sum_{k=1}^{10} 1^2 = \sum_{k=1}^{10} 1 = 10.$$

   (h) False. By Theorem 4.2, we have

$$\sum_{k=1}^{n} (e^x)^2 = (e^x)^2 \sum_{k=1}^{n} 1 = n(e^x)^2 = ne^{2x}.$$

3. It would be difficult because there is no obvious pattern to the terms, so that there is no obvious way to define a_k by a formula involving k.

5. It means the sum of the square of every integer from 3 to 87 inclusive.

7. (a) $b_p + b_{p+1} + b_{p+2} + \ldots + b_{q-1} + b_q$.

   (b) The index of the sum is the letter used to iterate the values, in this case i. The starting value is p, and the ending value is q. The term after the $\sum$ sign, in this case b_i, defines the form of each term of the sum.

   (c) Yes, p and q must be integers; see Definition 4.1. Again from Definition 4.1, p and q must be nonnegative (although it is possible to relax the definition to allow negative integers). There are no restrictions on b_i; the b_i may be positive, negative, or zero. Finally, we must have $p \leq q$ (again see Definition 4.1).

9. There are presumably many possibilities, but the most obvious one is that $a_k = k^2$. Assuming that, we have $m = 3$ since the first term is $3^2 = 9$. Also, $n = 7$ since the last term is $7^2 = 49$.

11. We have

$$\sum_{k=0}^{8} \frac{1}{k^2 + 1} = \frac{1}{0^2 + 1} + \frac{1}{1^2 + 1} + \frac{1}{2^2 + 1} + \frac{1}{3^2 + 1} + \frac{1}{4^2 + 1} + \frac{1}{5^2 + 1} + \frac{1}{6^2 + 1} + \frac{1}{7^2 + 1} + \frac{1}{8^2 + 1}$$

$$2\sum_{k=0}^{8} \frac{1}{2k^2 + 2} = 2\sum_{k=0}^{8} \frac{1}{2(k^2 + 1)}$$

$$= 2\left(\frac{1}{2(0^2 + 1)} + \frac{1}{2(1^2 + 1)} + \frac{1}{2(2^2 + 1)} + \frac{1}{2(3^2 + 1)} + \frac{1}{2(4^2 + 1)} + \frac{1}{2(5^2 + 1)} \right.$$

$$\left. + \frac{1}{2(6^2 + 1)} + \frac{1}{2(7^2 + 1)} + \frac{1}{2(8^2 + 1)} \right)$$

$$= \frac{2}{2(0^2 + 1)} + \frac{2}{2(1^2 + 1)} + \frac{2}{2(2^2 + 1)} + \frac{2}{2(3^2 + 1)} + \frac{2}{2(4^2 + 1)} + \frac{2}{2(5^2 + 1)}$$

$$+ \frac{2}{2(6^2 + 1)} + \frac{2}{2(7^2 + 1)} + \frac{2}{2(8^2 + 1)}$$

$$= \frac{1}{0^2 + 1} + \frac{1}{1^2 + 1} + \frac{1}{2^2 + 1} + \frac{1}{3^2 + 1} + \frac{1}{4^2 + 1} + \frac{1}{5^2 + 1} + \frac{1}{6^2 + 1} + \frac{1}{7^2 + 1} + \frac{1}{8^2 + 1}$$

13. (a) With a starting value of $k = 1$, the terms are $\frac{2}{k^2}$; the last term is for $k = 5$, so the formula is $\sum_{k=1}^{5} \frac{2}{k^2}$.

   (b) With a starting value of $k = 2$, the terms are $\frac{2}{(k-1)^2}$; the last term is for $k = 6$, so the formula is $\sum_{k=2}^{6} \frac{2}{(k-1)^2}$.

   (c) With a starting value of $k = 0$, the terms are $\frac{2}{(k+1)^2}$; the last term is for $k = 4$, so the formula is $\sum_{k=0}^{4} \frac{2}{(k+1)^2}$.

15. (a) $\sum_{k=1}^{2} k = 1 + 2 = 3 = \frac{2(2+1)}{2}$.

   (b) $\sum_{k=1}^{8} k = 1 + 2 + 3 + 4 + 5 + 6 + 7 + 8 = 36 = \frac{8(8+1)}{2}$.

   (c) $\sum_{k=1}^{9} k = 1 + 2 + 3 + 4 + 5 + 6 + 7 + 8 + 9 = 45 = \frac{9(9+1)}{2}$.

17. See Theorem 4.4:

   (a) $\sum_{k=1}^{n} 1 = n$.

(b) $\sum_{k=1}^{n} k = \frac{n(n+1)}{2}$.

(c) $\sum_{k=1}^{n} k^2 = \frac{n(n+1)(2n+1)}{6}$.

(d) $\sum_{k=1}^{n} k^3 = \frac{n^2(n+1)^2}{4}$.

19. The discussion following the example points out that the sum of the areas of rectangles under the velocity curve is both a good approximation of distance and a good approximation of the area under the velocity curve. Since velocity is the derivative of distance, we might expect that in general the area under the graph of a derivative function f' is related to the function f.

Skills

21. With $m = 1$, $n = 8$, and $a_k = 3$ for all k, the sum is $\sum_{k=1}^{8} a_k$.

23. With $m = 1$, $n = 5$, and $a_k = \frac{k+2}{k^3}$, the sum is $\sum_{k=1}^{5} a_k$.

25. With $m = 2$, $n = 10$, and $a_k = k^2 + 1$, the sum is $\sum_{k=2}^{10} a_k$.

27. With $m = 1$ and $a_k = \frac{k}{n}$, the sum is $\sum_{k=1}^{n} \frac{k}{n}$.

29. $\displaystyle\sum_{k=4}^{9} k^2 = 4^2 + 5^2 + 6^2 + 7^2 + 8^2 + 9^2 = 271$.

31. $\displaystyle\sum_{k=0}^{5} \left(\frac{1}{2}k\right)^2 \left(\frac{1}{2}\right) = \frac{1}{8}\sum_{k=0}^{5} k^2 = \frac{1}{8}\left(0^2 + 1^2 + 2^2 + 3^2 + 4^2 + 5^2\right) = \frac{55}{8}$.

33.

$$\sum_{k=0}^{9} \sqrt{3 + \frac{1}{10}k} \cdot \left(\frac{1}{10}\right) = \frac{1}{10}\sum_{k=0}^{9} \sqrt{3 + \frac{1}{10}k}$$

$$= \frac{1}{10}\left(\sqrt{3} + \sqrt{\frac{31}{10}} + \sqrt{\frac{32}{10}} + \sqrt{\frac{33}{10}} + \sqrt{\frac{34}{10}} + \sqrt{\frac{35}{10}}\right.$$

$$\left. + \sqrt{\frac{36}{10}} + \sqrt{\frac{37}{10}} + \sqrt{\frac{38}{10}} + \sqrt{\frac{39}{10}}\right).$$

35. Using Theorem 4.4,

$$\sum_{k=1}^{n}(3 - k) = \sum_{k=1}^{n} 3 - \sum_{k=1}^{n} k = 3\sum_{k=1}^{n} 1 - \sum_{k=1}^{n} k = 3n - \frac{n(n+1)}{2} = \frac{6n - n^2 - n}{2} = \frac{5n - n^2}{2}.$$

Thus

n	100	500	1000
$\sum_{k=1}^{n}(3-k)$	-4750	-123750	-497500

37.

$$\sum_{k=3}^{n}(k+1)^2 = \sum_{k=4}^{n+1} k^2 = -1^2 - 2^2 - 3^2 + \sum_{k=1}^{n+1} k^2$$

$$= -14 + \frac{(n+1)(n+2)(2(n+1)+1)}{6} = \frac{2n^3 + 9n^2 + 13n - 78}{6}.$$

Thus

n	100	500	1000
$\sum_{k=1}^{n}(k+1)^2$	348537	42042737	334835487

39.

$$\sum_{k=1}^{n}\frac{k^3-1}{n^4} = \sum_{k=1}^{n}\frac{k^3}{n^4} - \sum_{k=1}^{n}\frac{1}{n^4} = \frac{1}{n^4}\sum_{k=1}^{n}k^3 - \frac{1}{n^4}\sum_{k=1}^{n}1 = \frac{n^2(n+1)^2}{4n^4} - \frac{n}{n^4} = \frac{n^2+2n+1}{4n^2} - \frac{1}{n^3}$$

$$= \frac{n^3+2n^2+n-4}{4n^3}.$$

Thus

n	100	500	1000
$\sum_{k=1}^{n}\frac{k^3-1}{n^4}$	$\frac{15939}{62500}$	$\frac{7843781}{31250000}$	$\frac{250500249}{1000000000}$

41. Pulling out the terms for $k = 0, 1$, and 2 from the left-hand sum as well as the $k = 101$ term from the right-hand sum gives

$$2\sum_{k=0}^{100}a_k - \sum_{k=3}^{101}a_k = 2\sum_{k=0}^{3}a_k + 2\sum_{k=3}^{100}a_k - \sum_{k=3}^{100}a_k - a_{101}$$

$$= 2\sum_{k=0}^{3}a_k - a_{101} + \sum_{k=3}^{100}(2a_k - a_k)$$

$$= 2(a_0 + a_1 + a_2) - a_{101} + \sum_{k=3}^{100}a_k$$

This can also be written as $a_0 + a_1 + a_2 - a_{101} + \sum_{k=0}^{100}a_k$.

43. Pulling out the term for $k = 25$ from the first sum, and the terms for $k = 0$, $k = 1$, and $k = 25$ from the third sum gives

$$3\sum_{k=2}^{25}k^2 + 2\sum_{k=2}^{24}k - \sum_{k=0}^{25}1 = 3\cdot25^2 - 1 - 1 - 1 + 3\sum_{k=2}^{24}k^2 + 2\sum_{k=2}^{24}k - \sum_{k=2}^{24}1$$

$$= 1872 + \sum_{k=2}^{24}(3k^2 + 2k - 1).$$

45. Since $\sum_{k=0}^{4}a_k = a_0 + \sum_{k=1}^{4}a_k = 2 + 7 = 9$, we have

$$\sum_{k=0}^{4}(2a_k + 3b_k) = 2\sum_{k=0}^{4}a_k + 3\sum_{k=0}^{4}b_k = 2\cdot9 + 3\cdot10 = 48.$$

47. We know that

$$\sum_{k=1}^{n} \frac{k^2 + k + 1}{n^3} = \frac{1}{n^3} \sum_{k=1}^{n} (k^2 + k + 1)$$

$$= \frac{1}{n^3} \left(\sum_{k=1}^{n} k^2 + \sum_{k=1}^{n} k + \sum_{k=1}^{n} 1 \right)$$

$$= \frac{1}{n^3} \left(\frac{n(n+1)(2n+1)}{6} + \frac{n(n+1)}{2} + n \right)$$

$$= \frac{n^2 + 3n + 5}{3n^2}$$

Thus

$$\lim_{n \to \infty} \sum_{k=1}^{n} \frac{k^2 + k + 1}{n^3} = \lim_{n \to \infty} \frac{n^2 + 3n + 5}{3n^2} = \lim_{n \to \infty} \frac{1 + 3/n + 5/n^2}{3} = \frac{1}{3}.$$

49. We have

$$\sum_{k=1}^{n} \frac{(k+1)^2}{n^3 - 1} = \frac{1}{n^3 - 1} \sum_{k=1}^{n} (k^2 + 2k + 1)$$

$$= \frac{1}{n^3 - 1} \left(\sum_{k=1}^{n} k^2 + 2 \sum_{k=1}^{n} k + \sum_{k=1}^{n} 1 \right)$$

$$= \frac{1}{n^3 - 1} \left(\frac{n(n+1)(2n+1)}{6} + n(n+1) + n \right)$$

$$= \frac{2n^3 + 9n^2 + 13n}{6n^3 - 6}.$$

Therefore

$$\lim_{n \to \infty} \sum_{k=1}^{n} \frac{(k+1)^2}{n^3 - 1} = \lim_{n \to \infty} \frac{2n^3 + 9n^2 + 13n}{6n^3 - 6} = \lim_{n \to \infty} \frac{2 + 9/n + 13/n^2}{6 - 6/n^3} = \frac{1}{3}.$$

51. Simplifying,

$$\sum_{k=1}^{n} \left(1 + \frac{k}{n} \right)^2 \cdot \frac{1}{n} = \sum_{k=1}^{n} \left(\frac{1}{n}(n+k) \right)^2 \cdot \frac{1}{n}$$

$$= \sum_{k=1}^{n} \frac{(n+k)^2}{n^3}$$

$$= \frac{1}{n^3} \sum_{k=1}^{n} (n^2 + 2kn + k^2)$$

$$= \frac{1}{n^3} \left(n^2 \sum_{k=1}^{n} 1 + 2n \sum_{k=1}^{n} k + \sum_{k=1}^{n} k^2 \right)$$

$$= \frac{1}{n^3} \left(n^3 + 2n \frac{n(n+1)}{2} + \frac{n(n+1)(2n+1)}{6} \right)$$

$$= \frac{14n^2 + 9n + 1}{6n^2}.$$

So, taking limits,

$$\lim_{n \to \infty} \sum_{k=1}^{n} \left(1 + \frac{k}{n} \right)^2 \cdot \frac{1}{n} = \lim_{n \to \infty} \frac{14n^2 + 9n + 1}{6n^2} = \lim_{n \to \infty} \frac{14 + 9/n + 1/n^2}{6} = \frac{7}{3}.$$

Applications

53. Dividing the time interval into eight pieces, and using the middle point of each piece as the assumed constant velocity gives the following picture:

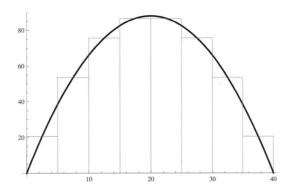

Since each piece is 5 seconds wide, the middle of the segments are at $t = 2.5 + 5n$ for $n \in \{0, 1, 2, 3, 4, 5, 6, 7\}$. Then the approximate area is given by the sum of the areas of the rectangles:

$$\sum_{n=0}^{7} \text{width} \times \text{height} = \sum_{n=0}^{7} 5v(2.5 + 5n) = 5\left(-0.22 \sum_{n=0}^{7} t^2 + 8.8 \sum_{n=0}^{7} t\right) \approx 2365.$$

Other answers are possible if a point other than the midpoint was used as the velocity in each subinterval.

55. (a) Since $I(t)$ is the change in value of your investments (measured by $B(t)$), it follows that $B(t)$ in the *next* year is $B(t)$ this year plus the value of $I(t)$. Thus $B(t)$ accumulates successive values of $I(t)$.

(b) A plot of $I(t)$ is

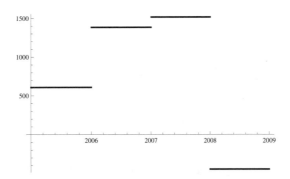

$B(t_0)$ is the area under the graph of $I(t)$ for $2005 \le t \le t_0$. Note that during 2008, the area "under" $I(t)$ is negative. (One could have a discussion about whether $B(t)$ changes during the year, and whether it changes at the constant rate implied by the fact that $I(t)$ is constant, but for the purposes of this exercise, we assume this is true).

(c) When $I(t)$ is positive, then $B(t)$ is increasing, since as t increases, we are adding *positive* area under $I(t)$ to $B(t)$. When $I(t)$ is negative, then $B(t)$ is decreasing, since as t increases, we are adding *negative* area under $I(t)$ to $B(t)$. Since the sign of $I(t)$ determines whether $B(t)$ is increasing or decreasing, it seems likely that $I(t) = B'(t)$ rather than the other way around.

57. (a) Let $b(n)$ be the amount of money spent at the second step, so that for example $b(1) = 10(0.7) = 7$ and $b(2) = 7(0.7)$. Since $b(n)$ for any $n > 1$ is just $0.7b(n-1)$, we see that

$A(n)$, which is the *total* amount spent after n steps as a result of the original 10 billion tax cut, is

$$A(n) = b(0) + b(1) + \cdots + b(n) = 7 + 7(0.7) + \cdots + 7(0.7)^{n-1} = \sum_{k=0}^{n-1} 7(0.7)^k.$$

(Note that $A(n) = \sum_{k=1}^{n} 10(0.7)^k$ is an equally possible, and equally valid, way of expressing this sum).

(b) We have

$$A(3) = 7 + 7(0.7) + 7(0.7)^2 = 7 + 4.9 + 3.43 = \$15.33\text{B},$$
$$A(4) = 7 + 7(0.7) + 7(0.7)^2 + 7(0.7)^3 = 7 + 4.9 + 3.43 + 2.401 = \$17.731\text{B},$$
$$A(5) = 7 + 7(0.7) + 7(0.7)^2 + 7(0.7)^3 + 7(0.7)^4 = 7 + 4.9 + 3.43 + 2.401 + 1.6807$$
$$= \$19.4117\text{B}.$$

(c) Calculations for n up to 15 are (rounded to the nearest million dollars):

n	1	2	3	4	5	6	7	8	9	10
$A(n)$	7	11.9	15.33	17.731	19.412	20.588	21.412	21.988	22.392	22.674

n	11	12	13	14	15
$A(n)$	22.872	23.010	23.107	23.175	23.223

(d) It appears that the total amount of spending generated is approaching a value between \$23 and \$24 billion. Looking at the graph, we see that it is plausible that the graph has a horizontal asymptote at about this value as $n \to \infty$. These two statements mean the same thing.

Proofs

59. Expanding gives

$$\sum_{k=0}^{n} 3a_k = 3a_0 + 3a_1 + \cdots + 3a_n$$

$$= (a_0 + a_0 + a_0) + (a_1 + a_1 + a_1) + \ldots (a_n + a_n + a_n)$$
$$= (a_0 + a_1 + \cdots + a_n) + (a_0 + a_1 + \cdots + a_n) + (a_0 + a_1 + \cdots + a_n)$$
$$= \sum_{k=0}^{n} a_k + \sum_{k=0}^{n} a_k + \sum_{k=0}^{n} a_k$$
$$= 3 \sum_{k=0}^{n} a_k.$$

61. If n is even, then group n with 1, $n-1$ with 2, and so forth. The last grouping is $\frac{n}{2} + 1$ and $\frac{n}{2}$. (For example, with $n = 6$, the pairs would be 6 and 1, 5 and 2, and 4 and 3). Note that the sum of each pair is $n+1$, and that there are exactly $\frac{n}{2}$ pairs. Thus

$$\sum_{k=1}^{n} k = \frac{n}{2}(n+1) = \frac{n(n+1)}{2}.$$

Thinking Forward

Functions defined by area accumulation

▷ Clearly $A(1) = 0$, since it is the area under the graph of $f(x)$ from $x = 1$ to $x = 1$. From the shaded area in the diagram, we can estimate $A(2)$ by noting that it contains two full squares, plus a tiny bit of another square between $x = 1$ and $x = \frac{3}{2}$, plus slightly more than one half of a square between $x = \frac{3}{2}$ and $x = 2$. As a rough estimate, then, $A(2) \approx 2 + \frac{1}{10} + \frac{1}{2} = 2.6$. Finally, $A(3) = A(2)$ plus the additional area under the curve between $x = 2$ and $x = 3$. This additional region contains five full squares plus more than one half of two other squares, so that $A(3) \approx A(2) + 6 \approx 8.6$.

▷ f is positive everywhere, so that $A(x)$ is increasing on $[1, \infty)$ (its domain of definition), since any time we increase x we are adding a positive area. Since $A(1) = 0$, we see that $A(x)$ is positive everywhere on $(1, \infty)$.

▷ If f were decreasing and positive, then $A(x)$ would still be increasing (since f is positive), but would be increasing at a decreasing rate. If f were increasing and negative, then $A(x)$ would be decreasing (since f is negative), but at a decreasing rate. This is illustrated in the following two diagrams:

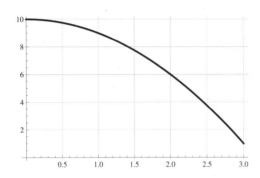
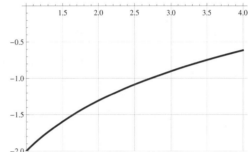

Approximating the area under a curve

▷

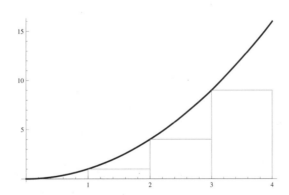

The width of each rectangle is 1, so the area of the rectangle between n and $n + 1$ is $f(n) = n^2$ (since we are choosing the left endpoint). Thus the approximate area is

$$\sum_{k=0}^{3} 1 \cdot f(k) = 0^2 + 1^2 + 2^2 + 3^2 = 14.$$

▷

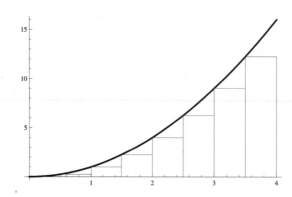

The width of each rectangle is $\frac{1}{2}$, so the area of the rectangle between x and $x + \frac{1}{2}$ is $f(x) = x^2$ (since we are choosing the left endpoint). Thus the approximate area is

$$\sum_{k=0}^{7} \frac{1}{2} \cdot \left(\frac{k}{2}\right)^2 = \frac{1}{8} \sum_{k=0}^{7} k^2 = \frac{1}{8} \left(\frac{7(8)(15)}{6}\right) = \frac{35}{2}.$$

This is clearly an underapproximation of the area, since each rectangle lies wholly below the curve.

Sequences of partial sums

▷ Computing values gives

n	1	2	3	4	5	6	7	8	9	10
$A(n)$	7	11.9	15.33	17.731	19.4117	20.5882	21.4117	21.9882	22.3917	22.6742

It appears from the data we have that the partial sums may not be increasing without bound, but may instead be approaching a value around 23.

▷ It appears that this sequence of partial sums converges.

4.2 Riemann Sums

Thinking Back

Simple area formulas

▷ πr^2.

▷ The area of a semicircle is one half the area of the circle of the same radius, so $\frac{1}{2}\pi r^2$.

▷ The area of a triangle is one half the base times the height; for a right triangle we can take the base to be one leg and the height the other, so the area is $\frac{1}{2}ab$.

▷ The area of a triangle is one half the base times the height, so the area is $\frac{1}{2}bh$.

▷ The area is wl.

▷ The area is $\frac{h_1 + h_2}{2} \cdot w$.

Concepts

1. (a) False. This is not true if the right endpoint of an interval is not the lowest point on the curve in that interval. For example, $f(x) = x^2$ on a subinterval $[1,2]$: the lower sum on that subinterval will generate a rectangle with area 1, while the right sum will generate a rectangle with area $1 \cdot 2^2 = 4$.

    (b) True. Since f is positive and increasing, the rectangle defined by the left sum on any subinterval will have its upper left corner touching the graph of f, and will be completely below f since f is increasing. Since this is true for each rectangle, the left sum is an underapproximation.

    (c) False. It depends on the behavior of f on the entire interval. For example, let $f(a) = 2$, $f(b) = 1$, and $f(x) = 0$ for $x \in (a,b)$. Then the area under the curve is zero, yet the right-sum approximation with one subinterval gives approximate area $1 \cdot 1 = 1$.

    (d) False. For example, if f is a constant function, the two are the same (and both give the exact area).

    (e) False. It may be either an underapproximation or an overapproximation. For example, let $f(x) = x^2$ with $a = -4$ and $b = 1$. The left-hand graph below shows the left-sum approximation with one subinterval; it is clearly an overapproximation. The right-hand graph shows the left-sum approximation with four subintervals; it is clearly an underapproximation.

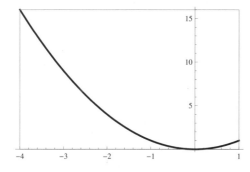 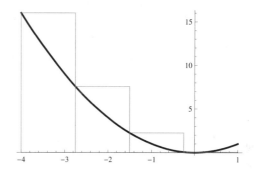

    (f) True. Since each trapezoid, say on the subinterval $[c,d]$, joins $(c, f(c))$ to $(d, f(d))$, the upper edge of the trapezoid lies totally above the graph of f since f is concave upwards. Since f is positive, the trapezoid has larger area than the area under the graph of f.

    (g) True. In general it will be an overapproximation, since the upper edge of the rectangle will always lie on or above the graph of f, but if f is constant, it will be exact.

    (h) False. Consider for example the curve below:

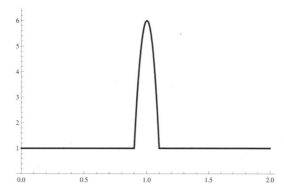

    Clearly here the left sum approximation with one rectangle is better than the left sum approximation with two rectangles, since in the second case one of the rectangles will be very high.

A similar graph can be constructed for the case of 5 versus 10 rectangles as in the problem statement.

3. (a) The region contains 45 complete squares; adding up the fractions in the remainder gives about 16 more, for about 61 squares. Each square is $\frac{1}{4}$ square unit, so the area is about $\frac{61}{4}$ square units.

   (b) If we decreased the side length of each square, we could get a better estimate since more complete squares would fit inside the blob, and the estimation error on the remainder would be smaller. Alternatively, we could be more careful about trying to count the areas of partial squares.

5. (a) The right sum with four rectangles in Example 1 is clearly an overestimate, since each rectangle lies above the curve. The left sum in Example 4 is clearly an underestimate, since each rectangle lies below the curve. The midpoint sum in Example 4 is probably an underestimate as well, since it appears that the areas of the regions left out exceed the areas of the extra regions included. The trapezoidal approximation is clearly an overestimate, since the curve is concave up, but not by much. Probably the trapezoidal approximation is closest to the true area, but it could also be the midpoint sum.

   (b) The right sum was 5.75, the left sum 3.75, the midpoint sum 4.625, and the trapezoidal sum 4.75. Since the true area is 4.6667, the midpoint sum was low by about 0.042 and the trapezoidal sum was high by about 0.09. The midpoint sum was closest.

7. (a) If the intervals are $[x_i, x_{i+1}]$ for $0 \leq i \leq n$, and they have width Δx, then the right sum is $\sum_{k=1}^{n} f(x_{k-1}) \Delta x$ and the left sum is $\sum_{k=1}^{n} f(x_k) \Delta x$. Now, the trapezoid sum is

$$\sum_{k=1}^{n} \frac{f(x_{k-1}) + f(x_k)}{2} \Delta x = \frac{1}{2} \Delta x \left(\sum_{k=1}^{n} f(x_{k-1}) + \sum_{k=1}^{n} f(x_k) \right)$$

$$= \frac{1}{2} \left(\sum_{k=1}^{n} f(x_{k-1}) \Delta x + \sum_{k=1}^{n} f(x_k) \Delta x \right)$$

   so that the trapezoid sum is the average of the left sum and the right sum. This average is $\frac{1}{2}(7.5 + 8.2) = 7.85$.

   (b) The midpoint sum need not be between 7.5 and 8.2. For example, suppose $f(x)$ had a large peak at the middle of each of the eight intervals. Then the midpoint sum would be larger than either the left or the right sum. An example where $a = 0$ and $b = 8$ is:

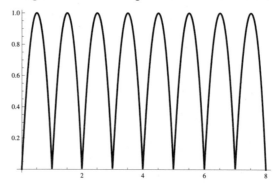

   Clearly both the left and right sums are zero, but the midpoint sum is not.

   (c) For the curve in part (b), since the midpoint is the largest point in each interval, for this example the midpoint sum is equal to the upper sum. Also clearly the lower sum is zero since each interval has a point at which the function is zero. For the curve in part (a), all we can say is that the upper sum is ≥ 8.2 and that the lower sum is ≤ 7.5.

(d) No, f need not be decreasing everywhere. An example where $a = 0$ and $b = 8$ is

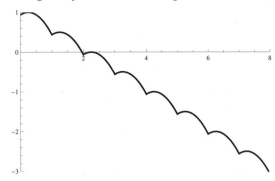

Here the left sum is clearly greater than the right sum since on each subinterval (there are eight of them) the value at the left endpoint exceeds that at the right endpoint. Yet the function is not decreasing on $[0, 8]$.

(e) No. If f is increasing on $[a, b]$, then for each subinterval $[x_{i-1}, x_i]$, we have $f(x_{i-1}) \leq f(x_i)$, so that each rectangle for the left sum is no greater than the corresponding right-sum rectangle. Thus the left sum is no greater than the right sum.

9. Any Riemann sum approximation for f on n subintervals chooses some point x_i^* in $[x_i, x_{i+1}]$ to use in the sum $\sum_{k=1}^{n} f(x_i^*)\Delta_x$. Let x_i^+ be a point in $[x_i, x_{i+1}]$ at which f reaches its maximum value (see the Extreme Value Theorem). Then

$$\sum_{k=1}^{n} f(x_i^*)\Delta_x \leq \sum_{k=1}^{n} f(x_i^+)\Delta_x,$$

but the sum on the right is just the upper sum.

11. Note that $f(x)$ is decreasing on $[0, 1]$ and increasing on $[1, 2]$.

(a) For $n = 2$, the subintervals are $[0, 1]$ and $[1, 2]$. The maximum value of $f(x)$ on $[0, 1]$ is thus $f(0) = 1$ (since f is decreasing on $[0, 1]$), and the maximum value on $[1, 2]$ is $f(2) = 1$ (since f is increasing on $[1, 2]$). Thus $M_1 = 0$ and $M_2 = 2$.

(b) For $n = 3$, the subintervals are $\left[0, \frac{2}{3}\right]$, $\left[\frac{2}{3}, \frac{4}{3}\right]$, and $\left[\frac{4}{3}, 2\right]$. For the first interval, the largest value occurs at $x = 0$, and for the third interval it occurs at $x = 2$. For the middle interval, f decreases from $\frac{2}{3}$ to 1 and increases from 1 to $\frac{4}{3}$. Since $f\left(\frac{2}{3}\right) = f\left(\frac{4}{3}\right) = \frac{1}{9}$, we can choose either point. Thus $M_1 = 0$, $M_2 = \frac{2}{3}$ or $\frac{4}{3}$, and $M_3 = 2$.

(c) For $n = 4$, the subintervals are $\left[0, \frac{1}{2}\right]$, $\left[\frac{1}{2}, 1\right]$, $\left[1, \frac{3}{2}\right]$, and $\left[\frac{3}{2}, 2\right]$. Again using the behavior of f we get $M_1 = 0$, $M_2 = \frac{1}{2}$, $M_3 = \frac{3}{2}$, and $M_4 = 2$.

Sketches of these three situations are:

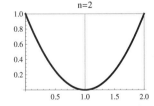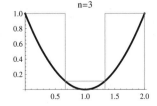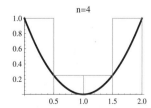

13. (a) Since $v(t_0)$ is measured in meters per second and time is measured in seconds, the units of $v(t_0)\Delta t$ are meters per second times seconds, or meters.

(b) $v(t_0)\Delta t$ represents an approximation to the change in position from $t = t_0$ to $t = t_0 + \Delta t$. If the particle were moving at velocity $v(t_0)$ for the entire time between t_0 and $t_0 + \Delta t$, its change in position would be exactly $v(t_0)\Delta t$.

(c) $v(t_0)\Delta t$ is the area of a small rectangle approximating the area under part of the velocity curve.

15. If this were a right sum, there are 20 subintervals on $[-3, 2]$, so each subinterval has width $\frac{1}{4}$. Thus successive right endpoints are separated by $\frac{1}{4}$. But two successive endpoints are $-3 + \frac{k}{2}$ and $-3 + \frac{k+1}{2}$; these differ by $\frac{1}{2}$.

17. There are clearly $n = 4$ subintervals of width $\Delta x = \frac{1}{2}$, so that the width of the entire interval is $4 \cdot \frac{1}{2} = 2$. The function appears to be $f(x) = x^2$, and from the form of the argument, the most likely possibility is that $a = 1$, so that $b = 3$, so that this is a right sum with $x_k = 1 + \frac{k}{2}$. Other answers are possible. A graph of this sum is

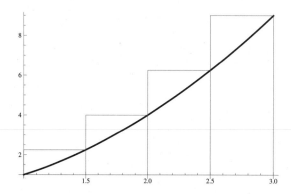

19. There are $n = 100$ subintervals each with a width of $\Delta x = 0.05$, so that the width of the entire interval is $100 \cdot 0.05 = 5$. The function appears to be $\sin x$, and from the form of the argument, this is most likely a trapezoid sum with $a = 0$, $b = 5$, and $x_k = 0.05k$. Other answers are possible. A graph of this sum is

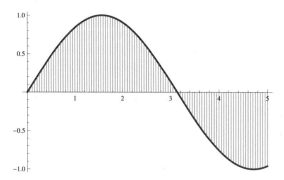

Skills

21. The actual area is $\frac{2}{3}(2\sqrt{2} - 1) \approx 1.22$; approximations will vary depending on the calculator and the method used.

23. The actual area is $e(e^3 - 1) \approx 51.88$; approximations will vary depending on the calculator and the method used.

25. The actual area is $\frac{8}{3} \approx 2.67$; approximations will vary depending on the calculator and the method used.

27. (a) With $n = 3$, since the width of the interval is 3, we have $\Delta x = 1$, and $x_k = k$ for $k \in \{0, 1, 2, 3\}$, so that the left sum is

$$\sum_{k=1}^{3} f(x_{k-1})\Delta x = \sum_{k=1}^{3} x_{k-1}^2 \cdot 1 = 0^2 + 1^2 + 2^2 = 5.$$

Since $f(x)$ is increasing on $[0, 3]$, each of the left rectangles lies completely below the curve, so that this is an underestimate.

(b) With $n = 6$, since the width of the interval is 3, we have $\Delta x = \frac{1}{2}$, and $x_k = \frac{k}{2}$ for $k \in \{0, 1, 2, 3, 4, 5, 6\}$, so that the left sum is

$$\sum_{k=1}^{6} f(x_{k-1})\Delta x = \sum_{k=1}^{6} x_{k-1}^2 \cdot \frac{1}{2} = \frac{1}{2}\left(0^2 + \left(\frac{1}{2}\right)^2 + 1^2 + \left(\frac{3}{2}\right)^2 + 2^2 + \left(\frac{5}{2}\right)^2\right) = \frac{55}{8} \approx 6.875.$$

Since $f(x)$ is increasing on $[0, 3]$, each of the left rectangles lies completely below the curve, so that this is an underestimate.

29. (a) With $n = 4$, since the width of the interval is 1, we have $\Delta x = \frac{1}{4}$, and $x_k = 2 + \frac{k}{4}$ for $k \in \{0, 1, 2, 3, 4\}$, so that the left sum is

$$\sum_{k=1}^{4} f(x_{k-1})\Delta x = \sum_{k=1}^{4} \sqrt{2 + \frac{k-1}{4} - 1} \cdot \frac{1}{4} = \frac{1}{4}\left(\sqrt{1} + \sqrt{\frac{5}{4}} + \sqrt{\frac{6}{4}} + \sqrt{\frac{7}{4}}\right)$$

$$= \frac{2 + \sqrt{5} + \sqrt{6} + \sqrt{7}}{8} \approx 1.166.$$

Since $\sqrt{x-1}$ is increasing, each left rectangle lies completely below the graph, so that this is an underestimate.

(b) n, Δx, and x_k are as in part (a), and the right sum is

$$\sum_{k=1}^{4} f(x_k)\Delta x = \sum_{k=1}^{4} \sqrt{2 + \frac{k}{4} - 1} \cdot \frac{1}{4} = \frac{1}{4}\left(\sqrt{\frac{5}{4}} + \sqrt{\frac{6}{4}} + \sqrt{\frac{7}{4}} + \sqrt{2}\right)$$

$$= \frac{\sqrt{5} + \sqrt{6} + \sqrt{7} + 2\sqrt{2}}{8} \approx 1.270.$$

Since $\sqrt{x-1}$ is increasing, each right rectangle lies completely above the graph, so that this is an overestimate.

31. (a) With $n = 6$, since the width of the interval is 3, we have $\Delta x = \frac{1}{2}$, and $x_k = 1 + \frac{k}{2}$ for $k \in \{0, 1, 2, 3, 4, 5, 6\}$. The midpoints of the intervals are thus $\frac{5}{4} + \frac{k}{2}$. Then the midpoint sum is

$$\sum_{k=1}^{6} f\left(\frac{x_{k-1} + x_k}{2}\right)\Delta x = \frac{1}{2}\left(e^{5/4} + e^{7/4} + e^{9/4} + e^{11/4} + e^{13/4} + e^{15/4}\right) \approx 51.343.$$

Looking at a graph, it is not obvious whether this is an overestimate or an underestimate:

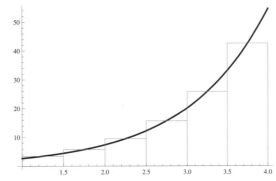

However, since the curve is concave up, it seems likely that in each subinterval, the area of the extra included area exceeds the area of the excluded area, so that this would be an overestimate — but not by much.

(b) n, Δx, and x_k are all as in part (a). The trapezoid sum is

$$\sum_{k=1}^{6} \frac{f(x_{k-1}) + f(x_k)}{2} \Delta x = \frac{1}{2} \sum_{k=1}^{6} \frac{e^{x_{k-1}} + e^{x_k}}{2}$$

$$= \frac{1}{2} \left(\frac{e + e^{3/2}}{2} + \frac{e^{3/2} + e^2}{2} + \frac{e^2 + e^{5/2}}{2} \right.$$

$$\left. + \frac{e^{5/2} + e^3}{2} + \frac{e^3 + e^{7/2}}{2} + \frac{e^{7/2} + e^4}{2} \right)$$

$$= \frac{e + 2e^{3/2} + 2e^2 + 2e^{5/2} + 2e^3 + 2e^{7/2} + e^4}{4} \approx 52.96$$

Since the curve is concave up, each trapezoid lies above the curve, so that this is an overestimate.

33. (a) With $n = 2$, since the width of the interval is 2, we have $\Delta x = 1$, and $x_k = k + 1$ for $k \in \{0, 1, 2\}$. Since $(x - 2)^2 + 1$ decreases on $[1, 2]$ and increases on $[2, 3]$, the lower sum is easy to find:

$$\sum_{k=1}^{2} f(m_k) \Delta x = ((2 - 2)^2 + 1) + ((2 - 2)^2 + 1) = 2.$$

The lower sum underestimates the true value, since for a nonconstant function, each rectangle in a lower sum has smaller area than the corresponding piece of the graph (it lies completely below the graph).

(b) With $n = 3$ and an interval width of 2, we have $\Delta x = \frac{2}{3}$, and $x_k = 1 + \frac{2}{3}k$ for $k \in \{0, 1, 2, 3\}$. Again using the increasing/decreasing behavior of $f(x)$, the lower sum is

$$\sum_{k=1}^{3} f(m_k) \Delta x = \frac{2}{3} \left(\left(\left(\frac{5}{3} - 2 \right)^2 + 1 \right) + \left((2 - 2)^2 + 1 \right) + \left(\left(\frac{7}{3} - 2 \right)^2 + 1 \right) \right)$$

$$= \frac{2}{3} \cdot \frac{29}{9} = \frac{58}{27}.$$

Again the lower sum underestimates the true value, for the same reason as in (a).

(c) With $n = 4$ and an interval width of 2, we have $\Delta x = \frac{1}{2}$, and $x_k = 1 + \frac{1}{2}k$ for $k \in \{0, 1, 2, 3, 4\}$. Again using the increasing/decreasing behavior of $f(x)$, the lower sum is

$$\sum_{k=1}^{4} f(m_k) \Delta x = \frac{1}{2} \left(\left(\left(\frac{3}{2} - 2 \right)^2 + 1 \right) + \left((2 - 2)^2 + 1 \right) + \left((2 - 2)^2 + 1 \right) \right.$$

$$\left. + \left(\left(\frac{5}{2} - 2 \right)^2 + 1 \right) \right) = \frac{1}{2} \cdot \frac{18}{4} = \frac{9}{4}.$$

Again the lower sum underestimates the true value, for the same reason as in (a).

35. From the graph below, the area under the curve consists of a rectangle with width 2 and height 10 together with a right triangle with base 2 and height 6, so the area is $2 \times 10 + \frac{1}{2}2 \times 6 = 26$.

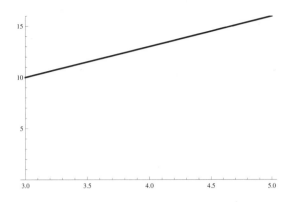

With $n = 4$ and an interval width of 2, we have $\Delta x = \frac{1}{2}$ so that $x_k = 3 + \frac{1}{2}k$ for $k \in \{0, 1, 2, 3, 4\}$. Then the various sums are

(left): $\displaystyle\sum_{k=1}^{4} f(x_{k-1}) \Delta x$

$$= \frac{1}{2}\left((3 \cdot 3 + 1) + \left(3 \cdot \frac{7}{2} + 1\right) + (3 \cdot 4 + 1) + \left(3 \cdot \frac{9}{2} + 1\right)\right) = \frac{1}{2}(49) = \frac{49}{2}$$

(right): $\displaystyle\sum_{k=1}^{4} f(x_k) \Delta x$

$$= \frac{1}{2}\left(\left(3 \cdot \frac{7}{2} + 1\right) + (3 \cdot 4 + 1) + \left(3 \cdot \frac{9}{2} + 1\right) + (3 \cdot 5 + 1)\right) = \frac{1}{2}(55) = \frac{55}{2}$$

(midpoint): $\displaystyle\sum_{k=1}^{4} f\left(\frac{x_{k-1} + x_k}{2}\right) \Delta x = \frac{1}{2}\sum_{k=1}^{4} f\left(\frac{2k + 11}{4}\right)$

$$= \frac{1}{2}\left(\left(3 \cdot \frac{13}{4} + 1\right) + \left(3 \cdot \frac{15}{4} + 1\right) + \left(3 \cdot \frac{17}{4} + 1\right) + \left(3 \cdot \frac{19}{4} + 1\right)\right)$$

$$= \frac{1}{2} \cdot \frac{104}{2} = 26$$

(upper): $\displaystyle\sum_{k=1}^{4} f(M_k) \Delta x$

$$= \frac{1}{2}\left(\left(3 \cdot \frac{7}{2} + 1\right) + (3 \cdot 4 + 1) + \left(3 \cdot \frac{9}{2} + 1\right) + (3 \cdot 5 + 1)\right) = \frac{1}{2} \cdot 55 = \frac{55}{2}$$

(lower): $\displaystyle\sum_{k=1}^{4} f(m_k) \Delta x$

$$= \frac{1}{2}\left((3 \cdot 3 + 1) + \left(3 \cdot \frac{7}{2} + 1\right) + (3 \cdot 4 + 1) + \left(3 \cdot \frac{9}{2} + 1\right)\right) = \frac{1}{2} \cdot 49 = \frac{49}{2}$$

(trapezoid): $\displaystyle\sum_{k=1}^{4} \frac{f(x_{k-1}) + f(x_k)}{2} \Delta x$

$$= \frac{1}{2}\left(\frac{(3 \cdot 3 + 1) + (3 \cdot \frac{7}{2} + 1)}{2} + \frac{(3 \cdot \frac{7}{2} + 1) + (3 \cdot 4 + 1)}{2}\right.$$

$$\left. + \frac{(3 \cdot 4 + 1) + (3 \cdot \frac{9}{2} + 1)}{2} + \frac{(3 \cdot \frac{9}{2} + 1) + (3 \cdot 5 + 1)}{2}\right) = \frac{1}{2} \cdot 52 = 26.$$

The midpoint and trapezoid sums provide exact answers.

37. From the graph below, the area under the curve consists of a semicircle with radius 1, so it has area $\frac{\pi}{2} \approx 1.5708$.

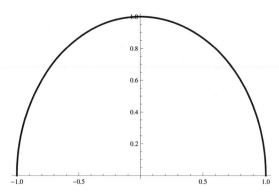

With $n = 4$ and an interval width of 2, we have $\Delta x = \frac{1}{2}$ so that $x_k = \frac{1}{2}k - 1$ for $k \in \{0, 1, 2, 3, 4\}$. Then the various sums are

$$\text{(left):} \quad \sum_{k=1}^{4} f(x_{k-1}) \Delta x = \frac{1}{2} \left(\sqrt{1 - (-1)^2} + \sqrt{1 - \left(-\frac{1}{2}\right)^2} + \sqrt{1 - 0^2} + \sqrt{1 - \left(\frac{1}{2}\right)^2} \right)$$

$$= \frac{1}{2} \left(\frac{\sqrt{3}}{2} + 1 + \frac{\sqrt{3}}{2} \right) = \frac{1 + \sqrt{3}}{2} \approx 1.366$$

$$\text{(right):} \quad \sum_{k=1}^{4} f(x_k) \Delta x = \frac{1}{2} \left(\sqrt{1 - \left(-\frac{1}{2}\right)^2} + \sqrt{1 - 0^2} + \sqrt{1 - \left(\frac{1}{2}\right)^2} + \sqrt{1 - 1^2} \right)$$

$$= \frac{1}{2} \left(\frac{\sqrt{3}}{2} + 1 + \frac{\sqrt{3}}{2} \right) = \frac{1 + \sqrt{3}}{2} \approx 1.366$$

$$\text{(midpoint):} \quad \sum_{k=1}^{4} f\left(\frac{x_{k-1} + x_k}{2} \right) \Delta x = \frac{1}{2} \sum_{k=1}^{4} f\left(\frac{2k - 5}{4} \right)$$

$$= \frac{1}{2} \left(\sqrt{1 - \left(-\frac{3}{4}\right)^2} + \sqrt{1 - \left(-\frac{1}{4}\right)^2} + \sqrt{1 - \left(\frac{1}{4}\right)^2} + \sqrt{1 - \left(\frac{3}{4}\right)^2} \right)$$

$$= \frac{1}{2} \left(\frac{\sqrt{7}}{4} + \frac{\sqrt{15}}{4} + \frac{\sqrt{15}}{4} + \frac{\sqrt{7}}{4} \right)$$

$$= \frac{\sqrt{7} + \sqrt{15}}{4} \approx 1.62968$$

$$\text{(upper):} \quad \sum_{k=1}^{4} f(M_k) \Delta x = \frac{1}{2} \left(\sqrt{1 - \left(-\frac{1}{2}\right)^2} + \sqrt{1 - 0^2} + \sqrt{1 - 0^2} + \sqrt{1 - \left(\frac{1}{2}\right)^2} \right)$$

$$= \frac{1}{2} \left(\frac{\sqrt{3}}{2} + 1 + 1 + \frac{\sqrt{3}}{2} \right) = \frac{2 + \sqrt{3}}{2} \approx 1.866$$

$$\text{(lower):} \quad \sum_{k=1}^{4} f(m_k) \Delta x = \frac{1}{2} \left(\sqrt{1 - (-1)^2} + \sqrt{1 - \left(-\frac{1}{2}\right)^2} + \sqrt{1 - \left(\frac{1}{2}\right)^2} + \sqrt{1 - 1^2} \right)$$

$$= \frac{1}{2} \left(\frac{\sqrt{3}}{2} + \frac{\sqrt{3}}{2} \right) = \frac{\sqrt{3}}{2} \approx 0.866$$

(trapezoid): $\displaystyle\sum_{k=1}^{4} \frac{f(x_{k-1}) + f(x_k)}{2} \Delta x$

$$= \frac{1}{2}\left(\frac{\sqrt{1-(-1)^2} + \sqrt{1-\left(-\frac{1}{2}\right)^2}}{2} + \frac{\sqrt{1-\left(-\frac{1}{2}\right)^2} + \sqrt{1-0^2}}{2}\right.$$

$$\left. + \frac{\sqrt{1-0^2} + \sqrt{1-\left(\frac{1}{2}\right)^2}}{2} + \frac{\sqrt{1-\left(\frac{1}{2}\right)^2} + \sqrt{1-1^2}}{2}\right)$$

$$= \frac{1}{4}\left(\frac{\sqrt{3}}{2} + \frac{\sqrt{3}}{2} + 1 + 1 + \frac{\sqrt{3}}{2} + \frac{\sqrt{3}}{2}\right) = \frac{1+\sqrt{3}}{2} \approx 1.366.$$

The midpoint sum is closest to the true value; it is off by only about 0.06.

39. Here $n = 4$ and the interval width is 1, so that $\Delta x = \frac{1}{4}$. Thus, $x_k = 2 + \frac{k}{4}$ for $k \in \{0,1,2,3,4\}$. Then the right sum is

$$\sum_{k=1}^{4} f(x_k)\Delta x = \sum_{k=1}^{4}\left(\frac{1}{4}\sqrt{2 + \frac{k}{4} - 1}\right).$$

41. $n = 6$ and the interval width is 3, so that $\Delta x = \frac{1}{2}$. Thus $x_k = 1 + \frac{k}{2}$ for $k \in \{0,1,2,3,4,5,6\}$. Then

$$\frac{x_{k-1} + x_k}{2} = \frac{1 + \frac{k-1}{2} + 1 + \frac{k}{2}}{2} = \frac{2k+3}{4},$$

so the midpoint sum is

$$\sum_{k=1}^{6} f\left(\frac{x_{k-1} + x_k}{2}\right)\Delta x = \sum_{k=1}^{6}\left(e^{(2k+3)/4} \cdot \frac{1}{2}\right).$$

43. $n = 4$ and the interval width is π, so that $\Delta x = \frac{\pi}{4}$. Thus $x_k = \frac{\pi}{4}k$ for $k \in \{0,1,2,3,4\}$. Then the trapezoid sum is

$$\sum_{k=1}^{4} \frac{f(x_{k-1}) + f(x_k)}{2}\Delta x = \sum_{k=1}^{4}\left(\frac{\sin\left(\frac{\pi}{4}(k-1)\right) + \sin\left(\frac{\pi}{4}k\right)}{2} \cdot \frac{\pi}{4}\right).$$

Applications

45. With $n = 4$ rectangles, we have $\Delta x = \frac{40}{4} = 10$, so that $x_k = 10k$ for $k \in \{0,1,2,3,4\}$. Note that v is increasing on $[0,20]$ and decreasing on $[20,40]$.

(a) The lower sum is

$$v(0) \cdot 10 + v(10) \cdot 10 + v(30) \cdot 10 + v(40) \cdot 10$$

$$= 10(-.22 \cdot 100 + 8.8 \cdot 10) + 10(-.22 \cdot 900 + 8.8 \cdot 30) = 1320.$$

(b) The upper sum is

$$v(10) \cdot 10 + v(20) \cdot 10 + v(20) \cdot 10 + v(30) \cdot 10$$
$$= 10(-.22 \cdot 100 + 8.8 \cdot 10) + 2 \cdot 10(-.22 \cdot 400 + 8.8 \cdot 20)$$
$$+ 10(-.22 \cdot 900 + 8.8 \cdot 30) = 3080.$$

47. (a) First, $T'(t)$ is always negative, so that $T(t)$ is decreasing everywhere (i.e., the casserole is cooling). Second, $T'(t)$ is increasing everywhere, so that the rate of decrease of $T(t)$ is decreasing over time (i.e., the casserole cools more slowly the cooler it gets).

(b) The area under the graph of $T'(t)$ between t_0 and t_1 has units of degrees. It measures the total amount of cooling in that number of minutes, so it measures the total change in $T(t)$ over that time. Thus rectangle approximations to the area under $T'(t)$ will approximate the change in temperature T of the casserole.

(c) Estimating $T'(t)$ gives (roughly)

$$T'(0) = -15 \quad T'(0.5) = -13 \quad T'(1) = -10 \quad T'(1.5) = -8 \quad T'(2) = -7 \quad T'(2.5) = -5$$
$$T'(3) = -4 \quad T'(3.5) = -3 \quad T'(4) = -2.75 \quad T'(4.5) = -2.5 \quad T'(5) = -2.$$

Since $n = 10$ and the width of the interval is 5, we have $\Delta x = 0.5$, so that the left sum is

$$\sum_{k=1}^{10} T'(0.5(k-1)) \cdot 0.5 = 0.5(-15 - 13 - 10 - 8 - 7 - 5 - 4 - 3 - 2.75 - 2.5) = -35.125.$$

The total change in temperature of the casserole over those 5 minutes was about $-35.125°$.

49. (a) The left Riemann sums are

$$c_1 \approx \frac{1}{105}(2100 \cdot (120 - 90) + 6300 \cdot (150 - 120) + 11000 \cdot (180 - 150) + 4000 \cdot (195 - 180))$$
$$\approx 6114.15$$

$$c_2 \approx \frac{1}{9.95}\left(2100\left(\sin\left(\frac{(90-90)\pi}{105}\right) - \frac{2}{\pi}\right) \cdot (120 - 90)\right.$$
$$+ 6300\left(\sin\left(\frac{(120-90)\pi}{105}\right) - \frac{2}{\pi}\right) \cdot (150 - 120)$$
$$+ 11000\left(\sin\left(\frac{(150-90)\pi}{105}\right) - \frac{2}{\pi}\right) \cdot (180 - 150)$$
$$\left. + 4000\left(\sin\left(\frac{(180-90)\pi}{105}\right) - \frac{2}{\pi}\right) \cdot (195 - 180)\right) \approx 8722.35$$

(b) Plugging in those coefficients gives for $g(t)$

$$g(t) = 6114.15 + 8722.35\left(\sin\left(\frac{(t-90)\pi}{105}\right) - \frac{2}{\pi}\right).$$

A plot of $g(t)$ together with the four data points between 90 and 195 days from the previous exercise gives

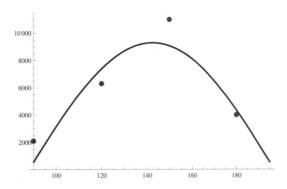

Proofs

51. Let f be defined on $[a, b]$, and assume f is positive and increasing there. Choose a positive integer n. Let $\Delta x = \frac{b-a}{n}$ and $x_k = a + k\Delta x$. Since f is increasing, clearly $f(x_{k-1}) \leq f(x_k)$ for all k. Then

$$\sum_{k=1}^{n} f(x_{k-1})\Delta x \leq \sum_{k=1}^{n} f(x_k)\Delta x,$$

so that the left Riemann sum on n rectangles is less than or equal to the right Riemann sum on n rectangles.

53. Recall that if a function f is concave up on an interval $[a, b]$, that means that its graph lies below any secant line between two points in $[a, b]$. But the trapezoid sum is the sum of the areas of trapezoids the slanted edge of which is a secant between x_{k-1} and x_k. That secant line lies above the graph of f, so that the area of the trapezoid on the base $[x_{k-1}, x_k]$ exceeds the area under the graph of f on $[x_{k-1}, x_k]$. Summing over all the intervals, we see that the trapezoid sum overestimates the area under the graph of f.

Thinking Forward

Approximating the length of a curve

Many answers are possible for all parts. For part (b), one way to accomplish this is to use smaller and smaller segments of roadway, treat each one as straight, and measure its length with a ruler. Using the scale on the map, one could get an approximation of the distance.

Limits of Riemann sums

▷ $A(n)$ is the right Riemann sum for $f(x)$ using n rectangles. As the number of rectangles increases, it appears that this value approaches a limit. We have

$$A(10) = \sum_{k=1}^{10} \left(1 + \frac{4k^2}{100}\right)\left(\frac{1}{5}\right)$$

$$= \frac{1}{5} \cdot \frac{1}{100} \sum_{k=1}^{10} (100 + 4k^2)$$

$$= 0.002 \left(\sum_{k=1}^{10} 100 + 4\sum_{k=1}^{10} k^2\right)$$

$$= 0.002 \left(1000 + 4\frac{10(11)(21)}{6}\right)$$

$$= 0.002 \left(1000 + 4 \cdot 385\right) = 5.08.$$

▷ The graph of $A(n)$ versus n is

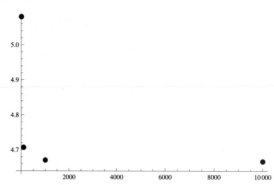

The graph of A seems to have a horizontal asymptote as $n \to \infty$, i.e., $A(n)$ converges to some finite limit.

▷ The right-sum approximations give successively better approximations to the area under $f(x)$ on $[1, 3]$.

Approximations and error

▷ The antiderivative of $\frac{1}{x}$ is $\ln x$, since $(\ln x)' = \frac{1}{x}$. Then by the given fact, we have

$$\int_1^4 \frac{1}{x}\, dx = \ln 4 - \ln 1 = \ln 4.$$

▷ With $n = 4$, we have $\Delta x = \frac{3}{4}$ and $x_k = 1 + \frac{3}{4}k$. The right-sum approximation is then

$$\int_1^4 \frac{1}{x}\, dx \approx \sum_{n=1}^4 \frac{1}{x_k} \cdot \frac{3}{4}$$

$$= \frac{1}{7/4} \cdot \frac{3}{4} + \frac{1}{10/4} \cdot \frac{3}{4} + \frac{1}{13/4} \cdot \frac{3}{4} + \frac{1}{16/4} \cdot \frac{3}{4}$$

$$= \frac{3}{7} + \frac{3}{10} + \frac{3}{13} + \frac{3}{16} \approx 1.147.$$

Since $\ln 4 \approx 1.386$, this right-sum approximation is an underestimate by about 0.239.

▷ Computing the right-sum approximation for $n = 5$ gives a value of ≈ 1.188, which is an underestimate by about $0.198 < 0.25$. Thus $n = 5$ is the value we want.

▷ With the same $n = 4$ and Δx as in part (b), the midpoints are $\frac{6k+5}{8}$, so that the midpoint approximation is

$$\int_1^4 \frac{1}{x}\, dx \approx \sum_{n=1}^4 \frac{1}{(6k+5)/8} \cdot \frac{3}{4}$$

$$= \frac{1}{11/8} \cdot \frac{3}{4} + \frac{1}{17/8} \cdot \frac{3}{4} + \frac{1}{23/8} \cdot \frac{3}{4} + \frac{1}{29/8} \cdot \frac{3}{4}$$

$$= \frac{6}{11} + \frac{6}{17} + \frac{6}{23} + \frac{6}{29} \approx 1.366.$$

This estimate is already within 0.25 of $\ln 4$; in fact, it differs by only ≈ 0.02. Even with $n = 1$, the approximation is within $\frac{1}{4}$: for $n = 1$, $\Delta x = 3$, the midpoint is $\frac{5}{2}$, and the midpoint sum is

$$\int_1^4 \frac{1}{x}\, dx \approx \frac{1}{5/2} \cdot 3 = \frac{6}{5} = 1.2,$$

which differs from $\ln 4$ by ≈ 0.186.

4.3   Definite Integrals

Thinking Back

Commuting with sums

Saying that derivatives commute with sums means that given two functions, you get the same answer if you add them and then differentiate, or if you differentiate and then add the results, i.e., $f'(x) + g'(x) = (f + g)'(x)$. Saying that limits commute with sums means that given two functions, you get the same answer if you add them and then take the limit at c, or if you first take each limit at c and then add the results, i.e., that $\lim_{x \to c} f(x) + \lim_{x \to c} g(x) = \lim_{x \to c}(f(x) + g(x))$. Saying that sums written in sigma notation commute with sums means that adding two sequences together can be done either by adding each sequence and then adding the results, or by adding each pair of terms together and then summing the resulting sequence, i.e., $\sum_{k=1}^{n} a_k + \sum_{k=1}^{n} b_k = \sum_{k=1}^{n}(a_k + b_k)$.

Commuting with constant multiples

Saying that derivatives commute with constant multiples means that given a function, you get the same answer if you multiply it by a constant and then differentiate, or if you differentiate and then multiply by the constant, i.e., $(cf)'(x) = c \cdot f'(x)$. Saying that limits commute with constant multiples means that given a function, you get the same answer if you multiply it by a constant and then take the limit at c, or if you first take each limit at c and then multiply by the constant, i.e., that $\lim_{x \to c}(kf)(x) = k \lim_{x \to c} f(x)$. Saying that sums written in sigma notation commute with constant multiples means that multiplying each term of a sequence and then adding them up gives the same answer as adding up the sequence and then multiplying by the constant, i.e., $\sum_{k=1}^{n}(ca_k) = c \sum_{k=1}^{n} a_k$.

Concepts

1. (a) False. For example, if the function is increasing everywhere, then the right-sum approximations will always be upper approximations, and left-sum approximations will always be lower approximations, so that the former will always be larger than the latter.

    (b) True. It can be approximated by a Riemann sum, and its value is the limit of that sum.

    (c) False. Since no limits were given on the integration, this is not a real number — it is an antiderivative, which is a function.

    (d) False. What is true is that $\int_{-3}^{-2} f(x)\,dx = -\int_{-2}^{-3} f(x)\,dx$.

    (e) False. For example, let $f(x) = \frac{3}{2}$ and $g(x) = 1$. Then $f(g(x)) = \frac{3}{2}$, so that $\int_0^2 f(g(x))\,dx = 3$.

    (f) False. For example, let $f(x) = \frac{3}{2}$ and $g(x) = 1$. Then $\int_0^2 f(x)g(x)\,dx = \int_0^2 f(x)\,dx = 3$.

    (g) True. See Theorem 4.12(b).

    (h) False. For example, f might not even be defined for $x \in (2, 4]$.

3. The signed area between the graph of a continuous function f and the x-axis on $[a, b]$ is represented by the notation $\int_a^b f(x)\,dx$ and is called the definite integral of $f(x)$ on $[a, b]$.

5. The definite integral of an integrable function f from $x = a$ to $x = b$ is defined to be

$$\int_a^b f(x)\,dx = \lim_{n \to \infty} \sum_{k=1}^{n} f(x_k^*)\Delta x,$$

where $\Delta x = \frac{b-a}{n}$, $x_k = a + k\Delta x$, and x_k^* is any choice of point in $[x_{k-1}, x_k]$.

7. The region under the curve $f(x)$ between $x = a$ and $x = a$ is a line at $x = a$. This looks like a rectangle with width 0 and height $f(a)$, so it has area zero. Thus it makes sense to say that $\int_a^a f(x)\, dx = 0$.

9. A graph of this function on the given interval is

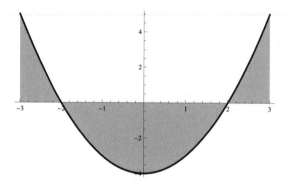

It appears that the shaded region below the axis has larger area than the shaded region above the axis, so that the integral is negative.

11. A graph of this function on the given interval is

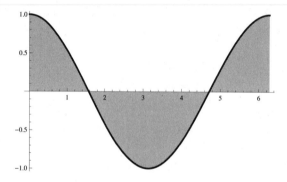

It appears that the shaded region below the axis has about the same area as the shaded region above the axis, so that the integral may be zero.

13. (a) The graph with the region above $[-2, 6]$ shaded is

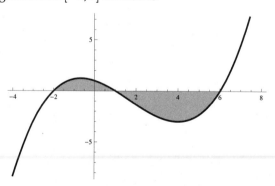

It appears that the area below the x axis, between 1 and 6, exceeds that above the x axis, between -2 and 1, so that the total signed area is negative.

(b) The graph with the region above $[-4, 2]$ shaded is

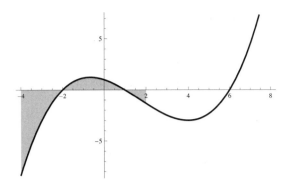

It appears that the area below the x axis, between -4 and -2 and also between 1 and 2, exceeds that above the x axis, between -2 and 1, so that the total signed area is negative.

15. (a)

$$\int_a^b (f(x) + g(x))\, dx = \int_a^b f(x)\, dx + \int_a^b g(x)\, dx,$$

but

$$\int_a^b f(x)g(x)\, dx \neq \left(\int_a^b f(x)\, dx\right)\left(\int_a^b g(x)\, dx\right).$$

(b) For example, let $f(x) = x$ and $g(x) = 3x$. Then $f(x) + g(x) = 4x$, so that by Theorem 4.11(b) and 4.13(b),

$$\int_0^1 (f(x) + g(x))\, dx = \int_0^1 4x\, dx = 4\int_0^1 x\, dx = 4 \cdot \frac{1}{2}(1^2 - 0^2) = 2$$

$$\int_0^1 f(x)\, dx + \int_0^1 g(x)\, dx = \int_0^1 x\, dx + 3\int_0^1 x\, dx = \frac{1}{2}(1^2 - 0^2) + 3 \cdot \frac{1}{2}(1^2 - 0^2) = 2,$$

and the two are equal.

(c) Let $f(x) = g(x) = x$, so that $f(x)g(x) = x^2$. Then by Theorem 4.13(b) and (c),

$$\int_0^1 f(x)g(x)\, dx = \int_0^1 x^2\, dx = \frac{1}{3}(1^3 - 0^3) = \frac{1}{3}$$

$$\left(\int_a^b f(x)\, dx\right)\left(\int_a^b g(x)\, dx\right) = \left(\int_0^1 x\, dx\right)^2 = \left(\frac{1}{2}(1^2 - 0^2)\right)^2 = \frac{1}{4},$$

so that the two are unequal.

17. (a) An n-rectangle right sum has $x_k = \frac{3k}{n}$ for $k = 0, 1, \ldots, n$, and $\Delta x = \frac{3}{n}$. Then the right sum is

$$\sum_{k=1}^n f(x_i)\Delta x = \sum_{k=1}^n \left(\frac{3k}{n}\right)^2 \left(\frac{3}{n}\right)$$

$$= \left(\frac{3}{n}\right)^3 \sum_{k=1}^n k^2$$

$$= \frac{27}{n^3} \cdot \frac{n(n+1)(2n+1)}{6}$$

$$= \frac{9(n+1)(2n+1)}{2n^2}.$$

(b) x_k and Δx are as in part (a), and the left sum is

$$\sum_{k=1}^{n} f(x_{i-1})\Delta x = \sum_{k=1}^{n} \left(\frac{3(k-1)}{n}\right)^2 \left(\frac{3}{n}\right)$$

$$= \frac{3}{n} \sum_{k=1}^{n} \frac{9k^2 - 18k + 9}{n^2}$$

$$= \frac{3}{n} \left(\frac{9}{n^2} \sum_{k=1}^{n} k^2 - \frac{18}{n^2} \sum_{k=1}^{n} k + \frac{9}{n^2} \sum_{k=1}^{n} 1\right)$$

$$= \frac{3}{n} \left(\frac{9}{n^2} \cdot \frac{n(n+1)(2n+1)}{6} - \frac{18}{n^2} \cdot \frac{n(n+1)}{2} + \frac{9}{n^2} \cdot n\right)$$

$$= \frac{9(n-1)(2n-1)}{2n^2}.$$

(c) For $n = 100$, the two sums are

$$\text{right:} \quad \frac{9(100+1)(2 \cdot 100 + 1)}{2 \cdot 100^2} \approx 9.13545$$

$$\text{left:} \quad \frac{9(100-1)(2 \cdot 100 - 1)}{2 \cdot 100^2} \approx 8.86545.$$

For $n = 1000$, the two sums are

$$\text{right:} \quad \frac{9(1000+1)(2 \cdot 1000 + 1)}{2 \cdot 1000^2} \approx 9.0135$$

$$\text{left:} \quad \frac{9(1000-1)(2 \cdot 1000 - 1)}{2 \cdot 1000^2} \approx 8.9865.$$

(d) We take the limits of the expressions in parts (a) and (b) as $n \to \infty$:

$$\lim_{n\to\infty} \frac{9(n+1)(2n+1)}{2n^2} = \lim_{n\to\infty} \frac{18n^2 + 27n + 9}{2n^2} = \lim_{n\to\infty} \frac{18 + 27/n + 9/n^2}{2} = 9.$$

$$\lim_{n\to\infty} \frac{9(n-1)(2n-1)}{2n^2} = \lim_{n\to\infty} \frac{18n^2 - 27n - 9}{2n^2} = \lim_{n\to\infty} \frac{18 - 27/n - 9/n^2}{2} = 9.$$

The common limit, 9, is the value of the integral $\int_0^3 x^2\,dx$. It is the limit of the areas of either the left or right sum as the number of rectangles gets larger and larger, and represents the area under the graph of $y = x^2$ between $x = 0$ and $x = 3$.

19. (a) A graph is

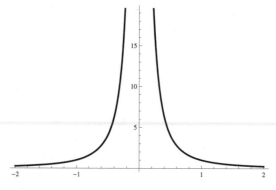

f has an infinite discontinuity at $x = 0$. Both one-sided limits are ∞.

(b) Since the function gets extremely large near $x = 0$, it is reasonable to think that the areas of rectangles near zero will get larger and larger as the chosen point in the subintervals get closer to zero.

(c) The Riemann sum approximation for this area approaches ∞ as $n \to \infty$.

(d) Graphs of both functions on the same ranges are:

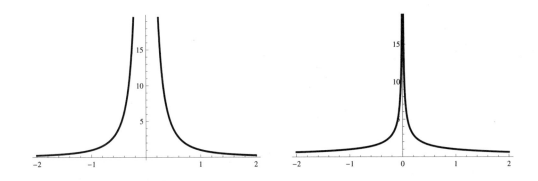

Clearly $\frac{1}{x^{2/3}}$ gets large much more slowly than $\frac{1}{x^2}$. Thus x must be much closer to zero in order for $\frac{1}{x^{2/3}}$ to reach the same value as for $\frac{1}{x^2}$. Thus the width of the rectangles for $\frac{1}{x^{2/3}}$ at a given height near zero is much smaller, so perhaps they get small rapidly enough that this integral is finite.

Skills

21. A graph of $2 - x$ on $[0, 2]$ is

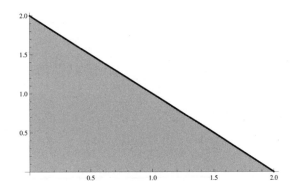

The area under the curve is a triangle with base and height each equal to 2, so that its area is $\frac{1}{2}2 \cdot 2 = 2$.

23. Since the function is a constant, the area under the graph from -3 to 8 is a rectangle with base $8 - (-3) = 11$ and height 24, so that the integral is $11 \cdot 24 = 264$.

25. A graph of $\sqrt{1-x^2}$ on $[-1,1]$ is

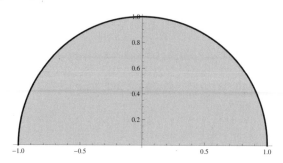

The area under the curve is a semicircle of radius 1, so the area, and thus the value of the integral, is $\frac{1}{2}\pi \cdot 1^2 = \frac{\pi}{2}$.

27. A graph of $1 - |x-1|$ on $[0,3]$ is

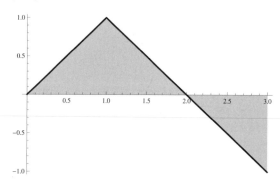

The area consists of two triangles above the x axis, each with base 1 and height 1 (one extends from $x = 0$ to $x = 1$, the other from $x = 1$ to $x = 2$). There is also a triangle below the x axis, with base 1 and height 1. This counts as a negative area. The end result is that the net area is one triangle with base 1 and height 1, so that the net area under the curve is $\frac{1}{2}1 \cdot 1 = \frac{1}{2}$.

Note that for Exercises 29-40, we will freely make use of Theorems 4.11 and 4.12 without explicit mention.

29. Since $\int_{-2}^{3} f(x)\,dx + \int_{3}^{6} f(x)\,dx = \int_{-2}^{6} f(x)\,dx$, we see that $4 + \int_{3}^{6} f(x)\,dx = 9$, so that $\int_{3}^{6} f(x)\,dx = 5$.

31.
$$\int_{-2}^{6}(f(x) + g(x))\,dx = \int_{-2}^{6} f(x)\,dx + \int_{-2}^{6} g(x)\,dx$$
$$= \int_{-2}^{6} f(x)\,dx + \int_{-2}^{3} g(x)\,dx + \int_{3}^{6} g(x)\,dx$$
$$= 9 + 2 + 3 = 14.$$

33. There is not enough information to evaluate this integral. There is no reason to believe that $\int_{-2}^{3} f(x)g(x)\,dx = \left(\int_{-2}^{3} f(x)\,dx\right)\left(\int_{-2}^{3} g(x)\,dx\right)$, and we have no other formulas that would allow us to determine a value.

35. There is not enough information to evaluate this integral. There is no reason to believe that $\int_{3}^{6} g(x)^2\,dx = \left(\int_{3}^{6} g(x)\,dx\right)^2$, and we have no other formulas that would allow us to determine a value.

37. $\int_6^3 (f(x) + g(x))\, dx = -\int_3^6 (f(x) + g(x))\, dx = -\int_3^6 f(x)\, dx - \int_3^6 g(x)\, dx = -5 - 3 = -8$ using the result of Exercise 29.

39.

$$\int_3^{-2} (2x^2 - 3g(x))\, dx = -\int_{-2}^3 (2x^2 - 3g(x))\, dx$$

$$= -2 \int_{-2}^3 x^2\, dx + 3 \int_{-2}^3 g(x)\, dx$$

$$= -2 \cdot \frac{1}{3}(3^3 - (-2)^3) + 3 \cdot 2$$

$$= -\frac{70}{3} + 6 = -\frac{52}{3}.$$

41. (a) The interval width is 3, so that $\Delta x = \frac{3}{n}$ and $x_k = 2 + \frac{3k}{n}$ for $k = 0, 1, \ldots, n$. Then the right sum is

$$\sum_{k=1}^n (5 - x_k)\Delta x = \sum_{k=1}^n \left(5 - \left(2 + \frac{3k}{n}\right)\right)\Delta x$$

$$= \frac{3}{n} \sum_{k=1}^n \left(3 - \frac{3k}{n}\right)$$

$$= \frac{3}{n}\left(\sum_{k=1}^n 3 - \frac{3}{n}\sum_{k=1}^n k\right)$$

$$= \frac{3}{n}\left(3n - \frac{3}{n} \cdot \frac{n(n+1)}{2}\right)$$

$$= 9 - \frac{9(n+1)}{2n}$$

(b) For $n = 100$ and $n = 1000$ we get

$$n = 100: \quad 9 - \frac{9(101)}{200} \approx 4.455 \qquad n = 1000: \quad 9 - \frac{9(1001)}{2000} \approx 4.4955.$$

(c) $\lim\limits_{n\to\infty}\left(9 - \frac{9(n+1)}{2n}\right) = 9 - \lim\limits_{n\to\infty}\frac{9+9/n}{2} = 9 - \frac{9}{2} = \frac{9}{2} = 4.5.$

43. (a) Since the interval width is 1, we get $\Delta x = \frac{1}{n}$ and $x_k = \frac{k}{n}$ for $k = 0, 1, \ldots, n$. Then

$$\sum_{k=1}^n 2x^2 \Delta x = \sum_{k=1}^n 2\frac{k^2}{n^2} \cdot \frac{1}{n}$$

$$= \frac{2}{n^3} \sum_{k=1}^n k^2$$

$$= \frac{2}{n^3} \cdot \frac{n(n+1)(2n+1)}{6}$$

$$= \frac{2n^3 + 3n^2 + n}{3n^3} = \frac{2n^2 + 3n + 1}{3n^2}.$$

(b) For $n = 100$ and $n = 1000$ we get

$$n = 100: \quad \frac{2 \cdot 100^2 + 3 \cdot 100 + 1}{3 \cdot 100^2} \approx 0.6767 \qquad n = 1000: \quad \frac{2 \cdot 1000^2 + 3 \cdot 1000 + 1}{3 \cdot 1000^2} \approx 0.667667.$$

(c) $\displaystyle\lim_{n\to\infty}\frac{2n^2+3n+1}{3n^2}=\frac{2+3/n+1/n^2}{3}=\frac{2}{3}.$

45. (a) Since the interval width is 1, we get $\Delta x=\frac{1}{n}$ and $x_k=2+\frac{k}{n}$ for $k=0,1,\ldots,n$. Then

$$\sum_{k=1}^{n}(x+1)^2\Delta x=\sum_{k=1}^{n}\left(2+\frac{k}{n}+1\right)^2\cdot\frac{1}{n}$$

$$=\frac{1}{n}\sum_{k=1}^{n}\left(\frac{k^2}{n^2}+\frac{6k}{n}+9\right)$$

$$=\frac{1}{n^3}\sum_{k=1}^{n}k^2+\frac{6}{n^2}\sum_{k=1}^{n}k+\frac{9}{n}\sum_{k=1}^{n}1$$

$$=\frac{1}{n^3}\cdot\frac{n(n+1)(2n+1)}{6}+\frac{6}{n^2}\cdot\frac{n(n+1)}{2}+\frac{9}{n}\cdot n$$

$$=\frac{74n^2+21n+1}{6n^2}.$$

(b) For $n=100$ and $n=1000$ we get

$$n=100:\qquad\frac{74\cdot100^2+21\cdot100+1}{6\cdot100^2}\approx12.3684$$

$$n=1000:\qquad\frac{74\cdot1000^2+21\cdot1000+1}{6\cdot1000^2}\approx12.3368.$$

(c) $\displaystyle\lim_{n\to\infty}\frac{74n^2+21n+1}{6n^2}=\lim_{n\to\infty}\frac{74+21/n+1/n^2}{6}=\frac{37}{3}.$

47. $\displaystyle\int_{2}^{4}(x^2+1)\,dx=\int_{2}^{4}x^2\,dx+\int_{2}^{4}1\,dx=\frac{1}{3}(4^3-2^3)+1(4-2)=\frac{56}{3}+2=\frac{62}{3}.$

49.

$$\int_{5}^{2}(9+10x-x^2)\,dx=-\int_{2}^{5}(-x^2+10x+9)\,dx$$

$$=\int_{2}^{5}(x^2-10x-9)\,dx$$

$$=\int_{2}^{5}x^2\,dx-\int_{2}^{5}10x\,dx-\int_{2}^{5}9\,dx$$

$$=\int_{2}^{5}x^2\,dx-10\int_{2}^{5}x\,dx-\int_{2}^{5}9\,dx$$

$$=\frac{1}{3}(5^3-2^3)-10\cdot\frac{1}{2}(5^2-2^2)-9(5-2)$$

$$=-93.$$

51. Expanding the integrand gives

$$((2x-3)^2+5)=4x^2-12x+9+5=4x^2-12x+14,$$

so that

$$\int_0^4 ((2x-3)^2 + 5)\, dx = \int_0^4 (4x^2 - 12x + 14)\, dx$$

$$= 4\int_0^4 x^2\, dx - 12\int_0^4 x\, dx + \int_0^4 14\, dx$$

$$= 4 \cdot \frac{1}{3}(4^3 - 0^3) - 12 \cdot \frac{1}{2}(4^2 - 0^2) + 14(4 - 0)$$

$$= \frac{136}{3}.$$

Applications

53. (a) The total distance traveled is the area under the velocity curve, which is

$$\int_0^{40} v(t)\, dt = \int_0^{40} (-0.22t^2 + 8.8t)\, dt.$$

(b) Use the right Riemann sum. If there are n subintervals, then $\Delta x = \frac{40}{n}$ and $x_k = \frac{40k}{n}$ for $k = 0, 1, \ldots, n$. The right Riemann sum is then

$$\sum_{k=1}^n f(x_k)\Delta x = \frac{40}{n} \sum_{k=1}^n \left(-0.22\left(\frac{40k}{n}\right)^2 + 8.8\frac{40k}{n} \right)$$

$$= \frac{40}{n} \sum_{k=1}^n \left(-0.22\left(\frac{40k}{n}\right)^2 \right) + \frac{40}{n} \sum_{k=1}^n 8.8\frac{40k}{n}$$

$$= -0.22\frac{64000}{n^3} \sum_{k=1}^n k^2 + 8.8\frac{1600}{n^2} \sum_{k=1}^n k$$

$$= -\frac{14080}{n^3} \cdot \frac{n(n+1)(2n+1)}{6} + \frac{14080}{n^2} \cdot \frac{n(n+1)}{2}$$

$$= -\frac{14080(n+1)(2n+1)}{6n^2} + \frac{7040(n+1)}{n}$$

$$= \frac{-28160n^2 - 42240n - 14080 + 42240n^2 + 42240n}{6n^2}$$

$$= \frac{14080n^2 - 14080}{6n^2}.$$

Then the limit of the Riemann sum as $n \to \infty$ is

$$\lim_{n\to\infty} \frac{14080n^2 - 14080}{6n^2} = \lim_{n\to\infty} \frac{14080 - 14080/n^2}{6} = \frac{14080}{6} \approx 2346.67.$$

The total distance traveled by the car in those 40 seconds is about 2346.67 feet.

Proofs

55. This proof is essentially the use of the constant multiplier rule for functions, sums, and limits. We have

$$\int_a^b cf(x)\, dx = \lim_{n\to\infty} \sum_{k=1}^n (cf)(x_k^*)\Delta x = \lim_{n\to\infty} \sum_{k=1}^n c(f(x_k^*))\Delta x = \lim_{n\to\infty} c \sum_{k=1}^n (f(x_k^*))\Delta x$$

$$= c \lim_{n\to\infty} \sum_{k=1}^n (f(x_k^*))\Delta x = c \int_a^b f(x)\, dx.$$

57. We have

$$\int_b^a f(x)\,dx = \lim_{n\to\infty} \sum_{k=1}^n f(x_k^*)\frac{a-b}{n}$$

$$= \lim_{n\to\infty} \sum_{k=1}^n \left(-f(x_k^*)\frac{b-a}{n}\right)$$

$$= \lim_{n\to\infty} -\left(\sum_{k=1}^n f(x_k^*)\frac{b-a}{n}\right)$$

$$= -\lim_{n\to\infty} \sum_{k=1}^n f(x_k^*)\frac{b-a}{n}$$

$$= -\int_a^b f(x)\,dx.$$

59. The graph of $f(x) = x$ between a and b is shown below (for $a = 2$ and $b = 5$), with the area under the graph shaded:

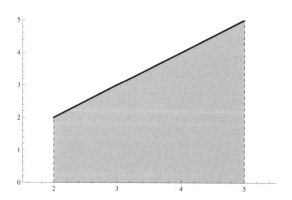

It is apparent that this area is (in the general case) a trapezoid with width $b - a$ and with heights $h_1 = a$ and $h_2 = b$, so by the formula for the area of a trapezoid, we get

$$\int_a^b x\,dx = \frac{1}{2}(a+b)\cdot(b-a) = \frac{1}{2}(b^2-a^2).$$

61. For each partition into n rectangles, use the right Riemann sum. Then $\Delta x = \frac{b-a}{n}$, and $x_k = a + \frac{b-a}{n}k$, so that

$$
\begin{aligned}
\int_a^b x^2\,dx &= \lim_{n\to\infty} \sum_{k=1}^n f(x_k)\Delta x \\
&= \lim_{n\to\infty} \sum_{k=1}^n \left(a + \frac{b-a}{n}k\right)^2 \frac{b-a}{n} \\
&= \lim_{n\to\infty} \sum_{k=1}^n \left(a^2 + 2a\frac{b-a}{n}k + \frac{(b-a)^2}{n^2}k^2\right)\frac{b-a}{n} \\
&= \lim_{n\to\infty} \left(a^2\frac{b-a}{n}\sum_{k=1}^n 1 + 2a\frac{(b-a)^2}{n^2}\sum_{k=1}^n k + \frac{(b-a)^3}{n^3}\sum_{k=1}^n k^2\right) \\
&= \lim_{n\to\infty} \left(a^2\frac{b-a}{n}\cdot n + 2a\frac{(b-a)^2}{n^2}\cdot\frac{n(n+1)}{2} + \frac{(b-a)^3}{n^3}\cdot\frac{n(n+1)(2n+1)}{6}\right) \\
&= \lim_{n\to\infty} \frac{(2b^3 - 2a^3)n^2 + (3a^3 - 3a^2b - 3ab^2 + 3b^3)n + (b^3 - 3ab^2 + 3a^2b - a^3)}{6n^2} \\
&= \lim_{n\to\infty} \frac{(2b^3 - 2a^3) + (3a^3 - 3a^2b - 3ab^2 + 3b^3)/n + (b^3 - 3ab^2 + 3a^2b - a^3)/n^2}{6} \\
&= \frac{1}{3}(b^3 - a^3).
\end{aligned}
$$

Thinking Forward

Functions defined by area accumulation

▷ We know that $\int_0^b 2x\,dx = 2\int_0^b x\,dx = 2\cdot\frac{1}{2}(b^2 - 0^2) = b^2$, so a table of integrals is

b	-3	-2	-1	0	1	2	3
$\int_0^b 2x\,dx$	9	4	1	0	1	4	9

▷ The word is "square".

▷ We know that $\int_1^b 2x\,dx = 2\int_1^b x\,dx = 2\cdot\frac{1}{2}(b^2 - 1^2) = b^2 - 1$, so a table of integrals is

b	-3	-2	-1	0	1	2	3
$\int_0^b 2x\,dx$	8	3	0	-1	0	3	8

▷ $1 + \int_1^b 2x\,dx = \int_0^b 2x\,dx$.

4.4   Indefinite Integrals

Thinking Back

Antiderivatives

Since h is the derivative of g, we see that g is the antiderivative of h.

Definite Integrals

(See Definition 4.9) If f is a function defined on $[a, b]$, then the definite integral of f from $x = a$ to $x = b$ is the number

$$\int_a^b f(x)\, dx = \lim_{x \to \infty} \sum_{k=1}^{n} f(x_k^*) \Delta x$$

if this limit exists, where $\Delta a = \frac{b-a}{n}$ and $x_k = a + \frac{b-a}{n}k$, for $x_k^* \in [x_{k-1}, x_k]$.

The Mean Value Theorem

(See Theorem 3.5) If f is continuous on $[a, b]$ and differentiable on (a, b), then there exists at least one value $c \in (a, b)$ such that $f'(c) = \frac{f(b) - f(a)}{b - a}$.

Derivatives and Antiderivatives

▷ Since $f(x)$ is a constant, its derivative is zero. An antiderivative is $\pi^2 x$, since $(\pi^2 x)' = \pi^2$.

▷ Here $f(x) = x^{-2/11}$, so that $f'(x) = -\frac{2}{11}x^{-13/11}$. Also, an antiderivative of $f(x)$ is $\frac{11}{9}x^{9/11}$.

▷ Since $f(x) = \frac{1}{5}x^{-1}$, we get $f'(x) = -\frac{1}{5}x^{-2}$. Since an antiderivative of x^{-1} is $\ln|x|$, an antiderivative of $f(x)$ is $\frac{1}{5}\ln|x|$.

▷ $f'(x) = 2\sec(3x + 1)\tan^2(3x + 1) \cdot 3 = 6\sec(3x + 1)\tan^2(3x + 1)$. Since $\tan'(x) = \sec^2 x$, we can start by guessing $\tan(3x + 1)$ for an antiderivative. That has derivative $\sec^2(3x + 1) \cdot 3 = 3\sec^2(3x + 1)$, which is too big by a factor of 3, so an antiderivative is $\frac{1}{3}\tan(3x + 1)$.

Concepts

1. (a) True. If $f(x) - g(x) = 2$, then $f(x) = g(x) + 2$, so that $f'(x) = (g(x) + 2)' = g'(x) + (2)' = g'(x)$, so that f and g have the same derivative.

   (b) False. For example, let $f(x) = x + 2$ and $g(x) = x$. Then an antiderivative of f is $\frac{1}{2}x^2 + 2x$, which is not an antiderivative of g (since its derivative is not equal to g).

   (c) False. $F(x) = x^2 + C$ for an arbitrary constant C.

   (d) True. An antiderivative of $x^3 + 1$ is $\frac{1}{4}x^4 + x$. The antiderivative given is $\frac{1}{4}x^4 + x + (2 + C)$; here $2 + C$ is an arbitrary constant since C is.

   (e) False. Taking the derivative of the right-hand side gives $\left(-\cos(x^2) + C\right)' = \sin(x^2) \cdot 2x = 2x\sin(x^2) \neq \sin(x^2)$.

   (f) False. Taking the derivative of the right-hand side gives $(e^x \sin x + C)' = e^x \sin x + e^x \cos x \neq e^x \cos x$.

   (g) False. Taking the derivative of the right-hand side (and noting that $|x^2 + 1| = x^2 + 1$ for all x) gives $\left(\ln(x^2 + 1) + C\right)' = \frac{1}{x^2+1} \cdot 2x = \frac{2x}{x^2+1} \neq \frac{1}{x^2+1}$.

   (h) False. Taking the derivative of the right-hand side (and noting that $|x^2 + 1| = x^2 + 1$ for all x) gives $\left(\frac{1}{2x}\ln(x^2 + 1) + C\right)' = \frac{1}{2x} \cdot \frac{1}{x^2+1} - \frac{1}{2x^2}\ln(x^2 + 1) \neq \frac{1}{x^2+1}$.

3. The indefinite integral of a function f is the collection of all antiderivatives of f. If $F(x)$ is any antiderivative, then the indefinite integral is $F(x) + C$ where C is an arbitrary constant.

5. The definite integral has as its value a real number; it has a unique answer; and it has a geometric meaning, which is the area under a curve between two specific points. The indefinite integral, on the other hand, has as its value a collection of functions; it does not have a unique answer; and it has (as yet) no obvious geometric meaning.

7. (a) $\int x^6\, dx = \frac{1}{7}x^7 + C$. This is because $\left(\frac{1}{7}x^7 + C\right) = x^6$.

   (b) $\frac{1}{7}x^7 + 33$ is an antiderivative of x^6 (as is, for example, $\frac{1}{7}x^7 + 257$).

   (c) The derivative of $\frac{1}{7}x^7$ is x^6.

9. We have an integration formula for $f(x) = \sin x$ because we know a function, namely $\cos x$, whose derivative is $\sin x$. We do not have an integration formula for $\sec x$ because we do *not* know a function whose derivative is $\sec x$.

11. These are encompassed in Theorems 4.16 through 4.23.

13. To show $F(x)$ is an antiderivative of $x^3 e^{(x^2)}$, we must show that its derivative is $x^3 e^{(x^2)}$. But

$$F'(x) = \frac{1}{2}\left(e^{(x^2)}(x^2 - 1)\right)'$$
$$= \frac{1}{2}\left(e^{(x^2)}(2x) + 2x \cdot e^{(x^2)}(x^2 - 1)\right)$$
$$= e^{(x^2)}(x + x(x^2 - 1))$$
$$= x^3 e^{(x^2)}.$$

15. To verify the equation it suffices to show that the derivative of the right-hand side is $\ln x$. But

$$(x(\ln x - 1) + C)' = (\ln x - 1) + x(\ln x - 1)' = \ln x - 1 + x\frac{1}{x} = \ln x - 1 + 1 = \ln x.$$

17. For example, let $f(x) = x$ and $g(x) = 1$. Then

$$\int f(x)g(x)\, dx = \int x\, dx = \frac{1}{2}x^2 + C,$$
$$\left(\int f(x)\, dx\right)\left(\int g(x)\, dx\right) = \left(\int x\, dx\right)\left(\int 1\, dx\right) = \left(\frac{1}{2}x^2 + C_1\right)(x + C_2),$$

and the two are clearly not equal since the second line contains an x^3 term but the first does not.

19. For $x < 0$, we have $F'(x) = -(\cot x)' = \csc^2 x$ while for $x > 0$ we have $F'(x) = (-\cot x + 100)' = -(\cot x)' + 100' = \csc^2 x$. Thus $F'(x) = \csc^2 x$ for all $x \neq 0$. Plots of $F(x)$ and $-\cot x$ are below:

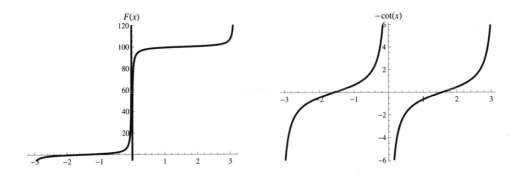

$F(x)$ differs from $-\cot x$ for $x > 0$ by a constant additive value, $C = 100$. By Theorem 4.16(b), this means that $F(x)$ and $-\cot x$ should have the same derivative, which they do.

Skills

21. Since the integral of x^k is $\frac{1}{k+1}x^{k+1}$ for $k \neq -1$, we get

$$\int (x^2 - 3x^5 - 7)\, dx = \frac{1}{3}x^3 - \frac{3}{6}x^6 - 7x + C = \frac{1}{3}x^3 - \frac{1}{2}x^6 - 7x + C.$$

Differentiating gives $3 \cdot \frac{1}{3}x^2 - 6 \cdot \frac{1}{2}x^5 - 7 = x^2 - 3x^5 - 7$

23. Expanding the integrand gives $3x^3 + 5x^2 - 3x - 5$. Since the integral of x^k is $\frac{1}{k+1}x^{k+1}$ for $k \neq -1$, we get

$$\int (x^2 - 1)(3x + 5)\, dx = \frac{3}{4}x^4 + \frac{5}{3}x^3 - \frac{3}{2}x^2 - 5x + C.$$

Differentiating gives $4 \cdot \frac{3}{4}x^3 + 3 \cdot \frac{5}{3}x^2 - 2 \cdot \frac{3}{2}x - 5 = 3x^3 + 5x^2 - 3x - 5 = (x^2 - 1)(3x + 5)$.

25. The integral of b^x is $\frac{1}{\ln b}b^x + C$, so that

$$\int (x^2 + 2^x + 2^2)\, dx = \frac{1}{3}x^3 + \frac{1}{\ln 2}2^x + 4x + C.$$

Differentiating gives $3 \cdot \frac{1}{3}x^2 + \frac{1}{\ln 2} \cdot 2^x \ln 2 + 4 = x^2 + 2^x + 4.$

27. Simplifying the integrand gives $\frac{x+1}{\sqrt{x}} = x^{1/2} + x^{-1/2}$. Then using the power rule, we have

$$\int \frac{x+1}{\sqrt{x}}\, dx = \frac{2}{3}x^{3/2} + 2x^{1/2} + C.$$

Differentiating gives $\frac{3}{2} \cdot \frac{2}{3}x^{1/2} + \frac{1}{2} \cdot 2x^{-1/2} = x^{1/2} + x^{-1/2} = \frac{x+1}{\sqrt{x}}.$

29. Using the rule for integrating exponentials gives

$$\int (3.2(1.43)^x - 50)\, dx = 3.2 \int 1.43^x\, dx - 50x + C = \frac{3.2}{\ln 1.43}1.43^x - 50x + C.$$

Differentiating gives $\frac{3.2}{\ln 1.43} \cdot 1.43^x \ln 1.43 - 50 = 3.2 \cdot 1.43^x - 50.$

31. Expanding the integrand gives $4(e^{x-3})^2 = 4e^{2x-6} = 4e^{-6}e^{2x}$. Now use the rule for integrating exponentials.

$$\int 4(e^{x-3})^2\, dx = 4e^{-6} \cdot \frac{1}{2}e^{2x} + C = 2e^{-6}e^{2x} + C = 2(e^{x-3})^2 + C.$$

Differentiating gives $2 \cdot 2e^{x-3} \cdot e^{x-3} = 4(e^{x-3})^2.$

33. We know that $(\sec x)' = \sec x \tan x$, so try $\sec 3x$. The derivative of $\sec 3x$ is $\sec 3x \tan 3x \cdot (3x)' = 3\sec 3x \tan 3x$, so this is three times too big. Thus

$$\int \sec(3x) \tan(3x)\, dx = \frac{1}{3}\sec 3x + C.$$

Differentiating gives $\frac{1}{3}\sec 3x \tan 3x \cdot 3 = \sec 3x \tan 3x.$

35. Since $\tan^2 x + 1 = \sec^2 x$, we have $3\tan^2 x = -3 + 3\sec^2 x$. The integral of $\sec^2 x$ is $\tan x$, so that

$$\int 3\tan^2 x\, dx = \int (-3 + 3\sec^2 x)\, dx = -3x + 3\tan x + C.$$

Differentiating gives $-3 + 3\sec^2 x = 3(\sec^2 x - 1) = 3\tan^2 x$ using a Pythagorean identity.

37. Since π is a constant, the integral is $\pi x + C$. Differentiating clearly gives π.

39. The integrand has the form of a derivative of $\sin^{-1}$. Since we have $(2x)^2$ instead of x^2, and an extra factor of 7, try $7\sin^{-1}(2x)$. The derivative of this function is $7\dfrac{1}{\sqrt{1-(2x)^2}} \cdot 2$, which is too large by a factor of 2. Thus

$$\int \frac{7}{\sqrt{1-(2x)^2}}\, dx = \frac{7}{2}\sin^{-1}(2x) + C.$$

The derivative of $\sin^{-1} x$ is $\dfrac{1}{\sqrt{1-x^2}}$, so differentiating gives

$$\frac{7}{2} \cdot \frac{1}{\sqrt{1-(2x)^2}} \cdot 2 = \frac{7}{\sqrt{1-(2x)^2}}.$$

41. The integrand has the form of a derivative of $\sec^{-1}$; since we have $3x$ instead of x, we try $\sec^{-1}(3x)$. Because the Chain Rule will give us an extra factor of 3 in the derivative, we will need to correct that by multiplying by $\frac{1}{3}$. Thus

$$\int \frac{1}{|3x|\sqrt{9x^2-1}}\, dx = \frac{1}{3}\sec^{-1}(3x) + C.$$

Since the derivative of $\sec^{-1} x$ is $\dfrac{1}{|x|\sqrt{x^2-1}}$, differentiating gives

$$\frac{1}{3} \cdot \frac{1}{|3x|\sqrt{9x^2-1}} \cdot 3 = \frac{1}{|3x|\sqrt{9x^2-1}}.$$

43. Since the numerator is equal to the derivative of the denominator, this is an application of the chain rule to $\ln(1+x^2)$. We get

$$\int \frac{2x}{1+x^2}\, dx = \ln(1+x^2) + C.$$

Note that we do not need absolute values on the log function since $x^2 + 1 > 0$ for all values of x, so that $\ln(x^2+1)$ is defined. Differentiating gives $\frac{1}{1+x^2} \cdot 2x = \frac{2x}{1+x^2}$.

45. This integrand looks like it must be an application of the product rule. With e^x as one factor, we see that $2x$ is the derivative of x^2, so that the whole thing is just the derivative of $x^2 e^x$. Thus

$$\int (x^2 e^x + 2xe^x)\, dx = x^2 e^x + C.$$

Differentiating gives $(x^2)'e^x + x^2(e^x)' = 2xe^x + x^2 e^x$.

47. The numerator is the derivative of the denominator, so that this is an application of the chain rule to $\ln(3x^2 + 1)$.

$$\int \frac{6x}{3x^2+1}\, dx = \ln(3x^2+1) + C.$$

Note that we do not need absolute values on the log function since $3x^2 + 1 > 0$ for all values of x, so that $\ln(3x^2 + 1)$ is defined. Differentiating gives $\frac{1}{3x^2+1}(6x) = \frac{6x}{3x^2+1}$.

49. Simplifying the integrand gives $e^x - 2e^{2x}$. Using the rule for integrating exponentials gives

$$\int \frac{e^{3x} - 2e^{4x}}{e^{2x}}\, dx = e^x - e^{2x} + C.$$

Differentiating gives $e^x - 2e^{2x}$, which is the simplified form of the integrand.

51. Since $3x^2$ is the derivative of $x^3 + 1$, we try $\frac{1}{3}(x^3 + 1)^6$ (using the $\frac{1}{3}$ to get rid of an unwanted factor of 3 in the result). This has as derivative $\frac{6}{3}(x^3 + 2)^5(3x^2) = 6x^2(x^3 + 1)^5$, which is too large by a factor of 6. Correcting for that gives

$$\int x^2(x^3 + 1)^5 \, dx = \frac{1}{18}(x^3 + 1)^6 + C.$$

Differentiating gives $6 \cdot \frac{1}{18}(x^3 + 1)^5(3x^2) = x^2(x^3 + 1)^5$.

53. This looks like an application of the quotient rule where the denominator was originally $\ln x$, so that presumably an antiderivative is $\frac{f(x)}{\ln x}$. Since $2x \ln x = f'(x) \ln x$, we should have $f(x) = x^2$, so that an antiderivative is $\frac{x^2}{\ln x}$. Check this by taking the derivative, then

$$\int \frac{2x \ln x - x}{(\ln x)^2} \, dx = \frac{x^2}{\ln x} + C.$$

Differentiating gives, using the product rule and the chain rule,

$$2x \cdot \frac{1}{\ln x} + x^2 \left(\frac{1}{\ln x} \right)' = \frac{2x}{\ln x} + x^2 \cdot \frac{-1}{(\ln x)^2} \cdot \frac{1}{x} = \frac{2x \ln x - x}{(\ln x)^2}.$$

55. This looks like it might be an application of the product rule, where one factor is e^x and the other factor is $x^{1/2}$ (so that the derivative of $x^{1/2}$ accounts for the $x^{-1/2}$). Differentiating $e^x x^{1/2}$, shows that this is correct, so

$$\int \left(e^x \sqrt{x} + \frac{e^x}{2\sqrt{x}} \right) dx = e^x \sqrt{x} + C.$$

Differentiating gives $e^x \left(\sqrt{x} \right)' + (e^x)' \sqrt{x} = \frac{e^x}{2\sqrt{x}} + e^x \sqrt{x}$.

57. This looks like an application of the product rule. Since $(\ln x)' = \frac{1}{x}$, one factor is $\ln x$, so the other factor should be $\cos x$. Checking the derivative of $\cos x \cdot \ln x$ shows that this is correct, so that

$$\int \left(\frac{1}{x} \cos x - \sin x \ln x \right) dx = \cos x \ln x + C.$$

Differentiating gives $\cos x (\ln x)' + (\cos x)' \ln x = \frac{\cos x}{x} - \sin x \ln x$.

59. Since the integral of $\cosh x$ is $\sinh x$, first try $\sinh 2x$. The derivative is $2 \cosh 2x$, which is off by a factor of $\frac{3}{2}$. Correcting gives

$$\int 3 \cosh 2x \, dx = \frac{3}{2} \sinh 2x + C.$$

Differentiating gives $\frac{3}{2} \cosh 2x \cdot 2 = 3 \cosh 2x$.

61. Since $(\tanh x)' = \text{sech}^2 x$, this is probably an application of the product rule where one factor is $\tanh x$. The other factor must be the coefficient of $\text{sech}^2 x$, which is x. Differentiating $x \tanh x$ shows that this is correct.

$$\int \left(\tanh x + x \, \text{sech}^2 x \right) dx = x \tanh x + C.$$

Differentiating gives $x' \tanh x + x(\tanh x)' = \tanh x + \text{sech}^2 x$.

63. Since $\left(e^{bx+c} \right)' = b e^{bx+c}$, this is almost right, but needs correction by a factor of $\frac{a}{b}$. Then $\frac{a}{b} e^{bx+c}$ has exactly the right derivative, so that

$$\int a e^{bx+c} \, dx = \frac{a}{b} e^{bx+c} + C.$$

65. Note that $(\sec(ax))' = a\sec(ax)\tan(ax)$, so that

$$\int \frac{\sec(ax)\tan(ax)}{b}\,dx = \frac{1}{ab}\int(a\sec(ax)\tan(ax))\,dx = \frac{1}{ab}\sec(ax) + C.$$

67. This looks like the derivative of $\tan^{-1}(bx)$, although the derivative of this function has an extra factor of b. Thus

$$\int \frac{a}{1+(bx)^2}\,dx = \frac{a}{b}\int \frac{b}{1+(bx)^2}\,dx = \frac{a}{b}\tan^{-1}(bx) + C.$$

Applications

69. (a) Since $a(t) = -32$, we get $\int a(t)\,dt = -\int 32\,dt = -32t + C$. Since the velocity at time zero was $v_0 = 42$, we must have $42 = -32\cdot 0 + C$, so that $C = 42$ and $v(t) = -32t + 42$.

(b) Integrating again gives

$$\int v(t)\,dx = \int(-32t + 42)\,dt = -\int 32t\,dt + \int 42\,dt = -16t^2 + 42t + C.$$

Since the initial position was $s_0 = 4$, we must have $4 = -16\cdot 0^2 + 42\cdot 0 + C$, so that $C = 4$. Hence $s(t) = -16t^2 + 42t + 4$.

(c) The acceleration graph is the constant line at $y = -32$. The velocity graph is the straight line, and the position graph is the curve. Acceleration is the derivative of velocity, which in turn is the derivative of position. Thought of another way, the velocity graph shows the area under the acceleration graph, adjusted so that its initial value is not zero but rather is $v_0 = 42$. Similarly, the position graph shows the area under the velocity graph, but adjusted so that its initial value is not zero but rather is $s_0 = 4$.

Proofs

71. (a) Assume $k \neq 1$. Then

$$\left(\frac{1}{k+1}x^{k+1} + C\right)' = \frac{1}{k+1}(k+1)x^k + C' = x^k,$$

so that $\int x^k\,dx = \frac{1}{k+1}x^{k+1} + C$.

(b) Note that both sides are defined for $x \neq 0$. For $x > 0$, $\ln|x| + C = \ln x + C$, while for $x < 0$, $\ln|x| + C = \ln(-x) + C$. Thus:

$$(x > 0:)\quad (\ln|x| + C)' = (\ln x)' + C' = \frac{1}{x}$$

$$(x < 0:)\quad (\ln|x| + C)' = (\ln(-x))' + C' = \frac{1}{-x}\cdot(-1) = \frac{1}{x}.$$

Thus for all x, the derivative relationship holds, so that $\int \frac{1}{x}\,dx = \ln|x| + C$.

73. (a) $\left(\frac{1}{k}e^{kx} + C\right)' = \left(\frac{1}{k}e^{kx}\right)' + C' = \frac{1}{k}\cdot ke^{kx} = e^{kx}$.

(b) $\left(\frac{1}{\ln b}b^x + C\right)' = \left(\frac{1}{\ln b}b^x\right)' + C' = \frac{1}{\ln b}\cdot(\ln b)b^x = b^x$.

75. (a) $(\sin^{-1}x + C)' = (\sin^{-1}x)' + C' = \frac{1}{\sqrt{1-x^2}}$.

(b) $(\tan^{-1}x + C)' = (\tan^{-1}x)' + C' = \frac{1}{1+x^2}$.

(c) $(\sec^{-1} x + C)' = (\sec^{-1} x)' + C' = \frac{1}{|x|\sqrt{x^2-1}}.$

77. Since addition commutes with taking the derivative, if F is an antiderivative of f and G is an antiderivative of g, then the derivative of $F + G$ is both the function $f + g$ and also the derivative of F plus the derivative of G.

79. Suppose F and G are differentiable functions, and that $G(x) = F(x) + C$. Then because differentiation commutes with addition, and because the derivative of a constant function is zero, we have
$$G'(x) = (F(x) + C)' = F'(x) + C' = F'(x),$$
so that $F'(x) = G'(x)$.

Thinking Forward

Initial-Value Problems

▷ Integrating gives
$$\int f'(x)\,dx = \int (x^8 - x^2 + 1)\,dx = \frac{1}{9}x^9 - \frac{1}{3}x^3 + x + C.$$
Then $f(0) = 0 - 0 + 0 + C = -1$, so that $C = -1$ and $f(x) = \frac{1}{9}x^9 - \frac{1}{3}x^3 + x - 1$.

▷ $f'(x) = 2x^{-1/3}$, so integrating gives
$$\int f'(x)\,dx = 3x^{2/3} + C.$$
Then $f(1) = 3 \cdot 1^{2/3} + C = 3 + C = 2$, so that $C = -1$ and $f(x) = 3x^{2/3} - 1$.

▷ Integrating gives
$$\int (3e^{2x} + 1)\,dx = \frac{3}{2}\int 2e^{2x}\,dx + \int 1\,dx = \frac{3}{2}e^{2x} + x + C.$$
Since $f(0) = \frac{3}{2}e^{2\cdot0} + 0 + C = \frac{3}{2} + C = 2$, we have $C = \frac{1}{2}$, so that $f(x) = \frac{3}{2}e^{2x} + x + \frac{1}{2}$.

▷ Integrating gives
$$\int 3\left(\frac{4}{5}\right)^x dx = 3\int \left(\frac{4}{5}\right)^x = \frac{3}{\ln(4/5)}\left(\frac{4}{5}\right)^x + C.$$
Since
$$f(1) = \frac{3}{\ln(4/5)}\left(\frac{4}{5}\right)^1 + C = \frac{3}{\ln(4/5)} \cdot \frac{4}{5} + C = 0,$$
we have $C = -\frac{3}{\ln(4/5)} \cdot \frac{4}{5}$, so that
$$f(x) = \frac{3}{\ln(4/5)}\left(\left(\frac{4}{5}\right)^x - \frac{4}{5}\right).$$

▷ This is the derivative of $\ln|3x - 2|$ except for constant factors. Thus
$$\int \frac{5}{3x-2}\,dx = \frac{5}{3}\int \frac{3}{3x-2}\,dx = \frac{5}{3}\ln|3x-2| + C.$$
Since $f(1) = \frac{5}{3}\ln 1 + C = 0 + C = C = 8$, we have $C = 8$, so that
$$f(x) = \frac{5}{3}\ln|3x-2| + 8.$$

▷ Note that $\sin(-2x) = -\sin(2x)$. Then integrating gives

$$\int (3\sin(-2x) - 5\cos(3x))\,dx = \int (-3\sin(2x) - 5\cos(3x))\,dx = \frac{3}{2}\cos(2x) - \frac{5}{3}\sin 3x + C.$$

Since $f(0) = \frac{3}{2}\cos 0 - \frac{5}{3}\sin 0 + C = \frac{3}{2} + C = 0$, we get $C = -\frac{3}{2}$, so that

$$f(x) = -\frac{3}{2}\cos(2x) - \frac{5}{3}\sin 3x - \frac{3}{2}.$$

Preview of Differential Equations

▷ We know that the derivative of e^x is e^x, so $f(x) = e^x$ works.

▷ Since the derivative of e^{2x} is $2e^{2x}$, we see that $f(x) = e^{2x}$ works.

▷ Since the derivative of $\cos(bt)$ is $-b\sin(bt)$, if we set $f(t) = -\frac{1}{b}\cos(bt)$, then $f'(t) = \sin(bt)$.

▷ Since $(\sin t)' = \cos t$ and $(\cos t)' = \sin t$, if we set $f(t) = -\cos t$, then

$$f''(t) = (-\cos t)'' = -(\cos t)'' = -(-\sin t)' = (\sin t)' = \cos t.$$

4.5   The Fundamental Theorem of Calculus

Thinking Back

Definite integrals

(See Definition 4.9) If f is a function defined on $[a, b]$, then the definite integral of f from $x = a$ to $x = b$ is the number

$$\int_a^b f(x)\,dx = \lim_{x \to \infty} \sum_{k=1}^n f(x_k^*)\Delta x$$

if this limit exists, where $\Delta a = \frac{b-a}{n}$ and $x_k = a + \frac{b-a}{n}k$, for $x_k^* \in [x_{k-1}, x_k]$.

Indefinite integrals

(See Definition 4.15) The *indefinite integral* of a continuous function f is defined to be the family of antiderivatives $\int f(x)\,dx = F(x) + C$ where F is an antiderivative of f — that is, a function for which $F' = f$.

Approximating with a Riemann sum

For $n = 10$, since the width of the interval is 5, we have $\Delta x = \frac{1}{2}$, and $x_k = \frac{k}{2}$. Using a right sum, we get for the approximate integral

$$\sum_{k=1}^n f(x_k)\Delta x = \sum_{k=1}^{10} \frac{1}{2}\left(\frac{k}{2}\right)^2$$

$$= \frac{1}{8}\sum_{k=1}^{10} k^2$$

$$= \frac{1}{8} \cdot \frac{10(11)(21)}{6} = \frac{385}{8}.$$

Limits of Riemann sums

Using the right sum as above, we have $\Delta x = \frac{5}{n}$ and $x_k = \frac{5k}{n}$, so that the area under the graph is

$$\lim_{n\to\infty}\sum_{k=1}^{n} f(x_k)\Delta x = \lim_{n\to\infty}\sum_{k=1}^{n}\left(\frac{5k}{n}\right)^2\cdot\frac{5}{n}$$

$$= \lim_{n\to\infty}\frac{125}{n^3}\sum_{k=1}^{n} k^2$$

$$= \lim_{n\to\infty}\frac{125}{n^3}\cdot\frac{n(n+1)(2n+1)}{6}$$

$$= \lim_{n\to\infty}\frac{250n^3+375n^2+125n}{6n^3}$$

$$= \lim_{n\to\infty}\frac{250+375/n+125/n^2}{6} = \frac{250}{6} = \frac{125}{3}.$$

Concepts

1. (a) False. Definite integrals compute area under a curve between two specific points and result in a real number. Indefinite integrals give a family of antiderivatives.

    (b) False. This equation holds only if F is an antiderivative of f.

    (c) True, since f is continuous on $[0, \pi]$ (see Theorem 4.24).

    (d) False, since f is discontinuous at $\frac{\pi}{2}$ (see Theorem 4.24).

    (e) False. For example, let $f(x) = x^2$; then $\int_{-1}^{1} x^2\,dx = \left[\frac{1}{3}x^3\right]_{-1}^{1} = \frac{2}{3}$.

    (f) True. Since f is an odd function, it is rotationally symmetric around the origin, so that the area under f between $-a$ and 0 is the negative of the area between 0 and a. Thus

    $$\int_{-a}^{a} f(x)\,dx = \int_{-a}^{0} f(x)\,dx + \int_{0}^{a} f(x)\,dx = -\int_{0}^{a} f(x)\,dx + \int_{0}^{a} f(x)\,dx = 0.$$

    (g) True. Since f is an even function, it is symmetric around the y axis, so that the area under f between $-a$ and 0 is simply the mirror image of the area under f between 0 and a, so that the two are equal.

    (h) False. Since f is an odd function, it is rotationally symmetric around the origin, so that the area under f between $-a$ and 0 is the negative of the area between 0 and a. For example, let $f(x) = 2x$; then

    $$\int_{0}^{a} 2x\,dx = \left[x^2\right]_{0}^{a} = a^2 - 0 = a^2, \qquad \int_{-a}^{0} 2x\,dx = \left[x^2\right]_{-a}^{0} = 0 - (-a)^2 = -a^2.$$

3. First, there is no reason a priori to expect a connection between a limiting process that defines the area under a curve and the values of the antiderivative of a function, so it is completely unexpected. Second, more pragmatically, it gives us a convenient way to compute definite integrals; the only other way we know is via a cumbersome limiting process that will not even be computable for some fairly simple integrands.

5. The Mean Value Theorem is used at a critical step in the proof, to make a good choice for $x_k^* \in [x_{k-1}, x_k]$.

7. The area under the velocity curve on $[a, b]$ is the net change in position from $t = a$ to $t = b$. If the velocity is positive, the particle is moving in a positive direction. Similarly, if the velocity is negative, then the particle is moving in a negative direction. Thus if the velocity is sometimes negative and sometimes positive, the negative pieces measure the change in position in the negative direction. So in total, it makes sense that the area under the velocity curve measures the net change in position. Since position is an antiderivative of velocity, the signed area under the velocity curve on $[a, b]$ is $s(b) - s(a)$, which says in a different way that the signed area under the velocity curve is the net change in position.

9. We have $\int_3^4 f'(x) \, dx = f(4) - f(3) = f(4) - 2$. Computing the area under the curve between 3 and 4 gives ≈ 2, so that $2 = f(4) - 2$ and $f(4) \approx 4$.

11. We have $\int_{-2}^2 f'(x) \, dx = f(2) - f(-2) = 3 - f(-2)$. Computing the area under the curve between -2 and 2 gives ≈ 0.4, so that $0.4 \approx 3 - f(-2)$ and thus $f(-2) \approx 2.6$ (note that the area consists of ≈ -1.8 below the x axis and ≈ 2.2 above).

13. (a) Using the right Riemann sum, we have $\Delta x = \frac{1}{n}$ and $x_k = 1 + \frac{k}{n}$. Then

$$\int_1^2 3x^2 \, dx = \lim_{n \to \infty} \sum_{k=1}^n f(x_k) \Delta x$$

$$= \lim_{n \to \infty} \frac{1}{n} \sum_{k=1}^n 3 \left(1 + \frac{k}{n} \right)^2$$

$$= \lim_{n \to \infty} \frac{3}{n} \left(\sum_{k=1}^n 1 + \frac{2}{n} \sum_{k=1}^n k + \frac{1}{n^2} \sum_{k=1}^n k^2 \right)$$

$$= \lim_{n \to \infty} \left(\frac{3}{n} \cdot n + \frac{6}{n^2} \cdot \frac{n(n+1)}{2} + \frac{3}{n^3} \cdot \frac{n(n+1)(2n+1)}{6} \right)$$

$$= 3 + \lim_{n \to \infty} \left(\frac{3n+3}{n} + \frac{2n^2 + 3n + 1}{2n^2} \right)$$

$$= 3 + \lim_{n \to \infty} \left(\frac{3 + 3/n}{1} + \frac{2 + 3/n + 1/n^2}{2} \right)$$

$$= 3 + 3 + 1 = 7.$$

(b) Using the definite integral formulas, we have

$$\int_1^2 3x^2 \, dx = 3 \int_1^2 x^2 \, dx = 3 \cdot \frac{1}{3} (2^3 - 1^3) = 7.$$

(c) Using the Fundamental Theorem of Calculus, note that x^3 is an antiderivative of $3x^2$, so that

$$\int_1^2 3x^2 \, dx = \left[x^3 \right]_1^2 = 2^3 - 1^3 = 7.$$

Clearly all three answers are the same.

15. (a) We have

$$\sum_{k=1}^{100} \left(\frac{1}{k} - \frac{1}{k+1} \right) = \left(\frac{1}{1} - \frac{1}{2} \right) + \left(\frac{1}{2} - \frac{1}{3} \right) + \cdots + \left(\frac{1}{k-1} - \frac{1}{k} \right)$$

$$+ \cdots + \left(\frac{1}{100} - \frac{1}{101} \right)$$

$$= \frac{1}{1} - \frac{1}{101} = \frac{101 + 1 - 1}{101} = \frac{100}{101}.$$

(b)

$$\sum_{k=1}^{100} \frac{k^2 - (k-1)^2}{2} = \sum_{k=1}^{100} \left(\frac{k^2}{2} - \frac{(k-1)^2}{2} \right)$$

$$= \left(\frac{1^2}{2} - \frac{0^2}{2} \right) + \left(\frac{2^2}{2} - \frac{1^2}{2} \right) + \cdots + \left(\frac{(k-1)^2}{2} - \frac{(k-2)^2}{2} \right)$$

$$+ \cdots + \left(\frac{100^2}{2} - \frac{99^2}{2} \right)$$

$$= \frac{100^2}{2} - \frac{0^2}{2} = \frac{100^2}{2} = 5000.$$

17. (a) True. Substitute $g'(x)$ for $f(x)$ to get $\int_a^b g'(x)\,dx = g(b) - g(a)$, which is the Net Change Theorem.

(b) False. Substituting gives $\int_a^b f(x)\,dx = [f'(x)]_b^a = f'(b) - f'(a)$, which is *not* the Net Change Theorem - the roles of f and f' are reversed.

(c) True. If $g(x)$ is an antiderivative of $h(x)$, then $h(x) = g'(x)$; substituting gives $\int_a^b g'(x)\,dx = g(b) - g(a)$, which is the Net Change Theorem.

(d) True. Write $f(x) = h'(x)$; then $f'(x) = h''(x)$; substituting in the given formula gives $\int_a^b f'(x)\,dx = [f(x)]_a^b$, which is the Net Change Theorem.

(e) True. This is just the Fundamental Theorem in Evaluation Notation.

Skills

19. An antiderivative of $x^4 + 3x + 1$ is $\frac{1}{5}x^5 + \frac{3}{2}x^2 + x$, so that

$$\int_{-1}^{1} (x^4 + 3x + 1)\,dx = \left[\frac{1}{5}x^5 + \frac{3}{2}x^2 + x \right]_{-1}^{1} = \left(\frac{1}{5} + \frac{3}{2} + 1 \right) - \left(-\frac{1}{5} + \frac{3}{2} - 1 \right) = \frac{12}{5}.$$

A graph is

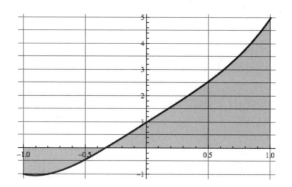

Counting rectangles each of area $\frac{1}{4}$ shows that this answer is reasonable.

21. An antiderivative of $3(2^x)$ is $\frac{3}{\ln 2}(2^x)$, so that

$$\int_{1}^{4} 3(2^x)\,dx = \left[\frac{3}{\ln 2}(2^x) \right]_{1}^{4} = \frac{3}{\ln 2} \cdot 16 - \frac{3}{\ln 2} \cdot 2 = \frac{42}{\ln 2} \approx 60.59.$$

A graph is

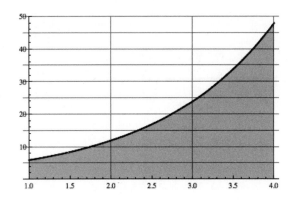

Counting rectangles each of area 5 shows that this answer is reasonable.

23. An antiderivative of $\frac{1}{\sqrt{x^5}} = x^{-5/2}$ is $-\frac{2}{3}x^{-3/2}$, so that

$$\int_2^5 \frac{1}{\sqrt{x^5}}\, dx = \left[-\frac{2}{3}x^{-3/2}\right]_2^5 = -\frac{2}{3}5^{-3/2} + \frac{2}{3}2^{-3/2} \approx 0.176$$

A graph is

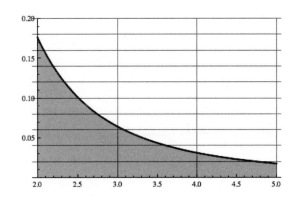

Counting rectangles each of area 0.02 shows that this answer is reasonable.

25. An antiderivative of $\frac{\sqrt{x}-1}{\sqrt{x}} = 1 - x^{-1/2}$ is $x - 2\sqrt{x}$, so that

$$\int_1^3 \frac{\sqrt{x}-1}{\sqrt{x}}\, dx = \left[x - 2\sqrt{x}\right]_1^3 = (3 - 2\sqrt{3}) - (1 - 2\sqrt{1}) = 4 - 2\sqrt{3} \approx 0.536.$$

A graph is

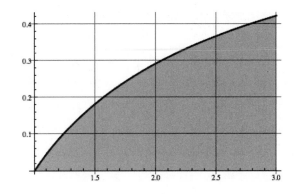

Counting rectangles each of area 0.05 shows that this answer is reasonable.

27. An antiderivative of $\sec x \tan x$ is $\sec x$, so that

$$\int_0^{\pi/4} \sec x \tan x \, dx = [\sec x]_0^{\pi/4} = \sec \frac{\pi}{4} - \sec 0 = \sqrt{2} - 1 \approx 0.414.$$

A graph is

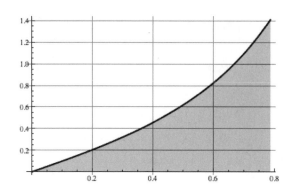

Counting rectangles each of area 0.04 shows that this answer is reasonable.

29. An antiderivative of $\frac{1}{\sqrt{1-4x^2}}$ is $\frac{1}{2} \sin^{-1} 2x$, so that

$$\int_0^{\sqrt{2}/4} \frac{1}{\sqrt{1-4x^2}} \, dx = \left[\frac{1}{2} \sin^{-1} 2x\right]_0^{\sqrt{2}/4} = \frac{1}{2} \sin^{-1} \frac{\sqrt{2}}{2} - \frac{1}{2} \sin^{-1} 0 = \frac{\pi}{8} \approx 0.393$$

A graph is

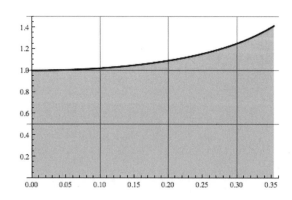

Counting rectangles each of area 0.05 shows that this answer is reasonable.

31. An antiderivative of $1 + \sin x$ is $x - \cos x$, so that

$$\int_{-\pi}^{\pi} (1 + \sin x) \, dx = [x - \cos x]_{-\pi}^{\pi} = (\pi - \cos \pi) - (-\pi - \cos(-\pi)) = 2\pi \approx 6.28.$$

A graph is

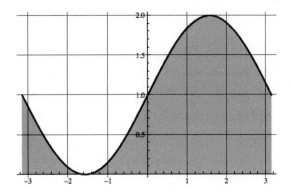

Counting rectangles each of area 0.5 shows that this answer is reasonable.

33. An antiderivative of $\frac{1}{2e^x} = \frac{1}{2}e^{-x}$ is $-\frac{1}{2}e^{-x}$, so that

$$\int_0^1 \frac{1}{2e^x}\,dx = \left[-\frac{1}{2}e^{-x}\right]_0^1 = -\frac{1}{2}e^{-1} + \frac{1}{2}e^0 = \frac{1}{2} - \frac{1}{2}e^{-1} \approx 0.316$$

A graph is

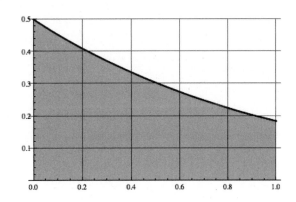

Counting rectangles each of area 0.02 shows that this answer is reasonable.

35. An antiderivative of $\frac{1}{1+x^2}$ is $\tan^{-1} x$, so that

$$\int_0^1 \frac{1}{1+x^2}\,dx = \left[\tan^{-1}\right]_0^1 = \tan^{-1} 1 - \tan^{-1} 0 = \frac{\pi}{4} \approx 0.785$$

A graph is

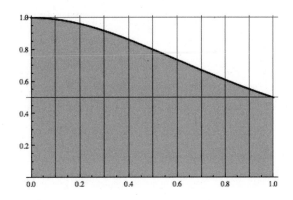

Counting rectangles each of area 0.05 shows that this answer is reasonable.

37. Since $\frac{2^x}{4^x} = \left(\frac{2}{4}\right)^x = \left(\frac{1}{2}\right)^2 = 2^{-x}$, an antiderivative is $-\frac{1}{\ln 2}2^{-x}$, so that

$$\int_{-1}^{1} \frac{2^x}{4^x}\, dx = \left[-\frac{1}{\ln 2}2^{-x}\right]_{-1}^{1} = \frac{1}{\ln 2}2^1 - \frac{1}{\ln 2}2^{-1} \approx 2.164.$$

A graph is

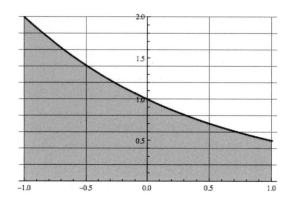

Counting rectangles each of area 0.1 shows that this answer is reasonable.

39. An antiderivative of $\frac{1}{x+5}$ is $\ln(x+5)$, so that

$$\int_{-3}^{2} \frac{1}{x+5}\, dx = [\ln(x+5)]_{-3}^{2} = \ln 7 - \ln 2 \approx 1.253$$

A graph is

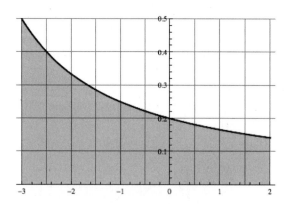

Counting rectangles each of area 0.05 shows that this answer is reasonable.

41. Since $2x$ is the derivative of $x^2 + 5$, an antiderivative of $\frac{2x}{x^2+5}$ is $\ln(x^2+5)$, so that

$$\int_{-3}^{2} \frac{2x}{x^2+5}\, dx = \left[\ln(x^2+5)\right]_{-3}^{2} = \ln 9 - \ln 14 \approx -0.44.$$

A graph is

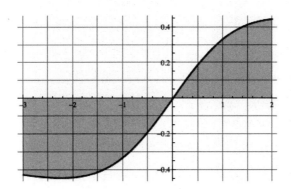

Counting signed rectangles each of area ± 0.05 shows that this answer is reasonable.

43. Since $2x + 3$ is the derivative of $x^2 + 3x + 4$, an antiderivative of $\frac{2x+3}{x^2+3x+4}$ is $\ln(x^2 + 3x + 4)$, so that

$$\int_1^4 \frac{2x+3}{x^2+3x+4}\,dx = \left[\ln(x^2 + 3x + 4)\right]_1^4 = \ln 32 - \ln 8 = \ln 4 \approx 1.386$$

A graph is

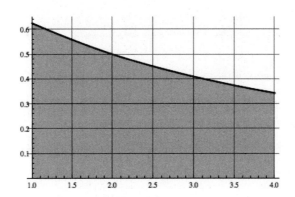

Counting rectangles each of area 0.05 shows that this answer is reasonable.

45. An antiderivative of $\frac{1}{\sqrt{9-x^2}} = \frac{1}{3} \cdot \frac{1}{\sqrt{1-(x/3)^2}}$ is $\sin^{-1}\frac{x}{3}$, so that

$$\int_0^{3/2} \frac{1}{\sqrt{9-x^2}}\,dx = \left[\sin^{-1}\frac{x}{3}\right]_0^{3/2} = \sin^{-1}\frac{1}{2} - \sin^{-1}0 = \frac{\pi}{6} \approx 0.524.$$

A graph is

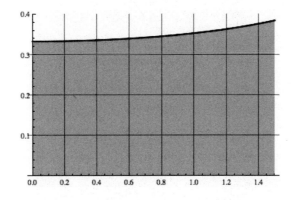

Counting rectangles each of area 0.02 shows that this answer is reasonable.

47. An antiderivative of $\frac{3\pi}{2x+6} = \frac{3\pi}{2} \cdot \frac{1}{x+3}$ is $\frac{3\pi}{2} \ln(x+3)$. Note also that $e - 3 \approx -0.28 < 0$, so that

$$\int_1^{e-3} \frac{3\pi}{2x+6}\, dx = -\int_{e-3}^1 \frac{3\pi}{2x+6} = -\left[\frac{3\pi}{2}\ln(x+3)\right]_{e-3}^1 = -\frac{3\pi}{2}\ln(4) + \frac{3\pi}{2}\ln e$$

$$= \frac{3\pi}{2}(1 - \ln 4) \approx -1.82.$$

A graph is

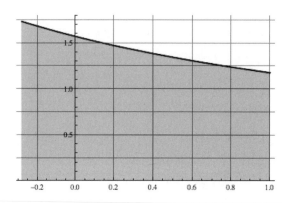

Counting rectangles each of which has area 0.05 shows that this answer is reasonable (don't forget that although this looks like a positive area, the bounds of integration are reversed, so that although the areas of the rectangles add up to ≈ 1.82, the result is really -1.82!)

49. An antiderivative of $x^{2/3} - x^{1/3}$ is $\frac{3}{5}x^{5/3} - \frac{3}{4}x^{4/3}$. Thus

$$\int_0^1 (x^{2/3} - x^{1/3})\, dx = \left[\frac{3}{5}x^{5/3} - \frac{3}{4}x^{4/3}\right]_0^1 = \frac{3}{5} - \frac{3}{4} = -\frac{3}{20} = -0.15.$$

A graph is

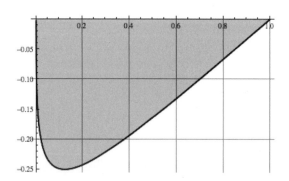

Counting signed rectangles each of which has area 0.02 shows that this answer is reasonable.

51. Since $\sin x = -(1 + \cos x)'$, an antiderivative of $\sin x(1 + \cos x)$ is $-\frac{1}{2}(1 + \cos x)^2$. Thus

$$\int_0^{\pi/2} \sin x(1 + \cos x)\, dx = \left[-\frac{1}{2}(1 + \cos x)^2\right]_0^{\pi/2} = -\frac{1}{2}(1 + 0)^2 + \frac{1}{2}(1 + 1)^2 = \frac{3}{2}.$$

A graph is

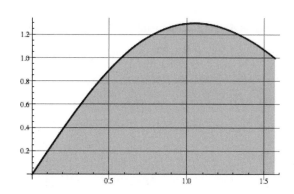

Counting rectangles each of which has area 0.1 shows that this answer is reasonable.

53. To find an antiderivative, note that this looks like an application of the quotient rule with denominator e^x and numerator x; differentiating $\frac{x}{e^x} = xe^{-x}$ reveals that this guess is correct. Thus

$$\int_{-1}^{1} \frac{e^x - xe^x}{e^{2x}}\, dx = \left[xe^{-x}\right]_{-1}^{1} = e^{-1} - (-1)e^1 = e + e^{-1} \approx 3.086.$$

A graph is

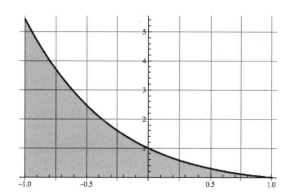

Counting rectangles each of which has area 0.25 shows that this answer is reasonable.

55. This looks like an application of the chain rule, since we see that the derivative of $\cot(x^2)$ is $-\csc(x^2) \cdot (2x) = -2x\csc(x^2)$; correcting for the factor of -2 gives $-\frac{1}{2}\cot(x^2)$ for an antiderivative. Thus

$$\int_{\pi/4}^{\pi/2} x\csc^2(x^2)\, dx = \left[-\frac{1}{2}\cot(x^2)\right]_{\pi/4}^{\pi/2} = -\frac{1}{2}\cot\frac{\pi^2}{4} + \frac{1}{2}\cot\frac{\pi^2}{16} \approx 1.331$$

A graph is

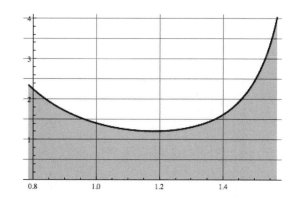

Counting signed rectangles each of which has area 0.1 shows that this answer is reasonable.

57. Since $-2x = (1 - x^2)'$, an antiderivative of $\frac{3x}{1-x^2} = -3\frac{x}{x^2-1}$ on $[2,4]$ is $-\frac{3}{2}\ln(x^2 - 1)$. Thus

$$\int_2^4 \frac{3x}{1-x^2}\, dx = \left[-\frac{3}{2}\ln(x^2-1)\right]_2^4 = -\frac{3}{2}\ln 15 + \frac{3}{2}\ln 3 = -\frac{3}{2}\ln 5 \approx -2.414.$$

A graph is

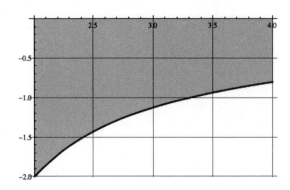

Counting signed rectangles each of which has area 0.25 shows that this answer is reasonable.

59. The integrand is

$$|x - 2| = \begin{cases} x - 2 & x \geq 2, \\ 2 - x & x < 2, \end{cases}$$

so that an antiderivative is

$$\begin{cases} \frac{1}{2}x^2 - 2x & x \geq 2, \\ 2x - \frac{1}{2}x^2 & x < 2. \end{cases}$$

Thus

$$\int_{-2}^5 |x - 2|\, dx = \int_{-2}^2 (2 - x)\, dx + \int_2^5 (x - 2)\, dx$$

$$= \left[2x - \frac{1}{2}x^2\right]_{-2}^2 + \left[\frac{1}{2}x^2 - 2x\right]_2^5$$

$$= (4 - 2) - (-4 - 2) + \left(\frac{25}{2} - 10 - (2 - 4)\right) = \frac{25}{2}$$

A graph is

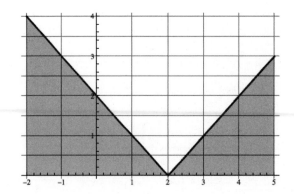

Counting rectangles each of which has area 0.5 shows that this answer is correct.

61. Since $2x^2 - 5x - 3 = (2x+1)(x-3)$, the integrand has roots at 3 and $-\frac{1}{2}$. Sign analysis shows that it is negative for $-\frac{1}{2} < x < 3$ and positive elsewhere (except obviously at the roots themselves). Thus the integrand is

$$\left|2x^2 - 5x - 3\right| = \begin{cases} 2x^2 - 5x - 3 & x \le -\frac{1}{2}, \\ -2x^2 + 5x + 3, & -\frac{1}{2} < x < 3, \\ 2x^2 - 5x - 3 & x \ge 3, \end{cases}$$

so that an antiderivative is

$$\begin{cases} \frac{2}{3}x^3 - \frac{5}{2}x^2 - 3x, & x \le -\frac{1}{2}, \\ -\frac{2}{3}x^3 + \frac{5}{2}x^2 + 3x, & -\frac{1}{2} < x < 3, \\ \frac{2}{3}x^3 - \frac{5}{2}x^2 - 3x, & x \ge 3. \end{cases}$$

Thus

$$\int_{-2}^{4} \left|2x^2 - 5x - 3\right| dx = \int_{-2}^{-1/2} (2x^2 - 5x - 3)\, dx + \int_{-1/2}^{3} (-2x^2 + 5x + 3)\, dx$$

$$+ \int_{3}^{4} (2x^2 - 5x - 3)\, dx$$

$$= \left[\frac{2}{3}x^3 - \frac{5}{2}x^2 - 3x\right]_{-2}^{-\frac{1}{2}} + \left[-\frac{2}{3}x^3 + \frac{5}{2}x^2 + 3x\right]_{-\frac{1}{2}}^{3} + \left[\frac{2}{3}x^3 - \frac{5}{2}x^2 - 3x\right]_{3}^{4}$$

$$= \frac{343}{12} \approx 28.583.$$

A graph is

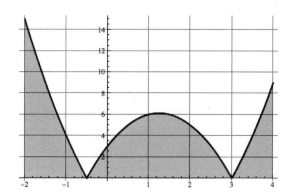

Counting rectangles each of which has area 2 shows that this answer is reasonable.

63. Since $\sin x$ is positive for $0 < x < \pi$ and negative for $\pi < x < \frac{5\pi}{4}$, we have

$$|\sin x| = \begin{cases} \sin x & 0 < x \le \pi, \\ -\sin x & \pi < x \le \frac{5\pi}{4}, \end{cases}$$

so that an antiderivative is

$$\begin{cases} -\cos x & 0 < x \le \pi, \\ \cos x & \pi < x \le \frac{5\pi}{4}. \end{cases}$$

Thus

$$\int_0^{5\pi/4} |\sin x|\, dx = \int_0^\pi (\sin x)\, dx + \int_\pi^{5\pi/4} (-\sin x)\, dx$$

$$= [-\cos x]_0^\pi + [\cos x]_\pi^{5\pi/4}$$

$$= (-\cos \pi + \cos 0) + \left(\cos \frac{5\pi}{4} - \cos \pi\right) = 3 - \frac{\sqrt{2}}{2} \approx 2.293$$

A graph is

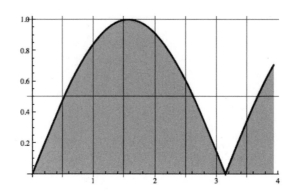

Counting rectangles each of which has area 0.25 shows that this answer is reasonable.

Applications

65. Since the rate of leakage is $0.05t$ gallons per hour, so after one day (which is 24 hours), the total amount of leakage is $\int_0^{24} 0.05t\, dt$ gallons. An antiderivative of $0.05t$ is $0.025t^2$, so that

$$\int_0^{24} 0.05t\, dt = \left[0.025t^2\right]_0^{24} = 0.025 \cdot 576 = 14.4$$

gallons of oil have leaked out after one day.

67. (a) For $t \leq 2$ approximately, $v(t)$ is positive and increasing, so that the figurine is moving right and speeding up. From 2 to 5, approximately, $v(t)$ is positive but decreasing, so that the figuring is moving right but slowing down. For the next 2.5 seconds, until $t \approx 7.5$, the velocity is negative and decreasing, so that the figuring is moving left and speeding up. Then until $t = 10$, the velocity is negative and increasing, so that the figuring is moving left but slowing down. From $t = 10$ to $t = 13$, it is moving right and speeding up; from $t = 13$ to $t = 15$, it is moving right but slowing down. Finally, it is moving left and speeding up from $t = 15$ to $t = 18$, and then moving left and slowing down from $t = 18$ to $t = 20$. Since the velocity at $t = 20$ is zero, the figurine is stopped after 20 seconds.

(b) We have

$$v(t) = 0.00015t(t-5)(t-10)(t-15)(t-20) = 0.00015(t^5 - 50t^4 + 875t^3 - 6250t^2 + 15000t),$$

so that after s seconds, the net distance traveled is the area under the velocity curve from $t = 0$ to $t = s$. Since an antiderivative of $v(t)$ is

$$0.00015\left(\frac{1}{6}t^6 - 10t^5 + \frac{875}{4}t^4 - \frac{6250}{3}t^3 + 7500t^2\right),$$

we have

$$(5 \text{ sec}:)\quad \left[0.00015\left(\frac{1}{6}t^6 - 10t^5 + \frac{875}{4}t^4 - \frac{6250}{3}t^3 + 7500t^2\right)\right]_0^5 \approx 5.273$$

$$(10 \text{ sec}:)\quad \left[0.00015\left(\frac{1}{6}t^6 - 10t^5 + \frac{875}{4}t^4 - \frac{6250}{3}t^3 + 7500t^2\right)\right]_0^{10} = 3.125$$

$$(20 \text{ sec}:)\quad \left[0.00015\left(\frac{1}{6}t^6 - 10t^5 + \frac{875}{4}t^4 - \frac{6250}{3}t^3 + 7500t^2\right)\right]_0^{20} = 0.$$

(c) The total distance traveled is the integral of the absolute value of $v(t)$, since we must count all distances traveled, whether rightwards or leftwards, as positive. From the graph, $v(t)$ is positive for $0 < t < 5$ and for $10 < t < 15$, and is negative elsewhere except at the roots. Thus

$$|v(t)| = \begin{cases} v(t), & 0 < t < 5 \text{ or } 10 < t < 15, \\ -v(t), & 5 < t < 10 \text{ or } 15 < t < 20. \end{cases}$$

So

$$\int_0^{20} |v(t)|\, dt = \int_0^5 v(t)\, dt - \int_5^{10} v(t)\, dt + \int_{10}^{15} v(t)\, dt - \int_{15}^{20} v(t)\, dt$$

$$= \left[0.00015\left(\frac{1}{6}t^6 - 10t^5 + \frac{875}{4}t^4 - \frac{6250}{3}t^3 + 7500t^2\right)\right]_0^5$$

$$- \left[0.00015\left(\frac{1}{6}t^6 - 10t^5 + \frac{875}{4}t^4 - \frac{6250}{3}t^3 + 7500t^2\right)\right]_5^{10}$$

$$+ \left[0.00015\left(\frac{1}{6}t^6 - 10t^5 + \frac{875}{4}t^4 - \frac{6250}{3}t^3 + 7500t^2\right)\right]_{10}^{15}$$

$$- \left[0.00015\left(\frac{1}{6}t^6 - 10t^5 + \frac{875}{4}t^4 - \frac{6250}{3}t^3 + 7500t^2\right)\right]_{15}^{20}$$

$$\approx 14.844.$$

The figurine travels a total distance of about 14.8 inches over the full 20 seconds.

69. Computing gives

$$\int_0^1 x f(x)\, dx = \int_0^1 6435 x^9 (1-x)^4\, dx$$

$$= 6435 \int_0^1 x^9 (1 - 4x + 6x^2 - 4x^3 + x^4)\, dx$$

$$= 6435 \int_0^1 (x^9 - 4x^{10} + 6x^{11} - 4x^{12} + x^{13})\, dx$$

$$= 6435 \left[\frac{1}{10}x^{10} - \frac{4}{11}x^{11} + \frac{1}{2}x^{12} - \frac{4}{13}x^{13} + \frac{1}{14}x^{14}\right]_0^1$$

$$= 6435 \left(\frac{1}{10} - \frac{4}{11} + \frac{1}{2} - \frac{4}{13} + \frac{1}{14}\right)$$

$$= \frac{9}{14} \approx 0.642857.$$

The probability that the next such message will be spam is thus about 64%.

71. (a) Noon corresponds to $t = 4$, since the shop opens at 8 A.M. Computing the area either by counting rectangles or using geometry to find the area of the triangle tells us that 80 people entered the store in the first 4 hours (the base of the triangle is 4; its height is 40, so its area is $\frac{1}{2} \cdot 4 \cdot 40 = 80$). Since fewer than 30 customers arrived in each of the first three hours, they have all been dealt with. However, in the last hour, the area under the curve is 35; since 35 customers arrived in the last hour, there is a line of about 5 people at noon.

   (b) At 1 P.M., customers are arriving at a rate of 50 per hour, so that the line is growing at a rate of $50 - 30 = 20$ per hour.

   (c) Seven hours after the store opened, at 3 P.M. Again customers arriving before 11 A.M. were all dealt with. Since 11 A.M. a total of 160 customers have arrived (this is the area under the curve from $t = 3$ to $t = 7$), but only 120 can be dealt with in that four hour window. Thus there is a line of about 40 people. After 3 P.M., the number of arrivals per hour dips below 30, so that the line starts to decrease.

   (d) There were 40 people in line at 3 P.M. Over the next three hours a total of 45 new customers arrive, making a total of 85 customers. Only 60 of them can be handled in the three hours, leaving a line of about 25 people at closing time.

Proofs

73. Since an antiderivative of x is $\frac{1}{2}x^2$, using the Fundamental Theorem of Calculus gives

$$\int_a^b x \, dx = \left[\frac{1}{2}x^2\right]_a^b = \frac{1}{2}b^2 - \frac{1}{2}a^2 = \frac{1}{2}(b^2 - a^2).$$

75. If $F(x)$ is an antiderivative of $f(x)$, then $kF(x)$ is an antiderivative of $kf(x)$ by the constant multiple rule for derivatives. Then

$$\int_a^b kf(x) \, dx = kF(b) - kF(a) = k(F(b) - F(a)) = k \int_a^b f(x) \, dx.$$

77. We must show that the integral is $F(b) - F(a)$ where F is any antiderivative of f. To do this, we must show that the limit of Riemann sums is $F(b) - F(a)$. So choose any antiderivative F of f, and choose a subdivision of $[a, b]$ into n subintervals. On each subinterval $[x_{k-1}, x_k]$, by the Mean Value Theorem, there is a point x_k^* such that $F'(x_k^*) = f(x_k^*)$ is equal to the average change in F on the interval $[x_{k-1}, x_k]$. Thus that term in the Riemann sum is $\frac{F(x_k) - F(x_{k-1})}{x_k - x_{k-1}}$. But the sum of the average change in F over any subinterval multiplied by the length of that subinterval is just the total change in F over that subinterval. Adding up over all subintervals shows that the Riemann sum is just the total change in F over $[a, b]$, which is $F(b) - F(a)$.

79. Suppose that $G(x) = F(x) + C$ where C is a constant. Then

$$[G(x)]_a^b = [F(x) + C]_a^b = (F(b) + C) - (F(a) + C) = F(b) - F(a) = [F(x)]_a^b.$$

Thinking Forward

Functions Defined by Integrals

▷ The area under the curve between two points is just the integral of the function between those two points. In this case, the two points are 0 and x, so that

$$A(x) = \int_0^x t^2 \, dt.$$

▷ Since $\frac{1}{3}t^3$ is an antiderivative of t^2, and since $\frac{1}{3}t^3\Big|_{t=0} = 0$, we have

$$A(0) = \left[\frac{1}{3}t^3\right]_0^0 = 0, \quad A(1) = \left[\frac{1}{3}t^3\right]_0^1 = \frac{1}{3}, \quad A(2) = \left[\frac{1}{3}t^3\right]_0^2 = \frac{8}{3}, \quad A(3) = \left[\frac{1}{3}t^3\right]_0^3 = 9,$$

$$A(4) = \left[\frac{1}{3}t^3\right]_0^4 = \frac{64}{3}.$$

▷ A graph, with $f(x)$ the solid line and $A(x)$ the dashed line, is

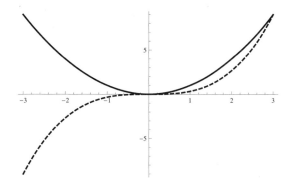

▷ See the first exercise in this group. $A(x) = \frac{1}{3}x^3$.

▷ $A'(x) = \frac{1}{3} \cdot 3x^2 = x^2$, so that $A'(x) = f(x)$.

▷ Since $A(x)$ is the area from $t = 0$ to $t = x$ under $f(x) = x^2$, we see that $A(x)$ is increasingly negative as x moves away from zero in the negative direction (recall that since this is the area *from 0 to x*, it is the negative of the signed area from x to 0). Also, $A(x)$ is increasingly positive as x moves away from 0 in the positive direction.

4.6   Areas and Average Values

Thinking Back

Finding sign intervals

With $f(x) = x^3 - 4x^2 - 12x = x(x-6)(x+2)$, we see that f can change sign only at $-2, 0$, and 6. Since $f(-3) = -27 < 0$, $f(-1) = 7 > 0$, $f(1) = -15 < 0$, and $f(7) = 63 > 0$, we see that f is positive on $(-2,0) \cup (6,\infty)$ and negative on $(-\infty, -2) \cup (0,6)$.

Finding intersection points

The curves $f(x) = x^3 - 9x^2$ and $g(x) = x - 9$ intersect where their difference is zero. Their difference is $f(x) - g(x) = x^3 - 9x^2 - x + 9 = (x-9)(x-1)(x+1)$. Thus the curves intersect at $x = -1, x = 1$, and $x = 9$.

The Intermediate Value Theorem

The Intermediate Value Theorem (Theorem 1.18) says that if f is continuous on $[a,b]$, then for any K strictly between $f(a)$ and $f(b)$, there is some $c \in (a,b)$ with $f(c) = K$. For example, in the graph

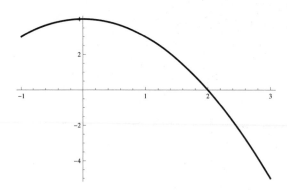

since $f(-1) = 3$ and $f(3) = -5$, we can choose $K = 0$, for example, and we see that there is some $c \in (a, b)$, namely $c = 2$, with $f(c) = K = 0$.

The Mean Value Theorem

The Mean Value Theorem (Theorem 3.5) says that if f is continuous on $[a, b]$ and differentiable on (a, b), then there exists at least one $c \in (a, b)$ with

$$f'(c) = \frac{f(b) - f(a)}{b - a}.$$

For example, for the graph from the previous part, the Mean Value Theorem says that there is at least one point $c \in (-1, 3)$ with $f'(c) = \frac{-5-3}{4} = -2$, and $c = 1$ is such a point:

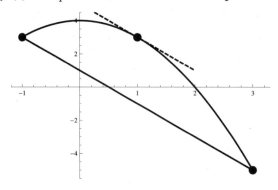

Absolute values as piecewise functions

▷ $f(x) = \begin{cases} -x, & x < 0, \\ x, & x \geq 0. \end{cases}$

▷ Since $2 - 3x < 0$ for $x < \frac{2}{3}$, we have

$$f(x) = \begin{cases} 3x - 2, & x < \frac{2}{3}, \\ 2 - 3x, & x \geq \frac{2}{3}. \end{cases}$$

▷ Since $x^2 - 3x - 4 = (x - 4)(x + 1)$, sign analysis shows that $x^2 - 3x - 4 < 0$ for $-1 < x < 4$ and $f(x) > 0$ for $x < -1$ and for $x > 4$. Thus

$$f(x) = \begin{cases} x^2 - 3x - 4, & x < -1, \\ -x^2 + 3x + 4, & -1 \leq x < 4, \\ x^2 - 3x - 4, & x \geq 4. \end{cases}$$

▷ $f(x) = 0$ when $|x - 1| = 1$, so when $x = 0$ or $x = 2$. Checking signs shows that $|x - 1| - 1 < 0$ for $0 < x < 2$, and $|x - 1| - 1 > 0$ for $x < 0$ and $x > 2$. Further, $x - 1 < 0$ for $x < 1$, and $x - 1 > 0$ for $x > 1$, so that

$$||x - 1| - 1| = |1 - x - 1| = |-x| = x \text{ on } [0, 1],$$
$$||x - 1| - 1| = |x - 1 - 1| = |x - 2| = 2 - x \text{ on } [1, 2].$$

Finally,

$$||x - 1| - 1| = |1 - x - 1| = |-x| = -x \text{ for } x < 0, \qquad ||x - 1| - 1| = x - 2 \text{ for } x > 2.$$

Putting this all together gives

$$f(x) = \begin{cases} -x & x \le 0, \\ x & 0 < x \le 1, \\ 2 - x & 1 < x \le 2 \\ x - 2 & x > 2. \end{cases}$$

Concepts

1. (a) False. It is equal to $\int_a^b |f(x)|\, dx$. If f is negative on part of the interval and positive on another part, these two are unequal.

   (b) False. Since the area between $f(x)$ and $g(x)$ is $\int_{-3}^3 |f(x) - g(x)|\, dx$, the area between the graphs is always nonnegative.

   (c) True. Since $f(x) \le |f(x)|$, we also have $\int_a^b f(x)\, dx \le \int_a^b |f(x)|\, dx$.

   (d) False. By Definition 4.29, it is given by $\int_a^b |f(x) - g(x)|\, dx$.

   (e) False. The average value of a function on an interval is $\frac{1}{b-a} \int_a^b f(x)\, dx$, not the average of its value at the endpoints of the interval.

   (f) False. The average value of a function on an interval is $\frac{1}{b-a} \int_a^b f(x)\, dx$, not the average of its value at the endpoints of the interval.

   (g) False. The average value of f on $[1, 5]$ is

   $$\frac{1}{4} \int_1^5 f(x)\, dx = \frac{\int_1^2 f(x)\, dx + \int_2^5 f(x)\, dx}{4},$$

   while the average of the two average values is

   $$\frac{1}{2} \left(\frac{1}{2-1} \int_1^2 f(x)\, dx + \frac{1}{5-2} \int_2^5 f(x)\, dx \right) = \frac{1}{2} \int_1^2 f(x)\, dx + \frac{1}{6} \int_2^5 f(x)\, dx,$$

   and these two values need not be equal.

   (h) True. The average on $[1, 5]$ is as in the previous part, and the average of the averages is

   $$\frac{1}{2} \left(\frac{1}{3-1} \int_1^2 f(x)\, dx + \frac{1}{5-3} \int_2^5 f(x)\, dx \right) = \frac{1}{4} \int_1^2 f(x)\, dx + \frac{1}{4} \int_2^5 f(x)\, dx,$$

   and the two are equal.

3. Since $\sin x$ is negative on $\left(-\frac{\pi}{2}, 0\right)$ and also on $(\pi, 2\pi)$, while it is positive on $(0, \pi)$, we would need three definite integrals, one on each of these intervals. The absolute area will be positive, since by definition it cannot be negative, and it is not zero since $\sin x$ is not everywhere zero. The signed area will be negative, since $-\cos x$ is an antiderivative of $\sin x$, so that

$$\int_{-\pi/2}^{2\pi} \sin x\, dx = [-\cos x]_{-\pi/2}^{2\pi} = -\cos 2\pi + \cos\left(-\frac{\pi}{2}\right) = -1.$$

5. (a) The signed area is simply the area under the curve, which is $\int_{-3}^{4} f(x)\,dx$.

    (b) Since f is negative on $[-1, 2]$, the absolute value is $-f$ on that interval, so that the absolute area between f and the x axis is

    $$\int_{-3}^{4} |f(x)|\,dx = \int_{-3}^{-1} f(x)\,dx - \int_{-1}^{2} f(x)\,dx + \int_{2}^{4} f(x)\,dx.$$

7. Since f is negative on $[-3, 2]$, the area of any rectangle in a Riemann sum is negative, so that the limit of the Riemann sums is negative. Thus $\int_{-3}^{2} f(x)\,dx < 0$. It follows that, multiplying through by -1, $-\int_{-3}^{2} f(x)\,dx > 0$.

9. For example, let $a = -1$, $b = 1$, and $f(x) = -x$. Then since f is an odd function, $\left| \int_{-1}^{1} f(x)\,dx \right| = |0| = 0$, but since $f(x)$ is zero at only one point, $\int_{-1}^{1} |f(x)|\,dx > 0$.

11. (a) The graph with the left sum for 8 rectangles is

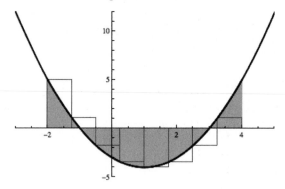

    (b) The signed area is just the integral: $\int_{-2}^{4} f(x)\,dx$.

    (c) Since $f(x)$ is negative on $[-1, 3]$ and positive elsewhere, the signed area is

    $$\int_{-2}^{-1} f(x)\,dx - \int_{-1}^{3} f(x)\,dx + \int_{3}^{4} f(x)\,dx.$$

13. (a) The graph with the left sum for 8 rectangles is

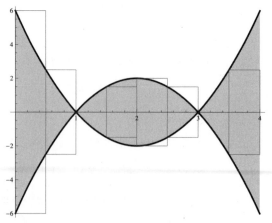

    (note that the rectangles on $\left[1, \frac{3}{2}\right]$ and $\left[2, \frac{4}{2}\right]$ have zero height).

(b) Since $f(x) \leq g(x)$ on $[0,1] \cup [3,4]$ and $f(x) \geq g(x)$ on $[1,3]$, the area between the graphs is

$$\int_0^1 (g(x) - f(x))\, dx + \int_1^3 (f(x) - g(x))\, dx + \int_3^4 (g(x) - f(x))\, dx.$$

15. Drawing a line at $y = -\frac{10}{3}$, the area excluded by that rectangle appears about equal to the extra area included, so the average value should be about $-\frac{10}{3}$.

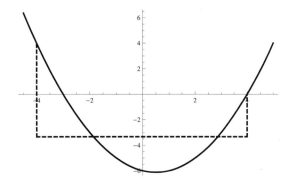

17. Since the average value of f on $[-2,5]$ is 10, we must have $\frac{1}{7} \int_{-2}^5 f(x)\, dx = 10$, so that $\int_{-2}^5 f(x)\, dx = 70$. Since the average rate of change of f on the same interval is -3, we must have $\frac{f(5)-f(-2)}{7} = -3$. A graph of such a function is:

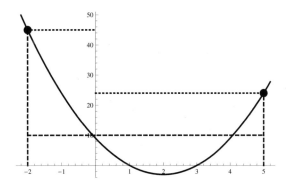

For example, note that $f(-2) = 45$ and $f(5) = 24$, so that $\frac{f(5)-f(-2)}{7} = 3$.

19. For the statement, see Theorem 4.31. The theorem tells us that for any continuous function on an interval, the function achieves its average value somewhere on that interval. For an example, consider the function from Exercise 17:

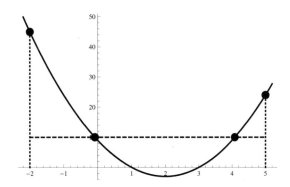

The average value of the function on $[-2,5]$ is 10, and the function achieves that value twice on $(-2,5)$.

21. The Mean Value Theorem for Integrals applies to any function that is continuous on an interval. Since $f(x)$ is a polynomial, it is continuous everywhere, thus continuous on $[1,5]$. The theorem tells us that there is at least one $c \in (1,5)$ such that

$$
\begin{aligned}
f(c) &= \frac{1}{5-1} \int_1^5 x(x-6)\,dx \\
&= \frac{1}{4} \int_1^5 (x^2 - 6x)\,dx \\
&= \frac{1}{4} \left[\frac{1}{3}x^3 - 3x^2 \right]_1^5 \\
&= \frac{1}{4} \left(\frac{1}{3} \cdot 125 - 3 \cdot 25 - \frac{1}{3} + 3 \right) \\
&= -\frac{23}{3}.
\end{aligned}
$$

To find the solutions to $f(x) = -\frac{23}{3}$, write $x(x-6) = -\frac{23}{3}$ and expand and simplify to get $3x^2 - 18x + 23 = 0$. The solutions to this quadratic are $x = 3 \pm \frac{2}{3}\sqrt{3}$. The marked points on the graph below indicate the values where $f(x) = -\frac{23}{3}$.

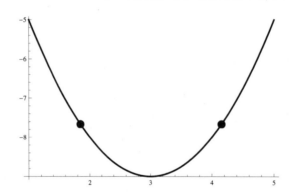

Skills

23. Use 10 rectangles, so that $\Delta x = 1$, and use right sums. Then $x_k = -5 + k$, and a plot of $f(x)$ together with the rectangles is

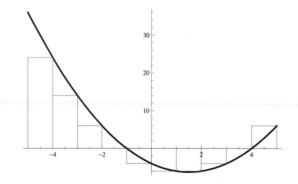

(a) For the signed area,

$$\int_{-5}^{5} f(x)\,dx \approx \sum_{n=1}^{10} f(x_k)\Delta x$$

$$= \sum_{n=1}^{10} \left((-5+k)^2 - 3(-5+k) - 4\right)$$

$$= \sum_{n=1}^{10} (k^2 - 13k + 36)$$

$$= \frac{10(11)(21)}{6} - 13\frac{10(11)}{2} + 360 = 30.$$

(b) For the unsigned area, we use $|f(x_k)|$ instead of $f(x_k)$.

$$\int_{-5}^{5} |f(x)|\,dx \approx \sum_{n=1}^{10} |f(x_k)|\,\Delta x$$

$$= |f(-4)| + |f(-4)| + \cdots + |f(4)| + |f(5)|$$

$$= 24 + 14 + 6 + 0 + 4 + 6 + 6 + 4 + 0 + 6 = 70.$$

25. Use 8 rectangles, so that $\Delta x = \frac{1}{2}$, and use right sums. Then $x_k = -1 + \frac{k}{2}$, and a plot of $f(x)$ together with the rectangles is

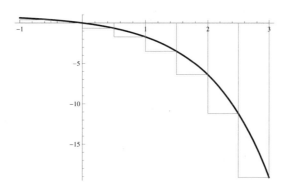

(a) For the signed area,

$$\int_{-1}^{3} f(x)\,dx \approx \sum_{n=1}^{8} f(x_k)\Delta x$$

$$= \sum_{n=1}^{8} (1 - e^{-1+k/2}) \cdot \frac{1}{2}$$

$$= \frac{1}{2}\left((1 - e^{-1/2}) + (1 - e^0) + (1 - e^{1/2}) + \cdots + (1 - e^{5/2}) + (1 - e^3)\right)$$

$$\approx -21.06.$$

(b) For the unsigned area, we use $|f(x_k)|$ instead of $f(x_k)$.

$$\int_{-1}^{3} f(x)\,dx \approx \sum_{n=1}^{8} |f(x_k)|\,\Delta x$$

$$= \sum_{n=1}^{8} \left|1 - e^{-1+k/2}\right| \cdot \frac{1}{2}$$

$$= \frac{1}{2}\left(\left|1 - e^{-1/2}\right| + \left|1 - e^{0}\right| + \left|1 - e^{1/2}\right| + \cdots + \left|1 - e^{5/2}\right| + \left|1 - e^{3}\right|\right)$$

$$\approx 21.45.$$

27. Note that an antiderivative of $3 - x$ is $3x - \frac{1}{2}x^2$. Then

(a) The signed area is

$$\int_{0}^{5} (3 - x)\,dx = \int_{0}^{5} 3\,dx - \int_{0}^{5} x\,dx = 3\,[x]_{0}^{5} - \left[\frac{1}{2}x^2\right]_{0}^{5} = 3(5 - 0) - \frac{1}{2}\cdot 25 = \frac{5}{2}.$$

(b) Since $3 - x > 0$ for $x < 3$ and $3 - x < 0$ for $x > 3$, the absolute area is

$$\int_{0}^{5} |3 - x|\,dx = \int_{0}^{3} (3 - x)\,dx + \int_{3}^{5} (x - 3)\,dx$$

$$= \left[3x - \frac{1}{2}x^2\right]_{0}^{3} + \left[\frac{1}{2}x^2 - 3x\right]_{3}^{5}$$

$$= 3\cdot 3 - \frac{1}{2}\cdot 3^2 + \frac{1}{2}\cdot 5^2 - 3\cdot 5 - \left(\frac{1}{2}\cdot 3^2 - 3\cdot 3\right)$$

$$= \frac{13}{2}.$$

29. Note that an antiderivative of $2x^2 - 7x + 3$ is $\frac{2}{3}x^3 - \frac{7}{2}x^2 + 3x$. Then

(a) The signed area is

$$\int_{0}^{4} (2x^2 - 7x + 3)\,dx = \left[\frac{2}{3}x^3 - \frac{7}{2}x + 3x\right]_{0}^{4} = \frac{2}{3}\cdot 4^3 - \frac{7}{2}\cdot 4^2 + 12 = -\frac{4}{3}.$$

(b) Since $2x^2 - 7x + 3 = (2x - 1)(x - 3)$, sign analysis shows that the integrand is negative for $\frac{1}{2} < x < 3$ and positive elsewhere, so that the absolute area is

$$\int_{0}^{4} \left|2x^2 - 7x + 3\right|\,dx = \int_{0}^{1/2} (2x^2 - 7x + 3)\,dx + \int_{1/2}^{3} (-2x^2 + 7x - 3)\,dx$$

$$+ \int_{3}^{4} (2x^2 - 7x + 3)\,dx$$

$$= \left[\frac{2}{3}x^3 - \frac{7}{2}x^2 + 3x\right]_{0}^{1/2} - \left[\frac{2}{3}x^3 - \frac{7}{2}x^2 + 3x\right]_{1/2}^{3} + \left[\frac{2}{3}x^3 - \frac{7}{2}x^2 + 3x\right]_{3}^{4}$$

$$= \frac{109}{12}.$$

31. Note that an antiderivative of $3x^2 + 5x - 2$ is $x^3 + \frac{5}{2}x^2 - 2x$. Then

(a) The signed area is

$$\int_{-3}^{2} (3x^2 + 5x - 2)\,dx = \left[x^3 + \frac{5}{2}x^2 - 2x\right]_{-3}^{2} = \frac{25}{2}.$$

(b) Since $3x^2 + 5x - 2 = (3x - 1)(x + 2)$, sign analysis shows that the integrand is negative for $-2 < x < \frac{1}{3}$ and positive elsewhere, so that the absolute area is

$$\int_{-3}^{2} \left|3x^2 + 5x - 2\right| dx = \int_{-3}^{-2} (3x^2 + 5x - 2) \, dx + \int_{-2}^{1/3} (-3x^2 - 5x + 2) \, dx$$

$$+ \int_{1/3}^{2} (3x^2 + 5x - 2) \, dx$$

$$= \left[x^3 + \frac{5}{2}x^2 - 2x\right]_{-3}^{-2} + \left[x^3 + \frac{5}{2}x^2 - 2x\right]_{-2}^{1/3} + \left[x^3 + \frac{5}{2}x^2 - 2x\right]_{1/3}^{2}$$

$$= \frac{1361}{54}.$$

33. Using the Chain Rule for the first term, we see that an antiderivative of $(2x - 1)^2 - 4$ is $\frac{1}{6}(2x - 1)^3 - 4x$. Then

(a) The signed area is

$$\int_{-2}^{4} ((2x - 1)^2 - 4) \, dx = \left[\frac{1}{6}(2x - 1)^3 - 4x\right]_{-2}^{4} = 54.$$

(b) Note that $f(x) > 0$ when $(2x - 1)^2 > 4$, which happens when $2x - 1 < -2$ or $2x - 1 > 2$. Thus $f(x)$ is positive for $x < -\frac{1}{2}$ and for $x > \frac{3}{2}$, and is negative elsewhere. Then the absolute area is

$$\int_{-2}^{4} \left|(2x - 1)^2 - 4\right| dx = \int_{-2}^{-1/2} ((2x - 1)^2 - 4) \, dx + \int_{-1/2}^{3/2} (4 - (2x - 1)^2) \, dx$$

$$+ \int_{3/2}^{4} ((2x - 1)^2 - 4) \, dx$$

$$= \left[\frac{1}{6}(2x - 1)^3 - 4x\right]_{-2}^{-1/2} + \left[4x - \frac{1}{6}(2x - 1)^3\right]_{-1/2}^{3/2}$$

$$+ \left[\frac{1}{6}(2x - 1)^3 - 4x\right]_{3/2}^{4}$$

$$= \frac{194}{3}.$$

35. An antiderivative of $\frac{1}{2x^3} = \frac{1}{2}x^{-3}$ is $-\frac{1}{4}x^{-2}$. Then

(a) The signed area is

$$\int_{1}^{3} \frac{1}{2x^3} \, dx = \left[-\frac{1}{4}x^{-2}\right]_{1}^{3} = \frac{2}{9}.$$

(b) Since $\frac{1}{2x^3}$ is positive everywhere on $[1, 3]$, the absolute area is the same as the signed area, which is $\frac{2}{9}$.

37. An antiderivative of $\frac{1}{1+x^2}$ is $\tan^{-1} x$. Then

(a) The signed area is

$$\int_{-1}^{1} \frac{1}{1 + x^2} \, dx = \left[\tan^{-1} x\right]_{-1}^{1} = \frac{\pi}{4} - \frac{-\pi}{4} = \frac{\pi}{2}.$$

(b) Since $\frac{1}{1+x^2} > 0$ for all x, the absolute area is the same as the signed area, which is $\frac{\pi}{2}$.

39. Note that $x < 2$ for $-4 < x < 2$, and $x > 2$ otherwise. Thus we are looking for the integral of the piecewise function

$$h(x) = \begin{cases} g(x) - f(x) = 2 - x, & -4 \le x < 2, \\ f(x) - g(x) = x - 2, & 2 \le x \le 4. \end{cases}$$

Use eight rectangles and the right Riemann sum. Then the interval width is $\Delta x = \frac{8}{8} = 1$, and $x_k = -4 + k$. The first plot below shows $f(x)$ as the solid line and $g(x)$ as the dashed line. The second plot shows $h(x)$ together with the rectangles used in the Riemann sum.

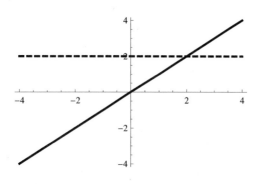
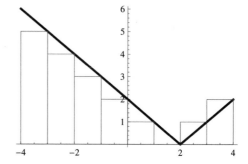

The Riemann sum is then

$$\int_0^\pi h(x)\,dx \approx \sum_{k=1}^8 h(x_k)\Delta x$$
$$= (2 - (-3)) + (2 - (-2)) + (2 - (-1)) + (2 - 0)$$
$$\qquad + (2 - 1) + (2 - 2) + (3 - 2) + (4 - 2)$$
$$= 18.$$

41. Since $f(x) - g(x) = 2x - 1$, we see that $f(x) - g(x) > 0$ for $x > \frac{1}{2}$ and $f(x) - g(x) < 0$ for $x < \frac{1}{2}$. Thus the area between the curves is

$$\int_0^3 |f(x) - g(x)|\,dx = \int_0^{1/2}(1 - 2x)\,dx + \int_{1/2}^3 (2x - 1)\,dx$$
$$= \left[x - x^2\right]_0^{1/2} + \left[x^2 - x\right]_{1/2}^3$$
$$= \frac{1}{2} - \frac{1}{4} + 9 - 3 - \frac{1}{4} + \frac{1}{2} = \frac{13}{2}.$$

43. We have $f(x) - g(x) = x^2 - x - 2 = (x - 2)(x + 1)$. Using sign analysis, we see that $f(x) - g(x) < 0$ for $-1 < x < 2$ and $f(x) - g(x) > 0$ elsewhere. Also, an antiderivative of $f(x) - g(x)$ is $\frac{1}{3}x^3 - \frac{1}{2}x^2 - 2x$. Thus the area between the curves is

$$\int_{-2}^2 |f(x) - g(x)|\,dx = \int_{-2}^{-1}(f(x) - g(x))\,dx + \int_{-1}^2 (g(x) - f(x))\,dx$$
$$= \left[\frac{1}{3}x^3 - \frac{1}{2}x^2 - 2x\right]_{-2}^{-1} + \left[-\frac{1}{3}x^3 + \frac{1}{2}x^2 + 2x\right]_{-1}^2$$
$$= \frac{19}{3}.$$

45. We have $f(x) - g(x) = x^2 - x - 2 = (x - 2)(x + 1)$. Using sign analysis, we see that $f(x) - g(x) < 0$ for $-1 < x < 2$ and $f(x) - g(x) > 0$ elsewhere. Also, an antiderivative of $f(x) - g(x)$ is $\frac{1}{3}x^3 - \frac{1}{2}x^2 - 2x$. Thus the area between the curves is

$$\int_{-3}^{3} |f(x) - g(x)| \, dx = \int_{-3}^{-1} (f(x) - g(x)) \, dx + \int_{-1}^{2} (g(x) - f(x)) \, dx$$
$$+ \int_{2}^{3} (f(x) - g(x)) \, dx$$
$$= \left[\frac{1}{3}x^3 - \frac{1}{2}x^2 - 2x \right]_{-3}^{-1} + \left[-\frac{1}{3}x^3 + \frac{1}{2}x^2 + 2x \right]_{-1}^{2}$$
$$+ \left[\frac{1}{3}x^3 - \frac{1}{2}x^2 - 2x \right]_{2}^{3}$$
$$= 15.$$

47. We have $f(x) - g(x) = x - 1 - (x^2 - 2x - 1) = -x^2 + 3x = -x(x - 3)$. Using sign analysis, we see that $f(x) - g(x) > 0$ for $0 < x < 3$ and $f(x) - g(x) < 0$ elsewhere. Also, an antiderivative of $f(x) - g(x)$ is $-\frac{1}{3}x^3 + \frac{3}{2}x^2$. Thus the area between the curves is

$$\int_{-1}^{3} |f(x) - g(x)| \, dx = \int_{-1}^{0} (g(x) - f(x)) \, dx + \int_{0}^{3} (f(x) - g(x)) \, dx$$
$$= \left[\frac{1}{3}x^3 - \frac{3}{2}x^2 \right]_{-1}^{0} + \left[-\frac{1}{3}x^3 + \frac{3}{2}x^2 \right]_{0}^{3}$$
$$= \frac{19}{3}.$$

49. We have $f(x) - g(x) = x^3 - x^2 + 4x - 4 = (x - 1)(x^2 + 4)$. Since $x^2 + 4 > 0$ everywhere, we see that $f(x) - g(x) < 0$ for $x < 1$ and $f(x) - g(x) > 0$ for $x > 1$. Also, an antiderivative of $f(x) - g(x)$ is $\frac{1}{4}x^4 - \frac{1}{3}x^3 + 2x^2 - 4x$. Thus the area between the curves is

$$\int_{-1}^{2} |f(x) - g(x)| \, dx = \int_{-1}^{1} (g(x) - f(x)) \, dx + \int_{1}^{2} (f(x) - g(x)) \, dx$$
$$= \left[-\frac{1}{4}x^4 + \frac{1}{3}x^3 - 2x^2 + 4x \right]_{-1}^{1} + \left[\frac{1}{4}x^4 - \frac{1}{3}x^3 + 2x^2 - 4x \right]_{1}^{2}$$
$$= \frac{145}{12}.$$

51. $f(x) - g(x) = \frac{2}{1+x^2} - 1 = \frac{1-x^2}{1+x^2}$, so that $f(x) - g(x) > 0$ for $0 < x < 1$ and $f(x) - g(x) < 0$ for $1 < x < \sqrt{3}$. Also, an antiderivative of $f(x) - g(x)$ is $2\tan^{-1} x - x$. Thus the area between the curves is

$$\int_{0}^{\sqrt{3}} |f(x) - g(x)| \, dx = \int_{0}^{1} (f(x) - g(x)) \, dx + \int_{1}^{\sqrt{3}} (g(x) - f(x)) \, dx$$
$$= \left[2\tan^{-1} x - x \right]_{0}^{1} + \left[-2\tan^{-1} x + x \right]_{1}^{\sqrt{3}}$$
$$= \sqrt{3} - 2 + \frac{\pi}{3}.$$

53. Use nine points equally spaced from -1 to 1, so that $x_k = -1 + \frac{k}{4}$ for $k \in \{0, 1, 2, 3, 4, 5, 6, 7, 8\}$. Then the average value of f on $[-1, 1]$ is approximately

$$\frac{2^{-1} + 2^{-3/4} + 2^{-1/2} + 2^{-1/4} + 2^0 + 2^{1/4} + 2^{1/2} + 2^{3/4} + 2^1}{9} \approx 1.103.$$

55. Use 9 points equally spaced on $[0, 2\pi]$, so that $x_k = \frac{k\pi}{4}$ for k an integer from 0 to 8. Then the average value of $\sin x$ on $[0, 2\pi]$ is approximately

$$\frac{1}{9}\left(\sin 0 + \sin\frac{\pi}{4} + \sin\frac{\pi}{2} + \sin\frac{3\pi}{4} + \sin\pi + \sin\frac{5\pi}{4} + \sin\frac{3\pi}{2} + \sin\frac{7\pi}{4} + \sin 2\pi\right) = 0.$$

57. Since an antiderivative of $3x + 1$ is $\frac{3}{2}x^2 + x$, the average value of $f(x)$ on $[0, 4]$ is

$$\frac{1}{4-0}\int_0^4 (3x+1)\,dx = \frac{1}{4}\left[\frac{3}{2}x^2 + x\right]_0^4 = \frac{1}{4}\left(\frac{3}{2}\cdot 16 + 4\right) = 7.$$

A graph of $f(x)$ on $[0, 4]$ is below, with a dashed line drawn at the average value:

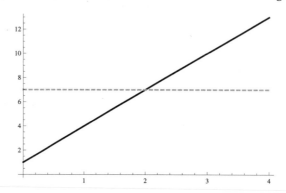

It is plausible that the area excluded by the line matches the extra area included by the line, so that this is a reasonable answer.

59. Since an antiderivative of 4 is $4x$, the average value of $f(x)$ on $[-37.2, 103.75]$ is

$$\frac{1}{103.75-(-37.2)}\int_{-37.2}^{103.75} 4\,dx = \frac{1}{103.75-(-37.2)}\left[4x\right]_{-37.2}^{103.75}$$

$$= \frac{1}{103.75-(-37.2)}\cdot 4(103.75-(-37.2)) = 4.$$

A graph of $f(x)$ on the given interval is below. Since f is constant, obviously its average value on any interval is 4.

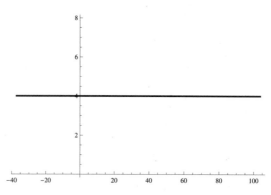

61. Since an antiderivative of $x^2 - 2x - 1$ is $\frac{1}{3}x^3 - x^2 - x$, the average value of $f(x)$ on $[0, 3]$ is

$$\frac{1}{3-0}\int_0^3 (x^2 - 2x - 1)\,dx = \frac{1}{3}\left[\frac{1}{3}x^3 - x^2 - x\right]_0^3 = -1.$$

A graph of $f(x)$ on the given interval is below, with a dashed line drawn at the average value:

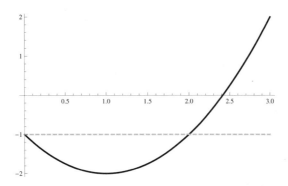

It is plausible that the area excluded by the line matches the extra area included by the line, so that this is a reasonable answer.

63. Since an antiderivative of $(e^x)^2 = e^{2x}$ is $\frac{1}{2}e^{2x}$, the average value of $f(x)$ on $[-1,1]$ is

$$\frac{1}{1-(-1)} \int_{-1}^{1} (e^x)^2 \, dx = \frac{1}{2} \left[\frac{1}{2}e^{2x} \right]_{-1}^{1} = \frac{1}{2} \left(e^2 - e^{-2} \right) \approx 1.813$$

A graph of $f(x)$ on the given interval is below, with a dashed line drawn at the average value:

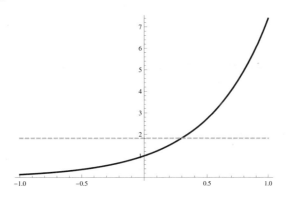

It is plausible that the area excluded by the line matches the extra area included by the line, so that this is a reasonable answer.

65. Since an antiderivative of $2 - \sqrt[3]{x} = 2 - x^{1/3}$ is $2x - \frac{3}{4}x^{4/3}$, the average value of $f(x)$ on $[1,8]$ is

$$\frac{1}{8-1} \int_{1}^{8} (2 - \sqrt[3]{x}) \, dx = \frac{1}{7} \left[2x - \frac{3}{4}x^{4/3} \right]_{1}^{8} = \frac{11}{28} \approx 0.393.$$

A graph of $f(x)$ on the given interval is below, with a dashed line drawn at the average value:

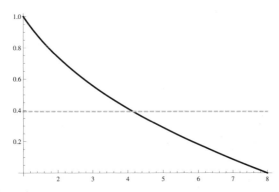

It is plausible that the area excluded by the line matches the extra area included by the line, so that this is a reasonable answer.

67. To find an antiderivative of $f(x)$, note that this looks like an application of the product rule to $x \sin x$; taking the derivative of that function shows that this guess is correct. Thus the average value of $f(x)$ on $[-\pi, \pi]$ is

$$\frac{1}{\pi - (-\pi)} \int_{-\pi}^{\pi} (\sin x + x \cos x)\, dx = \frac{1}{2\pi} [x \sin x]_{-\pi}^{\pi} = 0.$$

A graph of $f(x)$ on the given interval is below:

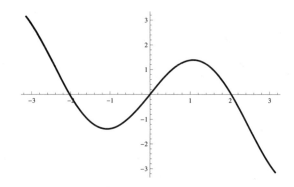

It is clear that the points above the average value of 0 exactly the balance the points below the average value, so that this answer is correct.

69. An antiderivative of $2x^2 + 1$ is $\frac{2}{3}x^3 + x$, so that the average value of $f(x)$ on $[-1, 2]$ is

$$\frac{1}{2 - (-1)} \int_{-1}^{2} (2x^2 + 1)\, dx = \frac{1}{3} \left[\frac{2}{3}x^3 + x\right]_{-1}^{2} = 3.$$

Solving $2x^2 + 1 = 3$ gives $x = \pm 1$. A graph of $f(x)$ on the given interval is below, with a dashed line drawn at the average value:

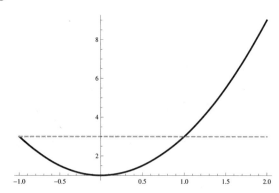

The average value line intersects the curve at $x = \pm 1$, as it should.

71. An antiderivative of $1 + x + 2x^2$ is $x + \frac{1}{2}x^2 + \frac{2}{3}x^3$, so that the average value of $f(x)$ on $[-2, 2]$ is

$$\frac{1}{2 - (-2)} \int_{-2}^{2} (1 + x + 2x^2)\, dx = \frac{1}{4} \left[x + \frac{1}{2}x^2 + \frac{2}{3}x^3\right]_{-2}^{2} = \frac{11}{3}.$$

Solving $2x^2 + x + 1 = \frac{11}{3}$ gives $x = \frac{1}{12}(-3 \pm \sqrt{201}) \approx -1.43, 0.93$. A graph of $f(x)$ on the given interval is below, with a dashed line drawn at the average value:

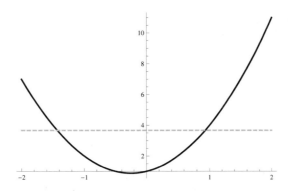

The average value line appears to intersect the curve at $x \approx -1.43, 0.93$, as it should.

73. An antiderivative of $(x+1)^2$ is $\frac{1}{3}(x+1)^3$, so that the average value of $f(x)$ on $[-1,2]$ is

$$\frac{1}{2-(-1)} \int_{-1}^{2} (x+1)^2 \, dx = \frac{1}{3} \left[\frac{1}{3}(x+1)^3 \right]_{-1}^{2} = 3.$$

Solving $(x+1)^2 = 3$ gives $x = -1 \pm \sqrt{3} \approx -2.732, 0.732$. On $[-1,2]$, we need only consider 0.732. A graph of $f(x)$ on the given interval is below, with a dashed line drawn at the average value:

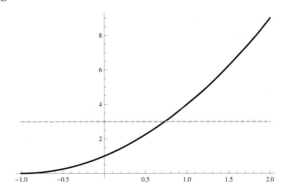

The average value line appears to intersect the curve at $x \approx 0.732$, as it should.

Applications

75. (a) Since the average rate of growth of the plant during those four days was $\frac{f(4)-f(0)}{4-0} = 1.44$, the Mean Value Theorem says that at some time during those four days, the plant was growing at 1.44 centimeters per day.

   (b) The Mean Value Theorem for Integrals tells us that at some point during those four days, the height of the plant was equal to its average height over those four days, which was

$$\frac{1}{4-0} \int_{0}^{4} 0.36t^2 \, dx = \frac{1}{4} \left[0.12t^3 \right]_{0}^{4} = 1.92 \text{ cm.}$$

77. (a) We have

$$f(x) = \frac{1}{2}(x-12)^2 = 55 \quad \Rightarrow \quad x = 12 \pm \sqrt{110} \approx 1.512, 22.488,$$

$$g(x) = \frac{1}{2}(x-24)^2 = 55 \quad \Rightarrow \quad x = 24 \pm \sqrt{110} \approx 13.512, 34.488,$$

$$r(x) = (x-12)^2 + 10 = 55 \quad \Rightarrow \quad x = 12 \pm \sqrt{45} \approx 5.292, 18.708,$$

$$s(x) = (x-24)^2 + 10 = 55 \quad \Rightarrow \quad x = 24 \pm \sqrt{45} \approx 17.292, 30.708.$$

From the picture, the values we are interested in are for $f(x)$, 1.512; for $g(x)$, 34.488; for $r(x)$, 5.292; and for $s(x)$, 30.708. Also, the graphs of r and s intersect where $(x - 12)^2 + 10 = (x - 24)^2 + 10$, which is for $x = 18$. We divide up the region as shown in the following diagram:

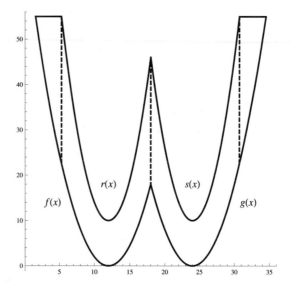

Then the area of the sign is the sum of four integrals:

$$A = \int_{12-\sqrt{110}}^{12-\sqrt{45}} (55 - f(x))\,dx + \int_{12-\sqrt{45}}^{18} (r(x) - f(x))\,dx$$

$$+ \int_{18}^{24+\sqrt{45}} (s(x) - g(x))\,dx + \int_{24+\sqrt{45}}^{24+\sqrt{110}} (55 - g(x))\,dx.$$

(b) Computing the integrals one at a time, we have

$$\int_{12-\sqrt{110}}^{12-\sqrt{45}} (55 - f(x))\,dx = \int_{12-\sqrt{110}}^{12-\sqrt{45}} \left(-\frac{1}{2}x^2 + 12x - 17\right)\,dx$$

$$= \left[-\frac{1}{6}x^3 + 6x^2 - 17x\right]_{12-\sqrt{110}}^{12-\sqrt{45}} \approx 65.924,$$

$$\int_{12-\sqrt{45}}^{18} (r(x) - f(x))\,dx = \int_{12-\sqrt{45}}^{18} \left(\frac{x^2}{2} - 12x + 82\right)\,dx$$

$$= \left[\frac{1}{6}x^3 - 6x^2 + 82x\right]_{12-\sqrt{45}}^{18} \approx 213.394,$$

$$\int_{18}^{24+\sqrt{45}} (s(x) - g(x))\,dx = \int_{18}^{24+\sqrt{45}} \left(\frac{1}{2}x^2 - 24x + 298\right)\,dx$$

$$= \left[\frac{1}{6}x^3 - 12x + 298x\right]_{18}^{24+\sqrt{45}} \approx 213.394,$$

$$\int_{24+\sqrt{45}}^{24+\sqrt{110}} (55 - g(x))\,dx = \int_{24+\sqrt{45}}^{24+\sqrt{110}} \left(-\frac{1}{2}x^2 + 24x - 233\right)\,dx$$

$$= \left[-\frac{1}{6}x^3 + 12x^2 - 233x\right]_{24+\sqrt{45}}^{24+\sqrt{110}} \approx 65.924.$$

Adding these up gives for the total area 558.635 square feet. The sign is too large by about 30 square feet.

Proofs

79. Recall the triangle inequality, $|r + s| \leq |r| + |s|$. Then

$$\left| \int_a^b f(x)\, dx \right| = \left| \lim_{n \to \infty} \sum_{k=1}^n f(x_k^*) \Delta x \right|$$

$$= \lim_{n \to \infty} \left| \sum_{k=1}^n f(x_k^*) \Delta x \right|$$

$$\leq \lim_{n \to \infty} \sum_{k=1}^n |f(x_k^*) \Delta x|$$

$$= \lim_{n \to \infty} \sum_{k=1}^n |f(x_k^*)| \, \Delta x$$

$$= \int_a^b |f(x)| \, dx.$$

81. (a) Since

$$\int_a^b f(x)\, dx = \lim_{n \to \infty} \sum_{k=1}^n f(x_k) \Delta x = \lim_{n \to \infty} \sum_{k=1}^n f(x_k) \cdot \frac{b-a}{n},$$

we divide both sides by $b - a$ and simplify to get

$$\frac{1}{b-a} \int_a^b f(x)\, dx = \lim_{n \to \infty} \frac{1}{b-a} \sum_{k=1}^n f(x_k) \cdot \frac{b-a}{n} = \lim_{n \to \infty} \sum_{k=1}^n f(x_k) \cdot \frac{1}{n} = \lim_{n \to \infty} \frac{\sum_{k=1}^n f(x_k)}{n}.$$

(b) The formula inside the limit is simply the average value computed using n specific points. It makes sense that the larger the number of points we use, the closer we will get to the true average value, and that if we use "all" of the points, then in the limit we will get the true average value.

(c) The right-hand side is the total area under the curve divided by the length of the interval, so it is the height of a rectangle whose total area is the area under the curve. This is the definition of average value.

Thinking Forward

The Second Fundamental Theorem of Calculus

▷ You come to a full stop when your velocity is zero. $v(t) = s'(t) = 9t^2 - 24t - 9$, which has zeros at $t = 3$ and $t = -\frac{1}{3}$. Rejecting the negative zero, we come to a full stop after 3 seconds.

▷ The average distance from the stop sign is the average value of $s(t)$ on $[0, 3]$, which is

$$\frac{1}{3} \int_0^3 s(t)\, dt = \frac{1}{3} \left[\frac{3}{4} t^4 - 4t^3 - \frac{9}{2} t^2 + 54t \right]_0^3 = \frac{99}{4} = 24.75 \text{ feet.}$$

▷ The average rate of change of position from $t = 0$ to $t = 3$ is

$$\frac{1}{3}(s(3) - s(0)) = \frac{1}{3}(0 - 54) = -18,$$

since our position changed from $s(0)$ to $s(3)$ in 3 seconds.

▷ Since $v(t) = s'(t) = 9t^2 - 24t - 9$, the average value of the velocity function over the three seconds is

$$\frac{1}{3} \int_0^3 v(t)\, dt = \frac{1}{3} \int_0^3 s'(t)\, dt = \frac{1}{3}(s(3) - s(0)) = -18.$$

▷ Since these two answers are the same, we have validated the Fundamental Theorem of Calculus in this particular case. The area under the graph of the velocity curve, which is three times the integral in the previous part, is the same as the total distance traveled, which is just $s(3) - s(0)$.

4.7   Functions Defined by Integrals

Thinking Back

The relationship between a function and its derivative

▷ If f has a zero, then f' *not applicable*.

▷ If f' has a zero, then f *has a critical point*.

▷ If f is increasing, then f' is *positive*.

▷ If f is positive, then f' is *not applicable*.

▷ If f' is negative, then f is *decreasing*.

▷ If f'' is negative, then f is *concave down*.

▷ If f'' is negative, then f' is *decreasing*.

Functions and their properties

▷ A *function f* from a set A to a set B is a rule that assigns to each element a of A a unique element $b \in B$; the rule is written $b = f(a)$.

▷ A function f from $[a, b]$ to $\mathbb{R}$ is *positive* if $f(x) > 0$ for all $x \in [a, b]$.

▷ A function f is *increasing* on $[a, b]$ if whenever $x < y$ we have $f(x) < f(y)$.

Properties of logarithms and exponents

▷ False. $\ln(ab) = \ln a + \ln b$.

▷ True.

▷ True.

▷ False. $e^{a/b} = (e^a)^{1/b}$.

▷ True. The fact that $e^{\ln x} = x$ is almost the definition of logarithms.

▷ False. $\ln(e^x) = x$, not 1.

Concepts

1. (a) False. $\int f(v)\,du = uf(v)$, so that $\int_2^u f(v)\,du = [uf(v)]_{u=2}^{u=u} = uf(v) - 2f(v)$. Compare with Theorem 4.35.

   (b) False. By Theorem 4.35, $\frac{d}{dx}\int_a^{x^2} \sin t\,dt = 2x \sin x^2$.

   (c) False. For example, let $f(x) = 4$ on any interval $[a,b]$. Then any $c \in [a,b]$ has $f(c)$ equal to the average value of f on the interval.

   (d) True. Since $3 - t$ is positive on $[0,3]$, the area accumulation function is increasing there.

   (e) False. $A'(x) = 3 - x$ by Theorem 4.35, so that $A''(x) = -1$ and A is concave down.

   (f) True, depending on the meaning of the terms. Every continuous function is integrable; however, not all functions have antiderivatives that can be expressed in terms of common functions. e^{-x^2} is an example of such a function, as is $\frac{\sin x}{x}$.

   (g) True, again depending on the meaning of the terms. Every continuous function is integrable, so has multiple antiderivatives differing by a constant, but the antiderivatives need not be expressible in terms of common functions

   (h) False. By Definition 4.36, $\ln 5 = \int_1^5 \frac{1}{t}\,dt$.

3. This function is just the area accumulation function for t^2 from $t = 2$; it is

$$\int_2^x t^2\,dt.$$

5. The independent variable is x. The dependent variable represents the signed area under the graph of $f(t)$ from $t = a$ to $t = x$. t is a dummy variable since it is meaningful only as part of the integral - it could equally well be u, v, or any letter except for a and x without changing the meaning of the expression.

7. (a) Since f is positive, we know that A is increasing by Exercise 6; however, since f is decreasing, it is clear geometrically that the amount that A increases from x to $x + \Delta$ is greater than the amount by which it increases from $x + \Delta$ to $x + 2\Delta$. Thus f increases more slowly for larger x, which means that the rate of change of the slope of f is decreasing, so A is concave down.

   (b) By Theorem 4.35, $A'(x) = f(x)$ so that $A''(x) = f'(x)$. Since f is decreasing, $f'(x) < 0$ so that $A''(x) < 0$. This means that A is concave down.

9. (a) By Theorem 4.35,

$$A'(x) = x^2 = B'(x),$$

   so that A and B have the same derivative. Thus they must differ by a constant.

   (b) We have

$$A(x) = \int_0^x t^2\,dt = \int_0^3 t^2\,dt + \int_3^x t^2\,dt = \left[\frac{1}{3}t^3\right]_0^3 + B(x) = B(x) + 9.$$

   Thus A and B differ by a constant.

   (c) Since $A(x)$ is the signed area under t^2 from $t = 0$ to $t = x$, and $B(x)$ is the signed area under t^2 from $t = 3$ to $t = x$, clearly the two are the same except for the area under t^2 from $t = 0$ to $t = 3$; this constant area is the difference between A and B.

11. Write $G(x) = \int_3^x \sin t\,dt$. Then $F(x) = G(x^2)$. The inside function is $f(x) = x^2$, and the outside function is $G(x)$.

13. $A(0) = 0$ since it is the area under the curve between 0 and 0. $A(-1)$ is positive, since $A(-1) = \int_0^{-1} f(t)\,dt = -\int_{-1}^0 f(t)\,dt$ and f is negative on $[-1, 0]$. Similarly, $A(-2) > A(-1)$. However, from the previous problem, $A(5) < 0$. Thus the correct order is (d), (a), (b), (c).

15. Since f is positive for $0 < x < 1$, $A(x)$ is positive and increasing there. Then f becomes negative, so that $A(x)$ decreases starting at $x = 1$. It continues to decrease, eventually becoming negative around $x = 2$, and remains negative and decreasing until $x = 5$, when f is zero again. At that point $A(x)$ starts increasing again, although it is still negative since the net signed area under the curve is still negative. A rough graph of $A(x)$ is

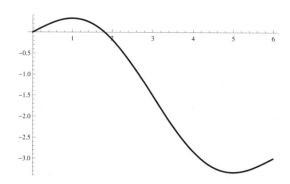

17. If we write $G(x) = \int_a^x f(t)\,dt$, then if $F(x) = \int_a^{u(x)} f(t)\,dt$, clearly $F(x) = G(u(x))$. Then the Chain Rule gives

$$\frac{d}{dx} \int_a^{u(x)} f(t)\,dt = F'(x) = G'(u(x))u'(x).$$

However, $G'(x) = f(x)$, so that this is simply equal to $f(u(x))u'(x)$.

19. (a) By the Net Change Theorem, $\int_a^b f'(x)\,dx = f(b) - f(a)$.

    (b) If F is an antiderivative of f, then $\int_a^b f(x)\,dx = F(b) - F(a)$, which is a constant, so that $\frac{d}{dx} \int_a^b f(x)\,dx = \frac{d}{dx}(F(b) - F(a)) = 0$.

    The two functions are definitely not inverses. In fact, their domains and ranges do not match up as they should for inverse functions, and they have completely different meanings.

21. (a) Since the integral $\ln x = \int_1^x \frac{1}{t}\,dt$ is the area under the curve $y = \frac{1}{t}$ between 1 and x, that area exists whenever $x > 0$. The area under $\frac{1}{t}$ between 0 and 1 is infinite (see Capstone Problem D for this chapter for details), so that 0 is not in the domain of $\ln x$. Thus $(0, \infty)$ is the domain of $\ln x$. Finally, note that the total area under $\frac{1}{x}$ on $[1, \infty)$ is bounded above by the sum $\frac{1}{2} + \frac{1}{3} + \ldots$, since $\sum_{k=2}^n \frac{1}{k}$ is a lower Riemann sum on $[2, n]$ for any n. But this sum is easily seen to diverge, since

$$\frac{1}{2} \geq \frac{1}{2},$$
$$\frac{1}{3} + \frac{1}{4} \geq 2\frac{1}{4} = \frac{1}{2}$$
$$\frac{1}{5} + \frac{1}{6} + \frac{1}{7} + \frac{1}{8} \geq 4\frac{1}{8} = \frac{1}{2}$$
$$\cdots$$

so that the sum can be made larger than any number of multiples of $\frac{1}{2}$ by choosing n large enough. Thus $\ln x$ can assume any positive real value, so that its range is $\mathbb{R}$.

(b) Since e^x is the inverse of $\ln x$, it has domain $\mathbb{R}$ and range $(0, \infty)$.

23. Since $\ln x = \int_1^x \frac{1}{t}\, dt$, the signed area under the graph of $\frac{1}{x}$ from $x = e$ to $x = 10$ is

$$\int_e^{10} \frac{1}{t}\, dt = \int_1^{10} \frac{1}{t}\, dt - \int_1^e \frac{1}{t}\, dt = \ln 10 - \ln e = -1 + \ln 10.$$

25. We have

$$\log_b x = \frac{1}{\ln b} \ln x = \frac{1}{\ln b} \int_1^x \frac{1}{t}\, dt = \int_1^x \frac{1}{t \ln b}\, dt.$$

Skills

27. (a) The shaded area in the graph below represents $A(2)$, the area under the curve between $t = 1$ and $t = 2$:

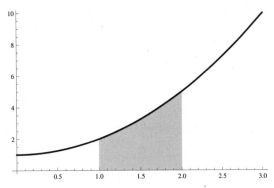

(b) We have

$$A(2) = \int_1^2 (t^2 + 1)\, dt = \left[\frac{1}{3} t^3 + t \right]_1^2 = \frac{10}{3},$$

$$A(5) = \int_1^5 (t^2 + 1)\, dt = \left[\frac{1}{3} t^3 + t \right]_1^5 = \frac{136}{3}.$$

(c) Integrating as above gives

$$A(x) = \int_1^x (t^2 + 1)\, dt = \left[\frac{1}{3} t^3 + t \right]_1^x = \frac{1}{3} x^3 + x - \frac{4}{3}.$$

29. (a) The shaded area in the graph below represents $A(2)$, the area under the curve between $t = 3$ and $t = 2$. However, note that the *signed* area is actually the negative of this area, since we are integrating from 3 to 2:

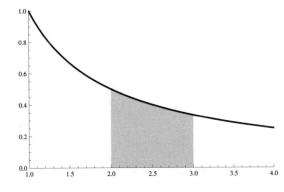

(b) We have

$$A(2) = \int_3^2 \frac{1}{t}\, dt = [\ln t]_3^2 = \ln 2 - \ln 3 = \ln\frac{2}{3},$$

$$A(5) = \int_3^5 \frac{1}{t}\, dt = [\ln t]_3^5 = \ln 5 - \ln 3 = \ln\frac{5}{3}.$$

(c) Integrating as above gives

$$A(x) = \int_3^x \frac{1}{t}\, dt = [\ln t]_3^x = \ln x - \ln 3 = \ln\frac{x}{3}.$$

31. For example, $\displaystyle\int_1^x \sin^2(3^t)\, dt, \qquad \int_7^x \sin^2(3^t)\, dt, \qquad 5 + \int_2^x \sin^2(3^t)\, dt.$

33. For example, $\displaystyle\int_0^x e^{(t^2)}\, dt, \qquad \int_{-4}^x e^{(t^2)}\, dt, \qquad 3 + \int_{100}^x e^{(t^2)}\, dt.$

35. $\displaystyle\frac{d}{dx}\int_x^4 e^{t^2+1}\, dt = \frac{d}{dx}\left(-\int_4^x e^{t^2+1}\, dt\right) = -\frac{d}{dx}\int_4^x e^{t^2+1}\, dt = -e^{x^2+1}.$

37. $\displaystyle\frac{d}{dx}\int_0^{x^2} \cos t\, dt = \cos(x^2)\cdot\frac{d}{dx}x^2 = 2x\cos(x^2).$

39. Note that since x is a constant with respect to integration by t, it can be pulled out of the integration:

$$\frac{d}{dx}\int_1^{\sqrt{x}} x\ln t\, dt = \frac{d}{dx}\left(x\int_1^{\sqrt{x}} \ln t\, dt\right)$$

$$= \int_1^{\sqrt{x}} \ln t\, dt + x\frac{d}{dx}\int_1^{\sqrt{x}} \ln t\, dt$$

$$= \int_1^{\sqrt{x}} \ln t\, dt + x\left(\ln(\sqrt{x})\cdot(\sqrt{x})'\right)$$

$$= \int_1^{\sqrt{x}} \ln t\, dt + x\left(\frac{1}{2}x^{-1/2}\ln(\sqrt{x})\right)$$

$$= \int_1^{\sqrt{x}} \ln t\, dt + \frac{1}{2}\sqrt{x}\ln(\sqrt{x}).$$

41. Since $\int_1^8 \ln t\, dt$ is a constant, its derivative is zero.

43.

$$\frac{d^2}{dx^2}\int_{x^2}^1 \ln|t|\, dt = -\frac{d^2}{dx^2}\int_1^{x^2} \ln|t|\, dt$$

$$= -\frac{d}{dx}\left(\ln\left|x^2\right|\cdot(x^2)'\right)$$

$$= -\frac{d}{dx}\left(2x\ln x^2\right)$$

$$= -\left(2\ln x^2 + 2x\cdot\frac{1}{x^2}\cdot(x^2)'\right)$$

$$= -\left(2\ln x^2 + 4\right).$$

45. $\displaystyle\frac{d^2}{dx^2}\int_2^{3x}(t^2+1)\, dt = \frac{d}{dx}\left(((3x)^2+1)\cdot(3x)'\right) = \frac{d}{dx}\left(3(9x^2+1)\right) = 54x.$

47.

$$\frac{d}{dx}\int_x^{x+2}\sin(t^2)\,dt = \frac{d}{dx}\left(\int_x^1 \sin(t^2)\,dt + \int_1^{x+2}\sin(t^2)\,dt\right)$$

$$= \frac{d}{dx}\int_1^{x+2}\sin(t^2)\,dt - \frac{d}{dx}\int_1^x \sin(t^2)\,dt$$

$$= \sin((x+2)^2) - \sin(x^2).$$

49. This looks like a derivative of a log; try $\ln(2x-1)$. This has derivative $\frac{2}{2x-1}$; correcting for the factor of 2 gives $f(x) = \frac{1}{2}\ln(2x-1) + C$. Then $f(1) = \frac{1}{2}\ln(2\cdot 1 - 1) + C = C = 3$, so that $C = 3$ and we get for an answer $f(x) = \frac{1}{2}\ln(2x-1) + 3$.

51. We do not know how to find an antiderivative of $f'(x)$ in terms of elementary functions, so we use the Second Fundamental Theorem to write

$$f(x) = \int_2^x \frac{1}{t^3+1}\,dt + C.$$

Then $f(2) = 0$ means that $\int_2^2 \frac{1}{t^3+1}\,dt + C = 0$, so that $C = 0$ and the answer is

$$f(x) = \int_2^x \frac{1}{t^3+1}\,dt.$$

53. We do not know how to find an antiderivative of e^{-x^2} in terms of elementary functions, so we use the Second Fundamental Theorem to write

$$f(x) = \int_1^x e^{-t^2}\,dt + C.$$

Then $f(1) = 0$ means that $\int_1^1 e^{-t^2}\,dt + C = 0$, so that $C = 0$ and the answer is

$$f(x) = \int_1^x e^{-t^2}\,dt.$$

55. The definition of ln tells us that $\ln 10 = \int_1^{10} \frac{1}{t}\,dt$. We approximate this integral using the upper and lower Riemann sums with nine rectangles. Then $x_k = k$ for $k \in \{1,2,3,4,5,6,7,8,9\}$. Since $\frac{1}{x}$ is a decreasing function, we see that its maximum on $[x_k, x_{k+1}]$ occurs at x_k and its minimum occurs at x_{k+1}. Then the upper Riemann sum is

$$\sum_{k=1}^9 \frac{1}{x_k}\Delta x = \sum_{k=1}^9 \frac{1}{k} = \frac{1}{1} + \frac{1}{2} + \frac{1}{3} + \frac{1}{4} + \frac{1}{5} + \frac{1}{6} + \frac{1}{7} + \frac{1}{8} + \frac{1}{9} = \frac{7129}{2520} \approx 2.829,$$

and the lower Riemann sum is

$$\sum_{k=1}^9 \frac{1}{k+1}\Delta x = \sum_{k=1}^9 \frac{1}{k+1} = \frac{1}{2} + \frac{1}{3} + \frac{1}{4} + \frac{1}{5} + \frac{1}{6} + \frac{1}{7} + \frac{1}{8} + \frac{1}{9} + \frac{1}{10} = \frac{4861}{2520} \approx 1.929,$$

Thus $1.929 \le \ln 10 \le 2.829$.

57. The definition of ln tells us that $\ln e = \int_1^e \frac{1}{t}\,dt$. We approximate this integral using upper and lower Riemann sums with four rectangles. Then $\Delta x = \frac{e-1}{4}$, and $x_k = 1 + \frac{e-1}{4}k$ for k an integer from 0

to 4. Since $\frac{1}{x}$ is a decreasing function, we see that its maximum on $[x_k, x_{k+1}]$ occurs at x_k and its minimum occurs at x_{k+1}. Then the upper Riemann sum is

$$\sum_{k=1}^{4} \frac{1}{x_{k-1}} \Delta x = \frac{e-1}{4} \sum_{k=1}^{4} \frac{1}{1 + \frac{e-1}{4}(k-1)}$$

$$= \frac{e-1}{4} \left(\frac{1}{1} + \frac{1}{1 + \frac{e-1}{4}} + \frac{1}{1 + \frac{e-1}{4} \cdot 2} + \frac{1}{1 + \frac{e-1}{4} \cdot 3} \right)$$

$$\approx 1.149,$$

and the lower Riemann sum is

$$\sum_{k=1}^{4} \frac{1}{x_k} \Delta x = \frac{e-1}{4} \sum_{k=1}^{4} \frac{1}{1 + \frac{e-1}{4}k}$$

$$= \frac{e-1}{4} \left(\frac{1}{1 + \frac{e-1}{4}} + \frac{1}{1 + \frac{e-1}{4} \cdot 2} + \frac{1}{1 + \frac{e-1}{4} \cdot 3} + \frac{1}{1 + \frac{e-1}{4} \cdot 4} \right)$$

$$\approx 0.877,$$

so that

$$0.877 \le \ln e \le 1.149.$$

Applications

59. (a) Since position is the integral of velocity, the position of the particle at time t is

$$s(t) = \int_0^t v(x)\, dx = \int_0^t \sin(0.1x^2)\, dx$$

(Note that since we want s to be a function of t, here we have used x as the dummy variable).

(b) Using a Riemann sum with 10 rectangles, we have $\Delta x = 1$ and $x_k = k$ for k an integer from 0 to 10, so that the left Riemann sum is

$$\sum_{k=1}^{10} v(x_{k-1})\Delta x = \sum_{k=1}^{10} \sin(0.1 x_{k-1}^2)$$

$$= \sin 0 + \sin 0.1 + \sin 0.4 + \sin 1.6 + \cdots + \sin 6.4 + \sin 8.1$$

$$\approx 2.532$$

The position of the particle after 10 seconds is about 2.5 feet to the right of the starting point.

(c) Since Δx is time, it is measured in seconds. Since v is the velocity, its units are feet per second. Thus $v(x_{k-1})\Delta x$ has units of feet, so that $s(10)$ does as well.

(d) As t gets larger, $v(t)$ looks like sin of some very large number, so that it oscillates very rapidly between -1 and $+1$. Thus after a long time, say 100 seconds, the particle oscillates very rapidly between a velocity of -1 and a velocity of $+1$.

61. (a) We want to integrate a function to find the number of cubic feet flowing past the gauge, but the measurement we have is in cubic feet per second. Integrating $f(t)$ from t_0 to t gives a value whose units are cubic feet per second times days, so to convert that to cubic feet, we must multiply by the number of seconds per day, or 86400. Thus we want to integrate $86400f(t)$.

Now, if F is an antiderivative of $86400f$, then

$$Ft_0(t) = F(t) - F(t_0) = \int_{t_0}^{t} 86400 f(x)\, dx$$

is the total amount of water flowing past the gauge between day t_0 and day t. Then

$$F_{t_0}(t) = \int_{t_0}^{t} 86400 f(x)\,dx$$

$$= 86400 \int_{t_0}^{t} (800 + 3.1(x - 90)(195 - x))\,dx$$

$$= 86400 \int_{t_0}^{t} (-3.1x^2 + 883.5x - 53605)\,dx$$

$$= 86400 \left[-1.033x^3 + 441.75x^2 - 53605x \right]_{t_0}^{t} \text{ cubic feet}$$

of water flow past the gauge between day t_0 and day t.

(b) With $t_0 = 95$ and $t = 195$, we have

$$F_{90}(195) = 86400 \left[-1.033x^3 + 441.75x^2 - 53605x \right]_{90}^{195} \approx 58.9 \text{ billion cubic feet.}$$

(c) For days from 0 to 89 and from 196 to 365, the flow is 800 cubic feet of water per second; in those $90 + 170 = 260$ days, then, the total number of cubic feet was $800 \cdot 260 \cdot 86400 \approx 17.971$ billion cubic feet. Thus the total amount of water that flows through the gauge in a year is $58.9 + 17.971 \approx 76.9$ billion cubic feet, of which $\frac{58.9}{76.9} \approx 76\%$ flows through between days 90 and 195.

Proofs

63. If G is an antiderivative of f, then by the Fundamental Theorem of Calculus, we have $A(x) = \int_0^x f(t)\,dt = G(x) - G(0)$. But then $A(x)$ and $G(x)$ differ by the constant $G(0)$, so that A is also an antiderivative of f.

65. The proof starts by showing that if $F(x) = \int_a^x f(t)\,dt$ is an antiderivative of f, then $F'(x) = \lim_{h \to 0^+} \frac{\int_x^{x+h} f(t)\,dt}{h}$. Then the proof bounds the area under the graph of f between x and $x + h$ by two rectangles whose heights are the maximum and minimum value of f on $[x, x + h]$. Dividing through by h taking limits, and using the Squeeze Theorem shows that the integral above, which is $F'(x)$, is bounded both below and above by $f(x)$, since both the minimum and maximum values of f on $[x, x + h]$ approach $f(x)$ as $h \to 0$.

For $h < 0$, we have

$$F'(x) = \lim_{h \to 0^-} \frac{F(x + h) - F(x)}{h}$$

$$= \lim_{h \to 0^-} \frac{\int_a^{x+h} f(t)\,dt - \int_a^x f(t)\,dt}{h}$$

$$= \lim_{h \to 0^-} \frac{\int_a^{x+h} f(t)\,dt - \left(\int_a^{x+h} f(t)\,dt + \int_{x+h}^x f(t)\,dt \right)}{h}$$

$$= \lim_{h \to 0^-} \frac{-\int_{x+h}^x f(t)\,dt}{h}.$$

Note that $h < 0$ since we are considering the left derivative. The quantity $\int_{x+h}^x f(t)\,dt$ represents the signed area between the graph of $f(t)$ and the t-axis from $t = x + h$ to $t = x$ (remember that $h < 0$). For any (negative) value of h, define m_h and M_h to be points in $[x + h, x]$ at which f attains its minimum and maximum values, respectively, in that interval. These points exist by the

Extreme Value Theorem. Then clearly $hf(m_h)$ is the area of a rectangle that is everywhere below the graph of $f(t)$ on $[x + h, x]$, and $hf(M_h)$ is the area of a rectangle that completely encloses the graph of $f(t)$ on $[x + h, x]$. Thus (since h is negative)

$$f(m_h)(-h) \le \int_{x+h}^{x} f(t)\, dt \le f(M_h)(-h).$$

Dividing through by $-h > 0$ gives

$$f(m_h) \le \frac{-\int_{x+h}^{x} f(t)\, dt}{h} \le f(M_h).$$

Now take limits of this inequality as $h \to 0$:

$$\lim_{h \to 0^-} f(m_h) \le \lim_{h \to 0^-} \frac{-\int_{x+h}^{x} f(t)\, dt}{h} \le \lim_{h \to 0^-} f(M_h).$$

Since f is continuous, both $f(m_h)$ and $f(M_h)$ approach $f(x)$ as $h \to 0$, while the middle term is, from the equation at the start of this proof, equal to $F'(x)$. Thus

$$f(x) \le F'(x) \le f(x),$$

so that $F'(x) = f(x)$.

67. Suppose f is continuous on $[a, b]$. Define $F(x) = \int_a^x f(t)\, dt$ for all $x \in [a, b]$. Then by the Second Fundamental Theorem, $F'(x) = f(x)$ for $x \in [a, b]$; in other words, for $x \in [a, b]$,

$$\frac{d}{dx} \int_a^x f(t)\, dt = f(x).$$

69. (a) If f is continuous on $[a, b]$, define $F(x) = \int_a^x f(t)\, dt$. Then by the Second Fundamental Theorem, F is continuous on $[a, b]$ and differentiable on (a, b), and also $F'(x) = f(x)$.

   (b) Since F is continuous on $[a, b]$ and differentiable on (a, b), it satisfies the conditions of the Mean Value Theorem (Theorem 3.5), so that there is at least one $c \in (a, b)$ such that

$$F'(c) = \frac{F(b) - F(a)}{b - a}.$$

   (c) $F'(c) = f(c)$, and

$$\frac{F(b) - F(a)}{b - a} = \frac{\int_a^b f(t)\, dt - \int_a^a f(t)\, dt}{b - a} = \frac{1}{b - a} \int_a^b f(t)\, dt$$

   since the second integral is zero. Putting these together gives

$$f(c) = \frac{1}{b - a} \int_a^b f(t)\, dt.$$

71. Given $r \in (0, \infty)$, we must show that $\ln x$ is continuous and differentiable at $x = r$. Let $F_r(x) = \int_{r/2}^x \frac{1}{t}\, dt$ on $\left[\frac{r}{2}, 2r\right]$; by the Second Fundamental Theorem, F_r is continuous on $\left[\frac{r}{2}, 2r\right]$ and differentiable at $\left(\frac{r}{2}, 2r\right)$, so in particular it is continuous and differentiable at r. However,

$$\ln x = \int_1^x \frac{1}{t}\, dt = \int_1^{r/2} \frac{1}{t}\, dt + \int_{r/2}^x \frac{1}{t}\, dt = F_r(x) + \int_1^{r/2} \frac{1}{t}\, dt.$$

Thus $\ln x = F_r(x)$ plus a constant, so that it too is continuous and differentiable at $x = r$.

73. To see that $\ln 1 = 0$, note that

$$\ln 1 = \int_1^1 \frac{1}{t}\, dt = 0$$

since the lower and upper bounds of integration are equal. If $0 < x < 1$, then

$$\ln x = \int_1^x \frac{1}{t}\, dt = -\int_x^1 \frac{1}{t}\, dt.$$

Since $0 < x < 1$ and $\frac{1}{t}$ is positive for $t > 0$, we see that $\int_x^1 \frac{1}{t}\, dt > 0$, so that $\ln x < 0$. Finally, if $x > 1$, then

$$\ln x = \int_1^x \frac{1}{t}\, dt.$$

Since $x > 1$ and $\frac{1}{t} > 0$ for $t > 0$, the integral is positive, so that $\ln x > 0$.

75. (a) By definition, $\ln ax = \int_1^{ax} \frac{1}{t}\, dt$. By Theorem 4.35, this integral is equal to $\frac{1}{ax} \cdot (ax)' = \frac{a}{ax} = \frac{1}{x}$.

    (b) Since $\ln ax$ and $\ln x$ have the same derivative, they must differ by a constant C, so that $\ln ax = \ln x + C$.

    (c) Evaluating at $x = 1$ gives $\ln a = \ln 1 + C = c$, so that $C = \ln a$ and thus $\ln ax = \ln a + \ln x$ for any x. So, given $a, b > 0$, we can substitute b for x into this equation (since it holds for any x) to get $\ln(ab) = \ln a + \ln b$.

77. Using the two previous exercises, we have

$$\ln\left(\frac{a}{b}\right) = \ln\left(ab^{-1}\right) \overset{75}{=} \ln a + \ln\left(b^{-1}\right) \overset{76}{=} \ln a + (-1)\ln b = \ln a - \ln b.$$

Thinking Forward

Differential Equations

▷ In order for f to exist, $g(x)$ must be continuous, for then we can define $f(x) = \int_a^x g(t)\, dt$ and use the Second Fundamental Theorem.

▷ Since $-\cos x$ is an antiderivative of $\sin x$, the solution is the family of functions $f(x) = C - \cos x$ where C is an arbitrary constant.

▷ The solution is the family of functions $C + \int_0^x \sin(e^t)\, dt$, where C is an arbitrary constant, by the Second Fundamental Theorem and using the fact that any two antiderivatives differ by a constant. An alternative way of expressing these solutions is $\int_C^x \sin(e^t)\, dt$ where C is an arbitrary constant.

▷ The general solution to this equation is $f(x) = Ce^x$ (see Theorem 2.14).

Initial-Value Problems

▷ Since f_1 and f_2 are both antiderivatives of f', they differ by a constant, say $f_2(x) = f_1(x) + C$. If $f_2(a) = f_1(a)$ for some a, then $f_2(a) = f_1(a) = f_1(a) + C$ so that $C = 0$ and $f_1 = f_2$.

▷ Since the family of solutions is $f(x) = C - \cos x$, the condition $f(\pi) = 0$ means that $C - \cos \pi = 0$, so that $C = -1$ and the solution is $f(x) = -1 - \cos x$.

▷ Using the form $f(x) = \int_C^x \sin(e^t)\, dt$ and applying the initial condition, we substitute π for x to get $f(\pi) = \int_C^\pi \sin(e^t)\, dt = 0$, so that $C = \pi$.

▷ Since $f(x) = Ce^x$, applying the initial condition gives $f(0) = Ce^0 = C = 3$, so that $C = 3$ and the solution is $f(x) = 3e^x$.

Chapter Review and Self-Test

1. $\displaystyle\sum_{k=1}^{10} \frac{1}{k^2} = 1 + \frac{1}{4} + \frac{1}{9} + \cdots + \frac{1}{81} + \frac{1}{100} \approx 1.5498.$

3. $\displaystyle\sum_{k=1}^{200}(k^2 + 1) = \sum_{k=1}^{200} k^2 + \sum_{k=1}^{200} 1 = \frac{200(201)(401)}{6} + 200 = 2686900.$

5.

$$\lim_{n\to\infty} \sum_{k=1}^{n} \frac{k^2 + 1}{n^3} = \lim_{n\to\infty}\left(\frac{1}{n^3} \sum_{k=1}^{n}(k^2 + 1)\right)$$

$$= \lim_{n\to\infty}\left(\frac{1}{n^3}\left(\sum_{k=1}^{n} k^2 + \sum_{k=1}^{n} 1\right)\right)$$

$$= \lim_{n\to\infty}\left(\frac{1}{n^3}\left(\frac{n(n+1)(2n+1)}{6} + n\right)\right)$$

$$= \lim_{n\to\infty} \frac{2n^3 + 3n^2 + 7n}{6n^3}$$

$$= \lim_{n\to\infty} \frac{2 + 3/n + 7/n^2}{6} = \frac{1}{3}.$$

7. With $n = 4$, since the interval width is 4, we have $\Delta x = 1$, and $x_k = k$ for $k \in \{0, 1, 2, 3, 4\}$. Then the left sum is

$$\sum_{k=1}^{4} \sqrt{x_{k-1}}\Delta x = 0 + 1 + \sqrt{2} + \sqrt{3} \approx 4.146.$$

9. With $n = 3$, since the interval width is 3, we have $\Delta x = 1$, and $x_k = k$ for $k \in \{0, 1, 2, 3\}$. Then the interval midpoints are $x_k^* = \frac{1}{2}, \frac{3}{2}, \frac{5}{2}$ for $k = 0, 1, 2$, and the midpoint sum is

$$\sum_{k=0}^{2}(9 - (x_k^*)^2)\Delta x = \left(9 - \frac{1}{4}\right) + \left(9 - \frac{9}{4}\right) + \left(9 - \frac{25}{4}\right) = \frac{73}{4}.$$

11. With $n = 3$, since the interval width is 3, we have $\Delta x = 1$, and $x_k = k$ for k an integer from 0 to 3. Since $f'(x) = 4 - 2x$ is positive on $[0, 2]$ and negative on $[2, 3]$, we see that f is increasing on $[0, 2]$ and decreasing on $[2, 3]$, so that have for the upper sum

$$\sum_{k=1}^{3} f(x_k^*)\Delta x = f(1) + f(2) + f(2) = 11.$$

13. Using a right Riemann sum, since the interval width is 2, we have $\Delta x = \frac{2}{n}$ and $x_k = 3 + \frac{2k}{n}$ for k an integer from 0 to n. Then the right Riemann sum is

$$
\begin{aligned}
\sum_{k=1}^{n} f(x_k)\Delta x &= \frac{2}{n} \sum_{k=1}^{n} \left(2 \cdot \left(3 + \frac{2k}{n} \right) + 4 \right) \\
&= \frac{2}{n} \sum_{k=1}^{n} \left(\frac{4}{n}k + 10 \right) \\
&= \frac{8}{n^2} \sum_{k=1}^{n} k + \frac{20}{n} \sum_{k=1}^{n} 1 \\
&= \frac{8n(n+1)}{2n^2} + 20 \\
&= \frac{48n^2 + 8n}{2n^2} = 24 + \frac{4}{n}.
\end{aligned}
$$

Thus

$$
\int_{3}^{5} (2x + 4)\, dx = \lim_{n\to\infty} \sum_{k=1}^{n} f(x_k)\Delta x = \lim_{n\to\infty} \left(24 + \frac{4}{n} \right) = 24.
$$

15. Using a right Riemann sum, since the interval width is 5, we have $\Delta x = \frac{5}{n}$ and $x_k = -2 + \frac{5k}{n}$ for k an integer from 0 to n. Then the right Riemann sum is

$$
\begin{aligned}
\sum_{k=1}^{n} f(x_k)\Delta x &= \frac{5}{n} \sum_{k=1}^{n} \left(4 - \left(-2 + \frac{5k}{n} \right)^2 \right) \\
&= \frac{5}{n} \sum_{k=1}^{n} \left(\frac{20}{n}k - \frac{25}{n^2}k^2 \right) \\
&= \frac{100}{n^2} \sum_{k=1}^{n} k - \frac{125}{n^3} \sum_{k=1}^{n} k^2 \\
&= \frac{100n(n+1)}{2n^2} - \frac{125n(n+1)(2n+1)}{6n^3} \\
&= \frac{300n^2(n+1) - 125n(n+1)(2n+1)}{6n^3} \\
&= \frac{50n^3 - 75n^2 - 125n}{6n^3}.
\end{aligned}
$$

Thus

$$
\int_{-2}^{3} (4 - x^2)\, dx = \lim_{n\to\infty} \sum_{k=1}^{n} f(x_k)\Delta x = \lim_{n\to\infty} \frac{50n^3 - 75n^2 - 125n}{6n^3} = \lim_{n\to\infty} \frac{50 - 75/n - 125/n^2}{6}
$$

$$
= \frac{25}{3}.
$$

17. We know how to antidifferentiate powers of x, so

$$
\int (3x^4 - x^3 + 2)\, dx = \frac{3}{5}x^5 - \frac{1}{4}x^4 + 2x + C.
$$

19. Simplifying gives $e^x(1 - e^{3x}) = e^x - e^{4x}$, and we know how to integrate e^{kx} for any k, so that

$$
\int e^x(1 - e^{3x})\, dx = \int (e^x - e^{4x})\, dx = e^x - \frac{1}{4}e^{4x} + C.
$$

21. This looks like the derivative of a log function; trying $\ln(1 + 4x)$ gives a derivative of $\frac{4}{1+4x}$, which is too large by a factor of $\frac{4}{3}$. Thus

$$\int \frac{3}{1 + 4x}\, dx = \frac{3}{4}\ln(1 + 4x) + C.$$

23. The integrand looks like the result of applying the product rule where one factor is x^2 and the other is 3^x. Differentiating $x^2 3^x$ shows that this gives the correct result, so that

$$\int \left(2x 3^x + x^2 (\ln 3) 3^x\right)\, dx = x^2 3^x + C.$$

25. Since the derivative of $\tanh x$ is $\operatorname{sech}^2 x$, we have

$$\int 3\operatorname{sech}^2 x\, dx = 3\tanh x + C.$$

27. Since an antiderivative of $x^2 - 1$ is $\frac{1}{3}x^3 - x$, we see that

$$\int_{-2}^{2} (x^2 - 1)\, dx = \left[\frac{1}{3}x^3 - x\right]_{-2}^{2} = \frac{4}{3}.$$

A graph of this function, with the relevant region shaded, is below:

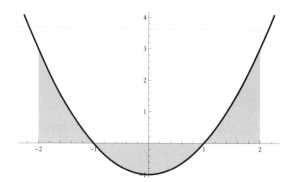

It appears that the areas above the x axis are slightly larger than the region below, so this answer is reasonable.

29. Since an antiderivative of $\sin 2x$ is $-\frac{1}{2}\cos 2x$, we see that

$$\int_{-\pi}^{\pi} \sin 2x\, dx = \left[-\frac{1}{2}\cos 2x\right]_{-\pi}^{\pi} = 0.$$

A graph of this function, with the relevant region shaded, is below:

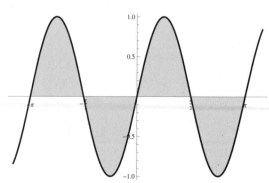

It is clear that the positive and negative regions balance, so that this is the correct answer.

31. Since an antiderivative of $(x-1)(x-2) = x^2 - 3x + 2$ is $\frac{1}{3}x^3 - \frac{3}{2}x^2 + 2x$, we see that

$$\int_0^4 (x-1)(x-2)\, dx = \left[\frac{1}{3}x^3 - \frac{3}{2}x^2 + 2x\right]_0^4 = \frac{16}{3}$$

A graph of this function, with the relevant region shaded, is below:

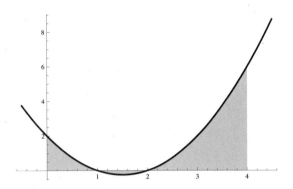

Judging from the graph, the answer seems reasonable.

33. Since $2x$ is the derivative of $1 + x^2$, this is the derivative of a log function, specifically $\frac{1}{2}\ln(1+x^2)$. Thus

$$\int_{-1}^{1} \frac{x}{1+x^2}\, dx = \left[\frac{1}{2}\ln(1+x^2)\right]_{-1}^{1} = \frac{1}{2}\ln 2 - \frac{1}{2}\ln 2 = 0.$$

A graph of this function, with the relevant region shaded, is below:

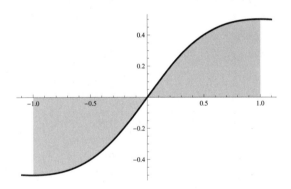

Judging from the graph, it seems clear that the positive and negative regions cancel.

35. The signed area is just the integral, so it is

$$\int_0^4 (3x^2 - 7x + 2)\, dx = \left[x^3 - \frac{7}{2}x^2 + 2x\right]_0^4 = 16.$$

37. Note that an antiderivative of $\sin 2x$ is $-\frac{1}{2}\cos 2x$, so that the average value of $\sin 2x$ on $[0, \pi]$ is

$$\frac{1}{\pi - 0}\int_0^\pi \sin 2x\, dx = \frac{1}{\pi}\left[-\frac{1}{2}\cos 2x\right]_0^\pi = \frac{1}{\pi}\left(-\frac{1}{2} - \frac{1}{2}\right) = -\frac{1}{\pi}.$$

39. Since $\frac{1}{4}x^4 + C$ is an antiderivative of x^3, we have

$$\frac{d}{dx} \int x^3 \, dx = \frac{d}{dx}\left(\frac{1}{4}x^4 + C\right) = x^3.$$

41. Since $\ln(x^2 + 1)$ is clearly an antiderivative of $\frac{d}{dx}\ln(x^2 + 1)$, we have

$$\int_{-2}^{2} \frac{d}{dx}\ln(x^2 + 1)\,dx = \left[\ln(x^2 + 1)\right]_{-2}^{2} = \ln 5 - \ln 5 = 0.$$

43. Since $\int_0^3 e^{-t^2}\,dt$ is a constant, its derivative is zero.

45. By Theorem 4.35,

$$\frac{d}{dx}\int_0^{\ln x} \sin^3 t\,dt = \sin^3(\ln x)\cdot(\ln x)' = \frac{1}{x}\sin^3(\ln x).$$

Chapter 5

Techniques of Integration

5.1   Integration by Substitution

Thinking Back

Algebra review

▷ $\sqrt{x}(x^{-2} - 5x^{2/3}) = x^{1/2}(x^{-2} - 5x^{2/3}) = x^{-2+1/2} - 5x^{2/3+1/2} = x^{-3/2} - 5x^{7/6}$.

▷ $\ln(5x^3) + 3\ln(2^x) = \ln(5x^3) + \ln((2^x)^3) = \ln(5x^3 \cdot 2^{3x}) = \ln(5x^3 2^{3x})$.

▷ $\frac{x+1}{2\sqrt[3]{x}} = \frac{1}{2}\left(x \cdot x^{-1/3} + x^{-1/3}\right) = \frac{1}{2}x^{2/3} + \frac{1}{2}x^{-1/3}$.

▷ $\frac{e^{3x} - 5e^{-2x}}{3e^{2x}} = \frac{1}{3} \cdot \frac{e^{3x}}{e^{2x}} - \frac{5}{3} \cdot \frac{e^{-2x}}{e^{2x}} = \frac{1}{3}e^x - \frac{5}{3}e^{-4x}$.

▷ $\sec x \tan x \cos^2 x = \frac{1}{\cos x} \cdot \frac{\sin x}{\cos x} \cdot \cos^2 x = \sin x$.

▷ $\sqrt{1 - \sin^2 x} = \sqrt{\cos^2 x} = |\cos x|$.

▷ $\frac{2\sqrt{x}-5}{3\sqrt{x}} = \frac{2}{3} - \frac{5}{3\sqrt{x}} = \frac{2}{3} - \frac{5}{3}x^{-1/2}$.

▷ $2^x(1 - 3^x) = 2^x - 2^x 3^x = 2^x - 6^x$.

Differentiation review

▷ $f'(x) = -\frac{1}{2}\left(-\csc^2 x^2\right)(2x) = x\csc^2 x^2$.

▷ $f'(x) = \frac{1}{3} \cdot \frac{1}{3x+1} \cdot (3x)' = \frac{1}{3x+1}$.

▷ $f'(x) = \frac{1}{2\pi}(2\sin \pi x)(\sin \pi x)' = \frac{1}{\pi}\sin \pi x (\cos \pi x \cdot \pi) = \sin \pi x \cos \pi x$.

▷ $f'(x) = \frac{2}{3} \cdot \frac{3}{2}(\ln x)^{1/2}(\ln x)' = \frac{\sqrt{\ln x}}{x}$.

▷ $f'(x) = \frac{1}{2}\left(\sin x^2\right)^{-1/2}(\sin x^2)' = \frac{1}{2}\left(\sin x^2\right)^{-1/2}\cos x^2 \cdot 2x = x\cos x^2 \left(\sin x^2\right)^{-1/2}$.

▷ $f'(x) = \frac{1}{2} \cdot 2\sin^{-1} x \cdot (\sin^{-1} x)' = \frac{\sin^{-1} x}{\sqrt{1-x^2}}$.

▷

$$f'(x) = -\sin(\sin^2 x^3) \cdot (\sin^2 x^3)' = -\sin(\sin^2 x^3) \cdot (2\sin x^3) \cdot (\sin x^3)'$$
$$= -2\sin x^3 \sin(\sin^2 x^3) \cdot \cos x^3 \cdot 3x^2 = -6x^2 \cos x^3 \sin x^3 \sin(\sin^2 x^3).$$

▷ $f'(x) = \frac{1}{\ln(\ln x)} \cdot (\ln(\ln x))' = \frac{1}{\ln(\ln x)} \cdot \frac{1}{\ln x} \cdot (\ln x)' = \frac{1}{x \ln x \ln(\ln x)}$.

Antidifferentiation review

▷ Using the power rule, an antiderivative of $\sqrt{x} + \sqrt[3]{x} = x^{1/2} + x^{1/3}$ is $\frac{2}{3}x^{3/2} + \frac{3}{4}x^{4/3}$.

▷ Since the derivative of tan is $\sec^2$, start with $\tan 5x$; this has derivative $5 \sec^2 5x$, which is too large. Correct for the extra factor to get an antiderivative of $\frac{1}{5}\tan 5x$.

▷ $\frac{2}{3x} = \frac{2}{3} \cdot \frac{1}{x}$, so that an antiderivative is $\frac{2}{3}\ln x$.

▷ Since the derivative of $\cos \pi x$ is $-\pi \sin \pi x$, an antiderivative of the given function is $-\frac{1}{2\pi}\cos \pi x$.

▷ $\frac{3}{x^2+1} = 3 \cdot \frac{1}{x^2+1}$, so an antiderivative is $3\tan^{-1} x$.

▷ $\frac{-2}{\sqrt{1-x^2}} = -2 \cdot \frac{1}{\sqrt{1-x^2}}$, so an antiderivative is $-2\sin^{-1} x$.

Expressing geometric quantities with integrals

▷ The signed area between the graph of a function and the x axis is the integral of that function, so the answer is $\int_0^{3\pi/2} \sin x \, dx$.

▷ The absolute area between the graph of a function and the x axis is the integral of the absolute value of that function, so the answer is $\int_0^{3\pi/2} |\sin x| \, dx$.

▷ Again the signed area between the graphs is just the integral of the difference, so the area between the curves is $\int_{-5}^{5}(2^x - x^2) \, dx$.

▷ The average value is $\frac{1}{10-1} \int_1^{10} \frac{1}{x} \, dx = \frac{1}{9} \int_1^{10} \frac{1}{x} \, dx$.

Concepts

1. (a) True, as long as g and h are continuous and differentiable, since $(g(h(x)))' = g'(h(x))h'(x)$ by the chain rule.

(b) False. If $v = u^2 + 1$, then $dv = 2u \, du$, so that $du = \frac{1}{2u} dv = \frac{1}{2\sqrt{v-1}} dv$, and then $\int \sqrt{u^2+1} \, du = \int \frac{\sqrt{v}}{2\sqrt{v-1}} \, dv$.

(c) False. The change of variables from x to u must also be applied to the factor of x, not only to $\sin(x^3)$.

(d) False. The left-hand side is the same thing as $\int_0^3 x^2 \, dx$, but that does not equal the right-hand side since on the right, the integral runs from $x = 0$ to $x = 3$, not from $u(x) = 0$ to $u(x) = 3$. For example, let $u(x) = x^2$, so that $du = 2x \, dx$; then the right-hand side is $\int_{x=0}^{x=3}(u(x))^2 \, du = \int_{x=0}^{x=3} x^4 \cdot 2x \, dx$, which clearly is not the same thing as $\int_0^3 x^2 \, dx$.

(e) True. x and u are both dummy variables; the name actually used does not matter.

(f) False. With the substitution $u = x^2 - 1$, we get $du = 2x \, dx$, so that $x \, dx = \frac{1}{2} du$, and then $\int_2^4 x e^{x^2-1} \, dx = \int_2^4 e^{x^2-1}(x \, dx) = \frac{1}{2} \int_{x=2}^{x=4} e^u \, du$. The statement is true except that the bounds of integration have not been changed to reflect the new variable, u.

(g) True. The right-hand side is $\int_{u(2)}^{u(3)} f(u) \, du = \int_{u(2)}^{u(3)} f(u(x)) \, du(x)$. But as u goes from $u(2)$ to $u(3)$, x goes from 2 to 3, so that this is in turn equal to $\int_2^3 f(u(x))u'(x) \, dx$.

(h) False. $\int_0^6 f(u(x))u'(x)\,dx = \left[\int_f(u)\,du\right]_{u(0)}^{u(6)}$.

3. Using the substitution $u = x^2 + 1$ in the first integral gives $du = 2x\,dx$, so that the integral becomes $\int \frac{1}{u}\,du$. Using the substitution $u = \ln x$ in the second integral gives $du = \frac{1}{x}\,dx$, so that the integral becomes $\int \frac{1}{u}\,du$, which is the same as the first integral after substitution.

5. For example,

$$\int 2x \sin x^2\,dx \quad \text{after the substitution} \quad u = x^2,\ du = 2x\,dx$$

$$\int e^x \sin(e^x) \quad \text{after the substitution} \quad u = e^x,\ du = e^x\,dx$$

$$\int \cos x \sin(\sin x)\,dx \quad \text{after the substitution} \quad u = \sin x,\ du = \cos x\,dx.$$

7. For example,

$$\int \frac{\cos x}{\sqrt{1 + \sin x}}\,dx \quad \text{after the substitution} \quad u = 1 + \sin x,\ du = \cos x\,dx$$

$$\int \frac{e^x}{\sqrt{e^x - 5}}\,dx \quad \text{after the substitution} \quad u = e^x - 5,\ du = e^x\,dx$$

$$\int \frac{3x^2}{\sqrt{x^3 + 4}}\,dx \quad \text{after the substitution} \quad u = x^3 + 4,\ du = 3x^2\,dx.$$

9. Since $\frac{du}{dx} = 2x + 1$, we have $du = (2x + 1)\,dx$.

11. Since $\frac{du}{dx} = \cos x$, we have $du = \cos x\,dx$.

13. The integral of u^2 with respect to u is $\frac{1}{3}u^3$, so that

$$\int_{-1}^5 u^2\,du = \left[\frac{1}{3}u^3\right]_{-1}^5 = \frac{125}{3} - \frac{-1}{3} = 42.$$

For the second integral, note that $x = -1$ corresponds to $u = 1$, and $x = 5$ corresponds to $u = 25$, so that

$$\int_{x=-1}^{x=5} u^2\,du = \int_1^{25} u^2\,du = \left[\frac{1}{3}u^3\right]_1^{25} = \frac{15625}{3} - \frac{1}{3} = 5208.$$

Finally, the third integral is the same as the second, since $u(-1) = 1$ and $u(5) = 25$, so that its value is also 5208.

15. For example, Exercises 21, 22, and 51. There are other examples.

17. (a) Use the substitution $u = x^2 - 1$, so that $du = 2x\,dx$ and then $x\,dx = \frac{1}{2}\,du$. Then

$$\int x(x^2 - 1)^2\,dx = \int \frac{1}{2}u^2\,du = \frac{1}{6}u^3 + C = \frac{1}{6}(x^2 - 1)^3 + C_1.$$

(b) The integrand is $x(x^2 - 1)^2 = x(x^4 - 2x^2 + 1) = x^5 - 2x^3 + x$, so that

$$\int x(x^2 - 1)^2\,dx = \int (x^5 - 2x^3 + x)\,dx = \frac{1}{6}x^6 - \frac{1}{2}x^4 + \frac{1}{2}x^2 + C_2.$$

(c) Since both answers are antiderivatives of $x(x^2-1)^2$, they must differ by a constant. And in fact

$$\frac{1}{6}(x^2-1)^3 + C_1 - \left(\frac{1}{6}x^6 - \frac{1}{2}x^4 + \frac{1}{2}x^2 + C_2\right)$$

$$= \frac{1}{6}(x^6 - 3x^4 + 3x^2 - 1) + C_1 - \left(\frac{1}{6}x^6 - \frac{1}{2}x^4 + \frac{1}{2}x^2 + C_2\right)$$

$$= \frac{1}{6}x^6 - \frac{1}{2}x^4 + \frac{1}{2}x^2 - \frac{1}{6} + C_1 - \left(\frac{1}{6}x^6 - \frac{1}{2}x^4 + \frac{1}{2}x^2 + C_2\right)$$

$$= -\frac{1}{6} + C_1 - C_2,$$

which is a constant.

19. (a) Use the substitution $u = 5 - 3x$, so that $du = -3\,dx$ and thus $dx = -\frac{1}{3}\,du$. Then

$$\int_{-2}^{2} e^{5-3x}\,dx = \int_{x=-2}^{x=2} \left(-\frac{1}{3}e^u\right) du = -\frac{1}{3}\left[e^u\right]_{x=-2}^{x=2} = -\frac{1}{3}\left[e^{5-3x}\right]_{x=-2}^{x=2} = -\frac{1}{3}e^{-1} + \frac{1}{3}e^{11}.$$

(b) Use the same substitution. When $x = -2$, we have $u = 5 - 3(-2) = 11$; when $x = 2$, we have $u = 5 - 3(2) = -1$. Then

$$\int_{-2}^{2} e^{5-3x}\,dx = \int_{11}^{-1} \left(-\frac{1}{3}e^u\right) du = -\frac{1}{3}\left[e^u\right]_{11}^{-1} = -\frac{1}{3}e^{-1} + \frac{1}{3}e^{11}.$$

Skills

21. This can be done either way. With substitution, use $u = 3x + 1$ so that $du = 3\,dx$. Or, expand the integrand:

$$\int (3x+1)^2\,dx = \int (9x^2 + 6x + 1)\,dx = 3x^3 + 3x^2 + x + C.$$

23. Use the substitution $u = x^2 + 1$, so that $du = 2x\,dx$. Then

$$\int \frac{8x}{x^2+1}\,dx = 4\int \frac{2x}{x^2+1}\,dx = 4\int \frac{1}{u}\,du = 4\ln|u| + C = 4\ln(x^2+1) + C.$$

25. Simplify the integrand and integrate using the power rule:

$$\int \frac{\sqrt{x}+3}{2\sqrt{x}}\,dx = \int \left(\frac{1}{2} + \frac{3}{2}x^{-1/2}\right) dx = \frac{1}{2}x + 3x^{1/2} + C.$$

27. Use the substitution $u = x^2$ so that $du = 2x\,dx$ and $\frac{1}{2}\,du = x\,dx$. Then

$$\int x\csc^2 x^2\,dx = \int \left(\frac{1}{2}\csc^2 u\right) du = -\frac{1}{2}\cot u + C = -\frac{1}{2}\cot x^2 + C.$$

29. Use the substitution $u = 3x + 1$, so that $du = 3\,dx$ and $dx = \frac{1}{3}\,du$. Then

$$\int \frac{1}{3x+1}\,dx = \frac{1}{3}\int \frac{1}{u}\,du = \frac{1}{3}\ln|u| + C = \frac{1}{3}\ln|3x+1| + C.$$

31. Use the substitution $u = \sin \pi x$, so that $du = \pi\cos \pi x\,dx$ and $\frac{1}{\pi}\,du = \cos \pi x\,dx$. Then

$$\int \sin \pi x \cos \pi x\,dx = \int \frac{1}{\pi}u\,du = \frac{1}{2\pi}u^2 + C = \frac{1}{2\pi}\sin^2 \pi x + C.$$

33. Let $u = 2x$, so that $du = 2\,dx$ and $\frac{1}{2}\,du = dx$. Then

$$\int \sec 2x \tan 2x\,dx = \frac{1}{2}\int \sec u \tan u\,du = \frac{1}{2}\sec u + C = \frac{1}{2}\sec 2x + C.$$

35. Since $\cot' = -\csc^2$, use the substitution $u = \cot x$. Then $du = -\csc^2 x\,dx$, so that

$$\int \cot^5 x \csc^2\,dx = -\int u^5\,du = -\frac{1}{6}u^6 + C = -\frac{1}{6}\cot^6 x + C.$$

37. Expand the integrand and use the power rule:

$$\int x^4(x^3+1)^2\,dx = \int (x^{10} + 2x^7 + x^4)\,dx = \frac{1}{11}x^{11} + \frac{1}{4}x^8 + \frac{1}{5}x^5 + C.$$

39. With $u = x^{5/4}$, we have $du = \frac{5}{4}x^{1/4}\,dx$, so that $x^{1/4}\,dx = \frac{4}{5}\,du$. Then

$$\int x^{1/4} \sin x^{5/4}\,dx = \frac{4}{5}\int \sin u\,du = -\frac{4}{5}\cos u + C = -\frac{4}{5}\cos x^{5/4} + C.$$

41. With $u = x^2 + 1$, we have $du = 2x\,dx$, so that $x\,dx = \frac{1}{2}\,du$. Then

$$\int x(2^{x^2+1})\,dx = \int \frac{1}{2}\cdot 2^u\,du = \frac{1}{2\ln 2}2^u + C = \frac{1}{2\ln 2}2^{x^2+1} + C = \frac{1}{\ln 2}2^{x^2} + C.$$

43. With $u = \ln x$, we have $du = \frac{1}{x}\,dx$. Then

$$\int \frac{\sqrt{\ln x}}{x}\,dx = \int \sqrt{u}\,du = \frac{2}{3}u^{3/2} + C = \frac{2}{3}(\ln x)^{3/2} + C.$$

45. With $u = \ln\sqrt{x} = \ln x^{1/2}$, we have $du = \frac{1}{x^{1/2}}\cdot\frac{1}{2}x^{-1/2}\,dx = \frac{1}{2}\cdot\frac{1}{x}\,dx$, so that $\frac{1}{x}\,dx = 2\,du$. Then

$$\int \frac{\ln\sqrt{x}}{x}\,dx = 2\int u\,du = 2\cdot\frac{1}{2}u^2 + C = u^2 + C = \ln^2\sqrt{x} + C.$$

An alternate approach is to simplify the numerator: $\ln\sqrt{x} = \frac{1}{2}\ln x$. Then use the substitution $u = \ln x$, so that $du = \frac{1}{x}\,dx$, and the integral is

$$\int \frac{\ln\sqrt{x}}{x}\,dx = \frac{1}{2}\int \frac{\ln x}{x}\,dx = \frac{1}{2}\int u\,du = \frac{1}{2}\cdot\frac{1}{2}u^2 + C = \frac{1}{4}\ln^2 x + C.$$

Note that these two answers are the same.

47. First simplify the integrand: $\frac{3x+1}{x^2+1} = \frac{3x}{x^2+1} + \frac{1}{x^2+1}$. The second term has an antiderivative of $\tan^{-1} x$. For the first term, use the substitution $u = x^2 + 1$ so that $du = 2x\,dx$ and $\frac{1}{2}\,du = x\,dx$. Then

$$\int \frac{3x+1}{x^2+1}\,dx = \tan^{-1}x + 3\int \frac{1}{2u}\,du + C = \tan^{-1}x + \frac{3}{2}\ln|u| + C = \tan^{-1}x + \frac{3}{2}\ln(x^2+1) + C.$$

49. Let $u = 2 - e^x$, so that $du = -e^x\,dx$. Then

$$\int \frac{e^x}{2-e^x}\,dx = -\int \frac{1}{u}\,du = -\ln|u| + C = -\ln|2 - e^x| + C.$$

51. Simplify the integrand and integrate:

$$\int \frac{2 - e^x}{e^x}\, dx = \int (2e^{-x} - 1)\, dx = -2e^{-x} - x + C.$$

53. Let $u = \cos x$, so that $du = -\sin x\, dx$. Then

$$\int \sin x \cos^7 x\, dx = -\int u^7\, du = -\frac{1}{8}u^8 + C = -\frac{1}{8}\cos^8 x + C.$$

55. Let $u = \sin x^2$. Then by the chain rule, $du = \cos x^2 \cdot 2x\, dx = 2x \cos x^2$, so that $\frac{1}{2}\, du = x \cos x^2$. Then

$$\int \frac{x \cos x^2}{\sqrt{\sin x^2}}\, dx = \frac{1}{2}\int \frac{1}{\sqrt{u}}\, du = u^{1/2} + C = \sqrt{\sin x^2} + C.$$

57. Let $u = \frac{1}{x}$, so that $du = -\frac{1}{x^2}\, dx$. Then

$$\int \frac{\sin\left(\frac{1}{x}\right)}{x^2}\, dx = -\int \sin u\, du = \cos u + C = \cos\left(\frac{1}{x}\right) + C.$$

59. Let $u = x^2 + 1$, so that $du = 2x\, dx$ and $\frac{1}{2}\, du = x\, dx$. Then

$$\int x\sqrt{x^2 + 1}\, dx = \frac{1}{2}\int u^{1/2}\, du = \frac{1}{3}u^{3/2} + C = \frac{1}{3}(x^2 + 1)^{3/2} + C.$$

61. Let $u = x + 1$, so that $du = dx$. Then also $x = u - 1$, and

$$\begin{aligned}
\int x\sqrt{x + 1}\, dx &= \int (u - 1)\sqrt{u}\, du \\
&= \int \left(u^{3/2} - u^{1/2}\right) du \\
&= \frac{2}{5}u^{5/2} - \frac{2}{3}u^{3/2} + C \\
&= \frac{2}{5}(x + 1)^{5/2} - \frac{2}{3}(x + 1)^{3/2} + C.
\end{aligned}$$

63. Let $u = 3x + 1$, so that $du = 3\, dx$. Also, $x = \frac{1}{3}(u - 1)$. Then

$$\begin{aligned}
\int \frac{x}{\sqrt{3x + 1}}\, dx &= \int \frac{\frac{1}{3}(u - 1)}{\sqrt{u}} \cdot \frac{1}{3}\, du \\
&= \frac{1}{9}\int (u - 1)u^{-1/2}\, du \\
&= \frac{1}{9}\int (u^{1/2} - u^{-1/2})\, du \\
&= \frac{2}{27}u^{3/2} - \frac{2}{9}u^{1/2} + C \\
&= \frac{2}{27}(3x + 1)^{3/2} - \frac{2}{9}(3x + 1)^{1/2} + C.
\end{aligned}$$

65. The integrand is $e^x \sin e^x$. Let $u = e^x$, so that $du = e^x\, dx$. Then

$$\int \frac{e^x}{\csc e^x}\, dx = \int (\sin e^x)(e^x\, dx) = \int \sin u\, du = -\cos u + C = -\cos e^x + C.$$

67. This could be done by expanding, but it would be pretty ugly. Instead, let $u = x - 1$, so that $du = dx$ and $x = u + 1$ so that $x^2 = u^2 + 2u + 1$. Then

$$
\begin{aligned}
\int x^2 (x-1)^{100}\, dx &= \int (u+1)^2 u^{100}\, du \\
&= \int (u^{102} + 2u^{101} + u^{100})\, du \\
&= \frac{1}{103} u^{103} + \frac{1}{51} u^{102} + \frac{1}{101} u^{101} + C \\
&= \frac{1}{103}(x-1)^{103} + \frac{1}{51}(x-1)^{102} + \frac{1}{101}(x-1)^{101} + C.
\end{aligned}
$$

69. Let $u = x^2$, so that $du = 2x\, dx$ and $\frac{1}{2}\, du = x\, dx$. Then

$$
\int x \sinh x^2\, dx = \frac{1}{2} \int \sinh u\, du = \frac{1}{2} \cosh u + C = \frac{1}{2} \cosh x^2 + C.
$$

71. Let $u = x^2 + 1$ so that $du = 2x\, dx$ and $\frac{1}{2}\, du = x\, dx$. Then

$$
\int_0^3 x(x^2+1)^{1/3}\, dx = \frac{1}{2} \int_{x=0}^{x=3} u^{1/3}\, du = \frac{1}{2}\left[\frac{3}{4} u^{4/3} \right]_{x=0}^{x=3} = \frac{1}{2}\left[\frac{3}{4}(x^2+1)^{4/3} \right]_{x=0}^{x=3} = \frac{3}{8}\left(10^{4/3} - 1 \right).
$$

73. Let $u = \sin x$, so that $du = \cos x\, dx$. Then

$$
\int_{-\pi}^{\pi} \sin x \cos x\, dx = \int_{x=-\pi}^{x=\pi} u\, du = \left[\frac{1}{2} u^2 \right]_{x=-\pi}^{x=\pi} = \left[\frac{1}{2} \sin^2 x \right]_{x=-\pi}^{x=\pi} = \frac{1}{2}\left(\sin^2 \pi - \sin^2(-\pi) \right) = 0.
$$

75. Let $u = \ln x$; then $du = \frac{1}{x}\, dx$ and

$$
\int_2^3 \frac{1}{x\sqrt{\ln x}}\, dx = \int_{x=2}^{x=3} \frac{1}{\sqrt{u}}\, du = \left[2\sqrt{u} \right]_{x=2}^{x=3} = \left[2\sqrt{\ln x} \right]_{x=2}^{x=3} = 2(\sqrt{\ln 3} - \sqrt{\ln 2}).
$$

77. Let $u = x^2$ so that $du = 2x\, dx$ and $\frac{1}{2}\, du = x\, dx$. Then

$$
\begin{aligned}
\int_0^1 \frac{x}{1+x^4}\, dx = \int_0^1 \frac{x}{1+(x^2)^2}\, dx = \frac{1}{2} \int_{x=0}^{x=1} \frac{1}{1+u^2}\, du &= \frac{1}{2}\left[\tan^{-1} u \right]_{x=0}^{x=1} \\
&= \frac{1}{2}\left[\tan^{-1} x^2 \right]_{x=0}^{x=1} = \frac{\pi}{8}.
\end{aligned}
$$

79. (a) The signed area under the curve is just the integral $\int_{-1}^{3} f(x)\, dx$. Let $u = x^2 + 1$, so that $du = 2x\, dx$ and $\frac{1}{2}\, du = x\, dx$. Then

$$
\begin{aligned}
\int_{-1}^{3} \frac{x}{x^2+1}\, dx = \frac{1}{2} \int_{x=-1}^{x=3} \frac{1}{u}\, du &= \frac{1}{2}\left[\ln |u| \right]_{x=-1}^{x=3} = \frac{1}{2}\left[\ln(x^2+1) \right]_{x=-1}^{x=3} \\
&= \frac{1}{2}(\ln 10 - \ln 2) = \frac{\ln 5}{2}.
\end{aligned}
$$

(We can remove the absolute value signs in the ln function since $x^2 + 1$ is always positive).

(b) Note that $g(x) \geq f(x)$ on $[-1,0]$ and on $[\sqrt{3},3]$, while $f(x) \geq g(x)$ on $[0,\sqrt{3}]$. Thus the area between the curves is

$$
\int_{-1}^{0} \left(\frac{x}{4} - \frac{x}{x^2+1} \right) dx - \int_{0}^{\sqrt{3}} \left(\frac{x}{4} - \frac{x}{x^2+1} \right) dx + \int_{\sqrt{3}}^{3} \left(\frac{x}{4} - \frac{x}{x^2+1} \right) dx.
$$

We proceed by finding an indefinite integral of the integrand, and then evaluating it. The first term of the integrand is easily integrable using the power rule, getting $\frac{1}{8}x^2$. For the second term, we get the same indefinite integral as in part (a), which is $\frac{1}{2}\ln(x^2+1)$. Thus the area between the curves is

$$\left[\frac{1}{8}x^2 - \frac{1}{2}\ln(x^2+1)\right]_{-1}^{0} - \left[\frac{1}{8}x^2 - \frac{1}{2}\ln(x^2+1)\right]_{0}^{\sqrt{3}} + \left[\frac{1}{8}x^2 - \frac{1}{2}\ln(x^2+1)\right]_{\sqrt{3}}^{3}$$

$$= 0 - \left(\frac{1}{8} - \frac{1}{2}\ln 2\right) - \left(\frac{1}{8}\cdot 3 - \frac{1}{2}\ln 4\right) + 0$$

$$+ \left(\frac{1}{8}\cdot 9 - \frac{1}{2}\ln 10\right) - \left(\frac{1}{8}\cdot 3 - \frac{1}{2}\ln 4\right)$$

$$= \frac{1}{4} + \frac{5}{2}\ln 2 - \frac{1}{2}\ln 10.$$

81.   (a) The average value of $f(x)$ on $\left[\frac{1}{2}, 2\right]$ is

$$\frac{1}{2-\frac{1}{2}}\int_{1/2}^{2} f(x)\,dx = \frac{2}{3}\int_{1/2}^{2} \frac{\ln x}{x}\,dx.$$

Let $u = \ln x$ so that $du = \frac{1}{x}\,dx$. Then

$$\frac{2}{3}\int_{1/2}^{2} \frac{\ln x}{x}\,dx = \frac{2}{3}\int_{x=1/2}^{x=2} u\,du$$

$$= \frac{2}{3}\left[\frac{1}{2}u^2\right]_{x=1/2}^{x=2}$$

$$= \frac{1}{3}\left[\ln^2 x\right]_{x=1/2}^{x=2}$$

$$= \frac{1}{3}\left(\ln^2 2 - \ln^2 \frac{1}{2}\right)$$

$$= \frac{1}{3}\left(\ln^2 2 - (-\ln 2)^2\right) = 0.$$

The average value of $f(x)$ is zero on $\left[\frac{1}{2}, 2\right]$.

(b) Since $\ln 1 = 0$, we see that $c = 1$ is such a point.

Applications

83. The amount of rainfall predicted is $\int_0^3 r(t)\, dt$. To integrate $r(t)$, let $u = 3 - t$, so that $du = -dt$. Further, $t = 3 - u$, so that $t^2 = 9 - 6u + u^2$. Then

$$\int_0^3 0.6t^2 \sqrt{3 - t}\, dt = -\int_{t=0}^{t=3} 0.6(9 - 6u + u^2)\sqrt{u}\, du$$

$$= -0.6 \int_{t=0}^{t=3} (9 - 6u + u^2)u^{1/2}\, du$$

$$= -0.6 \int_{t=0}^{t=3} \left(9u^{1/2} - 6u^{3/2} + u^{5/2}\right) du$$

$$= -0.6 \left[6u^{3/2} - \frac{12}{5}u^{5/2} + \frac{2}{7}u^{7/2}\right]_{t=0}^{t=3}$$

$$= -0.6 \left[6(3 - t)^{3/2} - \frac{12}{5}(3 - t)^{5/2} + \frac{2}{7}(3 - t)^{7/2}\right]_{t=0}^{t=3}$$

$$= 0.6 \left(6 \cdot 3^{3/2} - \frac{12}{5} \cdot 3^{5/2} + \frac{2}{7}3^{7/2}\right)$$

$$= 0.6 \left(18\sqrt{3} - \frac{108}{5}\sqrt{3} + \frac{54}{7}\sqrt{3}\right)$$

$$= 0.6 \cdot \frac{144}{35}\sqrt{3} \approx 4.276 \text{ inches.}$$

85. We must evaluate

$$\pi \int_a^b (f(x))^2\, dx = \pi \int_1^{10} \left(2\left(\frac{\ln x}{x}\right)^{1/2}\right)^2 dx = 4\pi \int_1^{10} \frac{\ln x}{x}\, dx.$$

To evaluate the integral, let $u = \ln x$, so that $du = \frac{1}{x}\, dx$. Then

$$4\pi \int_1^{10} \frac{\ln x}{x}\, dx = 4\pi \int_{x=1}^{x=10} u\, du = 4\pi \left[\frac{1}{2}u^2\right]_{x=1}^{x=10} = 2\pi \left[\ln^2 x\right]_{x=1}^{x=10} = 2\pi \ln^2 10 \text{ cu in.}$$

Proofs

87. (a) The signed area is

$$\int_0^{k\pi} \sin\left(2\left(x - \frac{\pi}{4}\right)\right) dx.$$

To evaluate this integral, let $u = 2\left(x - \frac{\pi}{4}\right)$, so that $du = 2\, dx$ and $\frac{1}{2}\, du = dx$. Then

$$\int_0^{k\pi} \sin\left(2\left(x - \frac{\pi}{4}\right)\right) dx = \frac{1}{2} \int_{x=0}^{x=k\pi} \sin u\, du$$

$$= -\frac{1}{2} \left[\cos u\right]_{x=0}^{x=k\pi}$$

$$= -\frac{1}{2} \left[\cos\left(2\left(x - \frac{\pi}{4}\right)\right)\right]_{x=0}^{x=k\pi}$$

$$= -\frac{1}{2} \left(\cos\left(2\left(k\pi - \frac{\pi}{4}\right)\right) - \cos\left(-\frac{\pi}{2}\right)\right)$$

$$= -\frac{1}{2} \left(\cos\left(2k\pi - \frac{\pi}{2}\right) - \cos\left(-\frac{\pi}{2}\right)\right)$$

$$= 0$$

since $\cos z = 0$ for z an odd multiple of $\frac{\pi}{2}$.

(b) The graph of $f(x)$ on $[0, 5\pi]$ is

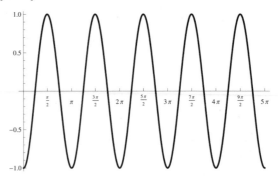

Note that between each multiple of π and the next, the positive area is exactly balanced by the two pieces of negative area, so that the net area under the curve is zero between any two integral multiples of π.

89. To prove this, we show that the integrand is the derivative of the right-hand side. But by the chain rule, $(f(u(x)) + C)' = (f(u(x)))' + C' = f'(u(x))u'(x)$. Since the integrand is the derivative of the right-hand side, by definition the right-hand side is an antiderivative of the integrand.

91. (a) Write $\tan x = \frac{\sin x}{\cos x}$, and then let $u = \cos x$ so that $du = -\sin x \, dx$. Then

$$\int \tan x \, dx = \int \frac{\sin x}{\cos x} \, dx = -\int \frac{1}{u} \, du = -\ln|u| + C = -\ln|\cos x| + C$$

$$= \ln \frac{1}{|\cos x|} + C = \ln|\sec x| + C.$$

(b) For $\sec x > 0$, we have

$$(\ln|\sec x|)' = (\ln \sec x)' = \frac{1}{\sec x}(\sec x)' = \frac{\sec x \tan x}{\sec x} = \tan x,$$

while for $\sec x < 0$, we have

$$(\ln|\sec x|)' = (\ln(-\sec x))' = \frac{1}{-\sec x}(-\sec x)' = \frac{-\sec x \tan x}{-\sec x} = \tan x.$$

Thinking Forward

Trigonometric integrals

▷ We have $\sin^5 x = (\sin^2 x)^2 \sin x = (1 - \cos^2 x)^2 \sin x$. Then writing $u = \cos x$, so that $du = -\sin x \, dx$, we have

$$\int \sin^5 x \, dx = \int (1 - \cos^2 x)^2 \sin x \, dx$$

$$= -\int (1 - u^2)^2 \, du$$

$$= -\int (1 - 2u^2 + u^4) \, du$$

$$= -\frac{1}{5}u^5 + \frac{2}{3}u^3 - u + C$$

$$= -\frac{1}{5}\cos^5 x + \frac{2}{3}\cos^3 x - \cos x + C.$$

▷ We have $\sin^3 x \cos^4 x = \sin^2 x \sin x \cos^4 x = (1 - \cos^2 x)\cos^4 x \sin x$. Then writing $u = \cos x$, so that $du = -\sin x\, dx$, we have

$$\int \sin^3 x \cos^4 x\, dx = \int (1 - \cos^2 x)\cos^4 x \sin x\, dx$$

$$= -\int (1 - u^2)u^4\, du$$

$$= \int (u^6 - u^4)\, du$$

$$= \frac{1}{7}u^7 - \frac{1}{5}u^5 + C$$

$$= \frac{1}{7}\cos^7 x - \frac{1}{5}\cos^5 x + C.$$

▷ We have $\sec^4 x \tan^3 x = \sec^2 x \sec^2 x \tan^3 x = (\tan^2 x + 1)\tan^3 x \sec^2 x$. Then writing $u = \tan x$, so that $du = \sec^2 x\, dx$, we have

$$\int \sec^4 x \tan^3 x\, dx = \int (\tan^2 x + 1)\tan^3 x \sec^2 x\, dx$$

$$= \int (u^2 + 1)u^3\, du$$

$$= \int (u^5 + u^3)\, du$$

$$= \frac{1}{6}u^6 + \frac{1}{4}u^4 + C$$

$$= \frac{1}{6}\tan^6 x + \frac{1}{4}\tan^4 x + C.$$

5.2   Integration by Parts

Thinking Back

Differentiation review

▷ $f'(x) = (xe^x)' - (e^x)' = e^x + xe^x - e^x = xe^x.$

▷ $f'(x) = \frac{1}{2}\left(x^2 \ln x\right)' - \frac{1}{4}(x^2)' = \frac{1}{2}\left(2x \ln x + x^2 \frac{1}{x}\right) - \frac{1}{4}2x = x \ln x.$

▷

$$f'(x) = -\left(e^{-x}(x^2 + 1)\right)' - 2(xe^{-x})' - 2(e^{-x})'$$

$$= -e^{-x}(2x) + e^{-x}(x^2 + 1) - 2(e^{-x} - xe^{-x}) + 2e^{-x}$$

$$= (x^2 + 1)e^{-x}.$$

▷ $f'(x) = \frac{1}{3}(x^3)' - 2(xe^x - e^x)' + \frac{1}{2}(e^{2x})' = x^2 - 2(e^x + xe^x - e^x) + \frac{1}{2}2e^{2x} = x^2 - 2xe^x + e^{2x}.$

▷ Using the result of Exercise 92 from the previous section,

$$f'(x) = -(x \cot x)' + (\ln |\sin x|)' = -\cot x + x \csc^2 x - (\ln |\csc x|)'$$

$$= -\cot x + x \csc^2 x + \cot x = x \csc^2 x.$$

▷

$$f'(x) = (x^3 \sin x)' + 3(x^2 \cos x)' - 6(x \sin x)' - 6(\cos x)'$$
$$= 3x^2 \sin x + x^3 \cos x + 3(2x \cos x - x^2 \sin x) - 6(\sin x + x \cos x) + 6 \sin x$$
$$= x^3 \cos x.$$

Review of integration by substitution

▷ Let $u = 3x + 1$, so that $du = 3\,dx$ and $\frac{1}{3}\,du = dx$. Then

$$\int e^{3x+1}\,dx = \frac{1}{3}\int e^u\,du = \frac{1}{3}e^u + C = \frac{1}{3}e^{3x+1} + C.$$

▷ Let $u = x^2 + 1$, so that $du = 2x\,dx$ and $\frac{1}{2}\,du = x\,dx$. Then

$$\int xe^{x^2+1}\,dx = \frac{1}{2}\int e^u\,du = \frac{1}{2}e^u + C = \frac{1}{2}e^{x^2+1} + C.$$

▷ Let $u = \ln x$, so that $du = \frac{1}{x}\,dx$. Then

$$\int \frac{\ln x}{x}\,dx = \int u\,du = \frac{1}{2}u^2 + C = \frac{1}{2}\ln^2 x + C.$$

▷ Let $u = \ln x$, so that $du = \frac{1}{x}\,dx$. Then

$$\int \frac{1}{x\ln x}\,dx = \int \frac{1}{u}\,du = \ln|u| + C = \ln|\ln x| + C.$$

▷ Let $u = \sec x$, so that $du = \tan x \sec x\,dx$. Then

$$\int \tan x \sec^2 x\,dx = \int \sec x(\tan x \sec x\,dx) = \int u\,du = \frac{1}{2}u^2 + C = \frac{1}{2}\sec^2 x + C.$$

An equally good approach is to let $u = \tan x$ so that $du = \sec^2 x$. This gives the equivalent solution $\frac{1}{2}\tan^2 x + C$; the two are equivalent since $\sec^2 x - \tan^2 x = 1$, which is a constant.

▷ No substitution is required; we recognize this as $\tan^{-1} x + C$.

▷ Let $u = \cos x$, so that $du = -\sin x\,dx$. Then

$$\int \sin x\sqrt{\cos x}\,dx = -\int \sqrt{u}\,du = -\frac{2}{3}u^{3/2} + C = -\frac{2}{3}(\cos x)^{3/2} + C.$$

▷ Let $u = e^x$, so that $du = e^x\,dx$. Then

$$\int e^x \sin e^x\,dx = \int \sin u\,du = -\cos u + C = -\cos e^x + C.$$

Concepts

1. (a) False. In fact $du = 2x\,dx$.

    (b) Technically this is false, since v is the antiderivative of x^2, so it must be $v = \frac{1}{3}x^3 + C$. However, in doing integration by parts, we choose a particular antiderivative, say $\frac{1}{3}x^3$, and use that as v.

(c) False. We would have to use $u = \frac{1}{x}$, since that is what appears in the integrand.

(d) False. The integral of dv is the integral of $\ln x$, which is more complicated than $\ln x$ itself.

(e) True. See Theorem 5.6.

(f) False. As a general rule, it is only useful when separating the integrand into two factors fg has the properties that f' is simpler than f and the integral of g is not more complicated than g.

(g) True. For example, when integrating $\ln x$, this is the correct procedure.

(h) False. The bounds on xe^x were not given. The correct answer is

$$\int_0^3 xe^x \, dx = [xe^x]_0^3 - \int_0^3 e^x \, dx.$$

3. (a) $\int u \, dv = uv - \int v \, du$.

   (b) $\int (u'(x)v(x) + u(x)v'(x)) \, dx = u(x)v(x) + C$.

5. For the first, consider $\int xe^x \, dx$. With $u = x$ and $dv = e^x \, dx$, the resulting integral is $\int e^x \, dx$. For the second, consider $\int x \ln x \, dx$. With $u = \ln x$ and $dv = x \, dx$, the resulting integral becomes $\int \frac{1}{2}x \, dx$.

7. For example, $\int \ln x \, dx$; see the text for details.

9. To apply integration by parts, we must be able to integrate whatever we choose for dv. If we choose that to be the entire integrand, then to apply integration by parts to the integrand, we must first be able to integrate the integrand. But this integration is exactly what we are trying to achieve in the first place.

11. For example, Exercises 27, 28, 30, 35, and 51, as well as the first step in Exercise 65.

13. If $u(x) = \sin 3x$, then $du = 3\cos 3x \, dx$. If $v(x) = x$, then $dv = 1 \, dx = dx$. Then $\int u \, dv = \int (\sin 3x \cdot dx) = \int \sin 3x \, dx$, while $\int v \, du = \int (x \cdot 3\cos 3x \, dx) = \int 3x \cos 3x \, dx$. Clearly the first integral is easier to find than the second. Thus if we are faced with something like the second integral, say $\int x \cos 3x \, dx$, it makes sense to apply integration by parts with $u = x$ and $dv = \cos 3x \, dx$.

15. We have

$$[g(x)]_a^b - [h(x)]_a^b = g(b) - g(a) - (h(b) - h(a)) = (g(b) - h(b)) - (g(a) - h(a))$$
$$= [g(x) - h(x)]_a^b.$$

Thus if we are faced with an integral like $\int (g(x) + h(x))k(x) \, dx$, and we can integrate $g(x)k(x)$ and $h(x)k(x)$ using integration by parts, we know that the original integral will be the sum of the two results.

17. (a) $\frac{d}{dx}(u(x)v(x)) = u(x)v'(x) + v(x)u'(x) = \ln x + x \cdot \frac{1}{x} = 1 + \ln x$.

    (b) $\int (1 + \ln x) \, dx = u(x)v(x) + C = x \ln x + C$.

    (c) $\int u \, dv = x \ln x - \int x \cdot \frac{1}{x} \, dx = x \ln x - \int 1 \, dx$.

19. Since uv contains a factor of x while $\int v \, du$ does not, we presume that $u = x$, so that $v = \frac{2^x}{\ln 2}$. Thus $du = dx$ and $dv = \frac{d}{dx} \frac{2^x}{\ln 2} = 2^x$. So the original integral is

$$\int x2^x \, dx = \frac{x2^x}{\ln 2} - \frac{1}{\ln 2} \int 2^x \, dx.$$

21. Since the $\ln x$ disappears between uv and $\int v\,du$, we must have $u = \ln x$ and $v = -x^{-2}$. Then $du = \frac{1}{x}\,dx$ and $dv = 2x^{-3}$, so that the original integral is

$$\int \left(2\frac{\ln x}{x^3} \right) dx = -\frac{\ln x}{x^2} - \int \left(-x^{-2} \cdot \frac{1}{x} \right) dx = -\frac{\ln x}{x^2} + \int \frac{1}{x^3}\,dx.$$

23. Clearly $u = \tan^{-1} x$ since then $du = \frac{1}{x^2+1}\,dx$. Then $v = x$ and $dv = dx$, so that the original integral was

$$\int \tan^{-1} x\,dx = x \tan^{-1} x - \int \frac{x}{x^2+1}\,dx.$$

25.   (a) Using integration by parts with $u = \ln(x^3)$ and $dv = x^2\,dx$ gives $du = \frac{1}{x^3} \cdot 3x^2\,dx = \frac{3}{x}\,dx$ while $v = \frac{1}{3}x^3$. Thus

$$\int x^2 \ln(x^3)\,dx = uv - \int v\,du = \frac{1}{3}x^3 \ln(x^3) - \frac{1}{3}\int \left(x^3 \cdot \frac{3}{x} \right) dx = \frac{1}{3}x^3 \ln(x^3) - \int x^2\,dx$$

$$= \frac{1}{3}x^3(\ln(x^3) - 1) + C.$$

(b) Using u-substitution with $u = x^3$ gives $du = 3x^2\,dx$ so that $\frac{1}{3}\,du = x^2\,dx$. Then

$$\int x^2 \ln(x^3)\,dx = \frac{1}{3}\ln u\,du = \frac{1}{3}(u \ln u - u) + C = \frac{1}{3}(x^3 \ln(x^3) - x^3) + C$$

$$= \frac{1}{3}x^3(\ln(x^3) - 1) + C.$$

(c) The two answers are identical up to a constant, as they must be since they are both antiderivatives of $x^2 \ln(x^3)$.

Skills

27. Use integration by parts with $u = x$ and $dv = e^x\,dx$, so that $du = dx$ and $v = e^x$. Then

$$\int xe^x\,dx = uv - \int v\,du = xe^x - \int e^x\,dx = xe^x - e^x + C.$$

29. Use integration by parts with $u = \ln x$ and $dv = x\,dx$, so that $du = \frac{1}{x}\,dx$ and $v = \frac{1}{2}x^2\,dx$. Then

$$\int x \ln x\,dx = uv - \int v\,du = \frac{1}{2}x^2 \ln x - \frac{1}{2}\int x^2 \cdot \frac{1}{x}\,dx = \frac{1}{2}x^2 \ln x - \frac{1}{2}\int x\,dx$$

$$= \frac{1}{2}x^2 \ln x - \frac{1}{4}x^2 + C.$$

31. Use u-substitution with $u = x^2$, so that $du = 2x\,dx$ and thus $\frac{1}{2}\,du = x\,dx$. Then

$$\int x \sin x^2\,dx = \frac{1}{2}\int \sin u\,du = -\frac{1}{2}\cos u + C = -\frac{1}{2}\cos x^2 + C.$$

33. Use integration by parts with $u = x^2$ and $dv = e^{3x}\,dx$, so that $du = 2x\,dx$ and $v = \frac{1}{3}e^{3x}$. Then

$$\int x^2 e^{3x}\,dx = uv - \int v\,du = \frac{1}{3}x^2 e^{3x} - \frac{2}{3}\int xe^{3x}\,dx.$$

To evaluate this integral, again use integration by parts, with $u = x$ and $dv = e^{3x}\,dx$, so that $du = dx$ and $v = \frac{1}{3}e^{3x}$. Then

$$\int x^2 e^{3x}\,dx = \frac{1}{3}x^2 e^{3x} - \frac{2}{3}\int xe^{3x}\,dx = \frac{1}{3}x^2 e^{3x} - \frac{2}{3}\left(\frac{1}{3}xe^{3x} - \frac{1}{3}\int e^{3x}\,dx\right)$$
$$= \frac{1}{3}x^2 e^{3x} - \frac{2}{9}xe^{3x} + \frac{2}{27}e^{3x} + C.$$

35. Use integration by parts with $u = x$ and $dv = e^{-x}\,dx$, so that $du = dx$ and $v = -e^{-x}$. Then

$$\int \frac{x}{e^x}\,dx = -xe^{-x} + \int e^{-x}\,dx = -xe^{-x} - e^{-x} + C.$$

37. Use u-substitution with $u = x^2$, so that $du = 2x\,dx$ and $\frac{1}{2}\,du = x\,dx$. Then

$$\int 3xe^{x^2}\,dx = \frac{3}{2}\int e^u\,du = \frac{3}{2}e^u + C = \frac{3}{2}e^{x^2} + C.$$

39. Rewrite this as $\ln(x^3) = 3\ln x$, and then use integration by parts with $u = \ln x$ and $dv = dx$, so that $du = \frac{1}{x}\,dx$ and $v = x$. Then

$$\int \ln(x^3)\,dx = 3\int \ln x\,dx = 3\left(x\ln x - \int x \cdot \frac{1}{x}\,dx\right) = 3(x\ln x - x) + C.$$

41. Rewrite the integrand as $\frac{x^2+1}{e^x} = \frac{x^2}{e^x} + e^{-x}$. Integrating the second term is easy. Integrating the first term is just Exercise 36, so we use integration by parts twice starting with $u = x^2$ and $dv = e^{-x}\,dx$, to finally get the answer

$$\int \frac{x^2+1}{e^x}\,dx = -e^{-x}(x^2 + 2x + 2) - e^{-x} + C = -e^{-x}(x^2 + 2x + 3) + C.$$

43. This can be done using algebra and basic integration rules:

$$\int \frac{3x+1}{x}\,dx = \int \left(3 + \frac{1}{x}\,dx\right) = \int 3\,dx + \int \frac{1}{x}\,dx = 3x + \ln|x| + C.$$

45. We have

$$\int (x - e^x)^2\,dx = \int (x^2 - 2xe^x + e^{2x})\,dx = \frac{1}{3}x^3 + \frac{1}{2}e^{2x} - 2\int xe^x\,dx + C.$$

To evaluate the remaining integral, use integration by parts with $u = x$ and $dv = e^x\,dx$, so that $du = 1$ and $v = e^x$. Then

$$\int xe^x\,dx = xe^x - \int e^x\,dx = xe^x - e^x + C,$$

so that the original integral is

$$\int (x - e^x)^2\,dx = \frac{1}{3}x^3 + \frac{1}{2}e^{2x} - 2(xe^x - e^x + C) = \frac{1}{3}x^3 + \frac{1}{2}e^{2x} - 2xe^x + 2e^x + C_1.$$

47. Use integration by parts with $u = \ln x$ and $dv = \sqrt{x}\,dx$, so that $du = \frac{1}{x}\,dx$ and $v = \frac{2}{3}x^{3/2}$. Then

$$\int \sqrt{x}\ln x\,dx = \frac{2}{3}x^{3/2}\ln x - \frac{2}{3}\int x^{1/2}\,dx = \frac{2}{3}\left(x^{3/2}\ln x - \frac{2}{3}x^{3/2}\right) + C.$$

49. Use integration by parts with $u = x^2$ and $dv = xe^{x^2}$, so that $du = 2x\, dx$ and $v = \frac{1}{2}e^{x^2}$. Then

$$\int x^3 e^{x^2}\, dx = \frac{1}{2}x^2 e^{x^2} - \int xe^{x^2}\, dx.$$

Now use a u-substitution with $u = x^2$, so that $\frac{1}{2}du = x\, dx$, and thus

$$\int x^3 e^{x^2}\, dx = \frac{1}{2}x^2 e^{x^2} - \int xe^{x^2}\, dx = \frac{1}{2}x^2 e^{x^2} - \frac{1}{2}\int e^u\, du = \frac{1}{2}(x^2 e^{x^2} - e^u) + C = \frac{1}{2}e^{x^2}(x^2 - 1) + C.$$

51. Use integration by parts with $u = x$ and $dv = \csc^2 x\, dx$, so that $du = dx$ and $v = -\cot x$. Then

$$\int x \csc^2 x\, dx = -x \cot x + \int \cot x\, dx = -x \cot x - \ln|\csc x| + C,$$

using the results of Exercise 92 in section 1.

53. Use integration by parts with $u = x^3$ and $dv = \cos x\, dx$, so that $du = 3x^2\, dx$ and $v = \sin x$. Then

$$\int x^3 \cos x\, dx = x^3 \sin x - 3 \int x^2 \sin x\, dx.$$

Use integration by parts again with $u = x^2$ and $dv = \sin x\, dx$, so that $du = 2x\, dx$ and $v = -\cos x$. Then

$$\int x^3 \cos x\, dx = x^3 \sin x - 3 \int x^2 \sin x\, dx$$

$$= x^3 \sin x - 3\left(-x^2 \cos x + 2 \int x \cos x\, dx\right)$$

$$= x^3 \sin x + 3x^2 \cos x - 6 \int x \cos x\, dx.$$

Now use integration by parts a third time with $u = x$ and $dv = \cos x\, dx$, so that $du = dx$ and $v = \sin x$. Then

$$\int x^3 \cos x\, dx = x^3 \sin x + 3x^2 \cos x - 6 \int x \cos x\, dx$$

$$= x^3 \sin x + 3x^2 \cos x - 6\left(x \sin x - \int \sin x\, dx\right)$$

$$= x^3 \sin x + 3x^2 \cos x - 6x \sin x - 6 \cos x + C.$$

55. Use integration by parts with $u = x^2$ and $dv = x \sin x^2\, dx$, so that $du = 2x\, dx$ and $v = -\frac{1}{2}\cos x^2$. Then

$$\int x^3 \sin x^2\, dx = -\frac{1}{2}x^2 \cos x^2 + \int x \cos x^2\, dx = -\frac{1}{2}x^2 \cos x^2 + \frac{1}{2}\sin x^2 + C.$$

57. Use integration by parts with $u = \tan^{-1} 3x$ and $dv = dx$, so that $du = \frac{3}{1+9x^2}\, dx$ and $v = x$. Then

$$\int \tan^{-1} 3x\, dx = x \tan^{-1} 3x - \int \frac{3x}{1+9x^2}\, dx.$$

Now use u-substitution with $u = 1 + 9x^2$, so that $du = 18x\, dx$ and then $\frac{1}{6}du = 3x\, dx$. Then we get

$$\int \tan^{-1} 3x\, dx = x \tan^{-1} 3x - \int \frac{3x}{1+9x^2}\, dx$$

$$= x \tan^{-1} 3x - \frac{1}{6}\frac{1}{u}\, du$$

$$= x \tan^{-1} 3x - \frac{1}{6}\ln|u| + C$$

$$= x \tan^{-1} 3x - \frac{1}{6}\ln(1+9x^2) + C.$$

59. Use integration by parts with $u = \sin x$ and $dv = e^{2x}\,dx$, so that $du = \cos x\,dx$ and $v = \frac{1}{2}e^{2x}$. Then

$$\int e^{2x} \sin x\,dx = \frac{1}{2}e^{2x}\sin x - \frac{1}{2}\int e^{2x}\cos x\,dx.$$

Now use integration by parts again with $u = \cos x$ and $dv = e^{2x}\,dx$, so that $du = -\sin x\,dx$ and $v = \frac{1}{2}e^{2x}$. Then

$$\int e^{2x} \cos x\,dx = \frac{1}{2}e^{2x}\cos x + \frac{1}{2}\int e^{2x}\sin x\,dx.$$

Plugging this result into the first equation gives

$$\int e^{2x} \sin x\,dx = \frac{1}{2}e^{2x}\sin x - \frac{1}{2}\left(\frac{1}{2}e^{2x}\cos x + \frac{1}{2}\int e^{2x}\sin x\,dx\right)$$

$$= \frac{1}{2}e^{2x}\sin x - \frac{1}{4}e^{2x}\cos x - \frac{1}{4}\int e^{2x}\sin x\,dx.$$

We can now solve for $\int e^{2x} \sin x\,dx$ to get

$$\int e^{2x} \sin x\,dx = \frac{2}{5}e^{2x}\sin x - \frac{1}{5}e^{2x}\cos x + C.$$

61. The integrand is $\frac{2^x}{3^x} = \left(\frac{2}{3}\right)^x$, so that

$$\int \frac{2^x}{3^x}\,dx = \int \left(\frac{2}{3}\right)^x dx = \frac{1}{\ln(2/3)}\left(\frac{2}{3}\right)^x + C.$$

63. Use integration by parts with $u = \cos x$ and $dv = e^{-x}\,dx$, so that $du = -\sin x\,dx$ and $v = -e^{-x}$. Then

$$\int e^{-x} \cos x\,dx = -e^{-x}\cos x - \int e^{-x}\sin x\,dx.$$

Now use integration by parts again with $u = \sin x$ and $dv = e^{-x}\,dx$, so that $du = \cos x\,dx$ and $v = -e^{-x}$. Then

$$\int e^{-x} \sin x\,dx = -e^{-x}\sin x + \int e^{-x}\cos x\,dx.$$

Plugging this result into the first equation gives

$$\int e^{-x} \cos x\,dx = -e^{-x}\cos x + e^{-x}\sin x - \int e^{-x}\cos x\,dx,$$

and solving for the integral gives

$$\int e^{-x} \cos x\,dx = \frac{1}{2}e^{-x}(\sin x - \cos x) + C.$$

65. The integrand is $x\sec^2 x$. Use integration by parts with $u = x$ and $dv = \sec^2 x\,dx$, so that $du = dx$ and $v = \tan x$. Then

$$\int \frac{x}{\cos^2 x}\,dx = x\tan x - \int \tan x\,dx = x\tan x - \ln|\sec x| + C$$

by Exercise 91 in section 1.

67. Use integration by parts with $u = \tan^{-1} x$ and $dv = x\,dx$, so that $du = \frac{1}{1+x^2}\,dx$ and $v = \frac{1}{2}x^2$. Then

$$\int x \tan^{-1} x\,dx = \frac{1}{2}x^2 \tan^{-1} x - \frac{1}{2}\int \frac{x^2}{1+x^2}\,dx$$

$$= \frac{1}{2}x^2 \tan^{-1} x - \frac{1}{2}\int \left(1 - \frac{1}{1+x^2}\right)\,dx$$

$$= \frac{1}{2}x^2 \tan^{-1} x - \frac{x}{2} + \frac{1}{2}\tan^{-1} x + C.$$

69. Use integration by parts with $u = x$ and $dv = \frac{x}{\sqrt{x^2+1}}\,dx$. Then $du = dx$, and (after a u-substitution if desired with $u = x^2 + 1$), $v = \sqrt{x^2+1}$. Thus

$$\int \frac{x^2}{\sqrt{x^2+1}}\,dx = x\sqrt{x^2+1} - \int \sqrt{x^2+1}\,dx$$

$$= x\sqrt{x^2+1} - \int \frac{x^2+1}{\sqrt{x^2+1}}\,dx$$

$$= x\sqrt{x^2+1} - \int \frac{1}{\sqrt{x^2+1}}\,dx - \int \frac{x^2}{\sqrt{x^2+1}}\,dx$$

$$= x\sqrt{x^2+1} - \sinh^{-1} x - \int \frac{x^2}{\sqrt{x^2+1}}\,dx.$$

Solving for the integral gives

$$\int \frac{x^2}{\sqrt{x^2+1}}\,dx = \frac{1}{2}x\sqrt{x^2+1} - \frac{1}{2}\sinh^{-1} x + C.$$

71. Use integration by parts with $u = \ln x$ and $dv = dx$, so that $du = \frac{1}{x}\,dx$ and $v = x$. Then

$$\int_1^2 \ln x\,dx = [x \ln x]_1^2 - \int_1^2 x \cdot \frac{1}{x}\,dx = 2\ln 2 - [x]_1^2 = \ln 4 - 1.$$

73. Use integration by parts with $u = x$ and $dv = e^{-x}\,dx$, so that $du = dx$ and $v = -e^{-x}$. Then

$$\int_{-1}^1 \frac{x}{e^x}\,dx = \left[-xe^{-x}\right]_{-1}^1 + \int_{-1}^1 e^{-x}\,dx = -\frac{1}{e} - e - \left[e^{-x}\right]_{-1}^1 = -e^{-1} - e - e^{-1} + e = -2e^{-1}$$

75. Use integration by parts with $u = \sin x$ and $dv = e^x\,dx$, so that $du = \cos x\,dx$ and $v = e^x$. Then

$$\int_0^\pi e^x \sin x\,dx = [e^x \sin x]_0^\pi - \int_0^\pi e^x \cos x\,dx = -\int_0^\pi e^x \cos x\,dx.$$

Now use integration by parts again with $u = \cos x$ and $dv = e^x\,dx$, so that $du = -\sin x\,dx$ and $v = e^x$. Then

$$\int_0^\pi e^x \sin x\,dx = -\int_0^\pi e^x \cos x\,dx = -\left([e^x \cos x]_0^\pi + \int_0^\pi e^x \sin x\,dx\right) = -\int_0^\pi e^x \sin x\,dx + e^\pi + 1.$$

Solving for the integral gives

$$\int_0^\pi e^x \sin x\,dx = \frac{1}{2}(e^\pi + 1).$$

77. Use integration by parts with $u = x$ and $dv = \csc^2 x\, dx$, so that $du = dx$ and $v = -\cot x$. Then (using Exercise 92 from section 1)

$$\int_{\pi/4}^{\pi/2} x \csc^2 x\, dx = -\left[x \cot x\right]_{\pi/4}^{\pi/2} + \int_{\pi/4}^{\pi/2} \cot x\, dx$$

$$= \frac{\pi}{4} - \left[\ln|\csc x|\right]_{\pi/4}^{\pi/2}$$

$$= \frac{\pi}{4} - \ln 1 + \ln \sqrt{2} = \frac{\pi}{4} + \frac{\ln 2}{2}.$$

79. We will need to compute the integral of xe^{-x}. To compute the indefinite integral, use integration by parts with $u = x$ and $dv = e^{-x}\, dx$, so that $du = dx$ and $v = -e^{-x}$. Then

$$\int xe^{-x}\, dx = -xe^{-x} + \int e^{-x}\, dx = -xe^{-x} - e^{-x} + C.$$

(a) The signed area is just the integral from -1 to 1:

$$\int_{-1}^{1} xe^{-x}\, dx = \left[-xe^{-x} - e^{-x}\right]_{-1}^{1} = -e^{-1} - e^{-1} - e + e = -\frac{2}{e}.$$

(b) Note that $xe^{-x} > 0$ for $x > 0$, and $xe^{-x} < 0$ for $x < 0$, so that the absolute area is

$$\int_{-1}^{0}(-xe^{-x})\, dx + \int_{0}^{1} xe^{-x}\, dx = \left[xe^{-x} + e^{-x}\right]_{-1}^{0} + \left[-xe^{-x} - e^{-x}\right]_{0}^{1}$$

$$= 1 + e - e - e^{-1} - e^{-1} + 1$$

$$= 2 - \frac{2}{e}.$$

81. We will need to integrate $x \cos 2x$. To find the indefinite integral, use integration by parts with $u = x$ and $dv = \cos 2x\, dx$, so that $du = dx$ and $v = \frac{1}{2} \sin 2x$. Then

$$\int x \cos 2x\, dx = \frac{1}{2} x \sin 2x - \frac{1}{2} \int \sin 2x\, dx = \frac{1}{2} x \sin 2x + \frac{1}{4} \cos 2x + C.$$

(a) The signed area is just the integral from 0 to $\frac{3\pi}{4}$, so is

$$\int_{0}^{3\pi/4} x \cos 2x\, dx = \left[\frac{1}{2} x \sin 2x + \frac{1}{4} \cos 2x\right]_{0}^{3\pi/4}$$

$$= \frac{1}{2} \cdot \frac{3\pi}{4} \sin \frac{3\pi}{4} + \frac{1}{4} \cos \frac{3\pi}{2} - 0 - \frac{1}{4} \cos 0$$

$$= -\frac{3\pi}{8} - \frac{1}{4}.$$

(b) Note that $f(x) \geq g(x)$ for $0 \leq x \leq \frac{\pi}{4}$, and $g(x) \geq f(x)$ otherwise. Thus the absolute area is

$$\int_{0}^{\pi/4} (x \cos 2x - (-\cos 2x))\, dx + \int_{\pi/4}^{3\pi/4} (-\cos 2x - x \cos 2x)\, dx$$

$$= \int_{0}^{\pi/4} (x \cos 2x + \cos 2x)\, dx - \int_{\pi/4}^{3\pi/4} (x \cos 2x + \cos 2x)\, dx.$$

Since the integral of $\cos 2x$ is $\frac{1}{2}\sin 2x$, we get for the absolute area

$$\left[\frac{1}{2}x\sin 2x + \frac{1}{4}\cos 2x + \frac{1}{2}\sin 2x\right]_0^{\pi/4} - \left[\frac{1}{2}x\sin 2x + \frac{1}{4}\cos 2x + \frac{1}{2}\sin 2x\right]_{\pi/4}^{3\pi/4}$$

$$= \frac{\pi}{8} + \frac{1}{2} - \frac{1}{4} - \left(-\frac{3\pi}{8} - \frac{1}{2} \cdot \frac{\pi}{8} - \frac{1}{2}\right)$$

$$= \frac{5\pi}{8} + \frac{5}{4}.$$

Applications

83. First use integration by parts with $u = 0.6t^2$ and $dv = \sqrt{3-t}\,dt$, so that (integrating dv using the power rule), $du = 1.2t\,dt$ and $v = -\frac{2}{3}(3-t)^{3/2}$. Then

$$\int_0^3 r(t)\,dt = -\left[\frac{2}{3}\cdot 0.6t^2(3-t)^{3/2}\right]_0^3 + \frac{2.4}{3}\int_0^3 t(3-t)^{3/2}\,dt$$

$$= -\frac{2}{5}\left[t^2(3-t)^{3/2}\right]_0^3 + \frac{4}{5}\int_0^3 t(3-t)^{3/2}\,dt$$

$$= \frac{4}{5}\int_0^3 t(3-t)^{3/2}\,dt.$$

Now integrate again using $u = t$ and $dv = (3-t)^{3/2}\,dt$, so that $du = dt$ and $v = -\frac{2}{5}(3-t)^{5/2}$, and then

$$\int_0^3 r(t)\,dt = \frac{4}{5}\int_0^3 t(3-t)^{3/2}\,dt$$

$$= \frac{4}{5}\left[\left[-\frac{2}{5}t(3-t)^{5/2}\right]_0^3 + \frac{2}{5}\int_0^3 (3-t)^{5/2}\,dt\right]$$

$$= 0 + \frac{8}{25}\left[-\frac{2}{7}(3-t)^{7/2}\right]_0^3$$

$$= \frac{16}{175}\cdot 3^{7/2} = \frac{432}{175}\sqrt{3}.$$

This is identical to the answer from Exercise 83 in Section 5.1, although the answer was expressed slightly differently.

85. We must calculate $\pi\int_{0.5}^5 \ln^2 x\,dx$. To do this, use integration by parts with $u = \ln^2 x$ and $dv = dx$, so that $du = 2\ln x \cdot \frac{1}{x}\,dx = 2\frac{\ln x}{x}\,dx$ and $v = x$. Then (using Theorem 5.8, and noting that the wine is held only in the portion of the goblet to the right of $x = 1$)

$$\pi\int_{0.5}^5 \ln^2 x\,dx = \pi\left(\left[x\ln^2 x\right]_1^5 - 2\int_1^5 \left(x\cdot\frac{\ln x}{x}\right)dx\right)$$

$$= \pi\left(5\ln^2 5 - 2\int_1^5 \ln x\,dx\right)$$

$$= \pi\left(5\ln^2 5 - 2\left[x\ln x - x\right]_1^5\right)$$

$$= \pi\left(5\ln^2 5 - 10\ln 5 + 10 - 2\right)$$

$$= \pi\left(5\ln^2 5 - 10\ln 5 + 8\right).$$

This is about 15.259 cubic inches. If you insist on putting wine in the bottom of the goblet, between $x = 0.5$ and $x = 1$, as well, this adds 0.209 cubic inches to the answer, for a total of 15.468.

87. (a) The position of the particle at time t, measured as feet to the left or the right of the starting position, is simply $\int_0^t v(x)\, dx = \int_0^t 5x \sin \pi x\, dx$. To integrate, use integration by parts with $u = 5x$ and $dv = \sin \pi x\, dx$. Then $du = 5\, dx$ and $v = -\frac{1}{\pi} \cos \pi x$, so that

$$
\begin{aligned}
\int_0^t v(x)\, dx &= -\frac{5}{\pi}\, [x \cos \pi x]_0^t + \frac{5}{\pi} \int_0^t \cos \pi x\, dx \\
&= -\frac{5}{\pi} t \cos \pi t + \frac{5}{\pi^2} \sin \pi t \\
&= \frac{5}{\pi^2} (\sin \pi t - \pi t \cos \pi t).
\end{aligned}
$$

(b) A graph of $s(t)$ is

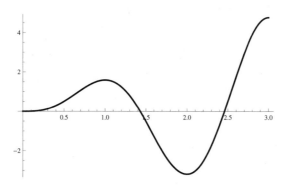

Until about 0.6 seconds, the particle is moving right and accelerating. At 0.6 seconds, it begins decelerating, but is still moving to the right until at $t = 1$ the velocity becomes negative and the particle starts moving back to the left. At about $t = 1.4$, the particle reaches its original starting point and continues moving to the left. At $t = 2$, the velocity becomes positive again, so that the particle starts moving back to the right; it reaches its original starting point again just before $t = 2.5$. It continues moving to the right until $t = 3$, although it is decelerating after about $t = 2.5$.

(c) At $t = 3$, the particle is $\frac{15}{\pi} \approx 4.77$ feet to the right of where it started. It has stopped moving to the right and is about to start moving back to the left.

(d) The particle moves back and forth with increasingly larger swings back and forth, due to the factor of t in the cosine term. Thus near $t = 100$, the amplitude of the swings will be approximately $\frac{5}{\pi^2} \cdot 100\pi \approx 159$.

Proofs

89. (a) Using integration by parts with $u = \ln x$ and $dv = dx$, we get $du = \frac{1}{x}\, dx$ and $v = x$, so that

$$
\int \ln x\, dx = x \ln x - \int x \cdot \frac{1}{x}\, dx = x \ln x - \int 1\, dx = x \ln x - x + C.
$$

(b) Differentiating $x \ln x - x$ gives

$$
(x \ln x - x)' = (x \ln x)' - x' = \ln x + x \cdot \frac{1}{x} - 1 = \ln x + 1 - 1 = \ln x.
$$

91. (a) Using integration by parts with $u = \tan^{-1} x$ and $dv = dx$ gives $du = \frac{1}{1+x^2}\, dx$ and $v = x$. Thus

$$
\int \tan^{-1} x\, dx = x \tan^{-1} x - \int \frac{x}{1 + x^2}\, dx.
$$

Now use a u-substitution with $u = 1 + x^2$, so that $du = 2x\,dx$ and $\frac{1}{2}\,du = x\,dx$. Then

$$\int \tan^{-1} x\,dx = x\tan^{-1} x - \int \frac{x}{1+x^2}\,dx$$
$$= x\tan^{-1} x - \frac{1}{2}\int \frac{1}{u}\,du$$
$$= x\tan^{-1} x - \frac{1}{2}\ln|u| + C$$
$$= x\tan^{-1} x - \frac{1}{2}\ln(x^2+1) + C.$$

(b) Differentiating gives

$$\left(x\tan^{-1} x - \frac{1}{2}\ln(x^2+1)\right)' = (x\tan^{-1} x)' - \frac{1}{2}(\ln(x^2+1))'$$
$$= \tan^{-1} x + \frac{x}{x^2+1} - \frac{1}{2}\frac{1}{x^2+1}\cdot 2x$$
$$= \tan^{-1} x.$$

Thinking Forward

Volumes of surfaces of revolution

▷ With $f(x) = \ln x$, we want to calculate $\pi \int_1^3 (f(x))^2\,dx = \pi \int_1^3 \ln^2 x\,dx$. To do this, use integration by parts with $u = \ln^2 x$ and $dv = dx$, so that $du = 2\ln x \cdot \frac{1}{x}\,dx = 2\frac{\ln x}{x}\,dx$ and $v = x$. Then (using Theorem 5.8)

$$\pi \int_1^3 \ln^2 x\,dx = \pi \left(\left[x\ln^2 x\right]_1^3 - 2\int_1^3 \left(x\cdot \frac{\ln x}{x}\right)\,dx\right)$$
$$= \pi \left(3\ln^2 3 - 2\int_1^3 \ln x\,dx\right)$$
$$= \pi \left(3\ln^2 3 - 2\,[x\ln x - x]_1^3\right)$$
$$= \pi \left(3\ln^2 3 - 6\ln 3 + 6 - 2\right)$$
$$= \pi \left(3\ln^2 3 - 6\ln 3 + 4\right).$$

▷ With $f^{-1}(y) = e^y$, we want to evaluate

$$2\pi \int_0^{\ln 3} y(3 - f^{-1}(y))\,dy = 2\pi \int_0^{\ln 3} (3y - ye^y)\,dy = 2\pi \left[\frac{3}{2}y^2\right]_0^{\ln 3} - 2\pi \int_0^{\ln 3} ye^y\,dy$$
$$= 3\pi \ln^2 3 - 2\pi \int_0^{\ln 3} ye^y\,dy.$$

To integrate ye^y, use integration by parts with $u = y$ and $dv = e^y\,dy$, so that $du = dy$ and $v = e^y$.

Then we get

$$2\pi \int_0^{\ln 3} y(3 - f^{-1}(y))\, dy = 3\pi \ln^2 3 - 2\pi \int_0^{\ln 3} y e^y\, dy$$

$$= 3\pi \ln^2 3 - 2\pi \left([y e^y]_0^{\ln 3} - \int_0^{\ln 3} e^y\, dy \right)$$

$$= 3\pi \ln^2 3 - 2\pi \left(3\ln 3 - [e^y]_0^{\ln 3} \right)$$

$$= 3\pi \ln^2 3 - 2\pi \left(3\ln 3 - 3 + 1 \right)$$

$$= \pi \left(3\ln^2 3 - 6\ln 3 + 4 \right).$$

▷ The two formulas are the same.

Centroid of a shape

▷ With $f(x) = x^2$, we have

$$\bar{x} = \frac{\int_0^3 x \cdot x^2\, dx}{\int_0^3 x^2\, dx} = \frac{\left[\frac{1}{4}x^4\right]_0^3}{\left[\frac{1}{3}x^3\right]_0^3} = \frac{81/4}{9} = \frac{9}{4},$$

$$\bar{y} = \frac{\frac{1}{2}\int_0^3 (x^2)^2\, dx}{\int_0^3 x^2\, dx} = \frac{\frac{1}{2}\int_0^3 x^4\, dx}{\left[\frac{1}{3}x^3\right]_0^3} = \frac{\frac{1}{2}\left[\frac{1}{5}x^5\right]_0^3}{9} = \frac{3^5/10}{9} = \frac{27}{10}.$$

Thus the centroid of this region is at $\left(\frac{9}{4}, \frac{27}{10}\right)$.

▷ For the denominator, we have

$$\int_{-\pi/2}^{\pi/2} \cos x\, dx = [\sin x]_{-\pi/2}^{\pi/2} = 2.$$

The numerator of $\bar{x}$ is $\int_{-\pi/2}^{\pi/2} x \cos x\, dx$, which we can integrate by parts with $u = x$ and $dv = \cos x\, dx$, so that $du = dx$ and $v = \sin x$. Then

$$\int_{-\pi/2}^{\pi/2} x \cos x\, dx = [x \sin x]_{-\pi/2}^{\pi/2} - \int_{-\pi/2}^{\pi/2} \sin x\, dx$$

$$= 0 + [\cos x]_{-\pi/2}^{\pi/2} = 0.$$

Thus $\bar{x} = \frac{0}{2} = 0$. The numerator of $\bar{y}$ is $\frac{1}{2}\int_{-\pi/2}^{\pi/2} \cos^2 x\, dx$, which we integrate using the double-angle formula:

$$\frac{1}{2}\int_{-\pi/2}^{\pi/2} \cos^2 x\, dx = \frac{1}{2}\int_{-\pi/2}^{\pi/2} \frac{1}{2}(1 + \cos 2x)\, dx$$

$$= \frac{1}{4}\int_{-\pi/2}^{\pi/2} 1 + \cos 2x\, dx$$

$$= \frac{1}{4}\left[x + \frac{1}{2}\sin 2x\right]_{-\pi/2}^{\pi/2}$$

$$= \frac{1}{4}\left(\frac{\pi}{2} + 0 - \left(-\frac{\pi}{2} + 0\right)\right) = \frac{\pi}{4}.$$

Thus $\bar{y} = \frac{\pi/4}{2} = \frac{\pi}{8}$.

The centroid is at $\left(0, \frac{\pi}{8}\right)$.

5.3   Partial Fractions and Other Algebraic Techniques

Thinking Back

Factoring polynomials

▷ Since 2 is a root, $x - 2$ is a factor, and $x^3 - 2x^2 - x + 2 = (x-2)(x^2-1) = (x-2)(x-1)(x+1)$.

▷ 1 is a root of $x^3 - 1$, and $x^3 - 1 = (x-1)(x^2+x+1)$.

▷ By experimentation, we find that 2 is a root, so that $x - 2$ is a factor, and $x^3 - 4x^2 + x + 6 = (x-2)(x^2-2x-3) = (x-2)(x-3)(x+1)$.

▷ By experimentation, we find that -1 is a root, so that $x + 1$ is a factor, and $x^3 - x^2 - 5x - 3 = (x+1)(x^2-2x-3) = (x+1)^2(x-3)$.

Integration

▷ Use a u-substitution with $u = x - 1$, so that $du = dx$, and

$$\int \frac{1}{(x-1)^2}\, dx = \int \frac{1}{u^2}\, du = -u^{-1} + C = -\frac{1}{x-1} + C.$$

▷ Rewrite the integrand as $\frac{x-1}{x^2-1} = \frac{1}{x+1}$. Then

$$\int \frac{x-1}{x^2-1}\, dx = \int \frac{1}{x+1}\, dx = \ln|x+1| + C.$$

▷ $\int \frac{1}{x^2+1}\, dx = \tan^{-1} x + C.$

▷ Use a u-substitution with $u = x^2 + 1$ so that $du = 2x\, dx$ and $\frac{1}{2}\, du = x\, dx$. Then

$$\int \frac{x}{x^2+1}\, dx = \frac{1}{2} \int \frac{1}{u}\, du = \frac{1}{2} \ln|u| + C = \frac{1}{2} \ln(x^2+1).$$

▷ Simplify the integrand and use the power rule:

$$\int \frac{x^2+1}{x^3}\, dx = \int \left(x^{-1} + x^{-3} \right) dx = \ln|x| - \frac{1}{2}x^{-2} + C.$$

▷ Rewrite the integrand as $\frac{x+1}{x^2+1} = \frac{x}{x^2+1} + \frac{1}{x^2+1}$. We recognize the second of these as the derivative of $\tan^{-1} x$; for the first, use the u-substitution $u = x^2 + 1$, so that $du = 2x\, dx$ and $\frac{1}{2}\, du = x\, dx$. Then

$$\begin{aligned}
\int \frac{x+1}{x^2+1}\, dx &= \int \frac{x}{x^2+1}\, dx + \int \frac{1}{x^2+1}\, dx \\
&= \frac{1}{2} \int \frac{1}{u}\, du + \tan^{-1} x + C \\
&= \frac{1}{2} \ln|u| + \tan^{-1} x + C \\
&= \frac{1}{2} \ln(x^2+1) + \tan^{-1} x + C.
\end{aligned}$$

Concepts

1. (a) False. A proper rational function is one in which the degree of the numerator is strictly less than that of the denominator.

   (b) True. See Theorem 5.11.

   (c) True. See the discussion preceding Theorem 5.11.

   (d) True. See Theorem 5.13 and the related discussion.

   (e) True. As shown in the text, an improper rational function must first be reduced to a polynomial and a proper rational function before applying partial fractions.

   (f) False. It is of the form $\frac{A}{x} + \frac{B}{x^2} + \frac{C}{x-3}$.

   (g) False. From Theorem 5.12, it is of the form $\frac{A}{x} + \frac{B}{x^2} + \frac{C}{x-3}$.

   (h) True. See Theorem 5.14.

3. A rational function is a function of the form $\frac{p(x)}{q(x)}$, where p and q are both polynomials with $q(x) \neq 0$. It is proper if $\deg p < \deg q$, and improper otherwise.

5. Since $p(x)$ is $q(x)$ times the quotient, plus the remainder, we have

$$p(x) = (x^2 - x + 3)q(x) + (3x - 1),$$
$$\frac{p(x)}{q(x)} = x^2 - x + 3 + \frac{3x - 1}{q(x)}.$$

7. Since $q(x)$ is nonzero, we may divide the original equation through by $q(x)$ to get

$$\frac{p(x)}{q(x)} = \frac{q(x)m(x) + R(x)}{q(x)} = m(x) + \frac{R(x)}{q(x)}.$$

9. If the degree of $p(x)$ is two greater than the degree of $q(x)$, then with $p(x) = q(x)m(x) + R(x)$, and $\deg R(x) < \deg q(x)$, we must have $\deg m(x) = \deg p(x) - \deg q(x) = 2$ to make the degrees of the polynomials on both sides equal. So $m(x)$ is quadratic, and $\deg R(x) < \deg q(x)$. Polynomial long division works only after reducing $\frac{p(x)}{q(x)}$ to $m(x) + \frac{R(x)}{q(x)}$ first.

11. Clearly $\deg R(x) < \deg q(x)$, since that holds whenever the division algorithm is applied. Now, if $m(x)$ is not the zero polynomial, then

$$\deg(q(x)m(x)) = \deg q(x) + \deg m(x) \geq \deg q(x) > \deg p(x).$$

But this is impossible, since in that case $\deg(q(x)m(x) + \deg R(x))$ would equal $\deg(q(x)m(x))$ since the terms in $R(x)$ are not of high enough degree to cancel the leading term in $q(x)m(x)$. But this means that

$$\deg p(x) = \deg(q(x)m(x) + R(x)) = \deg(q(x)m(x)) > \deg p(x),$$

a contradiction. So we must have $m(x)$ the zero polynomial, so that $R(x) = p(x)$.

13. The partial fraction decomposition of $\frac{p(x)}{q(x)}$ if q is an irreducible quadratic is $\frac{Ax+B}{q(x)}$. If q is a reducible quadratic, say $q(x) = (x - c)(x - d)$ then the partial fraction decomposition of $\frac{p(x)}{q(x)}$ is $\frac{A}{x-c} + \frac{B}{x-d}$ if $c \neq d$, and $\frac{A}{x-c} + \frac{B}{(x-c)^2}$ if $c = d$. Thus for $q(x) = x^2 + 1$ we get $\frac{Ax+B}{x^2+1}$, and for $q(x) = (x - 1)(x - 2)$ we get $\frac{A}{x-1} + \frac{B}{x-2}$.

15. Completing the square, we have $f(x) = x^2 - 3x + 5 = x^2 - 3x + \frac{9}{4} + 5 - \frac{9}{4} = \left(x - \frac{3}{2}\right)^2 + \frac{11}{4}$. Now, this is the graph of $g(x) = x^2$ shifted right by $\frac{3}{2}$ and up by $\frac{11}{4}$. Since the vertex of x^2 is at $(0,0)$, the vertex of $f(x)$ must be at $\left(\frac{3}{2}, \frac{11}{4}\right)$.

Skills

17. Expand the integrand using partial fractions:

$$\frac{3}{(x-2)(x+1)} = \frac{A}{x-2} + \frac{B}{x+1} = \frac{(A+B)x + A - 2B}{(x-2)(x+1)}.$$

Thus $A + B = 0$ and $A - 2B = 3$, so that $A = 1$ and $B = -1$. Hence

$$\int \frac{3}{(x-2)(x+1)}\, dx = \int \left(\frac{1}{x-2} - \frac{1}{x+1}\right)\, dx = \ln|x-2| - \ln|x+1| + C = \ln\left|\frac{x-2}{x+1}\right| + C.$$

19. Expand the integrand using partial fractions:

$$\frac{x+1}{(x-1)^2} = \frac{A}{x-1} + \frac{B}{(x-1)^2} = \frac{A(x-1) + B}{(x-1)^2} = \frac{Ax + (B-A)}{(x-1)^2}.$$

Thus $A = 1$ and $B - A = 1$, so that $B = 2$. Hence

$$\int \frac{x+1}{(x-1)^2}\, dx = \int \left(\frac{1}{x-1} + \frac{2}{(x-1)^2}\right)\, dx = \ln|x-1| - 2(x-1)^{-1} + C.$$

21. This is an improper rational function, so we first divide $(x+3)(x-2) = x^2 + x - 6$ by x^2 to get $x^2 + x - 6 = x^2(1) + x - 6$. Thus

$$\int \frac{(x+3)(x-2)}{x^2}\, dx = \int \left(1 + \frac{x-6}{x^2}\right)\, dx = \int 1\, dx + \int \frac{1}{x}\, dx - \int \frac{6}{x^2}\, dx = x + \ln|x| + \frac{6}{x} + C.$$

23. This is an improper rational function, so we first divide x^3 by $(x+3)(x-2) = x^2 + x - 6$ to get $x^3 = (x^2 + x - 6)(x-1) + (7x - 6)$. Thus

$$\int \frac{5x^3}{(x+3)(x-2)}\, dx = 5\left(\int (x-1)\, dx + \int \frac{7x-6}{(x+3)(x-2)}\, dx\right).$$

For the second integral, expand the integrand using partial fractions:

$$\frac{7x-6}{(x+3)(x-2)} = \frac{A}{x+3} + \frac{B}{x-2} = \frac{(A+B)x + (3B - 2A)}{(x+3)(x-2)},$$

so that $A + B = 7$ and $-2A + 3B = -6$. Thus $A = \frac{27}{5}$ and $B = \frac{8}{5}$, so that

$$\int \frac{5x^3}{(x+3)(x-2)}\, dx = 5\left(\int (x-1)\, dx + \int \frac{7x-6}{(x+3)(x-2)}\, dx\right)$$

$$= 5\left(\frac{1}{2}x^2 - x + \frac{27}{5}\int \frac{1}{x+3}\, dx + \frac{8}{5}\int \frac{1}{x-2}\, dx\right)$$

$$= 5\left(\frac{1}{2}x^2 - x + \frac{27}{5}\ln|x+3| + \frac{8}{5}\ln|x-2|\right) + C.$$

25. Expand using partial fractions; $x^2 + 3$ is an irreducible quadratic:

$$\frac{x}{(x^2+3)(x-2)} = \frac{Ax + B}{x^2 + 3} + \frac{C}{x-2} = \frac{(A+C)x^2 + (B - 2A)x + (3C - 2B)}{(x^2+3)(x-2)}.$$

Thus $A + C = 0$, $B - 2A = 1$, and $3C - 2B = 0$. Substituting $A = -C$ in the second equation gives $B + 2C = 1$; solving that together with the third equation gives $C = \frac{2}{7}$ so that $A = -\frac{2}{7}$ and $B = \frac{3}{7}$. Hence (with $u = x^2 + 3$ and $v = \frac{x}{\sqrt{3}}$)

$$\int \frac{x}{(x^2 + 3)(x - 2)}\,dx = \frac{1}{7}\int \frac{-2x + 3}{x^2 + 3}\,dx + \frac{2}{7}\int \frac{1}{x - 2}\,dx$$

$$= \frac{1}{7}\left(-\int \frac{2x}{x^2 + 3}\,dx + \int \frac{3}{x^2 + 3}\,dx + 2\int \frac{1}{x - 2}\,dx\right)$$

$$= \frac{1}{7}\left(-\int \frac{1}{u}\,du + \int \frac{1}{(x/\sqrt{3})^2 + 1}\,dx + 2\ln|x - 2|\right)$$

$$= \frac{1}{7}\left(-\ln|u| + \sqrt{3}\int \frac{1}{v^2 + 1}\,dv + 2\ln|x - 2|\right)$$

$$= \frac{1}{7}\left(-\ln(x^2 + 3) + \sqrt{3}\tan^{-1}v + 2\ln|x - 2|\right) + C$$

$$= \frac{1}{7}\left(\ln \frac{(x - 2)^2}{x^2 + 3} + \sqrt{3}\tan^{-1}\frac{x}{\sqrt{3}}\right) + C.$$

27. The denominator factors as $x^3 - 1 = (x - 1)(x^2 + x + 1)$; now expand using partial fractions:

$$\frac{3x + 3}{(x - 1)(x^2 + x + 1)} = \frac{A}{x - 1} + \frac{Bx + C}{x^2 + x + 1} = \frac{(A + B)x^2 + (A - B + C)x + (A - C)}{(x - 1)(x^2 + x + 1)}.$$

Thus $A + B = 0$, $A - B + C = 3$, and $A - C = 3$. Solving gives $A = 2$, $B = -2$ and $C = -1$. Hence (with $u = x^2 + x + 1$ so that $du = (2x + 1)\,dx$),

$$\int \frac{3x + 3}{(x - 1)(x^2 + x + 1)}\,dx = 2\int \frac{1}{x - 1}\,dx - \int \frac{2x + 1}{x^2 + x + 1}\,dx$$

$$= 2\ln|x - 1| - \int \frac{1}{u}\,du$$

$$= 2\ln|x - 1| - \ln|u| + C$$

$$= 2\ln|x - 1| - \ln\left|x^2 + x + 1\right| + C.$$

29. Since the denominator is irreducible, this integrand is its own partial fraction representation. The derivative of the denominator is $2x - 4$, so we rewrite the integral:

$$\int \frac{2(x - 1)}{x^2 - 4x + 5}\,dx = \int \frac{(2x - 4) + 2}{x^2 - 4x + 5}\,dx$$

$$= \int \frac{2x - 4}{x^2 - 4x + 5}\,dx + 2\int \frac{1}{(x^2 - 4x + 4) + 1}\,dx$$

$$= \int \frac{2x - 4}{x^2 - 4x + 5}\,dx + 2\int \frac{1}{(x - 2)^2 + 1}\,dx.$$

Now make the substitutions $u = x^2 - 4x + 5$ in the first integral and $v = x - 2$ in the second to get

$$\int \frac{2(x - 1)}{x^2 - 4x + 5}\,dx = \int \frac{1}{u}\,du + 2\int \frac{1}{v^2 + 1}\,dv$$

$$= \ln|u| + 2\tan^{-1}v + C$$

$$= \ln\left|x^2 - 4x + 5\right| + 2\tan^{-1}(x - 2) + C.$$

31. Expand the integrand using partial fractions:

$$\frac{3x^2 - 13x + 13}{(x-2)^3} = \frac{A}{x-2} + \frac{B}{(x-2)^2} + \frac{C}{(x-2)^3} = \frac{A(x-2)^2 + B(x-2) + C}{(x-2)^3}$$

$$= \frac{Ax^2 + (-4A+B)x + (4A-2B+C)}{(x-2)^3}.$$

Thus $A = 3$, $-4A + B = -13$, and $4A - 2B + C = 13$. Solving gives $B = -1$ and $C = -1$. Thus

$$\int \frac{3x^2 - 13x + 13}{(x-2)^3}\, dx = 3\int \frac{1}{x-2}\, dx - \int \frac{1}{(x-2)^2}\, dx - \int \frac{1}{(x-2)^3}\, dx$$

$$= 3\ln|x-2| + (x-2)^{-1} + \frac{1}{2}(x-2)^{-2} + C.$$

33. The denominator factors as $3x^2 - 8x - 3 = (3x+1)(x-3)$. Expand using partial fractions:

$$\frac{-5x-5}{3x^2 - 8x - 3} = \frac{A}{3x+1} + \frac{B}{x-3} = \frac{(A+3B)x + (B-3A)}{3x^3 - 8x - 3}.$$

Thus $A + 3B = -5$ and $B - 3A = -5$, so that $A = 1$ and $B = -2$. Hence

$$\int \frac{-5x-5}{3x^2 - 8x - 3}\, dx = \int \frac{1}{3x+1}\, dx - 2\int \frac{1}{x-3}\, dx = \frac{1}{3}\ln|3x+1| - 2\ln|x-3| + C.$$

35. Note that the denominator factors as $(x^2+1)(x^2+3)$. Now expand using partial fractions:

$$\frac{x^2 + 2x + 3}{x^4 + 4x^2 + 3} = \frac{Ax+B}{x^2+1} + \frac{Cx+D}{x^2+3}$$

$$= \frac{(Ax+B)(x^2+3) + (Cx+D)(x^2+1)}{x^4 + 4x^2 + 3}$$

$$= \frac{(A+C)x^3 + (B+D)x^2 + (3A+C)x + (D+3B)}{x^4 + 4x^2 + 3}.$$

Thus $A + C = 0$, $3A + C = 2$, $B + D = 1$, and $3B + D = 3$. Solving gives $A = 1$, $C = -1$, $B = 1$, and $D = 0$. Hence (using u-substitutions of $u = x^2 + 1$ and $v = x^2 + 3$)

$$\int \frac{x^2 + 2x + 3}{x^4 + 4x^2 + 3}\, dx = \int \frac{x+1}{x^2+1}\, dx - \int \frac{x}{x^2+3}\, dx$$

$$= \frac{1}{2}\int \frac{2x}{x^2+1}\, dx + \int \frac{1}{x^2+1}\, dx - \frac{1}{2}\int \frac{2x}{x^2+3}\, dx$$

$$= \frac{1}{2}\int \frac{1}{u}\, du + \tan^{-1} x - \frac{1}{2}\int \frac{1}{v}\, dv$$

$$= \frac{1}{2}(\ln|u| - \ln|v|) + \tan^{-1} x + C$$

$$= \frac{1}{2}\left(\ln\left|x^2+1\right| - \ln\left|x^2+3\right|\right) + \tan^{-1} x + C$$

$$= \frac{1}{2}\ln\left|\frac{x^2+1}{x^2+3}\right| + \tan^{-1} x + C.$$

37. Expanding using partial fractions gives

$$\frac{1-2x}{x^2(x^2+1)} = \frac{A}{x} + \frac{B}{x^2} + \frac{Cx+D}{x^2+1} = \frac{Ax(x^2+1) + B(x^2+1) + (Cx+D)x^2}{x^2(x^2+1)}$$

$$= \frac{(A+C)x^3 + (B+D)x^2 + Ax + B}{x^2(x^2+1)}.$$

Thus $A = -2$, $B = 1$, $C = 2$, and $D = -1$, so that

$$\int \frac{1-2x}{x^2(x^2+1)}\, dx = -2\int \frac{1}{x}\, dx + \int \frac{1}{x^2}\, dx + \int \frac{2x-1}{x^2+1}\, dx$$

$$= -2\ln|x| - \frac{1}{x} + \int \frac{2x}{x^2+1}\, dx - \int \frac{1}{x^2+1}\, dx$$

$$= -2\ln|x| - \frac{1}{x} + \ln(x^2+1) - \tan^{-1}x + C$$

$$= \ln\frac{x^2+1}{x^2} - \frac{1}{x} - \tan^{-1}x + C.$$

39. Note that -1 is a root of the denominator, which then factors as $(x+1)(x^2+1)$. Now expand the integrand using partial fractions:

$$\frac{2x^2+4x}{x^3+x^2+x+1} = \frac{A}{x+1} + \frac{Bx+C}{x^2+1} = \frac{(A+B)x^2+(B+C)x+(A+C)}{x^3+x^2+x+1}.$$

Then $A+B = 2$, $B+C = 4$, and $A+C = 0$, so that $A = -1$, $B = 3$, and $C = 1$. Hence

$$\int \frac{2x^2+4x}{x^3+x^2+x+1}\, dx = -\int \frac{1}{x+1}\, dx + \int \frac{3x+1}{x^2+1}\, dx$$

$$= -\ln|x+1| + \frac{3}{2}\int \frac{2x}{x^2+1}\, dx + \int \frac{1}{x^2+1}\, dx$$

$$= -\ln|x+1| + \frac{3}{2}\ln(x^2+1) + \tan^{-1}x + C.$$

41. Since the degree of the numerator exceeds that of the denominator, use polynomial long division first to get $2x(x^2-1) = 2x^3 - 2x = (x^2-4)(2x) + 6x$. Then

$$\int \frac{2x(x^2-1)}{x^2-4}\, dx = \int \left(2x + \frac{6x}{x^2-4}\right) dx = x^2 + 3\int \frac{2x}{x^2-4}\, dx = x^2 + 3\ln\left|x^2-4\right| + C.$$

43. Checking $x = -\frac{1}{2}$, we see that it is a root of the numerator as well, which factors as $(x-3)(x-2)(2x+1)$. Thus

$$\int \frac{2x^3-9x^2+7x+6}{1+2x}\, dx = \int (x^2-5x+6)\, dx = \frac{1}{3}x^3 - \frac{5}{2}x^2 + 6x + C.$$

45. Use polynomial long division first: $2x^4 - 8x^3 + 9x - 4 = (x^2-4x+1)(2x^2-2) + (x-2)$, so that

$$\int \frac{2x^4-8x^3+9x-4}{x^2-4x+1}\, dx = \int (2x^2-2)\, dx + \int \frac{x-2}{x^2-4x+1}\, dx$$

$$= \frac{2}{3}x^3 - 2x + \frac{1}{2}\int \frac{2x-4}{x^2-4x+1}\, dx$$

$$= \frac{2}{3}x^3 - 2x + \frac{1}{2}\ln\left|x^2-4x+1\right| + C.$$

47. This looks like a form of the derivative of $\tanh^{-1}$; some algebraic manipulations will put it in that form:

$$\int_3^4 \frac{4}{x^2-4}\, dx = -\int_3^4 \frac{1}{1-(x/2)^2}\, dx.$$

Now use the substitution $u = \frac{x}{2}$, so that $2\,du = dx$, and

$$-\int_3^4 \frac{1}{1 - (x/2)^2}\,dx = -2\int_{x=3}^{x=4} \frac{1}{1 - u^2}\,du = -2\left[\tanh^{-1} u\right]_{x=3}^{x=4} = -2\left[\tanh^{-1}\frac{x}{2}\right]_3^4$$

$$= -2\left(\tanh^{-1} 2 - \tanh^{-1}\frac{3}{2}\right).$$

Note that you can also use partial fractions, writing $\frac{4}{x^2 - 4} = \frac{4}{(x-2)(x+2)}$ as a pair of fractions; the indefinite integral will be $\ln\left|\frac{x-2}{x+2}\right|$, which looks different but is actually equivalent to the indefinite integral above.

49. Expand using partial fractions:

$$\frac{1}{x(x+1)} = \frac{A}{x} + \frac{B}{x+1} = \frac{A(x+1) + Bx}{x(x+1)}.$$

Setting $x = 0$ gives $A = 1$, while setting $x = -1$ gives $B = -1$. Thus

$$\int_1^3 \frac{1}{x(x+1)}\,dx = \int_1^3 \frac{1}{x}\,dx - \int_1^3 \frac{1}{x+1}\,dx$$

$$= \left[\ln|x|\right]_1^3 - \left[\ln|x+1|\right]_1^3$$

$$= \ln 3 - \ln 4 + \ln 2 = \ln\frac{3}{2}.$$

51. Use long division first to get an improper rational function: $x^2 - 2x - 4 = (x-3)(x+1) - 1$. Then

$$\int_{-1}^1 \frac{x^2 - 2x - 4}{x - 3}\,dx = \int_{-1}^1 \left(x + 1 - \frac{1}{x-3}\right)\,dx$$

$$= \left[\frac{1}{2}x^2 + x - \ln|x-3|\right]_{-1}^1$$

$$= \frac{1}{2} + 1 - \ln 2 - \left(\frac{1}{2} - 1 - \ln 4\right)$$

$$= 2 + \ln 2.$$

53. First substitute $u = \cos x$, so that $du = -\sin x\,dx$. Then

$$\int \frac{\sin x}{\cos x + \cos^2 x}\,dx = \int \frac{-1}{u + u^2}\,du = \int \frac{-1}{u(u+1)}\,du.$$

Expand using partial fractions:

$$\frac{-1}{u(u+1)} = \frac{A}{u} + \frac{B}{u+1} = \frac{A(u+1) + Bu}{u(u+1)}.$$

Substituting $u = 0$ gives $A = -1$; substituting $u = -1$ gives $B = 1$. Thus

$$\int \frac{\sin x}{\cos x + \cos^2 x}\,dx = \int \frac{-1}{u(u+1)}\,du$$

$$= -\int \frac{1}{u}\,du + \int \frac{1}{u+1}\,du$$

$$= -\ln|u| + \ln|u+1| + C$$

$$= \ln\left|\frac{u+1}{u}\right| + C$$

$$= \ln\left|\frac{1 + \cos x}{\cos x}\right| + C.$$

55. Substitute $u = \ln x$, so that $du = \frac{1}{x}\,dx$; then the integral becomes

$$\int \frac{\ln x}{x(1+\ln x)^2}\,dx = \int \frac{u}{(1+u)^2}\,du.$$

Now expand using partial fractions:

$$\frac{u}{(1+u)^2} = \frac{A}{1+u} + \frac{B}{(1+u)^2} = \frac{Au + (A+B)}{(1+u)^2}.$$

Thus $A = 1$ and $A + B = 0$, so that $B = -1$. Then

$$\int \frac{\ln x}{x(1+\ln x)^2}\,dx = \int \frac{u}{(1+u)^2}\,du$$

$$= \int \frac{1}{1+u}\,du - \int \frac{1}{(1+u)^2}\,du$$

$$= \ln|1+u| + \frac{1}{1+u} + C$$

$$= \ln|1+\ln x| + \frac{1}{1+\ln x} + C.$$

57. (a) Expand $f(x)$ using partial fractions:

$$\frac{1}{x^2-4} = \frac{A}{x-2} + \frac{B}{x+2} = \frac{A(x+2) + B(x-2)}{x^2-4}.$$

Setting $x = 2$ gives $A = \frac{1}{4}$; setting $x = -2$ gives $B = -\frac{1}{4}$. Then the area under $f(x)$ on $[-1, 1]$ is

$$\int_{-1}^{1} \frac{1}{x^2-4}\,dx = \frac{1}{4}\int_{-1}^{1}\left(\frac{1}{x-2} - \frac{1}{x+2}\right)dx$$

$$= \frac{1}{4}\left[\ln|x-2| - \ln|x+2|\right]_{-1}^{1}$$

$$= \frac{1}{4}\left(\ln 1 - \ln 3 - (\ln 3 - \ln 1)\right) = -\frac{\ln 3}{2}.$$

(b) Note that $f(x) \geq g(x)$ on $[-1, 0]$, and $f(x) \leq g(x)$ on $[0, -1]$, so that the area between the graphs is

$$\int_{-1}^{0}(f(x) - g(x))\,dx + \int_{0}^{1}(g(x) - f(x))\,dx$$

$$= \int_{-1}^{0} f(x)\,dx - \int_{0}^{1} f(x)\,dx - \int_{-1}^{0} g(x)\,dx + \int_{0}^{1} g(x)\,dx.$$

Since $\int g(x)\,dx = \int \frac{1}{4}(x-1)\,dx = \frac{1}{8}(x-1)^2$, and we computed $\int f(x)\,dx$ in part (a), we get for the answer

$$\frac{1}{4}\left([\ln|x-2| - \ln|x+2|]_{-1}^{0} - [\ln|x-2| - \ln|x+2|]_{0}^{1}\right) - \left[\frac{1}{8}(x-1)^2\right]_{-1}^{0} + \left[\frac{1}{8}(x-1)^2\right]_{0}^{1},$$

which evaluates to

$$\frac{1}{4}(\ln 2 - \ln 2 - \ln 3 + \ln 1 - (\ln 1 - \ln 3 - \ln 2 + \ln 2)) - \frac{1}{8} + \frac{1}{2} - \frac{1}{8} = \frac{1}{4}.$$

59. Expand using partial fractions:

$$\frac{2x^3 + x}{(x^2 + 1)^2} = \frac{Ax + B}{x^2 + 1} + \frac{Cx + D}{(x^2 + 1)^2} = \frac{(Ax + B)(x^2 + 1) + Cx + D}{(x^2 + 1)^2}$$

$$= \frac{Ax^3 + Bx^2 + (A + C)x + (B + D)}{(x^2 + 1)^2}.$$

Then $A = 2$, $B = 0$, $C = -1$, and $D = 0$, so that

$$\int \frac{2x^3 + x}{(x^2 + 1)^2}\, dx = \int \frac{2x}{x^2 + 1}\, dx - \int \frac{x}{(x^2 + 1)^2}\, dx.$$

For each of these integrals, use the substitution $u = x^2 + 1$ with $du = 2x\, dx$. Then we get

$$\int \frac{2x^3 + x}{(x^2 + 1)^2}\, dx = \int \frac{1}{u}\, du - \frac{1}{2}\int \frac{1}{u^2}\, du = \ln|u| + \frac{1}{2u} + C = \ln\left|x^2 + 1\right| + \frac{1}{2(x^2 + 1)} + C.$$

(a) The average value of f on $[0, 2]$ is

$$\frac{1}{2 - 0}\int_0^2 f(x)\, dx = \frac{1}{2}\left[\ln\left|x^2 + 1\right| + \frac{1}{2(x^2 + 1)}\right]_0^2 = \frac{1}{2}\left(\ln 5 + \frac{1}{10} - \ln 1 - \frac{1}{2}\right)$$

$$= \frac{1}{2}\left(\ln 5 - \frac{2}{5}\right) \approx 0.605.$$

(b) From the graph

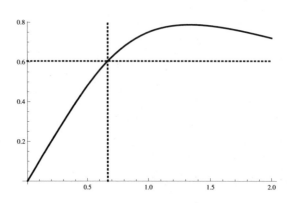

we see that f achieves its average value at approximately $x = 0.67$ (a more precise value is $x \approx 0.668487$).

Applications

61. Rewriting the integrand gives

$$\frac{1}{q\left(1 - \frac{q}{K}\right)} = \frac{K}{q(K - q)} = \frac{-K}{q(q - K)}.$$

To integrate, expand using partial fractions:

$$\frac{-K}{q(q - K)} = \frac{A}{q} + \frac{B}{q - K} = \frac{A(q - K) + Bq}{q(q - K)}.$$

Substituting $q = 0$ gives $-K = -AK$, so that $A = 1$. Substituting $q = K$ gives $BK = -K$, so that $B = -1$. Thus we have

$$
\begin{aligned}
r(t) &= \int_{p_0}^{p(t)} \frac{1}{q\left(1 - \frac{q}{K}\right)}\, dq \\
&= \int_{p_0}^{p(t)} \left(\frac{1}{q} - \frac{1}{q - K}\right) dq \\
&= \left[\ln |q| - \ln |q - K|\right]_{p_0}^{p(t)} \\
&= \ln p(t) - \ln(K - p(t)) - \ln p_0 + \ln(K - p_0) \\
&= \ln \frac{p(t)(K - p_0)}{p_0(K - p(t))}.
\end{aligned}
$$

Solving this equation for $p(t)$ gives

$$
p(t) = \frac{p_0 K e^{r(t)}}{K - p_0 + p_0 e^{r(t)}}.
$$

Proofs

63. Expand $(x - k)^2 + C$ with $k = -\frac{b}{2}$ and $C = c - \frac{b^2}{4}$:

$$
\left(x - \frac{-b}{2}\right)^2 + c - \frac{b^2}{4} = x^2 + 2 \cdot \frac{b}{2} x + \frac{b^2}{4} + c - \frac{b^2}{4} = x^2 + bx + c.
$$

Thinking Forward

Separable differential equations

▷ Expanding the integrand using partial fractions gives

$$
\frac{1}{P(100 - P)} = \frac{A}{P} + \frac{B}{100 - P} = \frac{A(100 - P) + BP}{P(100 - P)}.
$$

Substituting $P = 0$ gives $100A = 1$; substituting $P = 100$ gives $100B = 1$, so that $A = B = \frac{1}{100}$. Then

$$
\int \frac{1}{P(100 - P)}\, dP = \frac{1}{100} \int \left(\frac{1}{P} + \frac{1}{100 - P}\right) dP = \frac{1}{100}\left(\ln |P| - \ln |100 - P|\right) + C_1.
$$

▷ Since $\ln a - \ln b = \ln \frac{a}{b}$, we get

$$
\int \frac{1}{P(100 - P)}\, dP = \frac{1}{100} \ln \left|\frac{P}{100 - P}\right| + C_1.
$$

▷ Setting the two sides equal gives

$$\frac{1}{100}\ln\left|\frac{P}{100-P}\right| + C_1 = t + C_2 \quad\Rightarrow$$

$$\ln\left|\frac{P}{100-P}\right| = 100(t + C_3) \quad\Rightarrow$$

$$\left|\frac{P}{100-P}\right| = e^{100t+100C_3} = e^{100C_3}e^{100t} = C_4e^{100t} \quad\Rightarrow$$

$$\frac{P}{100-P} = \pm C_4e^{100t} = Ae^{100t} \quad\Rightarrow$$

$$P = 100Ae^{100t} - PAe^{100t} \quad\Rightarrow$$

$$P(1 + Ae^{100t}) = 100Ae^{100t} \quad\Rightarrow$$

$$P(t) = \frac{100Ae^{100t}}{1 + Ae^{100t}}.$$

▷ We have

$$\lim_{t\to\infty} P(t) = \lim_{t\to\infty}\frac{100Ae^{100t}}{1 + Ae^{100t}} = \lim_{t\to\infty}\frac{100A}{e^{-100t} + A} = \frac{100A}{A} = 100.$$

Thus over time, the population approaches 100 individuals.

5.4 Trigonometric Integrals

Thinking Back

Pythagorean identities

▷ Suppose the terminal ray of an acute angle θ intersects the unit circle at (x, y). Then (x, y) is in the first quadrant, since the angle is acute. Thus the three points $(0, 0)$, $(x, 0)$, and (x, y) form a right triangle with hypotenuse from $(0, 0)$ to (x, y). Since (x, y) lies on the unit circle, we know that $y = \sin\theta$ and $x = \cos\theta$, and that $x^2 + y^2 = 1$ by the Pythagorean theorem. Substituting gives $\sin^2\theta + \cos^2\theta = 1$. This is called a Pythagorean identity because it is simply the Pythagorean Theorem in different clothes.

▷ Now let θ be an arbitrary angle, intersecting the unit circle at (x, y). The three points $(0, 0)$, $(x, 0)$, and $(y, 0)$ again form a right triangle, although here x and y may be either positive or negative. But again the Pythagorean Theorem tells us that $x^2 + y^2 = 1$ since (x, y) lies on the unit circle, and $x = \sin\theta$, $y = \cos\theta$. Substituting again gives $\sin^2\theta + \cos^2\theta = 1$.

▷ Divide the identity $\sin^2 x + \cos^2 x = 1$ through by $\cos^2 x$ to get

$$\frac{\sin^2 x + \cos^2 x}{\cos^2 x} = \frac{1}{\cos^2 x}, \quad\text{or}\quad \frac{\sin^2 x}{\cos^2 x} + 1 = \frac{1}{\cos^2 x}.$$

Since $\frac{\sin x}{\cos x} = \tan x$ and $\frac{1}{\cos x} = \sec x$, this is the same as $1 + \tan^2 x = \sec^2 x$.

▷ Divide the identity $\sin^2 x + \cos^2 x = 1$ through by $\sin^2 x$ to get

$$\frac{\sin^2 x + \cos^2 x}{\sin^2 x} = \frac{1}{\sin^2 x}, \quad\text{or}\quad 1 + \frac{\cos^2 x}{\sin^2 x} = \frac{1}{\sin^2 x}.$$

Since $\frac{\cos x}{\sin x} = \cot x$ and $\frac{1}{\sin x} = \csc x$, this is the same as $1 + \cot^2 x = \csc^2 x$.

Double-angle identities

▷ Expanding the right-hand side gives

$$\frac{1}{2}(1 - \cos(2x)) = \frac{1}{2}(1 - \cos(x + x)) = \frac{1}{2}(1 - (\cos^2 x - \sin^2 x)) = \frac{1}{2}((1 - \cos^2 x) + \sin^2 x).$$

By the identity $\sin^2 x + \cos^2 x = 1$, we may write $\sin^2 x$ for $1 - \cos^2 x$, so that we get

$$\frac{1}{2}(1 - \cos(2x)) = \frac{1}{2}((1 - \cos^2 x) + \sin^2 x) = \frac{1}{2}(\sin^2 x + \sin^2 x) = \sin^2 x.$$

▷ Expanding the right-hand side gives

$$\frac{1}{2}(1 + \cos(2x)) = \frac{1}{2}(1 + \cos(x + x)) = \frac{1}{2}(1 + (\cos^2 x - \sin^2 x)) = \frac{1}{2}((1 - \sin^2 x) + \cos^2 x).$$

By the identity $\sin^2 x + \cos^2 x = 1$, we may write $\cos^2 x$ for $1 - \sin^2 x$, so that we get

$$\frac{1}{2}(1 + \cos(2x)) = \frac{1}{2}((1 - \sin^2 x) + \cos^2 x) = \frac{1}{2}(\cos^2 x + \cos^2 x) = \cos^2 x.$$

Concepts

1. (a) False. Taking the derivative of the right-hand side gives $\frac{1}{9} \cdot 9 \cos^8 x \cdot (\cos x)' = -\sin x \cos^8 x$. (To see that this is not equal to $\cos^8 x$, evaluate at $x = 0$, for example). The integrand is not in the form $\int u^9 \, du$.

    (b) False. Taking the derivative of the right-hand side gives $\frac{1}{9} \cdot 9 \sin^8 x \cdot (\sin x)' = \cos x \sin^8 x$. (To see that this is not equal to $\cos^8 x$, evaluate at $\frac{\pi}{4}$, for example).

    (c) False. Taking the derivative of the right-hand side gives $-\frac{1}{9} \cdot 9 \sin^8 x \cdot (\sin x)' = -\cos x \sin^8 x$. (To see that this is not equal to $\cos^8 x$, evaluate at $\frac{\pi}{4}$, for example).

    (d) False. It can be rewritten so that integration by substitution is possible with $u = \cos x$. If k is odd, then $\sin^k x = \sin x(\sin^{2r} x) = \sin x(\sin^2 x)^r = \sin x(1 - \cos^2 x)^r$; the substitution $u = \cos x$ turns this into a power integral.

    (e) False. Substitute $1 - \cos^2 x$ for $\sin^2 x$ to get $\int \sin x(1 - \cos^2 x) \cos^2 x \, dx$ and integrate using $u = \cos x$.

    (f) False. Since $\sec x = \frac{1}{\cos x}$, this integral is simply $\int \cos x \, dx = \sin x + C$.

    (g) True. Since $(\cot x)' = -\csc^2 x$, this integral is of the form $-\int u^3 \, du$ after that substitution.

    (h) True. Theorem 4.18 gives the integrals of $\sin x$ and $\cos x$; Exercises 91 and 92 in Section 5.1 give the integrals of $\tan x$ and $\cot x$; and Theorem 5.17 gives the integrals of $\sec x$ and $\csc x$.

3. Differentiating gives

$$\left(-\frac{1}{5}\cos^5 x + \frac{1}{7}\cos^7 x + C\right)' = -\frac{1}{5} \cdot 5 \cos^4 x(-\sin x) + \frac{1}{7} \cdot 7 \cos^6 x(-\sin x)$$

$$= \sin x \cos^4 x - \sin x \cos^6 x$$

$$= \sin x \cos^4 x(1 - \cos^2 x)$$

$$= \sin x \cos^4 x(\sin^2 x)$$

$$= \sin^3 x \cos^4 x.$$

5. Multiply by $1 = \frac{\csc x + \cot x}{\csc x + \cot x}$. See Exercise 83.

7. For example, Exercises 22, 34, and 41.

9. If $k = 2r$, use the double-angle identities to write $\cos^k x = (\cos^2 x)^r = \left(\frac{1}{2}(1 + \cos 2x)\right)^r$. Expanding gives powers of $\cos 2x$ no higher than r. For odd exponents, we can use the method in Exercise 8; for even exponents, apply the method of this problem again.

11. Write $\cot^k x = \cot^{k-2} x(\csc^2 x - 1)$. If $k = 2$, this is $\csc^2 x - 1$, which has the antiderivative $-\cot x - x$. If $k > 2$, write the expression as $\cot^{k-2} x \csc^2 x - \cot^{k-2} x$. The first term can be integrated using the substitution $u = \cot x$; for the second, apply this procedure again to further reduce the power.

13. If $k > 2$ is even, say $k = 2r$, write $\csc^k x = \csc^2 x(\csc^2 x)^{r-1} = \csc^2 x(\cot^2 x + 1)^r$. Now integrate using $u = \cot x$. If $k > 2$ is odd, use integration by parts with $u = \csc^{k-2} x$ and $dv = \csc^2 x\, dx$, so that

$$\int \csc^k x\, dx = -\csc^{k-2} x \cot x + \int (k-2)\csc^{k-3} x \cdot (-\cot x \csc x) \cot x\, dx$$

$$= -\csc^{k-2} x \cot x - (k-2)\int \csc^{k-2} x \cot^2 x\, dx$$

$$= -\csc^{k-2} x \cot x - (k-2)\int \csc^{k-2} x(\csc^2 x - 1)\, dx$$

$$= -\csc^{k-2} x \cot x - (k-2)\int \csc^k x\, dx + (k-2)\int \csc^{k-2} x\, dx.$$

The final integral above can be evaluated by repeating this procedure. The remaining integral on the right is just the original integral, so solving the equation for this integral gives the answer.

15. Use the double-angle identities to reduce $\sin^2 x$ and $\cos^2 x$ to expressions involving $\cos 2x$. Then the entire integrand becomes a polynomial in $\cos 2x$. Odd powers of $\cos 2x$ may be integrated using the method of Exercise 8, and even powers by the method of Exercise 9.

17. Since n is odd, say $n = 2k + 1$, use the identity $\tan^2 x = \sec^2 x - 1$ to reduce $\tan^n x$ to $\tan x(\sec^2 x - 1)^k$; then the integrand is $\sec^m x(\sec^2 x - 1)^k \tan x$. We may assume $m > 0$, since otherwise we can use Exercise 16, so that this is $\sec^{m-1} x(\sec^2 x - 1)^k(\sec x \tan x)$; this is now integrable using $u = \sec x$.

19. (a) We have

$$\int \sin^2 x \cos^2 x\, dx = \int (1 - \cos^2 x)\cos^2 x\, dx = \int (\cos^2 x - \cos^4 x)\, dx.$$

Now apply the double-angle identity to get

$$\int \left(\frac{1}{2}(1 + \cos 2x) - \left(\frac{1}{2}(1 + \cos 2x)\right)^2\right) dx$$

$$= \int \left(\frac{1}{2}(1 + \cos 2x) - \frac{1}{4}\left(1 + 2\cos 2x + \cos^2 2x\right)\right) dx = \int \left(-\frac{1}{4}\cos^2 2x + \frac{1}{4}\right) dx.$$

Apply the double-angle identity again to get

$$\int \left(-\frac{1}{4}\left(\frac{1}{2}(1 + \cos 4x)\right) + \frac{1}{4}\right) dx = \frac{1}{8}\int (1 - \cos 4x)\, dx = \frac{1}{8}\left(x - \frac{1}{4}\sin 4x\right) + C.$$

(b) Using the identity $\sin x \cos x = \frac{1}{2}\sin 2x$, we get

$$\int \sin^2 x \cos^2 x\, dx = \frac{1}{4}\sin^2 2x.$$

Now apply the double-angle identity to get

$$\frac{1}{4}\int \frac{1}{2}(1 - \cos 4x)\, dx = \frac{1}{8}\left(x - \frac{1}{4}\sin 4x\right) + C.$$

Skills

21. Use the double-angle identity:

$$\int \sin^2 3x \, dx = \int \frac{1}{2}(1 - \cos 6x) \, dx = \frac{x}{2} - \frac{1}{12} \sin 6x + C.$$

23. Use a Pythagorean substitution:

$$\int \cos^5 x \, dx = \int \cos x (\cos^2 x)^2 \, dx = \int \cos x (1 - \sin^2 x)^2 \, dx.$$

Now substitute $u = \sin x$, so that $du = \cos x \, dx$:

$$\int \cos x (1 - \sin^2 x)^2 \, dx = \int (1 - u^2)^2 \, du$$

$$= \int (1 - 2u^2 + u^4) \, du$$

$$= u - \frac{2}{3}u^3 + \frac{1}{5}u^5 + C$$

$$= \sin x - \frac{2}{3} \sin^3 x + \frac{1}{5} \sin^5 x + C.$$

25. Use the substitution $u = 2x$, so that $\frac{1}{2} du = dx$. Then

$$\int \tan 2x \, dx = \frac{1}{2} \int \tan u \, du = \frac{1}{2} \ln |\sec u| = \frac{1}{2} \ln |\sec 2x| + C.$$

27. Use the substitution $u = 4x$, so that $\frac{1}{4} du = dx$. Then

$$\int \sec 4x \, dx = \frac{1}{4} \int \sec u \, du = \frac{1}{4} \ln |\sec u + \tan u| + C = \frac{1}{4} \ln |\sec 4x + \tan 4x| + C.$$

29. Start with

$$\cot^5 x = \cot^3 x (\csc^2 x - 1)$$
$$= \cot^3 x \csc^2 x - \cot^3 x$$
$$= \cot^3 x \csc^2 x - \cot x (\csc^2 x - 1)$$
$$= \cot^3 x \csc^2 x - \cot x \csc^2 x + \cot x$$
$$= (\cot^3 x - \cot x) \csc^2 x + \cot x.$$

Then using the substitution $u = \cot x$ so that $du = -\csc^2 x \, dx$, we get

$$\int \cot^5 x \, dx = \int \left((\cot^3 x - \cot x) \csc^2 x \right) dx + \int \cot x \, dx$$

$$= -\int (u^3 - u) \, du + \int \cot x \, dx$$

$$= -\frac{1}{4}u^4 + \frac{1}{2}u^2 - \ln |\csc x| + C$$

$$= -\frac{1}{4} \cot^4 x + \frac{1}{2} \cot^2 x - \ln |\csc x| + C.$$

31. As in the text, use integration by parts with $u = \sec 2x$ and $dv = \sec^2 2x\, dx$. Then we get $du = 2 \sec 2x \tan 2x$ and $v = \frac{1}{2} \tan 2x$, so that

$$\int \sec^3 2x\, dx = \frac{1}{2} \sec 2x \tan 2x - \int \sec 2x \tan^2 2x\, dx$$

$$= \frac{1}{2} \sec 2x \tan 2x - \int \sec 2x (\sec^2 2x - 1)\, dx$$

$$= \frac{1}{2} \sec 2x \tan 2x - \int \sec^3 2x\, dx + \int \sec 2x\, dx.$$

Use the substitution $u = 2x$ for the last integral, so the result is $\frac{1}{2} \ln |\sec 2x + \tan 2x|$. Add the remaining integral to both sides and divide by 2 to get

$$\int \sec^3 2x\, dx = \frac{1}{4} (\sec 2x \tan 2x + \ln |\sec 2x + \tan 2x|) + C.$$

33. We have, using the double-angle formula twice,

$$\cos^4 x = (\cos^2 x)^2 = \left(\frac{1}{2}(1 + \cos 2x) \right)^2$$

$$= \frac{1}{4}(1 + 2\cos 2x + \cos^2 2x)$$

$$= \frac{1}{4} \left(1 + 2\cos 2x + \frac{1}{2}(1 + \cos 4x) \right)$$

$$= \frac{3}{8} + \frac{1}{2} \cos 2x + \frac{1}{8} \cos 4x.$$

Thus

$$\int \cos^4 x\, dx = \int \left(\frac{3}{8} + \frac{1}{2} \cos 2x + \frac{1}{8} \cos 4x \right) dx = \frac{3}{8}x + \frac{1}{4} \sin 2x + \frac{1}{32} \sin 4x + C.$$

35. Using the substitution $u = \cot x$ so that $du = -\csc^2 x\, dx$, we have

$$\int \csc^4 x\, dx = \int \csc^2 x (\csc^2 x)\, dx$$

$$= \int \csc^2 x (\cot^2 x + 1)\, dx$$

$$= \int \csc^2 x \cot^2 x\, dx + \int \csc^2 x\, dx$$

$$= -\int u^2\, du + \int \csc^2 x\, dx + C$$

$$= -\frac{1}{3} u^3 - \cot x + C$$

$$= -\frac{1}{3} \cot^3 x - \cot x + C.$$

37. Use the identity $\cos^2 x = 1 - \sin^2 x$ followed by the substitution $u = \sin x$:

$$\begin{aligned}
\int \cos^3 x \sin^4 x \, dx &= \int \cos x (\cos^2 x) \sin^4 x \, dx \\
&= \int \cos x (1 - \sin^2 x) \sin^4 x \, dx \\
&= \int (1 - u^2) u^4 \, du \\
&= \frac{1}{5} u^5 - \frac{1}{7} u^7 + C \\
&= \frac{1}{5} \sin^5 x - \frac{1}{7} \sin^7 x + C.
\end{aligned}$$

39. Use the identity $\tan^2 x = \sec^2 x - 1$ followed by the substitution $u = \sec x$, $du = \sec x \tan x \, dx$:

$$\begin{aligned}
\int \sec x \tan^3 x \, dx &= \int \sec x \tan x (\tan^2 x) \, dx \\
&= \int (\sec^2 x - 1) \sec x \tan x \, dx \\
&= \int (u^2 - 1) \, du \\
&= \frac{1}{3} u^3 - u + C \\
&= \frac{1}{3} \sec^3 x - \sec x + C.
\end{aligned}$$

41. Using the double angle formula gives

$$\sin^2 3x \cos^2 3x = \frac{1}{2}(1 - \cos 6x) \cdot \frac{1}{2}(1 + \cos 6x) = \frac{1}{4}(1 - \cos^2 6x) = \frac{1}{4}\left(1 - \frac{1}{2}(1 + \cos 12x)\right)$$
$$= \frac{1}{8} - \frac{1}{8}\cos 12x.$$

Thus

$$\int \sin^2 3x \cos^2 3x \, dx = \frac{1}{8} \int (1 - \cos 12x) \, dx = \frac{1}{8}x - \frac{1}{96}\sin 12x + C.$$

43. Make the substitution $u = \sec x$, so that $du = \sec x \tan x \, dx$:

$$\int \sec^8 x \tan x \, dx = \int \sec^7 x (\sec x \tan x) \, dx = \int u^7 \, du = \frac{1}{8}u^8 + C = \frac{1}{8}\sec^8 x + C.$$

45. Use the Pythagorean identity $\tan^2 x = \sec^2 x - 1$ to remove all but one copy of $\tan x$, and then use the substitution $u = \sec x$, $du = \sec x \tan x \, dx$:

$$\begin{aligned}
\int \sec^3 x \tan^5 x \, dx &= \int \sec^3 x \tan x (\tan^2 x)^2 \, dx \\
&= \int \sec^2 x (\sec^2 x - 1)^2 (\sec x \tan x) \, dx \\
&= \int u^2 (u^2 - 1)^2 \, du \\
&= \int (u^6 - 2u^4 + u^2) \, du \\
&= \frac{1}{7}u^7 - \frac{2}{5}u^5 + \frac{1}{3}u^3 + C \\
&= \frac{1}{7}\sec^7 x - \frac{2}{5}\sec^5 x + \frac{1}{3}\sec^3 x + C.
\end{aligned}$$

47. Use the identity $\cos^2 x = 1 - \sin^2 x$ to remove all but one of the copies of $\cos x$, and then substitute $u = \sin x$, $du = \cos x\, dx$:

$$\int \sin^{13} x \cos^5 x\, dx = \int \sin^{13} x (\cos^2 x)^2 \cos x\, dx$$

$$= \int \sin^{13} x (1 - \sin^2 x)^2 \cos x\, dx$$

$$= \int u^{13} (1 - u^2)^2\, du$$

$$= \int (u^{17} - 2u^{15} + u^{13})\, du$$

$$= \frac{1}{18} u^{18} - \frac{1}{8} u^{16} + \frac{1}{14} u^{14} + C$$

$$= \frac{1}{18} \sin^{18} x - \frac{1}{8} \sin^{16} x + \frac{1}{14} \sin^{14} x + C.$$

49. Start with

$$\int \csc x \cot^2 x\, dx = \int \csc x (\csc^2 x - 1)\, dx = \int \csc^3 x\, dx - \int \csc x\, dx.$$

To integrate $\csc^3 x$, use integration by parts with $u = \csc x$ and $dv = \csc^2 x\, dx$, so that $du = -\cot x \csc x\, dx$ and $v = -\cot x$. Then

$$\int \csc x \cot^2 x\, dx = \int \csc^3 x\, dx - \int \csc x\, dx$$

$$= -\csc x \cot x - \int \csc x \cot^2 x\, dx - \int \csc x\, dx$$

$$= -\csc x \cot x - \int \csc x \cot^2 x\, dx + \ln|\csc x + \cot x|.$$

Solving for the remaining integral algebraically gives

$$\int \csc x \cot^2 x = \frac{1}{2} \left(\ln|\csc x + \cot x| - \csc x \cot x \right) + C.$$

51. The integrand is equal to $\tan x \sec^3 x$. Write this as $\sec^2 x (\sec x \tan x)$ and then use the substitution $u = \sec x$, $du = \sec x \tan x\, dx$:

$$\int \frac{\tan x}{\cos^3 x}\, dx = \int \sec^2 x (\sec x \tan x)\, dx = \int u^2\, du = \frac{1}{3} u^3 + C = \frac{1}{3} \sec^3 x + C.$$

53. $\displaystyle\int \sin x \sec x\, dx = \int \frac{\sin x}{\cos x}\, dx = \int \tan x\, dx = \ln|\sec x| + C.$

55. $\displaystyle\int \cos^3 x \sec^2 x\, dx = \int \frac{\cos^3 x}{\cos^2 x}\, dx = \int \cos x\, dx = \sin x + C.$

57. $\displaystyle\int \tan^2 x \csc x\, dx = \int \frac{\sin^2 x}{\cos^2 x} \cdot \frac{1}{\sin x}\, dx = \int \frac{\sin x}{\cos^2 x}\, dx = \int \sec x \tan x\, dx = \sec x + C.$

59. First rewrite the integrand as $\sin x \cos x \ln(\cos x)$. Then use the u-substitution $u = \cos x$, $du = -\sin x\, dx$:

$$\int \frac{\cos x \ln(\cos x)}{\csc x}\, dx = \int \sin x \cos x \ln(\cos x)\, dx = -\int u \ln u\, du.$$

To integrate this, use integration by parts with $v = \ln u$ and $dw = u\,du$, so that $dv = \frac{1}{u}\,du$ and $w = \frac{1}{2}u^2$. Then

$$
\begin{aligned}
- \int u \ln u\,du &= -\frac{1}{2}u^2 \ln u + \int \frac{1}{2}u^2 \cdot \frac{1}{u}\,du \\
&= -\frac{1}{2}u^2 \ln u + \frac{1}{2}\int u\,du \\
&= -\frac{1}{2}u^2 \ln u + \frac{1}{4}u^2\,du + C \\
&= -\frac{1}{2}\cos^2 x \ln \cos x + \frac{1}{4}\cos^2 x + C.
\end{aligned}
$$

61. Use the Pythagorean identity $\sin^2 x = 1 - \cos^2 x$ followed by the substitution $u = \cos x$:

$$
\begin{aligned}
\int (\sin x \sqrt{\cos x})^3\,dx &= \int \sin^3 x (\cos x)^{3/2}\,dx \\
&= \int \sin x (1 - \cos^2 x)(\cos x)^{3/2}\,dx \\
&= -\int (1 - u^2) u^{3/2}\,du \\
&= \int (u^2 - 1) u^{3/2}\,du \\
&= \frac{2}{9}u^{9/2} - \frac{2}{5}u^{5/2} + C \\
&= \frac{2}{9}\cos^{9/2} x - \frac{2}{5}\cos^{5/2} x + C.
\end{aligned}
$$

63. Use the substitution $u = \cosh x$, $du = \sinh x\,dx$:

$$
\int \sinh x \cosh^2 x\,dx = \int u^2\,du = \frac{1}{3}u^3 + C = \frac{1}{3}\cosh^3 x + C.
$$

65. Use the substitution $u = \cosh x$, $du = \sinh x\,dx$:

$$
\begin{aligned}
\int \tanh x \cosh^7 x\,dx &= \int \frac{\sinh x}{\cosh x} \cosh^7 x\,dx \\
&= \int \cosh^6 x \sinh x\,dx \\
&= \int u^6\,du \\
&= \frac{1}{7}u^7 + C \\
&= \frac{1}{7}\cosh^7 x + C.
\end{aligned}
$$

67. Use the Pythagorean identity $\sin^2 x = 1 - \cos^2 x$ followed by the substitution $u = \cos x$:

$$
\begin{aligned}
\int_0^\pi \sin^5 x \, dx &= \int_0^\pi (\sin^2 x)^2 \sin x \, dx \\
&= \int_0^\pi (1 - \cos^2 x)^2 \sin x \, dx \\
&= -\int_{x=0}^{x=\pi} (1 - u^2)^2 \, du \\
&= \left[-\frac{1}{5}u^5 + \frac{2}{3}u^3 - u \right]_{x=0}^{x=\pi} \\
&= \left[-\frac{1}{5}\cos^5 x + \frac{2}{3}\cos^3 x - \cos x \right]_0^\pi \\
&= \left(\frac{1}{5} - \frac{2}{3} + 1 \right) - \left(-\frac{1}{5} + \frac{2}{3} - 1 \right) = \frac{16}{15}.
\end{aligned}
$$

69. Use the Pythagorean identity $\cos^2 x = 1 - \sin^2 x$ followed by the substitution $u = \sin x$:

$$
\begin{aligned}
\int_{-\pi/2}^{\pi/2} \cos^3 x \, dx &= \int_{-\pi/2}^{\pi/2} \cos^2 x \cos x \, dx \\
&= \int_{-\pi/2}^{\pi/2} (1 - \sin^2 x) \cos x \, dx \\
&= \int_{x=-\pi/2}^{x=\pi/2} (1 - u^2) \, du \\
&= \left[u - \frac{1}{3}u^3 \right]_{x=-\pi/2}^{x=\pi/2} \\
&= \left[\sin x - \frac{1}{3}\sin^3 x \right]_{-\pi/2}^{\pi/2} \\
&= \left(1 - \frac{1}{3} \right) - \left(-1 + \frac{1}{3} \right) = \frac{4}{3}.
\end{aligned}
$$

71. Use the Pythagorean identity $\sin^2 x = 1 - \cos^2 x$ followed by the substitution $u = \cos x$:

$$
\begin{aligned}
\int_0^\pi \sin^3 x \cos^2 x \, dx &= \int_0^\pi \sin^2 x \cos^2 x \sin x \, dx \\
&= \int_0^\pi (1 - \cos^2 x) \cos^2 x \sin x \, dx \\
&= \int_{x=0}^{x=\pi} (u^2 - 1) u^2 \, du \\
&= \left[\frac{1}{5}u^5 - \frac{1}{3}u^3 \right]_{x=0}^{x=\pi} \\
&= \left[\frac{1}{5}\cos^5 x - \frac{1}{3}\cos^3 x \right]_0^\pi \\
&= \left(-\frac{1}{5} + \frac{1}{3} \right) - \left(\frac{1}{5} - \frac{1}{3} \right) = \frac{4}{15}.
\end{aligned}
$$

73. Use the identity $\sec^2 x = \tan^2 x + 1$ followed by the substitution $u = \tan x$, $du = \sec^2 x\, dx$:

$$\int_0^{\pi/4} \sec^4 x \tan^2 x\, dx = \int_0^{\pi/4} \sec^2 x \tan^2 x \sec^2 x\, dx$$

$$= \int_0^{\pi/4} (\tan^2 x + 1) \tan^2 x \sec^2 x\, dx$$

$$= \int_{x=0}^{x=\pi/4} (u^2 + 1)u^2\, du$$

$$= \left[\frac{1}{5}u^5 + \frac{1}{3}u^3\right]_{x=0}^{x=\pi/4}$$

$$= \left[\frac{1}{5}\tan^5 x + \frac{1}{3}\tan^3 x\right]_0^{\pi/4}$$

$$= \frac{1}{5} + \frac{1}{3} = \frac{8}{15}.$$

75. Since the integrand is equal to $\tan^2 x \sec^2 x$, use the substitution $u = \tan x$, $du = \sec^2 x\, dx$:

$$\int_0^{\pi/4} \frac{\sin^2 x}{\cos^4 x}\, dx = \int_0^{\pi/4} \tan^2 x \sec^2 x\, dx$$

$$= \int_{x=0}^{x=\pi/4} u^2\, du$$

$$= \left[\frac{1}{3}u^3\right]_{x=0}^{x=\pi/4}$$

$$= \left[\frac{1}{3}\tan^3 x\right]_0^{\pi/4}$$

$$= \frac{1}{3}.$$

77. Start by finding an antiderivative for $f(x)$. Use the identity $\sin^2 x = 1 - \cos^2 x$ followed by the substitution $u = \cos x$:

$$F(x) = \int f(x)\, dx = 4\int \sin^3 x \cos^2 x\, dx$$

$$= 4\int \sin^2 x \cos^2 x \sin x\, dx$$

$$= 4\int (1 - \cos^2 x) \cos^2 x \sin x\, dx$$

$$= -4\int (1 - u^2)u^2\, du$$

$$= \frac{4}{5}u^5 - \frac{4}{3}u^3$$

$$= \frac{4}{5}\cos^5 x - \frac{4}{3}\cos^3 x.$$

(a) We want $\int_{-\pi}^{\pi} f(x)\, dx$, which is just $F(\pi) - F(-\pi)$, or

$$\frac{4}{5}\cos^5 \pi - \frac{4}{3}\cos^3 \pi - \left(\frac{4}{5}\cos^5(-\pi) - \frac{4}{3}\cos^3(-\pi)\right) = 0.$$

(b) Since $f(x)$ is negative for $x < 0$ and positive for $x > 0$, we want

$$-\int_{-\pi}^{0} f(x)\, dx + \int_0^{\pi} f(x)\, dx = -(F(0) - F(-\pi)) + F(\pi) - F(0) = F(\pi) + F(-\pi) - 2F(0).$$

But this is simply

$$\frac{4}{5}\cos^5 \pi - \frac{4}{3}\cos^3 \pi + \left(\frac{4}{5}\cos^5(-\pi) - \frac{4}{3}\cos^3(-\pi)\right) - 2\left(\frac{4}{5} - \frac{4}{3}\right) = \frac{32}{15}.$$

(c) The average value of $f(x)$ on $\left[0, \frac{\pi}{2}\right]$ is

$$\frac{1}{\frac{\pi}{2} - 0}\int_0^{\pi/2} f(x)\,dx = \frac{2}{\pi}\left(F\left(\frac{\pi}{2}\right) - F(0)\right) = \frac{2}{\pi}\left(\frac{4}{3} - \frac{4}{5}\right) = \frac{16}{15\pi}.$$

79. (a) The area between $f(x)$ and $f(x) + \frac{1}{2}$ on $[0,3]$ is

$$\int_0^3 \left(f(x) + \frac{1}{2}\right) - f(x)\,dx = \int_0^3 \frac{1}{2}\,dx = \left[\frac{x}{2}\right]_0^3 = \frac{3}{2}.$$

(b) The curves $f(x)$ and $-f(x) + \frac{1}{2}$ cross where $2f(x) = 2\sin^2 \pi x = \frac{1}{2}$, or where $\sin^2 \pi x = \frac{1}{4}$. This happens when $\sin \pi x = \pm\frac{1}{2}$; on $[0,3]$, these values are $\frac{1}{6}, \frac{5}{6}, \frac{7}{6}, \frac{11}{6}, \frac{13}{6}$, and $\frac{17}{6}$. Now,

$$\int \left(2\sin^2 \pi x - \frac{1}{2}\right)dx = \int \left(1 - \cos 2\pi x - \frac{1}{2}\right)dx = \frac{x}{2} - \frac{1}{2\pi}\sin 2\pi x.$$

Note that $f(x) = \sin^2(\pi x) = \frac{1}{2}(1 - \cos(2\pi x))$, so that $f(x)$ is periodic with period 1; hence, $-f(x) + \frac{1}{2}$ is as well. Thus the area between the graphs on $[0,3]$ is three times the area between them on $[0,1]$. Now, on $[0,1]$, we see that $f(x) \geq -f(x) + \frac{1}{2}$ on $\left[\frac{1}{6}, \frac{5}{6}\right]$ and $f(x) \leq -f(x) + \frac{1}{2}$ elsewhere, so that the area between the curves on $[0,3]$ is

$$3\int_0^1 \left|f(x) - \left(-f(x) + \frac{1}{2}\right)\right|dx = 3\left(-\int_0^{1/6}\left(2f(x) - \frac{1}{2}\right)dx + \int_{1/6}^{5/6}\left(2f(x) - \frac{1}{2}\right)dx\right.$$

$$\left. - \int_{5/6}^1 \left(2f(x) - \frac{1}{2}\right)dx\right)$$

$$= 3\left(-\left[\frac{x}{2} - \frac{1}{2\pi}\sin 2\pi x\right]_0^{1/6} + \left[\frac{x}{2} - \frac{1}{2\pi}\sin 2\pi x\right]_{1/6}^{5/6}\right.$$

$$\left. - \left[\frac{x}{2} - \frac{1}{2\pi}\sin 2\pi x\right]_{5/6}^1\right)$$

$$= \frac{1}{2} + \frac{3\sqrt{3}}{\pi}.$$

Applications

81. With $f(x) = \ln(2\cos x)$, we have $f'(x) = \frac{1}{2\cos x}(2\cos x)' = -\frac{\sin x}{\cos x} = -\tan x$, so that

$$\sqrt{1 + f'(x)^2} = \sqrt{1 + \tan^2 x} = \sqrt{\sec^2 x} = \sec x.$$

Thus the length of the filament is

$$\int_{-\pi/3}^{\pi/3} \sec x\,dx = [\ln|\sec x + \tan x|]_{-\pi/3}^{\pi/3}$$

$$= \ln(2 + \sqrt{3}) - \ln(2 - \sqrt{3}) = \ln\frac{2 + \sqrt{3}}{2 - \sqrt{3}}.$$

Proofs

83. (a) Multiply $\csc x$ by $1 = \frac{\csc x + \cot x}{\csc x + \cot x}$ and simplify to get

$$\int \csc x \, dx = \int \csc x \cdot \frac{\sec x + \cot x}{\csc x + \cot x} \, dx = \int \frac{\csc^2 x + \csc x \cot x}{\csc x + \cot x} \, dx.$$

Now use the substitution $u = \csc x + \cot x$, so that $du = -(\csc x \cot x + \csc^2 x) \, dx$, which is the negative of numerator of the above integrand. Then

$$\int \csc x \, dx = -\int \frac{1}{u} \, du = -\ln|u| = -\ln|\csc x + \cot x| + C.$$

(b) If $\csc x + \cot x > 0$, then

$$(-\ln|\csc x + \cot x|)' = (-\ln(\csc x + \cot x))' = \frac{-1}{\csc x + \cot x} \cdot (\csc x + \cot x)'$$

$$= \frac{1}{-\csc x - \cot x} \cdot (-\csc x \cot x - \csc^2 x) = \frac{\csc x(-\csc x - \cot x)}{-\csc x - \cot x} = \csc x.$$

If $\csc x + \cot x < 0$, then

$$(-\ln|\csc x + \cot x|)' = (-\ln(-\csc x - \cot x))' = \frac{-1}{-\csc x - \cot x} \cdot (-\csc x - \cot x)'$$

$$= \frac{1}{\csc x + \cot x} \cdot (\csc x \cot x + \csc^2 x) = \frac{\csc x(\csc x + \cot x)}{\csc x + \cot x} = \csc x.$$

85. (a) Note first that, using the double-angle formula,

$$\int \sin^2 x \, dx = \frac{1}{2} \int (1 - \cos 2x) \, dx = \frac{x}{2} - \frac{1}{4} \sin 2x.$$

Thus, using the reduction formula,

$$\int \sin^4 x \, dx = -\frac{1}{4} \sin^3 x \cos x + \frac{3}{4} \int \sin^2 x \, dx$$

$$= -\frac{1}{4} \sin^3 x \cos x + \frac{3}{4} \left(\frac{x}{2} - \frac{1}{4} \sin 2x \right)$$

$$= -\frac{1}{4} \sin^3 x \cos x + \frac{3x}{8} - \frac{3}{16} \sin 2x$$

$$\int \sin^8 x \, dx = -\frac{1}{8} \sin^7 x \cos x + \frac{7}{8} \int \sin^6 x \, dx$$

$$= -\frac{1}{8} \sin^7 x \cos x + \frac{7}{8} \left(-\frac{1}{6} \sin^5 x \cos x + \frac{5}{6} \int \sin^4 x \, dx \right)$$

$$= -\frac{1}{8} \sin^7 x \cos x - \frac{7}{48} \sin^5 x \cos x + \frac{35}{48} \left(-\frac{1}{4} \sin^3 x \cos x + \frac{3x}{8} - \frac{3}{16} \sin 2x \right)$$

$$= -\frac{1}{8} \sin^7 x \cos x - \frac{7}{48} \sin^5 x \cos x - \frac{35}{192} \sin^3 x \cos x + \frac{35x}{128} - \frac{35}{256} \sin 2x.$$

(b) It is called a reduction formula because it reduces the power of $\sin x$ under the integration sign by two, until it gets to 2, where we know how to integrate it.

(c) Let $u = \sin^{k-1} x$ and $dv = \sin x\, dx$, so that $v = -\cos x\, dx$ and $du = (k-1)\sin^{k-2} x \cos x\, dx$. Then

$$\int \sin^k x\, dx = -\sin^{k-1} x \cos x + (k-1)\int \sin^{k-2} x \cos^2 x\, dx$$

$$= -\sin^{k-1} x \cos x + (k-1)\int \sin^{k-2} x (1 - \sin^2 x)\, dx$$

$$= -\sin^{k-1} x \cos x + (k-1)\int \sin^{k-2} x\, dx - (k-1)\int \sin^k x\, dx.$$

Now add the final term to the both sides and divide through by k, giving

$$\int \sin^k x\, dx = -\frac{1}{k}\sin^{k-1} x \cos x + \frac{k-1}{k}\int \sin^{k-2} x\, dx.$$

Thinking Forward

Trigonometric Substitution

▷ Since $u = \sin^{-1} x$, we have $du = \frac{1}{\sqrt{1-x^2}}\, dx$. Alternatively, since $x = \sin u$, we also have $dx = \cos u\, du$.

▷ If we substitute $\sin u$ for x, and $\cos u\, du$ for dx, then we get

$$\int \frac{1}{x\sqrt{1-x^2}}\, dx = \int \frac{\cos u}{\sin u \sqrt{1 - \sin^2 u}}\, du.$$

▷ Since $1 - \sin^2 u = \cos^2 u$, this is equivalent to

$$\int \frac{\cos u}{\sin u \cos u}\, du = \int \csc u\, du.$$

▷ We know how to integrate $\csc u$:

$$\int \frac{1}{x\sqrt{1-x^2}}\, dx = \int \csc u\, du = -\ln|\csc u + \cot u| + C$$

$$= -\ln\left|\csc(\sin^{-1} x) + \cot(\sin^{-1} x)\right| + C.$$

5.5   Trigonometric Substitution

Thinking Back

Domains and ranges of inverse trigonometric functions

▷ (a) The domain of $\sin^{-1} x$ is the range of $\sin x$, which is $[-1, 1]$. The range of $\sin^{-1} x$ is defined to be $\left[-\frac{\pi}{2}, \frac{\pi}{2}\right]$.

(b) A graph of $f(x) = \sin^{-1} x$ is

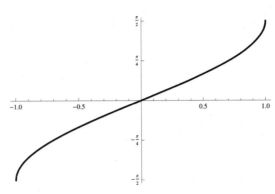

(c) The domain of $\sin^{-1} x$ is the set of y-values covered by the unit circle, which is $[-1, 1]$. The range of $\sin^{-1} x$ is defined to be the set of angles between (x, y) and the positive x axis as the point moves from $y = -1$ to $y = 1$ along the right half of the unit circle.

▷ (a) The domain of $\tan^{-1} x$ is the range of $\tan x$, which is $(-\infty, \infty)$. The range of $\tan^{-1} x$ is defined to be $\left(-\frac{\pi}{2}, \frac{\pi}{2}\right)$.

(b) A graph of $f(x) = \tan^{-1} x$ is

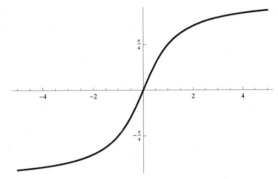

(c) The domain of $\tan^{-1} x$ is the set of y-to-x ratios covered by the unit circle, which is $(-\infty, \infty)$. The range of $\tan^{-1} x$ is defined to be the set of angles between (x, y) and the positive x axis as the point moves from $(0, -1)$ (which has a ratio of $-\infty$) to $(0, 1)$ (which has a ratio of ∞) along the right half of the unit circle.

▷ (a) The domain of $\sec^{-1} x$ is the range of $\sec x$, which is $(-\infty, -1] \cup [1, \infty)$. The range of $\sec^{-1} x$ is defined to be $\left[0, \frac{\pi}{2}\right) \cup \left(\frac{\pi}{2}, \pi\right]$.

(b) A graph of $f(x) = \sec^{-1} x$ is

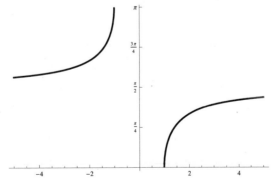

(c) The domain of $\sec^{-1} x$ is the set of inverses of x coordinates covered by the unit circle, since $\sec x = \frac{1}{\cos x}$. This set is $(-\infty, -1] \cup [1, \infty)$. The range of $\sec^{-1} x$ is defined to be values that come from one such set of coordinates, namely $\left[0, \frac{\pi}{2}\right) \cup \left(\frac{\pi}{2}, \pi\right]$.

▷ (a) The domain of $y = \sin^{-1}\frac{x}{3}$ is the range of $x = 3\sin y$, which is $[-3,3]$. The range of $\sin^{-1}\frac{x}{3}$ is still $\left[-\frac{\pi}{2}, \frac{\pi}{2}\right]$, since as x goes from -3 to 3, $\frac{x}{3}$ goes from -1 to 1, so that this function covers all possible arguments to $\sin^{-1} x$.

(b) A graph of $f(x) = \sin^{-1}\frac{x}{3}$ is

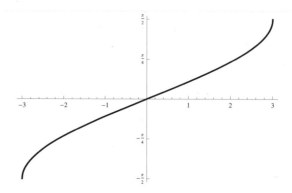

(c) The domain of $\sin^{-1}\frac{x}{3}$ is the set of y-values covered by the unit circle, which is $[-1,1]$, expanded by a factor of 3, so it is $[-3,3]$. The range of $\sin^{-1}\frac{x}{3}$ is defined to be values that come from choosing an angle between $-\frac{\pi}{2}$ and $\frac{\pi}{2}$ for any given argument (as opposed to any of the other angles that have the same sine, such as that plus multiples of 2π).

Working with inverse trigonometric functions

▷ From the diagram

whether $x > 0$ (first quadrant) or $x < 0$ (fourth quadrant), $\sin u = \frac{x}{\sqrt{1-x^2}}$.

▷ If $x = 3 \sin u$, then $\sin u = \frac{x}{3}$. Then from the diagram

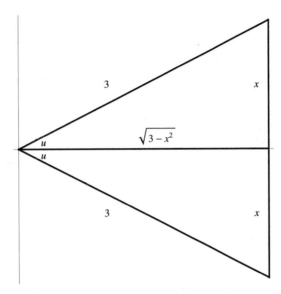

whether $x > 0$ (first quadrant) or $x < 0$ (fourth quadrant), $\cot u = \frac{\sqrt{3-x^2}}{x}$.

▷ From the diagram

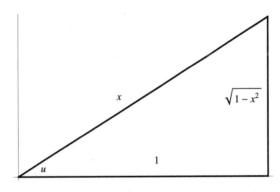

since $x = \sec u$ we see that $\tan u = \sqrt{1 - x^2}$.

▷ From the diagram

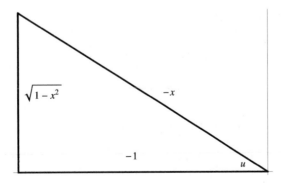

since $x = \sec u$ we see that $\tan u = -\sqrt{1 - x^2}$. (Note that we must have $x < 0$ since u is in the second quadrant, so the hypotenuse, $-x$, is positive).

Review of trigonometric integrals

▷ We have

$$\int \sec x\, dx = \int \sec x \cdot \frac{\sec x + \tan x}{\sec x + \tan x}\, dx$$

$$= \int \frac{\sec^2 x + \sec x \tan x}{\sec x + \tan x}\, dx$$

$$= \int \frac{d(\sec x + \tan x)}{\sec x + \tan x}$$

$$= \ln|\sec x + \tan x| + C.$$

▷ $\int \sec^2 x\, dx = \tan x + C.$

▷ Integrate by parts with $u = \sec x$ and $dv = \sec^2 x\, dx$, so that $du = \sec x \tan x\, dx$ and $v = \tan x$. Then

$$\int \sec^3 x\, dx = \sec x \tan x - \int \tan x \cdot \sec x \tan x\, dx$$

$$= \sec x \tan x - \int \sec x (\sec^2 x - 1)\, dx$$

$$= \sec x \tan x + \int \sec x\, dx - \int \sec^3 x\, dx$$

$$= \sec x \tan x + \ln|\sec x + \tan x| - \int \sec^3 x\, dx.$$

Add the final integral to both sides and divide through by 2 to get

$$\int \sec^3 x\, dx = \frac{1}{2}(\sec x \tan x + \ln|\sec x + \tan x|) + C.$$

▷ We have, using the results immediately above together with the identity $\tan^2 x = \sec^2 x - 1$,

$$\int \sec x \tan^2 x\, dx = \int \sec x (\sec^2 x - 1)\, dx$$

$$= \int \sec^3 x\, dx - \int \sec x\, dx$$

$$= \frac{1}{2}(\sec x \tan x - \ln|\sec x + \tan x|) + C.$$

Review of other integration methods

▷ $\int \frac{1}{\sqrt{1-x^2}}\, dx = \sin^{-1} x + C$ since we recognize the integrand as the derivative of $\sin^{-1} x$.

▷ Use the substitution $u = x^2 - 9$ so that $du = 2x\, dx$; then

$$\int \frac{x}{\sqrt{x^2 - 9}}\, dx = \frac{1}{2}\int \frac{1}{\sqrt{u}}\, du = \sqrt{u} + C = \sqrt{x^2 - 9} + C.$$

▷ We have $\int \frac{9 + x^2}{x}\, dx = \int \frac{9}{x}\, dx + \int x\, dx = 9\ln|x| + \frac{1}{2}x^2 + C.$

▷ First reduce the fraction using long division: $x^4 + 1 = (x^2 + 1)(x^2 - 1) + 2$. Then we recognize $\frac{2}{1+x^2}$ as the derivative of $2\tan^{-1} x$, so that

$$\int \frac{x^4 + 1}{x^2 + 1}\, dx = \int (x^2 - 1)\, dx + \int \frac{2}{x^2 + 1}\, dx = \frac{1}{3}x^3 - x + 2\tan^{-1} x + C.$$

▷ Use the substitution $u = 2x$, so that $\frac{1}{2}\,du = dx$; then

$$\int \frac{1}{1+4x^2}\,dx = \int \frac{1}{1+(2x)^2}\,dx = \frac{1}{2}\int \frac{1}{1+u^2}\,du = \frac{1}{2}\tan^{-1}u + C = \frac{1}{2}\tan^{-1}2x + C.$$

▷ Use the substitution $u = x^2 + 1$, so that $du = 2x\,dx$ and also $x^2 = u - 1$. Then

$$\int \frac{x^3}{\sqrt{x^2+1}}\,dx = \frac{1}{2}\int \frac{(u-1)\,du}{\sqrt{u}}$$
$$= \frac{1}{2}\int (u^{1/2} - u^{-1/2})\,du$$
$$= \frac{1}{3}u^{3/2} - u^{1/2} + C$$
$$= \frac{1}{3}(x^2+1)^{3/2} - (x^2+1)^{1/2} + C.$$

Concepts

1. (a) This substitution will work, giving $dx = 2\sec u \tan u\,du$ and $\frac{1}{x^2-4} = \frac{1}{4\sec^2 u - 4} = \frac{1}{4}\cot^2 u$, so that

$$\int \frac{1}{x^2-4}\,dx = \frac{2}{4}\int \sec u \tan u \cot^2 u\,du = \frac{1}{2}\int \csc u\,du = -\frac{1}{2}\ln|\csc u + \cot u|.$$

However, the back-substitution is complicated. It is easier to recognize this as similar to the derivative of $\tanh^{-1}x$, which is $\frac{1}{1-x^2}$, perform algebraic manipulations to get this integrand into that form, and work from there.

(b) True. See Example 4.

(c) As with part (a), this substitution will work, giving $dx = 2\sec^2 u\,du$ and $\frac{1}{x^2+4} = \frac{1}{4\tan^2 u + 4} = \frac{1}{4}\cos^2 u$, so that

$$\int \frac{1}{x^2+4}\,dx = \frac{2}{4}\int \sec^2 u \cos^2 u\,du = \frac{1}{2}\int 1\,du = \frac{1}{2}u + C = \frac{1}{2}\tan^{-1}\frac{x}{2} + C.$$

This can be done algebraically as well, by recognizing the integrand as similar to the derivative of $\tan^{-1}x$, performing algebraic manipulations to get it into that form, and proceeding from there.

(d) False. The denominator becomes $(x^2+4)^{5/2} = (4\sin^2 u + 4)^{5/2} = 2^5(\sin^2 u + 1)^5$, which is not easily reducible further. It would be better to try $x = 2\tan u$.

(e) False. For example, it is of no help if trying to integrate $\ln(\sqrt[3]{1+\sqrt{x^2-a^2}})$.

(f) True. The result of a trigonometric substitution is another integral in the variable u.

(g) False. In fact, possible values for x are in the range $[-a, a]$; this range is connected and the integral will have a single solution there.

(h) True. Since the range of $\sec u$ is $(-\infty, -1] \cup [1, \infty)$, the range of $a\sec u$ is $(-\infty, -a] \cup [a, \infty)$. Since this range is disconnected, the integral may have different solutions on the two halves.

3. (a) If u is a function of x, then using implicit differentiation on $x = \tan u$ with $u = u(x)$ gives

$$1 = \sec^2 u \frac{du}{dx}, \quad \text{so} \quad dx = \sec^2 u\,du$$

after multiplying through by dx.

(b) Differentiating gives $\frac{dx}{du} = \sec^2 u$; multiplying through by du gives $dx = \sec^2 u\, du$.

5. (a) With $u = \sin^{-1} x$, we have $du = \frac{1}{\sqrt{1-x^2}}\, dx$, and also $x = \sin u$. Then

$$\int \frac{1}{x^2\sqrt{1-x^2}}\, dx = \int \frac{1}{\sin^2 u}\, du = \int \csc^2 u\, du = -\cot u + C = -\cot(\sin^{-1} x) + C.$$

(b) Using instead the substitution $x = \sin u$ with $dx = \cos u\, du$, we get

$$\int \frac{1}{x^2\sqrt{1-x^2}}\, dx = \int \frac{1}{\sin^2 u\sqrt{1-\sin^2 u}}\cos u\, du$$

$$= \int \frac{1}{\sin^2 u \cos u}\cos^2 u\, du$$

$$= \int \csc^2 u\, du$$

$$= -\cot u + C = -\cot(\sin^{-1} x) + C.$$

(c) The second method is more straightforward in that no back-substitution is involved, but the two methods both require that the substitution $x = \sin u$ or its inverse be recognized as appropriate for the problem at hand.

7. We could use it, but we choose instead to use $x = \sin u$. This is applicable when $a^2 - x^2$ appears in the integrand; both substitutions give similar results, allowing expressions like $\sqrt{a^2 - x^2}$ to be simplified. There is no substantial difference between them, except that using $x = \cos u$ means that $dx = -\sin u\, du$, introducing an unnecessary minus sign.

9. The substitution $x = a\sin u$ makes sense only for integrals whose domain is contained in $[-a, a]$ (see the previous exercise), and that is precisely the domain of $\sqrt{a^2 - x^2}$, while the domain of $a^2 - x^2$ is $\mathbb{R}$. The substitution $x = a\tan u$, on the other hand, makes sense for any domain of x values, since the range of $a\tan u$ is $\mathbb{R}$.

11. (a) No. Simplify the integrand algebraically to get $4x^{-1} + x$ and integrate each term.

(b) No. Since the derivative of the denominator is twice the numerator, use the substitution $u = 4 + x^2$.

(c) Yes. The substitution $x = 2\tan u$ results in the integral $\int (2\tan^2 u)\, du$, which can be solved using the methods of the previous section. This integral can also be solved using integration by parts with $u = x$, but it is more complicated.

(d) No. The numerator can be written $16 - x^4 = (4 - x^2)(4 + x^2)$, so that $\int \frac{16-x^4}{4+x^2}\, dx = \int (4 - x^2)\, dx$, which can be integrated term by term.

13. If $x = a\sin u$, then (since we assume $a > 0$, see Theorem 5.18)

$$\sqrt{a^2 - x^2} = \sqrt{a^2 - a^2\sin^2 u} = \sqrt{a^2(1 - \sin^2 u)} = \sqrt{a^2\cos^2 u} = a\,|\cos u|.$$

Since $x = a\sin u$, and $\sin u \in [-1, 1]$, we see that $a\sin u \in [-a, a]$, so that $a^2 - x^2$ makes sense. Since u is in the range of $\sin^{-1}$, which is $\left(-\frac{\pi}{2}, \frac{\pi}{2}\right)$, we see that u is in the first or fourth quadrants. But $\cos u$ is positive for u in the first or fourth quadrants, so that $a\,|\cos u| = a\cos u$.

15. Simplifying gives

$$\sqrt{a^2\sec^2 u - a^2} = \sqrt{a^2(\sec^2 u - 1)} = \sqrt{a^2\tan^2 u} = a\,|\tan u|.$$

If $x < -a$, then $\sec u < -1$, so that $u \in \left(\frac{\pi}{2}, \pi\right]$, which is a second quadrant angle. Since $\tan u$ is negative in the second quadrant, we see that here $a \left|\tan u\right| = -a \tan u$. If $x > a$, then $\sec u > 1$, so that $u \in \left[0, \frac{\pi}{2}\right)$, which is a first quadrant angle. Since $\tan u$ is positive in the first quadrant, we see that here $a \left|\tan u\right| = a \tan u$.

17. Since $1 - x^2 < 0$ for $x \in [2, 3]$, the integrand is not defined there, so the integral does not make sense.

19. For example, Exercises 40, 42, and 46. Exercises 40 and 42 can be solved using the substitution $u = x^2 + 1$, and Exercise 46 can be solved by simplifying algebraically and integrating term by term. Other possible choices include Exercises 41, 54, 60, 67, 69, 70, 71, and 73.

21. From the figure, we see that whether x is positive or negative, $\cos(\sin^{-1} x) = \sqrt{1 - x^2}$.

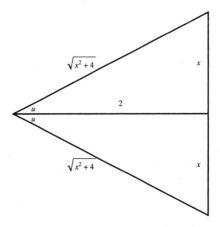

23. From the figure, we see that whether x is positive or negative, $\sin\left(\tan^{-1} \frac{x}{2}\right) = \frac{x}{\sqrt{x^2 + 4}}$.

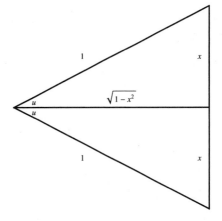

25. Since $x - 5 = 3 \sin u$, we have $\sin u = \frac{x-5}{3}$. From the figure, whether x is positive or negative, $\tan u = \frac{x-5}{\sqrt{9 - (x-5)^2}}$, so that

$$\tan^2 u = \frac{(x-5)^2}{9 - (x-5)^2}.$$

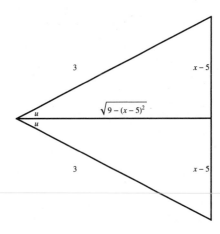

27. Since $\sin(2\theta) = 2\sin\theta\cos\theta$, we have

$$\sin(2\cos^{-1}x) = 2\sin(\cos^{-1}x)\cos(\cos^{-1}x) = 2\sqrt{1-x^2}\,x = 2x\sqrt{1-x^2}.$$

Note that since $\cos^{-1}x$ is in the first or second quadrant, $\sin(\cos^{-1}x)$ is always positive, so is equal to $\sqrt{1-x^2}$, not to $-\sqrt{1-x^2}$.

29. $x^2 + 6x - 2 = (x^2 + 6x + 9) - 11 = (x+3)^2 - 11$. This is of the form $x^2 - a^2$, so the substitution $x + 3 = \sqrt{11}\sec u$ would be appropriate.

31. $x^2 - 5x + 1 = \left(x^2 - 5x + \frac{25}{4}\right) - \frac{21}{4} = \left(x - \frac{5}{2}\right)^2 - \frac{21}{4}$. This is of the form $x^2 - a^2$, so the substitution $x - \frac{5}{2} = \frac{\sqrt{21}}{2}\sec u$ would be appropriate.

33. $4 - 8x - 2x^2 = 2\left(-x^2 - 4x + 2\right) = 2\left(-x^2 - 4x - 4 + 6\right) = 2\left(6 - (x+2)^2\right)$. This is of the form $a^2 - x^2$, so the substitution $x + 2 = \sqrt{6}\sin u$ would be appropriate.

35. (a) With the substitution $x = 3\tan u$, we get $dx = 3\sec^2 u\,du$ and $u = \tan^{-1}\frac{x}{3}$, so that

$$\int \frac{1}{x^2 + 9}\,dx = \int \frac{1}{9\tan^2 u + 9}\cdot 3\sec^2 u\,du$$

$$= \int \frac{1}{9\sec^2 u}\cdot 3\sec^2 u\,du$$

$$= \int \frac{1}{3}\,du$$

$$= \frac{1}{3}u + C = \frac{1}{3}\tan^{-1}\frac{x}{3} + C.$$

(b) Divide numerator and denominator by 9 and then use the substitution $u = \frac{x}{3}$, so that $3\,du = dx$:

$$\int \frac{1}{x^2 + 9}\,dx = \int \frac{1/9}{\frac{x^2}{9} + 1}\,dx$$

$$= \frac{1}{9}\int \frac{1}{(x/3)^2 + 1}\,dx$$

$$= \frac{1}{3}\int \frac{1}{u^2 + 1}\,du$$

$$= \frac{1}{3}\tan^{-1}u + C$$

$$= \frac{1}{3}\tan^{-1}\frac{x}{3} + C.$$

37. (a) Using the substitution $u = 4 + x^2$ so that $du = 2x\,dx$ and $\frac{1}{2}du = x\,dx$, we get

$$\int \frac{x}{4 + x^2}\,dx = \frac{1}{2}\int \frac{1}{u}\,du = \frac{1}{2}\ln|u| + C = \frac{1}{2}\ln(x^2 + 4).$$

(b) Using the trigonometric substitution $x = 2\tan u$, we get $dx = 2\sec^2 u\,du$, and then

$$\int \frac{x}{4 + x^2}\,dx = \int \frac{2\tan u}{4 + 4\tan^2 u}\cdot 2\sec^2 u\,du$$

$$= \int \frac{4\tan u\sec^2 u}{4\sec^2 u}\,du$$

$$= \int \tan u\,du$$

$$= \ln|\sec u| + C = \ln\left|\sec\left(\tan^{-1}\frac{x}{2}\right)\right| + C.$$

As a unit circle diagram shows, $\sec\left(\tan^{-1}\frac{x}{2}\right) = \pm\frac{1}{2}\sqrt{x^2+4}$, so that we get for the integral

$$\ln\left|\sec\left(\tan^{-1}\frac{x}{2}\right)\right| + C = \ln\frac{\sqrt{x^2+4}}{2} + C = \ln\sqrt{x^2+4} - \ln 2 + C = \frac{1}{2}\ln(x^2+4) + C_1.$$

Skills

39. Let $x = 2\sin u$, so that $dx = 2\cos u$. Then

$$\begin{aligned}
\int \frac{\sqrt{4-x^2}}{x^2}\,dx &= \int \frac{\sqrt{4-4\sin^2 u}}{4\sin^2 u} \cdot 2\cos u\,du \\
&= \int \frac{\sqrt{4\cos^2 u}}{4\sin^2 u} \cdot 2\cos u\,du \\
&= \int \frac{2\cos u \cdot 2\cos u}{4\sin^2 u}\,du \\
&= \int \cot^2 u\,du \\
&= \int (\csc^2 u - 1)\,du \\
&= -\cot u - u + C \\
&= -\cot\left(\sin^{-1}\frac{x}{2}\right) - \sin^{-1}\frac{x}{2} + C \\
&= -\frac{\sqrt{4-x^2}}{x} - \sin^{-1}\frac{x}{2} + C.
\end{aligned}$$

41. Let $u = x^2 - 1$, so that $\frac{1}{2}\,du = x\,dx$. Then

$$\int \frac{x}{\sqrt{x^2-1}}\,dx = \frac{1}{2}\int \frac{1}{\sqrt{u}}\,du = \sqrt{u} + C = \sqrt{x^2-1} + C.$$

43. Use the substitution $x = \sqrt{3}\sin u$, so that $dx = \sqrt{3}\cos u\,du$. Then

$$\begin{aligned}
\int \frac{1}{\sqrt{3-x^2}}\,dx &= \int \frac{1}{\sqrt{3-3\sin^2 u}} \cdot \sqrt{3}\cos u\,du \\
&= \int \frac{1}{\sqrt{3}\cos u} \cdot \sqrt{3}\cos u\,du \\
&= \int 1\,du \\
&= u + C = \sin^{-1}\frac{x}{\sqrt{3}} + C.
\end{aligned}$$

45. Use the substitution $x = 3 \sin u$, so that $dx = 3 \cos u \, du$. Then

$$
\begin{aligned}
\int \frac{\sqrt{9 - x^2}}{x} \, dx &= \int \frac{\sqrt{9(1 - \sin^2 u)}}{3 \sin u} \cdot 3 \cos u \, du \\
&= \int \frac{3 \cos u}{3 \sin u} \cdot 3 \cos u \, du \\
&= 3 \int \frac{\cos^2 u}{\sin u} \, du \\
&= 3 \int \frac{1 - \sin^2 u}{\sin u} \, du \\
&= 3 \int (\csc u - \sin u) \, du \\
&= 3 \left(-\ln |\csc u + \cot u| + \cos u \right) + C \\
&= 3 \left(-\ln \left| \csc \left(\sin^{-1} \frac{x}{3} \right) + \cot \left(\sin^{-1} \frac{x}{3} \right) \right| + \cos \left(\sin^{-1} \frac{x}{3} \right) \right) + C \\
&= -3 \ln \left| \frac{3}{x} + \frac{\sqrt{9 - x^2}}{x} \right| + \sqrt{9 - x^2} + C.
\end{aligned}
$$

47. Since $(3 - x)(x - 1) = -x^2 + 4x - 3 = -(x^2 - 4x + 4) + 1 = 1 - (x - 2)^2$, we use the substitution $x - 2 = \sin u$, so that $dx = \cos u \, du$. Then

$$
\begin{aligned}
\int \sqrt{3 - x} \sqrt{x - 1} \, dx = \int \sqrt{-x^2 + 4x - 3} \, dx &= \int \sqrt{1 - (x - 2)^2} \, dx \\
&= \int \sqrt{1 - \sin^2 u} \cos u \, du \\
&= \int \cos^2 u \, du \\
&= \frac{1}{2} \int (1 + \cos 2u) \, du \\
&= \frac{1}{2} u + \frac{1}{4} \sin 2u + C \\
&= \frac{1}{2} \sin^{-1}(x - 2) + \frac{1}{4} \sin(2 \sin^{-1}(x - 2)) + C \\
&= \frac{1}{2} \sin^{-1}(x - 2) + \frac{1}{2} \left(\sin(\sin^{-1}(x - 2)) \cos(\sin^{-1}(x - 2)) \right) + C \\
&= \frac{1}{2} \sin^{-1}(x - 2) + \frac{x - 2}{2} \sqrt{1 - (x - 2)^2} + C \\
&= \frac{1}{2} \sin^{-1}(x - 2) + \frac{x - 2}{2} \sqrt{3 - x} \sqrt{x - 1} + C.
\end{aligned}
$$

49. Use the substitution $x = 3 \sec u$, so that $dx = 3 \sec u \tan u \, du$. The integrand is defined on the domain $(-\infty, -3] \cup [3, \infty)$. Taking first the case where $x \geq 3$, note that $\sec u > 0$, so that u is a

first-quadrant angle and thus $\tan u$ is positive as well. Then we get for the integral

$$
\begin{aligned}
\int \frac{1}{x^2 \sqrt{x^2 - 9}} \, dx &= \int \frac{1}{9 \sec^2 u \sqrt{9(\sec^2 u - 1)}} \cdot 3 \sec u \tan u \, du \\
&= \int \frac{\tan u}{3 \sec u \sqrt{9 \tan^2 u}} \, du \\
&= \int \frac{\tan u}{9 \sec u \tan u} \, du \qquad\qquad (*) \\
&= \frac{1}{9} \int \cos u \, du \\
&= \frac{1}{9} \sin u + C = \frac{1}{9} \sin \left(\sec^{-1} \frac{x}{3} \right) \\
&= \frac{\sqrt{x^2 - 9}}{9x} + C. \qquad\qquad (**)
\end{aligned}
$$

Note that we used the fact that $\tan u$ was positive at $(*)$ and the fact that x is positive together with the fact that $\sin u$ is positive for u a first quadrant angle at $(**)$. If instead we have $x \leq -3$, then $\sec u < 0$, so that u is a second-quadrant angle and thus $\tan u$ is negative. Then in the computation above, a minus sign is introduced at $(*)$. At $(**)$, we would get $\frac{\sqrt{x^2-9}}{-9x}$, since $\sin u$ is positive in the second quadrant as well, but $x < 0$. These minus signs cancel, so that the result is the same as above in this case as well. So $(**)$ is the value of the integral.

51. Use the substitution $x = \sin u$, so that $dx = \cos u \, du$. Then

$$
\begin{aligned}
\int (1 - x^2)^{-3/2} \, dx &= \int (1 - \sin^2 u)^{-3/2} \cos u \, du \\
&= \int (\cos^2 u)^{-3/2} \cos u \, du \\
&= \int \sec^2 u \, du \\
&= \tan u + C = \tan(\sin^{-1} x) + C = \frac{x}{\sqrt{1 - x^2}} + C.
\end{aligned}
$$

53. After completing the square, use the substitution $x - 2 = 3 \tan u$, so that $dx = 3 \sec^2 u \, du$:

$$
\begin{aligned}
\int \frac{1}{x^2 - 4x + 13} &= \int \frac{1}{(x^2 - 4x + 4) + 9} \, dx = \int \frac{1}{(x - 2)^2 + 9} \, dx \\
&= \int \frac{1}{9 \tan^2 u + 9} \cdot 3 \sec^2 u \, du \\
&= \frac{1}{3} \int \frac{\sec^2 u}{\tan^2 u + 1} \, du \\
&= \frac{1}{3} \int 1 \, du = \frac{1}{3} u + C = \frac{1}{3} \tan^{-1} \frac{x - 2}{3} + C.
\end{aligned}
$$

55. Since $9 - 4x^2 = 9 - (2x)^2$, use the substitution $2x = 3\sin u$, or $x = \frac{3}{2}\sin u$. Then $dx = \frac{3}{2}\cos u$, and

$$
\begin{aligned}
\int \frac{3+x}{\sqrt{9-4x^2}}\,dx &= \int \frac{3+\frac{3}{2}\sin u}{\sqrt{9-9\sin^2 u}} \cdot \frac{3}{2}\cos u\,du \\
&= \frac{1}{2}\int \frac{3+\frac{3}{2}\sin u}{\cos u}\cos u\,du \\
&= \frac{1}{2}\int \left(3+\frac{3}{2}\sin u\right)\,du \\
&= \frac{1}{2}\left(3u - \frac{3}{2}\cos u\right) + C = \frac{3}{2}\sin^{-1}\frac{2x}{3} - \frac{3}{4}\cos\left(\sin^{-1}\frac{2x}{3}\right) + C \\
&= \frac{3}{2}\sin^{-1}\frac{2x}{3} - \frac{3}{4}\cdot\frac{\sqrt{9-4x^2}}{3} + C \\
&= \frac{3}{2}\sin^{-1}\frac{2x}{3} - \frac{\sqrt{9-4x^2}}{4} + C.
\end{aligned}
$$

57. Completing the square gives $\sqrt{x^2 - 8x + 25} = \sqrt{(x^2 - 8x + 16) + 9} = \sqrt{(x-4)^2 + 9}$. Use the substitution $x - 4 = 3\tan u$, so that $dx = 3\sec^2 u\,du$. Then (using Example 3 from the previous section or any of several earlier exercises)

$$
\begin{aligned}
\int \sqrt{x^2 - 8x + 25}\,dx &= \int \sqrt{(x-4)^2 + 9}\,dx = \int \sqrt{9(\tan^2 u + 1)}\cdot 3\sec^2 u\,du \\
&= 9\int \sec^3 u\,du \\
&= \frac{9}{2}(\sec u\tan u + \ln|\sec u + \tan u|) + C \\
&= \frac{9}{2}\left(\sec\left(\tan^{-1}\frac{x-4}{3}\right)\tan\left(\tan^{-1}\frac{x-4}{3}\right)\right. \\
&\qquad \left. + \ln\left|\sec\left(\tan^{-1}\frac{x-4}{3}\right) + \tan\left(\tan^{-1}\frac{x-4}{3}\right)\right|\right) + C \\
&= \frac{9}{2}\left(\frac{\sqrt{x^2 - 8x + 25}}{3}\cdot\frac{x-4}{3} + \ln\left|\frac{\sqrt{x^2 - 8x + 25}}{3} + \frac{x-4}{3}\right|\right) + C \\
&= \frac{(x-4)\sqrt{x^2 - 8x + 25}}{2} + \frac{9}{2}\ln\left|\frac{\sqrt{x^2 - 8x + 25}}{3} + \frac{x-4}{3}\right| + C.
\end{aligned}
$$

59. Use the substitution $x = \sqrt{3}\sin u$, so that $dx = \sqrt{3}\cos u\,du$. Then (using the formula $\sin 4\theta = 4\cos^3\theta\sin\theta - 4\cos\theta\sin^3\theta$),

$$\int (3 - x^2)^{3/2}\,dx = \int (3 - 3\sin^2 u)^{3/2}\sqrt{3}\cos u\,du$$

$$= 9\int (\cos^2 u)^{3/2}\cos u\,du$$

$$= 9\int \cos^4 u\,du$$

$$= 9\int \left(\frac{1}{2}(1 + \cos 2u)\right)^2 du$$

$$= \frac{9}{4}\int (1 + 2\cos 2u + \cos^2 2u)\,du$$

$$= \frac{9}{4}\int \left(1 + 2\cos 2u + \frac{1}{2}(1 + \cos 4u)\right) du$$

$$= \frac{9}{4}\left(u + \sin 2u + \frac{u}{2} + \frac{1}{8}\sin 4u\right) + C.$$

Now substitute back to get

$$\int (3 - x^2)^{3/2}\,dx = \frac{27}{8}\sin^{-1}\frac{x}{\sqrt{3}} + \frac{9}{4}\sin\left(2\sin^{-1}\frac{x}{\sqrt{3}}\right) + \frac{9}{32}\sin\left(4\sin^{-1}\frac{x}{\sqrt{3}}\right) + C$$

$$= \frac{27}{8}\sin^{-1}\frac{x}{\sqrt{3}} + \frac{9}{4}\left(2\sin\left(\sin^{-1}\frac{x}{\sqrt{3}}\right)\cos\left(\sin^{-1}\frac{x}{\sqrt{3}}\right)\right)$$

$$\qquad + \frac{9}{32}\left(4\cos^3\left(\sin^{-1}\frac{x}{\sqrt{3}}\right)\sin\left(\sin^{-1}\frac{x}{\sqrt{3}}\right)\right.$$

$$\qquad\qquad\left. -4\cos\left(\sin^{-1}\frac{x}{\sqrt{3}}\right)\sin^3\left(\sin^{-1}\frac{x}{\sqrt{3}}\right)\right) + C$$

$$= \frac{27}{8}\sin^{-1}\frac{x}{\sqrt{3}} + \frac{9}{2}\cdot\frac{x}{\sqrt{3}}\cdot\frac{\sqrt{3 - x^2}}{\sqrt{3}}$$

$$\qquad + \frac{9}{32}\left(4\cdot\left(\frac{\sqrt{3 - x^2}}{\sqrt{3}}\right)^3\cdot\frac{x}{\sqrt{3}} - 4\frac{\sqrt{3 - x^2}}{\sqrt{3}}\cdot\left(\frac{x}{\sqrt{3}}\right)^3\right) + C$$

$$= \frac{27}{8}\sin^{-1}\frac{x}{\sqrt{3}} + \frac{3x\sqrt{3 - x^2}}{2} + \frac{9}{8}\left(\frac{x(3 - x^2)\sqrt{3 - x^2}}{9} - \frac{x^3\sqrt{3 - x^2}}{9}\right) + C$$

$$= \frac{27}{8}\sin^{-1}\frac{x}{\sqrt{3}} + \frac{3x\sqrt{3 - x^2}}{2} + \frac{(3x - 2x^3)\sqrt{3 - x^2}}{8} + C$$

$$= \frac{27}{8}\sin^{-1}\frac{x}{\sqrt{3}} + \frac{15x - 2x^3}{8}\sqrt{3 - x^2} + C.$$

61. Use the substitution $x = \sqrt{2}\sin u$ so that $dx = \sqrt{2}\cos u\,du$. Then

$$\int \sqrt{2 - x^2}\,dx = \int \sqrt{2(1 - \sin^2 u)} \cdot \sqrt{2}\cos u\,du$$

$$= 2\int \sqrt{1 - \sin^2 u}\,\cos u\,du$$

$$= 2\int \cos^2 u\,du$$

$$= \int (1 + \cos 2u)\,du$$

$$= u + \frac{1}{2}\sin 2u + C = u + \frac{1}{2}\sin\left(2\sin^{-1}\frac{x}{\sqrt{2}}\right)$$

$$= \sin^{-1}\frac{x}{\sqrt{2}} + \frac{1}{2}\left(2\sin\left(\sin^{-1}\frac{x}{\sqrt{2}}\right)\cos\left(\sin^{-1}\frac{x}{\sqrt{2}}\right)\right)$$

$$= \sin^{-1}\frac{x}{\sqrt{2}} + \frac{x}{\sqrt{2}} \cdot \frac{\sqrt{2 - x^2}}{\sqrt{2}} + C$$

$$= \sin^{-1}\frac{x}{\sqrt{2}} + \frac{x\sqrt{2 - x^2}}{2} + C.$$

63. Use the substitution $x = \sec u$, so that $dx = \sec u \tan u\,du$. The integrand is defined on $(-\infty, -1] \cup [1, \infty)$. Taking first the case where $x \geq 1$, note that $\sec u > 0$, so that u is a first-quadrant angle and thus $\tan u$ is positive as well. Then we get for the integral

$$\int \frac{\sqrt{x^2 - 1}}{x}\,dx = \int \frac{\sqrt{\sec^2 u - 1}}{\sec u}\,\sec u \tan u\,du$$

$$= \int \frac{\tan u}{\sec u}\,\sec u \tan u\,du$$

$$= \int \tan^2 u\,du$$

$$= \int (\sec^2 u - 1)\,du$$

$$= \tan u - u + C = \tan(\sec^{-1} x) - \sec^{-1} x + C$$

$$= \sqrt{x^2 - 1} - \sec^{-1} x + C.$$

In the other case, where $x \leq -1$, then since $\sec u < 0$, we see that u is a second-quadrant angle so that $\tan u$ is negative. Then we get for the integral

$$\int \frac{\sqrt{x^2 - 1}}{x}\,dx = \int \frac{\sqrt{\sec^2 u - 1}}{\sec u}\,\sec u \tan u\,du$$

$$= -\int \frac{\tan u}{\sec u}\,\sec u \tan u\,du$$

$$= -\int \tan^2 u\,du$$

$$= -\int (\sec^2 u - 1)\,du$$

$$= -\tan u + u + C = -\tan(\sec^{-1} x) + \sec^{-1} x + C$$

$$= \sqrt{x^2 - 1} + \sec^{-1} x + C.$$

Note that in the last step, we end up with the positive square root rather than the negative square root because $\tan u < 0$ and there is an additional minus sign.

So in this case, the integrals are different depending on the range of x values used.

65. Use the substitution $x = \sqrt{2} \sec u$, so that $dx = \sqrt{2} \sec u \tan u \, du$. The integrand is defined on $(-\infty, -\sqrt{2}] \cup [\sqrt{2}, \infty)$. Taking first the case where $x \geq \sqrt{2}$, note that $\sec u > 0$, so that u is a first-quadrant angle and thus $\tan u$ is positive as well. Then we get for the integral

$$\int \frac{1}{\sqrt{x^2 - 2}} \, dx = \int \frac{1}{\sqrt{2(\sec^2 u - 1)}} \cdot \sqrt{2} \sec u \tan u \, du$$

$$= \int \frac{1}{\sqrt{\sec^2 u - 1}} \cdot \sec u \tan u \, du$$

$$= \int \sec u \, du$$

$$= \ln|\sec u + \tan u| + C = \ln \left| \sec \left(\sec^{-1} \frac{x}{\sqrt{2}} \right) + \tan \left(\sec^{-1} \frac{x}{\sqrt{2}} \right) \right|$$

$$= \ln \left| \frac{x}{\sqrt{2}} + \frac{\sqrt{x^2 - 2}}{\sqrt{2}} \right| + C.$$

In the other case, where $x \leq -\sqrt{2}$, then since $\sec u < 0$, we see that u is a second-quadrant angle so that $\tan u$ is negative. Then we get for the integral

$$\int \frac{1}{\sqrt{x^2 - 2}} \, dx = \int \frac{1}{\sqrt{2(\sec^2 u - 1)}} \cdot \sqrt{2} \sec u \tan u \, du$$

$$= \int \frac{1}{\sqrt{\sec^2 u - 1}} \cdot \sec u \tan u \, du$$

$$= - \int \sec u \, du$$

$$= - \ln|\sec u + \tan u| + C = - \ln \left| \sec \left(\sec^{-1} \frac{x}{\sqrt{2}} \right) + \tan \left(\sec^{-1} \frac{x}{\sqrt{2}} \right) \right|$$

$$= - \ln \left| \frac{x}{\sqrt{2}} - \frac{\sqrt{x^2 - 2}}{\sqrt{2}} \right| + C.$$

Note that in the last step, we end up with the negative square root because $\tan u < 0$.

So in this case, the integrals are different depending on the range of x values used.

67. Use integration by parts with $u = x^2$ and $dv = x\sqrt{x^2 + 1}$. Then $du = 2x \, dx$ and $v = \frac{1}{3}(x^2 + 1)^{3/2}$, and we have

$$\int x^3 \sqrt{x^2 + 1} \, dx = \frac{1}{3}x^2(x^2 + 1)^{3/2} - \frac{2}{3} \int x\sqrt{x^2 + 1}^{3/2} \, dx = \frac{1}{3}x^2(x^2 + 1)^{3/2} - \frac{2}{15}(x^2 + 1)^{5/2} + C.$$

69. Use integration by parts with $u = \ln(x^2 + 1)$ and $dv = dx$, so that $du = \frac{2x}{x^2 + 1} \, dx$ and $v = x$. Then

$$\int \ln(x^2 + 1) \, dx = x \ln(x^2 + 1) - 2 \int \frac{x^2}{x^2 + 1} \, dx = x \ln(x^2 + 1) - 2 \int \left(1 - \frac{1}{x^2 + 1} \right) dx$$

$$= x \ln(x^2 + 1) - 2(x - \tan^{-1} x) + C.$$

71. Make the substitution $u = e^{2x} + 1$, so that $du = 2e^{2x} \, dx$ and $\frac{1}{2} du = e^{2x} \, dx$. Then

$$\int e^{2x} \sqrt{e^{2x} + 1} \, dx = \frac{1}{2} \int \sqrt{u} \, du = \frac{1}{3} u^{3/2} + C = \frac{1}{3}(e^{2x} + 1)^{3/2} + C.$$

73. First substitute $u = e^x$ so that $du = e^x\,dx$ and thus $dx = \frac{1}{u}\,du$. Then

$$\int \frac{1}{e^x\sqrt{e^{2x}+3}}\,dx = \frac{1}{u\sqrt{u^2+3}} \cdot \frac{1}{u}\,du = \int \frac{1}{u^2\sqrt{u^2+3}}\,du.$$

Now substitute $u = \sqrt{3}\tan v$, so that $du = \sqrt{3}\sec^2 v\,dv$. Note that since $u = e^x$ that $u > 0$ always, so that v is a first-quadrant angle (because $v = \tan^{-1}\frac{u}{\sqrt{3}}$) and thus $\tan v$ and $\sec v$ are both positive. Then

$$\int \frac{1}{u^2\sqrt{u^2+3}}\,dx = \int \frac{1}{3\tan^2 v\sqrt{3(\tan^2 v+1)}} \cdot \sqrt{3}\sec^2 v\,dv$$

$$= \frac{1}{3}\int \frac{\sec^2 v}{\tan^2 v\sec v}\,dv$$

$$= \frac{1}{3}\int \frac{\cos v}{\sin^2 v}\,dv$$

$$= \frac{1}{3}\int \frac{1}{w^2}\,dw \text{ after } w = \sin v$$

$$= -\frac{1}{3}w^{-1} + C = -\frac{1}{3}\csc v + C$$

$$= -\frac{1}{3}\csc\left(\tan^{-1}\frac{u}{\sqrt{3}}\right) + C$$

$$= -\frac{1}{3} \cdot \frac{\sqrt{u^2+3}}{u} + C$$

$$= -\frac{\sqrt{e^{2x}+3}}{3e^x} + C.$$

75. Using long division gives $x^3 = (x^2+4)(x) - 4x$, so that (using the substitution $u = x^2+4$)

$$\int \frac{x^3}{x^2+4}\,dx = \int\left(x - \frac{4x}{x^2+4}\right)dx = \frac{1}{2}x^2 - 2\int \frac{1}{u}\,du = \frac{1}{2}x^2 - 2\ln|u| + C$$

$$= \frac{1}{2}x^2 - 2\ln(x^2+4) + C.$$

77. Using polynomial long division gives $x^4 - 3 = (3x^2+2)\left(\frac{1}{3}x^2 - \frac{2}{9}\right) - \frac{23}{9}$, so that

$$\int \frac{x^4-3}{3x^2+2}\,dx = \int\left(\frac{1}{3}x^2 - \frac{2}{9} - \frac{23}{9}\cdot\frac{1}{3x^2+2}\right)dx.$$

For $\frac{1}{3x^2+2}$, use the substitution $x\sqrt{3} = \sqrt{2}\tan u$, or $x = \sqrt{\frac{2}{3}}\tan u$. Then $dx = \sqrt{\frac{2}{3}}\sec^2 u\,du$, and

$$\int \frac{x^4-3}{3x^2+2}\,dx = \int\left(\frac{1}{3}x^2 - \frac{2}{9} - \frac{23}{9}\cdot\frac{1}{3x^2+2}\right)dx$$

$$= \frac{1}{9}x^3 - \frac{2}{9}x - \frac{23}{9}\int \frac{1}{2(\tan^2 u+1)} \cdot \sqrt{\frac{2}{3}}\sec^2 u\,du$$

$$= \frac{x^3-2x}{9} - \frac{23\sqrt{2}}{18\sqrt{3}}\int 1\,du$$

$$= \frac{x^3-2x}{9} - \frac{23\sqrt{6}}{54}u + C$$

$$= \frac{x^3-2x}{9} - \frac{23\sqrt{6}}{54}\tan^{-1}\left(x\sqrt{\frac{3}{2}}\right) + C.$$

79. Use the substitution $u = x^2 + 4$, so that $\frac{1}{2} du = x \, dx$. Then

$$
\begin{aligned}
\int_0^4 x\sqrt{x^2 + 4} \, dx &= \frac{1}{2} \int_{x=0}^{x=4} u^{1/2} \, du \\
&= \frac{1}{2} \left[\frac{2}{3} u^{3/2} \right]_{x=0}^{x=4} \\
&= \frac{1}{3} \left[(x^2 + 4)^{3/2} \right]_0^4 \\
&= \frac{1}{3} (20^{3/2} - 4^{3/2}) = \frac{1}{3}(40\sqrt{5} - 8).
\end{aligned}
$$

81. Use the substitution $x = 3 \tan u$, so that $dx = 3 \sec^2 u \, du$. Note that since the bounds of integration are both positive, we may assume that u is a first-quadrant angle. Then

$$
\begin{aligned}
\int_4^5 \frac{1}{x\sqrt{x^2 + 9}} \, dx &= \int_{x=4}^{x=5} \frac{1}{3 \tan u \sqrt{9(\tan^2 u + 1)}} \cdot 3 \sec^2 u \, du \\
&= \frac{1}{3} \int_{x=4}^{x=5} \frac{\sec^2 u}{\tan u \sec u} \, du \\
&= \frac{1}{3} \int_{x=4}^{x=5} \csc u \, du \\
&= -\frac{1}{3} \left[\ln |\csc u + \cot u| \right]_{x=4}^{x=5} \\
&= -\frac{1}{3} \left[\ln \left| \csc \left(\tan^{-1} \frac{x}{3} \right) + \cot \left(\tan^{-1} \frac{x}{3} \right) \right| \right]_4^5 \\
&= -\frac{1}{3} \left[\ln \left| \frac{3 + \sqrt{x^2 + 9}}{x} \right| \right]_4^5 \\
&= -\frac{1}{3} \left(\ln \left(\frac{3 + \sqrt{34}}{5} \right) - \ln 2 \right) \\
&= \frac{1}{3} \ln \left(\frac{10}{3 + \sqrt{34}} \right).
\end{aligned}
$$

83. Use the substitution $x = \sin u$, so that $dx = \cos u \, du$. Then

$$
\begin{aligned}
\int_{1/4}^{1/2} \frac{1}{x^2 \sqrt{1 - x^2}} \, dx &= \int_{x=1/4}^{x=1/2} \frac{1}{\sin^2 u \sqrt{1 - \sin^2 u}} \cos u \, du \\
&= \int_{x=1/4}^{x=1/2} \frac{\cos u}{\sin^2 u \cos u} \, du \\
&= \int_{x=1/4}^{x=1/2} \csc^2 u \, du \\
&= -\left[\cot u \right]_{x=1/4}^{x=1/2} = -\left[\cot(\sin^{-1} x) \right]_{1/4}^{1/2} = -\left[\frac{\sqrt{1 - x^2}}{x} \right]_{1/4}^{1/2} \\
&= -\left(\frac{\sqrt{3/4}}{1/2} - \frac{\sqrt{15/16}}{1/4} \right) = \sqrt{15} - \sqrt{3}.
\end{aligned}
$$

85. Let $2x^2 = 9 \sin^2 u$, so that $x = \frac{3}{\sqrt{2}} \sin u$ and $dx = \frac{3}{\sqrt{2}} \cos u \, du$. Then

$$\int_1^2 \frac{3}{(9 - 2x^2)^{3/2}} \, dx = \int_{x=1}^{x=2} \frac{3}{(9(1 - \sin^2 u))^{3/2}} \cdot \frac{3}{\sqrt{2}} \cos u \, du$$

$$= \frac{9}{\sqrt{2}} \int_{x=1}^{x=2} \frac{\cos u}{9^{3/2}(1 - \sin^2 u)^{3/2}} \, du$$

$$= \frac{1}{3\sqrt{2}} \int_{x=1}^{x=2} \frac{\cos u}{\cos^3 u} \, du$$

$$= \frac{1}{3\sqrt{2}} \int_{x=1}^{x=2} \sec^2 u \, du$$

$$= \frac{1}{3\sqrt{2}} \left[\tan u \right]_{x=1}^{x=2} = \frac{1}{3\sqrt{2}} \left[\tan \left(\sin^{-1} \frac{\sqrt{2}}{3} x \right) \right]_1^2 = \frac{1}{3\sqrt{2}} \left[\frac{x\sqrt{2}}{\sqrt{9 - 2x^2}} \right]_1^2$$

$$= \frac{1}{3\sqrt{2}} \left(2\sqrt{2} - \sqrt{\frac{2}{7}} \right) = \frac{2}{3} - \frac{1}{3\sqrt{7}}.$$

87. Let $x = 2 \sinh u$, so that $dx = 2 \cosh u$. Then

$$\int \frac{1}{\sqrt{x^2 + 4}} \, dx = \int \frac{1}{\sqrt{4 \sinh^2 u + 4}} \cdot 2 \cosh u \, du$$

$$= \int \frac{1}{2\sqrt{\cosh^2 u}} \cdot 2 \cosh u \, du$$

$$= \int 1 \, du$$

$$= u + C = \sinh^{-1} \frac{x}{2} + C.$$

89. Let $x = \sinh u$, so that $dx = \cosh u \, du$. Then

$$\int \frac{x^2}{(x^2 + 1)^{3/2}} \, dx = \int \frac{\sinh^2 u}{(\sinh^2 u + 1)^{3/2}} \cdot \cosh u \, du$$

$$= \int \frac{\cosh u \sinh^2 u}{(\cosh^2 u)^{3/2}} \, du$$

$$= \int \tanh^2 u \, du$$

$$= \int (1 - \operatorname{sech}^2 u) \, du$$

$$= u - \tanh u + C = \sinh^{-1} x - \tanh(\sinh^{-1} x) + C.$$

To further simplify the term $\tanh(\sinh^{-1} x)$, recall from Exercise 97 in Section 2.6 that $\sinh^{-1} x =$

$\ln(x + \sqrt{x^2 + 1})$, so that

$$\tanh(\sinh^{-1} x) = \frac{e^{\ln(x+\sqrt{x^2+1})} - e^{-\ln(x+\sqrt{x^2+1})}}{e^{\ln(x+\sqrt{x^2+1})} + e^{-\ln(x+\sqrt{x^2+1})}}$$

$$= \frac{x + \sqrt{x^2 + 1} - \frac{1}{x+\sqrt{x^2+1}}}{x + \sqrt{x^2 + 1} + \frac{1}{x+\sqrt{x^2+1}}}$$

$$= \frac{(x + \sqrt{x^2 + 1})^2 - 1}{(x + \sqrt{x^2 + 1})^2 + 1}$$

$$= \frac{x^2 + 2x\sqrt{x^2 + 1} + x^2 + 1 - 1}{x^2 + 2x\sqrt{x^2 + 1} + x^2 + 1 + 1}$$

$$= \frac{2x(x + \sqrt{x^2 + 1})}{2x^2 + 2 + 2x\sqrt{x^2 + 1}}$$

$$= \frac{2x(x + \sqrt{x^2 + 1})(2x^2 + 2 - 2x\sqrt{x^2 + 1})}{(2x^2 + 2 + 2x\sqrt{x^2 + 1})(2x^2 + 2 - 2x\sqrt{x^2 + 1})}$$

$$= \frac{4x\sqrt{x^2 + 1}}{4(x^2 + 1)}$$

$$= \frac{x}{\sqrt{x^2 + 1}}.$$

Thus the result of the integral is $\sinh^{-1} x - \frac{x}{\sqrt{x^2+1}} + C.$

Applications

91. (a) The graph of $f(x)$ intersects the x-axis at $x = \pm\sqrt{42}$, so that the volume of the solid is given by

$$\pi \int_a^b (f(x))^2 \, dx = \pi \int_{-\sqrt{42}}^{\sqrt{42}} \left(2(42 - x^2)^{1/4}\right)^2 \, dx = 4\pi \int_{-\sqrt{42}}^{\sqrt{42}} (42 - x^2)^{1/2} \, dx.$$

(b) To integrate this expression, use the substitution $x = \sqrt{42} \sin u$, so that $dx = \sqrt{42} \cos u \, du.$

Then

$$\pi \int_a^b (f(x))^2 \, dx = 4\pi \int_{-\sqrt{42}}^{\sqrt{42}} (42 - x^2)^{1/2} \, dx$$

$$= 4\pi \int_{x=-\sqrt{42}}^{x=\sqrt{42}} (42(1 - \sin^2 u))^{1/2} \cdot \sqrt{42} \cos u \, du$$

$$= 168\pi \int_{x=-\sqrt{42}}^{x=\sqrt{42}} (\cos^2 u)^{1/2} \cos u \, du$$

$$= 168\pi \int_{x=-\sqrt{42}}^{x=\sqrt{42}} \cos^2 u \, du$$

$$= 168\pi \int_{x=-\sqrt{42}}^{x=\sqrt{42}} \frac{1}{2}(1 + \cos 2u) \, du$$

$$= 84\pi \left[u + \frac{1}{2} \sin 2u \right]_{x=-\sqrt{42}}^{x=\sqrt{42}}$$

$$= 84\pi \left[\sin^{-1} \frac{x}{\sqrt{42}} + \frac{1}{2} \sin \left(2 \sin^{-1} \frac{x}{\sqrt{42}} \right) \right]_{x=-\sqrt{42}}^{x=\sqrt{42}}$$

$$= 84\pi \left[\sin^{-1} \frac{x}{\sqrt{42}} + \sin \left(\sin^{-1} \frac{x}{\sqrt{42}} \right) \cos \left(\sin^{-1} \frac{x}{\sqrt{42}} \right) \right]_{-\sqrt{42}}^{\sqrt{42}}$$

$$= 84\pi \left[\sin^{-1} \frac{x}{\sqrt{42}} + \frac{x\sqrt{42 - x^2}}{42} \right]_{-\sqrt{42}}^{\sqrt{42}}$$

$$= 84\pi \left(\sin^{-1} 1 + 0 - \sin^{-1}(-1) - 0 \right)$$

$$= 84\pi^2 \text{ cubic mm.}$$

Proofs

93. (a) The area of the circle of radius r centered at the origin is twice the area of the upper semicircle. Since the equation of the upper semicircle is $y = \sqrt{r^2 - x^2}$, the total area of the circle is

$$2 \int_{-r}^{r} \sqrt{r^2 - x^2} \, dx.$$

(b) To solve this integral, let $x = r \sin u$, so that $dx = r \cos u$; then

$$2 \int_{-r}^{r} \sqrt{r^2 - x^2} \, dx = 2 \int_{x=-r}^{x=r} \sqrt{r^2(1 - \sin^2 u)} \cdot r \cos u \, du$$

$$= 2r^2 \int_{x=-r}^{x=r} \sqrt{\cos^2 u} \cos u \, du$$

$$= 2r^2 \int_{x=-r}^{x=r} \cos^2 u \, du$$

$$= r^2 \int_{x=-r}^{x=r} (1 + \cos 2u) \, du$$

$$= r^2 \left[u + \frac{1}{2} \sin 2u \right]_{x=-r}^{x=r}$$

$$= r^2 \left[\sin^{-1} \frac{x}{r} + \frac{1}{2} \sin \left(2 \sin^{-1} \frac{x}{r} \right) \right]_{x=-r}^{x=r}$$

$$= r^2 \left[\sin^{-1} \frac{x}{r} + \sin \left(\sin^{-1} \frac{x}{r} \right) \cos \left(\sin^{-1} \frac{x}{r} \right) \right]_{-r}^{r}$$

$$= r^2 \left[\sin^{-1} \frac{x}{r} + \frac{x\sqrt{r^2 - x^2}}{r^2} \right]_{-r}^{r}$$

$$= r^2 \left(\sin^{-1} 1 - \sin^{-1}(-1) \right) = \pi r^2.$$

Thinking Forward

Arc length of a curve

$\triangleright$ $\sqrt{(\Delta x)^2 + (\Delta y)^2} = \Delta x \sqrt{\frac{1}{(\Delta x)^2}} \sqrt{(\Delta x)^2 + (\Delta y)^2} = \Delta x \sqrt{\frac{(\Delta x)^2 + (\Delta y)^2}{(\Delta x)^2}} = \sqrt{1 + \left(\frac{\Delta y}{\Delta x} \right)^2} \, \Delta x.$

$\triangleright$ Using four line segments, the endpoints of the line segments have x-coordinates $-3, -\frac{3}{2}, 0, \frac{3}{2}$, and 3. Then $\Delta x = \frac{3}{2}$. We now compute Δy and $\frac{\Delta y}{\Delta x}$ for each of the segments:

Segment	Δy	$\frac{\Delta y}{\Delta x}$	$\sqrt{1 + \left(\frac{\Delta y}{\Delta x} \right)^2} \, \Delta x$
1	$f\left(-\frac{3}{2}\right) - f(-3) = \frac{3}{2}\sqrt{3}$	$\sqrt{3}$	3
2	$f(0) - f\left(-\frac{3}{2}\right) = 3 - \frac{3}{2}\sqrt{3}$	$2 - \sqrt{3}$	$\frac{3}{2}\sqrt{8 - 4\sqrt{3}}$
3	$f\left(\frac{3}{2}\right) - f(0) = \frac{3}{2}\sqrt{3} - 3$	$\sqrt{3} - 2$	$\frac{3}{2}\sqrt{8 - 4\sqrt{3}}$
4	$f(3) - f\left(\frac{3}{2}\right) = -\frac{3}{2}\sqrt{3}$	$-\sqrt{3}$	3

The approximation is the sum over all four segments of $\sqrt{1 + \left(\frac{\Delta y}{\Delta x} \right)^2} \Delta x$, which is

$$6 + 3\sqrt{8 - 4\sqrt{3}} \approx 9.1058.$$

With eight line segments, the endpoints of the line segments have x-coordinates $\pm 3, \pm\frac{9}{4}, \pm\frac{3}{2}, \pm\frac{3}{4}$, and 0. Now $\Delta x = \frac{3}{4}$. Again we compute Δy and $\frac{\Delta y}{\Delta x}$ for each of these segments.

Segment	Δy	$\frac{\Delta y}{\Delta x}$	$\sqrt{1 + \left(\frac{\Delta y}{\Delta x}\right)^2}\,\Delta x$
1	$f\left(-\frac{9}{4}\right) - f(-3) = \frac{3\sqrt{7}}{4}$	$\sqrt{7}$	$\frac{3}{\sqrt{2}}$
2	$f\left(-\frac{3}{2}\right) - f\left(-\frac{9}{4}\right) = \frac{3\sqrt{3}}{2} - \frac{3\sqrt{7}}{4}$	$2\sqrt{3} - \sqrt{7}$	$\frac{3}{2}\sqrt{5 - \sqrt{21}}$
3	$f\left(-\frac{3}{4}\right) - f\left(-\frac{3}{2}\right) = -\frac{3\sqrt{3}}{2} + \frac{3\sqrt{15}}{4}$	$\sqrt{3}(\sqrt{5} - 2)$	$\frac{3}{2}\sqrt{7 - 3\sqrt{5}}$
4	$f(0) - f\left(-\frac{3}{4}\right) = 3 - \frac{3\sqrt{15}}{4}$	$4 - \sqrt{15}$	$3\sqrt{2 - \frac{\sqrt{15}}{2}}$
5	$f\left(\frac{3}{4}\right) - f(0) = \frac{3\sqrt{15}}{4} - 3$	$\sqrt{15} - 4$	$3\sqrt{2 - \frac{\sqrt{15}}{2}}$
6	$f\left(\frac{3}{2}\right) - f\left(-\frac{3}{4}\right) = \frac{3\sqrt{3}}{2} - \frac{3\sqrt{15}}{4}$	$-\sqrt{3}(\sqrt{5} - 2)$	$\frac{3}{2}\sqrt{7 - 3\sqrt{5}}$
7	$f\left(\frac{9}{4}\right) - f\left(\frac{3}{2}\right) = -\frac{3\sqrt{3}}{2} + \frac{3\sqrt{7}}{4}$	$-2\sqrt{3} + \sqrt{7}$	$\frac{3}{2}\sqrt{5 - \sqrt{21}}$
8	$f(3) - f\left(\frac{9}{4}\right) = -\frac{3\sqrt{7}}{4}$	$-\sqrt{7}$	$\frac{3}{\sqrt{2}}$

The approximation is the sum over all eight segments of $\sqrt{1 + \left(\frac{\Delta y}{\Delta x}\right)}\,\Delta x$, which is

$$3\sqrt{2} + 3\sqrt{5 - \sqrt{21}} + 3\sqrt{7 - 3\sqrt{5}} + 6\sqrt{2 - \frac{\sqrt{15}}{2}} \approx 9.3135.$$

▷ This integral is most likely the limit of the Riemann sum of $\sqrt{1 + \left(\frac{\Delta y_k}{\Delta x}\right)^2}\,\Delta x$ as $\Delta x \to 0$. But as $\Delta x \to 0$, $\frac{\Delta y_k}{\Delta x}$ is approximately $f'(x)$ at the point corresponding to y_k. So it is reasonable to expect that this sum approaches $\int_a^b \sqrt{1 + f'(x)^2}\,dx$, and that the integral is the length of the curve.

▷ With $f(x) = \sqrt{9 - x^2} = (9 - x^2)^{1/2}$, we get $f'(x) = -x(9 - x^2)^{-1/2}$, so that

$$\sqrt{1 + f'(x)^2} = \sqrt{1 + \frac{x^2}{9 - x^2}} = \sqrt{\frac{9}{9 - x^2}} = \frac{3}{\sqrt{9 - x^2}}.$$

Then the arc length of $f(x)$ on $[-3, 3]$ is

$$\int_{-3}^{3} \sqrt{1 + f'(x)^2}\,dx = \int_{-3}^{3} \frac{3}{\sqrt{9 - x^2}}\,dx.$$

To evaluate this integral, use the substitution $x = 3\sin u$, so that $dx = 3\cos u$. Then

$$\int_{-3}^{3} \frac{3}{\sqrt{9 - x^2}}\,dx = \int_{x=-3}^{x=3} \frac{3}{\sqrt{9(1 - \sin^2 u)}} \cdot 3\cos u\,du$$

$$= 3\int_{x=-3}^{x=3} \frac{\cos u}{\sqrt{\cos^2 u}}\,du$$

$$= 3\int_{x=-3}^{x=3} 1\,du$$

$$= 3\left[u\right]_{x=-3}^{x=3} = 3\left[\sin^{-1}\frac{x}{3}\right]_{-3}^{3} = 3\pi.$$

▷ The circumference of a circle with radius 3 is $2\pi r = 6\pi$, so the circumference of the upper half-circle, which is what was measured by the integral above, is half that, or 3π.

Surface area of a solid of revolution

▷ With $f(x) = 2x^3$, we have $f'(x) = 6x^2$, so that the surface area over $[0,3]$ is

$$S = 2\pi \int_0^3 2x^3 \sqrt{1 + (6x^2)^2}\, dx = 2\pi \int_0^3 2x^3 \sqrt{36x^4 + 1}\, dx.$$

With the substitution $u = 36x^4 + 1$, we have $du = 144x^3\, dx$, so that $2x^3\, dx = \frac{1}{72}\, du$. Then

$$S = 2\pi \int_0^3 2x^3 \sqrt{36x^4 + 1}\, dx$$

$$= \frac{\pi}{36} \int_{x=0}^{x=3} \sqrt{u}\, du$$

$$= \frac{\pi}{36} \left[\frac{2}{3} u^{3/2} \right]_{x=0}^{x=3}$$

$$= \frac{\pi}{54} \left[(36x^4 + 1)^{3/2} \right]_{x=0}^{x=3}$$

$$= \frac{\pi}{54} \left(2917^{3/2} - 1 \right).$$

▷ We get a sphere of radius r by rotating the region under the graph of $f(x) = \sqrt{r^2 - x^2}$ around the x-axis, since that graph encloses a (semi-)circle of radius r. Then $f'(x) = -\frac{x}{\sqrt{r^2 - x^2}}$, and thus by the given formula, the surface area of the sphere is given by

$$S = 2\pi \int_{-r}^r \sqrt{r^2 - x^2} \sqrt{1 + \left(-\frac{x}{\sqrt{r^2 - x^2}} \right)^2}\, dx$$

$$= 2\pi \int_{-r}^r \sqrt{r^2 - x^2} \sqrt{1 + \frac{x^2}{r^2 - x^2}}\, dx$$

$$= 2\pi \int_{-r}^r \sqrt{r^2 - x^2 + x^2}\, dx$$

$$= 2\pi \int_{-r}^r r\, dx$$

$$= 2\pi \left[rx \right]_{-r}^r = 4\pi r^2.$$

5.6 Improper Integrals

Thinking Back

Limits

▷ $\lim\limits_{x \to \infty} x^{-2/3} = \lim\limits_{x \to \infty} \frac{1}{x^{2/3}} = \frac{1}{\infty} = 0.$

▷ $\lim\limits_{x \to \infty} x^{-4/3} = \lim\limits_{x \to \infty} \frac{1}{x^{4/3}} = \frac{1}{\infty} = 0.$

▷ $\lim\limits_{x \to 0^+} x^{-2/3} = \lim\limits_{x \to 0^+} \frac{1}{x^{2/3}} = \frac{1}{0^+} = \infty.$

▷ $\lim\limits_{x \to 0^+} x^{-4/3} = \lim\limits_{x \to 0^+} \frac{1}{x^{4/3}} = \frac{1}{0^+} = \infty.$

▷ $\lim\limits_{x \to 1^-} \frac{1}{x \ln x} = \frac{1}{1 \cdot 0^-} = -\infty.$

▷ $\lim\limits_{x \to 1^+} \dfrac{1}{x \ln x} = \dfrac{1}{1 \cdot 0^+} = \infty.$

▷ $\lim\limits_{x \to \infty} e^{-x^2} = \lim\limits_{x \to \infty} \dfrac{1}{e^{x^2}} = \dfrac{1}{\infty} = 0.$

▷ $\lim\limits_{x \to \infty} x e^{-x^2} = \lim\limits_{x \to \infty} \dfrac{x}{e^{x^2}} \overset{L'H}{=} \lim\limits_{x \to \infty} \dfrac{1}{2x e^{x^2}} = 0.$

▷ $\lim\limits_{x \to \frac{\pi}{2}^-} \sec x = \lim\limits_{x \to \frac{\pi}{2}^-} \dfrac{1}{\cos x} = \dfrac{1}{0^+} = \infty.$

Integration

▷ Use the substitution $u = 3x - 1$, so that $\frac{1}{3} du = dx$. Then

$$\int (3x - 1)^{-2/3} dx = \frac{1}{3} \int u^{-2/3} du = u^{1/3} + C = (3x - 1)^{1/3} + C.$$

▷ Use the substitution $u = \ln x$, so that $du = \frac{1}{x} dx$. Then

$$\int \frac{1}{x \ln x} dx = \int \frac{1}{u} du = \ln |u| + C = \ln |\ln x| + C.$$

▷ Use the substitution $u = x^2$, so that $\frac{1}{2} du = x \, dx$. Then

$$\int x e^{-x^2} dx = \frac{1}{2} \int e^{-u} du = -\frac{1}{2} e^{-u} + C = -\frac{1}{2} e^{-x^2} + C.$$

▷ Use the substitution $u = \ln x$, so that $du = \frac{1}{x} dx$. Then

$$\int \frac{\ln x}{x} dx = \int u \, du = \frac{1}{2} u^2 + C = \frac{1}{2} \ln^2 x + C.$$

▷ Decompose the integrand using partial fractions:

$$\frac{1}{x(x-1)} = \frac{A}{x} + \frac{B}{x-1} = \frac{(A+B)x - A}{x(x-1)},$$

so that $A = -1$ and $B = 1$. Then

$$\int \frac{1}{x(x-1)} dx = \int \left(\frac{1}{x-1} - \frac{1}{x} \right) dx = \ln |x - 1| - \ln |x| + C = \ln \left| \frac{x-1}{x} \right| + C.$$

▷ By Theorem 5.17, which also presents the derivation of the formula,

$$\int \sec x \, dx = \ln |\sec x + \tan x| + C.$$

The Fundamental Theorem of Calculus

▷ $\displaystyle\int_1^{100} \frac{1}{\sqrt[3]{x^4}} dx = \int_1^{100} x^{-4/3} dx = \left[-3x^{-1/3} \right]_1^{100} = 3 - \frac{3}{\sqrt[3]{100}}.$

▷ $\displaystyle\int_1^{100} \frac{1}{\sqrt[4]{x^3}} dx = \int_1^{100} x^{-3/4} dx = \left[4x^{1/4} \right]_1^{100} = 4\sqrt{10} - 4.$

▷ $\displaystyle\int_{0.01}^1 \frac{1}{\sqrt[3]{x^4}} dx = \int_{0.01}^1 x^{-4/3} dx = \left[-3x^{-1/3} \right]_{0.01}^1 = \frac{3}{\sqrt[3]{10^{-2}}} - 3 = 3\sqrt[3]{100} - 3.$

▷ $\displaystyle\int_{0.01}^1 \frac{1}{\sqrt[4]{x^3}} dx = \int_{0.01}^1 x^{-3/4} dx = \left[4x^{1/4} \right]_{0.01}^1 = 4 - 4\sqrt{0.1} = 4 - 2\frac{\sqrt{10}}{5}.$

Concepts

1. (a) False. The Fundamental Theorem of Calculus requires that the integrand be continuous on the interval of integration, but $\frac{1}{1-x}$ is undefined at $x = 1$.

   (b) False. The Fundamental Theorem of Calculus applies on an interval $[a, b]$, where a and b are real numbers.

   (c) False. For example, let $f(x) = \frac{1}{x}$ (see the text).

   (d) False. For example, let $f(x) = \frac{1}{x^2}$ (see the text).

   (e) True. See Theorem 5.21.

   (f) True. See Theorem 5.22.

   (g) False. For example, let $f(x) = \frac{1}{x^2}$ and $g(x) = 2f(x)$. Then $\int_1^\infty g(x)\,dx = 2\int_1^\infty f(x)\,dx = 2$.

   (h) True. See Theorem 5.23.

3. Either one or both bounds of integration will be infinite, or else the integrand has an infinite discontinuity somewhere in the interval of integration (including possibly at one or both endpoints).

5. Because e^x is continuous and nonzero over the entire range of integration.

7. It does not apply since $(x - 3)^{-4/3}$ has an infinite discontinuity at $x = 3$, which is in the range of integration. Using the Fundamental Theorem of Calculus anyway gives

$$\int_0^5 (x - 3)^{-4/3}\,dx = \left[-3(x - 3)^{-1/3} \right]_0^5 = -3(2^{-1/3} - 3^{1/3}).$$

   This is not the proper answer: Since $x - 3$ is x shifted right by 3 units, the integral is the same as $\int_{-3}^2 x^{-4/3}$, which diverges by Theorem 5.22 since here $p = \frac{4}{3} \geq 1$.

9. For $p \geq 1$, we have $x^p \leq x$ since $x \leq 1$, so that $\frac{1}{x^p} \geq \frac{1}{x}$. Thus the graph of $\frac{1}{x^p}$ lies completely above the graph of $\frac{1}{x}$, so the area under its graph is at least as great as the area under $\frac{1}{x}$ from 0 to 1, which is itself infinite.

11. On $[1, \infty)$, we have $\frac{1}{x+1} < \frac{1}{x}$. Since $\int_1^\infty \frac{1}{x}\,dx$ diverges, the comparison test does not tell us anything about the convergence of $\frac{1}{x+1}$ — it could be *much* smaller than $\frac{1}{x}$ and converge, or it could be about the same size and diverge.

13. The comparison test cannot draw any conclusions about the divergence of $\int_1^\infty \frac{1}{x+1}\,dx$ for the reasons given in the solution to Exercise 11. Graphically, if $f(x) = \frac{1}{x}$, then $g(x) = \frac{1}{x+1} = f(x + 1)$ is simply f shifted left by one unit. Thus

$$\int_1^\infty g(x)\,dx = \int_1^\infty f(x + 1)\,dx = \int_2^\infty f(x)\,dx,$$

   and the latter integral diverges by the power rule.

15. Pick an arbitrary point in $[1, \infty)$, say 2; then

$$\int_1^\infty \frac{1}{x - 1}\,dx = \lim_{c \to 1^+} \int_c^2 \frac{1}{x - 1}\,dx + \lim_{c \to \infty} \int_2^c \frac{1}{x - 1}\,dx.$$

17. Since $2x^2 - 10x + 12 = 2(x^2 - 5x + 6) = 2(x - 2)(x - 3)$, the integrand is discontinuous at $x = 2$ and at $x = 3$. Choose a point, say $\frac{5}{2}$, in $(2, 3)$. Then

$$\int_1^5 \frac{1}{2x^2 - 10x + 12}\,dx = \lim_{c \to 2^-} \int_1^c \frac{1}{2x^2 - 10x + 12}\,dx + \lim_{c \to 2^+} \int_c^{5/2} \frac{1}{2x^2 - 10x + 12}\,dx$$

$$+ \lim_{c \to 3^-} \int_{5/2}^c \frac{1}{2x^2 - 10x + 12}\,dx + \lim_{c \to 3^+} \int_c^5 \frac{1}{2x^2 - 10x + 12}\,dx.$$

19. The integrand is discontinuous at $x = 0$. Choose two points, say ± 1, in $(-\infty, 0)$ and in $(0, \infty)$. Then

$$\int_{-\infty}^{\infty} \frac{e^{-x^2}}{x}\, dx = \lim_{c \to -\infty} \int_{c}^{-1} \frac{e^{-x^2}}{x}\, dx + \lim_{c \to 0^-} \int_{-1}^{c} \frac{e^{-x^2}}{x}\, dx + \lim_{c \to 0^+} \int_{c}^{1} \frac{e^{-x^2}}{x}\, dx + \lim_{c \to \infty} \int_{1}^{c} \frac{e^{-x^2}}{x}\, dx.$$

21. $\displaystyle \int_{1}^{\infty} \frac{1}{\sqrt[3]{x^4}}\, dx = \lim_{c \to \infty} \int_{1}^{c} x^{-4/3}\, dx = \lim_{c \to \infty} \left[-3x^{-1/3} \right]_{1}^{c} = \lim_{c \to \infty} \left(-3c^{-1/3} + 3 \right) = 3.$

23. $\displaystyle \int_{0}^{1} \frac{1}{\sqrt[3]{x^4}}\, dx = \lim_{c \to 0^+} \int_{c}^{1} x^{-4/3}\, dx = \lim_{c \to 0^+} \left[-3x^{-1/3} \right]_{c}^{1} = \lim_{c \to 0^+} \left(3c^{-1/3} - 3 \right) = -\infty.$

25. $\displaystyle \int_{0}^{1} x^{-0.99}\, dx = \lim_{c \to 0^+} \int_{c}^{1} x^{-0.99}\, dx = \lim_{c \to 0^+} \left[100x^{0.01} \right]_{c}^{1} = \lim_{c \to 0^+} (100 - 100c^{0.01}) = 100.$

27. $\displaystyle \int_{1}^{\infty} x^{-0.99}\, dx = \lim_{c \to \infty} \int_{1}^{c} x^{-0.99}\, dx = \lim_{c \to \infty} \left[100x^{0.01} \right]_{1}^{c} = \lim_{c \to \infty} (100c^{0.01} - 100) = \infty.$

29.

$$\int_{0}^{1} (3x - 1)^{-2/3}\, dx = \lim_{c \to \frac{1}{3}^-} \int_{0}^{c} (3x - 1)^{-2/3}\, dx + \lim_{c \to \frac{1}{3}^+} \int_{c}^{1} (3x - 1)^{-2/3}\, dx$$

$$= \lim_{c \to \frac{1}{3}^-} \left[(3x - 1)^{1/3} \right]_{0}^{c} + \lim_{c \to \frac{1}{3}^+} \left[(3x - 1)^{1/3} \right]_{c}^{1}$$

$$= \lim_{c \to \frac{1}{3}^-} \left((3c - 1)^{1/3} + 1 \right) + \lim_{c \to \frac{1}{3}^+} \left(2^{1/3} - (3c - 1)^{1/3} \right)$$

$$= 1 + 2^{1/3}.$$

31.

$$\int_{1}^{\infty} \frac{1}{(x - 2)^2}\, dx = \lim_{c \to 2^-} \int_{1}^{c} \frac{1}{(x - 2)^2}\, dx + \lim_{c \to 2^+} \int_{c}^{3} \frac{1}{(x - 2)^2}\, dx + \lim_{c \to \infty} \int_{3}^{c} \frac{1}{(x - 2)^2}\, dx$$

$$= \lim_{c \to 2^-} \left[-\frac{1}{x - 2} \right]_{1}^{c} + \lim_{c \to 2^+} \left[-\frac{1}{x - 2} \right]_{c}^{3} + \lim_{c \to \infty} \left[-\frac{1}{x - 2} \right]_{3}^{c}$$

$$= \lim_{c \to 2^-} \left(-\frac{1}{c - 2} + 1 \right) + \lim_{c \to 2^+} \left(-1 + \frac{1}{c - 2} \right) + \lim_{c \to \infty} \left(-\frac{1}{c - 2} + 1 \right)$$

$$= \infty + 1 - 1 + \infty - 0 + 1$$

so the integral is undefined.

33. Note that $2x - 4$ is the derivative of the denominator, so that

$$\int \frac{2x - 4}{x^2 - 4x + 3}\, dx = \ln \left| x^2 - 4x + 3 \right|.$$

Also, $x^2 - 4x + 3$ has zeroes at $x = 1$ and at $x = 3$, so that

$$\int_0^4 \frac{2x-4}{x^2-4x+3}\,dx = \lim_{c\to 1^-} \int_0^c \frac{2x-4}{x^2-4x+3}\,dx + \lim_{c\to 1^+} \int_c^2 \frac{2x-4}{x^2-4x+3}\,dx$$

$$+ \lim_{c\to 3^-} \int_2^c \frac{2x-4}{x^2-4x+3}\,dx + \lim_{c\to 3^+} \int_c^4 \frac{2x-4}{x^2-4x+3}\,dx$$

$$= \lim_{c\to 1^-} \left[\ln\left|x^2-4x+3\right|\right]_0^c + \lim_{c\to 1^+} \left[\ln\left|x^2-4x+3\right|\right]_c^2$$

$$+ \lim_{c\to 3^-} \left[\ln\left|x^2-4x+3\right|\right]_2^c + \lim_{c\to 3^+} \left[\ln\left|x^2-4x+3\right|\right]_c^4$$

$$= \lim_{c\to 1^-} \left(\ln\left|c^2-4c+3\right| - \ln 3\right) + \lim_{c\to 1^+} \left(\ln 1 - \ln\left|c^2-4c+3\right|\right)$$

$$+ \lim_{c\to 3^-} \left(\ln\left|c^2-4c+3\right| - \ln 1\right) + \lim_{c\to 3^+} \left(\ln 3 - \ln\left|c^2-4c+3\right|\right)$$

and the integral is undefined.

35. Using the substitution $u = -x^2$, we see that an antiderivative of xe^{-x^2} is $-\frac{1}{2}e^{-x^2}$, so that

$$\int_0^\infty xe^{-x^2}\,dx = \lim_{c\to\infty} \int_0^c xe^{-x^2}\,dx = \lim_{c\to\infty}\left(-\frac{1}{2}\left[e^{-x^2}\right]_0^c\right) = -\frac{1}{2}\lim_{c\to\infty}(e^{-c^2}-1) = \frac{1}{2}.$$

37. Use integration by parts with $u = x^2$ and $dv = e^{-x}\,dx$, so that $du = 2x\,dx$ and $v = -e^{-x}$. Then

$$\int x^2 e^{-x}\,dx = -x^2 e^{-x} + 2\int xe^{-x}\,dx.$$

Now use integration by parts again, with $u = x$ and $dv = e^{-x}\,dx$, so that $du = dx$ and $v = -e^{-x}$; then

$$\int x^2 e^{-x}\,dx = -x^2 e^{-x} - 2e^{-x}(x+1) = -e^{-x}(x^2+2x+2).$$

Thus

$$\int_0^\infty x^2 e^{-x}\,dx = \lim_{c\to\infty}\int_0^c x^2 e^{-x}\,dx = \lim_{c\to\infty}\left[-e^{-x}(x^2+2x+2)\right]_0^c = \lim_{c\to\infty}\left(-\frac{x^2+2x+2}{e^x}+2\right) = 2,$$

where the first term in the limit may be evaluated using L'Hôpital's Rule or by remembering that exponential functions dominate polynomial functions.

39. Let $u = \sqrt{x}$ so that $2\,du = x^{-1/2}\,dx$. Then

$$\int x^{-1/2}e^{-\sqrt{x}}\,dx = 2\int e^{-u}\,du = -2e^{-u} = -2e^{-\sqrt{x}}.$$

Hence

$$\int_0^\infty \frac{1}{\sqrt{x}e^{\sqrt{x}}}\,dx = \lim_{c\to 0^+}\int_c^1 \frac{1}{\sqrt{x}e^{\sqrt{x}}}\,dx + \lim_{c\to\infty}\int_1^c \frac{1}{\sqrt{x}e^{\sqrt{x}}}\,dx$$

$$= \lim_{c\to 0^+}\left[-2e^{-\sqrt{x}}\right]_c^1 + \lim_{c\to\infty}\left[-2e^{-\sqrt{x}}\right]_1^c$$

$$= \lim_{c\to 0^+}\left(-2e^{-1}+2e^{-\sqrt{c}}\right) + \lim_{c\to\infty}\left(-2e^{-\sqrt{c}}+2e^{-1}\right)$$

$$= -2e^{-1}+2+0+2e^{-1} = 2.$$

41. Use the substitution $u = x^2 + 1$, so that $\frac{1}{2} du = x\,dx$. Then

$$\int \frac{x}{(x^2+1)^2}\,dx = \frac{1}{2}\int \frac{1}{u^2}\,du = -\frac{1}{2}u^{-1} = -\frac{1}{2(x^2+1)}.$$

Hence

$$\int_{-\infty}^{\infty} \frac{x}{(x^2+1)^2}\,dx = \lim_{c\to-\infty}\int_c^0 \frac{x}{(x^2+1)^2}\,dx + \lim_{c\to\infty}\int_0^c \frac{x}{(x^2+1)^2}\,dx$$

$$= \lim_{c\to-\infty}\left[-\frac{1}{2(x^2+1)}\right]_c^0 + \lim_{c\to\infty}\left[-\frac{1}{2(x^2+1)}\right]_0^c$$

$$= -\frac{1}{2} + \lim_{c\to-\infty}\frac{1}{2(c^2+1)} + \lim_{c\to\infty}\left(-\frac{1}{2(c^2+1)}\right) + \frac{1}{2}$$

$$= -\frac{1}{2} + 0 + 0 + \frac{1}{2} = 0.$$

43. $\sec x$ is discontinuous at $x = \frac{\pi}{2}$, so that

$$\int_0^{\pi/2} \sec x\,dx = \lim_{c\to\frac{\pi}{2}^-}\int_0^c \sec x\,dx = \lim_{c\to\frac{\pi}{2}^-}\left[\ln|\sec x + \tan x|\right]_0^c$$

$$= \lim_{c\to\frac{\pi}{2}^-}\left(\ln|\sec c + \tan c| - 0\right) = \infty$$

since both $\sec c$ and $\tan c$ approach ∞ as $c \to \frac{\pi}{2}^-$.

45. With $u = \ln x$ we get $du = \frac{1}{x}\,dx$, so that $\int \frac{\ln x}{x}\,dx = \frac{1}{2}\ln^2 x$. Then

$$\int_0^1 \frac{\ln x}{x}\,dx = \lim_{c\to0^+}\int_c^1 \frac{\ln x}{x}\,dx$$

$$= \lim_{c\to0^+}\left[\frac{1}{2}\ln^2 x\right]_c^1$$

$$= \lim_{c\to0^+}\left(\frac{1}{2}(\ln 1)^2 - \frac{1}{2}(\ln c)^2\right)$$

$$= -\infty.$$

47. With $u = \ln x$ we get $du = \frac{1}{x}\,dx$, so that $\int \frac{1}{x\ln x}\,dx = \ln|\ln x|$. Then

$$\int_0^1 \frac{1}{x\ln x}\,dx = \lim_{c\to0^+}\int_c^{1/2} \frac{1}{x\ln x}\,dx + \lim_{c\to1^-}\int_{1/2}^c \frac{1}{x\ln x}\,dx$$

$$= \lim_{c\to0^+}\left[\ln|\ln x|\right]_c^{1/2} + \lim_{c\to1^-}\left[\ln|\ln x|\right]_{1/2}^c$$

$$= \lim_{c\to0^+}\left(\ln\ln 2 - \ln|\ln c|\right) + \lim_{c\to1^-}\left(\ln|\ln c| - \ln\ln 2\right).$$

The first of these limits goes to $-\infty$ as $c \to 0$, since $\ln c \to -\infty$, so that $\ln|\ln c|$ goes to $\ln\infty = \infty$. The second limit goes to $-\infty$ as well, since $|\ln c| \to 0$, so that $\ln|\ln c| \to -\infty$. Thus this limit is $-\infty$.

49. With $u = \ln x$ we get $du = \frac{1}{x}\,dx$, so that

$$\int \frac{1}{x(\ln x)^2}\,dx = \int \frac{1}{u^2}\,du = -u^{-1} = -\frac{1}{\ln x}.$$

Thus

$$\int_e^\infty \frac{1}{x(\ln x)^2}\, dx = \lim_{c\to\infty} \int_e^c \frac{1}{x(\ln x)^2}\, dx$$

$$= \lim_{c\to\infty} \left[-\frac{1}{\ln x} \right]_e^c$$

$$= \lim_{c\to\infty} \left(-\frac{1}{\ln c} + \frac{1}{\ln e} \right)$$

$$= 1 - \lim_{c\to\infty} \frac{1}{\ln c} = 1.$$

51. Let $u = \frac{1}{x}$, so that $du = -\frac{1}{x^2}\, dx$. Then

$$\int \frac{\sin\left(\frac{1}{x}\right)}{x^2}\, dx = -\int \sin u\, du = \cos u = \cos\left(\frac{1}{x}\right).$$

Then

$$\int_{1/\pi}^\infty \frac{\sin\left(\frac{1}{x}\right)}{x^2}\, dx = \lim_{c\to\infty} \int_{1/\pi}^c \frac{\sin\left(\frac{1}{x}\right)}{x^2}\, dx$$

$$= \lim_{c\to\infty} \left[\cos\left(\frac{1}{x}\right) \right]_{1/\pi}^c$$

$$= \lim_{c\to\infty} \left(\cos\frac{1}{c} - \cos\pi \right)$$

$$= 1 - (-1) = 2.$$

53. To find an antiderivative of the integrand, decompose the integrand using partial fractions:

$$\frac{1}{x(x-1)} = \frac{A}{x} + \frac{B}{x-1} = \frac{(A+B)x - A}{x(x-1)},$$

so that $A = -1$ and $B = 1$. Then

$$\int \frac{1}{x(x-1)}\, dx = \int \left(\frac{1}{x-1} - \frac{1}{x} \right)\, dx = \ln|x-1| - \ln|x| = \ln\left| \frac{x-1}{x} \right|.$$

Hence

$$\int_1^2 \frac{1}{x(x-1)}\, dx = \lim_{c\to1^+} \int_c^2 \frac{1}{x(x-1)}\, dx$$

$$= \lim_{c\to1^+} \left[\ln\left| \frac{x-1}{x} \right| \right]_c^2$$

$$= \ln\frac{1}{2} - \lim_{c\to1^+} \ln\left| \frac{c-1}{c} \right|$$

$$= \infty.$$

55. To find an antiderivative of the integrand, expand using partial fractions:

$$\frac{x-2}{x(x-1)} = \frac{A}{x} + \frac{B}{x-1} = \frac{(A+B)x - A}{x(x-1)},$$

so that $A + B = 1$ and $A = 2$; thus $B = -1$. So

$$\int \frac{x-2}{x(x-1)}\,dx = \int \left(\frac{2}{x} - \frac{1}{x-1}\right)\,dx = 2\ln|x| - \ln|x-1| = \ln\left|\frac{x^2}{x-1}\right|.$$

So

$$\int_0^2 \frac{x-2}{x(x-1)}\,dx = \lim_{c\to 0^+}\int_c^{1/2}\frac{x-2}{x(x-1)}\,dx + \lim_{c\to 1^-}\int_{1/2}^c\frac{x-2}{x(x-1)}\,dx + \lim_{c\to 1^+}\int_c^2\frac{x-2}{x(x-1)}\,dx$$

$$= \lim_{c\to 0^+}\left[\ln\left|\frac{x^2}{x-1}\right|\right]_c^{1/2} + \lim_{c\to 1^-}\left[\ln\left|\frac{x^2}{x-1}\right|\right]_{1/2}^c + \lim_{c\to 1^+}\left[\ln\left|\frac{x^2}{x-1}\right|\right]_c^2$$

$$= \ln\frac{1}{2} - \lim_{c\to 0^+}\ln\left|\frac{c^2}{c-1}\right| + \lim_{c\to 1^-}\ln\left|\frac{c^2}{c-1}\right| - \ln\frac{1}{2} + \ln 4 - \lim_{c\to 1^+}\ln\left|\frac{c^2}{c-1}\right|,$$

which is undefined.

57. Since $|\sin x| \le 1$, we have $\frac{x}{1+\sin^2 x} \ge \frac{x}{2}$, so that $\int_1^\infty \frac{x}{1+\sin^2 x}\,dx \ge \int_1^\infty \frac{x}{2}\,dx = \infty$, so that the integral diverges by comparison with $\frac{x}{2}$.

59. Since $\frac{1}{x^2+1} < \frac{1}{x^2}$, it follows that $\int_1^\infty \frac{1}{x^2+1}\,dx \le \int_1^\infty \frac{1}{x^2}$, which converges by the power rule. Thus the integral converges.

61. Since $\frac{1+\ln x}{x} \ge \frac{1}{x}$ for $x \ge 1$ (where $\ln x \ge 0$), it follows that $\int_1^\infty \frac{1+\ln x}{x}\,dx \ge \int_1^\infty \frac{1}{x}\,dx = \infty$, so that the integral diverges.

63. Since for $x \ge 1$ we have $\ln x < x$, it follows that $\frac{1+\ln x}{x^3} < \frac{1+x}{x^3} = \frac{1}{x^3} + \frac{1}{x^2}$. Thus

$$\int_1^\infty \frac{1+\ln x}{x^3}\,dx \le \int_1^\infty \left(\frac{1}{x^3} + \frac{1}{x^2}\right)\,dx.$$

Since both of the terms on the right-hand side have convergent integrals by the power rule, the left-hand side converges as well.

65. Let $f(x) = \frac{1}{(x-1)^2}$ and $g(x) = \frac{1}{x^2}$. Then

$$\lim_{x\to\infty}\frac{f(x)}{g(x)} = \lim_{x\to\infty}\frac{x^2-2x+1}{x^2} = \lim_{x\to\infty}\frac{1-2/x+1/x^2}{1} = 1.$$

Since the ratio converges to a real number, $f(x)$ and $g(x)$ either both converge or both diverge on $[2,\infty)$. But $\int_2^\infty g(x)\,dx$ converges by the power rule, so that $\int_2^\infty f(x)\,dx$ converges as well.

67. Let $f(x) = \frac{1}{3x^2-1}$ and $g(x) = \frac{1}{x^2}$. Then

$$\lim_{x\to\infty}\frac{f(x)}{g(x)} = \lim_{x\to\infty}\frac{x^2}{3x^2-1} = \lim_{x\to\infty}\frac{1}{3-1/x^2} = \frac{1}{3}.$$

Thus the integrals of f and g either both converge or both diverge. Since $\int_1^\infty g(x)\,dx$ converges by the power rule, $\int_1^\infty f(x)\,dx$ converges as well.

69. Let $f(x) = \frac{x}{\sqrt[4]{x^3}+100}$ and $g(x) = x^{1/4}$. Then

$$\lim_{x\to\infty}\frac{f(x)}{g(x)} = \lim_{x\to\infty}\frac{x}{x^{1/4}(\sqrt[4]{x^3}+100)} = \lim_{x\to\infty}\frac{x}{x+100x^{1/4}} = \lim_{x\to\infty}\frac{1}{1+100x^{-3/4}} = 1.$$

Thus the integrals of f and g either both converge or both diverge on $[1,\infty)$. Since the integral of g diverges since $g(x) \to \infty$ as $x \to \infty$, it follows that the integral of f diverges as well.

Applications

71. First evaluate the integral:

$$\int_T^\infty ke^{-0.023t}\, dt = \lim_{c\to\infty} \int_T^c ke^{-0.023t}\, dt$$

$$= \lim_{c\to\infty} \left[-\frac{1}{0.023} ke^{-0.023t} \right]_T^c$$

$$= \frac{1}{0.023} ke^{-0.023T} - \lim_{c\to\infty} \left(\frac{1}{0.023} ke^{-0.023c} \right)$$

$$= \frac{1}{0.023} ke^{-0.023T} \approx 43.478 ke^{-0.023T}.$$

We are given that $Q(12) = 0.95$, so

$$0.95 = Q(12) = \int_{12}^\infty ke^{-0.023t}\, dt \approx 43.478 ke^{-0.023\cdot 12} = 43.478 ke^{-0.276}.$$

Solving gives $k \approx 0.0288$. Thus the equation for the amount of cesium in the tank at time T is

$$Q(T) = \int_T^\infty 0.0288 e^{-0.023t}\, dt = 0.0288 \cdot 43.478 e^{-0.023T} \approx 1.25915 e^{-0.023T}.$$

Thus the amount in the tank at time $T = 0$ is ≈ 1.25 lb.

73. Evaluate the integral using integration by parts with $u = t$, $dv = 0.31e^{-0.31t}$. Then $du = dt$ and $v = -e^{-0.31t}$. Thus

$$\int_0^\infty tf(t)\, dt = \lim_{c\to\infty} \int_0^c tf(t)\, dt$$

$$= \lim_{c\to\infty} \left(\left[-te^{-0.31t} \right]_0^c + \int_0^c e^{-0.31t}\, dt \right)$$

$$= \lim_{c\to\infty} \left(-ce^{-0.31c} - \frac{1}{0.31} \left[e^{-0.31t} \right]_0^c \right)$$

$$= \lim_{c\to\infty} \left(-\frac{c}{e^{0.31c}} - \frac{1}{0.31e^{0.31c}} + \frac{1}{0.31} \right)$$

$$\approx 3.2258$$

since exponential functions dominate power functions. The pump can be expected to last 3.2 years.

Proofs

75. If $p > 1$, then

$$\int \frac{1}{x^p}\, dx = \int x^{-p}\, dx = -\frac{1}{p-1} x^{-p+1} = \frac{1}{1-p} x^{1-p}.$$

Thus in this case

$$\int_1^\infty \frac{1}{x^p}\, dx = \lim_{c\to\infty} \int_1^c \frac{1}{x^p}\, dx = \lim_{c\to\infty} \left[\frac{1}{1-p} x^{1-p} \right]_1^c = \lim_{c\to\infty} \left(\frac{c^{1-p}}{1-p} - \frac{1}{1-p} \right).$$

Now, since $p > 1$, we have $\lim_{c\to\infty} \frac{c^{1-p}}{1-p} = 0$, so that the integral has the value $-\frac{1}{1-p} = \frac{1}{p-1}$.

77. If $p > 1$, then

$$\int \frac{1}{x^p}\, dx = \int x^{-p}\, dx = -\frac{1}{p-1} x^{-p+1} = \frac{1}{1-p} x^{1-p}.$$

Thus in this case

$$\int_0^1 \frac{1}{x^p}\, dx = \lim_{c \to 0^+} \int_c^1 \frac{1}{x^p}\, dx = \lim_{c \to 0^+} \left[\frac{1}{1-p} x^{1-p} \right]_c^1 = \lim_{c \to 0^+} \left(\frac{1}{1-p} - \frac{c^{1-p}}{1-p} \right).$$

Now, since $p > 1$, we have $1 - p < 0$, so that $\lim\limits_{c \to 0^+} \frac{c^{1-p}}{1-p} = -\infty$, so that the integral diverges.

If $p = 1$, then $\int \frac{1}{x^p}\, dx = \int \frac{1}{x}\, dx = \ln|x|$, so in this case

$$\int_0^1 \frac{1}{x}\, dx = \lim_{c \to 0^+} \int_c^1 \frac{1}{x}\, dx = \lim_{c \to 0^+} \left[\ln|x| \right]_c^1 = \lim_{c \to 0^+} (-\ln|c|) = \infty.$$

Thus if $p \geq 1$, $\int_0^1 \frac{1}{x^p}\, dx$ diverges.

79. Since $0 \leq g(x) \leq f(x)$ for all $x \in (a, b]$, it follows that for any $c \in (a, b]$, $0 \leq \int_c^b g(x)\, dx \leq \int_c^b f(x)\, dx$. Thus

$$0 \leq \int_a^b g(x)\, dx = \lim_{c \to a^+} \int_c^b g(x)\, dx \leq \lim_{c \to a^+} \int_c^b f(x)\, dx = \int_a^b f(x)\, dx.$$

Thus if the final integral converges, the first must as well.

81. Suppose without loss of generality that $c < d$. Then

$$\int_{-\infty}^c f(x)\, dx + \int_c^\infty f(x)\, dx = \lim_{r \to -\infty} \int_r^c f(x)\, dx + \lim_{r \to \infty} \int_c^r f(x)\, dx$$

$$= \lim_{r \to -\infty} \int_r^c f(x)\, dx + \lim_{r \to \infty} \left(\int_c^d f(x)\, dx + \int_d^r f(x)\, dx \right)$$

$$= \left(\lim_{r \to -\infty} \int_r^c f(x)\, dx \right) + \int_c^d f(x)\, dx + \lim_{r \to \infty} \left(\int_d^r f(x)\, dx \right)$$

$$= \lim_{r \to -\infty} \left(\int_r^c f(x)\, dx + \int_c^d f(x)\, dx \right) + \lim_{r \to \infty} \left(\int_d^r f(x)\, dx \right)$$

$$= \lim_{r \to -\infty} \int_r^d f(x)\, dx + \lim_{r \to \infty} \int_d^r f(x)\, dx$$

$$= \int_{-\infty}^d f(x)\, dx + \int_d^\infty f(x)\, dx.$$

Thinking Forward

Gabriel's Horn

▷ Since in this case $f(x) = \frac{1}{x} = x^{-1}$, the volume, by the given formula, is

$$\pi \int_1^\infty (f(x))^2\, dx = \pi \int_1^\infty \frac{1}{x^2}\, dx.$$

Evaluating gives

$$\pi \int_1^\infty \frac{1}{x^2}\, dx = \pi \lim_{c \to \infty} \int_1^c \frac{1}{x^2}\, dx = \pi \lim_{c \to \infty} \left[-x^{-1} \right]_1^c = \pi \lim_{c \to \infty} \left(-\frac{1}{c} + 1 \right) = \pi.$$

Thus the volume of the solid is π.

▷ With $f(x) = x^{-1}$, we have $f'(x) = -x^{-2}$, so that the surface area is

$$2\pi \int_1^\infty f(x)\sqrt{1+f'(x)^2}\,dx = 2\pi \int_1^\infty x^{-1}\sqrt{1+(-x^{-2})^2}\,dx$$

$$= 2\pi \int_1^\infty x^{-1}\sqrt{1+x^{-4}}\,dx$$

$$= 2\pi \int_1^\infty \frac{\sqrt{1+x^{-4}}}{x}\,dx.$$

The integrand is greater than $\frac{1}{x}$ for $x \geq 1$, so that since $\int_1^\infty \frac{1}{x}$ diverges, so does the surface area integral. Thus the surface area is infinite.

5.7   Numerical Integration

Thinking Back

Monotonicity and concavity

▷ Since $f'(x) = 3x^2 - 4x$, we see that $f'(1) = -1 < 0$ while $f'(2) = 4 > 0$, so that f is decreasing on portions of $[1,3]$ and increasing on other parts. Since $f''(x) = 6x - 4$, we have $f''(x) > 0$ for $x > \frac{2}{3}$, so that in particular $f''(x) > 0$ everywhere on $[1,3]$. Thus f is concave up everywhere on $[1,3]$.

▷ Since $f'(x) = 3x^2 - 4x$, we see that $f'(x) = 0$ at $x = 0$ and at $x = \frac{4}{3}$. Since $f'(1) < 0$, it follows that f is decreasing everywhere on $[0,1]$. Since $f''(x) = 6x - 4$, we have $f''(0) < 0$ while $f''(1) > 0$, so that f is sometimes concave up and sometimes concave down on $[0,1]$ — it has an inflection point where $f''(x) = 0$, at $x = \frac{2}{3}$.

▷ Since $f'(x) = \frac{\pi}{2}\cos\left(\frac{\pi}{2}x\right)$, we see that $f'(x) \geq 0$ for $x \in [-1,1]$, and $f'(x) > 0$ except at $x = \pm 1$. Thus f is increasing everywhere on $[-1,1]$. Now, $f''(x) = -\frac{\pi^2}{4}\sin\left(\frac{\pi}{2}x\right)$, so that $f''(x)$ changes sign at $x = 0$, and f is concave up for $x < 0$ and concave down for $x > 0$.

Bounding the second derivative

▷ Since $f(x) = x^2(x-3) = x^3 - 3x^2$, we have $f'(x) = 3x^2 - 6x$ and $f''(x) = 6x - 6$. Thus $f''(x)$ is increasing on $[0,2]$, so it achieves its maximum magnitude at either 0 or 2. Since $|f''(0)| = |-6| = 6$ while $|f''(2)| = |6| = 6$, we can take $M = 6$.

▷ Again $f''(x)$ achieves its maximum magnitude at one endpoint or the other. Since $|f''(4)| = |18| = 18$, we take $M = 18$.

Determining quadratics

▷ Let $p(x) = ax^2 + bx + c$. Then

$$p(1) = a + b + c = f(1) = 2, \quad p(0) = c = f(0) = 1, \quad p(-1) = a - b + c = f(-1) = 0.$$

So $a + b = 1$ and $a - b = -1$; it follows that $a = 0$ and $b = 1$. Thus the unique "parabola" is in fact the line $p(x) = x + 1$ (the three points $(-1, f(-1))$, $(0, f(0))$, and $(1, f(1))$ are collinear).

▷ Let $p(x) = ax^2 + bx + c$. Then

$$p(1) = a + b + c = f(1) = 1, \quad p(2) = 4a + 2b + c = f(2) = \frac{1}{2}, \quad p(3) = 9a + 3b + c = f(3) = \frac{1}{3},$$

which has the solution $a = \frac{1}{6}$, $b = -1$, and $c = \frac{11}{6}$, so that $p(x) = \frac{1}{6}x^2 - x + \frac{11}{6}$.

▷ Let $p(x) = ax^2 + bx + c$. Then

$$p(0) = c = f(0) = 0, \; p\left(\frac{\pi}{2}\right) = \frac{\pi^2}{4}a + \frac{\pi}{2}b + c = f\left(\frac{\pi}{2}\right) = 1, \; p(\pi) = \pi^2 a + \pi b + c = f(\pi) = 0.$$

Solving these equations gives $a = -\frac{4}{\pi^2}$, $b = \frac{4}{\pi}$, and $c = 0$, so that $p(x) = -\frac{4}{\pi^2}x^2 + \frac{4}{\pi}x$.

Concepts

1. (a) True. Since the error is the difference between the Riemann sum and the true value of the integral, if we know the error and the Riemann sum, we can get the value of the integral.

    (b) True. For example, the magnitude of an error bound of a left sum is given by the difference between the left sum and the right sum.

    (c) False. For example, if the function is a constant function, the two left sums will be equal to the value of the integral. In most cases, however, the statement is true.

    (d) True, since each of the ten rectangles on the same base as a given rectangle in the $n = 10$ sum will be inside the given rectangle; since the function is monotonically decreasing, all rectangles except the first will have a lower height than the given rectangle. Thus the sum of those ten areas will be less than the area of the given rectangle.

    (e) True. If f is monotonically increasing, then the right sum is always larger than the integral and the left sum is always smaller, so that the error from the left sum is smaller than the difference of the right and left sums.

    (f) True. Since every trapezoid joins two points on the curve, the line joining the points will lie below the curve if it is concave down.

    (g) True. See Theorem 5.26.

    (h) False. For example, consider a piecewise linear curve consisting of ten lines. A trapezoid sum with $n = 10$ will exactly compute the area under this curve, but five parabolas clearly will not be exact.

3. Consider the figure

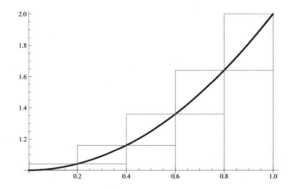

The left sum is the sum of the areas of the lower rectangles; the right sum is the sum of the area of the upper rectangles. The error incurred by the left sum is the sum of the five triangle-like regions lying between the curve and the lower rectangles, while the difference between the right and left sums is the sum of the areas of the five rectangles lying between the two sets of rectangles. Clearly the first is less than the second.

5. From the figures

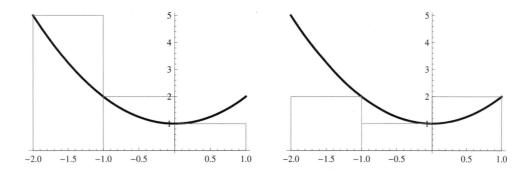

it is clear from the first figure that while the left approximation lies below the curve for $x > 0$, it lies above the curve for $x < 0$, and the extra included area there far exceeds the excluded area to the right of $x = 0$. Thus in this case $\text{LEFT}(n) > \int_a^b f(x)\,dx$. Similarly it is clear from the second figure that while the right approximation lies above the curve for $x > 0$, it lies below the curve for $x < 0$, and the excluded area there far exceeds the extra included area to the right of $x = 0$. Thus in this case $\text{RIGHT}(n) < \int_a^b f(x)\,dx$.

7. Consider the figure

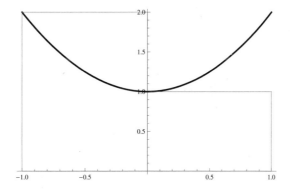

which shows a left sum approximation to $f(x) = x^2 + 1$ on $[-1, 1]$. The left sum approximation is $2 + 1 = 3$, and the actual integral is

$$\int_{-1}^{1} (x^2 + 1)\,dx = \left[\frac{1}{3}x^3 + x\right]_{-1}^{1} = \frac{8}{3},$$

so that the error $E(x)$ is nonzero. However, since $f(b) = f(a)$, the expression $(f(b) - f(a))\Delta x = 0$, so that the statement $|E(x)| \leq |(f(b) - f(a))\Delta x|$ is false.

9. The left sum is not guaranteed to be an overapproximation; for example, if f is increasing but concave down, the left sum will be an underapproximation by Theorem 5.24(a). Similarly, if f is decreasing and concave down, the right sum will be an underapproximation by Theorem 5.24(b). Since f is concave down, the trapezoid sum is an underapproximation and the midpoint sum is an overapproximation, by Theorem 5.26(b). Thus only (d) is guaranteed to be an overapproximation.

11. Approximate the integral by two separate sums, one on $[a, c]$ and the other on $[c, b]$. Since f is concave down on $[a, c]$, Theorem 5.27 applies there; similarly, since f is concave up on $[c, b]$, Theorem 5.27 applies there. If $|f''(x)| \leq M_a$ for $x \in [a, c]$ and $|f''(x)| \leq M_b$ for $x \in [c, b]$, and we use

approximations with n_a rectangles on $[a, c]$ and n_b rectangles on $[c, b]$, then the total error of the composite trapezoid or midpoint approximations are bounded by

$$\frac{M_a(c-a)^3}{12n_a^2} + \frac{M_b(b-c)^3}{12n_b^2} \quad \text{and} \quad \frac{M_a(c-a)^3}{24n_a^2} + \frac{M_b(b-c)^3}{24n_b^2}$$

respectively. See the figures below, where $a = -2$, $b = 1$, $c = 0$, $n_a = 3$ and $n_b = 2$:

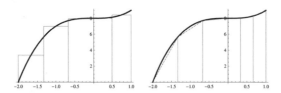

13. See the figure following Theorem 5.26. The midpoint trapezoid is formed as follows: given the rectangle resulting from a midpoint sum on some subinterval, draw the tangent line to the curve at the midpoint. This cuts off a piece of the rectangle on one side or the other of the midpoint. Remove that rectangle and glue it onto the triangle on the other side of the midpoint. The result is a trapezoid whose diagonal edge is along the tangent line to the curve, and whose area is equal to the area of the original rectangle. Thus, unless there is an inflection point in the subinterval, the diagonal edge of the midpoint trapezoid will lie wholly above or wholly below the curve. So in the case of a curve with constant concavity, all midpoint trapezoids have area smaller than (in the case of upwards concavity) or larger than (in the case of downwards concavity) the area under the curve. Thus for concave up curves, the midpoint approximation is an underapproximation, while for concave down curves, it is an overapproximation.

15. (a) From the given conditions, we know that f is positive, decreasing, and concave down on $[a, b]$. Since it is decreasing, we have by Theorem 5.24(b)

$$\text{RIGHT}(n) \leq \int_a^b f(x)\, dx \leq \text{LEFT}(n).$$

Since it is concave down, we have by Theorem 5.26(b)

$$\text{TRAP}n \leq \int_a^b f(x)\, dx \leq \text{MID}(n).$$

Since f is decreasing, the value at the midpoint of an interval lies between the values at the left and right ends, so that $\text{RIGHT}(n) \leq \text{MID}(n) \leq \text{LEFT}(n)$. Since f is concave down, the trapezoid on any subinterval completely includes the rectangle drawn at the height of the smaller (right-hand) endpoint, so that $\text{RIGHT}(n) \leq \text{TRAP}(n)$. Putting this all together gives

$$\text{RIGHT}(n) \leq \text{TRAP}(n) \leq \int_a^b f(x)\, dx \leq \text{MID}(n) \leq \text{LEFT}(n).$$

The figure below shows all four of these approximations, in increasing order, for such a function.

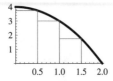 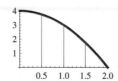 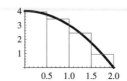 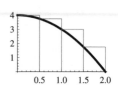

Note that in the second graph, the trapezoids approximate the curve very closely, so they are hard to see.

(b) From the discussion in part (a), the true value lies between the trapezoidal and midpoint approximations.

17. The formula in Exercise 63 gives the formula

$$\text{SIMP}(2n) = \frac{1}{3}\text{TRAP}(n) + \frac{2}{3}\text{MID}(n) = \frac{1}{3}(1.09861258) + \frac{2}{3}(1.09861214) = 1.09861229.$$

This should be the result of applying Simpson's rule to approximate the integral.

19. Computing the integral gives

$$\int_1^5 (3x^2 - x + 4)\, dx = \left[x^3 - \frac{1}{2}x^2 + 4x \right]_1^5 = 125 - \frac{25}{2} + 20 - 1 + \frac{1}{2} - 4 = 128.$$

The right-hand side of Theorem 5.28 evaluates in this case to

$$\frac{B-A}{6}\left(p(A) + 4p\left(\frac{A+B}{2}\right) + p(B) \right) = \frac{4}{6}\left(3\cdot 1^2 - 1 + 4 + 4\left(3\left(\frac{1+5}{2}\right)^2 - \frac{1+5}{2} + 4 \right) \right.$$

$$\left. + 3\cdot 5^2 - 5 + 4 \right)$$

$$= \frac{2}{3}\left(6 + 4\left(27 - 3 + 4\right) + 74 \right)$$

$$= \frac{2}{3}(192) = 128.$$

21. Suppose that on some interval $[a, b]$ we approximate the integral of $f(x)$ using Simpson's rule with 2 parabolas, so that $n = 4$. To do this, we choose equally spaced points $x_0 = a, x_1, x_2, x_3$, and $x_4 = b$ on $[a, b]$, with Δx the distance between consecutive points. We then construct one parabola $p_1(x)$ going through the points $(x_0, f(x_0))$, $(x_1, f(x_1))$, and $(x_2, f(x_2))$, and another, $p_2(x)$, through the points $(x_2, f(x_2))$, $(x_3, f(x_3))$, and $(x_4, f(x_4))$. Then the area under these parabolas is, by Theorem 5.28,

$$\int_{x_0}^{x_2} p_1(x)\, dx = \frac{x_2 - x_0}{6}\left(p_1(x_0) + 4p_1\left(\frac{x_0 + x_2}{2}\right) + p_1(x_2) \right)$$

$$\int_{x_2}^{x_4} p_2(x)\, dx = \frac{x_4 - x_2}{6}\left(p_2(x_2) + 4p_2\left(\frac{x_2 + x_4}{2}\right) + p_2(x_4) \right)$$

Note that $x_1 = \frac{x_0 + x_2}{2}$ and $x_3 = \frac{x_2 + x_4}{2}$ since the points are equally spaced. Also, by construction, $p_j(x_i) = f(x_i)$ for $i = 0, 1, 2, 3, 4$ and $j = 1, 2$. Thus the above equations become

$$\int_{x_0}^{x_2} p_1(x)\, dx = \frac{x_2 - x_0}{6}\left(f(x_0) + 4f(x_1) + f(x_2) \right) = \frac{2\Delta x}{6}\left(f(x_0) + 4f(x_1) + f(x_2) \right)$$

$$\int_{x_2}^{x_4} p_2(x)\, dx = \frac{x_4 - x_2}{6}\left(f(x_2) + 4f(x_3) + f(x_4) \right) = \frac{2\Delta x}{6}\left(f(x_2) + 4f(x_3) + f(x_4) \right).$$

Adding these equations gives on the left the approximate integral with $n = 4$:

$$\text{SIMP}(4) = \frac{\Delta x}{3}\left(f(x_0) + 4f(x_1) + 2f(x_2) + 4f(x_3) + f(x_4) \right).$$

Skills

23. With $n = 4$ we have $\Delta x = 1$ and $x_i = i$ for $i = 0, 1, 2, 3, 4$. Then the right sum is

$$\sum_{k=1}^{4} f(x_k)\Delta x = \sum_{k=1}^{4} (k^2 + k) = 40.$$

The error bound for this estimate is $|f(b) - f(a)|\,\Delta x = (4^2 + 4) - (0^2 + 0) = 20$, so the integral should be in the range $[20, 60]$. Since the true value of the integral is

$$\int_0^4 (x^2 + x)\,dx = \left[\frac{1}{3}x^3 + \frac{1}{2}x^2\right]_0^4 = \frac{64}{3} + 8 = \frac{88}{3} \approx 29.33,$$

the error bounds are accurate.

25. With $n = 6$ we have $\Delta x = \frac{3-1}{6} = \frac{1}{3}$, and $x_i = 1 + \frac{i}{3}$ for $i \in \{0, 1, 2, 3, 4, 5, 6\}$. Then the left sum is

$$\sum_{k=1}^{6} f(x_{k-1})\Delta x = \frac{1}{3}\sum_{k=1}^{6} e^{-(1+(k-1)/3)} = \frac{1}{3}\left(e^{-1} + e^{-4/3} + e^{-5/3} + e^{-2} + e^{-7/3} + e^{-8/3}\right) \approx 0.374.$$

The error bound for this estimate is $|f(b) - f(a)|\,\Delta x = \frac{1}{3}\left|e^{-3} - e^{-1}\right| \approx 0.106$, so the integral should be in the range $[0.268, 0.480]$. Since the true value of the integral is

$$\int_1^3 e^{-x}\,dx = \left[-e^{-x}\right]_1^3 = e^{-1} - e^{-3} \approx 0.318,$$

the error bounds are accurate.

27. With $n = 6$ we have $\Delta x = \frac{3-1}{6} = \frac{1}{3}$, and $x_i = 1 + \frac{i}{3}$ for $i \in \{0, 1, 2, 3, 4, 5, 6\}$. Then the trapezoid sum is

$$\sum_{k=1}^{6} \frac{f(x_{k-1}) + f(x_k)}{2}\Delta x = \frac{1}{6}\sum_{k=1}^{6}\left(e^{1+(k-1)/3} + e^{1+k/3}\right)$$

$$= \frac{1}{6}\left((e^{-1} + e^{-4/3}) + (e^{-4/3} + e^{-5/3}) + \cdots + (e^{-8/3} + e^{-3})\right)$$

$$\approx 0.321.$$

Now, $f''(x) = e^{-x}$, which is decreasing, so that it achieves its maximum at the left end of the interval; that maximum is e^{-1}. Thus by Theorem 5.27, the error estimate is

$$\left|E_{\text{TRAP}(6)}\right| \le \frac{e^{-1}(3 - 1)^3}{12 \cdot 6^2} \approx 0.0068.$$

Thus the true value of the integral should be in the interval $[0.3142, 0.3278]$. Since from Exercise 25 the true value is ≈ 0.318, the error bounds are accurate.

29. With $n = 12$ we have $\Delta x = \frac{7-1}{12} = \frac{1}{2}$, and $x_k = 1 + \frac{k}{2}$ for k from 0 to 12. Thus the midpoint values are

$$x_k^* = \frac{x_{k-1} + x_k}{2} = \frac{1 + \frac{k-1}{2} + 1 + \frac{k}{2}}{2} = \frac{3 + 2k}{4}$$

for k from 1 to 12. Then the approximation is

$$\sum_{k=1}^{12} f(x_k^*)\Delta x = \frac{1}{2}\sum_{k=1}^{12} \ln\frac{3 + 2k}{4} = \frac{1}{2}\left(\ln\frac{5}{4} + \ln\frac{7}{4} + \ln\frac{9}{4} + \cdots + \ln\frac{25}{4} + \ln\frac{27}{4}\right) \approx 7.630.$$

Now, $f''(x) = (x^{-1})' = -x^{-2}$, which is negative but increasing, so that it achieves its maximum magnitude at the left end of the interval; that maximum magnitude is $|f''(1)| = 1$. So by Theorem 5.27, the error estimate is

$$\left|E_{\text{MID}(12)}\right| \le \frac{1 \cdot (7-1)^3}{24 \cdot 12^2} = 0.0625,$$

so that the integral should lie in the interval $[7.5675, 7.6925]$. Since the true value of the integral is

$$\int_1^7 \ln x \, dx = [x \ln x - x]_1^7 = 7 \ln 7 - 7 + 1 = 7 \ln 7 - 6 \approx 7.621,$$

the error bounds are accurate.

31. With $n = 8$ we have $\Delta x = \frac{2-0}{8} = \frac{1}{4}$, so that $x_i = \frac{i}{4}$ for i from 0 to 8. Then the trapezoid approximation is

$$\sum_{k=1}^8 \frac{f(x_{k-1}) + f(x_k)}{2} \Delta x = \frac{1}{8} \sum_{k=1}^8 \sqrt{1 + \left(\frac{k-1}{4}\right)^3} + \sqrt{1 + \left(\frac{k}{4}\right)^3}$$

$$= \frac{1}{8} \left(\left(\sqrt{1} + \sqrt{1 + \frac{1}{64}} \right) + \left(\sqrt{1 + \frac{1}{64}} + \sqrt{1 + \frac{8}{64}} \right) + \cdots \right.$$

$$\left. + \left(\sqrt{1 + \frac{7^3}{64}} + \sqrt{1 + \frac{8^3}{64}} \right) \right)$$

$$\approx 3.2517.$$

Now,

$$f'(x) = \left((1 + x^3)^{1/2} \right)' = \frac{1}{2}(1 + x^3)^{-1/2} \cdot 3x^2 = \frac{3}{2}x^2(1 + x^3)^{-1/2}$$

$$f''(x) = \left(\frac{3}{2}x^2(1 + x^3)^{-1/2} \right)'$$

$$= 3x(1 + x^3)^{-1/2} - \frac{3}{4}x^2(1 + x^3)^{-3/2} \cdot 3x^2$$

$$= (1 + x^3)^{-3/2} \left(3x(1 + x^3) - \frac{9}{4}x^4 \right)$$

$$= (1 + x^3)^{-3/2} \left(\frac{3}{4}x^4 + 3x \right).$$

The first factor is ≤ 1 on $[0,2]$, and the second factor is $\le \frac{3}{4} \cdot 2^4 + 3 \cdot 2 = 18$, so that $|f''(x)| \le 18$ on $[0,2]$. (Note that this is a substantial overestimate; in fact $|f''(x)| \le 1.5$ on $[0,2]$, but this is harder to come by). Then by Theorem 5.27, the error bounds are

$$\left|E_{\text{TRAP}(8)}\right| \le \frac{18 \cdot (2-0)^3}{12 \cdot 8^2} = \frac{3}{16} = 0.1875.$$

We do not know how to find an antiderivative of this function, but integrating numerically using software gives an integral of ≈ 3.241, which is inside the error range.

33. With $n = 9$ we have $\Delta x = \frac{3-0}{9} = \frac{1}{3}$, so that $x_i = \frac{i}{3}$ for i from 0 to 9. Then the left sum approximation is

$$\sum_{k=1}^9 f(x_{k-1})\Delta x = \frac{1}{3} \sum_{k=1}^9 \cos\left(\frac{k-1}{3}\right)^2$$

$$= \frac{1}{3} \left(\cos 0 + \cos \frac{1}{9} + \cos \frac{4}{9} + \cdots + \cos \frac{64}{9} \right)$$

$$\approx 0.996.$$

The error estimate for this sum is $|f(b) - f(a)| \, \Delta x = \frac{1}{3} |\cos 9 - \cos 0| \approx 0.637$, so that the true value should lie in $[0.359, 1.633]$. We do not know how to find an antiderivative of this function, but integrating numerically using software gives an integral of ≈ 0.703, which is inside the error range.

35. With $n = 12$, we have $\Delta x = \frac{4-(-2)}{12} = \frac{1}{2}$, so that $x_i = -2 + \frac{i}{2}$ for i from 0 to 12. Then the Simpson's rule sum is (with $f(x) = x(x+1)(x-2) = x^3 - x^2 - 2x$)

$$\mathrm{SIMP}(12) = \frac{1/2}{3} \left(f(x_0) + 4f(x_1) + 2f(x_2) + \cdots + 4f(x_{11}) + f(x_{12}) \right)$$
$$= \frac{1}{6} \left(((-2)^3 - (-2)^2 - 2(-2)) + \cdots + (4^3 - 4^2 - 2 \cdot 4) \right)$$
$$= 24.$$

Since the function is a cubic, its fourth derivative is zero, so that the error estimate is $\left| E_{\mathrm{SIMP}(12)} \right| = 0$ and this is an exact result.

37. With $n = 4$, we have $\Delta x = \frac{2-(-1)}{4} = \frac{3}{4}$, so that $x_i = -1 + \frac{3i}{4} = \frac{3i-4}{4}$ for i from 0 to 4. Then the Simpson's rule sum is

$$\mathrm{SIMP}(4) = \frac{3/4}{3} \left(f(x_0) + 4f(x_1) + 2f(x_2) + 4f(x_3) + f(x_4) \right)$$
$$= \frac{1}{4} \left(((-1)^4 - 4(-1)^3 + 4(-1)^2) + \cdots + 4 \left(\left(\frac{5}{4}\right)^4 - 4\left(\frac{5}{4}\right)^3 + 4\left(\frac{5}{4}\right)^2 \right) \right.$$
$$\left. + (2^4 - 4 \cdot 2^3 + 4 \cdot 2^2) \right)$$
$$= \frac{477}{128} \approx 3.7266.$$

Since $f^{(4)}(x) = 24$, we may take $M = 24$ in the error estimate to get

$$\left| E_{\mathrm{SIMP}(4)} \right| \leq \frac{24(2-(-1))^5}{180 \cdot 4^4} = \frac{81}{640} \approx 0.1266.$$

Thus the true value should lie in $[3.600, 3.8532]$. Integrating gives

$$\int_{-1}^{2} (x^4 - 4x^3 + 4x^2) \, dx = \left[\frac{1}{5}x^5 - x^4 + \frac{4}{3}x^3 \right]_{-1}^{2}$$
$$= \frac{32}{5} - 16 + \frac{32}{3} + \frac{1}{5} + 1 + \frac{4}{3}$$
$$= \frac{18}{5} = 3.6,$$

so that the error estimate is valid.

39. With $n = 8$, we have $\Delta x = \frac{2\pi}{8} = \frac{\pi}{4}$, so that $x_i = -\pi + \frac{i\pi}{4} = \frac{(i-4)\pi}{4}$. Then the Simpson's rule sum is

$$\mathrm{SIMP}(8) = \frac{\pi/4}{3} \left(f(x_0) + 4f(x_1) + 2f(x_2) + \cdots + 4f(x_7) + f(x_8) \right)$$
$$= \frac{\pi}{12} \left(\cos(-\pi) + 4\cos\left(-\frac{3\pi}{4}\right) + 2\cos\left(-\frac{2\pi}{4}\right) + \cdots + 4\cos\left(\frac{3\pi}{4}\right) + \cos \pi \right)$$
$$= 0.$$

Since $f^{(4)}(x) = -\cos x$, we have $\left|f^{(4)}(x)\right| \le 1$, so we take $M = 1$, and then

$$\left|E_{\text{SIMP}(8)}\right| \le \frac{1 \cdot (2\pi)^5}{180 \cdot 8^4} \approx 0.013.$$

Since the true value of the integral is

$$\int_{-\pi}^{\pi} \cos x \, dx = [\sin x]_{-\pi}^{\pi} = 0,$$

obviously the error estimate is valid.

41. For the left sum to be accurate within 0.5, we must have

$$\left|E_{\text{LEFT}(n)}\right| = |f(b) - f(a)|\,\Delta x = |f(b) - f(a)|\frac{b-a}{n} \le 0.5.$$

Here $a = -1$, $b = 1$, and $f(a) = -1e^{-1}$, $f(b) = e$. So we want to find n so that

$$(e + e^{-1})\frac{2}{n} \le 0.5, \quad \text{or} \quad n \ge 4(e + e^{-1}) \approx 12.35,$$

so that $n = 13$ works. For $n = 13$, we have $\Delta x = \frac{2}{13}$ and $x_k = -1 + \frac{2k}{13}$; then the left sum is

$$\sum_{k=1}^{13} f(x_{k-1})\Delta x = \frac{2}{13} \sum_{k=1}^{13} \left(-1 + \frac{2(k-1)}{13}\right) e^{-1+\frac{2(k-1)}{13}} \approx 0.5091.$$

To calculate the integral, use integration by parts with $u = x$ and $dv = e^x\,dx$, so that $du = dx$ and $v = e^x$. Then

$$\int_{-1}^{1} xe^x\,dx = [xe^x]_{-1}^{1} - \int_{-1}^{1} e^x\,dx = e + e^{-1} - [e^x]_{-1}^{1} = e + e^{-1} - e + e^{-1} = 2e^{-1} \approx 0.736.$$

The actual value is within 0.5 of the left sum estimate.

43. For the right sum to be accurate to within 0.05, we must have

$$\left|E_{\text{RIGHT}(n)}\right| = |f(b) - f(a)|\,\Delta x = |f(b) - f(a)|\frac{b-a}{n} \le 0.05.$$

Here $a = 2$, $b = 4$, and $f(a) = \frac{1}{2}$, $f(b) = \frac{1}{4}$. So we want to find n so that

$$\frac{1}{4} \cdot \frac{2}{n} \le 0.05, \quad \text{or} \quad n \ge 10,$$

so that $n = 10$ works. For $n = 10$, we have $\Delta x = \frac{2}{10} = \frac{1}{5}$ and $x_k = 2 + \frac{k}{5} = \frac{k+10}{5}$; then the right sum is

$$\sum_{k=1}^{10} f(x_k)\Delta x = \frac{1}{5} \sum_{k=1}^{10} \frac{1}{(k+10)/5} = \frac{1}{5} \sum_{k=1}^{10} \frac{5}{k+10} \approx 0.669.$$

The true value of the integral is

$$\int_{2}^{4} \frac{1}{x}\,dx = [\ln x]_{2}^{4} = \ln 4 - \ln 2 \approx 0.693,$$

so that $\left|\text{RIGHT}(10) - \int_{2}^{4} \frac{1}{x}\,dx\right| \approx 0.024 < 0.05$.

45. For the midpoint sum to be accurate within 0.1, we must have

$$\left| E_{\text{MID}(n)} \right| \le \frac{M(b-a)^3}{24n^2} \le 0.1.$$

Since $f''(x) = 2$, we can take $M = 2$, and $a = 0$, $b = 3$, so that the error bound is

$$\frac{2(3-0)^3}{24n^2} = \frac{54}{24n^2} = \frac{9}{4n^2} \le 0.1.$$

This holds if $n^2 \ge \frac{90}{4} = 22.5$, so that $n = 5$ is sufficient. For $n = 5$, we have $\Delta x = \frac{3}{5}$, so that $x_i = \frac{3i}{5}$, and the midpoints are $x_i^* = \frac{x_{i-1}+x_i}{2} = \frac{\frac{3(i-1)}{5} + \frac{3i}{5}}{2} = \frac{6i-3}{10}$. Then the midpoint sum is

$$\sum_{k=1}^{5} f(x_k^*)\Delta x = \frac{3}{5} \sum_{k=1}^{5} f\left(\frac{6k-3}{10}\right) = \frac{3}{5} \sum_{k=1}^{5} \frac{(6k-3)^2}{100} = \frac{3}{500} \sum_{k=1}^{5} (6k-3)^2 \approx 8.91.$$

The true value of the integral is

$$\int_0^3 x^2 \, dx = \left[\frac{1}{3} x^3 \right]_0^3 = 9,$$

and the magnitude of the difference between the estimate and the true value is $0.09 < 0.1$.

47. For the trapezoid sum to be accurate within 0.005, we must have

$$\left| E_{\text{TRAP}(n)} \right| \le \frac{M(b-a)^3}{12n^2} \le 0.005.$$

Now, $f''(x) = -\frac{2}{9} x^{-4/3}$, so that $|f''(x)| = \frac{2}{9} x^{-4/3}$, which is a decreasing function. Thus it takes its maximum on $[1,4]$ at $x = 1$, and that maximum is $M = \frac{2}{9}$. Also, $a = 1$ and $b = 4$, so that the error bound is

$$\frac{\frac{2}{9}(4-1)^3}{12n^2} = \frac{54}{108n^2} = \frac{1}{2n^2} \le 0.005.$$

This holds if $n^2 \ge 100$, so that $n = 10$ is sufficient. For $n = 10$, we have $\Delta x = \frac{3}{10}$, so that $x_i = 1 + \frac{3i}{10} = \frac{3i+10}{10}$. Then the trapezoid sum is

$$\sum_{k=1}^{10} \frac{f(x_{k-1}) + f(x_k)}{2} \Delta x = \frac{3}{20} \sum_{k=1}^{10} \left(\left(\frac{3(k-1)+10}{10}\right)^{2/3} + \left(\frac{3k+10}{10}\right)^{2/3} \right)$$

$$= \frac{3}{2 \cdot 10^{5/3}} \sum_{k=1}^{10} \left((3k+7)^{2/3} + (3k+10)^{2/3} \right)$$

$$\approx 5.44577.$$

The true value of the integral is

$$\int_1^4 x^{2/3} \, dx = \left[\frac{3}{5} x^{5/3} \right]_1^4 = \frac{3}{5} 4^{5/3} - \frac{3}{5} \approx 5.44762,$$

and the magnitude of the difference between the estimate and the true value is $\approx 0.00205 < 0.005$.

49. For the Simpson sum to be accurate to within 0.01, we must have

$$\left| E_{\text{SIMP}(n)} \right| \le \frac{M(b-a)^5}{180n^4} \le 0.01,$$

where M is an upper bound for $f^{(4)}(x)$ on $[0, 2]$. Now, $f^{(4)}(x) = 10e^{-x}$, which is a decreasing function, so it takes its maximum on $[0, 2]$ at $x = 0$; that maximum is $M = 10$. Then

$$\frac{10(2-0)^5}{180n^4} = \frac{320}{180n^4} = \frac{16}{9n^4} \leq 0.01.$$

This holds for $n^4 \geq \frac{1600}{9} \approx 177.78$, so the smallest integer n is $n = 4$, since n must be even. For $n = 4$, we have $\Delta x = \frac{2}{4} = \frac{1}{2}$ so that $x_i = \frac{i}{2}$. Then the Simpson sum is

$$\begin{aligned} \text{SIMP}(4) &= \frac{\Delta x}{3}(f(x_0) + 4f(x_1) + 2f(x_2) + 4f(x_3) + f(x_4)) \\ &= \frac{10}{6}\left(e^{-0} + 4e^{-1/2} + 2e^{-1} + 4e^{-3/2} + e^{-2}\right) \\ &\approx 8.64956. \end{aligned}$$

The true value of the integral is

$$\int_0^2 10e^{-x}\, dx = -10\left[e^{-x}\right]_0^2 = 10 - 10e^{-2} \approx 8.64665,$$

which differs from the approximation by $\approx 0.00291 < 0.01$.

51. For the Simpson sum to be accurate to within 0.0005, we must have

$$\left|E_{SIMP(n)}\right| \leq \frac{M(b-a)^5}{180n^4} \leq 0.0005,$$

where M is an upper bound for $f^{(4)}(x)$ on $[0, \pi]$. Now, $f^{(4)}(x) = \sin x$, which has a maximum magnitude of $M = 1$. Then

$$\frac{1(\pi - 0)^5}{180n^4} = \frac{\pi^5}{180n^4} \leq 0.0005.$$

This holds for $n^4 \geq \frac{\pi^5}{180} \cdot 2000 = 6.6874$, so the smallest integer n is $n = 8$, since n must be even. For $n = 8$, we have $\Delta x = \frac{\pi}{8}$ so that $x_i = \frac{i\pi}{8}$. Then the Simpson sum is

$$\begin{aligned} \text{SIMP}(8) &= \frac{\Delta x}{3}(f(x_0) + 4f(x_1) + 2f(x_2) + 4f(x_3) + 2f(x_4) + 4f(x_5) + 2f(x_6) + 4f(x_7) + f(x_8)) \\ &= \frac{\pi}{24}\left(\sin 0 + 4\sin\frac{\pi}{8} + 2\sin\frac{\pi}{4} + 4\sin\frac{3\pi}{8} + 2\sin\frac{\pi}{2} + 4\sin\frac{5\pi}{8}\right. \\ &\qquad\qquad \left. + 2\sin\frac{3\pi}{4} + 4\sin\frac{7\pi}{8} + \sin\pi\right) \\ &\approx 2.00027. \end{aligned}$$

The true value of the integral is

$$\int_0^\pi \sin x\, dx = -\left[\cos x\right]_0^\pi = 2,$$

which differs from the approximation by $\approx 0.00027 < 0.0005$.

Applications

53. (a) On $[0, 4]$, with $n = 4$, we have $\Delta x = 1$, so that $x_i = i$ for $i = 0, 1, 2, 3, 4$. The interval midpoints are $x_i^* = i - \frac{1}{2}$ for $i = 1, 2, 3, 4$. Then the various approximations are:

$$\text{LEFT}(4) = \sum_{k=1}^{4} f(x_{k-1})\Delta x = 15e^0 + 15e^{-0.5} + 15e^{-1} + 15e^{-1.5} \approx 32.9631$$

$$\text{RIGHT}(4) = \sum_{k=1}^{4} f(x_k)\Delta x = 15e^{-0.5} + 15e^{-1} + 15e^{-1.5} + 15e^{-2} \approx 19.9931$$

$$\text{MID}(4) = \sum_{k=1}^{4} f(x_k^*)\Delta x = 15e^{-0.25} + 15e^{-0.75} + 15e^{-1.25} + 15e^{-1.75} \approx 25.6717$$

$$\text{TRAP}(4) = \sum_{k=1}^{4} \frac{f(x_{k-1}) + f(x_k)}{2} = \frac{1}{2}(15e^0 + 30e^{-0.5} + 30e^{-1} + 30e^{-1.5} + 15e^{-2}) \approx 26.4781$$

$$\text{SIMP}(4) = \frac{1}{3}(15e^0 + 60e^{-0.5} + 30e^{-1} + 60e^{-1.5} + 15e^{-2}) \approx 25.9487.$$

(b) Since $|f''(x)| = 3.75e^{-0.5t}$ and $\left|f^{(4)}(x)\right| = 0.9375e^{-0.5t}$, these are both decreasing functions, so they take their maximum value on $[0, 4]$ at $x = 0$, where the maxima are 3.75 and 0.9375 respectively. So for the midpoint and trapezoidal approximations, we take $M = 3.75$; for Simpson's rule we take $M = 0.9375$. The errors are:

$$\left|E_{\text{LEFT}(4)}\right| \leq |f(b) - f(a)|\, \Delta x = \left|15e^{-2} - 15\right| \approx 12.97$$

$$\left|E_{\text{RIGHT}(4)}\right| \leq |f(b) - f(a)|\, \Delta x = \left|15e^{-2} - 15\right| \approx 12.97$$

$$\left|E_{\text{MID}(4)}\right| \leq \frac{M(b-a)^3}{24n^2} = \frac{3.75 \cdot 4^3}{24 \cdot 4^2} \approx 0.625$$

$$\left|E_{\text{TRAP}(4)}\right| \leq \frac{M(b-a)^3}{12n^2} = \frac{3.75 \cdot 4^3}{12 \cdot 4^2} \approx 1.25$$

$$\left|E_{\text{SIMP}(4)}\right| \leq \frac{M(b-a)^5}{180n^4} = \frac{0.9375 \cdot 4^5}{180 \cdot 4^4} \approx 0.0208.$$

55. (a) Using Simpson's rule, we have $\Delta x = 1$, and the approximation is

$$\frac{1}{3}(2.7 + 4 \cdot 4.0 + 2 \cdot 2.7 + 4 \cdot 3.6 + 2 \cdot 4.4 + 4 \cdot 4.3 + 2 \cdot 4.1 + 4 \cdot 3.8 + 0.3) = 29.4.$$

(b) The best estimate is given by multiplying the year-on-year percentage increases: the GDP on December 31, 2001 is

$$6880(1.027)(1.04)(1.027)(1.036)(1.044)(1.043)(1.041)(1.038)(1.003) \approx 9226.9$$

so the estimate is \$9226.9 billion.

(c) The first derivative can be approximated by the slopes of secants. Since each pair of numbers represents one year, the slopes are just the differences in the values. So the first derivative can be approximated by the set of values $1.3, -1.3, 0.9, 0.8, -0.1, -0.2, -0.3, -3.5$. Continuing the process, we approximate the second derivative as the slopes of the secants for the first derivative, or $-2.6, 2.2, -0.1, -0.9, -0.1, -0.1, -3.2$; the third derivative as $4.8, -2.3, -0.8, 0.8, 0$, and -3.1, and the fourth derivative as $-7.1, 1.5, 1.6, -0.8, -3.1$. So the maximum estimated magnitude for the fourth derivative is 7.1. Then the error bound for Simpson's rule would be

$$\left|E_{\text{SIMP}(8)}\right| \leq \frac{7.1(2001 - 1993)^5}{180 \cdot 8^4} \approx 0.3156 \text{ percent.}$$

57. Assume at first that a year has 360 days in it. Then we are using Simpson's rule with $n = 6$ and $\Delta x = 60$, so that the approximation is

$$\text{SIMP}(6) = \frac{60}{3}(700 + 4 \cdot 1000 + 2 \cdot 6300 + 4 \cdot 4000 + 2 \cdot 500 + 4 \cdot 650 + 700) = 752000 \text{ cubic feet.}$$

To account for the missing five days, note that the flow at day 360 and the flow at day 0 are the same. So assume that the flow on each of the missing five days is equal to those, namely 700. This gives an additional 3500 cubic feet, for a total of 755500 cubic feet of water. However, this entire computation was based on cubic feet per day; there are 86400 seconds in a day, so we must multiply by 86400 to get 6.52752×10^{10} cubic feet.

Proofs

59. Consider any subinterval of $[a, b]$, say $[x_{k-1}, x_k]$. Since f is concave down, the line joining the points $(x_{k-1}, f(x_{k-1}))$ and $(x_k, f(x_k))$ lies below the graph of f, so that on this subinterval the area of the trapezoid is enclosed in the graph of f. So on each subinterval, the trapezoidal summand is less than or equal to the area under the curve, so that adding up over all subintervals gives $\text{TRAP}(n) \leq \int_a^b f(x)\,dx$. For the midpoint sum, note that since f is concave down, it increases less and less rapidly as x increases. Thus at the midpoint x_k^* between x_{k-1} and x_k, $f(x_k^*)$ is more than halfway from $f(x_{k-1})$ to $f(x_k)$, so that $(x_k - x_{k-1})f(x_k^*)$ exceeds the area under the curve. Adding up over all subintervals gives $\int_a^b f(x)\,dx \leq \text{MID}(n)$.

61. Let the points used in computing $\text{SIMP}(2n)$ be $x_0, x_1, \ldots, x_{2n}$, and note that $x_i = \frac{x_{i-1} + x_{i+1}}{2}$ for i not equal to 0 or $2n$. Then

$$\begin{aligned}
\text{SIMP}(2n) &= \frac{\Delta x}{3}(f(x_0) + 4f(x_1) + 2f(x_2) + \cdots + 4f(x_{2n-1}) + f(x_{2n})) \\
&= \frac{\Delta x}{3}(f(x_0) + 2f(x_2) + \cdots + 2f(x_{2n-2}) + f(x_{2n})) \\
&\quad + \frac{4\Delta x}{3}(f(x_1) + f(x_3) + \cdots + f(x_{2n-1})) \\
&= \frac{1}{3} \cdot 2\Delta x \left(\frac{f(x_0) + f(x_2)}{2} + \cdots + \frac{f(x_{2n-2}) + f(x_{2n})}{2} \right) \\
&\quad + \frac{2}{3} \cdot 2\Delta x \left(f\left(\frac{x_0 + x_2}{2}\right) + \cdots + \frac{x_{2n-2} + x_{2n}}{2} \right).
\end{aligned}$$

The first sum in the above formula is clearly the trapezoidal sum on the $n + 1$ equally spaced points $x_0, x_2, \ldots, x_{2n}$, so it is $\text{TRAP}(n)$. The second sum is the midpoint sum on the same $n + 1$ points, so it is $\text{MID}(n)$. Making these substitutions gives

$$\text{SIMP}(2n) = \frac{1}{3}\text{TRAP}(n) + \frac{2}{3}\text{MID}(n).$$

63. Suppose $x_0 = a, x_1, \ldots, x_{2n} = b$ are $2n + 1$ equally spaced points. Then on the interval $[x_{2i}, x_{2i+2}]$, we construct a parabola $p_i(x)$ which has the same values as $f(x)$ at x_{2i}, x_{2i+2}, and $\frac{x_{2i} + x_{2i+2}}{2} = x_{2i+1}$. Then by Theorem 5.28, the area under such a quadratic is

$$\begin{aligned}
\int_{x_{2i}}^{x_{2i+2}} p_i(x)\,dx &= \frac{x_{2i+2} - x_{2i}}{6}\left(p(x_{2i}) + 4p\left(\frac{x_{2i} + x_{2i+2}}{2}\right) + p(x_{2i+2}) \right) \\
&= \frac{x_{2i+2} - x_{2i}}{6}(p(x_{2i}) + 4p(x_{2i+1}) + p(x_{2i+2})) \\
&= \frac{x_{2i+2} - x_{2i}}{6}(f(x_{2i}) + 4f(x_{2i+1}) + f(x_{2i+2})).
\end{aligned}$$

Now add these up for $i = 0, 1, \ldots, n-1$; we see that the approximation is

$$\sum_{i=0}^{n-1} \frac{x_{2i+2} - x_{2i}}{6} \left(f(x_{2i}) + 4f(x_{2i+1}) + f(x_{2i+2})\right) = \sum_{i=0}^{n-1} \frac{2\Delta x}{6} \left(f(x_{2i}) + 4f(x_{2i+1}) + f(x_{2i+2})\right)$$

$$= \frac{\Delta x}{3}\left(f(x_0) + 4f(x_1) + 2f(x_2) + \cdots + 4f(x_{2n-2}) + f(x_{2n})\right).$$

Thinking Forward

Approximating with power series

▷ We have

$$\sum_{k=1}^{5} \left(\frac{1}{2}\right)^k \approx 0.96875, \quad \sum_{k=1}^{10} \left(\frac{1}{2}\right)^k \approx 0.999023, \quad \sum_{k=1}^{20} \left(\frac{1}{2}\right)^k \approx 0.999999.$$

Since $f\left(\frac{1}{2}\right) = \frac{1}{1-\frac{1}{2}} = 2$, the errors are $0.03125, 0.000977$, and 0.000001.

▷ For $x = 2$ we get

$$\sum_{k=1}^{5} 2^k = 62, \quad \sum_{k=1}^{10} 2^k = 2046, \quad \sum_{k=1}^{20} 2^k = 2097150.$$

Since $f(2) = \frac{1}{1-2} = -1$, these are clearly not good approximations.

Chapter Review and Self-Test

1. Use the Pythagorean identity $\cos^2 x = 1 - \sin^2 x$ followed by the substitution $u = \sin x$:

$$\int \cos^7 x \, dx = \int (\cos^2 x)^3 \cos x \, dx$$

$$= \int (1 - \sin^2 x)^3 \cos x \, dx$$

$$= \int (1 - u^2)^3 \, du$$

$$= \int (1 - 3u^2 + 3u^4 - u^6) \, du$$

$$= u - u^3 + \frac{3}{5}u^5 - \frac{1}{7}u^7 + C$$

$$= \sin x - \sin^3 x + \frac{3}{5}\sin^5 x - \frac{1}{7}\sin^7 x + C.$$

3. Use integration by parts with $u = x$ and $dv = \sec x \tan x \, dx$. Then $du = dx$ and $v = \sec x$, so that

$$\int x \sec x \tan x \, dx = x \sec x - \int \sec x \, dx = x \sec x - \ln|\sec x + \tan x| + C.$$

5. Use integration by parts with $u = \ln 2x$ and $dv = \frac{1}{x^2}\, dx = x^{-2} \, dx$, so that $du = \frac{1}{x}\, dx$ and $v = -x^{-1}$. Then

$$\int \frac{\ln 2x}{x^2} \, dx = -\frac{\ln 2x}{x} + \int \frac{1}{x^2} \, dx = -\frac{\ln 2x}{x} - \frac{1}{x} + C.$$

7. Use the identity $\cot^2 x = \csc^2 x - 1$:

$$\int \cot^6 x \, dx = \int \cot^4 x (\csc^2 x - 1) \, dx$$

$$= \int \cot^4 x \csc^2 x \, dx - \int \cot^4 x \, dx$$

$$= -\frac{1}{5} \cot^5 x - \int \cot^2 x (\csc^2 x - 1) \, dx$$

$$= -\frac{1}{5} \cot^5 x - \int \cot^2 x \csc^2 x \, dx + \int \cot^2 x \, dx$$

$$= -\frac{1}{5} \cot^5 x + \frac{1}{3} \cot^3 x + \int (\csc^2 x - 1) \, dx$$

$$= -\frac{1}{5} \cot^5 x + \frac{1}{3} \cot^3 x - \cot x - x + C.$$

9. Expand algebraically and integrate:

$$\int 2\pi x (8 - x^{3/2}) \, dx = \int (16\pi x - 2\pi x^{5/2}) \, dx = 8\pi x^2 - \frac{4}{7} \pi x^{7/2} + C.$$

11. Use integration by parts with $u = \ln(1 + 3x)$ and $dv = dx$, so that $du = \frac{3}{1+3x} \, dx$ and $v = x$. Then

$$\int_0^1 \ln(1 + 3x) \, dx = [x \ln(1 + 3x)]_0^1 - \int_0^1 \frac{3x}{1 + 3x} \, dx$$

$$= \ln 4 - \int_0^1 \left(1 - \frac{1}{1 + 3x} \right) dx$$

$$= \ln 4 + \int_0^1 \left(-1 + \frac{1}{1 + 3x} \right) dx$$

$$= \ln 4 + \left[-x + \frac{1}{3} \ln(1 + 3x) \right]_0^1$$

$$= \ln 4 + \left(-1 + \frac{1}{3} \ln 4 - \left(-0 + \frac{1}{3} \ln 1 \right) \right)$$

$$= \frac{4}{3} \ln 4 - 1.$$

13. Note that the derivative of the denominator is $(1 - x^{3/2})' = -\frac{3}{2} x^{1/2}$, which is a constant multiple of the numerator. So use the substitution $u = 1 - x\sqrt{x}$, so that $du = -\frac{3}{2} \sqrt{x} \, dx$ and $-\frac{2}{3} \, du = \sqrt{x} \, dx$. Then

$$\int \frac{\sqrt{x}}{1 - x\sqrt{x}} \, dx = -\frac{2}{3} \int \frac{1}{u} \, du = -\frac{2}{3} \ln |u| + C = -\frac{2}{3} \ln \left| 1 - x\sqrt{x} \right| + C.$$

15. Use integration by parts with $u = \sin(\ln x)$ and $dv = dx$. Then $du = \frac{1}{x} \cos(\ln x) \, dx$ and $v = x$. Thus

$$\int \sin(\ln x) \, dx = x \sin(\ln x) - \int x \cdot \frac{1}{x} \cos(\ln x) \, dx = x \sin(\ln x) - \int \cos(\ln x) \, dx.$$

Use integration by parts again, with $u = \cos(\ln x)$ and $dv = dx$, so that $du = -\frac{1}{x} \sin(\ln x)$ and

$v = x$. Thus

$$\int \sin(\ln x)\,dx = x\sin(\ln x) - \int \cos(\ln x)\,dx$$

$$= x\sin(\ln x) - \left(x\cos(\ln x) + \int \sin(\ln x)\,dx \right)$$

$$= x\sin(\ln x) - x\cos(\ln x) - \int \sin(\ln x)\,dx.$$

Collect the integrals on the left side and divide through by 2 to get

$$\int \sin(\ln x)\,dx = \frac{1}{2}(x\sin(\ln x) - x\cos(\ln x)).$$

17. Use integration by parts with $u = \ln\left(\frac{x^3}{2x+1}\right)$ and $dv = dx$, so that $v = x$ and

$$du = \frac{1}{\frac{x^3}{2x+1}} \cdot \left(\frac{x^3}{2x+1}\right)'$$

$$= \frac{2x+1}{x^3} \cdot \frac{(2x+1)(3x^2) - x^3(2)}{(2x+1)^2}\,dx$$

$$= \frac{6x^3 + 3x^2 - 2x^3}{x^3(2x+1)}\,dx$$

$$= \frac{x^2(4x+3)}{x^3(2x+1)} = \frac{4x+3}{x(2x+1)}\,dx.$$

Then

$$\int \ln\left(\frac{x^3}{x^2+1}\right) dx = x\ln\left(\frac{x^3}{x^2+1}\right) - \int \frac{x(4x+3)}{x(2x+1)}\,dx$$

$$= x\ln\left(\frac{x^3}{x^2+1}\right) - \int \frac{4x+3}{2x+1}\,dx$$

$$= x\ln\left(\frac{x^3}{x^2+1}\right) - \int \left(2 + \frac{1}{2x+1}\right) dx$$

$$= x\ln\left(\frac{x^3}{x^2+1}\right) - 2x - \frac{1}{2}\ln|2x+1| + C.$$

19. Use integration by parts three times:

$$\int (\ln x)^3\,dx = x(\ln x)^3 - 3\int (\ln x)^2\,dx \quad (u = (\ln x)^3, dv = dx)$$

$$= x(\ln x)^3 - 3\left(x(\ln x)^2 - 2\int \ln x\,dx \right) \quad (u = (\ln x)^2, dv = dx)$$

$$= x(\ln x)^3 - 3x(\ln x)^2 + 6\int \ln x\,dx$$

$$= x(\ln x)^3 - 3x(\ln x)^2 + 6(x\ln x - x) + C \quad (u = \ln x, dv = dx)$$

$$= x(\ln x)^3 - 3x(\ln x)^2 + 6x\ln x - 6x + C.$$

21. Use the identity $\tan^2 x = \sec^2 x - 1$ followed by the substitution $u = \sec x$:

$$\int \tan^3 x \sec^3 x \, dx = \int \tan^2 x \sec^2 x (\sec x \tan x) \, dx$$

$$= \int (\sec^2 x - 1) \sec^2 x (\sec x \tan x) \, dx$$

$$= \int (u^2 - 1) u^2 \, du$$

$$= \frac{1}{5} u^5 - \frac{1}{3} u^3 + C$$

$$= \frac{1}{5} \sec^5 x - \frac{1}{3} \sec^3 x + C.$$

23. Simplify algebraically and integrate:

$$\int_1^2 \frac{x^2 + 3x + 7}{\sqrt{x}} \, dx = \int_1^2 (x^{3/2} + 3x^{1/2} + 7x^{-1/2}) \, dx$$

$$= \left[\frac{2}{5} x^{5/2} + 2x^{3/2} + 14x^{1/2} \right]_1^2$$

$$= \frac{2}{5} 2^{5/2} + 2 \cdot 2^{3/2} + 14 \cdot 2^{1/2} - \frac{2}{5} - 2 - 14$$

$$= 2^{3/2} \left(\frac{4}{5} + 2 + 7 \right) - \frac{82}{5}$$

$$= \frac{49}{5} 2^{3/2} - \frac{82}{5}.$$

25. With the substitution $u = 3x$, we get

$$\int \csc 3x \, dx = \frac{1}{3} \int \csc u \, du = -\frac{1}{3} \ln |\csc u + \cot u| + C = -\frac{1}{3} \ln |\csc 3x + \cot 3x| + C.$$

27. Use the substitution $u = \frac{1}{x}$, so that $du = -\frac{1}{x^2}, dx$. Then

$$\int_1^3 \frac{e^{1/x}}{x^2} \, dx = -\int_{x=1}^{x=3} e^u \, du = -\left[e^u \right]_{x=1}^{x=3} = -\left[e^{1/x} \right]_1^3 = e - e^{1/3}.$$

29. Use the substitution $u = x + 5$, so that $du = dx$ and $x = u - 5$. Then

$$\int x(x+5)^{7/2} \, dx = \int (u - 5) u^{7/2} \, du$$

$$= \int (u^{9/2} - 5u^{7/2}) \, du$$

$$= \frac{2}{11} u^{11/2} - \frac{10}{9} u^{9/2} + C$$

$$= \frac{2}{11} (x+5)^{11/2} - \frac{10}{9} (x+5)^{9/2} + C.$$

31. Expand using partial fractions:

$$\frac{x+1}{x^4 + 3x^2} = \frac{x+1}{x^2(x^2+3)} = \frac{A}{x} + \frac{B}{x^2} + \frac{Cx+D}{x^2+3} = \frac{Ax(x^2+3) + B(x^2+3) + (Cx+D)x^2}{x^2(x^2+3)}$$

$$= \frac{(A+C)x^3 + (B+D)x^2 + 3Ax + 3B}{x^2(x^2+3)}.$$

This gives $A + C = B + D = 0, 3A = 1$, and $3B = 1$, so that $A = B = \frac{1}{3}$ and thus $C = D = -\frac{1}{3}$, so that

$$
\begin{aligned}
\int \frac{x+1}{x^4 + 3x^2}\, dx &= \frac{1}{3} \int \frac{1}{x}\, dx + \frac{1}{3} \int \frac{1}{x^2}\, dx - \frac{1}{3} \int \frac{x+1}{x^2+3}\, dx \\
&= \frac{1}{3}\left(\ln|x| - \frac{1}{x} \right) - \frac{1}{3} \int \frac{x}{x^2+3}\, dx - \frac{1}{3} \int \frac{1}{x^2+3}\, dx \\
&= \frac{1}{3}\left(\ln|x| - \frac{1}{x} \right) - \frac{1}{6}\ln(x^2+3) - \frac{1}{3} \int \frac{1}{x^2+3}\, dx.
\end{aligned}
$$

For the remaining integral, multiply top and bottom by $\frac{1}{3}$ to turn it into a derivative of a $\tan^{-1}$ function:

$$
\int \frac{1}{x^2+3}\, dx = \int \frac{1/3}{\left(\frac{x}{\sqrt{3}}\right)^2 + 1}\, dx = \frac{1}{\sqrt{3}}\tan^{-1}\frac{x}{\sqrt{3}} + C.
$$

Putting everything together gives for the answer:

$$
\frac{1}{3}\left(\ln|x| - \frac{1}{x} - \frac{1}{2}\ln(x^2+3) - \frac{1}{\sqrt{3}}\tan^{-1}\frac{x}{\sqrt{3}} \right) + C.
$$

33. Use integration by parts with $u = x^2$ and $dv = xe^{-x^2}\, dx$, so that $du = 2x\, dx$ and $v = -\frac{1}{2}e^{-x^2}$, followed by the substitution $u = x^2$. Then

$$
\begin{aligned}
\int \frac{x^3}{e^{x^2}}\, dx &= -\frac{1}{2}x^2 e^{-x^2} + \int xe^{-x^2}\, dx \\
&= -\frac{1}{2}x^2 e^{-x^2} + \frac{1}{2}\int e^{-u}\, du \\
&= -\frac{1}{2}x^2 e^{-x^2} - \frac{1}{2}e^{-u} + C \\
&= -\frac{1}{2}e^{-x^2}(x^2+1) + C.
\end{aligned}
$$

35. $\displaystyle \int 2^x e^x\, dx = \int (2e)^x\, dx = \frac{1}{\ln(2e)}(2e)^x + C = \frac{1}{1+\ln 2}(2e)^x + C.$

37. Noting that 1 is a root of the denominator, it factors as $(x-1)(x^2+2x+3)$. Expand using partial fractions:

$$
\begin{aligned}
\frac{7x^2+6x+5}{x^3+x^2+x-3} &= \frac{A}{x-1} + \frac{Bx+C}{x^2+2x+3} = \frac{A(x^2+2x+3) + (Bx+C)(x-1)}{x^3+x^2+x-3} \\
&= \frac{(A+B)x^2 + (2A-B+C)x + (3A-C)}{x^3+x^2+x-3}.
\end{aligned}
$$

Thus $A + B = 7, 2A - B + C = 6$, and $3A - C = 5$. Solving gives $A = 3$ and $B = C = 4$. Then

$$
\begin{aligned}
\int \frac{7x^2+6x+5}{x^3+x^2+x-3}\, dx &= 3 \int \frac{1}{x-1}\, dx + 4 \int \frac{x+1}{x^2+2x+3}\, dx \\
&= 3\ln|x-1| + 2 \int \frac{2x+2}{x^2+2x+3}\, dx \\
&= 3\ln|x-1| + 2\ln\left|x^2+2x+3\right| + C.
\end{aligned}
$$

39. Use the substitution $x = \frac{3}{2}\sin u$, so that $dx = \frac{3}{2}\cos u\, du$. Then

$$
\begin{aligned}
\int \frac{1}{(9 - 4x^2)^{3/2}}\, dx &= \frac{3}{2}\int \frac{\cos u}{(9 - 9\sin^2 u)^{3/2}}\, du \\
&= \frac{3}{2}\int \frac{\cos u}{27\cos^3 u}\, du \\
&= \frac{1}{18}\int \sec^2 u\, du \\
&= \frac{1}{18}\tan u + C \\
&= \frac{1}{18}\tan\left(\sin^{-1}\frac{2x}{3}\right) + C \\
&= \frac{1}{18}\cdot\frac{2x}{\sqrt{9 - 4x^2}} + C \\
&= \frac{x}{9\sqrt{9 - 4x^2}} + C.
\end{aligned}
$$

41. Let $u = \tan^{-1} x$ so that $du = \frac{1}{x^2+1}\, dx$, and then use integration by parts. Then

$$
\begin{aligned}
\int \frac{\ln(\tan^{-1} x)}{x^2 + 1}\, dx &= \int \ln u\, du \\
&= u\ln u - u + C \qquad (v = \ln u,\, dw = du) \\
&= \tan^{-1} x(\ln(\tan^{-1} x) - 1) + C.
\end{aligned}
$$

43. Let $u = x^2$, so that $\frac{1}{2}\, du = x\, dx$. Then

$$
\int x\,\mathrm{sech}^2\, x^2\, dx = \frac{1}{2}\int \mathrm{sech}^2\, u\, du = \frac{1}{2}\tanh u + C = \frac{1}{2}\tanh x^2 + C.
$$

45. Since the derivative of $\sinh^{-1} x$ is $\frac{1}{\sqrt{x^2+1}}$, use the substitution $u = \sinh^{-1} x$; then

$$
\int \frac{\sinh^{-1} x}{\sqrt{x^2 + 1}}\, dx = \int u\, du = \frac{1}{2}u^2 + C = \frac{1}{2}(\sinh^{-1} u)^2 + C.
$$

47. The signed area is simply the integral of $f(x)$ on $[-4, 4]$. To integrate, first use polynomial long division:

$$
\frac{x^2 - 1}{x^2 + 1} = 1 - \frac{2}{x^2 + 1}.
$$

Then

$$
\begin{aligned}
\int_{-4}^4 \frac{x^2 - 1}{x^2 + 1}\, dx &= \int_{-4}^4 \left(1 - \frac{2}{x^2 + 1}\right) dx \\
&= \left[x - 2\tan^{-1} x\right]_{-4}^4 \\
&= (4 - 2\tan^{-1} 4) - (-4 - 2\tan^{-1}(-4)) \\
&= 8 - 2\tan^{-1} 4 - 2\tan^{-1} 4 \\
&= 8 - 4\tan^{-1} 4.
\end{aligned}
$$

49. On $[0, \pi]$, we have $\sin^2 x \geq \frac{1}{4}$ on $\left[\frac{\pi}{6}, \frac{5\pi}{6}\right]$ and $\sin^2 x \leq \frac{1}{4}$ elsewhere. Thus

$$\int_0^\pi \left|\sin^2 x - \frac{1}{4}\right| dx = \int_0^{\pi/6} \left(\frac{1}{4} - \sin^2 x\right) dx - \int_{\pi/6}^{5\pi/6} \left(\frac{1}{4} - \sin^2 x\right) dx + \int_{5\pi/6}^\pi \left(\frac{1}{4} - \sin^2 x\right) dx.$$

Now, using the double-angle formula,

$$\int \left(\frac{1}{4} - \sin^2 x\right) dx = \int \left(\frac{1}{4} - \frac{1}{2} + \frac{1}{2}\cos 2x\right) dx$$

$$= \int \left(-\frac{1}{4} + \frac{1}{2}\cos 2x\right) dx$$

$$= \frac{1}{4}(-x + \sin 2x),$$

so that

$$\int_0^\pi \left|\sin^2 x - \frac{1}{4}\right| dx = \left[\frac{1}{4}(-x + \sin 2x)\right]_0^{\pi/6} - \left[\frac{1}{4}(-x + \sin 2x)\right]_{\pi/6}^{5\pi/6} + \left[\frac{1}{4}(-x + \sin 2x)\right]_{5\pi/6}^\pi$$

$$= \frac{\sqrt{3}}{2} + \frac{\pi}{12}.$$

51. To integrate xe^{-x} use integration by parts with $u = x$ and $dv = e^{-x}\,dx$, so that $du = dx$ and $v = -e^{-x}$. Then the average value is

$$\frac{1}{3 - (-1)} \int_{-1}^3 xe^{-x}\,dx = \frac{1}{4}\left(\left[-xe^{-x}\right]_{-1}^3 + \int_{-1}^3 e^{-x}\,dx\right)$$

$$= \frac{1}{4}\left(-3e^{-3} - e - \left[e^{-x}\right]_{-1}^3\right)$$

$$= \frac{1}{4}\left(-3e^{-3} - e - e^{-3} + e\right)$$

$$= -e^{-3}.$$

53. Since $x^{-1/3}$ is discontinuous at $x = 0$, we have

$$\int_0^8 x^{-1/3}\,dx = \lim_{c \to 0^+} \int_c^8 x^{-1/3}\,dx = \lim_{c \to 0} \left[\frac{3}{2}x^{2/3}\right]_c^8 = \lim_{c \to 0}\left(\frac{3}{2}\cdot 4 - \frac{3}{2}c^{2/3}\right) = 6.$$

55. To integrate $x \ln x$, use integration by parts with $u = \ln x$ and $dv = x\,dx$, so that $du = \frac{1}{x}\,dx$ and $v = \frac{1}{2}x^2$. Then

$$\int x \ln x\,dx = \frac{1}{2}x^2 \ln x - \frac{1}{2}\int x\,dx = \frac{1}{2}x^2 \ln x - \frac{1}{4}x^2.$$

Since $x \ln x$ is discontinuous at $x = 0$, we have

$$\int_0^\infty x \ln x\,dx = \lim_{c \to 0^+} \int_c^1 x \ln x\,dx + \lim_{c \to \infty} \int_1^c x \ln x\,dx$$

$$= \lim_{c \to 0^+} \left[\frac{1}{2}x^2 \ln x - \frac{1}{4}x^2\right]_c^1 + \lim_{c \to \infty} \left[\frac{1}{2}x^2 \ln x - \frac{1}{4}x^2\right]_1^c$$

$$= -\frac{1}{4} - \lim_{c \to 0^+}\left(\frac{1}{2}c^2 \ln c - \frac{1}{4}c^2\right) + \lim_{c \to \infty}\left(\frac{1}{2}c^2 \ln c - \frac{1}{4}c^2\right) + \frac{1}{4}.$$

Since the first term of the second limit is infinite, the integral is divergent.

57. To integrate $\sec x \tan^3 x$, use the identity $\tan^2 x = \sec^2 x - 1$ and then the substitution $u = \sec x$:

$$
\begin{aligned}
\int \sec x \tan^3 x \, dx &= \int (\tan^2 x)(\sec x \tan x) \, dx \\
&= \int (\sec^2 x - 1) \sec x \tan x \, dx \\
&= \int (u^2 - 1) \, du \\
&= \frac{1}{3} u^3 - u \\
&= \frac{1}{3} \sec^3 x - \sec x.
\end{aligned}
$$

Since $\sec x \tan^3 x$ is discontinuous at $x = \frac{\pi}{2}$, we have

$$
\begin{aligned}
\int_0^{\pi/2} \sec x \tan^3 x \, dx &= \lim_{c \to \frac{\pi}{2}^-} \int_0^c \sec x \tan^3 x \, dx \\
&= \lim_{c \to \frac{\pi}{2}^-} \left[\frac{1}{3} \sec^3 x - \sec x \right]_0^c \\
&= -\frac{2}{3} + \lim_{c \to \frac{\pi}{2}^-} \left(\frac{1}{3} \sec^3 c - \sec c \right) \\
&= \infty.
\end{aligned}
$$

59. Note that e^{-x^2} is an even function, and the domain of integration is $[-1, 1]$, so that the desired integral is in fact $2 \int_0^1 e^{-x^2} \, dx$. For the right sum approximation to this integral to be within 0.5, we must have the right sum approximation to $\int_0^1 e^{-x^2} \, dx$ within 0.25. For this to hold, we must have

$$
\left| E_{\text{RIGHT}(n)} \right| \le |f(b) - f(a)| \, \Delta x = |f(1) - f(0)| \frac{1 - 0}{n} = \frac{1 - e^{-1}}{n} \le 0.25.
$$

This happens if $n \ge 4(1 - e^{-1}) \approx 2.528$, so we choose $n = 3$. Then on $[-1, 1]$, we have $\Delta x = \frac{2}{3}$, and $x_i = -1 + \frac{2i}{3}$ for $i = 0, 1, 2, 3$, and the right sum is

$$
\begin{aligned}
\sum_{k=1}^3 f(x_k) \Delta x &= \frac{2}{3} \sum_{k=1}^3 e^{-\left(-1 + \frac{2k}{3}\right)^2} \\
&= \frac{2}{3} \left(e^{-1/9} + e^{-1/9} + e^{-1} \right) \approx 1.438.
\end{aligned}
$$

61. For Simpson's rule to be within 0.01, we must have

$$
\left| E_{\text{SIMP}(n)} \right| \le \frac{M(b - a)^5}{180 n^4} \le 0.01,
$$

where M is a bound on the magnitude of $f^{(4)}(x)$ on $[0, 3]$. A graph of $f^{(4)}(x)$ shows that it is bounded in magnitude by 8 on $[0, 3]$, so choose $M = 8$. Then we want

$$
\frac{8(3 - 0)^5}{180 n^4} = \frac{54}{5 n^4} \le 0.01,
$$

which happens for $n^4 \geq 1080$. Then $n \geq 5.73$, so that $n = 6$ is sufficient. For $n = 6$, we have $\Delta x = \frac{1}{2}$ and $x_k = \frac{k}{2}$. Thus the Simpson's rule sum is

$$\mathrm{SIMP}(6) = \frac{\Delta x}{3}(f(x_0) + 4f(x_1) + 2f(x_2) + 4f(x_3) + 2f(x_4) + 4f(x_5) + f(x_6))$$

$$= \frac{1}{6}\left(\sqrt{1} + 4\sqrt{1 + \frac{1}{8}} + 2\sqrt{1 + 1} + 4\sqrt{1 + \frac{27}{8}} + 2\sqrt{1 + 8} + 4\sqrt{1 + \frac{5^3}{8}} + \sqrt{1 + \frac{6^3}{8}}\right)$$

$$\approx 7.3398.$$

Chapter 6

Applications of Integration

6.1 Volumes by Slicing

Thinking Back

Definite integrals

▷ $\int_{-3}^{3} (9 - x^2)\, dx = \left[9x - \frac{1}{3}x^3 \right]_{-3}^{3} = 36.$

▷ $\int_{0}^{1} (x^4 + 2x^2)\, dx = \left[\frac{1}{5}x^5 + \frac{2}{3}x^3 \right]_{0}^{1} = \frac{1}{5} + \frac{2}{3} = \frac{13}{15}.$

▷ $\int_{0}^{1} 4\, dx = [4x]_{0}^{1} = 4.$

▷ To integrate $(\ln x)^2$, use integration by parts twice, each time with u equal to the entire integrand and $dv = dx$:

$$\int_{1}^{e^{2/3}} (4 - 9(\ln x)^2)\, dx = \int_{1}^{e^{2/3}} 4\, dx - 9 \int_{1}^{e^{2/3}} (\ln x)^2\, dx$$

$$= [4x]_{1}^{e^{2/3}} - 9 \left(\left[x(\ln x)^2 \right]_{1}^{e^{2/3}} - 2 \int_{1}^{e^{2/3}} \ln x\, dx \right)$$

$$= 4e^{2/3} - 4 - 9 \left(e^{2/3}(\ln e^{2/3})^2 - 0 - 2 \left([x \ln x]_{1}^{e^{2/3}} - \int_{1}^{e^{2/3}} 1\, dx \right) \right)$$

$$= 4e^{2/3} - 4 - 9 \left(\frac{4}{9}e^{2/3} - 2e^{2/3} \ln e^{2/3} + 2 [x]_{1}^{e^{2/3}} \right)$$

$$= 4e^{2/3} - 4 - 9 \left(\frac{4}{9}e^{2/3} - \frac{4}{3}e^{2/3} + 2e^{2/3} - 2 \right)$$

$$= -6e^{2/3} + 14.$$

▷ $\int_{0}^{2} (4y^2 - y^4)\, dy = \left[\frac{4}{3}y^3 - \frac{1}{5}y^5 \right]_{0}^{2} = \frac{32}{3} - \frac{32}{5} = \frac{64}{15}.$

▷ $\int_{0}^{1} (2 - \sqrt{y})^2\, dy = \int_{0}^{1} (4 - 4\sqrt{y} + y)\, dy = \left[4y - \frac{8}{3}y^{3/2} + \frac{1}{2}y^2 \right]_{0}^{1} = 4 - \frac{8}{3} + \frac{1}{2} = \frac{11}{6}.$

Definite integrals for geometric quantities

▷ The area under the curve is just the integral: $\int_0^4 \sqrt{x}\,dx$.

▷ Since $g(x) \leq 0$ on $[2,4]$ and $g(x) \geq 0$ on $[0,2]$, the absolute area is

$$\int_0^4 |g(x)|\,dx = \int_0^2 g(x)\,dx - \int_2^4 g(x)\,dx.$$

▷ The graphs of f and g cross where $\sqrt{x} = 2 - x$, which is at $x = 1$. Now, $g(x) \geq f(x)$ for $x \leq 1$ and $g(x) \leq f(x)$ for $x \geq 1$. Thus the area between the graphs is

$$\int_0^4 |f(x) - g(x)|\,dx = \int_0^1 (2 - x - \sqrt{x})\,dx + \int_1^4 (\sqrt{x} - (2 - x))\,dx.$$

▷ The average value is the integral divided by the interval length: $\dfrac{1}{4} \int_0^4 \sqrt{x}\,dx$.

Concepts

1. (a) False. Riemann sums have a specific form: $\sum f(x_i^*)\Delta x$, where each x_i^* is in the ith interval of width Δx.

   (b) False, for the same reason as in part (a).

   (c) False. The volume is the thickness times the *area* of a circle of the same radius.

   (d) True. See part (c).

   (e) True. The volume of a cylinder is the area of the base times the height.

   (f) True. If the outer radius is R and the inner radius is r, with height h, then you can think of the washer as a disc with an inner disc of radius r removed, so the remaining volume is $\pi R^2 h - \pi r^2 h = \pi(R^2 - r^2)h$.

   (g) True. The volume of such a cone is $\frac{1}{3}\pi r^2 h$, which is one third the volume of the corresponding cylinder.

   (h) True. See the introductory material to this section.

3. (a) When rotated around the x axis, the solid formed is a cylinder with height $2.25 - 2 = 0.25$ and radius $3 - 0 = 3$. Thus its volume is $\pi \cdot 3^2 \cdot 0.25 = \frac{9}{4}\pi$.

   (b) When rotated around the line $y = 5$, the solid formed is a washer with height $2.25 - 2 = 0.25$, outer radius $R = 5 - 0 = 5$, and inner radius $r = 5 - 3 = 2$. Thus its volume is $\pi(5^2 - 2^2) \cdot 0.25 = \frac{21}{4}\pi$.

5. A plot of the region together with the Riemann sum rectangles is

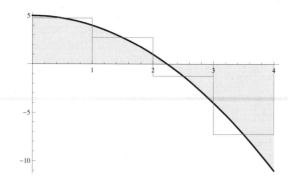

A four-disc approximation of this area would consist of the volumes of the cylinders which result when each of these four rectangles is rotated around the x axis.

7. A plot of the region with each x_i marked, with x_k^* at the left end of each interval, as well as marking the function values at each x_i^*, is:

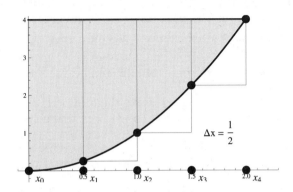

(a) The region from $y = 0$ to $y = 4$ corresponds to the region over $[0, 2]$. Since the interval width is 2 and there are four washers, $\Delta x = \frac{1}{2}$. Then $x_k = \frac{k}{2}$ for $k = 0, 1, 2, 3, 4$.

(b) Since x_k^* is the left end, we have $x_k^* = \frac{k-1}{2}$ for $k = 1, 2, 3, 4$.

(c) Computing gives

k	x_k^*	$f(x_k^*)$
1	0	0
2	$\frac{1}{2}$	$\frac{1}{4}$
3	1	1
4	$\frac{3}{2}$	$\frac{9}{4}$

(d) The width of the second washer is $\Delta x = \frac{1}{2}$, its outer radius is 4, and its inner radius is $\frac{1}{4}$, so its volume is $\pi \left(4^2 - \left(\frac{1}{4} \right)^2 \right) \cdot \frac{1}{2} = \frac{255}{32} \pi$.

9. Since $x_0 = 2$ and $x_n = 2 + n \left(\frac{1}{n} \right) = 3$, this is an integral on $[2, 3]$. The integrand is $(1 + x)^2$, so that this limit represents $\pi \int_2^3 (1 + x)^2 \, dx$. This is the volume of the solid of revolution gotten by revolving the area under the graph of $f(x) = 1 + x$ around the x axis from $x = 2$ to $x = 3$.

11. Since $x_0 = 0$ and $x_n = n \left(\frac{2}{n} \right) = 2$, this is an integral from $x = 0$ to $x = 2$. The integrand is $4 - x^2$, so the resulting integral is $\pi \int_0^2 (4 - x^2) \, dx$. This is probably the volume of the solid of revolution gotten by revolving the area between the graphs of $y = x$ and $y = 2$ between $x = 0$ and $x = 2$ around the x-axis. It could also be the volume of the solid gotten by revolving the area under the curve $y = \sqrt{4 - x^2}$ between $x = 0$ and $x = 2$ around the x-axis.

13. Since $\cos(\pi - x) = \cos \pi \cos x - \sin \pi \sin x = -\cos x$, we see that $\cos^2(\pi - x) = \cos^2 x$, so $\int_0^{\pi/2} \cos^2 x \, dx = \int_{\pi/2}^{\pi} \cos^2 x \, dx$, so the two integrals are equal.

15. Since x^4 is concave up, the line connecting $[0, 0]$ and $[2, 16]$ lies above the graph of $y = x^4$, so that the area between $y = x^4$ and $y = 16$ on $[0, 2]$ exceeds the area under $y = x^4$ on $[0, 2]$. Thus the second integral is larger.

17. Since $f(x) = x^4$ is concave up, it is clear that the region under $f(x)$ from $x = 0$ to $x = 2$ has smaller area than the region bounded by $f(x)$, $y = 4$, and the y axis. Since the first integral is the first area rotated around the x axis, and the second integral is the second area rotated around the y axis, the first volume will be smaller than the second.

19. Since the integral has the form $\pi \int_a^b (R^2 - r^2)\, dx$, the region could be the region bounded by $y = 4$ and $y = x + 1$ from $x = 1$ to $x = 3$, in which case the integral is the volume using washers of the solid of revolution around the x axis. Since it has the form $\pi \int_a^b R^2\, dx$, it could also be the region bounded by the x axis and $y = \sqrt{16 - (x+1)^2}$ from $x = 1$ to $x = 3$, in which case the integral is the volume using disks of the solid of revolution around the x axis.

21. Since $x^2 - 2x + 1 = (x-1)^2$, the integral takes the form $\pi \int_a^b R^2\, dx$, so that the region is the area under the graph of $y = x - 1$ between $x = 1$ and $x = 3$, rotated around the x axis.

23. The integral takes the form $\pi \int_a^b R^2\, dy$, so that the region is the area between the graph of $x = \frac{y-1}{2}$ (i.e., $y = 2x + 1$) and the y axis between $y = 1$ and $y = 5$ rotated around the y axis.

25. For $-3 \leq y \leq 0$, the solid of revolution is a disk whose radius is the curve $x = 2$, so its volume is $\pi \int_{-3}^0 2^2\, dy$. For $0 \leq y \leq 3$, use the washer method to get $\pi \int_0^3 (2^2 - f^{-1}(y)^2)\, dy$. Thus the total volume is

$$\pi \int_{-3}^0 2^2\, dy + \pi \int_0^3 (2^2 - f^{-1}(y)^2)\, dy.$$

Skills

27. When rotated, the rectangles turn into discs, each of which has thickness 1. The height of the disc on $[i-1, i]$ is $\sqrt{i}$, so that the approximate volume is

$$\pi \sum_{i=1}^4 (\sqrt{i})^2 \cdot 1 = \pi(1 + 2 + 3 + 4) = 10\pi.$$

29. When rotated, the rectangles turn into washers, each of which has thickness 1. The outer radius of the ith rectangle is $\sqrt{i} + 1$, and the inner radius is 1. Thus the approximate volume is

$$V = \pi \sum_{i=1}^4 \left((\sqrt{i} + 1)^2 - 1^2 \right) \cdot 1$$
$$= \pi \left((4 - 1) + \left(3 + 2\sqrt{2} - 1 \right) + \left(4 + 2\sqrt{3} - 1 \right) + (9 - 1) \right)$$
$$= (16 + 2\sqrt{2} + 2\sqrt{3})\pi.$$

31. When rotated, we use the disk method with radius equal to $f(x) = \frac{3}{x}$, so the volume of the solid is

$$\pi \int_1^3 f(x)^2\, dx = \pi \int_1^3 \left(\frac{3}{x} \right)^2 dx = -9\pi \left[x^{-1} \right]_1^3 = 6\pi.$$

33. When rotated, we will have to use the washer method with an inner radius of 1. However, the outer radius is constant at $3 - 0 = 3$ for $0 \leq y \leq 1$, and is $x = \frac{3}{y}$ for $1 \leq y \leq 3$. You could recognize the first of these as a rectangle and avoid integration altogether for this piece (since the rotated solid is just a washer from $y = 0$ to $y = 1$), or just integrate:

$$\pi \int_0^1 (3^2 - 1^2)\, dy + \pi \int_1^3 \left(\left(\frac{3}{y} \right)^2 - 1^2 \right) dy = 8\pi + \pi \left[-\frac{9}{y} - y \right]_1^3 = 12\pi.$$

35. When rotated, we get washers with outer radius 5 and inner radius $x = y + 2$. Noting that $f(5) = 3$, the integration is from $y = 0$ to $y = 3$. Thus the volume is

$$\pi \int_0^3 (5^2 - (y+2)^2)\, dy = \pi \int_0^3 (21 - 4y - y^2)\, dy = \pi \left[21y - 2y^2 - \frac{1}{3}y^3 \right]_0^3 = 36\pi.$$

37. When rotated, we get washers with outer radius 3, since the outer radius is delimited by the bottom of the region. The inner radius is $3 - (x - 2) = 5 - x$. Thus the volume of the solid is

$$\pi \int_2^5 (3^2 - (5-x)^2)\, dx = \pi \left[9x + \frac{1}{3}(5-x)^3 \right]_2^5 = 18\pi.$$

39. When rotated, we get washers with outer radius $5 - 2 = 3$. Since $y = x - 2$, we have $x = y + 2$, so that the inner radius is $(y+2) - 2 = y$. Further, the top of the region is at $x = 5$, which corresponds to $y = 3$. Then the volume is

$$\pi \int_0^3 (3^2 - y^2)\, dy = \pi \left[9y - \frac{1}{3}y^3 \right]_0^3 = 18\pi.$$

41. When rotated, we get washers from $y = 0$ to $y = 4$ and a disc from $y = 4$ to $y = 5$. The outer radius of the washers is 2. Since $y = 4 - x^2$, we have $x = \sqrt{4 - y}$, so that the inner radius of the washers is $\sqrt{4 - y}$. For the disc, the radius is 2. Thus the volume of the solid of revolution is

$$V = \pi \int_0^4 (2^2 - (\sqrt{4-y})^2)\, dy + \pi \int_4^5 2^2\, dy$$
$$= \pi \int_0^4 y\, dy + \pi \int_4^5 4\, dy$$
$$= \pi \left(\left[\frac{1}{2}y^2 \right]_0^4 + [4y]_4^5 \right)$$
$$= 12\pi.$$

43. When rotated, we get washers over the entire range $[0,5]$, with inner radius $3 - 2 = 1$. Since $y = 4 - x^2$, we have $x = \sqrt{4 - y}$. Thus on $[0,4]$, the outer radius is $3 - \sqrt{4 - y}$, while on $[4,5]$, the outer radius is $3 - 0 = 3$. Thus the volume of the solid is

$$V = \pi \int_0^4 ((3 - \sqrt{4-y})^2 - 1^2)\, dy + \pi \int_4^5 (3^2 - 1^2)\, dy$$
$$= \pi \int_0^4 (12 - y - 6\sqrt{4-y})\, dy + \pi\, [8y]_4^5$$
$$= 8\pi + \pi \left[12y - \frac{1}{2}y^2 + 4(4 - y)^{3/2} \right]_0^4$$
$$= 16\pi.$$

45. When rotated, we get washers along the entire interval $[1,4]$. However, since $f(x)$ and $g(x)$ cross at $x = \frac{5}{3}$, the outer radius is $5 - x$ on $\left[1, \frac{5}{3}\right]$ and $2x$ on $\left[\frac{5}{3}, 4\right]$. Similarly, the inner radius is $2x$ on

$\left[1, \frac{5}{3}\right]$ and $5 - x$ on $\left[\frac{5}{3}, 4\right]$. Thus the volume is

$$V = \pi \int_1^{5/3} \left((5-x)^2 - (2x)^2\right) dx + \pi \int_{5/3}^4 \left((2x)^2 - (5-x)^2\right) dx$$

$$= \pi \int_1^{5/3} \left(-3x^2 - 10x + 25\right) dx + \pi \int_{5/3}^4 \left(3x^2 + 10x - 25\right) dx$$

$$= \pi \left[-x^3 - 5x^2 + 25x\right]_1^{5/3} + \pi \left[x^3 + 5x^2 - 25x\right]_{5/3}^4$$

$$= \frac{112}{27}\pi + \frac{1813}{27}\pi = \frac{1925}{27}\pi.$$

47. Since $y = f(x) = x^2$, we have $x = f^{-1}(y) = \sqrt{y}$, and since $y = g(x) = 2x$, we have $x = g^{-1}(y) = \frac{y}{2}$. When rotated, therefore, we get washers with an outer radius of $\sqrt{y}$ and an inner radius of $\frac{y}{2}$. Since the range of x values is $[0, 2]$, the range of y values is $[0, 4]$, and the volume is

$$\pi \int_0^4 \left((\sqrt{y})^2 - \left(\frac{y}{2}\right)^2\right) dy = \pi \int_0^4 \left(y - \frac{y^2}{4}\right) dy = \pi \left[\frac{1}{2}y^2 - \frac{1}{12}y^3\right]_0^4 = \frac{8}{3}\pi.$$

49. When rotated, we get washers with an outer radius of $g(x) + 1 = 2x + 1$ and an inner radius of $f(x) + 1 = x^2 + 1$. Thus the volume is

$$V = \pi \int_0^2 \left((2x+1)^2 - (x^2+1)^2\right) dx$$

$$= \pi \int_0^2 \left(-x^4 + 2x^2 + 4x\right) dx$$

$$= \pi \left[-\frac{1}{5}x^5 + \frac{2}{3}x^3 + 2x^2\right]_0^2$$

$$= \frac{104}{15}\pi.$$

51. Since $x \geq (x-2)^2$ on $[1, 4]$, rotation gives washers with an outer radius of x and an inner radius of $(x-2)^2$. Thus the volume is

$$\pi \int_1^4 \left(x^2 - (x-2)^4\right) dx = \pi \left[\frac{1}{3}x^3 - \frac{1}{5}(x-2)^5\right]_1^4 = \frac{72}{5}\pi.$$

53. Since $x \geq (x-2)^2$ on $[1, 4]$, rotation gives washers with an outer radius of $x + 1$ and an inner radius of $(x-2)^2 + 1$. Thus the volume is

$$V = \pi \int_1^4 \left((x+1)^2 - ((x-2)^2 + 1)^2\right) dx$$

$$= \pi \int_1^4 \left(x^2 + 2x + 1 - 25 + 40x - 26x^2 + 8x^3 - x^4\right) dx$$

$$= \pi \int_1^4 \left(-x^4 + 8x^3 - 25x^2 + 42x - 24\right) dx$$

$$= \pi \left[-\frac{1}{5}x^5 + 2x^4 - \frac{25}{3}x^3 + 21x^2 - 24x\right]_1^4$$

$$= \frac{117}{5}\pi.$$

55. When rotated, we get discs with a radius of $1 - \cos x$, so that the volume is

$$
\begin{aligned}
V &= \pi \int_0^\pi (1 - \cos x)^2 \, dx \\
&= \pi \int_0^\pi (1 - 2\cos x + \cos^2 x) \, dx \\
&= \pi \int_0^\pi \left(1 - 2\cos x + \frac{1}{2} + \frac{1}{2}\cos 2x \right) dx \\
&= \pi \left[\frac{3}{2}x - 2\sin x + \frac{1}{4}\sin 2x \right]_0^\pi \\
&= \frac{3}{2}\pi^2.
\end{aligned}
$$

57. Since $y = 1 - \cos x$, we see that $x = \cos^{-1}(1 - y)$, so that when rotated, we get discs with a radius of $\pi - \cos^{-1}(1 - y)$. Further, $x = \pi$ corresponds to $y = 2$, so that the volume is $\pi \int_0^2 (\pi - \cos^{-1}(1 - y))^2 \, dy$. Approximating the integral numerically gives ≈ 18.4399.

Applications

59. (a) Since the perimeter is $4 \times 756 = 3024$ feet, the radius of the circle with perimeter 3024 is given by $2\pi r = 3024$. Solving for r gives $r \approx 481.285$, which is 481 feet to the nearest foot. The height of the Great Pyramid is thus 481 feet.

    (b) The volume of the Great Pyramid by this formula is $\frac{1}{3} \cdot 756^2 \cdot 481 = 91636272$ cubic feet.

    (c) Let $f(x)$ be the area of the horizontal cross-section Great Pyramid at height x. Each cross-section is a square, and using similar triangles we see that the side length of the square at height x is $756 - \frac{756}{481}x$. Thus $f(x) = \left(756 - \frac{756}{481}x \right)^2$. Now, choose n equally spaced points $x_0 = 0, x_1, \ldots, x_n = 481$ from 0 to 481, and use these points to divide the pyramid into horizontal slices each of which has thickness Δx. Approximate each slide as a parallelepiped with thickness Δx and a square base whose area is $f(x_k)$. Thus the volume of the kth slice is $\approx f(x_k)\Delta x$. The total approximate volume is thus

$$
\sum_{k=1}^n f(x_k)\Delta x,
$$

which corresponds to the integral

$$
\begin{aligned}
\int_0^{481} f(x) \, dx &= \int_0^{481} \left(756 - \frac{756}{481}x \right)^2 dx \\
&= \int_0^{481} \left(756^2 - \frac{2 \cdot 756^2}{481}x + \frac{756^2}{481^2}x^2 \right) dx \\
&= \left[756^2 x - \frac{756^2}{481}x^2 + \frac{756^2}{3 \cdot 481^2}x^3 \right]_0^{481} \\
&= 91636272.
\end{aligned}
$$

Proofs

61. Consider $f(x) = \frac{3}{5}x$ between $x = 0$ and $x = 5$. Rotating this graph around the x axis produces a cone of radius 3 and height 5, so the volume of that cone is

$$
V = \pi \int_0^5 \left(\frac{3}{5}x \right)^2 dx = \frac{9}{25}\pi \left[\frac{1}{3}x^3 \right]_0^5 = \frac{9}{25}\pi \cdot \frac{125}{3} = 15\pi.
$$

With $r = 3$ and $h = 5$, the given formula is $V = \frac{1}{3}\pi \cdot 3^2 \cdot 5 = 15\pi$, and the two are equal.

63. If either r or h is zero, the $V = 0$ and the formula holds. So assume neither is zero, and consider $f(x) = \frac{r}{h}x$ between $x = 0$ and $x = h$. Rotating this graph around the x axis produces a cone of radius r and height h, so the volume of that cone is

$$V = \pi \int_0^h \left(\frac{r}{h}x\right)^2 dx = \frac{r^2}{h^2}\pi \left[\frac{1}{3}x^3\right]_0^h = \frac{r^2}{h^2}\pi \cdot \frac{1}{3}h^3 = \frac{1}{3}\pi r^2 h.$$

Thinking Forward

Slicing to obtain a definite integral for work

▷ Since the work is weight times distance, and every point in the cylinder must be lifted 10 feet, the work in this case is $50 \times 10 = 500$ foot-pounds.

▷ The slice at y_k feet above the bottom must be raised $4 - y_k$ feet. Let $f(y)$ be the cross-sectional area of the tank at height y; then $f(y) = 10^2\pi = 100\pi$. Water weight 62.4 pounds per cubic foot (see Section 6.4), and the volume of a cylindrical slice of thickness Δy_k is $f(y)\Delta y$, so its weight is $62.4f(y)\Delta y$. Thus the work to lift all the water out of the tank is approximated by

$$\sum_{k=1}^n 62.4f(y_k)\Delta y,$$

which corresponds to the integral

$$\int_0^4 62.4f(y)\,dy = \int_0^4 62.4 \cdot 100\pi\,dy = [6240y]_0^4 = 24960 \text{ foot-pounds}.$$

6.2   Volumes by Shells

Thinking Back

Definite integrals

▷ Expand and integrate:

$$\int_0^1 (3-x)(x^2+1)\,dx = \int_0^1 (-x^3 + 3x^2 - x + 3)\,dx = \left[-\frac{1}{4}x^4 + x^3 - \frac{1}{2}x^2 + 3x\right]_0^1 = \frac{13}{4}.$$

▷ Expand and integrate:

$$\int_0^4 x(2 - \sqrt{x})\,dx = \int_0^4 (2x - x^{3/2})\,dx = \left[x^2 - \frac{2}{5}x^{5/2}\right]_0^4 = \frac{16}{5}.$$

▷ Expand and integrate:

$$\int_0^6 x((x-3)^2 + 2)\,dx = \int_0^6 (x^3 - 6x^2 + 11x)\,dx = \left[\frac{1}{4}x^4 - 2x^3 + \frac{11}{2}x^2\right]_0^6 = 90.$$

▷ Use integration by parts with $u = x$ and $dv = e^{x/3}\, dx$:

$$\int_1^2 xe^{x/3}\, dx = \left[3xe^{x/3}\right]_1^2 - 3\int_1^2 e^{x/3}\, dx$$

$$= 6e^{2/3} - 3e^{1/3} - 9\left[e^{x/3}\right]_1^2$$

$$= 6e^{2/3} - 3e^{1/3} - 9e^{2/3} + 9e^{1/3}$$

$$= 6e^{1/3} - 3e^{2/3}.$$

▷ $\int_0^1 (e^x)^2\, dx = \int_0^1 e^{2x}\, dx = \left[\frac{1}{2}e^{2x}\right]_0^1 = \frac{1}{2}e^2 - \frac{1}{2}.$

▷ Use integration by parts with $u = \ln x$ and $dv = x$:

$$\int_2^4 x \ln x\, dx = \left[\frac{1}{2}x^2 \ln x\right]_2^4 - \frac{1}{2}\int_2^4 x^2 \cdot \frac{1}{x}\, dx$$

$$= 8\ln 4 - 2\ln 2 - \frac{1}{2}\int_2^4 x\, dx$$

$$= 14\ln 2 - \frac{1}{4}\left[x^2\right]_2^4$$

$$= 14\ln 2 - 3.$$

▷ Use integration by parts with $u = y$ and $dv = e^y\, dy$:

$$\int_1^5 ye^y\, dy = [ye^y]_1^5 - \int_1^5 e^y\, dy = 5e^5 - e - [e^y]_1^5 = 4e^5.$$

▷ Use integration by parts twice, the first time with $u = (\ln y)^2$ and $dv = dy$; the second time with $u = \ln y$ and $dv = dy$:

$$\int_1^e (\ln y)^2\, dx = \left[y(\ln y)^2\right]_1^e - 2\int_1^e \ln y\, dy$$

$$= e - 2\left([y \ln y]_1^e - \int_1^e 1\, dy\right)$$

$$= e - 2\left(e - [y]_1^e\right)$$

$$= e - 2.$$

▷ Use integration by parts with $u = y$ and $dv = \sin y\, dy$:

$$\int_0^\pi y \sin y\, dy = -[y \cos y]_0^\pi + \int_0^\pi \cos y\, dy = \pi + [\sin y]_0^\pi = \pi.$$

▷ Use the substitution $u = 6 - y$, so that $-du = dy$ and $y = 6 - u$. Then

$$\int_2^4 y\sqrt{6-y}\,dy = -\int_{y=2}^{y=4}(6-u)\sqrt{u}\,du$$

$$= -\int_{y=2}^{y=4}(6u^{1/2} - u^{3/2})\,du$$

$$= \left[-4u^{3/2} + \frac{2}{5}u^{5/2}\right]_{y=2}^{y=4}$$

$$= \left[-4(6-y)^{3/2} + \frac{2}{5}(6-y)^{5/2}\right]_2^4$$

$$= \frac{96}{5} - \frac{32\sqrt{2}}{5}.$$

Finding volumes geometrically

▷ This is the union of a cylinder with radius 2 and height 1 and a cone with radius 2 and height $5 - 2 = 3$, so its volume is

$$\pi \cdot 2^2 \cdot 1 + \frac{1}{3}\pi \cdot 2^2 \cdot 3 = 8\pi.$$

▷ This is the union of two cylinders with height $\frac{1}{2}$ and radius 2 and two cones with height $\frac{3}{2}$ and radius 2. Thus its volume is

$$2\pi \cdot 2^2 \cdot \frac{1}{2} + 2 \cdot \frac{1}{3}\pi \cdot 2^2 \cdot \frac{3}{2} = 8\pi.$$

▷ The volume of this solid is the volume of the cylinder with height 6 and radius 3 less the volume of the "hole", which is a cylindrical hole with height 6 and radius 2. The volume is thus

$$\pi \cdot 3^2 \cdot 6 - \pi \cdot 2^2 \cdot 6 = 30\pi.$$

▷ The solid pictured is the frustum of a cone less the volume of the removed cylinder. To find the volume of the frustum, note that the top and bottom edges of the cone have equation $y = \frac{1}{3}x + 2$ and $y = -\frac{1}{3}x$; these two lines meet at the apex of the cone, which is therefore where $\frac{1}{3}x + 2 = -\frac{1}{3}x$, i.e., at $x = -3$. So the volume of the frustum is the volume of a cone with radius 2 and height 6 less the volume of a cone with radius 1 and height 3. The removed cylinder has height 3 and radius 1. Thus the volume of the solid is

$$\frac{1}{3}\pi \cdot 2^2 \cdot 6 - \frac{1}{3}\pi \cdot 1^2 \cdot 3 - \pi \cdot 1^2 \cdot 3 = 4\pi.$$

Concepts

1. (a) True. See Theorem 6.4.

   (b) True. See Theorem 6.4.

   (c) False. The volume is $2\pi rh\Delta x$. See Theorem 6.4.

   (d) False. The volume is $2\pi rh\Delta x$. See Theorem 6.4.

   (e) False. The volume is $2\pi rh\Delta x$ where r is the *average radius* of the shell.

   (f) True. See the picture preceding Theorem 6.4.

   (g) False. Such a rotation gives a cylinder or a washer.

   (h) True. Look at the second picture following Theorem 6.5.

3. (a) When rotated around the x axis, this becomes a washer with thickness $1.5 - 1 = 0.5$, inner radius 1, and outer radius 4. Thus its volume is $\pi(4^2 - 1^2)(0.5) = 7.5\pi$.

   (b) When rotated around the y axis, this becomes a washer with thickness $4 - 1 = 3$, inner radius 1, and outer radius 1.5. Thus its volume is $\pi(1.5^2 - 1^2)(3) = 3.75\pi$.

   (c) When rotated around the line $x = 1$, this becomes a cylinder (disc) with thickness $4 - 1 = 3$ and radius $1.5 - 1 = 0.5$. Thus its volume is $\pi \cdot 0.5^2 \cdot 3 = 0.75\pi$.

   (d) When rotated around the line $y = 1$, this becomes a cylinder (disc) with thickness $1.5 - 1 = 0.5$ and radius $4 - 1 = 3$. Thus its volume is $\pi \cdot 3^2 \cdot 0.5 = 4.5\pi$.

   (e) When rotated around the line $x = 2$, this becomes a washer with thickness $4 - 1 = 3$, inner radius $2 - 1.5 = 0.5$, and outer radius $2 - 1 = 1$. Thus its volume is $\pi(1^2 - 0.5^2) \cdot 3 = 2.25\pi$.

   (f) When rotated around the line $y = -1$, this becomes a washer with thickness $1.5 - 1 = 0.5$, inner radius $1 - (-1) = 2$, and outer radius $4 - (-1) = 5$. Thus its volume is $\pi(5^2 - 2^2) \cdot 0.5 = 10.5\pi$.

5. Shells and washers are, roughly speaking, the same geometrically. However, considering a shell and a washer of small thickness Δx, the shell will have fixed radii which differ by Δx and a thickness that may vary over the interval, while the washer will have a fixed thickness Δx and radii that may vary over the interval. This distinction gives rise to different integral formulas, one (washers) involving areas of circles, the other (shells) involving circumferences.

7. A graph of the approximation, with each x_i and x_i^* marked as well as $f(x_i^*) = (x_i^*)^2$, is:

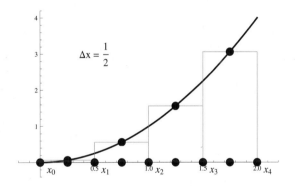

   (a) $\Delta x = \frac{2-0}{4} = \frac{1}{2}$, so that $x_k = \frac{k}{2}$.

   (b) The midpoint x_k^* of each subinterval is $x_k^* = \frac{x_{k-1}+x_k}{2} = \frac{1}{4} + \frac{k}{2} = \frac{1+2k}{4}$.

   (c) For k from 1 to 4, the values of the function at x_k^* are:

k	x_k^*	$f(x_k^*)$
1	$\frac{1}{4}$	$\frac{1}{16}$
2	$\frac{3}{4}$	$\frac{9}{16}$
3	$\frac{5}{4}$	$\frac{25}{16}$
4	$\frac{7}{4}$	$\frac{49}{16}$

   (d) The second shell has thickness $\Delta x = \frac{1}{2}$ and height $f(x_2^*) = \frac{9}{16}$, and radius $x_2^* = \frac{3}{4}$, so its volume is

$$2\pi r h \Delta x = 2\pi \cdot \frac{3}{4} \cdot \frac{9}{16} \cdot \frac{1}{2} = \frac{27}{64}\pi.$$

9. In the formula

$$2\pi rh\Delta y = 2\pi (y_k^*)^3 \left(\frac{2}{n}\right),$$

since $y_k = \frac{2k}{n}$, we see that Δy corresponds to $\frac{2}{n}$. Also, y_k^* is the radius of the kth shell. Thus $(y_k^*)^2$ is the height of the shell, so that the function is $x = y^2$ or $y = \sqrt{x}$. The y-range of the integral is from $y_0 = 0$ to $y_n = 2$. Since the shells are horizontal, we have a solid that is being rotated horizontally; that rotation is around $y = 0$ (the x axis). Thus the integral is

$$2\pi \int_0^2 y^3 \, dy,$$

and it represents the volume of the solid obtained by rotating the region bounded by $y = \sqrt{x}$, the y axis, and the line $y = 2$ around the x axis.

11. This is an x integral, so that $r = x$, the shells are vertical and this is a solid of revolution around the y axis. Comparing $2\pi rh$ to $2\pi x \cos x$, we see that $h = f(x) = \cos x$. So this is the volume of the solid obtained by rotating the region bounded by the x axis, the curve $y = \cos x$, and the line $x = \frac{\pi}{4}$ around the y axis.

13. This is a y integral, so that $r = y$, the shells are horizontal and this is a solid of revolution around the x axis. Comparing $2\pi rh$ to $2\pi y\sqrt{y-4}$, we see that $h = f^{-1}(y) = \sqrt{y-4}$, so that $y = x^2 + 4$. So this is the volume of the solid obtained by rotating the region bounded by the y axis, the line $y = 5$, and the curve $y = x^2 + 4$ around the x axis.

15. This is a y integral, so that $r = y$, the shells are horizontal and this is a solid of revolution around the x axis. Comparing $2\pi rh$ to $2\pi y \ln y$, we see that $h = f^{-1}(y) = \ln y$, so that $y = f(x) = e^x$. So this is the volume of the solid obtained by rotating the region bounded by the y axis, the line $y = e$, and the curve $y = e^x$ around the x axis.

17. Since the rotation is around the x axis, the shells will be horizontal. The height of the shells is given by $3 - f^{-1}(y)$, and the radius is y. Thus the integral is

$$2\pi \int_0^2 y(3 - f^{-1}(y)) \, dy.$$

19. Since the rotation is around the y axis, the shells will be vertical. The height of the shells is given by $f(x) - (-3) = f(x) + 3$, and the radius is x. Thus the integral is

$$2\pi \int_0^2 x(f(x) + 3) \, dx.$$

21. Note that the volume is twice the volume of the solid generated by rotating the region between the graph of $f(x) = |x|$ on $[0, 2]$ around the x axis, and here $|x| = x$. Then using the disc/washer method, we get discs with height x, so the volume is

$$2\pi \int_0^2 x^2 \, dx.$$

Using the shell method, we want horizontal shells. $x = 0$ corresponds to $y = 0$, and $x = 2$ to $y = 2$. Also $f^{-1}(y) = y$. The radius of the shells is y and the height is $2 - y$, so that the integral using the shell method is

$$2 \cdot 2\pi \int_0^2 y(2 - y) \, dy.$$

The first method is slightly easier, since the integrand is simpler. However, both integrands are pretty simple.

23. Using the washer method, we solve $y = x^2 + 1$ for x, giving $x = \sqrt{y-1}$. Then for $0 \le y \le 1$, we get washers of outer radius $3 - 0 = 3$ and inner radius $3 - 1 = 2$; for $1 \le y \le 2$, we get washers of outer radius $3 - \sqrt{y-1}$ and inner radius $3 - 1 = 2$. Thus the volume of the solid is

$$\pi \int_0^1 (3^2 - 2^2) \, dy + \pi \int_1^2 ((3 - \sqrt{y-1})^2 - 2^2) \, dy.$$

Using the shell method, we want vertical shells since the solid is a vertical rotation. The shells have radius $3 - x$ and height $x^2 + 1$, so that the volume is

$$2\pi \int_0^1 (3 - x)(x^2 + 1) \, dx.$$

The shell integral was easier to set up, and is clearly easier to evaluate.

Skills

25. This is a midpoint approximation, so $y_k^* = \frac{2k-1}{4}$ for $k = 1, 2, 3, 4$. There are four rectangles on $[0, 2]$, so that $\Delta y = \frac{1}{2}$. Since $y = f(x) = \sqrt{x}$, we have $x = f^{-1}(y) = y^2$. So the kth shell has radius y_k^* and height $4 - f^{-1}(y) = 4 - (y_k^*)^2$. The approximate volume is thus

$$V \approx \sum_{k=1}^4 2\pi y_k^* (4 - (y_k^*)^2) \cdot \frac{1}{2} = \pi \sum_{k=1}^4 \left(4y_k^* - (y_k^*)^3 \right)$$

$$= \pi \left(\left(1 - \frac{1^3}{4^3} \right) + \left(3 - \frac{3^3}{4^3} \right) + \left(5 - \frac{5^3}{4^3} \right) + \left(7 - \frac{7^3}{4^3} \right) \right)$$

$$= \pi \left(16 - \frac{496}{64} \right) = \frac{33}{4} \pi = 8.25\pi.$$

27. This is a midpoint approximation, so that $y_k^* = \frac{2k-1}{4}$ for $k = 1, 2, 3, 4$. There are four rectangles on $[0, 2]$, so that $\Delta y = \frac{1}{2}$. Since $y = f(x) = \sqrt{x}$, we have $x = f^{-1}(y) = y^2$. So the kth shell has radius $y_k^* + 1$ and height $4 - f^{-1}(y) = 4 - (y_k^*)^2$. The approximate volume is thus

$$V \approx \sum_{k=1}^4 2\pi (y_k^* + 1)(4 - (y_k^*)^2) \cdot \frac{1}{2} = \pi \sum_{k=1}^4 \frac{2k+3}{4} \left(4 - \frac{(2k-1)^2}{16} \right)$$

$$= \pi \left(\frac{5}{4} \left(4 - \frac{1^2}{16} \right) + \frac{7}{4} \left(4 - \frac{3^2}{16} \right) + \frac{9}{4} \left(4 - \frac{5^2}{16} \right) + \frac{11}{4} \left(4 - \frac{7^2}{16} \right) \right) = 19\pi.$$

29. Using the shell method, we have vertical shells with radius x and height $f(x) = 4 - x^2$, so the volume of the solid is

$$2\pi \int_0^2 x(4 - x^2) \, dx = 2\pi \left[2x^2 - \frac{1}{4}x^4 \right]_0^2 = 8\pi.$$

31. Using the shell method, we have vertical shells with radius $2 - x$ and height $4 - x^2$, so the volume of the solid is

$$2\pi \int_0^2 (2 - x)(4 - x^2) \, dx = 2\pi \int_0^2 (x^3 - 2x^2 - 4x + 8) \, dx = 2\pi \left[\frac{1}{4}x^4 - \frac{2}{3}x^3 - 2x^2 + 8x \right]_0^2 = \frac{40}{3}\pi.$$

33. Using the shell method, we have vertical shells with radius x and height $x - 2$, so that the volume of the solid is

$$2\pi \int_2^5 x(x - 2) \, dx = 2\pi \left[\frac{1}{3}x^3 - x^2 \right]_2^5 = 36\pi.$$

35. Using the shell method, we have horizontal shells with radius $3 - y$. To determine the height, note that $f(x) = x - 2$ means $f^{-1}(y) = y + 2$, so that the height is $5 - (y + 2) = 3 - y$. Then the volume is

$$2\pi \int_0^3 (3 - y)^2 \, dy = -\frac{2}{3}\pi \left[(3 - y)^3\right]_0^3 = 18\pi.$$

37. Using the shell method, we have vertical shells with radius $x - 2$ and height $x - 2$, so the volume is

$$2\pi \int_2^5 (x - 2)^2 \, dx = 2\pi \left[\frac{1}{3}(x - 2)^3\right]_2^5 = 18\pi.$$

39. Using the shell method, we have vertical shells with radius x. The height of the shells is $5 - (4 - x^2) = x^2 + 1$. Thus the volume is

$$2\pi \int_0^2 x(x^2 + 1) \, dx = 2\pi \left[\frac{1}{4}x^4 + \frac{1}{2}x^2\right]_0^2 = 12\pi.$$

41. Using the shell method, we have vertical shells with radius $3 - x$. The height of these shells is $5 - (4 - x^2) = x^2 + 1$. Thus the volume is

$$2\pi \int_0^2 (3 - x)(x^2 + 1) \, dx = 2\pi \int_0^2 (-x^3 + 3x^2 - x + 3) \, dx = 2\pi \left[-\frac{1}{4}x^4 + x^3 - \frac{1}{2}x^2 + 3x\right]_0^2 = 16\pi.$$

43. Using the shell method, we have vertical shells with radius $5 - x$. The height is the difference between the two curves which is $x + 1 - (-2) = x + 3$. Thus the volume of the solid is

$$2\pi \int_1^5 (5 - x)(x + 3) \, dx = 2\pi \int_1^5 (-x^2 + 2x + 15) \, dx = 2\pi \left[-\frac{1}{3}x^3 + x^2 + 15x\right]_1^5 = \frac{256}{3}\pi.$$

45. Using the shell method, we have horizontal shells with radius y. The height must be determined separately for $0 \le y \le 1$ and for $1 \le y \le 4$. Solving $y = (x - 2)^2$ for x gives $x = 2 \pm \sqrt{y}$, so that the left edge of the shell for $0 \le y \le 1$ is $2 - \sqrt{y}$ while the right edge is $2 + \sqrt{y}$, so that the height is $(2 + \sqrt{y}) - (2 - \sqrt{y}) = 2\sqrt{y}$. For $1 \le y \le 4$, the height is $2 + \sqrt{y} - y$. Thus the volume of the solid is

$$V = 2\pi \left(\int_0^1 2y\sqrt{y} \, dy + \int_1^4 y(2 + \sqrt{y} - y) \, dy\right)$$

$$= 2\pi \left(\left[\frac{4}{5}y^{5/2}\right]_0^1 + \left[y^2 + \frac{2}{5}y^{5/2} - \frac{1}{3}y^3\right]_1^4\right)$$

$$= \frac{72}{5}\pi.$$

47. Using the shell method, we have horizontal shells with radius $y - (-1) = y + 1$. The height must be determined separately for $0 \le y \le 1$ and for $1 \le y \le 4$. Solving $y = (x - 2)^2$ for x gives $x = 2 \pm \sqrt{y}$, so that the left edge of the shell for $0 \le y \le 1$ is $2 - \sqrt{y}$ while the right edge is $2 + \sqrt{y}$, so that the height is $(2 + \sqrt{y}) - (2 - \sqrt{y}) = 2\sqrt{y}$. For $1 \le y \le 4$, the height is $2 + \sqrt{y} - y$. Thus

the volume of the solid is

$$V = 2\pi \left(\int_0^1 2(y+1)\sqrt{y}\, dy + \int_1^4 (y+1)(2+\sqrt{y}-y)\, dy \right)$$

$$= 2\pi \left(\int_0^1 (2y^{3/2} + 2y^{1/2})\, dy + \int_1^4 (2 + y^{1/2} + y + y^{3/2} - y^2)\, dy \right)$$

$$= 2\pi \left[\frac{4}{5}y^{5/2} + \frac{4}{3}y^{3/2} \right]_0^1 + 2\pi \left[2y + \frac{2}{3}y^{3/2} + \frac{1}{2}y^2 + \frac{2}{5}y^{5/2} - \frac{1}{3}y^3 \right]_1^4$$

$$= \frac{117}{5}\pi.$$

49. A plot of the given region, with the axis of rotation shown as a dotted line, is

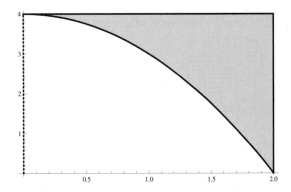

Use the washer method: the outer radius is 2 and the inner radius is found by solving $y = 4 - x^2$ for x, so it is $\sqrt{4-y}$. Then the volume is

$$\pi \int_0^4 (2^2 - (\sqrt{4-y})^2)\, dy = \pi \int_0^4 y\, dy = \pi \left[\frac{1}{2}y^2 \right]_0^4 = 8\pi.$$

51. A plot of the given region, with the axis of rotation shown as a dotted line, is

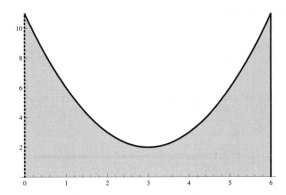

Use the shell method with vertical shells; the radius is x and the height is $(x-3)^2 + 2$, so the volume of the solid is

$$2\pi \int_0^6 x((x-3)^2 + 2)\, dx = 2\pi \int_0^6 (x^3 - 6x^2 + 11x)\, dx = 2\pi \left[\frac{1}{4}x^4 - 2x^3 + \frac{11}{2}x^2 \right]_0^6 = 180\pi.$$

53. A plot of the given region, with the axis of rotation shown as a dotted line, is

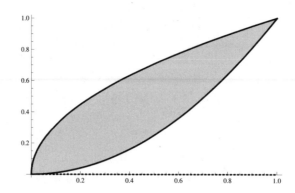

Since $\sqrt{x} \geq x^2$ on $[0, 1]$, using the washer method gives for the volume

$$\pi \int_0^1 \left((\sqrt{x})^2 - (x^2)^2\right) dx = \pi \int_0^1 (x - x^4) \, dx = \pi \left[\frac{1}{2}x^2 - \frac{1}{5}x^5\right]_0^1 = \frac{3}{10}\pi.$$

55. A plot of the given region, with the axis of rotation shown as a dotted line, is

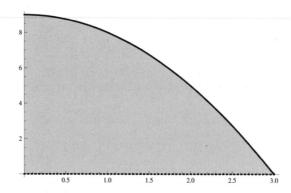

Using the disc method, the radius is $9 - x^2$, so that the volume is

$$\pi \int_0^3 (9 - x^2)^2 \, dx = \pi \int_0^3 (81 - 18x^2 + x^4) \, dx = \pi \left[81x - 6x^3 + \frac{1}{5}x^5\right]_0^3 = \frac{648}{5}\pi.$$

57. A plot of the given region, with the axis of rotation shown as a dotted line, is

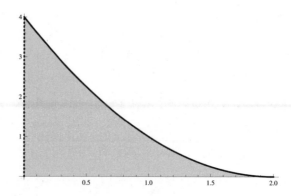

Using the shell method, we have vertical shells with radius x and height $(x-2)^2$, so that the volume is

$$2\pi \int_0^2 x(x-2)^2 \, dx = 2\pi \int_0^2 (x^3 - 4x^2 + 4x) \, dx = 2\pi \left[\frac{1}{4}x^4 - \frac{4}{3}x^3 + 2x^2\right]_0^2 = \frac{8}{3}\pi.$$

59. A plot of the given region, with the axis of rotation shown as a dotted line, is

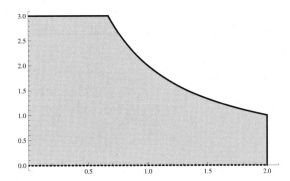

Either disc/washer or shell requires two separate integrals; we choose disc/washer. For the first integral, on $\left[0, \frac{2}{3}\right]$, the radius is 3; for the second integral, on $\left[\frac{2}{3}, 2\right]$, the radius is $\frac{2}{x}$. Then the volume of the solid is

$$\pi \left(\int_0^{2/3} 9 \, dx + \int_{2/3}^2 \frac{4}{x^2} \, dx\right) = \pi \left([9x]_0^{2/3} - 4\left[\frac{1}{x}\right]_{2/3}^2\right) = 6\pi - 4\pi \left(\frac{1}{2} - \frac{3}{2}\right) = 10\pi.$$

61. A plot of the given region, with the axis of rotation shown as a dotted line, is

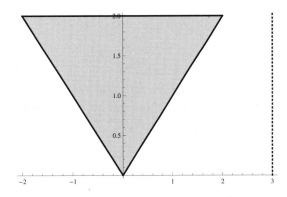

Use the washer method. The outer radius is $3 - (-y) = y + 3$, and the inner radius is $3 - y$. Thus the volume is

$$\pi \int_0^2 ((y+3)^2 - (3-y)^2) \, dy = \pi \int_0^2 12y \, dy = \pi \left[6y^2\right]_0^2 = 24\pi.$$

Applications

63. Since $f(4) = 3 - 0.15 \cdot 16 = 0.6$, the horizontal line in the diagram is at $y = 0.6$. Thus a solid bead is the solid obtained by rotating the area between the curve $f(x) = 3 - 0.15x^2$ and $g(x) = 0.6$ from $x = -4$ to $x = 4$ around the x axis. Using the shell method, the radius is y. To determine the height, solve $y = 3 - 0.15x^2$ for x to get $x = f^{-1}(y) = \pm\sqrt{\frac{3-y}{0.15}}$, so that the height is $2\sqrt{\frac{3-y}{0.15}}$

and thus the volume of the solid of revolution is (using the substitution $u = 3 - y$, $du = -dy$, and $y = 3 - u$)

$$2\pi \int_{0.6}^{3} 2y \sqrt{\frac{3-y}{0.15}} \, dy = -\frac{4\pi}{\sqrt{0.15}} \int_{0.6}^{3} y\sqrt{3-y} \, dy$$

$$= -\frac{4\pi}{\sqrt{0.15}} \int_{y=0.6}^{y=3} (3-u)\sqrt{u} \, du$$

$$= -\frac{4\pi}{\sqrt{0.15}} \left[2u^{3/2} - \frac{2}{5}u^{5/2} \right]_{y=0.6}^{y=3}$$

$$= -\frac{4\pi}{\sqrt{0.15}} \left[2(3-y)^{3/2} - \frac{2}{5}(3-y)^{5/2} \right]_{0.6}^{3}$$

$$= \frac{4\pi}{\sqrt{0.15}} \left(2 \cdot 2.4^{3/2} - \frac{2}{5} \cdot 2.4^{5/2} \right)$$

$$\approx 125.463 \text{ cubic centimeters.}$$

Since each bead uses 125.463 cubic centimeters, 100 beads will use about 12546 cubic centimeters of material.

Proofs

65. If the average radius of the shell is r, assume the inner radius is r_1 and the outer radius r_2; then $\frac{r_1+r_2}{2} = r$. Note that $r_2 - r_1 = \Delta x$, since this is the thickness. Now, the volume of the cylinder with height h and radius r_2 is $\pi r_2^2 h$; the volume of the cylinder with height h and radius r_1 is $\pi r_1^2 h$, so that the difference of these volumes, which is the volume of the shell with thickness $r_2 - r_1 = \Delta x$ and radii r_1 and r_2 (and thus average radius r) is:

$$\pi r_2^2 h - \pi r_1^2 h = \pi(r_2^2 - r_1^2)h = \pi(r_1 + r_2)(r_1 - r_2)h = 2\pi \left(\frac{r_1+r_2}{2} \right) h\Delta x = 2\pi r h \Delta x.$$

67. A sphere of radius 3 is the solid of revolution around the x axis of $y = \sqrt{9 - x^2}$ on $[-3,3]$. Using the shell method with horizontal shells, the radius is y and the height is $\sqrt{9-y^2} - (-\sqrt{9-y^2}) = 2\sqrt{9-y^2}$. Thus the volume of the solid of revolution, which is the cone, is (using the substitution $u = 9 - y^2$):

$$2\pi \int_0^3 y \left(2\sqrt{9-y^2} \right) dy = -2\pi \int_{y=0}^{y=3} \sqrt{u} \, du = -2\pi \left[\frac{2}{3}u^{3/2} \right]_{y=0}^{y=3} = -2\pi \left[\frac{2}{3}(9-y^2)^{3/2} \right]_0^3 = 36\pi,$$

which is equal to $\frac{4}{3}\pi r^3 = \frac{4}{3}\pi \cdot 3^3$.

69. A sphere of radius r is the solid of revolution around the x axis of $y = \sqrt{r^2 - x^2}$ on $[-r,r]$. Using the shell method with horizontal shells, the radius is y and the height is $\sqrt{r^2-y^2} - (-\sqrt{r^2-y^2}) = 2\sqrt{r^2-y^2}$. Thus the volume of the solid of revolution, which is the cone, is (using the substitution $u = r^2 - y^2$):

$$2\pi \int_0^r y \left(2\sqrt{r^2-y^2} \right) dy = -2\pi \int_{y=0}^{y=r} \sqrt{u} \, du = -2\pi \left[\frac{2}{3}u^{3/2} \right]_{y=0}^{y=r} = -2\pi \left[\frac{2}{3}(r^2-y^2)^{3/2} \right]_0^r$$

$$= -2\pi \left(-\frac{2}{3}r^3 \right) = \frac{4}{3}\pi r^3.$$

Thinking Forward

More "subdivide, approximate, and add"

▷ The length of a curve $y = f(x)$ should be the limit of the lengths of secants connecting $(x_k, f(x_k))$ and $(x_{k+1}, f(x_{k+1}))$ where $x_0 = a, x_1, \ldots, x_{n=1}, x_n = b$ is a subdivision of $[a, b]$ into equal-length subintervals. Instead of the areas of rectangles, we would be looking at the lengths of secants. The length of the secant over $[x_k, x_{k+1}]$ clearly depends on the slope of $f(x)$ over that interval, so in the end, the formula we come up with should be an integral (since we are adding things together) of some expression involving $f'(x)$.

▷ The solid of revolution obtained by rotating the region between the graph of $f(x)$ and the x axis around the x axis is bounded by the surface obtained by rotating the graph itself around the x axis. Thus if we have an approximation to the length of the graph, using some subdivision x_k as in the previous part, an approximation to the area of the surface over $[x_k, x_{k+1}]$ should be the lateral surface area of the frustum of a cone whose diagonal side is the secant. Thus the total volume is approximated by the sums of the lateral surface areas of the frustums. Here you can think of each frustum as a "disk" lying over $[x_k, x_{k+1}]$.

▷ Use the same idea as above for arc length, except that for the piece between x_k and x_{k+1} you need to weight its contribution by the density at some point $x^*_{k+1} \in [x_k, x_{k+1}]$, so that the total mass is the sum of the length of each secant multiplied by the density in that interval. The things we are adding up are thus masses.

6.3   Arc Length and Surface Area

Thinking Back

Approximating area with rectangles

▷ Since the length of the interval is 2 and $n = 4$, we have $\Delta x = \frac{2}{4} = \frac{1}{2}$, and $x_k = \frac{k}{2}$ for $k = 0, 1, 2, 3$. Then the right sum is

$$\sum_{k=1}^{4} f(x_k)\Delta x = \frac{1}{2}\sum_{k=1}^{4}\left(\sqrt{\frac{k}{2}} + 1\right) = \frac{1}{2}\left(4 + \sqrt{\frac{1}{2}} + 1 + \sqrt{\frac{3}{2}} + \sqrt{2}\right) = \frac{1}{2\sqrt{2}}(3 + 5\sqrt{2} + \sqrt{3}).$$

▷ Since the length of the interval is 1 and $n = 4$, we have $\Delta x = \frac{1}{4}$, and $x_k = 3 + \frac{k}{4} = \frac{12+k}{4}$ for $k = 0, 1, 2, 3$. Then the right sum is

$$\sum_{k=1}^{4} f(x_k)\Delta x = \frac{1}{4}\sum_{k=1}^{4}\frac{1}{\frac{12+k}{4} - 2} = \frac{1}{4}\sum_{k=1}^{4}\frac{4}{4+k} = \frac{1}{4}\left(\frac{4}{5} + \frac{2}{3} + \frac{4}{7} + \frac{1}{2}\right) = \frac{533}{840}.$$

Riemann sums

▷ The interval width is $e - 1$, so that $\Delta x = \frac{e-1}{n}$ and $x_k = 1 + \frac{e-1}{n}k = \frac{n+k(e-1)}{n}$ for k from 0 to n. Then the Riemann sum is
$$\lim_{n\to\infty}\sum_{k=1}^{n}\ln\left(\frac{n+k(e-1)}{n}\right)\cdot\frac{e-1}{n}.$$

▷ The interval width is 2π, so that $\Delta x = \frac{2\pi}{n}$ and $x_k = -\pi + \frac{2\pi k}{n} = \frac{(2k-n)\pi}{n}$ for k from 0 to n. Then the Riemann sum is
$$\lim_{n\to\infty}\sum_{k=1}^{n}\sin\left(\frac{(2k-n)\pi}{n}\right)\cdot\frac{2\pi}{n}.$$

Review of integration techniques

▷ Use the u-substitution $u = 3 + 16x$, so that $\frac{1}{16}\, du = dx$. Then

$$\int_0^4 \sqrt{3+16x}\, dx = \frac{1}{16} \int_{x=0}^{x=4} \sqrt{u}\, du = \frac{1}{16}\left[\frac{2}{3}u^{3/2}\right]_{x=0}^{x=4} = \frac{1}{24}\left[(3+16x)^{3/2}\right]_0^4 = \frac{1}{24}\left(67^{3/2} - 3^{3/2}\right).$$

▷ $\int_{-2}^2 \sqrt{26}\, dx = \left[\sqrt{26}\, x\right]_{-2}^2 = 4\sqrt{26}.$

▷ Divide top and bottom by 3 and then use the substitution $u = \frac{x}{3}$, so that $3\, du = dx$:

$$\int_0^3 \frac{3}{\sqrt{9-x^2}}\, dx = \int_0^3 \frac{1}{\sqrt{1-\left(\frac{x}{3}\right)^2}}$$

$$= 3\int_{x=0}^{x=3} \frac{1}{\sqrt{1-u^2}}\, du$$

$$= 3\left[\sin^{-1} u\right]_{x=0}^{x=3}$$

$$= 3\left[\sin^{-1}\frac{x}{3}\right]_0^3$$

$$= \frac{3}{2}\pi.$$

▷ $\int_{\pi/4}^{\pi/2} \csc x\, dx = -\left[\ln|\csc x + \cot x|\right]_{\pi/4}^{\pi/2} = \ln(1+\sqrt{2}).$

Approximating definite integrals

▷ $\int_{-1}^3 \sqrt{1+e^{2x}}\, dx \approx 20.7205.$

▷ $\int_0^{2\pi} \sqrt{1+\sin^2 x}\, dx \approx 7.6404.$

Concepts

1. (a) True. We have been using this concept since Chapter 4.

   (b) False. Since $\Delta x = \frac{b-a}{n}$, the kth division point is given by $a + \frac{b-a}{n}k$.

   (c) True. This is the Pythagorean Theorem.

   (d) False. $\sqrt{a^2 + b^2} = |a|\sqrt{1 + \left(\frac{b}{a}\right)^2}.$

   (e) True; Theorem 6.7 applies only if the function is continuous. However, for discontinuities such as jump discontinuities, the method can be adapted to produce a reasonable arc length by splitting the interval into two intervals at the point of discontinuity and computing each arc length separately.

   (f) False. The arc length approximations are always sums of secant lengths between two points on the graph, and a straight line is the shortest distance between two points, so this secant never has greater length than the actual arc length of the curve between those points.

   (g) False. Since the surface area of the frustums is computed using a secant of the underlying curve, the surface area of each frustum is always no greater than the actual surface area for the same reason as in part (f).

(h) True. The area is $2\pi rs$, where r is the average radius and s the slant length. See Theorem 6.8.

3. See Definition 6.6. The limit and the notation are both explained in the definition.

5. In the proof of Theorem 6.7, the Mean Value Theorem is used to convert the ratio $\frac{\Delta y_k}{\Delta x}$, which is the slope of a secant, into $f'(x_k^*)$ where x_k is some point in $[x_{k-1}, x_k]$. This is the critical step since then as $\Delta x \to 0$, $x_k^* \to x_k$.

7. The area under the graph of $y = \sqrt{1 + e^{2x}}$ on $[0, 2]$ is equal to the length of the graph of the function $y = e^x$ on the same interval, since $f'(x) = (e^x)' = e^x$.

9. A graph of $\tan x$ together with the five-segment approximation is

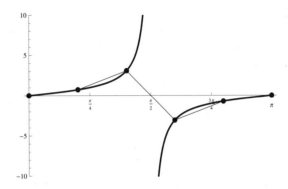

Clearly this is a poor approximation. Since $\tan x$ is discontinuous on $[0, \pi]$, there is no reason to expect these approximations to converge to the arc length (which is infinite, in fact).

11. Since $f'(x) = -g'(x) = 2x$, we see that $f'(x)^2 = g'(x)^2$, so that $\sqrt{1 + f'(x)^2} = \sqrt{1 + g'(x)^2}$, so that they have the same arc length on any interval. This makes sense, since $f(x)$ is x^2 shifted down by three units, while $g(x)$ is x^2 reflected in the x axis and shifted up by 5 units; none of these transformations affect arc length.

13. The circumference of a circle of radius 5 is twice the arc length of the upper half semicircle, which has the equation $f(x) = \sqrt{25 - x^2}$. Then $f'(x) = -x(25 - x^2)^{-1/2}$, so that the circumference of the circle is

$$2 \int_{-5}^{5} \sqrt{1 + (-x(25 - x^2))^{-1/2})^2} \, dx = 2 \int_{-5}^{5} \sqrt{1 + \frac{x^2}{25 - x^2}} \, dx.$$

15. (a) Using four line segments, the points are $x_0 = 0$, $x_1 = \frac{\pi}{4}$, $x_2 = \frac{\pi}{2}$, $x_3 = \frac{3\pi}{4}$, and $x_4 = \pi$, so that the segments and their lengths are

k	Segment	Length
$0, 1$	$(0, 1) - \left(\frac{\pi}{4}, \frac{\sqrt{2}}{2}\right)$	$\sqrt{\frac{\pi^2}{16} + \left(1 - \frac{\sqrt{2}}{2}\right)^2}$
$1, 2$	$\left(\frac{\pi}{4}, \frac{\sqrt{2}}{2}\right) - \left(\frac{\pi}{2}, 0\right)$	$\sqrt{\frac{\pi^2}{16} + \frac{1}{2}}$
$2, 3$	$\left(\frac{\pi}{2}, 0\right) - \left(\frac{3\pi}{4}, -\frac{\sqrt{2}}{2}\right)$	$\sqrt{\frac{\pi^2}{16} + \frac{1}{2}}$
$3, 4$	$\left(\frac{3\pi}{4}, -\frac{\sqrt{2}}{2}\right) - (\pi, -1)$	$\sqrt{\frac{\pi^2}{16} + \left(1 - \frac{\sqrt{2}}{2}\right)^2}$

The sum of the segment lengths is ≈ 3.7901.

(b) We have $(\cos x)' = -\sin x$. The x_k are as above; with four intervals, $\Delta x = \frac{\pi}{4}$. Thus the right approximation is

$$\frac{\pi}{4}\left(\sqrt{1+\sin^2\frac{\pi}{4}}+\sqrt{1+\sin^2\frac{\pi}{2}}+\sqrt{1+\sin^2\frac{3\pi}{4}}+\sqrt{1+\sin^2\pi}\right) \approx 3.81994.$$

(c) Since $(\cos x)' = -\sin x$, the arc length of $\cos x$ from 0 to π is $\int_0^\pi \sqrt{1+((\cos x)')^2}\,dx = \int_0^\pi \sqrt{1+\sin^2 x}\,dx$, which is just the area under the curve $\sqrt{1+\sin^2 x}$ from 0 to π. There seems no particular reason to expect either approximation to be better; they are different methods.

17. For a drawing, see the figure immediately following Theorem 6.8. Since the slant length s_k is the hypotenuse of a right triangle whose legs are Δx and Δy_k, its length is, by the Pythagorean Theorem, $s_k = \sqrt{(\Delta x)^2 + (\Delta y_k)^2}$.

Skills

19. A plot of $f(x)$ with the line segments drawn is

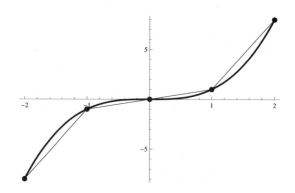

The points $(x_k, f(x_k)) = (x_k, x_k^3)$ for $k = 0, 1, 2, 3, 4$ are

k	0	1	2	3	4
x_k	-2	-1	0	1	2
$f(x_k)$	-8	-1	0	1	8

so that the approximate arc length is

$$\sqrt{(-1-(-2))^2+(-1-(-8))^2}+\sqrt{(0-(-1))^2+(0-(-1))^2}$$
$$+\sqrt{(1-0)^2+(1-0)^2}+\sqrt{(2-1)^2+(8-1)^2}$$
$$= 2\sqrt{50}+2\sqrt{2} = 12\sqrt{2} \approx 16.971.$$

21. A plot of $f(x)$ with the line segments drawn is

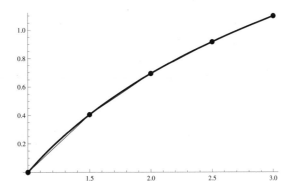

The points $(x_k, f(x_k)) = (x_k, \ln x_k)$ for $k = 0, 1, 2, 3, 4$ are

k	0	1	2	3	4
x_k	1	$\frac{3}{2}$	2	$\frac{5}{2}$	3
$f(x_k)$	0	$\ln 3 - \ln 2$	$\ln 2$	$\ln 5 - \ln 2$	$\ln 3$

so that the approximate arc length is

$$\sqrt{\frac{1}{4} + (\ln 3 - \ln 2)^2} + \sqrt{\frac{1}{4} + (\ln 3 - 2\ln 2)^2} + \sqrt{\frac{1}{4} + (\ln 5 - 2\ln 2)^2} + \sqrt{\frac{1}{4} + (\ln 5 - \ln 2 - \ln 3)^2}$$
$$\approx 2.3003.$$

23. A plot of $f(x)$ with the line segments drawn is

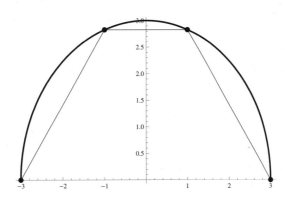

The points $(x_k, f(x_k)) = \left(x_k, \sqrt{9 - x_k^2}\right)$ for $k = 0, 1, 2, 3$ are

k	0	1	2	3
x_k	-3	-1	1	3
$f(x_k)$	0	$2\sqrt{2}$	$2\sqrt{2}$	0

so that the approximate arc length is

$$\sqrt{4 + (2\sqrt{2})^2} + \sqrt{4 + 0} + \sqrt{4 + (2\sqrt{2})^2} = 2 + 4\sqrt{3} \approx 8.9282.$$

25. The points $(x_k, f(x_k)) = (x_k, x_k^3)$ for $k = 0, 1, 2, 3, 4$ are shown in Exercise 19. $\Delta y_k = x_k^3 - x_{k-1}^3$ for $k = 1, 2, 3, 4$ are

k	1	2	3	4
Δy_k	7	1	1	7

Further, $\Delta x = 1$, so that the approximate arc length is

$$\sqrt{1 + \left(\frac{7}{1}\right)^2} + \sqrt{1 + \left(\frac{1}{1}\right)^2} + \sqrt{1 + \left(\frac{1}{1}\right)^2} + \sqrt{1 + \left(\frac{7}{1}\right)^2} = 2\sqrt{50} + 2\sqrt{2} \approx 16.971.$$

27. The points $(x_k, f(x_k)) = (x_k, x_k^3)$ for $k = 0, 1, 2, 3, 4$ are shown in Exercise 21. $\Delta y_k = \ln x_k - \ln x_{k-1}$ for $k = 1, 2, 3, 4$ are

k	1	2	3	4
Δy_k	$\ln 3 - \ln 2$	$2\ln 2 - \ln 3$	$\ln 5 - 2\ln 2$	$\ln 2 + \ln 3 - \ln 5$

Further, $\Delta x = \frac{1}{2}$, so that the approximate arc length is

$$\frac{1}{2}\left(\sqrt{1 + \left(\frac{\ln 3 - \ln 2}{1/2}\right)^2} + \sqrt{1 + \left(\frac{2\ln 2 - \ln 3}{1/2}\right)^2} + \sqrt{1 + \left(\frac{\ln 5 - 2\ln 2}{1/2}\right)^2}\right.$$
$$\left. + \sqrt{1 + \left(\frac{\ln 2 + \ln 3 - \ln 5}{1/2}\right)^2}\right)$$

$$\approx 2.3003.$$

29. The points $(x_k, f(x_k)) = (x_k, x_k^3)$ for $k = 0, 1, 2, 3$ are shown in Exercise 23. $\Delta y_k = \sqrt{9 - x_k^2} = \sqrt{9 - x_{k-1}^2}$ for $k = 1, 2, 3$ are

k	1	2	3
Δy_k	$2\sqrt{2}$	0	$-2\sqrt{2}$

Further, $\Delta x = 2$, so that the approximate arc length is

$$2\left(\sqrt{1 + \left(\frac{2\sqrt{2}}{2}\right)^2} + \sqrt{1 + \left(\frac{0}{2}\right)^2} + \sqrt{1 + \left(\frac{-2\sqrt{2}}{2}\right)^2}\right) = 2(1 + 2\sqrt{3}) \approx 8.9282.$$

31. Since $f'(x) = 3$, so that $f'(x)^2 = 9$, we have

$$L = \int_{-1}^{4} \sqrt{1 + 9}\, dx = \left[\sqrt{10}\, x\right]_{-1}^{4} = 5\sqrt{10}.$$

33. Since $f'(x) = 3x^{1/2}$, so that $f'(x)^2 = 9x$, we have (using the substitution $u = 1 + 9x$)

$$L = \int_{1}^{3} \sqrt{1 + 9x}\, dx = \frac{1}{9}\int_{x=1}^{x=3} \sqrt{u}\, du = \frac{2}{27}\left[u^{3/2}\right]_{x=1}^{x=3} = \frac{2}{27}\left[(1 + 9x)^{3/2}\right]_{1}^{3} = \frac{2}{27}(28^{3/2} - 10^{3/2}).$$

35. Since $f'(x) = \frac{3}{2}(2x + 3)^{1/2} \cdot 2 = 3\sqrt{2x + 3}$, so that $f'(x)^2 = 18x + 27$, we have (using the substitution $u = 18x + 28$)

$$L = \int_{-1}^{1} \sqrt{1 + (18x + 27)}\, dx = \frac{1}{18}\int_{x=-1}^{x=1} \sqrt{u}\, du = \frac{1}{27}\left[u^{3/2}\right]_{x=-1}^{x=1} = \frac{1}{27}\left[(18x + 28)^{3/2}\right]_{x=-1}^{x=1}$$

$$= \frac{1}{27}\left(46^{3/2} - 10^{3/2}\right).$$

37. We have $f'(x) = \frac{1}{2}(9 - x^2)^{-1/2} \cdot (-2x) = -\frac{x}{\sqrt{9-x^2}}$, so that $f'(x)^2 = \frac{x^2}{9-x^2}$ and $1 + f'(x)^2 = \frac{9}{9-x^2} = \frac{1}{1-(x/3)^2}$. Then (using the substitution $u = \frac{x}{3}$)

$$L = \int_{-3}^{3} \frac{1}{\sqrt{1-(x/3)^2}}\,dx = 3\int_{x=-3}^{x=3} \frac{1}{\sqrt{1-u^2}}\,du = 3\left[\sin^{-1} u\right]_{x=-3}^{x=3} = 3\left[\sin^{-1}\frac{x}{3}\right]_{-3}^{3} = 3\pi.$$

39. We have

$$f'(x) = \frac{1}{2}x^{1/2} - \frac{1}{2}x^{-1/2} = \frac{1}{2}(x^{1/2} - x^{-1/2}),$$

so that

$$1 + f'(x)^2 = 1 + \frac{1}{4}\left(x - 2 + \frac{1}{x}\right) = \frac{1}{4}x + \frac{1}{4x} + \frac{1}{2} = \left(\frac{1}{2}\sqrt{x} + \frac{1}{2\sqrt{x}}\right)^2.$$

Then the arc length is

$$L = \int_0^1 \sqrt{\left(\frac{1}{2}\sqrt{x} + \frac{1}{2\sqrt{x}}\right)^2}\,dx = \int_0^1 \left(\frac{1}{2}\sqrt{x} + \frac{1}{2\sqrt{x}}\right)\,dx = \left[\frac{1}{3}x^{3/2} + \sqrt{x}\right]_0^1 = \frac{4}{3}.$$

41. We have

$$f'(x) = -\frac{3}{2}\sqrt{4 - x^{2/3}} \cdot \frac{2}{3}x^{-1/3} = -x^{-1/3}\sqrt{4 - x^{2/3}},$$

so that

$$1 + f'(x)^2 = 1 + x^{-2/3}(4 - x^{2/3}) = 4x^{-2/3}.$$

Then the arc length is

$$L = \int_0^2 \sqrt{4x^{-2/3}}\,dx = \int_0^2 2x^{-1/3}\,dx = \left[3x^{2/3}\right]_0^2 = 3 \cdot 2^{2/3}.$$

43. Since $f'(x) = 2x - \frac{1}{8x}$, we get

$$1 + f'(x)^2 = \frac{1}{2} + 4x^2 + \frac{1}{64x^2} = \left(2x + \frac{1}{8x}\right)^2.$$

Then

$$L = \int_1^2 \sqrt{\left(2x + \frac{1}{8x}\right)^2}\,dx = \int_1^2 \left(2x + \frac{1}{8x}\right)\,dx = \left[x^2 + \frac{1}{8}\ln x\right]_1^2 = 3 + \frac{1}{8}\ln 2.$$

45. Since $f'(x) = \cot x$, we get $1 + f'(x)^2 = 1 + \cot^2 x = \csc^2 x$, so that

$$L = \int_{\pi/4}^{3\pi/4} \sqrt{\csc^2 x}\,dx = \int_{\pi/4}^{3\pi/4} \csc x\,dx = -\left[\ln|\csc x + \cot x|\right]_{\pi/4}^{3\pi/4} = \ln\frac{\sqrt{2}+1}{\sqrt{2}-1}.$$

47. Since $9e^{6x} = f'(x)^2$, we get $f'(x) = \pm 3e^{3x}$, so that $f(x) = \pm e^{3x}$. The interval $[a,b]$ is the domain of integration, which is $[-2,5]$.

49. Simplifying gives $\sqrt{\frac{x^4+1}{x^4}} = \sqrt{1 + x^{-4}}$, so that $f'(x)^2 = x^{-4}$. Thus $f'(x) = \pm x^{-2}$, so that $f(x) = \pm\frac{1}{x}$. The interval $[a,b]$ is the domain of integration, which is $[2,3]$.

51. Since $\sec x = \sqrt{\sec^2 x} = \sqrt{1 + \tan^2 x}$, we get $\tan^2 x = f'(x)^2$, so that $f'(x) = \pm\tan x$ and thus $f(x) = \pm\ln|\cos x|$. The interval $[a,b]$ is the domain of integration, which is $[0, \frac{\pi}{4}]$.

53. Since $f'(x) = 3x^2$, we have $L = \int_{-1}^{1} \sqrt{1 + 9x^4}\, dx \approx 3.0957$.

55. Since $f'(x) = 6x$, we have $L = \int_{1}^{2} \sqrt{1 + 36x^2}\, dx \approx 9.0576$.

57. With $n = 2$, we have $\Delta x = 2$ as well. A table of the various values we need is:

k	0	1	2
x_k	0	2	4
$f(x_k) = x_k^2$	0	4	16
$\Delta y_k = f(x_k) - f(x_{k-1})$		4	12
$r_k = \frac{f(x_k) + f(x_{k-1})}{2}$		2	10

Then the approximation is

$$2\pi \left(2\sqrt{1 + \left(\frac{4}{2}\right)^2} \cdot 2 + 10\sqrt{1 + \left(\frac{12}{2}\right)^2} \cdot 2 \right) = 8\pi\sqrt{5} + 40\pi\sqrt{37}.$$

59. With $n = 4$, we have $\Delta x = \frac{1}{2}$. A table of the various values we need is

k	0	1	2	3	4
x_k	1	$\frac{3}{2}$	2	$\frac{5}{2}$	3
$f(x_k) = x_k^3$	1	$\frac{27}{8}$	8	$\frac{125}{8}$	27
$\Delta y_k = f(x_k) - f(x_{k-1})$		$\frac{19}{8}$	$\frac{37}{8}$	$\frac{61}{8}$	$\frac{91}{8}$
$r_k = \frac{f(x_k) + f(x_{k-1})}{2}$		$\frac{35}{16}$	$\frac{91}{16}$	$\frac{189}{16}$	$\frac{341}{16}$

Then the approximation is

$$2\pi \cdot \frac{1}{2} \left(\frac{35}{16}\sqrt{1 + \left(\frac{19}{4}\right)^2} + \frac{91}{16}\sqrt{1 + \left(\frac{37}{4}\right)^2} + \frac{189}{16}\sqrt{1 + \left(\frac{61}{4}\right)^2} + \frac{341}{16}\sqrt{1 + \left(\frac{91}{4}\right)^2} \right)$$

$$= \frac{\pi}{64} \left(35\sqrt{377} + 91\sqrt{1385} + 189\sqrt{3737} + 341\sqrt{8297} \right).$$

61. With $n = 2$, we have $\Delta x = \frac{\pi}{2}$. A table of the various values we need is

k	0	1	2
x_k	0	$\frac{\pi}{2}$	π
$f(x_k) = \sin x_k$	0	1	0
$\Delta y_k = f(x_k) - f(x_{k-1})$		1	-1
$r_k = \frac{f(x_k) + f(x_{k-1})}{2}$		$\frac{1}{2}$	$\frac{1}{2}$

Then the approximation is

$$2\pi \cdot \frac{\pi}{2} \left(\frac{1}{2}\sqrt{1 + \left(\frac{2}{\pi}\right)^2} + \frac{1}{2}\sqrt{1 + \left(-\frac{2}{\pi}\right)^2} \right) = \frac{\pi^2}{2} \left(\frac{1}{\pi}\sqrt{\pi^2 + 4} + \frac{1}{\pi}\sqrt{\pi^2 + 4} \right) = \pi\sqrt{\pi^2 + 4}.$$

63. We have

$$A = 2\pi \int_0^2 f(x)\sqrt{1 + f'(x)^2}\,dx = 2\pi \int_0^2 x\sqrt{2}\,dx = \pi\sqrt{2}\left[x^2\right]_0^2 = 4\sqrt{2}\,\pi.$$

65. We have $f(x) = x^{1/2}$, so that $1 + f'(x)^2 = 1 + \frac{1}{4x}$. Then the area of the surface of revolution is

$$2\pi \int_0^4 f(x)\sqrt{1 + f'(x)^2} = 2\pi \int_0^4 \sqrt{x}\sqrt{1 + \frac{1}{4x}}\,dx$$

$$= 2\pi \int_0^4 \sqrt{x + \frac{1}{4}}\,dx$$

$$= 2\pi \left[\frac{2}{3}\left(x + \frac{1}{4}\right)^{3/2}\right]_0^4$$

$$= 2\pi \cdot \frac{2}{3}\left(\left(\frac{17}{4}\right)^{3/2} - \left(\frac{1}{4}\right)^{3/2}\right)$$

$$= \frac{\pi}{6} \cdot (17\sqrt{17} - 1).$$

67. We have $f'(x) = \frac{3}{2}(3x + 1)^{-1/2}$, so that

$$f(x)\sqrt{1 + f'(x)^2} = (3x + 1)^{1/2}\sqrt{1 + \frac{9}{4(3x+1)}} = \sqrt{3x + 1 + \frac{9}{4}} = \frac{1}{2}\sqrt{12x + 13}.$$

Then using the substitution $u = 12x + 13$, we get

$$A = 2\pi \int_0^3 f(x)\sqrt{1 + f'(x)^2}\,dx = 2\pi\frac{1}{2}\int_0^3 \sqrt{12x + 13}\,dx$$

$$= \frac{1}{12}\pi \int_{x=0}^{x=3} \sqrt{u}\,du$$

$$= \frac{1}{18}\pi \left[u^{3/2}\right]_{x=0}^{x=3}$$

$$= \frac{1}{18}\pi \left[(12x + 13)^{3/2}\right]_0^3$$

$$= \frac{1}{18}\pi \left(49^{3/2} - 13^{3/2}\right) = \frac{1}{18}\pi(343 - 13^{3/2}).$$

69. We have

$$A = 2\pi \int_{-1}^1 e^{4x}\sqrt{1 + (4e^{4x})^2}\,dx = 2\pi \int_{-1}^1 e^{4x}\sqrt{1 + 16e^{8x}}\,dx.$$

With the substitution $u = 4e^{4x}$ we have $du = 16e^{4x}\,dx$ and also $u^2 = 16e^{8x}$. Then (using the further substitution $u = \tan v$, $du = \sec^2 v\,dx$)

$$
\begin{aligned}
A &= 2\pi \cdot \frac{1}{16} \int_{x=-1}^{x=1} \sqrt{1+u^2}\,du \\
&= \frac{\pi}{8} \int_{x=-1}^{x=1} \sqrt{1+\tan^2 v}\,\sec^2 v\,dv \\
&= \frac{\pi}{8} \int_{x=-1}^{x=1} \sec^3 v\,dv \\
&= \frac{\pi}{8} \left[\frac{1}{2}\left(\sec v \tan v + \ln|\sec v + \tan v|\right) \right]_{x=-1}^{x=1} \\
&= \frac{\pi}{16} \left[\sec(\tan^{-1} u)\tan(\tan^{-1} u) + \ln\left|\sec(\tan^{-1} u) + \tan(\tan^{-1} u)\right| \right]_{x=-1}^{x=1} \\
&= \frac{\pi}{16} \left[u\sqrt{u^2+1} + \ln\left|u + \sqrt{u^2+1}\right| \right]_{x=-1}^{x=1} \\
&= \frac{\pi}{16} \left[4e^{4x}\sqrt{16e^{8x}+1} + \ln\left|4e^{4x} + \sqrt{16e^{8x}+1}\right| \right]_{-1}^{1} \\
&= \frac{\pi}{16} \left(4e^4\sqrt{16e^8+1} + \ln(4e^4 + \sqrt{16e^8+1}) - 4e^{-4}\sqrt{16e^{-8}+1} - \ln(4e^{-4} + \sqrt{16e^{-8}+1}) \right).
\end{aligned}
$$

71. We have (with the substitution $u = \cos x$ followed by the substitution $u = \tan v$)

$$
\begin{aligned}
A &= 2\pi \int_0^\pi \sin x \sqrt{1+\cos^2 x}\,dx \\
&= -2\pi \int_{x=0}^{x=\pi} \sqrt{1+u^2}\,du \\
&= -2\pi \int_{x=0}^{x=\pi} \sqrt{1+\tan^2 v}\,\sec^2 v\,dv \\
&= -2\pi \int_{x=0}^{x=\pi} \sec^3 v\,dv \\
&= -2\pi \left[\frac{1}{2}(\sec v \tan v + \ln|\sec v + \tan v|) \right]_{x=0}^{x=\pi} \\
&= -\pi \left[\sec(\tan^{-1} u)\tan(\tan^{-1} u) + \ln\left|\sec(\tan^{-1} u) + \tan(\tan^{-1} u)\right| \right]_{x=0}^{x=\pi} \\
&= -\pi \left[u\sqrt{u^2+1} + \ln\left|u + \sqrt{u^2+1}\right| \right]_{x=0}^{x=\pi} \\
&= -\pi \left[\cos x \sqrt{1+\cos^2 x} + \ln\left|\cos x + \sqrt{1+\cos^2 x}\right| \right]_{0}^{\pi} \\
&= -\pi \left(-\sqrt{2} + \ln(\sqrt{2}-1) - \sqrt{2} - \ln(\sqrt{2}+1) \right) \\
&= 2\pi\sqrt{2} + \pi \ln\left(\frac{\sqrt{2}+1}{\sqrt{2}-1} \right).
\end{aligned}
$$

Applications

73. The distance it travels is the arc length of $f(x)$ from $x = 0$ to when it hits the ground. It hits the ground when $f(x) = 0$, i.e., when $3 = \frac{1}{2}x^2$, so that $x = \sqrt{6}$. Then since $f'(x) = -x$, the arc length

is (using the substitution $x = \tan u$, $dx = \sec^2 u\, du$)

$$L = \int_0^{\sqrt{6}} \sqrt{1 + x^2}\, dx$$

$$= \int_{x=0}^{x=\sqrt{6}} \sqrt{1 + \tan^2 u}\, \sec^2 u\, du$$

$$= \int_{x=0}^{x=\sqrt{6}} \sec^3 u\, du$$

$$= \frac{1}{2} \left[\sec u \tan u + \ln |\sec u + \tan u| \right]_{x=0}^{x=\sqrt{6}}$$

$$= \frac{1}{2} \left[\sec(\tan^{-1} x) \tan(\tan^{-1} x) + \ln \left| \sec(\tan^{-1} x) + \tan(\tan^{-1} x) \right| \right]_0^{\sqrt{6}}$$

$$= \frac{1}{2} \left[x\sqrt{x^2 + 1} + \ln \left| x + \sqrt{x^2 + 1} \right| \right]_0^{\sqrt{6}}$$

$$= \frac{1}{2} \left(\sqrt{6} \cdot \sqrt{7} + \ln \left| \sqrt{6} + \sqrt{7} \right| \right)$$

$$= \frac{1}{2} \sqrt{42} + \frac{1}{2} \ln(\sqrt{6} + \sqrt{7}).$$

Proofs

75. From Definition 6.6, $x_k = a + k\Delta x$, so that $x_k - x_{k-1} = \Delta x$. Also from Definition 6.6, $\Delta y_k = f(x_k) - f(x_{k-1})$. Substituting these values gives

$$\sum_{k=1}^{n} \sqrt{(x_k - x_{k-1})^2 + (f(x_k) - f(x_{k-1}))^2} = \sum_{k=1}^{n} \sqrt{(\Delta x)^2 + (\Delta y_k)^2}$$

$$= \sum_{k=1}^{n} \sqrt{(\Delta x)^2 \left(1 + \left(\frac{\Delta y_k}{\Delta x} \right)^2 \right)}$$

$$= \sum_{k=1}^{n} \sqrt{1 + \left(\frac{\Delta y_k}{\Delta x} \right)^2}\, \Delta x.$$

77. (a) If $f(x) = mx + c$, then the arc length of $f(x)$ from $x = a$ to $x = b$ is the length of the line joining $(a, f(a))$ and $(b, f(b))$ (since $f(x)$ is itself a line), which is

$$L = \sqrt{(b - a)^2 + (f(b) - f(a))^2}$$

$$= \sqrt{(b - a)^2 + (mb + c - (ma + c))^2}$$

$$= \sqrt{(b - a)^2 + m^2(b - a)^2}$$

$$= (b - a)\sqrt{1 + m^2}.$$

(b) Using Theorem 6.7, since $f'(x) = m$, the arc length of the curve is

$$\int_a^b \sqrt{1 + f'(x)^2}\, dx = \int_a^b \sqrt{1 + m^2}\, dx = \left[x\sqrt{1 + m^2} \right]_a^b = (b - a)\sqrt{1 + m^2}.$$

79. The circumference of a circle of radius r is twice the circumference (i.e., arc length) of the upper half-circle, which has the equation $f(x) = \sqrt{r^2 - x^2}$. That graph meets the x axis at $\pm r$, so the

arc length integral will be from $-r$ to r. We have $f'(x) = -x(r^2 - x^2)^{-1/2}$, so that $1 + f'(x)^2 = \frac{r^2}{r^2 - x^2} = \frac{1}{1 - (x/r)^2}$. Then (with the substitution $u = \frac{x}{r}$)

$$L = 2 \int_{-r}^{r} \frac{1}{\sqrt{1 - (x/r)^2}} \, dx = 2r \int_{x=-r}^{x=r} \frac{1}{\sqrt{1 - u^2}} \, du = 2r \left[\sin^{-1} u \right]_{x=-r}^{x=r} = 2r \left[\sin^{-1} \frac{x}{r} \right]_{-r}^{r} = 2\pi r.$$

81. (a) Consider the following cross-section of the given diagram:

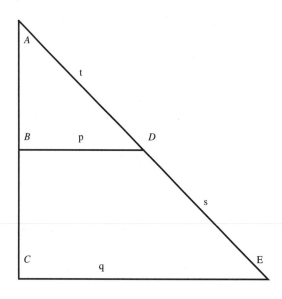

Since BD and CE are parallel, $\angle BDA = \angle CEA$; since the angles at B and C are right angles, $\triangle ABD$ and $\triangle ACE$ are similar. Thus $\frac{AB}{CE} = \frac{AD}{BD}$, or $\frac{s+t}{q} = \frac{t}{p}$. Clearing fractions gives $qt = ps + pt$.

(b) If the radius of the cone is A and the height is B, then the slant height is the hypotenuse of a right triangle with legs A and B, so it has length $\sqrt{A^2 + B^2}$. Thus the surface area of the cone is $\pi A \sqrt{A^2 + B^2} = \pi AC$ where C is the slant height.

(c) The surface area of the frustum is the surface area of the entire cone less the surface area of the upper cone (with radius p and slant height t. But this area is, by part (b),

$$S = \pi q(s + t) - \pi pt = \pi qs + \pi qt - \pi pt = (\pi ps + \pi pt) + \pi qs - \pi pt = \pi(p + q)s.$$

(d) If r is the average radius, then $r = \frac{p+q}{2}$, so that the surface area of the frustum is

$$S = \pi(p + q)s = 2\pi \frac{p + q}{2} s = 2\pi rs.$$

83. The surface area of the sphere is the surface area of the solid formed by rotating the equation of the circle of radius r, which is $f(x) = \sqrt{r^2 - x^2}$, around the x axis (from $-r$ to r). Using Theorem 6.10, we have $f'(x) = -x(r^2 - x^2)^{-1/2}$, so that $1 + f'(x)^2 = \frac{r^2}{r^2 - x^2}$. Then $f(x) \sqrt{1 + f'(x)^2} = r$, and thus

$$A = 2\pi \int_{-r}^{r} f(x) \sqrt{1 + f'(x)^2} \, dx = 2\pi \int_{-r}^{r} r \, dx = 2\pi \left[rx \right]_{-r}^{r} = 4\pi r^2.$$

Thinking Forward

Arc length of parametric curves

▷ The coordinates of the points are

$$(\cos 0, \sin 0) = (1, 0), \qquad \left(\cos \frac{\pi}{4}, \sin \frac{\pi}{4}\right) = \left(\frac{\sqrt{2}}{2}, \frac{\sqrt{2}}{2}\right), \qquad \left(\cos \frac{5\pi}{6}, \sin \frac{5\pi}{6}\right) = \left(-\frac{\sqrt{3}}{2}, \frac{1}{2}\right),$$

$$\left(\cos \frac{3\pi}{2}, \sin \frac{3\pi}{2}\right) = (0, -1), \qquad (\cos 2\pi, \sin 2\pi) = (1, 0).$$

▷ Clearly we should define $\Delta x_k = x(t_k) - x(t_{k-1})$ and $\Delta y_k = y(t_k) - y(t_{k-1})$, since these are the changes in the coordinates on the graph as t moves from t_{k-1} to t_k.

▷ Since the length of the segment from $(x_{k-1}(t), y_{k-1}(t))$ to $(x_k(t), y_k(t))$ is

$$\sqrt{(x_k(t) - x_{k-1}(t))^2 + (y_k(t) - y_{k-1}(t))^2} = \sqrt{(\Delta x_k)^2 + (\Delta y_k)^2}$$

by the Pythagorean Theorem, the approximation for a given division into n pieces is

$$\sum_{k=1}^{n} \sqrt{(\Delta x_k)^2 + (\Delta y_k)^2}.$$

As $n \to \infty$, the lengths of the segment approximations come closer and closer to the actual lengths of the pieces of the curve, so it is reasonable to define the arc length to be

$$\lim_{n \to \infty} \sum_{k=1}^{n} \sqrt{(\Delta x_k)^2 + (\Delta y_k)^2}.$$

▷ Rewriting gives

$$\lim_{n \to \infty} \sum_{k=1}^{n} \sqrt{(\Delta x_k)^2 + (\Delta y_k)^2} = \lim_{n \to \infty} \sum_{k=1}^{n} \sqrt{\left(\frac{x(t_k) - x(t_{k-1})}{t_k - t_{k-1}}\right)^2 + \left(\frac{y(t_k) - y(t_{k-1})}{t_k - t_{k-1}}\right)^2} \cdot (t_k - t_{k-1})$$

$$= \lim_{n \to \infty} \sum_{k=1}^{n} \sqrt{\left(\frac{x(t_k) - x(t_{k-1})}{t_k - t_{k-1}}\right)^2 + \left(\frac{y(t_k) - y(t_{k-1})}{t_k - t_{k-1}}\right)^2} \, \Delta t.$$

As $n \to \infty$, $\frac{x(t_k) - x(t_{k-1})}{t_k - t_{k-1}}$ approaches $x'(t_k^*)$ for some $t_k^* \in [t_{k-1}, t_k]$, and similarly for $\frac{y(t_k) - y(t_{k-1})}{t_k - t_{k-1}}$. Therefore, this can be thought of as a Riemann sum for the integral

$$\int_a^b \sqrt{(x'(t))^2 + (y'(t))^2} \, dt.$$

▷ Parametrize the unit circle by $(\cos t, \sin t)$ for $t \in [0, 2\pi]$. Then by the formula in the preceding part, the circumference of the unit circle is

$$\int_0^{2\pi} \sqrt{((\cos t)')^2 + ((\sin t)')^2} \, dt = \int_0^{2\pi} \sqrt{\sin^2 t + \cos^2 t} \, dt = \int_0^{2\pi} 1 \, dt = [t]_0^{2\pi} = 2\pi.$$

6.4   Real-World Applications of Integration

Thinking Back

Definite Integrals

▷ $\int_0^{10} 4\pi(11.2 + 0.072x^2)\,dx = 4\pi\left[11.2x + 0.024x^3\right]_0^{10} = 4\pi \cdot 136 = 544\pi.$

▷ $\int_0^4 (62.4\pi)(3.5)^2(4 - y)\,dy = 764.4\pi \int_0^4 (4 - y)\,dy = 764.4\pi\left[4y - \dfrac{1}{2}y^2\right]_0^4 = 6115.2\pi.$

▷

$$\int_0^4 62.4\left(10 + \frac{5}{2}y\right)(8 - y)\,dy = 31.2 \int_0^4 (20 + 5y)(8 - y)\,dy$$

$$= 31.2 \int_0^4 (-5y^2 + 20y + 160)\,dy$$

$$= 31.2\left[-\frac{5}{3}y^3 + 10y^2 + 160y\right]_0^4$$

$$= 21632.$$

▷

$$\pi \int_0^{100} (0.75x + 250)(200 - x)\,dx = \pi \int_0^{100} (-0.75x^2 - 100x + 50000)\,dx$$

$$= \pi\left[-0.25x^3 - 50x^2 + 50000x\right]_0^{100}$$

$$= 4.25 \times 10^6\,\pi.$$

▷

$$\pi \int_0^6 (-2y + 1.12)\left(100 - \frac{25}{2}y\right)\,dy = \pi \int_0^6 (25y^2 - 214y + 112)\,dy$$

$$= \pi\left[\frac{25}{3}y^3 - 107y^2 + 112y\right]_0^6$$

$$= -1380\pi.$$

Setting up and solving a word problem involving arc length

Place the origin at the apex of the parabola, 3 feet above the ceiling. Then the parabola passes through the points $(0,0)$, $(6, -3)$, and $(-6, -3)$. Plugging those values into $f(x) = ax^2 + bx + c$ gives for the equation of the parabola $f(x) = -\frac{1}{12}x^2$. Then $f'(x) = -\frac{1}{6}x$, so that the length of the string of lights is

(using the substitution $x = 6 \tan u$, so that $dx = 6 \sec^2 u \, du$)

$$\int_{-6}^{6} \sqrt{1 + \left(-\frac{1}{6}x\right)^2} \, dx = \int_{-6}^{6} \sqrt{1 + \frac{1}{36}x^2} \, dx$$

$$= 6 \int_{x=-6}^{x=6} \sqrt{1 + \tan^2 u} \sec^2 u \, du$$

$$= 6 \int_{x=-6}^{x=6} \sec^3 u \, du$$

$$= 6 \cdot \frac{1}{2} \left[\sec u \tan u + \ln|\sec u + \tan u| \right]_{x=-6}^{x=6}$$

$$= 3 \left[\sec\left(\tan^{-1}\frac{x}{6}\right) \tan\left(\tan^{-1}\frac{x}{6}\right) + \ln\left|\sec\left(\tan^{-1}\frac{x}{6}\right) + \tan\left(\tan^{-1}\frac{x}{6}\right)\right| \right]_{-6}^{6}$$

$$= 3 \left[\frac{x\sqrt{x^2 + 36}}{36} + \ln\left|\frac{x + \sqrt{x^2+36}}{6}\right| \right]_{-6}^{6}$$

$$= 3 \left[\sqrt{2} + \ln\left|\sqrt{2}+1\right| - \left(-\sqrt{2} + \ln\left|\sqrt{2}-1\right|\right) \right]$$

$$= 6\sqrt{2} + 3\ln\left(\frac{\sqrt{2}+1}{\sqrt{2}-1}\right).$$

Setting up and solving a word problem involving volume

The flask looks like this:

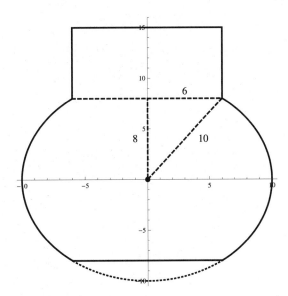

Since the radius of the sphere is 10, and 2 centimeters were cut off, the vertical dashed line is 8 long, so that the horizontal dashed line (the diameter of the cylindrical piece) is 12. Thus the radius of the cylinder is 6 and its height is 7, so its volume is $\pi r^2 h = 252\pi$ cubic centimeters. It remains to find the volume of the truncated sphere. The volume is the volume of the solid obtained by rotating the right-hand piece of the truncated circle in the picture around the y axis. The equation of the circular piece of the boundary, as a function of y, is $x = \sqrt{100 - y^2}$, so using the disc method, we get for the volume

$$\pi \int_{-8}^{8} \left(\sqrt{100 - y^2}\right)^2 dy = \pi \int_{-8}^{8} (100 - y^2) \, dy = \pi \left[100y - \frac{1}{3}y^3 \right]_{-8}^{8} = \frac{3776}{3}\pi.$$

Thus the total volume is

$$\left(252 + \frac{3776}{3}\right)\pi = \frac{4532}{3}\pi \text{ cubic centimeters.}$$

Concepts

1. (a) False. Mass is an attribute of the object; it measures the amount of material in the object. Weight is a force that depends not only on mass but also on the acceleration due to gravity.

   (b) True. The mass depends on the density. For example, a large foam ball may well have less mass than a small iron ball.

   (c) False. Mass is an attribute of the object that measures the amount of material in the object. This does not depend on gravity.

   (d) True. Since the acceleration due to gravity on the moon is different from that on the earth, objects of the same mass will have different forces acting on them on the earth than on the moon.

   (e) False. The units are pounds per cubic foot. Foot-pounds is a unit of work (a force through a distance).

   (f) True. Since the object is moving uphill, the work is force times distance, and the force required to lift a heavy object is greater than the force required to lift a light object.

   (g) False. Hydrostatic force measures the force exerted by a liquid on some surface.

   (h) True. See Definition 6.13. Since the force $F = \omega a d$ varies directly with the depth, it is stronger at greater depths than at lesser depths.

3. See the beginning of this section. Mass is an attribute of the object; it measures the amount of material in the object. Weight is a force that depends not only on mass but also on the acceleration due to gravity.

5. On the vertical wall, the depth varies depending on where you are on the wall, so an integral is used to compute the area. On the bottom, the depth is constant, so computing the hydrostatic force simply involves multiplication.

7. The limits of integration are the first and last x_k (namely, x_0 and x_n).

9. Lifting the hot tub four feet in the air requires that all the water be lifted 4 feet, which takes an amount of work equal to four times the total weight of the water. Pumping all the water out of the hot tub (which we assume to be 4 feet high) requires the water to be lifted varying amounts, from 0 to 4 feet. So this is a different (and smaller) amount of work.

11. The integers $1, 2, \ldots, n$ simply index the y_k, which are the y coordinates of the function being integrated. Since $y_0 = 0$ and $y_n = 4.5$, these are the limits of integration.

13. Divide the contents of the pool up into equal-sized cubes. The only force acting on each cube is the force due to the weight of water above the cube and the forces from adjacent cubes. But the force on a cube from an adjacent cube exactly balances the force on the adjacent cube from this one, so there is no net force between adjacent cubes. The only place this does not hold is at the edge of the pool, and here the force transmitted to the wall is due to the only force on the cube, which is the weight of water above it.

15. Let the weight-density be ω, volume be V, distance traveled be d, and work be W. Then the first sentence can be written $W = \omega V d$, and the second can be written $W = \int_0^h \omega A(y) d \, dy$, where the height of the object is h and $A(y)$ is the cross-sectional area at height y; here $\omega = \omega(y)$ may depend on y. The reason volume does not appear is that this is a weighted volume, as in the previous exercise: if ω is constant, then the formula becomes $\omega d \int_0^h A(y) \, dy = \omega d V$ since volume is the integral of the cross-sectional areas.

17. Considering the x coordinate, we see that the numerator is a sum of centroids, where centroids of larger areas "count" more than those for smaller areas, i.e., are "weighted" more. Dividing by the sum of the areas creates an average, which is thus a "weighted" average. Note that if all areas were the same, say equal to A, then the x coordinate would be

$$\frac{\sum_{k=1}^{n} \overline{x_k} A}{\sum_{k=1}^{n} A} = \frac{A \sum_{k=1}^{n} \overline{x_k}}{nA} = \frac{1}{n} \sum_{k=1}^{n} \overline{x_k}.$$

Thus in the case of equal areas this in fact becomes the average of the centroids. It makes sense to count centroids of larger areas more, since that area has more effect on the total center of mass than does a smaller area.

Skills

19. Since mass is volume times density, we have $m = \pi r^2 h \cdot \rho = \pi \cdot 4^2 \cdot 10 \cdot 12 = 1920\pi$ grams.

21. Since work is force times distance (or, in this case, weight times distance), we have $W = 20 \cdot 100 = 2000$ foot-pounds.

23. The hydrostatic force is $F = \omega A d$, where ω is the weight-density, A the area, and d the depth. So in this case we have $F = 62.4 \cdot (3 \cdot 4) \cdot 2 = 1497.6$ pounds.

25. The cylindrical rod is the solid obtained by rotating the region

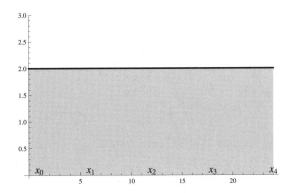

around the x axis. Here $\Delta x = \frac{24}{4} = 6$. We choose $x_k^* = x_k = 6k$ for $k = 1, 2, 3, 4$, so that the mass is approximated by

$$m \approx \sum_{k=1}^{4} \pi \cdot 2^2 \cdot (10.5 + 0.01527(x_k^*)^2) \cdot \Delta x$$

$$= 24\pi \left((10.5 + 0.01527 \cdot 36) + (10.5 + 0.01527 \cdot 12^2) + (10.5 + 0.01527 \cdot 18^2) + (10.5 + 0.01527 \cdot 24^2) \right)$$

$$\approx 4410.16 \text{ grams}$$

27. The solid is obtained by rotating the region

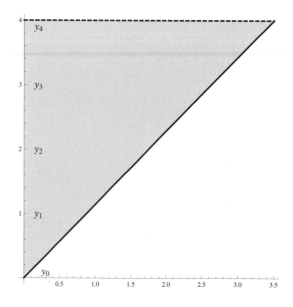

around the y axis. Here $\Delta y = \frac{4}{4} = 1$. We choose $y_k^* = y_k = k$ for $k = 1, 2, 3, 4$, and use the disc method. The radius of each disc is $\frac{3.5}{4} y_k^*$. The work required to pump out the tank can thus be approximated by

$$
\begin{aligned}
W &\approx \sum_{k=1}^{4} 62.4(4 - y_k^*) \cdot \pi \left(\frac{3.5}{4} y_k^* \right)^2 \cdot \Delta y \\
&= \frac{49}{64} \pi \cdot 62.4 \sum_{k=1}^{4} (y_k^*)^2 (4 - y_k^*) \\
&= \frac{49}{64} \pi \cdot 62.4 (1^2 \cdot 4 + 2^2 \cdot 2 + 3^2 \cdot 1 + 4^2 \cdot 0) \\
&\approx 3151.88 \text{ foot-pounds.}
\end{aligned}
$$

29. A side wall of the pool is a 4 foot by 4 foot square. The hydrostatic force is given by $F = \omega A d$ when A and d are constant; here $\omega = 62.4$ is the weight-density of water. Since there are 4 slices, $\Delta y = \frac{4}{4} = 1$. We choose $y_k^* = y_k = k$ for $k = 1, 2, 3, 4$. Then the area of each strip is $4\Delta y$, since the wall is 4 feet wide. Further, the strip that is y_k^* feet above the bottom is $4 - y_k^*$ feet of water deep. The hydrostatic force is thus

$$
F \approx \sum_{k=1}^{4} 62.4(4\Delta y)(4 - y_k^*) = 249.6 \sum_{k=1}^{4} (4 - k) = 249.6(3 + 2 + 1 + 0) = 1497.6 \text{ foot-pounds.}
$$

31. A diagram of the situation is

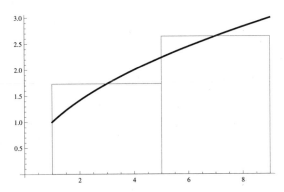

With $n = 2$, the centers of the two rectangles are at $(x_1^*, y_1^*) = \left(3, \frac{1}{2}f(3)\right) = \left(3, \frac{1}{2}\sqrt{3}\right)$ and $(x_2^*, y_2^*) = \left(7, \frac{1}{2}f(7)\right) = \left(7, \frac{1}{2}\sqrt{7}\right)$. The areas of the two rectangles are $A_1 = 4 \cdot f(3) = 4\sqrt{3}$ and $A_2 = 4 \cdot f(7) = 4\sqrt{7}$. Then the approximation is

$$(\overline{x}, \overline{y}) \approx \left(\frac{x_1^* A_1 + x_2^* A_2}{A_1 + A_2}, \frac{y_1^* A_1 + y_2^* A_2}{A_1 + A_2}\right)$$
$$= \left(\frac{3 \cdot 4\sqrt{3} + 7 \cdot 4\sqrt{7}}{4\sqrt{3} + 4\sqrt{7}}, \frac{\frac{1}{2}\sqrt{3} \cdot 4\sqrt{3} + \frac{1}{2}\sqrt{7} \cdot 4\sqrt{7}}{4\sqrt{3} + 4\sqrt{7}}\right)$$
$$= \left(\frac{3\sqrt{3} + 7\sqrt{7}}{\sqrt{3} + \sqrt{7}}, \frac{5}{\sqrt{3} + \sqrt{7}}\right)$$
$$\approx (5.417, 1.142).$$

33. A diagram of the situation is

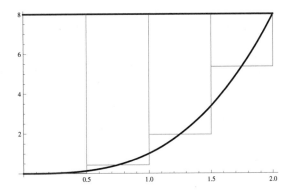

With $n = 4$, we have $\Delta x = \frac{2}{4} = \frac{1}{2}$, and

k	x_k^*	$\left| f(x_k^*) - g(x_k^*) \right|$	$\left| f(x_k^*)^2 - g(x_k^*)^2 \right|$	A_k
1	$\frac{1}{4}$	$\frac{511}{64}$	$\frac{262143}{4096}$	$\frac{511}{128}$
2	$\frac{3}{4}$	$\frac{485}{64}$	$\frac{261415}{4096}$	$\frac{485}{128}$
3	$\frac{5}{4}$	$\frac{387}{64}$	$\frac{246519}{4096}$	$\frac{387}{128}$
4	$\frac{7}{4}$	$\frac{169}{64}$	$\frac{144495}{4096}$	$\frac{169}{128}$

Then

$$(\bar{x}, \bar{y}) \approx \left(\frac{\sum_{k=1}^{4} x_k^* A_k}{\sum_{k=1}^{4} A_k}, \frac{\sum_{k=1}^{4} \frac{1}{2} \left| f(x_k^*)^2 - g(x_k^*)^2 \right| \Delta x}{\sum_{k=1}^{4} A_k} \right) = \left(\frac{1271}{1552}, \frac{228643}{49664} \right) \approx (0.8189, 4.6038).$$

35. For a graph, see Exercise 31. Using the equation of the centroid from Theorem 6.14, we have

$$\int_1^9 \sqrt{x}\, dx = \left[\frac{2}{3} x^{3/2} \right]_1^9 = \frac{52}{3}$$

$$\int_1^9 x\sqrt{x}\, dx = \left[\frac{2}{5} x^{5/2} \right]_1^9 = \frac{484}{5}$$

$$\int_1^9 \frac{1}{2} \left(\sqrt{x} \right)^2 dx = \frac{1}{2} \int_1^9 x\, dx = \frac{1}{2} \left[\frac{1}{2} x^2 \right]_1^9 = 20.$$

Thus

$$(\bar{x}, \bar{y}) = \left(\frac{\int_1^9 x\sqrt{x}\, dx}{\int_1^9 \sqrt{x}\, dx}, \frac{\int_1^9 \frac{1}{2} \left(\sqrt{x} \right)^2 dx}{\int_1^9 \sqrt{x}\, dx} \right) = \left(\frac{363}{65}, \frac{15}{13} \right) \approx (5.585, 1.154).$$

From the graph, this looks reasonable (the test is whether there seems to be as much material above $\bar{y}$ as below it, and as much to the left of $\bar{x}$ as to the right of it).

37. For a graph, see Exercise 33. Using the equation of the centroid from Theorem 6.14, we have

$$\int_0^2 \left| 8 - x^3 \right| dx = \int_0^2 (8 - x^3)\, dx = \left[8x - \frac{1}{4} x^4 \right]_0^2 = 12$$

$$\int_0^2 x \left| 8 - x^3 \right| dx = \int_0^2 (8x - x^4)\, dx = \left[4x^2 - \frac{1}{5} x^5 \right]_0^2 = \frac{48}{5}$$

$$\int_0^2 \frac{1}{2} \left| 64 - x^6 \right| dx = \frac{1}{2} \int_0^2 (64 - x^6)\, dx = \frac{1}{2} \left[64x - \frac{1}{7} x^7 \right]_0^2 = \frac{384}{7}.$$

Thus

$$(\bar{x}, \bar{y}) = \left(\frac{\int_0^2 x \left| 8 - x^3 \right| dx}{\int_0^2 \left| 8 - x^3 \right| dx}, \frac{\int_0^2 \frac{1}{2} \left| 64 - x^6 \right| dx}{\int_0^2 \left| 8 - x^3 \right| dx} \right) = \left(\frac{4}{5}, \frac{32}{7} \right) \approx (0.8, 4.57143).$$

From the graph, this looks reasonable (the test is whether there seems to be as much material above $\bar{y}$ as below it, and as much to the left of $\bar{x}$ as to the right of it).

39. A graph, with $f(x)$ solid black and $g(x)$ the dotted line, is

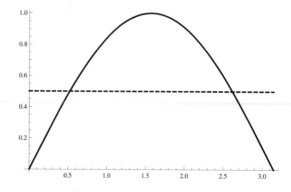

The curves cross at both $x = \frac{\pi}{6}$ and $x = \frac{5\pi}{6}$. We will need the facts that

$$\int x \sin x \, dx = \sin x - x \cos x + C, \qquad \text{(parts, with } u = x, dv = \sin x \, dx)$$

$$\int \sin^2 x \, dx = \frac{x}{2} - \frac{1}{4} \sin 2x + C, \qquad \text{(double-angle formula)}.$$

Then

$$\int_0^\pi \left| \sin x - \frac{1}{2} \right| dx = \int_0^{\pi/6} \left(\frac{1}{2} - \sin x \right) dx + \int_{\pi/6}^{5\pi/6} \left(\sin x - \frac{1}{2} \right) dx + \int_{5\pi/6}^\pi \left(\frac{1}{2} - \sin x \right) dx$$

$$= \left[\frac{1}{2} x + \cos x \right]_0^{\pi/6} + \left[-\cos x - \frac{1}{2} x \right]_{\pi/6}^{5\pi/6} + \left[\frac{1}{2} x + \cos x \right]_{5\pi/6}^\pi$$

$$= 2\sqrt{3} - 2 - \frac{\pi}{6}$$

$$\int_0^\pi x \left| \sin x - \frac{1}{2} \right| dx = \int_0^{\pi/6} \left(\frac{1}{2} x - x \sin x \right) dx + \int_{\pi/6}^{5\pi/6} \left(x \sin x - \frac{1}{2} x \right) dx$$

$$+ \int_{5\pi/6}^\pi \left(\frac{1}{2} x - x \sin x \right) dx$$

$$= \left[\frac{1}{4} x^2 + x \cos x - \sin x \right]_0^{\pi/6} + \left[\sin x - x \cos x - \frac{1}{4} x^2 \right]_{\pi/6}^{5\pi/6}$$

$$+ \left[\frac{1}{4} x^2 + x \cos x - \sin x \right]_{5\pi/6}^\pi$$

$$= \pi \left(\sqrt{3} - 1 - \frac{1}{12} \pi \right)$$

$$\int_0^\pi \frac{1}{2} \left| \sin^2 x - \frac{1}{4} \right| dx = \frac{1}{2} \int_0^{\pi/6} \left(\frac{1}{4} - \sin^2 x \right) dx + \frac{1}{2} \int_{\pi/6}^{5\pi/6} \left(\sin^2 x - \frac{1}{4} \right) dx$$

$$+ \frac{1}{2} \int_{5\pi/6}^\pi \left(\frac{1}{4} - \sin^2 x \right) dx$$

$$= \frac{1}{2} \left[-\frac{1}{4} x + \frac{1}{4} \sin 2x \right]_0^{\pi/6} + \frac{1}{2} \left[\frac{1}{4} x - \frac{1}{4} \sin 2x \right]_{\pi/6}^{5\pi/6}$$

$$+ \frac{1}{2} \left[-\frac{1}{4} x + \frac{1}{4} \sin 2x \right]_{5\pi/6}^\pi$$

$$= \frac{\pi}{24} + \frac{\sqrt{3}}{4}$$

Thus

$$(\bar{x}, \bar{y}) = \left(\frac{\pi}{2}, \frac{\pi + 6\sqrt{3}}{48(\sqrt{3} - 1) - 4\pi} \right) \approx (1.5708, 0.5996).$$

From the graph, this looks reasonable (the test is whether there seems to be as much material above $\bar{y}$ as below it, and as much to the left of $\bar{x}$ as to the right of it).

Applications

41. The volume of the block of copper is $20 \cdot 20 \cdot 13 = 5200$ cubic centimeters, so its mass is $8.93 \cdot 5200 = 46436$ grams.

43. The cross-sectional area of the rod is $1.5^2 = 2.25$ square inches, so the mass of the rod is

$$\int_0^{12} 2.25(4.2 + 0.4x - 0.03x^2)\, dx = \int_0^{12} (9.45 + 0.9x - 0.0675x^2)\, dx$$

$$= \left[9.45x + 0.45x^2 - 0.0225x^3 \right]_0^{12} = 139.32 \text{ grams.}$$

45. The cross-sectional area of the rod is $1^2 = 1$ square cm, so the mass of the rod is

$$\int_0^{10} 1(15 - 0.5x)\, dx = \left[15x - 0.25x^2 \right]_0^{10} = 125 \text{ grams.}$$

47. (a) Place the origin at the center of the circular base of the solid of revolution. Then the parabola goes through $(0,8)$ and $(\pm 5, 0)$, so its equation is $-\frac{8}{25}x^2 + 8 = 0$. The pudding is the solid of revolution obtained by rotating the region bounded by this parabola and the x and y axes around the y axis.

    (b) Since the density varies linearly, and the density at the bottom is $d(0) = 1.12$ while the density at the top is $d(8) = 0.15$, the slope of the density curve is $\frac{1.12 - 0.15}{0 - 8} = -\frac{0.97}{8} = -0.12125$, while its intercept is 1.12. Thus the equation is $d(y) = 1.12 - 0.12125y$.

    (c) To determine the cross-sectional area at height y we need to know the radius of the solid at height y. Thus we need to solve $y = -\frac{8}{25}x^2 + 8$ for x, giving $x = \frac{5}{4}\sqrt{16 - 2y}$, and so the cross-sectional area at height y is $\pi x^2 = \frac{25}{8}\pi(8 - y)$. Then the weight is the integral of cross-sectional area times weight-density:

$$\int_0^8 \frac{25}{8}\pi(8 - y)(1.12 - 0.12125y)\, dy = 3.125\pi \int_0^8 (0.12125y^2 - 2.09y + 8.96)\, dy$$

$$= 3.125\pi \left[8.96y - 1.045y^2 + 0.04042y^3 \right]_0^8 \approx 250.28 \text{ ounces.}$$

49. The weight-density of water is 62.4 pounds per cubic foot, and the volume of the container is $\pi \cdot 0.25^2 \cdot 0.5 = 0.03125\pi$ cubic feet, so its weight is 1.95π pounds. Since we are lifting it 1 foot in the air, the total work is $1.95\pi \approx 6.13$ foot-pounds.

51. The cross-sectional area of the tub at height y is $5 \cdot 8 = 40$ square feet, and the water y feet from the bottom must be lifted $3 - y$ feet. Thus the work required is the integral of the weight of a cross section times the height, or

$$\int_0^3 40 \cdot 62.4 \cdot (3 - y)\, dy = 2496 \left[3y - \frac{1}{2}y^2 \right]_0^3 = 11232 \text{ foot-pounds.}$$

53. Place the origin at the apex of the tank. Then the equation of the slant side of the tank is $y = \frac{5}{2}x$, so the radius of the circular cross-section at height y is $x = \frac{2}{5}y$. The water at height y must be raised $10 - y$ feet. Thus the total work required is

$$\int_0^{10} 62.4\pi \left(\frac{2}{5}y \right)^2 \cdot (10 - y)\, dy = 62.4\pi \cdot \frac{4}{25} \int_0^{10} (10y^2 - y^3)\, dy$$

$$= 9.984\pi \left[\frac{10}{3}y^3 - \frac{1}{4}y^4 \right]_0^{10} \approx 26138.1 \text{ foot-pounds.}$$

55. Place the origin at the bottom of the tank. The cross-sectional area of the tank is constant at $\pi r^2 = 16\pi$ square feet. The water y feet above the bottom of the tank must be raised $3 + (13 - y) = 16 - y$

feet, since the top of the tank is 3 feet below the surface. Since the tank is 13 feet high, the total work is

$$\int_0^{13} 62.4 \cdot 16\pi(16-y)\,dy = 998.4\pi\left[16y - \frac{1}{2}y^2\right]_0^{13} \approx 387366 \text{ foot-pounds.}$$

57. A diagram of the end-on view of the truck is:

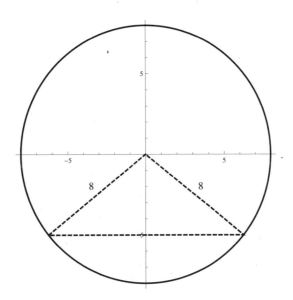

When the liquid is at the height of the horizontal dotted line, at height y, the surface of the liquid forms a rectangle with area 100 feet (the length of the tank) times the length of the horizontal dotted line. That length is, by the Pythagorean Theorem, $2\sqrt{64-y^2}$. Thus the cross-sectional area of the liquid at height y is $200\sqrt{64-y^2}$. The liquid at height y must be lifted $8-y$ feet, so the amount of work required is

$$\int_{-8}^{8} 42.3 \cdot 200\sqrt{64-y^2}(8-y)\,dy = 8460\int_{-8}^{8}\left(8\sqrt{64-y^2} - y\sqrt{64-y^2}\right)\,dy.$$

For the second term, use the substitution $u = 64 - y^2$ so that $-\frac{1}{2}\,du = y\,dy$. For the first term, use the substitution $y = 8 \sin v$, so that $8\sqrt{64 - y^2}\, dy$ becomes $512 \cos^2 v = 256 + 256 \cos(2v)$:

$$\int_{-8}^{8} 42.3 \cdot 200 \sqrt{64 - y^2}(8 - y)\, dy = 8460 \int_{-8}^{8} \left(8\sqrt{64 - y^2} - y\sqrt{64 - y^2} \right) dy$$

$$= 8460 \int_{y=-8}^{y=8} (256 + 256 \cos 2v)\, dv + 4230 \int_{y=-8}^{y=8} \sqrt{u}\, du$$

$$= 8460 \left[256v + 128 \sin 2v \right]_{y=-8}^{y=8} + 4230 \left[\frac{2}{3} u^{3/2} \right]_{y=-8}^{y=8}$$

$$= 8460 \left[256 \sin^{-1} \frac{y}{8} + 128 \sin \left(2 \sin^{-1} \frac{y}{8} \right) \right]_{-8}^{8}$$

$$+ 2820 \left[(64 - y^2)^{3/2} \right]_{-8}^{8}$$

$$= 8460 \left[256 \sin^{-1} \frac{y}{8} + 256 \sin \left(\sin^{-1} \frac{y}{8} \right) \cos \left(\sin^{-1} \frac{y}{8} \right) \right]_{-8}^{8} + 0$$

$$= 8460 \left[256 \sin^{-1} \frac{y}{8} + 256 y \frac{\sqrt{64 - y^2}}{8} \right]_{-8}^{8}$$

$$= 2165760\pi \text{ foot-pounds.}$$

59. The 20 pound mallet requires $20 \cdot 18 = 360$ foot-pounds of work to lift. There are initially 18 feet of rope weighing 0.25 pounds per foot, and the rope y feet off the ground must be lifted $18 - y$ feet, so the work required to lift the rope is

$$\int_0^{18} 0.25(18 - y)\, dy = \left[\frac{9}{2} y - \frac{1}{8} y^2 \right]_0^{18} = \frac{81}{2} = 40.5 \text{ foot-pounds.}$$

Thus the total amount of work is $40.5 + 360 = 400.5$ foot-pounds of work.

61. The hydrostatic force on the bottom is the area times the weight of the water above it. The volume of the cylinder is (in cubic feet) $\pi \cdot \left(\frac{2}{12} \right)^2 \cdot \frac{7}{12} = \frac{7}{432}\pi$; the weight-density of water is 62.4 pounds per cubic foot, so that the hydrostatic force on the bottom is $62.4 \cdot \frac{7}{432}\pi \approx 3.1765$ pounds.

63. The width of a horizontal slice along the side of the pool is 8, so its area is $8\Delta y$. Water at y feet from the bottom is at depth $3 - y$, so the hydrostatic force is

$$\int_0^3 62.4 \cdot 8(3 - y)\, dy = 499.2 \left[3y - \frac{1}{2} y^2 \right]_0^3 = 2246.4 \text{ pounds.}$$

65. The circumference of a horizontal slice is $2\pi \cdot 0.4 = 0.8\pi$ feet, so its area is $0.8\pi\Delta y$. Water at y feet from the bottom is at depth $1 - y$, so the hydrostatic force is

$$\int_0^1 62.4 \cdot 0.8\pi(1 - y)\, dy = 49.92\pi \int_0^1 (1 - y)\, dy = 49.92\pi \left[y - \frac{1}{2} y^2 \right]_0^1 \approx 78.4142 \text{ pounds.}$$

67. The width of the dam at height y is $250 + \frac{400-250}{200} y = 250 + \frac{3}{4} y$, so the area of a slice y feet from the bottom is $\left(250 + \frac{3}{4} y \right) \Delta y$. Water at y feet from the bottom is at depth $200 - y$, so the hydrostatic force is

$$\int_0^{200} 62.4 \cdot \left(250 + \frac{3}{4} y \right) (200 - y)\, dy = 62.4 \int_0^{200} \left(50000 - 100y - \frac{3}{4} y^2 \right) dy$$

$$= 62.4 \left[50000y - 50y^2 - \frac{1}{4} y^3 \right]_0^{200} = 3.744 \times 10^8 \text{ pounds.}$$

69. A diagram of the long side of the pool is

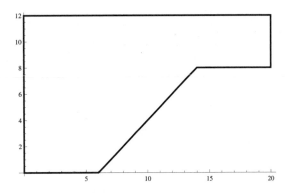

Between $y = 0$ and $y = 8$, the width is $y + 6$, and between $y = 8$ and $y = 12$, the width is 20. Thus the area of a horizontal slice at depth y is $(y + 6)\Delta y$ for $0 \leq y \leq 8$ and $20\Delta y$ for $8 \leq y \leq 12$. A slice at depth y is $12 - y$ feet from the surface, so the hydrostatic force is

$$\int_0^8 62.4(y+6)(12-y)\,dy + \int_8^{12} 62.4(20)(12-y)\,dy$$

$$= 62.4\left(\int_0^8 (72 + 6y - y^2)\,dy + \int_8^{12}(240 - 20y)\,dy\right)$$

$$= 62.4\left(\left[72y + 3y^2 - \frac{1}{3}y^3\right]_0^8 + \left[240y - 10y^2\right]_8^{12}\right)$$

$$= 62.4 \cdot \frac{2272}{3} = 47257.6 \text{ pounds.}$$

71. A diagram of the region is

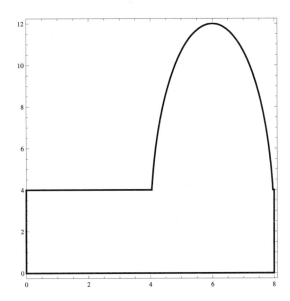

We wish to find the centroid of this region. The upper half of the ellipse has equation $y = 2 + 5\sqrt{-x^2 + 12x - 32}$. Since the line $y = 4$ intersects the ellipse at the two points $r_1 = \frac{2}{5}(15 - 2\sqrt{6})$ and $r_2 = \frac{2}{5}(15 + 2\sqrt{6})$, the region can be thought of as the region over $[0, 8]$ between the x axis and the curve

$$f(x) = \begin{cases} 4, & 0 \le x \le r_1, \\ 2 + 5\sqrt{-x^2 + 12x - 32}, & r_1 < x \le r_2, \\ 4, & r_2 < x \le 8. \end{cases}$$

The centroid of this region is (after a lengthy computation)

$$(\bar{x}, \bar{y}) = \left(\frac{\int_0^8 xf(x)\,dx}{\int_0^8 f(x)\,dx}, \frac{\int_0^8 \frac{1}{2}f(x)^2\,dx}{\int_0^8 f(x)\,dx} \right) \approx \left(\frac{268.818}{55.470}, \frac{236.353}{55.470} \right) \approx (4.846, 4.261).$$

Proofs

73. Choose equally spaced points $x_0 = a, x_1, x_2, \ldots, x_n = b$, so that $x_k - x_{k-1} = \Delta x$ for $k = 1, 2, \ldots, n$. Then the volume of the slice between x_{k-1} and x_k may be approximated by $A(x_k)\Delta x$, so that its mass is approximately $\rho(x_k)A(x_k)\Delta x$ by Definition 6.11. Then the mass m of the entire object is approximated by $\sum_{k=1}^n \rho(x_k)A(x_k)\Delta x$. Since both A and ρ are continuous, this represents a Riemann sum for the product ρA on $[a, b]$, so that

$$m = \lim_{n \to \infty} \sum_{k=1}^n \rho(x_k)A(x_k)\Delta x = \int_a^b \rho(x)A(x)\,dx.$$

75. Choose equally spaced points $y_0 = a, y_1, y_2, \ldots, y_n = b$, so that $y_k - y_{k-1} = \Delta y$ for $k = 1, 2, \ldots, n$. Then the area of the slice from y_{k-1} to y_k is approximated by $l(y_k)\Delta y$. The hydrostatic force on the wall on that slice is then approximated by $w(y_k)d(y_k)l(y_k)\Delta y$ by Definition 6.13. The total hydrostatic force on the vertical wall is thus approximated by $\sum_{k=1}^n w(y_k)l(y_k)d(y_k)\Delta y$. Since w, l, and d are all continuous, this represents a Riemann sum for the product wld on $[a, b]$, so that

$$F = \lim_{n \to \infty} \sum_{k=1}^n w(y_k)l(y_k)d(y_k)\Delta y = \int_a^b w(y)l(y)d(y)\,dy.$$

Thinking Forward

The volume of a torus

Solving the equation $y^2 + (x - 3)^2 = 1$ for x gives $x = 3 \pm \sqrt{1 - y^2}$. Now, divide the circle defined by this equation into horizontal slices on $y_0 = -1, y_1, y_2, \ldots, y_n = 1$. Then the height of each slice is Δy, and each slice when rotated becomes a washer with inner radius $\approx 3 - \sqrt{1 - y_k^2}$ and outer radius $\approx 3 + \sqrt{1 - y_k^2}$. Thus its volume when rotated is approximated by

$$\pi((3 + \sqrt{1 - y_k^2})^2 - (3 - \sqrt{1 - y_k^2})^2)\Delta y = 12\pi\sqrt{1 - y_k^2}\,\Delta y.$$

The volume of the entire solid is given by the sum of these volumes for k from 1 to n, which we recognize as a Riemann sum for

$$\int_{-1}^1 12\pi\sqrt{1 - y^2}\,dy.$$

To integrate, substitute $y = \sin u$ so that $du = \cos u$; then

$$\int_{-1}^{1} 12\pi \sqrt{1-y^2}\, dy = 12\pi \int_{y=-1}^{y=1} \sqrt{1-\sin^2 u}\,\cos u\, du$$

$$= 12\pi \int_{y=-1}^{y=1} \cos^2 u\, du$$

$$= 6\pi \int_{y=-1}^{y=1} (1 + \cos 2u)\, du$$

$$= 6\pi \left[u + \frac{1}{2}\sin 2u \right]_{y=-1}^{y=1}$$

$$= 6\pi \left[\sin^{-1} y + \frac{1}{2}\sin\left(2\sin^{-1} y\right) \right]_{-1}^{1}$$

$$= 6\pi \left[\sin^{-1} y + \sin(\sin^{-1} y)\cos(\sin^{-1} y) \right]_{-1}^{1}$$

$$= 6\pi \left[\sin^{-1} y + y\sqrt{1-y^2} \right]_{-1}^{1}$$

$$= 6\pi^2.$$

Pappus's Centroid Theorem

The centroid of the circle centered at $(3,0)$ with radius 1 is obviously $(3,0)$ by symmetry, so that by Pappus's Theorem, the volume of the solid of revolution is $2\pi r A$, where in this case r is the distance from $(3,0)$ to the axis of rotation, which is 3, and A is the area of the circle, which is π, so that $2\pi r A = 6\pi^2$.

Consider a division strategy in which we subdivide a region into a number of small tori each of which "goes around" the axis. The circumference of such a torus at its midpoint is proportional to the radius, so to its distance from the axis of rotation. Thus the sum of all the circumferences is the same as the number of tori times the circumference of the torus at the "average" of all the points in the region, which is the centroid.

6.5   Differential Equations

Thinking Back

Integration

▷ $\int \frac{1}{y^2}\, dy = \int y^{-2}\, dy = -y^{-1} + C.$

▷ $\int \frac{1}{1-y}\, dy = -\ln|1-y| + C.$

▷ $\int \frac{1}{y^2+1}\, dy = \tan^{-1} y + C.$

▷ Use the substitution $u = y^2 + 1$:

$$\int \frac{y}{y^2+1}\, dy = \frac{1}{2}\int \frac{2y}{y^2+1}\, dy = \frac{1}{2}\int \frac{1}{u}\, du = \frac{1}{2}\ln|u| + C = \frac{1}{2}\ln(y^2+1) + C.$$

▷ Use partial fractions:

$$\int \frac{1}{y(1-y)}\, dy = \int \left(\frac{1}{y} + \frac{1}{1-y} \right) dy = \ln|y| - \ln|1-y| + C.$$

▷ Use partial fractions:

$$\int \frac{1}{y(1-3y)}\, dy = \int \left(\frac{1}{y} + \frac{3}{1-3y} \right) dy = \ln|y| - \ln|1-3y| + C.$$

Local linearity

▷ We have $f(0+1) \approx f'(0)(1-0) + f(0) = 1 + 2 = 3$.

▷ We have $f(1+0.5) \approx f'(1)(0.5) + f(1) = -3 \cdot 0.5 + (-1) = -2.5$.

▷ We have $f(-2+0.25) \approx f'(-2)(0.25) + f(-2) = 2(0.25) - 3 = -2.5$.

▷ We have $f(1.5+0.1) \approx f'(1.5)(0.1) + f(1.5) = 2.2(0.1) + 6.2 = 6.42$.

Concepts

1. (a) True. See Theorem 6.16 and the material which follows.

    (b) False. For example, $y' = y$ has the solutions $y = Ce^x$; those solutions do not differ by a constant.

    (c) True. See the discussion preceding Theorem 6.16.

    (d) False. A differential equation is separable if it can be written in the form $\frac{dy}{dx} = p(x)q(y)$; see Theorem 6.17.

    (e) True. The slope at any point is $g(a,b)$; since $\frac{dy}{dx} = g(x,y)$, we see that $g(a,b) = \frac{dy}{dx}\big|_{(a,b)}$.

    (f) True. The slope at any point is $\frac{dy}{dx}$. Since $\frac{dy}{dx} = g(y)$, the slope depends only on y, so that the slope at $(2,b)$ and the slope at $(3,b)$ are the same since they have the same y value.

    (g) True. Successive points found using Euler's method form the points at which the piecewise linear function changes. See the discussion in the text.

    (h) True. If P is much smaller than K, then $1 - \frac{P}{K} \approx 1$, so that the logistic growth equation is $\frac{dP}{dt} \approx rP(1) = rP$, which is exponential growth.

3. A solution to a differential equation is in general a *family* of functions each of which traverses the slope field; adding initial conditions picks a particular curve out of the family that passes through a particular point.

5. We have $y(x) = (x+C)^{1/2}$, so that $y'(x) = \frac{1}{2}(x+C)^{-1/2} = \frac{1}{2\sqrt{x+C}} = \frac{1}{2y}$.

7. (a) Since acceleration is the time derivative of velocity, the differential equation giving velocity is $\frac{dv}{dt} = a = -9.8$. Since the object is dropped, its initial velocity is zero, so that the initial condition is $v(0) = 0$. The solution to the differential equation is $v(t) = -9.8t + C$; since $v(0) = 0$, we have $v(0) = -9.8 \cdot 0 + C$, so that $C = 0$ and $v(t) = -9.8t$.

    (b) Since velocity is the time derivative of position, the differential equation giving position is $\frac{ds}{dt} = v = -9.8t$. Since the initial height of the object is 100, the initial condition is $s(0) = 100$. The solution to the differential equation is, by antidifferentiation, $s(t) = -4.9t^2 + C$. Then $s(0) = -4.9 \cdot 0 + C = 100$, so that $C = 100$ and the solution is $s(t) = 100 - 4.9t^2$.

9. Dividing through by $q(y)$ gives $\frac{1}{q(y)} \frac{dy}{dx} = p(x)$. Now integrate both sides with respect to x, giving $\int \frac{1}{q(y)} \frac{dy}{dx}\, dx = \int p(x)\, dx$. Applying Exercise 8 gives $\int \frac{1}{q(y)}\, dy = \int p(x)\, dx$. Thus we can integrate the two functions separately with respect to their independent variables and set the two results equal to get a solution to the equation.

11. We have $-\ln|1-y| = x + C$, so that $\ln|1-y| = -x - C$ and $|1-y| = e^{-x-C} = e^{-C}e^{-x}$. Since we assume $y > 1$, this becomes $y - 1 = e^{-C}e^{-x}$. Solving for y gives $y = 1 + e^{-C}e^{-x} = 1 - Ae^{-x}$ where $A = -e^{-C}$.

13. Visually following the slope field lines can help us see what the solutions look like; doing so with a pencil will produce an approximation to the solution curves. See the discussion of slope fields in the text.

15. To compute the $(k+1)^{\text{st}}$ point (x_{k+1}, y_{k+1}), we set $x_{k+1} = x_k + \Delta x$ and then wish to compute y_{k+1}. Since the slope of the solution curve at (x_k, y_k) is $\frac{dy}{dx}\big|_{(x_k, y_k)}$ is known since we know (x_k, y_k), we use that slope as an approximation to the slope of the solution curve on $[x_k, x_{k+1}]$, so that the approximate change in y is $\Delta y_k = \Delta x \cdot \frac{dy}{dx}\big|_{(x_k, y_k)}$. Clearing fractions gives the equation in the problem statement.

17. (a) By Theorem 6.21, the carrying capacity is 500. Intuitively, as $P \to 500$, $1 - \frac{P}{500} \to 0$, so that $\frac{dP}{dt} \to 0$.

(b) If $0 < P(t) < 500$, then $\frac{P(t)}{500} < 1$ and $1 - \frac{P}{500} > 0$. Since $\frac{dP}{dt} > 0$ by assumption, and both $P(t) > 0$ and $1 - \frac{P}{500} > 0$, we also have $k > 0$.

(c) When P is small, then $\frac{P}{500} \approx 0$, so that $1 - \frac{P}{500} \approx 1$, so that $\frac{dP}{dt} \approx kP$.

(d) When $P \approx 500$, then $\frac{P}{500} \approx 1$, so that $1 - \frac{P}{500} \approx 0$, so that $\frac{dP}{dt} \approx 0$.

Skills

19. Simply antidifferentiate:
$$y = \int (x^3 + 4)^2 \, dx = \int (x^6 + 8x^3 + 16) \, dx = \frac{1}{7}x^7 + 2x^4 + 16x + C.$$

21. Separate the variables and use Theorem 6.17 to get $\int \frac{1}{y} \, dy = \int 2x \, dx$. Integrating gives $\ln|y| = x^2 + C$, so that $|y| = e^{x^2 + C} = e^C \cdot e^{x^2}$. Then $y = \pm e^C \cdot e^{x^2} = Ae^{x^2}$ where A is an arbitrary constant.

23. Separate the variables and use Theorem 6.17 to get $\int \frac{1}{y} \, dy = \int x^2 \, dx$. Integrating gives $\ln|y| = \frac{1}{3}x^3 + C$, so that $|y| = e^{\frac{1}{3}x^3 + C} = e^C \cdot e^{\frac{1}{3}x^3}$. Then $y = \pm e^C \cdot e^{\frac{1}{3}x^3} = Ae^{\frac{1}{3}x^3}$ where A is an arbitrary constant.

25. Separate the variables and use Theorem 6.17 to get $\int y \, dy = \int \frac{3x+1}{x} \, dx$. Integrating gives
$$\frac{1}{2}y^2 = \int \left(3 + \frac{1}{x}\right) dx = 3x + \ln|x| + C.$$

Thus $y^2 = 6x + 2\ln|x| + 2C$ and $y = \pm\sqrt{6x + 2\ln|x| + A}$ where A is an arbitrary constant.

27. Separate the variables and use Theorem 6.17 to get $\int e^y \, dy = \int x \, dx$. Integrating gives $e^y = \frac{1}{2}x^2 + C$, so that $y = \ln\left|\frac{1}{2}x^2 + C\right|$ where C is an arbitrary constant.

29. Separate the variables and use Theorem 6.17 to get $\int \frac{1}{y} \, dy = \int 3 \, dx$. Integrating gives $\ln|y| = 3x + C$, so that $|y| = e^{3x+C}$ and $y = \pm e^C \cdot e^{3x} = Ae^{3x}$. Then $y(0) = Ae^{3 \cdot 0} = A = 4$, so that $A = 4$ and $y = 4e^{3x}$.

31. Antidifferentiate to get $y = -2x^3 + C$. Then $y(1) = -2 \cdot 1^3 + C = 3$, so that $C = 5$ and $y = 5 - 2x^3$.

33. Separate variables and use Theorem 6.17 to get $\int \frac{1}{1-y}\,dy = \int 1\,dx$. Integrating gives $-\ln|1-y| = x+C$, so that $|1-y| = e^{-x-C} = e^{-C}\cdot e^{-x} = C_1 e^{-x}$. Thus $1-y = \pm C_1 e^{-x}$ so that $y = 1 \pm C_1 e^{-x} = 1 + Ae^{-x}$. Since $y(0) = 1 + Ae^{-0} = 4$, we have $A = 3$ and thus $y = 1 + 3e^{-x}$.

35. Separate variables and use Theorem 6.17 to get $\int y\,dy = \int 3\,dx$. Integrating gives $\frac{1}{2}y^2 = 3x+C$, so that $y^2 = 6x + 2C$ and $y = \pm\sqrt{6x+2C} = \pm\sqrt{6x+A}$. Since $y(0) = \pm\sqrt{6\cdot 0 + A} = 2$, we must have $6\cdot 0 + A = 4$, so that $A = 4$. Also we must choose the plus sign, since $2 > 0$, so that the solution is $y(0) = \sqrt{6x+4}$.

37. Separate variables and use Theorem 6.17 to get $\int \frac{1}{y}\,dy = \int x\,dx$. Integrating gives $\ln|y| = \frac{1}{2}x^2 + C$, so that $|y| = e^{\frac{1}{2}x^2+C} = e^C \cdot e^{\frac{1}{2}x^2}$. Then $y = \pm e^C \cdot e^{\frac{1}{2}x^2} = Ae^{\frac{1}{2}x^2}$. Since $y(0) = Ae^0 = -1$, we have $A = -1$ so that $y = -e^{\frac{1}{2}x^2}$.

39. Separate variables and use Theorem 6.17 to get $\int e^{-y}\,dy = \int e^x\,dx$. Integrating gives $-e^{-y} = e^x + C$, so that $e^{-y} = -e^x - C$. Now, the left-hand side is positive, so the right-hand side is as well, so that taking logs gives $-y = \ln(-e^x - C)$. Since $y(0) = -\ln(-e^0 - C) = -\ln(-1-C) = 2$, we have $\ln(-1-C) = -2$ so that $-1 - C = e^{-2}$ and $C = -1 - e^{-2}$. So we get the solution $y = -\ln(-e^x - (-1 - e^{-2})) = -\ln(-e^x + 1 + e^{-2})$.

41. Antidifferentiate, using the substitution $u = 1 + x^2$:

$$y = \int \frac{x}{1+x^2}\,dx = \frac{1}{2}\int \frac{1}{u}\,du = \frac{1}{2}\ln|u| + C = \frac{1}{2}\ln(x^2+1) + C.$$

Since $y(0) = \frac{1}{2}\ln(0^2+1) + C = 4$, we have $C = 4$ so that the solution is $y = \frac{1}{2}\ln(x^2+1) + 4$.

43. Separate variables and use Theorem 6.17 to get $\int \frac{1}{9.8 - \frac{0.3}{150}y}\,dy = \int 1\,dx$. Integrate both sides to get $-\frac{150}{0.3}\ln\left|9.8 - \frac{0.3}{150}y\right| = x + C$, or $-500\ln|9.8 - 0.002y| = x + C$. Thus $\ln|9.8 - 0.002y| = -\frac{1}{500}(x + C) = -\frac{1}{500}x + C_1$. Exponentiating gives $|9.8 - 0.002y| = e^{-\frac{1}{500}x + C_1} = C_2 e^{-x/500}$, so that $9.8 - 0.002y = \pm C_2 e^{-x/500}$. Simplifying gives $y = 9.8(500) + Ae^{-x/500} = 4900 + Ae^{-x/500}$, where $A = \pm 500 C_2$ is a constant. Since $y(0) = 4900 + A = 1000$, we have $A = -3900$ so that the solution is $y = 4900 - 3900e^{-x/500}$.

45. Separate variables and use Theorem 6.17 to get

$$\int \frac{1}{-2(1-3y)}\,dy = \int \frac{1}{6y-2}\,dy = \int 1\,dx.$$

Integrating gives $\frac{1}{6}\ln|6y-2| = x + C$, so that $\ln|6y-2| = 6x + C_1$. Exponentiating gives $|6y-2| = e^{6x+C_1} = C_2 e^{6x}$, so that $6y - 2 = \pm C_2 e^{6x}$. Simplifying gives $y = \frac{1}{3} + Ae^{6x}$, where $A = \pm\frac{1}{6}C_2$ is a constant. Since $y(0) = \frac{1}{3} + A = 2$, we have $A = \frac{5}{3}$, so that the solution is $y = \frac{1}{3}(1 + 5e^{6x})$.

47. Separate variables and use Theorem 6.17 to get

$$\int \frac{1}{y(1-y)}\,dy = \int 3\,dx.$$

To integrate the left-hand side, use partial fractions:

$$\int \frac{1}{y(1-y)}\,dy = \int \left(\frac{1}{y} + \frac{1}{1-y}\right)\,dy = \ln|y| - \ln|1-y| + C = \ln\left|\frac{y}{1-y}\right| + C.$$

Thus after integrating both sides we get

$$\ln\left|\frac{y}{1-y}\right| = 3x + C.$$

Exponentiate to get $\left|\frac{y}{1-y}\right| = e^{3x+C} = C_1 e^{3x}$, so that

$$\frac{y}{1-y} = \pm C_1 e^{3x} = A e^{3x}, \quad \text{so that} \quad y = (1-y)A e^{3x}.$$

Simplifying gives $y(1 + A e^{3x}) = A e^{3x}$, so that

$$y = \frac{A e^{3x}}{1 + A e^{3x}}.$$

Since $y(0) = \frac{A}{1+A} = \frac{2}{3}$, we get $A = 2$, and the solution is $y = \frac{2e^{3x}}{1+2e^{3x}}$.

49. Separate variables and apply Theorem 6.17 to get $\int \frac{1}{y^2} dy = \int \cos x\, dx$. Integrating both sides gives $-y^{-1} = \sin x + C$, so that $y^{-1} = -\sin x - C = A - \sin x$. Thus $y = \frac{1}{A - \sin x}$. Since $y(\pi) = \frac{1}{A - \sin \pi} = \frac{1}{A} = 1$, we have $A = 1$, so that the solution is $y = \frac{1}{1 - \sin x}$.

51. Separate variables and apply Theorem 6.17 to get $\int y^2 dy = \int (x^2 - 2)\, dx$. Integrating both sides gives $\frac{1}{3}y^3 = \frac{1}{3}x^3 - 2x + C$, so that $y^3 = x^3 - 6x + 3C = x^3 - 6x + A$. Then $y = \sqrt[3]{x^3 - 6x + A}$. Since $y(0) = \sqrt[3]{A} = 8$, we have $A = 512$, so that the solution is $y = \sqrt[3]{x^3 - 6x + 512}$.

53. Since we start with $y(0) = 3$, we have (with $\Delta x = 0.25$ and $g(x,y) = -2y$)

$$(x_0, y_0) = (0, 3)$$
$$(x_1, y_1) = (x_0 + \Delta x, y_0 + g(x_0, y_0)\Delta x) = (0.25, 3 + (-2 \cdot 3) \cdot 0.25) = (0.25, 1.5)$$
$$(x_2, y_2) = (x_1 + \Delta x, y_1 + g(x_1, y_1)\Delta x) = (0.5, 1.5 + (-2 \cdot 1.5) \cdot 0.25) = (0.5, 0.75)$$
$$(x_3, y_3) = (x_2 + \Delta x, y_2 + g(x_2, y_2)\Delta x) = (0.75, 0.75 + (-2 \cdot 0.75) \cdot 0.25) = (0.75, 0.375)$$
$$(x_4, y_4) = (x_3 + \Delta x, y_3 + g(x_3, y_3)\Delta x) = (1, 0.375 + (-2 \cdot 0.375) \cdot 0.25) = (1, 0.1875).$$

A plot of this approximate solution is

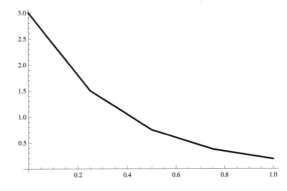

55. Since we start with $y(0) = 0$, we have (with $\Delta x = 0.5$ and $g(x,y) = x + y$)

$$(x_0, y_0) = (0, 0)$$
$$(x_1, y_1) = (x_0 + \Delta x, y_0 + g(x_0, y_0)\Delta x) = (0.5, 0 + (0 + 0)(0.5)) = (0.5, 0)$$
$$(x_2, y_2) = (x_1 + \Delta x, y_1 + g(x_1, y_1)\Delta x) = (1, 0 + (0.5 + 0)(0.5)) = (1, 0.25)$$
$$(x_3, y_3) = (x_2 + \Delta x, y_2 + g(x_2, y_2)\Delta x) = (1.5, 0.25 + (1 + 0.25)(0.5)) = (1.5, 0.875)$$
$$(x_4, y_4) = (x_3 + \Delta x, y_3 + g(x_3, y_3)\Delta x) = (2, 0.875 + (1.5 + 0.875)(0.5)) = (2, 2.0625).$$

A plot of this approximate solution is

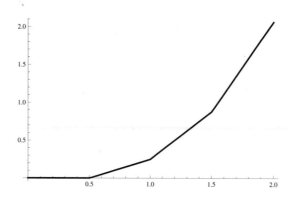

57. Since we start with $y(0) = 2$, we have (with $\Delta x = 0.5$ and $g(x, y) = \frac{x}{y}$)

$$(x_0, y_0) = (0, 2)$$

$$(x_1, y_1) = (x_0 + \Delta x, y_0 + g(x_0, y_0)\Delta x) = \left(0.5, 2 + \frac{0}{2} \cdot 0.5\right) = (0.5, 2)$$

$$(x_2, y_2) = (x_1 + \Delta x, y_1 + g(x_1, y_1)\Delta x) = \left(1, 2 + \frac{0.5}{2} \cdot 0.5\right) = (1, 2.125)$$

$$(x_3, y_3) = (x_2 + \Delta x, y_2 + g(x_2, y_2)\Delta x) = \left(1.5, 2.125 + \frac{1}{2.125} \cdot 0.5\right) \approx (1.5, 2.3603)$$

$$(x_4, y_4) = (x_3 + \Delta x, y_3 + g(x_3, y_3)\Delta x) = \left(2, 2.3605 + \frac{1.5}{2.3603} \cdot 0.5\right) \approx (2, 2.67805).$$

A plot of this approximate solution is

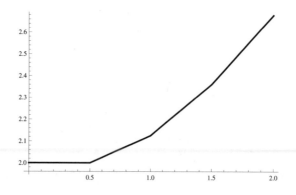

59. The slope field (in gray) together with solutions for $y(0) = -1, 0, 1, 2$ from bottom to top is

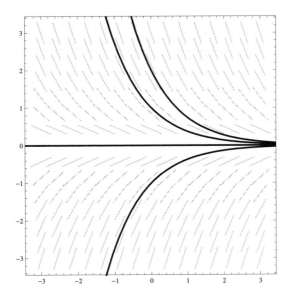

Note that the differential equation has solutions $\int \frac{1}{y} \, dy = \int (-1) \, dx$, so that $y = Ae^{-x}$.

61. The slope field (in gray) together with solutions for $y(0) = -2, -1, 0, 1$ from bottom to top is

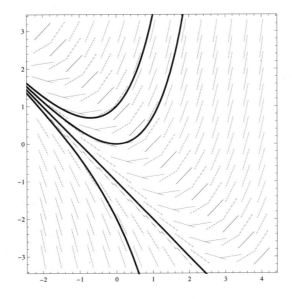

This differential equation is not separable; we do not know how to solve it yet.

63. The slope field (in gray) together with solutions for $y(0) = -2, -1, 0, 1$ from bottom to top is

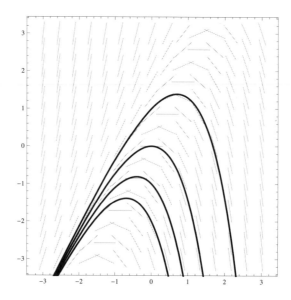

This differential equation is not separable; we do not know how to solve it yet.

Applications

65. (a) If the growth rate is 1.39%, that means that $\frac{dP}{dt} = 0.0139P$. The initial condition is that there are 1.08 million people now, at time $t = 0$, so that $P(0) = 1.08$. The differential equation is separable; using Theorem 6.17 gives

$$\int \frac{1}{P}\, dP = \int 0.0139\, dt, \quad \text{so that} \quad \ln|P| = 0.0139t + C.$$

$P > 0$, so we can drop the absolute values signs after exponentiation to get

$$P = e^{0.0139t + C} = Ae^{0.0139t}.$$

The initial condition gives $P(0) = Ae^{0.0139 \cdot 0} = A = 1.08$, so that $A = 1.08$ and the answer is $P(t) = 1.08e^{0.0139t}$.

(b) The population was half a million people when $P(t) = 0.5$. Solving $0.5 = 1.08e^{0.0139t}$ gives $e^{0.0139t} = \frac{0.5}{1.08} \approx 0.463$, so that $0.0139t = \ln 0.463 \approx -0.7701$. Dividing gives $t \approx -\frac{0.7701}{0.0139} \approx -55.404$, so between 55 and 56 years ago.

(c) The population will double when $P(t) = 2.16$. Solving $2.16 = 1.08e^{0.0139t}$ gives $0.0139t = \ln 2$, so that $t = \frac{\ln 2}{0.0139} \approx 49.867$, so between 49 and 50 years from now.

67. (a) Since the quantity of carbon 14 decreases at a rate proportional to the amount in the body, we must have $\frac{dC}{dt} = -kC$ where $k > 0$ is a constant. Separating variables and using Theorem 6.17 gives

$$\int \frac{1}{C}\, dC = \int (-k)\, dt, \quad \text{so that} \quad \ln|C| = -kt + K.$$

Since $C(t) \geq 0$ we can drop the absolute value sign after exponentiation to get

$$C(t) = e^{-kt + K} = Ae^{-kt}.$$

(b) Since the half-life is 5730 years, after 5730 years there will be $\frac{A}{2}$ of carbon-14, so that

$$\frac{A}{2} = Ae^{-5730k}, \quad \text{or} \quad -5730k = \ln\frac{1}{2}, \quad \text{or} \quad k = \frac{\ln 2}{5730} \approx 0.000121.$$

Thus

$$C(t) = Ae^{-0.000121t}.$$

(c) Solving $0.1A = Ae^{-0.000121t}$ gives $-0.000121t = \ln 0.1$, so that $t = \frac{\ln 0.1}{-0.000121} = \frac{\ln 10}{0.000121} \approx$ 19029.6, so about 19000 years.

69. (a) Using the general model from the instructions to this problem, with $k = 0.1$ and $L = 1000$, we get for the differential equation

$$\frac{dP}{dt} = 0.1P\left(1 - \frac{P}{1000}\right) = 0.0001P(1000 - P).$$

Separating variables and using Theorem 6.17 gives

$$\int \frac{1}{P(1000 - P)}\, dP = \int 0.0001\, dt.$$

Solve the left-hand integral using partial fractions:

$$\int \frac{1}{P(1000 - P)}\, dP = \int \left(\frac{1}{1000P} + \frac{1}{1000(1000 - P)}\right) dP = \frac{1}{1000}\left(\ln|P| - \ln|1000 - P|\right)$$

$$= \frac{1}{1000}\ln\left|\frac{P}{1000 - P}\right|.$$

Setting this equal to the integral on the right gives

$$\frac{1}{1000}\ln\left|\frac{P}{1000 - P}\right| = \frac{1}{10000}t + C, \quad \text{or} \quad \ln\left|\frac{P}{1000 - P}\right| = \frac{1}{10}t + C_1.$$

Then $\frac{P}{1000-P} = e^{t/10+C_1} = Ae^{t/10}$. Solving for P gives

$$P(t) = \frac{1000Ae^{t/10}}{1 + Ae^{t/10}} = \frac{1000}{1 + Be^{-t/10}}, \quad \text{where} \, B = A^{-1}.$$

Since $P(0) = 100 = \frac{1000}{1+B}$, we get $B = 9$, so that finally

$$P(t) = \frac{1000}{1 + 9e^{-t/10}}.$$

(b) A graph of $P(t)$ is

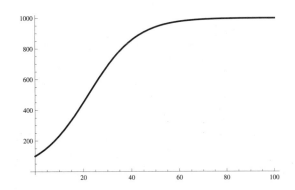

(c) Since the carrying capacity is 1000, we want to find t such that $P(t) = 500$, so we need to solve $500 = \frac{1000}{1+9e^{-t/10}}$ for t. Simplifying gives $1 + 9e^{-t/10} = 2$, so that $e^{-t/10} = \frac{1}{9}$ and thus $t = 10\ln 9 \approx 21.972$, so between 21 and 22 years from now.

71. (a) The basic equation is $\frac{dT}{dt} = k(A - T)$; here $A = 70$, so we have $\frac{dT}{dt} = k(70 - T)$. Separating gives $\int \frac{1}{70-T}\,dT = \int k\,dt$, so that $-\ln|70 - T| = kt + C$. Then $|70 - T| = e^{-kt-C} = Ae^{-kt}$. Since $T > 70$, we get $T - 70 = Ae^{-kt}$, so that $T = 70 + Ae^{kt}$. Let $t = 0$ correspond to 8 P.M. Then $T(0) = 88.8 = 70 + Ae^{-k \cdot 0} = 70 + A$, so that $A = 18.8$, and the equation is

$$T(t) = 70 + 18.8e^{-kt}.$$

(b) Since $T(0.5) = 87.5 = 70 + 18.8e^{-0.5k}$, we bet $e^{-0.5k} = \frac{17.5}{18.8}$. Taking logs and solving gives $k \approx 0.1433$.

(c) We want to solve $T(t) = 98.6 = 70 + 18.8e^{-0.1433t}$, or $\frac{28.6}{18.8} = e^{-0.1433t}$. Taking logs and solving gives $t \approx -2.927$, so the person died not quite 3 hours before 8 P.M., or just after 5 P.M.

73. (a) Since $\frac{dP}{dt}$ is proportional to $P^{1.1}$, we have $\frac{dP}{dt} = kP^{1.1}$. Separating variables gives $\int P^{-1.1}\,dP = \int k\,dt$, so that $-10P^{-0.1} = kt + C_1$. Thus $P^{-0.1} = -\frac{1}{10}kt + C_2$, so that

$$P = \left(-\frac{1}{\frac{1}{10}kt + C_2}\right)^{10} = \left(\frac{10}{kt + C}\right)^{10}.$$

(b) Let $t = 0$ correspond to the year 2005; then $P(0) = \left(\frac{10}{C}\right)^{10} = 6.451$. Taking roots and inverting gives $C = 10(6.451)^{-0.1} \approx 8.29922$, so that

$$P(t) = \left(\frac{10}{kt + 8.29922}\right)^{10}.$$

Since $P(5) = 6.775$, we have $6.775 = \left(\frac{10}{5k + 8.29922}\right)^{10}$; solving gives $k \approx -0.0081$.

(c) We want to solve $7 = \left(\frac{10}{-0.0081t + 8.29922}\right)^{10}$; simplifying and solving gives $t \approx 8.334$, so about one third of the way through the year 2013.

Proofs

75. To solve $\frac{dQ}{dt} = kQ$, separate variables to get $\int \frac{1}{Q}\,dQ = \int k\,dt$, so that $\ln|Q| = kt + C_1$. Exponentiating gives $|Q| = e^{kt+C_1}$, so that $Q = \pm e^{kt+C_1} = Ce^{kt}$. The initial condition gives $Q_0 = Q(0) = Ce^{k \cdot 0} = C$, so that $C = Q_0$ and then $Q(t) = Q_0 e^{kt}$.

77. From Theorem 6.21 and the previous exercise, the general solution to the logistic growth model is

$$P(t) = \frac{L}{1 + Ae^{-kt}}.$$

Now, $\lim_{t \to \infty} Ae^{-kt} = 0$ since we are assuming $k > 0$, so that

$$\lim_{t \to \infty} P(t) = \lim_{t \to \infty} \frac{L}{1 + Ae^{-kt}} = \frac{\lim_{t \to \infty} L}{\lim_{t \to \infty}(1 + Ae^{-kt})} = L.$$

79. From Theorem 6.22 and the previous exercise, the general solution to this equation is $T(t) = A - Ce^{-kt}$. Since we are assuming $k > 0$, $\lim_{t \to \infty} e^{-kt} = 0$, so that $\lim_{t \to \infty} T(t) = A - \lim_{t \to \infty}(Ce^{-kt}) = A$.

If the object is heating, its temperature approaches A from below; if it is cooling, it approaches A from above. Drawing some solution curves for a particular problem will make this behavior clear.

Thinking Forward

Implicit solution curves

▷ Separating variables gives $\int y\,dy = -\int x\,dx$, so that $\frac{1}{2}y^2 = -\frac{1}{2}x^2 + C_1$. Clearing fractions gives $y^2 = C - x^2$. This has two solutions, $y = \pm\sqrt{C - x^2}$. However, this equation also defines the function implicitly. If it passes through $(2, 0)$, then $0 = C - 2^2$, so that $C = 4$, and the equation is $y^2 = 4 - x^2$, which is the same as $x^2 + y^2 = 4$. This is the circle of radius 2 centered at the origin:

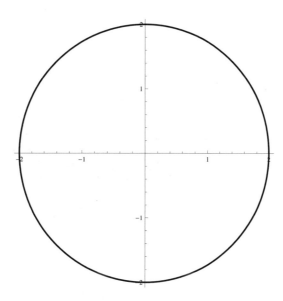

▷ Separating variables gives $\int y\,dy = \int x\,dx$ so that $\frac{1}{2}y^2 = \frac{1}{2}x^2 + C_1$. Clearing fractions gives $y^2 = x^2 + C$, or $y^2 - x^2 = C$. This equation defines a function implicitly; it is a hyperbola. If the hyperbola passes through $(2, 0)$, then $C = 0^2 - 2^2 = -4$, so that its equation is $y^2 - x^2 = -4$, or $x^2 - y^2 = 4$.

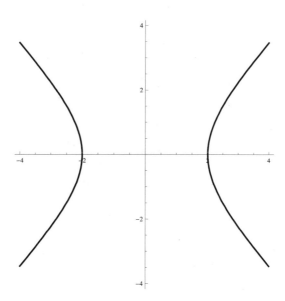

Chapter Review and Self-Test

1.

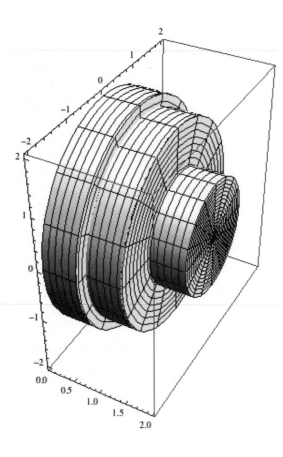

3.

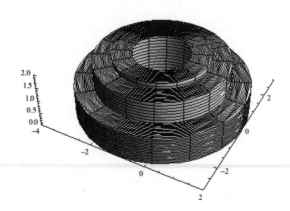

5. For example, for the disc between $x = 2$ and $x = 3$, we have

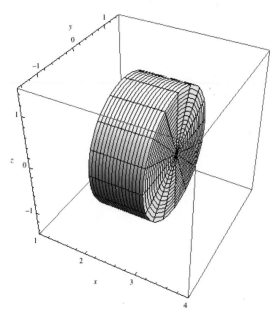

7. For example, for the shell between $x = 2$ and $x = 3$, we have

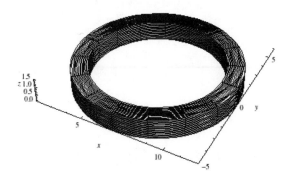

9. A graph of the region is

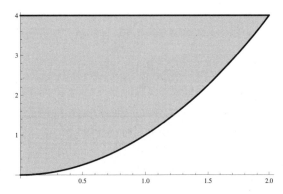

(a) Using the washer method, the washers have outer radius 4 and inner radius x^2, so that the volume is

$$\pi \int_0^2 (16 - x^4)\, dx = \pi \left[16x - \frac{1}{5}x^5 \right]_0^2 = \frac{128}{5}\pi.$$

Using horizontal shells, the shell radius is y, and the height is $x = \sqrt{y}$, so that the volume is

$$2\pi \int_0^4 y\sqrt{y}\, dy = 2\pi \left[\frac{2}{5}y^{5/2}\right]_0^4 = \frac{128}{5}\pi.$$

(b) Using the disc method, the discs have radius $x = \sqrt{y}$, so that the volume is

$$\pi \int_0^4 (\sqrt{y})^2\, dy = \pi \int_0^4 y\, dy = \pi \left[\frac{1}{2}y^2\right]_0^4 = 8\pi.$$

Using vertical shells, the shell radius is x, and the height is $4 - x^2$, so that the volume is

$$2\pi \int_0^2 x(4 - x^2)\, dx = 2\pi \int_0^2 (4x - x^3)\, dx = 2\pi \left[2x^2 - \frac{1}{4}x^4\right]_0^2 = 8\pi.$$

11. A graph of the region is

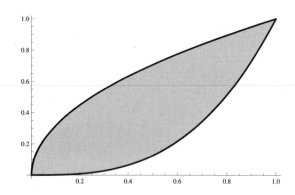

Note that $\sqrt{x} \geq x^3$ on $[0, 1]$.

(a) Using washers, we must solve $y = \sqrt{x}$ and $y = x^3$ for x, giving $x = y^2$ and $x = \sqrt[3]{y}$. The outer radius is $\sqrt[3]{y}$ and the inner radius is y^2. Thus the volume is

$$\pi \int_0^1 ((\sqrt[3]{y})^2 - (y^2)^2)\, dy = \pi \int_0^1 (y^{2/3} - y^4)\, dy = \pi \left[\frac{3}{5}y^{5/3} - \frac{1}{5}y^5\right]_0^1 = \frac{2}{5}\pi.$$

Using vertical shells, the radius of the shells is x and the height is $\sqrt{x} - x^3$, so that the volume is

$$2\pi \int_0^1 x(\sqrt{x} - x^3)\, dx = 2\pi \int_0^1 (x^{3/2} - x^4)\, dx = 2\pi \left[\frac{2}{5}x^{5/2} - \frac{1}{5}x^5\right]_0^1 = \frac{2}{5}\pi.$$

(b) Using washers, the outer radius is $2 - y^2$ and the inner radius is $2 - \sqrt[3]{y}$ (see part (a) for the solutions of the equations for x). Thus the volume is

$$\pi \int_0^1 ((2 - y^2)^2 - (2 - \sqrt[3]{y})^2)\, dy = \pi \int_0^1 (-4y^2 + y^4 + 4y^{1/3} - y^{2/3})\, dy$$

$$= \pi \left[-\frac{4}{3}y^3 + \frac{1}{5}y^4 + 3y^{4/3} - \frac{3}{5}y^{5/3}\right]_0^1$$

$$= \frac{19}{15}\pi.$$

Using shells, the radius of the shells is $2 - x$ and the height is $\sqrt{x} - x^3$, so that the volume is

$$2\pi \int_0^1 (2-x)(\sqrt{x} - x^3)\, dx = 2\pi \int_0^1 (2x^{1/2} - 2x^3 - x^{3/2} + x^4)\, dx$$

$$= 2\pi \left[\frac{4}{3}x^{3/2} - \frac{1}{2}x^4 - \frac{2}{5}x^{5/2} + \frac{1}{5}x^5 \right]_0^1$$

$$= \frac{19}{15}\pi.$$

13. Since $f(x) = \sin(\pi x)$, we get $f'(x) = \pi \cos(\pi x)$, so that $1 + f'(x)^2 = 1 + \pi^2 \cos^2(\pi x)$. The curve is rotationally symmetric, so the surface area of the solid of revolution is twice the surface area of the solid rotated over $[0,1]$. Thus the total surface area is

$$2 \cdot 2\pi \int_0^1 f(x) \sqrt{1 + f'(x)^2}\, dx = 4\pi \int_0^1 \sin(\pi x) \sqrt{1 + \pi^2 \cos^2 \pi x}\, dx.$$

Now use the substitution $u = \pi \cos(\pi x)$, so that $du = -\pi^2 \sin(\pi x)$; the integral becomes

$$4\pi \int_{x=0}^{x=1} \left(-\frac{1}{\pi^2} \sqrt{1 + u^2} \right) du = -\frac{4}{\pi} \int_{x=0}^{x=1} \sqrt{1 + u^2}\, du.$$

Next substitute $u = \tan v$, so that $du = \sec^2 v\, dv$; we get (recalling the integral of $\sec^3 v$ from prior work)

$$-\frac{4}{\pi} \int_{x=0}^{x=1} \sqrt{1 + \tan^2 v} \sec^2 v\, dv = -\frac{4}{\pi} \int_{x=0}^{x=1} \sec^3 v\, dv$$

$$= -\frac{4}{\pi} \cdot \frac{1}{2} \left[\sec v \tan v + \ln |\sec v + \tan v| \right]_{x=0}^{x=1}$$

$$= -\frac{2}{\pi} \left[\sec(\tan^{-1} u) \tan(\tan^{-1} u) + \ln \left| \sec(\tan^{-1} u) + \tan(\tan^{-1} u) \right| \right]_{x=0}^{x=1}$$

$$= -\frac{2}{\pi} \left[u\sqrt{1 + u^2} + \ln \left| u + \sqrt{1 + u^2} \right| \right]_{x=0}^{x=1}$$

$$= -\frac{2}{\pi} \left[\pi \cos(\pi x) \sqrt{1 + \pi^2 \cos^2 \pi x} + \ln \left| \pi \cos(\pi x) + \sqrt{1 + \pi^2 \cos^2 \pi x} \right| \right]_0^1$$

$$= -\frac{2}{\pi} \left(-\pi\sqrt{1 + \pi^2} + \ln \left| -\pi + \sqrt{1 + \pi^2} \right| - \left(\pi\sqrt{1 + \pi^2} + \ln \left| \pi + \sqrt{1 + \pi^2} \right| \right) \right)$$

$$= -\frac{2}{\pi} \left(-2\pi\sqrt{1 + \pi^2} + \ln \left| \frac{\sqrt{1 + \pi^2} - \pi}{\sqrt{1 + \pi^2} + \pi} \right| \right)$$

$$= 4\sqrt{1 + \pi^2} - \frac{2}{\pi} \ln \frac{\sqrt{1 + \pi^2} - \pi}{\sqrt{1 + \pi^2} + \pi}.$$

15. A graph of the region is

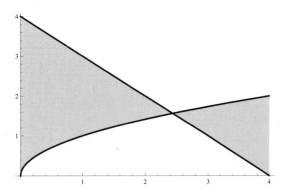

Note that $g(x) \geq f(x)$ on $\left[0, \frac{1}{2}(9 - \sqrt{17})\right]$ and $g(x) \leq f(x)$ elsewhere. Then

$$\int_0^4 |f(x) - g(x)|\, dx = \int_0^{(9-\sqrt{17})/2} (4 - x - x^{1/2})\, dx - \int_{(9-\sqrt{17})/2}^4 (x^{1/2} - 4 + x)\, dx$$

$$= \left[4x - \frac{1}{2}x^2 - \frac{2}{3}x^{3/2}\right]_0^{(9-\sqrt{17})/2} - \left[\frac{2}{3}x^{3/2} - 4x + \frac{1}{2}x^2\right]_{(9-\sqrt{17})/2}^4$$

$$\approx 5.8179$$

$$\int_0^4 x\,|f(x) - g(x)|\, dx = \int_0^{(9-\sqrt{17})/2} (4x - x^2 - x^{3/2})\, dx - \int_{(9-\sqrt{17})/2}^4 (x^{3/2} - 4x + x^2)\, dx$$

$$= \left[2x^2 - \frac{1}{3}x^3 - \frac{2}{5}x^{5/2}\right]_0^{(9-\sqrt{17})/2} - \left[\frac{2}{5}x^{5/2} - 2x^2 + \frac{1}{3}x^3\right]_{(9-\sqrt{17})/2}^4$$

$$\approx 8.8234$$

$$\int_0^4 \frac{1}{2}\left|f(x)^2 - g(x)^2\right| dx = \frac{1}{2}\left(\int_0^{(9-\sqrt{17})/2}\left((4-x)^2 - x\right) dx - \int_{(9-\sqrt{17})/2}^4\left(x - (4-x)^2\right) dx\right)$$

$$= \frac{1}{2}\left(\int_0^{(9-\sqrt{17})/2}\left(16 - 5x + x^2\right) dx - \int_{(9-\sqrt{17})/2}^4\left(-16 + 5x - x^2\right) dx\right)$$

$$= \frac{1}{2}\left(\left[16x - \frac{5}{2}x^2 + \frac{1}{3}x^3\right]_0^{(9-\sqrt{17})/2} - \left[-16x + \frac{5}{2}x^2 - \frac{1}{3}x^3\right]_{(9-\sqrt{17})/2}^4\right)$$

$$\approx 10.4244.$$

Thus

$$(\overline{x}, \overline{y}) = \left(\frac{\int_0^4 x\,|f(x) - g(x)|\, dx}{\int_0^2 |f(x) - g(x)|\, dx},\ \frac{\int_0^4 \frac{1}{2}|f(x)^2 - g(x)^2|\, dx}{\int_0^2 |f(x) - g(x)|\, dx}\right) \approx (1.5166, 1.7918).$$

17. The cross-sectional area of the tub is $\pi \cdot 4^2 = 16\pi$ square feet, so the weight of a slice is $62.4 \cdot 16\pi\Delta y$. A slice y feet from the bottom must be raised $3 - y$ feet, so the work required to raise that slice is $62.4 \cdot 16\pi(3 - y)\Delta y$. Thus the total work required is

$$\int_0^3 62.4 \cdot 16\pi(3 - y)\, dy = 998.4\pi \int_0^3 (3 - y)\, dy = 998.4\pi \left[3y - \frac{1}{2}y^2\right]_0^3$$

$$= 998.4 \times \frac{9}{2}\pi \approx 14114.5 \text{ foot-pounds.}$$

19. Separating variables gives $\int \frac{1}{y}\, dy = \int 5\, dx$; integrating then gives $\ln|y| = 5x + C_1$, so that $|y| = e^{5x+C_1}$. Then $y = \pm e^{5x+C_1} = Ce^{5x}$. Since $y(0) = Ce^{5\cdot 0} = C = 1$, we get $C = 1$ and the solution is $y = e^{5x}$.

21. Separating variables gives $\int \frac{1}{y(1-5y)}\, dy = \int 2\, dx$. To integrate the left-hand side, use partial fractions:

$$\int \frac{1}{y(1 - 5y)}\, dy = \int \left(\frac{1}{y} + \frac{5}{1 - 5y}\right) dy = \ln|y| - \ln|1 - 5y| = \ln\left|\frac{y}{1 - 5y}\right|.$$

Setting that equal to the right-hand side, which is $\int 2\, dx = 2x + C_1$ gives

$$\ln\left|\frac{y}{1 - 5y}\right| = 2x + C_1, \quad \text{so that} \quad \frac{y}{1 - 5y} = \pm e^{2x+C_1} = C_2 e^{2x}.$$

Solving for y gives

$$y = \frac{C_2 e^{2x}}{1 + 5C_2 e^{2x}} = \frac{1}{5 + Ce^{-2x}}.$$

Since $y(0) = \frac{1}{5+C} = 1$, we get $C = -4$, so that the solution is

$$y = \frac{1}{5 - 4e^{-2x}}.$$

Part III

Sequences and Series

Chapter 7

Sequences and Series

7.1 Sequences

Thinking Back

Functions

▷ A *function* f from a set A to a set B is an assignment f that associates to each element x of the *domain* A exactly one element of the *codomain* B.

▷ The *target* of a function is the same as its codomain.

▷ f is *increasing* on $[a, b]$ if $f(d) \geq f(c)$ for all $d > c$ in $[a, b]$.

▷ f is *strictly increasing* on $[a, b]$ if $f(d) > f(c)$ for all $d > c$ in $[a, b]$.

▷ f is *decreasing* on $[a, b]$ if $f(d) \leq f(c)$ for all $d > c$ in $[a, b]$.

▷ f is *strictly decreasing* on $[a, b]$ if $f(d) < f(c)$ for all $d > c$ in $[a, b]$.

▷ f is *constant* on $[a, b]$ if $f(c) = f(d)$ for all $c, d \in [a, b]$.

▷ f is *bounded* on $[a, b]$ if there exist $m, M \in \mathbb{R}$ such that $m \leq f(c) \leq M$ for all $c \in [a, b]$.

Concepts

1. (a) True. A sequence is a function from the set of indices (normally the positive integers) to the real numbers. See Definition 7.1.

   (b) True. Since the first term is for $k = 1$, the third term is when $k = 3$, and when $k = 3$, we see that $k + 1 = 4$.

   (c) False. Since the first term is for $k = 2$, the third term is for $k = 4$, and when $k = 4$, we have $k^2 = 16$. The *second* term of this sequence is 9.

   (d) False. For example, $\{1, 0, 1, 0, 1, 0, \dots\}$ is neither increasing nor decreasing.

   (e) False. For example, the sequence $\{0, -1, -2, -3, -4, \dots\}$ is unbounded below, and the sequence $\{1, \frac{1}{2}, \frac{1}{4}, \frac{1}{8}, \dots\}$ is bounded below but nonetheless has no smallest term.

   (f) False. For example, let $a_1 = 1$ and $a_k = 1 - a_{k-1}$ for all $k > 1$. Then $\{a_k\} = \{1, 0, 1, 0, 1, 0, \dots\}$.

   (g) False. For example, $\{1, -2, 3, -4, 5, -6, 7, -8, \dots\}$ has neither.

   (h) True. If the sequence is monotonically increasing, then it is bounded below by a_1, since $a_k \geq a_1$ for all $k \geq 1$. Similarly, if it is monotonically decreasing, then it is bounded above by a_1, since $a_k \leq a_1$ for all $k \geq 1$.

3. See Definition 7.1. A sequence $\{a_k\}$ of real numbers is a function f from positive integers to the reals, $f : k \mapsto a_k$. Alternatively, a sequence may be thought of as an ordered list of real numbers.

5. The index of a term of a sequence is its subscript, in the sense that the index of a_k is k.

7. We have

$$a_1 = 1, \quad a_2 = a_1 + 2 = 3, \quad a_3 = a_2 + 2 = 5, \quad a_4 = a_3 + 2 = 7, \quad a_5 = a_4 + 2 = 9.$$

A closed formula for a_k is $a_k = 2k - 1$.

9. With $a_1 = 1$, we have $a_k = k + 1$, or $a_k = a_{k-1} + 1$. Then $a_2 = a_1 + 1 = 2, a_3 = a_2 + 1 = 3$, and so on.

11. A sequence $\{a_k\}$ is eventually strictly increasing if there is some positive integer N such that whenever $k > N$ then $a_{k+1} > a_k$.

13. A sequence $\{a_k\}$ is eventually strictly decreasing if there is some positive integer N such that whenever $k > N$ then $a_{k+1} < a_k$.

15. A sequence $\{a_k\}$ is bounded above if there is some $M \in \mathbb{R}$ such that for all k we have $a_k \leq M$. It is bounded below if there is some $m \in \mathbb{R}$ such that for all k we have $a_k \geq m$. It is bounded if it is both bounded above and bounded below.

17. Let $a_k = \{e^1, -e^2, e^3, -e^4, \dots\}$. Since $\lim\limits_{k \to \infty} e^{2k+1} = \infty$ and $\lim\limits_{k \to \infty} (-e^{2k}) = -\infty$, this sequence is unbounded both above and below.

19. If we did not require the terms to be positive, then the sequence $\{-1, -2, -3, -4, \dots\}$ would be increasing, since $\frac{a_{k+1}}{a_k} = \frac{-(k+1)}{-k} = \frac{k+1}{k} \geq 1$. This is clearly undesirable. Essentially, it is because the quotient of two negative numbers is the quotients of their magnitudes.

21. Computing, we get

$$f_1 = \frac{(1+\sqrt{5})^1 - (1-\sqrt{5})^1}{2^1\sqrt{5}} = \frac{1+\sqrt{5}-1-\sqrt{5}}{2\sqrt{5}} = 1$$

$$f_2 = \frac{(1+\sqrt{5})^2 - (1-\sqrt{5})^2}{2^2\sqrt{5}} = \frac{1+2\sqrt{5}+5-1+2\sqrt{5}-5}{4\sqrt{5}} = \frac{4\sqrt{5}}{4\sqrt{5}} = 1$$

$$f_3 = \frac{(1+\sqrt{5})^3 - (1-\sqrt{5})^3}{2^3\sqrt{5}} = \frac{1+3\sqrt{5}+15+5\sqrt{5}-1+3\sqrt{5}-15+5\sqrt{5}}{8\sqrt{5}} = \frac{16\sqrt{5}}{8\sqrt{5}} = 2$$

$$f_4 = \frac{(1+\sqrt{5})^4 - (1-\sqrt{5})^4}{2^4\sqrt{5}} = \frac{1+4\sqrt{5}+30+20\sqrt{5}+25-1+4\sqrt{5}-30+20\sqrt{5}-25}{16\sqrt{5}}$$

$$= \frac{48\sqrt{5}}{16\sqrt{5}} = 3$$

$$f_5 = \frac{(1+\sqrt{5})^5 - (1-\sqrt{5})^5}{2^5\sqrt{5}}$$

$$= \frac{1+5\sqrt{5}+50+50\sqrt{5}+125+25\sqrt{5}-1+5\sqrt{5}-50+50\sqrt{5}-125+25\sqrt{5}}{32\sqrt{5}}$$

$$= \frac{160\sqrt{5}}{32\sqrt{5}} = 5.$$

23. The greatest lower bound property for nonempty subsets of real numbers is: every nonempty subset of $\mathbb{R}$ that is bounded below has a greatest lower bound.

Skills

25. Since it appears that $a_k = 0$ for k odd and $a_k = 1$ for k even, we could have $a_k = \frac{1}{2}(1 + (-1)^k)$ for $k \geq 1$.

27. The terms are in a consistent form except for the third term. However, $\frac{1}{9} = \frac{3}{27}$, and then it is clear that $a_k = \frac{k}{3^k}$ for $k \geq 1$ is a formula for the given terms.

29. The numerators increase by 1 each time, and the denominators by 2. So a plausible formula is $a_k = \frac{k+1}{2k+1}$ for $k \geq 1$.

31. Computing, we get

$$\frac{1 - (-1)^1}{1} = \frac{2}{1} = 2, \quad \frac{1 - (-1)^2}{2} = \frac{0}{2} = 0, \quad \frac{1 - (-1)^3}{3} = \frac{2}{3},$$
$$\frac{1 - (-1)^4}{4} = \frac{0}{4} = 0, \quad \frac{1 - (-1)^5}{5} = \frac{2}{5}.$$

33. Computing gives

$$a_1 = \frac{\cos(1 \cdot x)}{x^1 + 1^2} = \frac{\cos x}{x + 1}, \quad a_2 = \frac{\cos(2 \cdot x)}{x^2 + 2^2} = \frac{\cos 2x}{x^2 + 4}, \quad a_3 = \frac{\cos(3 \cdot x)}{x^3 + 3^2} = \frac{\cos 3x}{x^3 + 9},$$
$$a_4 = \frac{\cos(4 \cdot x)}{x^4 + 4^2} = \frac{\cos 4x}{x^4 + 16}, \quad a_5 = \frac{\cos(5 \cdot x)}{x^5 + 5^2} = \frac{\cos 5x}{x^5 + 25}.$$

35. Computing gives

$$\frac{\sqrt{1}}{1} = 1, \quad \frac{\sqrt{2}}{2}, \quad \frac{\sqrt{3}}{3}, \quad \frac{\sqrt{4}}{4} = \frac{1}{2}, \quad \frac{\sqrt{5}}{5}.$$

37. It is clear that 2 is an upper bound. Since $\frac{1}{(k+1)^2} < \frac{1}{k^2}$, we see that $2 - \frac{1}{k^2} < 2 - \frac{1}{(k+1)^2}$ so that the sequence is increasing. Since $\lim\limits_{k \to \infty} \frac{1}{k^2} = 0$, the terms of the sequence approach 2. Then 2 is the least upper bound, since for any $r < 2$, there is some k such that $r < 2 - \frac{1}{k^2} < 2$, so that r is not an upper bound.

39. The sequence is $\{-1, -2, -3, \dots\}$. By inspection, -1 is the least upper bound of this sequence: all terms are ≤ -1, and there is a term that is equal to -1.

41. The least upper bound of this sequence is 1. Note first that

$$a_k = \frac{1}{2} + \frac{1}{4} + \cdots + \frac{1}{2^k} = 1 - \frac{1}{2^k}.$$

To see this, note that

$$a_k + \frac{1}{2^k} = \frac{1}{2} + \frac{1}{4} + \cdots + \frac{1}{2^{k-1}} + \frac{1}{2^k} + \frac{1}{2^k}$$
$$= \frac{1}{2} + \frac{1}{4} + \cdots + \frac{1}{2^{k-2}} + \frac{1}{2^{k-1}} + \frac{1}{2^{k-1}}$$
$$= \frac{1}{2} + \frac{1}{4} + \cdots + \frac{1}{2^{k-2}} + \frac{1}{2^{k-2}}$$
$$= \cdots$$
$$= \frac{1}{2} + \frac{1}{2} = 1.$$

Then since $a_k < 1$ for all k, 1 is certainly an upper bound. Since $\lim\limits_{k \to \infty} \frac{1}{2^k} = 0$, the terms a_k approach 1 arbitrarily closely, so that no $r < 1$ can be an upper bound.

43. $\{c, cr, cr^2, cr^3, cr^4\} = \{3, \frac{3}{2}, \frac{3}{4}, \frac{3}{8}, \frac{3}{16}\}.$

45. $\{c, cr, cr^2, cr^3, cr^4\} = \{-2, -2 \cdot -3, -2 \cdot (-3)^2, -2 \cdot (-3)^3, -2 \cdot (-3)^4\} = \{-2, 6, -18, 54, -162\}.$

47. Since the first term is -3 and the difference between successive terms is 10, this sequence is $\{-3 + 10k\}_{k=0}^{\infty}$.

49. Since the first term is 1 and the difference between successive terms is -2, this sequence is $\{1 - 2k\}_{k=0}^{\infty}$.

51. Since

$$a_{k+1} - a_k = ((k+1)^2 - 5(k+1)) - (k^2 - 5k) = k^2 + 2k + 1 - 5k - 5 - k^2 + 5k = 2k - 4,$$

we see that $a_{k+1} - a_k \geq 0$ for $k \geq 2$. Thus this sequence is not monotonically increasing, since $a_2 < a_1$, but it is eventually monotonically increasing.

53. We have

$$a_{k+1} - a_k = \frac{k+1}{k+3} - \frac{k}{k+2} = \frac{(k+1)(k+2) - k(k+3)}{(k+2)(k+3)} = \frac{2}{(k+2)(k+3)} > 0.$$

Since $a_{k+1} - a_k > 0$ for all k, $\{a_k\}$ is strictly increasing.

55. Since all terms are positive, the ratio test applies. Then

$$\frac{a_{k+1}}{a_k} = \frac{(k+1)^2/(k+1)!}{k^2/k!} = \frac{(k+1)^2}{(k+1)!} \cdot \frac{k!}{k^2} = \frac{(k+1)^2}{(k+1)k^2} = \frac{k+1}{k^2}.$$

Now, $\frac{k+1}{k^2} \leq 1$ whenever $k + 1 \leq k^2$, i.e. whenever $k^2 - k - 1 \geq 0$. The roots of this quadratic are $\frac{1 \pm \sqrt{3}}{2}$, so for $k > \frac{1 + \sqrt{3}}{2} \approx 1.36$, the ratio is less than 1. Thus the sequence is not decreasing, since $a_1 = 1 < a_2 = 2$, but it is decreasing for $k \geq 2$ (and thus is eventually decreasing).

57. Since all terms are positive, the ratio test applies. Note that

$$2 \cdot 4 \cdot 6 \ldots (2k) = 2^k (1 \cdot 2 \cdot 3 \ldots k) = 2^k k!,$$

so that

$$\frac{a_{k+1}}{a_k} = \frac{\frac{3^{k+1}}{2^{k+1}(k+1)!}}{\frac{3^k}{2^k k!}} = \frac{3^{k+1}}{2^{k+1}(k+1)!} \cdot \frac{2^k k!}{3^k} = \frac{3}{2(k+1)} < 1.$$

Since the ratio is < 1 for $k \geq 1$, this is a strictly decreasing sequence.

59. Define $a(x) = \frac{\sqrt{x+1}}{x}$; then $a(k) = a_k$. Since

$$a'(x) = \frac{\frac{1}{2}x(x+1)^{-1/2} - (x+1)^{1/2}}{x^2} = \frac{\frac{1}{2}x - (x+1)}{x^2(x+1)^{1/2}} = \frac{-2 - x}{2x^2(x+1)^{1/2}},$$

we see that $a'(x) < 0$ for $x \geq 1$, so that $\{a_k\}$ is decreasing by the derivative test.

61. Define $a(x) = \frac{\sin x}{x}$; then $a(k) = a_k$. Since

$$a'(x) = \frac{x \cos x - \sin x}{x^2},$$

we see that $a'(x)$ changes sign infinitely often: for $x = \frac{\pi}{2} + 2n\pi$, the numerator is -1 while the denominator is positive, and for $x = 2n\pi$, the numerator is $2n\pi > 0$ while the denominator is still positive. Thus $a'(x)$ changes sign infinitely often, so we get no information from the derivative test.

63.

$$a_{k+1} - a_k = \left(2 - \frac{(k+1)-1}{10}\right) - \left(2 - \frac{k-1}{10}\right) = \frac{k-1}{10} - \frac{k}{10} = -\frac{1}{10} < 0.$$

Since the difference is negative, this sequence is decreasing. This sequence is unbounded: since $\frac{k-1}{10}$ is obviously unbounded, its difference with 2 is unbounded as well.

65. Since the terms of a_k alternate in sign, $\{a_k\}$ cannot be either increasing or decreasing. Since $|a_k| = \left|\frac{(-1)^k}{k}\right| = \frac{1}{k} \leq 1$, the sequence is bounded above by 1 and bounded below by -1, so it is bounded.

67. The sequence is neither increasing nor decreasing, by examination. It is bounded above by 1 and bounded below by -1, so it is bounded.

69. Since $\cos k$ changes sign infinitely often, so does $a_k = \frac{\cos k}{k}$. Thus $\{a_k\}$ cannot be either increasing or decreasing. Since $|a_k| = \left|\frac{\cos k}{k}\right| \leq \frac{1}{k} \leq 1$, the sequence is bounded below by -1 and bounded above by 1. Hence it is bounded.

71. Use the ratio test:

$$\frac{a_{k+1}}{a_k} = \frac{e^{k+1}}{(k+1)!} \cdot \frac{k!}{e^k} = \frac{e}{k+1}.$$

The ratio is thus ≤ 1 for $k \geq 2$, since $e \approx 2.718$. Thus this sequence is decreasing for $k \geq 2$; however, since $a_1 = e \approx 2.718 < a_2 = \frac{e^2}{2} \approx 3.695$, it is not a decreasing sequence. Since all terms are positive, the sequence is bounded below by 0. Since it is eventually decreasing, it is bounded above by $a_2 = \frac{e^2}{2}$. Thus the sequence is bounded.

73. Since this sequence alternates signs, it is neither increasing nor decreasing. It is unbounded above since $\lim_{k \to \infty} a_{2k} = \lim_{k \to \infty} (-1)^{2k}(2k) = \lim_{k \to \infty} 2k = \infty$, and unbounded below since $\lim_{k \to \infty} a_{2k+1} = \lim_{k \to \infty} (-1)^{2k+1}(2k+1) = \lim_{k \to \infty} (-2k) = -\infty$.

75. We have $f'(x) = 3x^2$, so that $x_{k+1} = x_k - \frac{x_k^3 - 2}{3x_k^2}$. Then

k	x_k	$|x_k - x_{k-1}|$
0	1	
1	1.33333	0.33333
2	1.26389	0.06944
3	1.25993	0.00396
4	1.25992	0.00001

77. We have $f'(x) = e^x + \cos x$, so that $x_{k+1} = x_k - \frac{e^{x_k} + \sin x_k}{e^{x_k} + \cos x_k}$. Then

k	x_k	$|x_k - x_{k-1}|$
0	-2	
1	-4.75616	2.75616
2	-24.0015	19.2453
3	-26.1281	2.12655
4	-24.5866	1.54146
5	-25.1944	0.60782
6	-25.1327	0.06175
7	-25.1327	0.00008

79. To find $\sqrt{a}$, we wish to find a root of $f(x) = x^2 - a$. Then $f'(x) = 2x$, so that Newton's method says that

$$x_{k+1} = x_k - \frac{x_k^2 - a}{2x_k} = x_k - \frac{1}{2}x_k + \frac{a}{2x_k} = \frac{1}{2}x_k + \frac{1}{2} \cdot \frac{a}{x_k} = \frac{1}{2}\left(x_k + \frac{a}{x_k}\right).$$

81. Using the formula from Exercise 79 with $a = 3$ gives

k	x_k	$|x_k - x_{k-1}|$
0	1	
1	2	1
2	1.75	0.25
3	1.73214	0.017857
4	1.73205	0.000092

83. Using the formula from Exercise 79 with $a = 101$ gives

k	x_k	$|x_k - x_{k-1}|$
0	1	
1	51	50
2	26.4902	24.5098
3	15.1515	11.3387
4	10.9087	4.24272
5	10.0837	0.82506
6	10.0499	0.03375
7	10.0499	0.00006

85. (a) With $f(x) = \sqrt[3]{x-2}$, we get $f'(x) = \frac{1}{3}(x-2)^{-2/3}$, so that $\frac{f(x)}{f'(x)} = \frac{(x-2)^{1/3}}{\frac{1}{3}(x-2)^{-2/3}} = 3(x-2)$.
Then

$$x_{k+1} = x_k - 3(x_k - 2) = -2x_k + 6.$$

Then in the limit, x_{k+1} is roughly $-2x_k$, so that the magnitude of the x_k increase without bound. For example,

$$x_1 = x_0 - 3(x_0 - 2) = 1 - 3(1 - 2) = 4$$
$$x_2 = x_1 - 3(x_1 - 2) = 4 - 3(4 - 2) = -2$$
$$x_3 = x_2 - 3(x_2 - 2) = -2 - 3(-2 - 2) = 10$$
$$x_4 = x_3 - 3(x_3 - 2) = 10 - 3(10 - 2) = -14$$

  (b) $\sqrt[3]{x-2} = 0$ when $x = 2$.

Applications

87. From Example 7, we have $L_1 = D$ and $L_{k+1} = \left(1 - \frac{p}{100}\right)L_k + D$ for $k \geq 1$. In this case, $D = 100$ and $p = 25$, so we get $L_1 = 100$ and

$$L_{k+1} = \left(1 - \frac{25}{100}\right)L_k + 100 = 100 + 0.75L_k.$$

Then

k	1	2	3	4	5	6	7
L_k	100	175	231.25	273.438	305.078	328.809	346.606

89. We have

$$a_0 = 100(1 + 0.05)^0 = 100, \quad a_1 = 100(1 + 0.05)^1 = 100(1.05) = 105,$$

$$a_2 = 100(1.05)^2 = 100(1.1025) = 110.25, \quad a_3 = 100(1.05)^3 = 100(1.15763) \approx 115.76,$$

$$a_4 = 100(1.05)^4 = 100(1.21551) \approx 121.55.$$

a_0 is the amount originally deposited. The sequence is not bounded, since $(1 + 0.05)^k = 1.05^k$ is unbounded as $k \to \infty$. Further, since $1.05^k > 1$ for all $k \geq 1$, the sequence is monotonically increasing (and strictly so).

Proofs

91. Suppose $\{a_k\}$ is both increasing and decreasing. Then for every k, we must have $a_k \leq a_{k+1}$ since the sequence is increasing, and also $a_k \geq a_{k+1}$ since it is decreasing. Thus $a_k = a_{k+1}$ for all k, so the sequence is constant.

93. To show that $\{S_n\}_{n=1}^\infty$ is increasing, it suffices to show, by the difference test, that $S_{k+1} - S_k \geq 0$ for all k. But

$$S_{k+1} - S_k = (a_1 + a_2 + \cdots + a_{k+1}) - (a_1 + a_2 + \cdots + a_k) = a_{k+1} > 0$$

since every a_k is positive by hypothesis.

95. (a) Since $\frac{a_{k+1}}{a_k} \geq 1$ for every $k \geq 1$, clearing fractions gives $a_{k+1} \geq a_k$, so that $a_k \leq a_{k+1}$, for every $k \geq 1$. This means that $\{a_k\}$ is increasing.

    (b) Since $\frac{a_{k+1}}{a_k} > 1$ for every $k \geq 1$, clearing fractions gives $a_{k+1} > a_k$, so that $a_k < a_{k+1}$, for every $k \geq 1$. This means that $\{a_k\}$ is strictly increasing.

    (c) Since $\frac{a_{k+1}}{a_k} \leq 1$ for every $k \geq 1$, clearing fractions gives $a_{k+1} \leq a_k$, so that $a_k \geq a_{k+1}$, for every $k \geq 1$. This means that $\{a_k\}$ is decreasing.

    (d) Since $\frac{a_{k+1}}{a_k} < 1$ for every $k \geq 1$, clearing fractions gives $a_{k+1} < a_k$, so that $a_k > a_{k+1}$, for every $k \geq 1$. This means that $\{a_k\}$ is strictly decreasing.

Thinking Forward

A sequence of sums

Evaluating the sum S_n for $n = 1, 2, \ldots, 10$ gives

n	S_n
1	2.0
2	2.5
3	2.6666666666666666667
4	2.7083333333333333333
5	2.7166666666666666667
6	2.7180555555555555556
7	2.7182539682539682540
8	2.7182787698412698413
9	2.7182815255731922399
10	2.7182818011463844797

Since $e \approx 2.7182818284590452354$, this approximation is correct to within 10^{-7}.

A sequence of polynomials

Graphs of p_1, p_2, p_5, and p_{10} along with e^x are (with the p_n in gray and e^x in black):

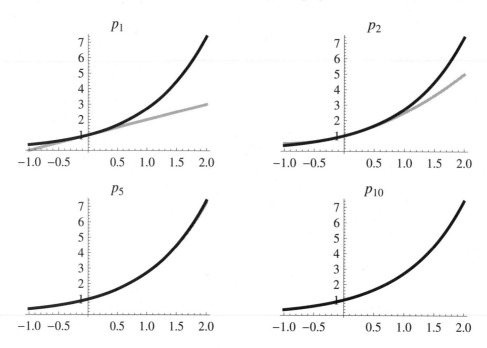

Clearly $p_n(x)$ approaches e^x very rapidly. Note that

$$p_n'(x) = \left(1 + x + \cdots + \frac{x^n}{n!}\right)' = 1 + x + \cdots + \frac{x^{n-1}}{(n-1)!} = p_{n-1}(x).$$

Then $p_n'(0) = 1 = (e(x))'\big|_{x=0}$. Thus each p_n is an approximation to e^x with matching first derivatives.

7.2   Limits of Sequences

Thinking Back

Analyzing the behavior of a continuous function

Since $f'(x) = \frac{e^x(3x^2) - x^3 e^x}{e^{2x}} = \frac{x^2(3-x)}{e^x}$, we see that $f(x)$ is increasing for $x < 3$ and decreasing for $x > 3$. The maxima and minima for f must occur at the critical point, $x = 3$, or at the endpoint, $x = 0$. Since $f'(x)$ changes sign from positive to negative at $x = 3$, that is a local maximum with value $\frac{27}{e^3}$. $x = 0$ is a local minimum, since $f'(x) > 0$ for x near zero, and $f(0) = 0$. Finally, since $f(x) > 0$ for $x > 0$, we see that 0 is a global minimum and 3 is a global maximum, and since exponential functions dominate power functions, $\lim\limits_{x \to \infty} \frac{x^3}{e^x} = 0$.

The asymptotic behavior of a continuous function

▷ Suppose f is increasing and has M as an upper bound. Then it has a least upper bound L, and since L is a *least* upper bound, it must be the case that values of $f(x)$ approach L arbitrarily closely as $x \to \infty$ (otherwise a smaller upper bound would exist). But this means that L is an asymptote as $x \to \infty$.

▷ Since f is increasing, if it has an asymptote (say $y = L$), then it approaches that asymptote arbitrarily closely from below, so that L is the least upper bound for $f(x)$. Thus if f has a horizontal asymptote, it has an upper bound, which shows that lack of an upper bound implies that there is no horizontal asymptote.

▷ If we define a sequence a_k by $a_k = f(k)$, then $\{a_k\}$ converges if and only if $\lim\limits_{x\to\infty} f(x)$ converges, which happens if and only if f has a horizontal asymptote as $x \to \infty$, which (by parts a and b) happens if and only if f has an upper bound. Finally, since both f and $\{a_k\}$ are increasing, f has an upper bound if and only if $\{a_k\}$ has an upper bound. We have thus shown, using parts a and b, that an increasing sequence converges if and only if it has an upper bound.

▷ For example, $f(x) = \frac{\sin x}{x}$. Since $f(x) > 0$ arbitrarily often (say, when $x = \frac{\pi}{2} + n\pi$ and $f(x) < 0$ arbitrarily often (say, when $x = \frac{3\pi}{2} + n\pi$), it cannot be either increasing or decreasing on any interval of the form $[a, \infty)$. However, $-\frac{1}{x} \le \frac{\sin x}{x} \le \frac{1}{x}$, so by the squeeze theorem,

$$0 = \lim_{x\to\infty}\left(-\frac{1}{x}\right) \le \lim_{x\to\infty}\frac{\sin x}{x} \le \lim_{x\to\infty}\frac{1}{x} = 0.$$

Concepts

1. (a) True. Suppose that $\{a_k\}_{k=1}^{\infty}$ is a sequence and that $\{a_k\}_{k=n}^{\infty}$ converges to L. This means that for any ϵ, there is some $N > n$ such that whenever $k > N$ then $|a_k - L| < \epsilon$. But then this statement holds for the original sequence as well, so that it also converges to L.

   (b) False. For example, $a_k = (-1)^k$ for all k, and $b_k = (-1)^{k+1}$ for all k. Then both $\{a_k\}$ and $\{b_k\}$ diverge, yet $a_k + b_k = 0$ for all k, so that $\{a_k + b_k\}$ converges.

   (c) False. For example, $3, 3.1, 3.14, 3.141, 3.1415, \ldots$ converges to π, which is irrational.

   (d) True. For suppose $\{a_k\}$ converges to L, and choose $\epsilon = 1$. Then there is some $N > 1$ such that whenever $n > N$ then $|a_n - L| < 1$, so that $L - 1 < a_n < L + 1$. Then $\{a_k\}$ is bounded above by $\max(a_0, a_1, \ldots, a_N, L + 1)$ and is bounded below by $\min(a_0, a_1, \ldots, a_N, L - 1)$.

   (e) False. For example, $\{1, 0, 1, 0, 1, 0, \ldots\}$ is bounded but does not converge.

   (f) True, since exponential functions dominate power functions.

   (g) True, by Theorem 7.19. The sequence is bounded above by 0, and is monotonic, so it converges.

   (h) False. For example, $\{1, 2, 3, 4, \ldots\}$ does not converge.

3. A sequence is convergent if for large enough indices, the terms of the sequence can be made arbitrarily close to the limit L. Since we are considering only "large enough" indices, any finite portion at the start of the sequence may safely be ignored.

5. Not possible. By Theorem 7.11 (a) and (b), if $\lim\limits_{k\to\infty} a_k = L$ and $\lim\limits_{k\to\infty} b_k = M$, then

$$\lim_{k\to\infty}(a_k - b_k) = \lim_{k\to\infty}a_k + \lim_{k\to\infty}(-b_k) = \lim_{k\to\infty}a_k - \lim_{k\to\infty}b_k = L - M.$$

7. Let $\{a_k\} = \{b_k\} = \{1, 0, 1, 0, 1, 0, \ldots\}$. Then $\{a_k \cdot b_k\} = \{1, 0, 1, 0, 1, 0, \ldots\}$ as well, and none of the three sequences converges.

9. Let $a_k = e^k$ and $b_k = k^2$. Then clearly both $\{a_k\}$ and $\{b_k\}$ diverge since they are unbounded, and $\{\frac{a_k}{b_k}\} = \{\frac{e^k}{k^2}\}$ diverges as well since exponential functions dominate power functions.

11. Let $a_k = 1$ and $b_k = \frac{1}{k}$. Then both $\{a_k\}$ and $\{b_k\}$ converge, but $\{\frac{a_k}{b_k}\} = \{k\}$ diverges.

13. For example, $a_k = \frac{\sin\left(\frac{\pi}{2}k\right)}{k}$. When $k = 1, 5, 9, 13, \ldots, a_k > 0$; when $k = 3, 7, 11, 15, \ldots, a_k < 0$. Thus a_k is not monotonic, since it changes sign infinitely often. However, by the Squeeze Theorem, Theorem 7.12(c), since

$$-\frac{1}{k} \leq \frac{\sin\left(\frac{\pi}{2}k\right)}{k} \leq \frac{1}{k},$$

and the sequences on the ends converge to 0, so does the sequence in the middle.

15. Such a sequence cannot exist. Since the sequence is decreasing, $a_1 \geq a_2 \geq a_3 \geq \ldots$, so that a_1 is an upper bound for the sequence.

17. For example, let $a_k = k!$; then $\frac{k!}{a_k} = 1$, and $\{1, 1, 1, \ldots\}$ converges. For a more complicated example, let $a_k = k^k$. Then with some work it can be seen that $\frac{k!}{a_k}$ is decreasing since the ratio of successive terms is < 1, and bounded below since all terms are positive, so it must converge.

19. Since exponential growth dominates power growth, this means that $\lim_{k\to\infty} \frac{k^3}{e^k} = 0$ since the numerator is a power function and the denominator is an exponential function.

21. By Theorem 7.11(a), if $\{r^k\}$ converges to L, then $\{cr^k\}$ converges to cL. Further, by Theorem 7.15, $\{r^k\}$ converges for $r > 0$ if and only if $r \leq 1$, in which case it converges to 1 for $r = 1$ and to 0 otherwise, and in particular it is bounded. If $r > 1$, then $\{r^k\}$ diverges and is unbounded. So:

   - If $r > 1$, then $\{cr^k\}$ diverges and is unbounded above since $c > 0$.
   - If $r = 1$, then $\{cr^k\} \to c$ since $\{r^k\} \to 1$. Thus $\{cr^k\}$ is bounded.
   - If $r < 1$, then $\{cr^k\} \to 0$ since $\{r^k\} \to 0$. Thus $\{cr^k\}$ is bounded.

Skills

23. This is the constant sequence, so it is bounded both above and below by 11, and is monotonic. It converges to 11.

25. Since $|\sin z| \leq 1$, this sequence is bounded above by 1 and below by -1. However, since

$$\sin\left(\frac{k\pi}{2}\right) = \begin{cases} 0, & k \text{ even}, \\ 1, & k = 1, 5, 9, 13, 17, \ldots, \\ -1, & k = 3, 7, 11, 15, 19, \ldots, \end{cases}$$

the sequence diverges. It is not monotonic nor eventually monotonic.

27. Since $\left|\frac{(-1)^k}{k+6}\right| = \frac{1}{k+6} \leq \frac{1}{7}$, the sequence is bounded in magnitude by $\frac{1}{7}$, so is bounded both above and below. Further, the absolute value of the sequence, $\{\frac{1}{k+6}\}$, converges to zero, so that by Theorem 7.12(d), so does the original sequence. However, since the signs of successive terms alternate, the sequence is not monotonic nor eventually monotonic.

29. All terms of this sequence are negative, since for $k \geq 1$, $3^{-k} < 2$. Thus the sequence is bounded above by zero. Since $3^{-k} > 0$ for all k, it is also bounded below by -2. Further,

$$\lim_{k\to\infty}\left(3^{-k} - 2\right) = \left(\lim_{k\to\infty} 3^{-k}\right) - 2 = -2.$$

Finally, since $3^{-(k+1)} = \frac{1}{3}3^{-k} < 3^{-k}$, we see that $a_{k+1} < a_k$, so that the sequence is monotonically decreasing.

31. We have $\frac{k-2^{-k}}{2^k} = \frac{k}{2^k} - 2^{-2k} = \frac{k}{2^k} - \left(\frac{1}{4}\right)^k$. The first term converges to zero since exponential functions dominate power functions. The second term converges to zero by Theorem 7.15. Thus by Theorem 7.12, the sequence converges to zero and thus is bounded both above and below. To determine monotonicity, note that

$$a_{k+1} - a_k = \frac{(k+1) - 2^{-(k+1)}}{2^{k+1}} - \frac{k - 2^{-k}}{2^k}$$

$$= \frac{(k+1) - 2^{-1-k} - 2(k - 2^{-k})}{2^{k+1}}$$

$$= \frac{1 - k - 2^{-1-k} + 2^{1-k}}{2^{k+1}}$$

$$= \frac{1 - k + 3 \cdot 2^{-1-k}}{2^{k+1}}.$$

For $k \geq 2$, we see that $1 - k < -1$ while $3 \cdot 2^{-1-k} < 1$, so that the numerator is negative while the denominator is always positive. Thus for $k \geq 2$, the sequence is decreasing, so it is eventually monotonically decreasing.

33. Since $k \leq 2 + k$ for all k, this sequence is bounded above by 1. Since all terms are positive, it is bounded below by 0. Additionally,

$$\lim_{k \to \infty} \frac{k}{2 + k} = \lim_{k \to \infty} \frac{1}{1 + 2/k} = 1.$$

To determine monotonicity, note that

$$a_{k+1} - a_k = \frac{k+1}{k+3} - \frac{k}{k+2} = \frac{(k+1)(k+2) - k(k+3)}{(k+2)(k+3)} = \frac{2}{(k+2)(k+3)} > 0,$$

so that the sequence is monotonically increasing.

35. We have

$$\lim_{k \to \infty} \left(1 + \frac{1}{k}\right)^2 = \lim_{k \to \infty} \left(1 + \frac{2}{k} + \frac{1}{k^2}\right) = 1$$

since both $\frac{1}{k}$ and $\frac{1}{k^2}$ converge to zero. Thus the sequence converges, and thus is also bounded both above and below. To determine monotonicity,

$$\frac{a_{k+1}}{a_k} = \frac{\left(1 + \frac{1}{k+1}\right)^2}{\left(1 + \frac{1}{k}\right)^2} = \left(\frac{\frac{k+2}{k+1}}{\frac{k+1}{k}}\right)^2 = \left(\frac{k(k+2)}{(k+1)^2}\right)^2 = \left(\frac{k^2 + 2k}{k^2 + 2k + 1}\right)^2,$$

which is clearly < 1. Thus the sequence is monotonically decreasing.

37. By Theorem 7.16, factorials dominate exponential functions, so this sequence converges to zero (or, simply apply Theorem 7.17). Hence it is also bounded both above and below. To determine monotonicity, note that

$$\frac{a_{k+1}}{a_k} = \frac{2^{k+1}}{(k+1)!} \cdot \frac{k!}{2^k} = \frac{2}{k+1}.$$

Since this is ≤ 1 when $k \geq 2$, this sequence is eventually monotonically decreasing.

39. By Theorem 7.18, $\{3^{1/k}\} \to 1$, so that the sequence converges and thus is also bounded both above and below. Since

$$\frac{a_{k+1}}{a_k} = \frac{3^{1/(k+1)}}{3^{1/k}} = 3^{1/(k+1) - 1/k} = 3^{-1/(k(k+1))} < 1,$$

the sequence is monotonically decreasing.

41. Writing out terms of this sequence gives $\{42 - 1, 42 + 1, 42 - 1, 42 + 1, \dots\} = \{41, 43, 41, 43, \dots\}$. This does not converge since its terms alternate between 41 and 43. However, it is bounded above by 43 and below by 41. Since the terms alternate between two values, the sequence is neither monotonic nor eventually monotonic.

43. Since

$$\lim_{k \to \infty} \left(\frac{2k^2}{k^2 + 2k + 2} \right)^3 = \lim_{k \to \infty} \left(\frac{2}{1 + 2/k + 2/k^2} \right)^3 = \left(\lim_{k \to \infty} \frac{2}{1 + 2/k + 2/k^2} \right)^3 = 2^3 = 8,$$

this sequence converges to 8, and thus also is bounded both above and below. To see that it is monotonically increasing, note that

$$\frac{2(k+1)^2}{(k+1)^2 + 2(k+1) + 2} - \frac{2k^2}{k^2 + 2k + 2} = \frac{4(k^2 + 3k + 1)}{(k^2 + 2k + 2)(k^2 + 4k + 5)},$$

which is > 0 for $k \geq 1$. Thus $\sqrt[3]{a_{k+1}} > \sqrt[3]{a_k}$, and thus $a_{k+1} > a_k$ so that the sequence is monotonically increasing.

45. Since

$$\lim_{k \to \infty} \frac{k}{k^2 - 2k + 50} = \lim_{k \to \infty} \frac{1}{k - 2 + 50/k} = 0,$$

the sequence converges to $42 - 0 = 42$, so is also bounded both above and below. To check monotonicity,

$$a_{k+1} - a_k = \left(42 - \frac{(k+1)}{(k+1)^2 - 2(k+1) + 50} \right) - \left(42 - \frac{k}{k^2 - 2k + 50} \right)$$

$$= \frac{k}{k^2 - 2k + 50} - \frac{(k+1)}{(k+1)^2 - 2(k+1) + 50}$$

$$= \frac{k^2 + k - 50}{(k^2 + 49)(k^2 - 2k + 50)}.$$

$k^2 + k - 50 > 0$ for $k > 6$. In the denominator, $k^2 + 49$ is always positive, and $k^2 - 2k + 50 = k^2 - 2k + 1 + 49 = (k-1)^2 + 49$ is always positive as well, so the denominator is always positive. Thus $a_{k+1} - a_k > 0$ for $k > 6$ and the sequence is eventually monotonically increasing.

47. Note that

$$\frac{1 \cdot 3 \cdot 5 \cdots (2k-1)}{10^k} = \frac{1 \cdot 2 \cdot 3 \cdot 4 \cdots (2k-1) \cdot (2k)}{10^k (2 \cdot 4 \cdot 6 \cdots (2k))} = \frac{(2k)!}{10^k \cdot 2^k \cdot (1 \cdot 2 \cdot 3 \cdots k)} = \frac{(2k)!}{20^k \cdot k!}.$$

Then

$$\lim_{k \to \infty} \frac{a_{k+1}}{a_k} = \lim_{k \to \infty} \frac{(2k+2)!}{20^{k+1} \cdot (k+1)!} \cdot \frac{20^k \cdot k!}{(2k)!} = \lim_{k \to \infty} \frac{(2k+2)(2k+1)}{20(k+1)} = \lim_{k \to \infty} \frac{2k+1}{10} = \infty.$$

Since the ratio of successive terms increases without bound, and the terms are positive, the terms themselves increase without bound, so the sequence diverges. Since the ratio of successive terms is eventually greater than 1, the sequence is eventually monotonically increasing. Finally, the sequence is bounded below by 0 and is unbounded above.

49. Let $\{m_k\} = \{-\frac{1}{k}\}$ and $\{M_k\} = \{\frac{1}{k}\}$. Then $m_k \leq \frac{\sin k}{k} \leq M_k$, so that

$$0 = \lim_{k \to \infty} m_k \leq \lim_{k \to \infty} \frac{\sin k}{k} \leq \lim_{k \to \infty} M_k = 0,$$

and thus by Theorem 7.12(c), $\lim_{k \to \infty} \frac{\sin k}{k} = 0$. However, since $\sin k$ changes sign infinitely often, the sequence is neither monotonic nor eventually monotonic.

51. We have

$$\frac{a_{k+1}}{a_k} = \frac{\frac{((k+1)!)^2}{(2(k+1))!}}{\frac{(k!)^2}{(2k)!}} = \frac{(k+1)!(k+1)!}{(2k+2)!} \cdot \frac{(2k)!}{k!k!} = \frac{(k+1)^2}{(2k+1)(2k+2)} = \frac{k+1}{2(2k+1)}$$

Now, this ratio is clearly < 1 for $k \geq 1$, so that the sequence is monotonically decreasing. Also, since

$$\frac{a_{k+1}}{a_k} = \frac{k+1}{2(2k+1)} < \frac{k+1}{2(k+1)} = \frac{1}{2},$$

and $a_1 = \frac{1}{2}$, we see that $0 \leq a_k \leq \left(\frac{1}{2}\right)^k$, so by Theorem 7.12(c), $\lim\limits_{k\to\infty} a_k = 0$. Thus $\{a_k\}$ converges, so also is bounded both above and below.

53. Since factorials dominate exponential functions, it follows that for any $b > 1$, $\lim\limits_{k\to\infty} (k!)^{1/k} \geq$ $\lim\limits_{k\to\infty} (b^k)^{1/k} = \lim\limits_{k\to\infty} b = b$. Hence the limit of the sequence is question is $> b$ for any $b > 1$, so it is infinite. The sequence diverges and is unbounded above; it is bounded below by zero.

55. Since $-1 \leq \sin x \leq 1$, it follows that $-\frac{1}{2^k} \leq \frac{\sin k}{2^k} \leq \frac{1}{2^k}$. Since $\lim\limits_{k\to\infty} \left(-\frac{1}{2^k}\right) = \lim\limits_{k\to\infty} \frac{1}{2^k} = 0$, the Squeeze Theorem for Sequences (Theorem 7.12(c)) tells us that $\lim\limits_{k\to\infty} \frac{\sin k}{2^k} = 0$.

57. Using Theorem 7.11 (d),

$$\lim_{k\to\infty} \frac{k-7}{3k+5} = \lim_{k\to\infty} \frac{1-7/k}{3+5/k} = \frac{\lim\limits_{k\to\infty}(1-7/k)}{\lim\limits_{k\to\infty}(3+5/k)} = \frac{1}{3}.$$

59. Using algebra together with Theorem 7.11 (d),

$$\begin{aligned}
\lim_{k\to\infty}\left(\sqrt{k^2+k}-k\right) &= \lim_{k\to\infty} \frac{(\sqrt{k^2+k}-k)(\sqrt{k^2+k}+k)}{\sqrt{k^2+k}+k} \\
&= \lim_{k\to\infty} \frac{k^2+k-k^2}{\sqrt{k^2+k}+k} \\
&= \lim_{k\to\infty} \frac{k}{\sqrt{k^2+k}+k} \\
&= \lim_{k\to\infty} \frac{1}{\frac{1}{k}\sqrt{k^2+k}+1} \\
&= \lim_{k\to\infty} \frac{1}{\sqrt{1+1/k}+1} \\
&= \frac{\lim\limits_{k\to\infty} 1}{\lim\limits_{k\to\infty}\left(\sqrt{1+1/k}+1\right)} \\
&= \frac{1}{2}.
\end{aligned}$$

Applications

61. Your bank balance after one year, which is a_k for each of the given values of k, will be

k	1	2	6	12	52	365
a_k	\$106	\$106.09	\$106.15	\$106.17	\$106.18	\$106.18

It appears that the sequence is bounded above by approximately \$106.18.

Proofs

63. It suffices to show that the sequence is unbounded. Consider the even-numbered terms; these terms are $r^{2k} = |r|^{2k}$. But for $|r| > 1$, the sequence $\{|r|^k\}$ is monotonically increasing and unbounded. Thus the subsequence $\{|r|^{2k}\}$ is also unbounded, so the original sequence is as well and thus diverges.

65. $\lim\limits_{k\to\infty} k^{1/k}$ has the form ∞^0, which is an indeterminate form. Define $f(x) = x^{1/x}$. If we show $\lim\limits_{x\to\infty} x^{1/x} = 1$, then by Theorem 7.9, we also have $\lim\limits_{k\to\infty} k^{1/k} = 1$. But taking logs gives $\frac{1}{x}\ln x$, and

$$\lim_{x\to\infty} \frac{\ln x}{x} \overset{\text{L'H}}{=} \lim_{x\to\infty} \frac{1/x}{1} = \lim_{x\to\infty} \frac{1}{x} = 0.$$

Since $\lim\limits_{x\to\infty} \ln\left(x^{1/x}\right) = 0$, it follows that $\lim\limits_{x\to\infty} x^{1/x} = e^0 = 1$ as desired, using the Composition Rule from Theorem 1.20.

67. To prove that $\{ca_k\} \to cL$, we must show that by taking k large enough, we can get ca_k as close to cL as desired. More precisely, for $\epsilon > 0$, we must show that there exists some $N > 0$ such that if $k > N$, then $|ca_k - cL| < \epsilon$. First, if $c = 0$, then $\{ca_k\} = \{0,0,0,\dots\} \to 0 = 0L$. So assuming $c \neq 0$, since $\{a_k\} \to L$, choose N such that for $k > N$ we have $|a_k - L| < \frac{\epsilon}{|c|}$. Then for $k > N$,

$$|ca_k - cL| = |c(a_k - L)| = |c|\,|a_k - L| < |c|\cdot\frac{\epsilon}{|c|} = \epsilon,$$

and we are done.

69. To prove that $\{a_k b_k\} \to LM$, we must show that by taking k large enough, we can get $a_k b_k$ as close to LM as desired. More precisely, for $\epsilon > 0$, we must show that there exists some $N > 0$ such that if $k > N$ then $|a_k b_k - LM| < \epsilon$. Assume that $L > 0$ and $M > 0$; other cases are similar. Choose $\epsilon' > 0$ so that $\epsilon' < L$ and $(L+M)\epsilon' + (\epsilon')^2 < \epsilon$, and choose N_1, N_2 such that if $k > N_1$ then $|a_k - L| < \epsilon'$ and if $k > N_2$ then $|b_k - M| < \epsilon'$. Then let $N = \max(N_1, N_2)$. Note that for $k > N$, we have $|a_k - L| < \epsilon'$, so that $L - \epsilon' < a_k < L + \epsilon'$ and thus (since $L > \epsilon'$) $|a_k| < L + \epsilon'$. Then (using the triangle inequality) for $k > N$ we have

$$\begin{aligned}
|a_k b_k - LM| &= |a_k b_k - a_k M + a_k M - LM| \\
&= |a_k(b_k - M) + M(a_k - L)| \\
&\leq |a_k(b_k - M)| + |M(a_k - L)| \\
&= |a_k|\,|b_k - M| + M\,|a_k - L| \\
&< (L+\epsilon')\epsilon' + M\epsilon' = (L+M)\epsilon' + (\epsilon')^2 < \epsilon,
\end{aligned}$$

and we are done.

71. Suppose by way of contradiction that $\{a_k\} \to L$ and $\{a_k\} \to M$ with $L \neq M$. We may assume that $L > M$ (otherwise reverse L and M). Then $L = M + r$ for some real number $r > 0$. Let $\epsilon = \frac{r}{2}$, and note that $(L - \epsilon, L + \epsilon) \cap (M - \epsilon, M + \epsilon)$ is empty. Since $\{a_k\} \to L$, we can find $N_1 > 0$ such that for $k > N_1$ we have $|a_k - L| < \epsilon$; since $\{a_k\} \to M$, we can find $N_2 > 0$ such that for $k > N_2$ we have $|a_k - M| < \epsilon$. Now let $N = \max(N_1, N_2)$. Then for $k > N$, we have both $|a_k - L| < \epsilon$, so that $a_k \in (L - \epsilon, L + \epsilon)$ and also $|a_k - M| < \epsilon$, so that $a_k \in (M - \epsilon, M + \epsilon)$. But the intersection of these two intervals is empty, so this is impossible. Thus $L = M$.

73. To show that $\{a_k\} \to L$, we must show that for any $\epsilon > 0$ there exists an $N > 0$ such that for $k > N$ we have $|a_k - L| < \epsilon$. Choose $\epsilon > 0$. Since $\{m_k\} \to L$, we can choose N_1 such that for $k > N_1$ we

have $|m_k - L| < \epsilon$, or $L - \epsilon < m_k < L + \epsilon$. Similarly we can choose N_2 such that for $k > N_2$ we have $|M_k - L| < \epsilon$, or $L - \epsilon < M_k < L + \epsilon$. Let $N = \max(N_1, N_2)$. Then for $k > N$ we have

$$L - \epsilon < m_k \le a_k \le M_k < L + \epsilon,$$

so that $|a_k - L| < \epsilon$, which is what was to be shown.

75. To show that $\{a_k\} \to L$ we need to show that for any $\epsilon > 0$ there is an N such that for $k > N$ we have $|a_k - L| < \epsilon$. Choose $\epsilon > 0$. Since $\lim\limits_{x \to \infty} a(x) = L$, we can choose N such that whenever $x > N$ we have $|a(x) - L| < \epsilon$. But then for $k > N$,

$$|a_k - L| = |a(k) - L| < \epsilon,$$

which is what was to be shown.

77. $\lim\limits_{k \to \infty} a_k = \infty$ means that for any $M > 0$ we can choose $N > 0$ such that if $k > N$ then $a_k > M$. We want to show that $\{\frac{1}{a_k}\} \to 0$; this means that for any $\epsilon > 0$ we can choose $N > 0$ such that if $k > N$ then $\left|\frac{1}{a_k} - 0\right| = \left|\frac{1}{a_k}\right| < \epsilon$. So choose $\epsilon > 0$ and let $M = \frac{1}{\epsilon}$. Choose $N > 0$ such that if $k > N$ then $a_k > M > 0$. Then $0 < \frac{1}{a_k} < \frac{1}{M}$, so that $\left|\frac{1}{a_k}\right| < \frac{1}{M} = \epsilon$, as was to be shown.

79. Let $f(x) = x^{-p}$. Then $f(k) = k^{-p}$, and $\lim\limits_{x \to \infty} f(x) = 0$, so that $\{k^{-p}\} \to 0$ as well.

81. The fact that $\{r^k\}$ diverges is the content of Theorem 7.15(a). For the second part, there are many possible proofs. For example, note that $\{|r|^k\} \gg \{k\}$ by Theorem 7.16; since $\{k\} \to \infty$, we must have $\{|r|^k\} \to \infty$ as well.

83. From Theorem 7.16, $k^a \ll b^k$ for $a > 0$ and $b > 1$. If $r > 0$, then since $p > 0$, we have $k^r \ll (1+p)^k$, so that $\{\frac{k^r}{(1+p)^k}\} \to 0$ by definition of $\ll$. If $r \le 0$, then

$$\frac{k^r}{(1+p)^k} = \frac{1}{k^{|r|}(1+p)^k} \le \frac{1}{(1+p)^k} = \left(\frac{1}{1+p}\right)^k.$$

By Theorem 7.15(d), this sequence converges to zero. So the sequence converges to zero in all cases.

85. Suppose that $\{a_k\} \to L$ since it is convergent. Choose $\epsilon > 0$, and choose an integer N such that if $k > N$ then $|a_k - L| < \epsilon$. This means that $L - \epsilon < a_k < L + \epsilon$ whenever $k > N$, so that a_k is bounded in absolute value by $\max(|L - \epsilon|, |L + \epsilon|)$ for $k > N$. Now let K be the maximum of the earlier elements of a_k: $K = \max(|a_1|, |a_2|, \ldots, |a_{N-1}|)$, and let $M = \max(K, |L - \epsilon|, |L + \epsilon|)$. Then clearly $|a_k| \le M$ for all k.

87. Choose $\epsilon > 0$. The statement that $\{a_{2n}\} \to L$ means that there exists an N_1 such that whenever $2n > N_1$ then $|a_{2n} - L| < \epsilon$. The statement that $\{a_{2n+1}\} \to L$ means that there exists an N_2 such that whenever $2n + 1 > N_1$ then $|a_{2n+1} - L| < \epsilon$. Let $N = \max(N_1, N_2)$. Then whenever $k > N$, if k is even then $|a_k - L| = \left|a_{2(k/2)} - L\right| < \epsilon$ since $\{a_{2n}\} \to L$, while if k is odd, then $|a_k - L| = \left|a_{2((k-1)/2)+1} - L\right| < \epsilon$ since $\{a_{2n+1}\} \to L$. Thus $|a_k - L| < \epsilon$, and $\{a_k\} \to L$.

Thinking Forward

A sequence of sums

Computing the first few terms gives $a_0 = 1$, $a_1 = \frac{3}{2}$, $a_2 = \frac{7}{4}$, $a_3 = \frac{15}{8}$, $a_4 = \frac{31}{16}$, so that it appears that $a_n = \frac{2^{n+1}-1}{2^n} = 2 - \frac{1}{2^n}$. To prove this, use induction on n. We just showed that the formula holds for $0 \le n \le 4$. Suppose it holds for $n = k$. Then

$$a_{k+1} = a_k + \frac{1}{2^{k+1}} = 2 - \frac{1}{2^k} + \frac{1}{2^{k+1}} = 2 + \frac{-2+1}{2^{k+1}} = 2 - \frac{1}{2^{k+1}}$$

as desired. The sequence is obviously bounded below by 1 since each term is a sum of positive numbers and the first term is 1. From the above closed form, it is also bounded above by 2, since the value of a term is 2 minus a positive number. Finally, since $a_{k+1} - a_k = \frac{1}{2^{k+1}} > 0$, it is monotonically increasing. Thus it is bounded and monotone, so that it converges. In fact it converges to 2 since $\lim\limits_{k \to \infty} \frac{1}{2^{k+1}} = 0$.

Another sequence of sums

The graph below shows the graph of $f(x) = \frac{1}{x}$ together with the function $a(x)$ which is equal to $\frac{1}{k}$ if $x \in [k, k+1)$:

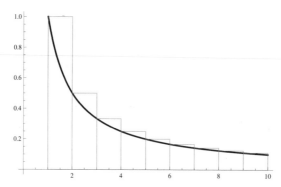

Since the area of each rectangle is $1 \cdot \frac{1}{k} = \frac{1}{k}$, it is clear that the sum of the areas of the first k rectangles (which is in fact a left sum approximation to $\int_1^{k+1} f(x)\,dx$) exceeds the value of $\int_1^{k+1} f(x)\,dx$ — that is

$$1 + \frac{1}{2} + \cdots + \frac{1}{k} = a_k \ge \int_1^{k+1} \frac{1}{x}\,dx.$$

But

$$\lim_{k \to \infty} \int_1^{k+1} \frac{1}{x}\,dx = \lim_{k \to \infty} [\ln x]_1^{k+1} = \lim_{k \to \infty} \ln(k+1) = \infty,$$

so that $\{a_k\} \to \infty$.

7.3   Series

Thinking Back

Working with sequences

▷ Since $\{r^{n+1}\} \to 0$ for $0 < r < 1$ by Theorem 7.15, $\{1 - r^{n+1}\} \to 1$ for $0 < r < 1$.

▷ Since $\{r^{n+1}\} \to \infty$ for $r > 1$ by Theorem 7.15, $\{1 - r^{n+1}\} \to -\infty$ for $r > 1$.

▷ Since $\{r^{n+1}\} \to 0$ for $-1 < r < 0$ by Theorem 7.15, $\{1 - r^{n+1}\} \to 1$ for $-1 < r < 0$.

▷ Since $\{r^{n+1}\}$ diverges for $r < -1$ and is unbounded both above and below by Theorem 7.15, $\{1 - r^{n+1}\}$ diverges and is unbounded both above and below for $r < -1$.

▷ Let $f(x) = x^r$. Then $f(n) = n^r$, and $\lim_{x\to\infty} f(x) = \infty$, so that $\{n^r\} \to \infty$ as well.

▷ Let $f(x) = x^{-r}$. Then $f(x) = n^{-r}$, and $\lim_{x\to\infty} f(x) = 0$, so that $\{n^{-r}\} \to 0$ as well.

Convergence of an improper integral

The integral converges to L if $\lim_{t\to\infty} \int_1^t f(x)\,dx = L$. The integral diverges if $\lim_{t\to\infty} \int_1^t f(x)\,dx$ does not exist.

Convergence of a sequence

A sequence converges to a limit L if the terms of the sequence are arbitrarily close to L if you go far enough out in the sequence. More precisely, $\{a_n\} \to L$ means that for any $\epsilon > 0$, there is some $N > 0$ such that if $k > N$ then $|a_k - L| < \epsilon$. A sequence diverges if it does not converge to any finite limit. Thus for example it could diverge to infinity, or it could alternate between two values, or any of many other possibilities.

Concepts

1. (a) False. For example, $\{1, 1, 1, \dots\}$ converges to 1, but $\sum_{n=1}^k a_n = k$, and $\{k\}$ diverges.

    (b) False. For example, let $a_k = (-1)^k$. Then its sequence of partial sums is $S_1 = -1$, $S_2 = 0$, $S_3 = -1$, $S_4 = 0, \dots$. Thus the sequence of partial sums alternates between 0 and -1.

    (c) False. For example, let $\sum_{k=1}^\infty a_k$ be any divergent series, and let $b_k = -a_k$. Then $\sum_{k=1}^\infty b_k$ is divergent as well, but since $a_k + b_k = 0$ for all k, we see that $\sum_{k=1}^\infty (a_k + b_k) = \sum_{k=1}^\infty 0 = 0$ is convergent.

    (d) True. By Theorem 7.24 (a) and (b) we have

    $$\sum_{k=1}^\infty (ca_k + db_k) = \sum_{k=1}^\infty ca_k + \sum_{k=1}^\infty db_k = c\sum_{k=1}^\infty a_k + d\sum_{k=1}^\infty b_k.$$

    (e) False. Suppose $a_k = -1$ for $1 \le k < 100$. Then

    $$\sum_{k=100}^\infty a_k = -a_1 - a_2 - \cdots - a_{100} + \sum_{k=1}^{100} a_k = 105.$$

    (f) True. The left-hand side is the sum of $a_N, a_{N+1}, a_{N+2}, \dots$, while the right-hand side is the sum of $a_{N+0}, a_{N+1}, a_{N+2}, \dots$, which is the same as $a_N, a_{N+1}, a_{N+2}, \dots$.

    (g) True. $\sum_{k=0}^\infty cr^k = c\sum_{k=0}^\infty r^k$, and by Theorem 7.26, this converges if and only if $|r| < 1$. But if $|r| < 1$, then
    $$\lim_{k\to\infty} cr^k = c\lim_{k\to\infty} r^k = c\cdot 0 = 0.$$

    (h) False. This is a geometric series with $a_1 = c = 4$ and $r = \frac{1}{2}$, so that $\sum_{k=1}^\infty a_k = \frac{4}{1-\frac{1}{2}} = 8$.

3. The sequence of partial sums for $\sum_{k=1}^\infty a_k$ is the sequence $\{S_n\}$ where $S_n = \sum_{k=1}^n a_k$. Since intuitively we want to say that $\sum_{k=1}^\infty a_k = \lim_{n\to\infty} \sum_{k=1}^n a_k$, the concept of partial sums naturally falls out of this intuition.

5. We have

$$\sum_{k=1}^{\infty} a_k = \sum_{k=1}^{9} a_k + \sum_{k=10}^{\infty} a_k.$$

Thus, if we knew that $\sum_{k=10}^{\infty} a_k = 42$, we would need to know $\sum_{k=1}^{9} a_k$ in order to know $\sum_{k=1}^{\infty} a_k$.

7. Suppose we want the index of the sum to start at N. We will write the new sum using j as the index letter, so we want the sum to look like $\sum_{j=N}^{\infty} b_j$. Thus we want $b_N = a_1$, so let $b_j = a_{j-(N-1)} = a_{j-N+1}$. Then $\sum_{k=1}^{\infty} a_k = \sum_{j=N}^{\infty} a_{j-N+1}$.

9. A geometric series is a series whose terms form a geometric progression. By Theorem 7.26, a geometric series $\sum_{k=1}^{\infty} cr^k$ converges if and only if $|r| < 1$ (or if $c = 0$).

11. Assume that some term a_n in the sum $\sum_{k=1}^{\infty} a_k$ is zero. Then since the series is geometric, all subsequent terms in the underlying geometric sequence, being a_n times some power of the ratio r, are also zero. Thus $\sum_{k=1}^{\infty} a_k = \sum_{k=1}^{n-1} a_k$, which converges. Thus if the series diverges, no term can be zero.

13. A geometric series with first term c and ratio $0 < r < 1$ converges to $\frac{c}{1-r}$. Since we want $\frac{c}{1-r} = 1$, we must have $c = 1 - r$. So, for example, choose $c = r = \frac{1}{2}$ to get the series

$$\sum_{k=0}^{\infty} cr^k = \sum_{k=0}^{\infty} \frac{1}{2} \left(\frac{1}{2} \right)^k = \sum_{k=0}^{\infty} \left(\frac{1}{2} \right)^{k+1} = \sum_{k=1}^{\infty} \left(\frac{1}{2} \right)^k = 1.$$

15. Let $a_k = k$ and $b_k = -k + \frac{1}{2^k}$. Then $\sum_{k=1}^{\infty} a_k$ does not converge — the kth partial sum is at least k (in fact, it is exactly $\frac{k(k+1)}{2}$), so the partial sums are unbounded. Also $\sum_{k=1}^{\infty} b_k$ does not converge since its kth partial sum is always less than $-k + 1$, so its partial sums are unbounded (negatively) as well. However, $a_k + b_k = \frac{1}{2^k}$. For $\sum_{k=1}^{\infty} (a_k + b_k) = \sum_{k=1}^{\infty} \frac{1}{2^k}$, the nth partial sum is

$$S_n = \frac{1}{2} + \frac{1}{4} + \cdots + \frac{1}{2^n}.$$

Note that $S_1 + \frac{1}{2} = 1$ and $S_2 + \frac{1}{4} = 1$. Claim that in general $S_n + \frac{1}{2^n} = 1$. Use induction on n: assume the statement is true for $n = k$. Then

$$S_{k+1} + \frac{1}{2^{k+1}} = \left(S_k + \frac{1}{2^{k+1}} \right) + \frac{1}{2^{k+1}} = S_k + \frac{1}{2^k} = 1.$$

Since $S_n + \frac{1}{2^n} = 1$, we see that $S_n = 1 - \frac{1}{2^n}$. Then the sequence of partial sums converges to 1, since

$$\lim_{n \to \infty} S_n = \lim_{n \to \infty} \left(1 - \frac{1}{2^n} \right) = 1 - \lim_{n \to \infty} \frac{1}{2^n} = 1.$$

17. Let $a_k = \left(\frac{1}{2} \right)^k$ and $b_k = \left(\frac{1}{3} \right)^k$. Then

$$\sum_{k=0}^{\infty} a_k = \frac{1}{1 - \frac{1}{2}} = 2, \qquad \sum_{k=0}^{\infty} b_k = \frac{1}{1 - \frac{1}{3}} = \frac{3}{2}.$$

Now, $a_k b_k = \left(\frac{1}{2} \right)^k \cdot \left(\frac{1}{3} \right)^k = \left(\frac{1}{6} \right)^k$, so that

$$\sum_{k=0}^{\infty} a_k b_k = \frac{1}{1 - \frac{1}{6}} = \frac{6}{5} \neq 2 \cdot \frac{3}{2},$$

so that $\sum_{k=0}^{\infty} a_k b_k \neq \left(\sum_{k=0}^{\infty} a_k \right) \left(\sum_{k=0}^{\infty} b_k \right)$.

19. Let $a_k = 2^k$ and $b_k = 3^k$. Since both $\sum_{k=0}^{\infty} a_k$ and $\sum_{k=0}^{\infty} b_k$ are geometric series with $r > 1$ ($r = 2$ for a_k and $r = 3$ for b_k), they both diverge. However,

$$\sum_{k=0}^{\infty} \frac{a_k}{b_k} = \sum_{k=0}^{\infty} \frac{2^k}{3^k} = \sum_{k=0}^{\infty} \left(\frac{2}{3}\right)^k$$

is a geometric series with ratio $r = \frac{2}{3}$ of absolute value less than 1, so it converges.

Skills

21. We have

$$a_1 = \frac{1}{1^2 + 2} = \frac{1}{3}, \qquad a_2 = \frac{1}{2^2 + 2} = \frac{1}{6}, \qquad a_3 = \frac{1}{3^2 + 2} = \frac{1}{11},$$
$$a_4 = \frac{1}{4^2 + 2} = \frac{1}{18}, \qquad a_5 = \frac{1}{5^2 + 2} = \frac{1}{27}.$$

23. We have

$$a_0 = (-1)^{0+1} = -1, \qquad a_1 = (-1)^{1+1} = 1, \qquad a_2 = (-1)^{2+1} = -1,$$
$$a_3 = (-1)^{3+1} = 1, \qquad a_4 = (-1)^{4+1} = -1.$$

25. Note that $\frac{i!}{(i+1)!} = \frac{1}{i+1}$; then

$$a_0 = \frac{1}{0+1} = 1, \qquad a_1 = \frac{1}{1+1} = \frac{1}{2}, \qquad a_2 = \frac{1}{2+1} = \frac{1}{3},$$
$$a_3 = \frac{1}{3+1} = \frac{1}{4}, \qquad a_4 = \frac{1}{4+1} = \frac{1}{5}.$$

27. We have

$$a_0 = \frac{(-1)^0}{(2 \cdot 0)!} = \frac{1}{1} = 1, \qquad a_1 = \frac{(-1)^1}{(2 \cdot 1)!} = -\frac{1}{2}, \qquad a_2 = \frac{(-1)^2}{(2 \cdot 2)!} = \frac{1}{24},$$
$$a_3 = \frac{(-1)^3}{(2 \cdot 3)!} = -\frac{1}{720}, \qquad a_4 = \frac{(-1)^4}{(2 \cdot 4)!} = \frac{1}{40320}.$$

29. We have

$$S_1 = \frac{1}{3}, \qquad S_2 = \frac{1}{3} + \frac{1}{6} = \frac{1}{2}, \qquad S_3 = \frac{1}{3} + \frac{1}{6} + \frac{1}{11} = \frac{13}{22},$$
$$S_4 = \frac{1}{3} + \frac{1}{6} + \frac{1}{11} + \frac{1}{18} = \frac{64}{99}, \qquad S_5 = \frac{1}{3} + \frac{1}{6} + \frac{1}{11} + \frac{1}{18} + \frac{1}{27} = \frac{203}{297}.$$

31. We have

$$S_0 = -1, \qquad S_1 = -1 + 1 = 0, \qquad S_2 = -1 + 1 - 1 = -1,$$
$$S_3 = -1 + 1 - 1 + 1 = 0, \qquad S_4 = -1 + 1 - 1 + 1 - 1 = -1.$$

33. Note that $\frac{i!}{(i+1)!} = \frac{1}{i+1}$; then

$$S_0 = 1, \qquad S_1 = 1 + \frac{1}{2} = \frac{3}{2}, \qquad S_2 = 1 + \frac{1}{2} + \frac{1}{3} = \frac{11}{6},$$
$$S_3 = 1 + \frac{1}{2} + \frac{1}{3} + \frac{1}{4} = \frac{25}{12}, \qquad S_4 = 1 + \frac{1}{2} + \frac{1}{3} + \frac{1}{4} + \frac{1}{5} = \frac{137}{60}.$$

35. We have

$$S_0 = 1, \qquad S_1 = 1 - \frac{1}{2} = \frac{1}{2}, \qquad S_2 = 1 - \frac{1}{2} + \frac{1}{24} = \frac{13}{24},$$

$$S_3 = 1 - \frac{1}{2} + \frac{1}{24} - \frac{1}{720} = \frac{389}{720}, \qquad S_4 = 1 - \frac{1}{2} + \frac{1}{24} - \frac{1}{720} + \frac{1}{40320} = \frac{4357}{8064}.$$

For the solutions to Exercises 37-42, note that from the proof of Theorem 7.26, if $\sum_{k=0}^{\infty} r^k$ is a geometric series, then the Nth partial sum is

$$S_N = \sum_{k=0}^{N} r^k = \frac{1 - r^{N+1}}{1 - r}.$$

37. This is a geometric series with ratio $r = \frac{1}{2}$, so that

$$S_{100} = \sum_{k=0}^{100} \frac{1}{2^k} = \frac{1 - \left(\frac{1}{2}\right)^{101}}{1 - \frac{1}{2}} = \frac{2^{101} - 1}{2^{101} - 2^{100}} = \frac{2^{101} - 1}{2^{100}} = 2 - \frac{1}{2^{100}}.$$

39. This is a geometric series with ratio $r = 2$, so that

$$S_{1000} = \frac{1 - 2^{1001}}{1 - 2} = 2^{1001} - 1.$$

41. This is a geometric series with ratio $r = \frac{2}{3}$. Thus

$$\sum_{k=0}^{150} \left(\frac{2}{3}\right)^k = S_{150} = \frac{1 - \left(\frac{2}{3}\right)^{151}}{1 - \frac{2}{3}} = 3 - 3\left(\frac{2}{3}\right)^{151}$$

$$\sum_{k=0}^{5} \left(\frac{2}{3}\right)^k = S_5 = \frac{1 - \left(\frac{2}{3}\right)^6}{1 - \frac{2}{3}} = 3 - 3\left(\frac{2}{3}\right)^6.$$

Then

$$\sum_{k=6}^{150} \left(\frac{2}{3}\right)^k = S_{150} - S_5 = \left(3 - 3\left(\frac{2}{3}\right)^{151}\right) - \left(3 - 3\left(\frac{2}{3}\right)^6\right) = 3\left(\frac{2}{3}\right)^6 - 3\left(\frac{2}{3}\right)^{151}.$$

43. $\sum_{k=0}^{\infty} a_k = a_0 + a_1 + \sum_{k=2}^{\infty} a_k = -3 + 5 + 7 = 9.$

45. $\sum_{k=3}^{\infty} a_k = -a_2 + \sum_{k=2}^{\infty} a_k = -(-4) + 7 = 11.$

47. This is a geometric series with first term equal to $c = 3$ and ratio $r = \frac{1}{2}$. Since the ratio is less than 1 in magnitude, the series converges, and it converges to

$$\frac{c}{1 - r} = \frac{3}{1 - \frac{1}{2}} = 6.$$

49. This is a geometric series with first term equal to $c = \frac{2^2}{3} = \frac{4}{3}$ and ratio $r = 2$. Since the ratio is greater than 1 in magnitude, the series diverges.

51. This is a geometric series with first term equal to $c = \pi \left(\frac{e}{3}\right)^0 = \pi$ and ratio $r = \frac{e}{3}$. Since the ratio is less than 1 in magnitude, the series converges, and it converges to

$$\frac{c}{1 - r} = \frac{\pi}{1 - (e/3)} = \frac{3\pi}{3 - e}.$$

53. This is a geometric series with first term equal to $c = \left(-\frac{3}{5}\right)^2 = \frac{9}{25}$ and ratio $r = -\frac{3}{5}$. Since the ratio is less than 1 in magnitude, the series converges, and it converges to

$$\frac{c}{1-r} = \frac{9/25}{1-(-3/5)} = \frac{9/25}{8/5} = \frac{9}{40}.$$

55. Rewrite this as

$$\sum_{k=0}^{\infty} \frac{2^{k+2}}{5^{k-1}} = \sum_{k=0}^{\infty} 20\frac{2^k}{5^k} = \sum_{k=0}^{\infty} 20\left(\frac{2}{5}\right)^k.$$

Then this is a geometric series with first term equal to $c = 20$ and ratio $r = \frac{2}{5}$. Since the ratio is less than 1 in magnitude, the series converges, and it converges to

$$\frac{c}{1-r} = \frac{20}{1-2/5} = \frac{20}{3/5} = \frac{100}{3}.$$

57. Rewrite this as

$$\sum_{k=0}^{\infty} \frac{(-3)^{k+1}}{4^{k-2}} = \sum_{k=0}^{\infty} \left(-3 \cdot 4^2 \frac{(-3)^k}{4^k}\right) = \sum_{k=0}^{\infty} \left(-48\left(\frac{-3}{4}\right)^k\right).$$

Then this is a geometric series with first term equal to $c = -48$ and ratio $r = -\frac{3}{4}$. Since the ratio is less than 1 in magnitude, the series converges, and it converges to

$$\frac{c}{1-r} = \frac{-48}{1-(-3/4)} = \frac{-48}{7/4} = -\frac{192}{7}.$$

59. Since each of the terms shown, after the first, is $-\frac{1}{4}$ times the previous term, this appears to be a geometric series with first term equal to $c = \frac{80}{3}$ and ratio $r = -\frac{1}{4}$. Since the ratio is less than 1 in magnitude, the series converges, and it converges to

$$\frac{c}{1-r} = \frac{80/3}{1-(-1/4)} = \frac{80/3}{5/4} = \frac{64}{3}.$$

61. The term $-\frac{1}{k+2}$ in the kth summand will cancel with the term $\frac{1}{k+2}$ in the $(k+2)^{\text{nd}}$ summand. Thus

$$\begin{aligned}
S_n &= \sum_{k=1}^{n} \left(\frac{1}{k} - \frac{1}{k+2}\right) \\
&= \left(\frac{1}{1} - \frac{1}{3}\right) + \left(\frac{1}{2} - \frac{1}{4}\right) + \left(\frac{1}{3} - \frac{1}{5}\right) + \left(\frac{1}{4} - \frac{1}{6}\right) + \dots \\
&\quad + \left(\frac{1}{n-3} - \frac{1}{n-1}\right) + \left(\frac{1}{n-2} - \frac{1}{n}\right) + \left(\frac{1}{n-1} - \frac{1}{n+1}\right) + \left(\frac{1}{n} - \frac{1}{n+2}\right) \\
&= \frac{1}{1} + \frac{1}{2} - \frac{1}{n+1} - \frac{1}{n+2} \\
&= \frac{3}{2} - \frac{2n+3}{n^2+3n+2}.
\end{aligned}$$

Since

$$\lim_{n\to\infty} \frac{2n+3}{n^2+3n+2} = \lim_{n\to\infty} \frac{2/n+3/n^2}{1+3/n+2/n^2} = 0,$$

we see that $\lim_{n\to\infty} S_n = \frac{3}{2}$, so that is the sum of the series.

63. The positive term in any summand will cancel with the negative term in the previous summand, so:

$$S_N = \sum_{k=0}^{N} \left(\frac{1}{3^k} - \frac{1}{3^{k+1}} \right) = \left(\frac{1}{3^0} - \frac{1}{3^1} \right) + \left(\frac{1}{3^1} - \frac{1}{3^2} \right) + \left(\frac{1}{3^2} - \frac{1}{3^3} \right) + \cdots + \left(\frac{1}{3^N} - \frac{1}{3^{N+1}} \right)$$

$$= \frac{1}{3^0} - \frac{1}{3^{N+1}} = 1 - \frac{1}{3^{N+1}}.$$

Since $\lim\limits_{N \to \infty} \frac{1}{3^{N+1}} = 0$, we see that $\lim\limits_{N \to \infty} S_N = 1$, so that is the sum of the series.

65. We have

$$S_n = \sum_{k=1}^{n} \ln \frac{k}{k+1}$$

$$= \sum_{k=1}^{n} (\ln k - \ln(k+1))$$

$$= (\ln 1 - \ln 2) + (\ln 2 - \ln 3) + \cdots + (\ln(n-1) - \ln n) + (\ln n - \ln(n+1))$$

$$= \ln 1 - \ln(n+1)$$

$$= -\ln(n+1).$$

Since $\lim\limits_{n \to \infty} \ln(n+1) = \infty$, we see that S_n diverges, so that the series diverges.

67. This is a geometric series with ratio x, so it converges only for $|x| < 1$.

69. This is a geometric series with ratio $\sin x$, so it converges for $|\sin x| < 1$. But $|\sin x| = 1$ only for x an odd multiple of $\frac{\pi}{2}$, so that the series converges for all x except for $x = \frac{\pi}{2} + n\pi$ for n an integer.

71. Since 237 is the repeating part, we have

$$0.237237237\ldots = 0.237(1 + 0.001 + 0.000001 + \ldots) = 0.237(1 + 10^{-3} + 10^{-6} + 10^{-9} + \ldots)$$

$$= \frac{0.237}{1 - 10^{-3}} = \frac{0.237}{1 - 0.001} = \frac{0.237}{0.999} = \frac{237}{999} = \frac{79}{333}.$$

73. Here 3051 is the repeating part, so we get

$$0.30513051305\ldots = 0.3051(1 + 0.0001 + 0.00000001 + \ldots)$$

$$= 0.3051(1 + 10^{-4} + 10^{-8} + 10^{-12} + \ldots) = \frac{0.3051}{1 - 10^{-4}} = \frac{0.3051}{0.9999} = \frac{3051}{9999} = \frac{339}{1111}$$

75. Since 27 is the repeating part, we have

$$1.272727\ldots = 1 + 0.27(1 + 0.01 + 0.0001 + \ldots) = 1 + \frac{0.27}{1 - 0.01} = 1 + \frac{27}{99} = 1 + \frac{3}{11} = \frac{14}{11}.$$

77. Since 9 is the repeating part, we have

$$0.199999\ldots = \frac{1}{10}(1 + 0.99999\ldots) = \frac{1}{10}(1 + 0.9(1 + 0.1 + 0.01 + \ldots))$$

$$= \frac{1}{10} \left(1 + \frac{0.9}{1 - 0.1} \right) = \frac{1}{10} \left(1 + \frac{0.9}{0.9} \right) = \frac{2}{10} = \frac{1}{5}.$$

79. The initial drop accounts for 1 meter of travel. Between the first and second bounces, then, it travels $2 \cdot 0.6 \cdot 1 = 1.2$ meters. Between each subsequent pair of bounces, it travels 0.6 times its distance between the preceding two bounces. Thus the total distance traveled is

$$1 + 1.2 + 1.2(0.6) + 1.2(0.6)^2 + \cdots = 1 + 1.2(1 + 0.6 + (0.6)^2 + \dots)$$

$$= 1 + \frac{1.2}{1 - 0.6} = 1 + \frac{1.2}{0.4} = 1 + 3 = 4 \text{ m.}$$

81. For reference, label the squares with horizontal sides $H_1, H_2, H_3, \ldots$ starting with the original square, and label the other squares $D_1, D_2, D_3, \ldots$ starting with the square immediately inside the original square. Then if the side length of H_{i-1} is s_{i-1}, the Pythagorean Theorem tells us that the side length of D_{i-1} is

$$\sqrt{\left(\frac{s_{i-1}}{2}\right)^2 + \left(\frac{s_{i-1}}{2}\right)^2} = \sqrt{\frac{s_{i-1}^2}{2}} = s_{i-1}\frac{\sqrt{2}}{2};$$

applying this argument a second time shows that the side length of H_i is

$$s_i = s_{i-1}\frac{\sqrt{2}}{2} \cdot \frac{\sqrt{2}}{2} = \frac{s_{i-1}}{2}.$$

Thus $s_i = \frac{1}{2^{i-1}}s_1$, so that $A(H_i)$, the area of H_i, is

$$A(H_i) = \frac{1}{2^{2i-2}}s_1^2.$$

Now, in H_i, the shaded area consists of four right triangles of side length $\frac{s_i}{2}$, so that the shaded area in H_i is

$$4\left(\frac{1}{2} \cdot \frac{s_i}{2} \cdot \frac{s_i}{2}\right) = \frac{s_i^2}{2} = \frac{1}{2}A(H_i) = \frac{1}{2^{2i-1}}s_1^2.$$

Thus the total shaded area is the sum of these areas, which is

$$\sum_{i=1}^{\infty}\left(\frac{1}{2^{2i-1}}s_1^2\right) = s_1^2\sum_{i=1}^{\infty}\frac{1}{2^{2i-1}}$$

$$= s_1^2\frac{1/2}{1 - 1/4} = s_1^2\frac{1/2}{3/4} = \frac{2}{3}s_1^2.$$

Since s_1, the side length of the original square is 1, the total area covered is $\frac{2}{3}$.

83. We will need the fact that the area of an equilateral triangle with side length s is $\frac{\sqrt{3}}{4}s^2$. At each step, we whiten the middle triangle from each of the existing black triangles. Thus at the first step, we whiten the middle triangle from the original triangle; at the second step, we whiten the middle triangle in each of the three remaining black triangles; at the third step, in each of the nine remaining black triangles, and so on. At each step, the number of black triangles increases by a factor of three, and the side length of each triangle decreases by a factor of 2. The white area added at any step is clearly $\frac{1}{4}$ of the area of each triangle at that step.

Putting this together, we see that at the kth step we have 3^{k-1} triangles of side length $\frac{1}{2^{k-1}}$, and we add a white area equal to

$$3^{k-1} \cdot \frac{1}{4} \cdot \frac{\sqrt{3}}{4}\left(\frac{1}{2^{k-1}}\right)^2 = \frac{\sqrt{3}}{16}\left(\frac{3}{4}\right)^{k-1}.$$

So the total white area added is the sum of these terms for k from 1 to ∞; this is a geometric series with first term $\frac{\sqrt{3}}{16}$ and ratio $\frac{3}{4}$, so its sum is

$$\frac{\sqrt{3}/16}{1-(3/4)} = \frac{\sqrt{3}/16}{1/4} = \frac{\sqrt{3}}{4}.$$

Note that this is the area of the original triangle! So the white areas fill the triangle.

Another, simpler way to see this answer is as follows: at each step, the amount of black area is $\frac{3}{4}$ of what it was at the previous step, since $\frac{1}{4}$ of each black triangle has been taken away. Thus the amount of black area is

$$\lim_{n\to\infty} \left(\frac{\sqrt{3}}{4} \cdot \left(\frac{3}{4}\right)^n \right) = \frac{\sqrt{3}}{4} \lim_{n\to\infty} \left(\frac{3}{4}\right)^n = 0,$$

so that no black area remains.

Proofs

85. Let $S_n = \sum_{k=1}^{n} a_k$ and $T_n = \sum_{k=1}^{n} b_k$ be the partial sums of the two series. Then $\lim_{n\to\infty} S_n = L$ and $\lim_{n\to\infty} T_n = M$. By Theorem 7.11, $\lim_{n\to\infty} (S_n + T_n) = L + M$. But

$$S_n + T_n = \sum_{k=1}^{n} a_k + \sum_{k=1}^{n} b_k = \sum_{k=1}^{n} (a_k + b_k),$$

so that $S_n + T_n$ is the nth partial sum for $\sum_{k=1}^{n} (a_k + b_k)$. Thus $\sum_{k=1}^{n} (a_k + b_k)$ converges to $L + M$.

87. Since $\sum_{k=0}^{\infty} cr^k$ converges, we have $|r| < 1$. Since $\sum_{k=0}^{\infty} bv^k$ converges, we also have $|v| < 1$. Hence $|rv| < 1$ as well, so that

$$\sum_{k=0}^{\infty} (cr^k \cdot bv^k) = \sum_{k=0}^{\infty} bc(rv)^k$$

is a geometric series with ratio rv; since $|rv| < 1$, it is convergent. Then

$$\left(\sum_{k=0}^{\infty} cr^k\right) \left(\sum_{k=0}^{\infty} bv^k\right) = \frac{c}{1-r} \cdot \frac{b}{1-v} = \frac{bc}{1-r-v+rv},$$

$$\sum_{k=0}^{\infty} bc(rv)^k = \frac{bc}{1-rv},$$

so that the two are equal, assuming that neither b nor c is zero, if and only if $1 - rv = 1 - r - v + rv$, i.e. if $2rv = r + v$ and both are less than 1 in magnitude. Thus, for example, let $r = -\frac{1}{2}$ and $v = \frac{1}{4}$. Then

$$\left(\sum_{k=0}^{\infty} cr^k\right) \left(\sum_{k=0}^{\infty} bv^k\right) = \frac{c}{1-(-1/2)} \cdot \frac{b}{1-(1/4)} = \frac{bc}{(3/2) \cdot (3/4)} = \frac{8bc}{9},$$

$$\sum_{k=0}^{\infty} bc(rv)^k = \frac{bc}{1-(-1/8)} = \frac{bc}{9/8} = \frac{8bc}{9},$$

and the two are equal. Thus it is possible.

89. Since b and v are both nonzero, $\frac{cr^k}{bv^k}$ makes sense for all $k \geq 0$, and

$$\sum_{k=0}^{\infty} \frac{cr^k}{bv^k} = \sum_{k=0}^{\infty} \frac{c}{b} \cdot \left(\frac{r}{v}\right)^k,$$

so that it is a geometric series with ratio $\frac{r}{v}$. It will be convergent if $\left|\frac{r}{v}\right| < 1$, i.e., if $|r| < |v|$.

Thinking Forward

Improper Integrals

▷ $\int_1^\infty \frac{1}{x}\, dx = \lim_{t\to\infty} \int_1^t \frac{1}{x}\, dx = \lim_{t\to\infty} [\ln x]_1^t = \lim_{t\to\infty} (\ln t - \ln 1) = \lim_{t\to\infty} \ln t = \infty$, so this integral diverges.

▷ $\int_1^\infty \frac{1}{x^2}\, dx = \lim_{t\to\infty} \int_1^t \frac{1}{x^2}\, dx = \lim_{t\to\infty} \left[-\frac{1}{x}\right]_1^t = \lim_{t\to\infty} \left(1 - \frac{1}{t}\right) = 1 - \lim_{t\to\infty} \frac{1}{t} = 1$, so this integral converges.

An Improper Integral and Infinite Series

The graph below shows the graph of $f(x) = \frac{1}{x}$ together with the function $a(x)$ which is equal to $\frac{1}{k}$ if $x \in [k, k+1)$:

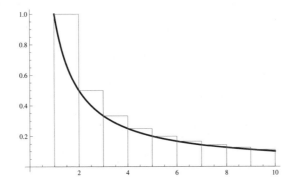

Since the area of each rectangle is $1 \cdot \frac{1}{k} = \frac{1}{k}$, it is clear that the sum of the areas of the first k rectangles (which is in fact a left sum approximation to $\int_1^{k+1} f(x)\, dx$) exceeds the value of $\int_1^{k+1} f(x)\, dx$ — that is

$$1 + \frac{1}{2} + \cdots + \frac{1}{k} = S_k \geq \int_1^{k+1} \frac{1}{x}\, dx.$$

But the integral diverges by the immediately preceding exercise, so that $\{S_k\} \to \infty$ and the series diverges.

7.4   Introduction to Convergence Tests

Thinking Back

The contrapositive

The contrapositive of "If A, then B" is "If not B, then not A".

▷ The contrapositive is "If a quadrilateral is not a rectangle, then it is not a square".

▷ The contrapositive is "If a positive integer does not have exactly two positive divisors, then it is not a prime".

▷ The contrapositive is "If $|r| \neq -r$, then $r \geq 0$".

▷ The contrapositive is "If a does not divide c, and b is any number, then either a does not divide b or b does not divide c".

Concepts

1. (a) False. Theorem 7.27 is *not* an if and only if test. For example, $\{\frac{1}{k}\} \to 0$, but $\sum_{k=1}^{\infty} \frac{1}{k}$ diverges.

   (b) True. See Theorem 7.27.

   (c) False. $f(x)$ must satisfy the hypotheses of Theorem 7.28 — it must be continuous, positive, and decreasing on $[1, \infty)$.

   (d) False. See Theorem 7.29 or the Thinking Forward problems for the previous section.

   (e) True. See Theorem 7.29 (a).

   (f) False. For example, if $f(x) = \frac{1}{x}$, then $\sum_{k=1}^{\infty} f(k)$ is the harmonic series, which diverges.

   (g) False. While $f(k)$ must go to zero, this need not be true for $f(x)$. For example, let

   $$f(x) = \begin{cases} \frac{1}{k^2}, & k \text{ a nonzero integer,} \\ 1, & \text{otherwise.} \end{cases}$$

   Then while $\lim_{k \to \infty} f(k) = 0$ when k is restricted to the integers, $\lim_{x \to \infty} f(x)$ does not exist when x is real.

   (h) True. If $\sum_{k=1}^{n} a_k = L$, then $\{S_n\} \to L$, so that $\{L - S_n\} \to 0$.

3. A *p*-series is a series of the form

   $$\sum_{k=1}^{\infty} k^{-p} = \sum_{k=1}^{\infty} \frac{1}{k^p},$$

   where p is a real number.

5. If the limit of the terms cr^k of a geometric series is zero, then we must have $|r| < 1$, since if $|r| > 1$ then cr^k grows without bound in magnitude, while if $|r| = 1$ then $|cr^k| = |c|$, so in both cases the series diverges by the divergence test. But for $|r| < 1$, a geometric series with ratio r converges.

7. Since $\lim_{x \to \infty} f(x) = \alpha > 0$, we can find $N > 0$ such that $f(x) > \frac{\alpha}{2}$ for $x > N$. Thus

   $$\int_{1}^{\infty} f(x)\,dx = \int_{1}^{N} f(x)\,dx + \int_{N}^{\infty} f(x)\,dx \geq \int_{1}^{N} f(x)\,dx + \int_{N}^{\infty} \frac{\alpha}{2}\,dx = \infty$$

   since the second integral is clearly infinite. Then by Theorem 7.28, since the integral diverges, and since $f(x)$ satisfies the hypotheses of the theorem, $\sum_{k=1}^{\infty} f(k)$ diverges.

9. The statement of the corresponding theorem is: If $a(x)$ is a function that is continuous on $[1, \infty)$, and is eventually positive and decreasing, and if $\{a_k\}$ is the sequence defined by $a_k = a(k)$ for every positive integer k, then

   $$\sum_{k=1}^{\infty} a_k \quad \text{and} \quad \int_{1}^{\infty} a(x)\,dx$$

   either both converge or both diverge.

   To prove this version of the theorem, suppose that $a(x)$ is continuous and is eventually positive and decreasing. Choose an integer $N > 0$ such that a is positive and decreasing on $[N, \infty)$, and let $g(x) = a(x + (N - 1))$. Then $g(x)$ is continuous since a is, and is positive and decreasing for $x \geq 1$ since a is positive and decreasing for $x \geq N$. Hence by the integral test, $\sum_{k=1}^{\infty} g(k)$ and $\int_{1}^{\infty} g(x)\,dx$ either both converge or both diverge. However,

   $$\sum_{k=1}^{\infty} g(k) = \sum_{k=1}^{\infty} a_{k+N-1} = \sum_{k=N}^{\infty} a_k,$$

   $$\int_{1}^{\infty} g(x)\,dx = \int_{1}^{\infty} a(x + N - 1)\,dx = \int_{N}^{\infty} a(x)\,dx,$$

so that $\sum_{k=N}^{\infty} a_k$ and $\int_N^{\infty} a(x)\,dx$ either both converge or both diverge. However,

$$\int_1^{\infty} a(x)\,dx = \int_1^N a(x)\,dx + \int_N^{\infty} a(x)\,dx,$$

$$\sum_{k=1}^{\infty} a_k = \sum_{k=1}^{N-1} a_k + \sum_{k=N}^{\infty} a_k.$$

Since $a(x)$ is continuous, we know that $\int_1^N a(x)\,dx$ converges, so that $\int_1^{\infty} a(x)\,dx$ and $\int_N^{\infty} a(x)\,dx$ either both converge or both diverge. Similarly, $\sum_{k=1}^{N-1} a_k$ exists, so that $\sum_{k=1}^{\infty} a_k$ and $\sum_{k=N}^{\infty} a_k$ either both converge or both diverge. Thus we have shown the string of convergence/divergence equivalences

$$\sum_{k=1}^{\infty} a_k \Leftrightarrow \sum_{k=N}^{\infty} a_k \Leftrightarrow \int_N^{\infty} a(x)\,dx \Leftrightarrow \int_1^{\infty} a(x)\,dx,$$

finishing the proof.

11. If the convergent series $\sum_{k=1}^{\infty} a_k$ satisfies the conditions of the integral test, there must be a continuous, positive, and decreasing function $a(x)$ with $a_k = a(k)$. Since $a(x)$ is positive, the sequence $\{a_k\}$ must also be positive. But this means that the remainder, which is the sum of the "tail", must also be positive since it is a sum of positive numbers:

$$R_n = L - S_n = \sum_{k=1}^{\infty} a_k - \sum_{k=1}^{n} a_k = \sum_{k=n+1}^{\infty} a_k > 0$$

In this case, by Theorem 7.31, the remainder R_n is bounded above by $\int_n^{\infty} a(x)\,dx$ (and below by 0), which, if the integral can be evaluated, gives us a bound on the size of the error in the nth partial sum.

13. If the function $a(x)$ is not continuous, then the Riemann sum approximation that is used as a basis for the proof of the integral test may not converge to the integral of the function. For example, let $a_k = \frac{1}{k}$ and let $a(x) = a_x$ for x an integer but $a(x) = 0$ otherwise. Then $\int_1^{\infty} a(x)\,dx = 0$ but $\sum_{k=1}^{\infty} a_k$ diverges. As another example, let $a_k = \frac{1}{2^k}$ and let $a(x) = a_x$ for x an integer but $a(x) = 1$ otherwise. Then $\int_1^{\infty} a(x)\,dx$ diverges but $\sum_{k=1}^{\infty} a_k$ converges.

15. Let $f(x) = \sin(\pi x)$. Then $f(k) = 0$ for k any integer, so that $\sum_{k=1}^{\infty} f(k) = 0$ converges. However,

$$\int_1^{\infty} f(x)\,dx = \lim_{t \to \infty} \int_1^t \sin(\pi x)\,dx = -\lim_{t \to \infty} \left[\frac{1}{\pi}\cos(\pi x)\right]_1^t = -\frac{1}{\pi} - \lim_{t \to \infty} \frac{1}{\pi}\cos(\pi t),$$

which diverges since $\cos(\pi t)$ oscillates between 1 and -1.

Skills

17. Since $\{\frac{1}{k^2}\} \to 0$, the divergence test fails — the divergence test provides information only if the limit of the terms of the series *do not* approach zero.

19. Since

$$\lim_{k \to \infty} \frac{k}{3k + 100} = \lim_{k \to \infty} \frac{1}{3 + 100/k} = \frac{1}{3},$$

the series diverges by the divergence test since the terms of the sequence do not approach zero.

21. Note that
$$\frac{k!}{k^k} = \frac{1 \cdot 2 \cdot 3 \ldots k}{k \cdot k \cdot k \ldots k} \leq \frac{1}{k},$$
so that $\{\frac{k!}{k^k}\} \to 0$. Thus the divergence test fails — the divergence test provides information only if the limit of the terms of the series *do not* approach zero.

23. By Theorem 7.16, $\{\frac{k^2+1}{k!}\} = \{\frac{k^2}{k!} + \frac{1}{k!}\} \to 0$ since factorials dominate power functions. Thus the divergence test fails — the divergence test provides information only if the limit of the terms of the series *do not* approach zero.

25. Let $f(x) = e^{-x}$; then $f(x)$ is continuous and positive everywhere, and $f'(x) = -e^{-x}$ is negative, so that f is decreasing everywhere. Thus the integral test applies, and
$$\int_0^\infty e^{-x}\,dx = \lim_{t\to\infty} \int_0^t e^{-x}\,dx = \lim_{t\to\infty} \left[-e^{-x}\right]_0^t = \lim_{t\to\infty} \left(-e^{-t} + e^{-0}\right) = 1.$$
Since the integral converges, so does the series.

27. Let $f(x) = \frac{1}{x\sqrt{\ln x}}$. Then for $x \geq 2$, $f(x)$ is positive and continuous. Also, since $x\sqrt{\ln x}$ is obviously increasing, $f(x)$ is decreasing. Thus $f(x)$ satisfies the conditions of the integral test on $[2,\infty)$, so by the integral test, $\sum_{k=2}^\infty \frac{1}{k\sqrt{\ln k}}$ converges if and only if $\int_2^\infty f(x)\,dx$ does. But (with the substitution $u = \ln x$, $du = \frac{1}{x}\,dx$, we get
$$\int_2^\infty \frac{1}{x\sqrt{\ln x}}\,dx = \lim_{t\to\infty}\int_{x=2}^{x=t} \frac{1}{u^{1/2}}\,du = \lim_{t\to\infty}\left[2u^{1/2}\right]_{x=2}^{x=t} = \lim_{t\to\infty}\left[2\sqrt{\ln x}\right]_2^t$$
$$= \lim_{t\to\infty}\left(2\sqrt{\ln t} - 2\sqrt{\ln 2}\right),$$
which diverges. Thus the series $\sum_{k=2}^\infty \frac{1}{k\sqrt{\ln k}}$ diverges as well.

29. Let $f(x) = \frac{2x-4}{x^2-4x+3}$. For $x \geq 4$, we have $2x - 4 > 0$. Since $x^2 - 4x + 3 = (x-3)(x-1)$, it is nonzero for $x \geq 4$, and is also positive there. Thus $f(x)$ is continuous and positive on $[4,\infty)$. Further,
$$f'(x) = \frac{(x^2-4x+3)(2) - (2x-4)^2}{(x^2-4x+3)^2} = -\frac{2(x^2-4x+5)}{(x^2-4x+3)^2}.$$
Since $x^2 - 4x + 5 > 0$ for all x (its discriminant is negative, so it has no roots, and it is positive at $x = 0$), and since the denominator is positive for $x \geq 4$, we see that $f'(x) < 0$ for $x \geq 4$, so that f is decreasing. So by the integral test, $\sum_{k=4}^\infty \frac{2k-4}{k^2-4k+3}$ and $\int_4^\infty f(x)\,dx$ either both converge or both diverge. But (with the substitution $u = x^2 - 4x + 3$, $du = 2x - 4$)
$$\int_4^\infty f(x)\,dx = \lim_{t\to\infty}\int_4^t \frac{2x-4}{x^2-4x+3}\,dx$$
$$= \lim_{t\to\infty}\int_{x=4}^{x=t} \frac{1}{u}\,du$$
$$= \lim_{t\to\infty}\left[\ln u\right]_{x=4}^{x=t}$$
$$= \lim_{t\to\infty}\left[\ln\left|x^2-4x+3\right|\right]_4^t$$
$$= \lim_{t\to\infty}\left(\ln(t^2-4t+3) - \ln 3\right),$$
which diverges. Thus $\sum_{k=4}^\infty \frac{2k-4}{k^2-4k+3}$ diverges as well.

31. Let $f(x) = \frac{1}{(x-2)^2}$. Then for $x \geq 3$, $f(x)$ is clearly continuous, positive, and decreasing. So by the integral test, $\sum_{k=3}^{\infty} \frac{1}{(k-2)^2}$ and $\int_3^{\infty} f(x)\,dx$ either both converge or both diverge. However,

$$\int_3^{\infty} f(x)\,dx = \lim_{t \to \infty} \int_3^t \frac{1}{(x-2)^2}\,dx = \lim_{t \to \infty} \left[-(x-2)^{-1} \right]_3^t = \lim_{t \to \infty} \left(-\frac{1}{t-2} + 1 \right) = 1,$$

so that the integral converges, and thus $\sum_{k=3}^{\infty} \frac{1}{(k-2)^2}$ converges as well.

33. Since $\{k^{-3/2}\} \to 0$, the divergence test will not be useful. Define $f(x) = x^{-3/2}$; then for $x \geq 1$, we see that $f(x)$ is continuous and positive. Also $f(x)$ is decreasing since $f'(x) = -\frac{3}{2}x^{-5/2} < 0$ for $x \geq 1$. Since

$$\int_1^{\infty} x^{-3/2}\,dx = \lim_{t \to \infty} \int_1^t x^{-3/2}\,dx = \lim_{t \to \infty} \left[-2x^{-1/2} \right]_1^t = \lim_{t \to \infty} \left(-2t^{-1/2} + 2 \right) = 2,$$

converges, so does the series $\sum_{k=1}^{\infty} k^{-3/2}$. Note that since this is a p-series with $p = \frac{3}{2}$, we could have simply applied Theorem 7.29 (which is derived from the integral test).

35. Since $\lim_{k \to \infty} \frac{k}{k^2+2} = 0$, the divergence test will not be useful. Let $f(x) = \frac{x}{x^2+2}$; then $f(x)$ is clearly continuous and positive for $x \geq 1$. Further,

$$f'(x) = \frac{x^2 + 2 - x(2x)}{(x^2+2)^2} = \frac{2 - x^2}{(x^2+2)^2},$$

so that f is eventually decreasing. So the generalization of the integral test proved in Exercise 9 applies. Use the substitution $u = x^2 + 2$, $du = 2x\,dx$ to evaluate the integral:

$$\int_1^{\infty} \frac{x}{x^2+2}\,dx = \lim_{t \to \infty} \int_1^t \frac{x}{x^2+2}\,dx = \lim_{t \to \infty} \frac{1}{2} \int_{x=1}^{x=t} \frac{1}{u}\,du = \lim_{t \to \infty} \frac{1}{2} \left[\ln u \right]_{x=1}^{x=t}$$

$$= \frac{1}{2} \lim_{t \to \infty} \left[\ln(x^2+2) \right]_1^t = \frac{1}{2} \lim_{t \to \infty} \left(\ln(t^2+2) - \ln 3 \right) = \infty.$$

Thus $\sum_{k=1}^{\infty} \frac{k}{k^2+2}$ diverges as well.

37. Since power functions dominate log functions, $\{ \frac{\ln k}{k} \} \to 0$, so the divergence test will not be useful. Define $f(x) = \frac{\ln x}{x}$; then for $x > 1$, $f(x)$ is positive and continuous. Also, since $f'(x) = \frac{1 - \ln x}{x^2}$, we see that $f'(x) < 0$ for $x > e$, so that f is eventually decreasing. So the generalization of the integral test proved in Exercise 9 applies. We have, using the substitution $u = \ln x$,

$$\int_1^{\infty} \frac{\ln x}{x}\,dx = \lim_{t \to \infty} \int_1^t \frac{\ln x}{x}\,dx = \lim_{t \to \infty} \int_{x=1}^{x=t} u\,du = \lim_{t \to \infty} \left[\frac{1}{2}u^2 \right]_{x=1}^{x=t} = \lim_{t \to \infty} \left[\frac{1}{2}\ln^2 x \right]_1^t$$

$$= \lim_{t \to \infty} \left(\frac{1}{2}\ln^2 t \right),$$

which diverges. Thus $\sum_{k=1}^{\infty} \frac{\ln k}{k}$ diverges as well.

39. Since exponential functions dominate power functions, $\{k^3 e^{-k^4}\} = \{\frac{k^3}{e^{k^4}}\} \to 0$, so the divergence test will not be useful. Define $f(x) = x^3 e^{-x^4}$. Then $f(x)$ is clearly continuous and positive on $[1, \infty)$, and since $f'(x) = x^3 e^{-x^4} - 4x^3(x^3 e^{-x^4}) = (x^3 - 4x^6)e^{-x^4}$, we see that $f'(x) < 0$ for $x \geq 1$

and thus f is decreasing. Then (with the substitution $u = x^4$, $du = 4x^3$)

$$\int_1^\infty x^3 e^{-x^4} \, dx = \lim_{t\to\infty} \int_1^t x^3 e^{-x^4} \, dx = \lim_{t\to\infty} \int_{x=1}^{x=t} \frac{1}{3} e^{-u} \, du = \lim_{t\to\infty} \left[-\frac{1}{3} e^{-u} \right]_{x=1}^{x=t}$$

$$= \lim_{t\to\infty} \left[-\frac{1}{3} e^{-x^4} \right]_1^t = \lim_{t\to\infty} \left(\frac{1}{3} e^{-1/4} - \frac{1}{3} e^{-t^4} \right),$$

which converges. Thus $\sum_{k=1}^\infty k^3 e^{-k^4}$ converges as well.

41. Since the terms clearly tend to zero, the divergence test will not be useful. Let $f(x) = \frac{1}{x(x-1)}$. Then for $x \geq 2$, clearly $f(x)$ is continuous and positive; it is also clear that it is decreasing since the denominator is increasing. Using partial fractions, we have

$$\int_2^\infty \frac{1}{x(x-1)} \, dx = \lim_{t\to\infty} \int_2^t \frac{1}{x(x-1)} \, dx$$

$$= \lim_{t\to\infty} \int_2^t \left(\frac{1}{x-1} - \frac{1}{x} \right) dx$$

$$= \lim_{t\to\infty} \left[\ln(x-1) - \ln x \right]_2^t$$

$$= \lim_{t\to\infty} \left[\ln \frac{x-1}{x} \right]_2^t$$

$$= \lim_{t\to\infty} \ln \frac{t-1}{t} - \ln \frac{1}{2} = \ln 1 - \ln \frac{1}{2} = \ln 2.$$

Since the integral converges, it follows from the integral test that $\sum_{k=2}^\infty \frac{1}{k(k-1)}$ converges as well.

43. Since exponentials dominate power functions, $\left\{ \frac{k}{3^k} \right\} \to 0$, so the divergence test is not useful. Let $f(x) = \frac{x}{3^x} = x3^{-x}$. Then clearly $f(x)$ is positive and continuous on $[1, \infty)$. Since $f'(x) = 3^{-x} - x3^{-x} \ln 3 = (1 - x \ln 3)3^{-x}$, which is negative for $x > 1$. Thus f is decreasing on $[1, \infty)$, so the integral test applies. We have, using integration by parts with $u = x$ and $dv = 3^{-x} \, dx$,

$$\int_1^\infty x3^{-x} \, dx = \lim_{t\to\infty} \int_1^t x3^{-x} \, dx$$

$$= \lim_{t\to\infty} \left(\left[-\frac{1}{\ln 3} x3^{-x} \right]_1^t + \frac{1}{\ln 3} \int 1 \cdot 3^{-x} \, dx \right)$$

$$= \lim_{t\to\infty} \left(-\frac{1}{\ln 3} (t3^{-t} - 3^{-1}) - \frac{1}{(\ln 3)^2} \left[3^{-x} \right]_1^t \right)$$

$$= \lim_{t\to\infty} \left(-\frac{t3^{-t} - 3^{-1}}{\ln 3} - \frac{3^{-t} - 3^{-1}}{(\ln 3)^2} \right)$$

$$= \frac{1}{3\ln 3} + \frac{1}{3(\ln 3)^2}$$

since both $\lim_{t\to\infty} 3^{-t}$ and $\lim_{t\to\infty} t3^{-t}$ are zero. Since the integral converges, the series $\sum_{k=1}^\infty \frac{k}{3^k}$ does as well, by the integral test.

45. (a) For a proof using the integral test that $\sum_{k=1}^\infty e^{-k}$ converges, see Exercise 25.

(b) The tenth term in the sequence of partial sums is

$$S_{10} = \sum_{k=0}^9 e^{-k} = 1 + \frac{1}{e} + \frac{1}{e^2} + \frac{1}{e^3} + \cdots + \frac{1}{e^9} \approx 1.58190.$$

(c) Integrating gives

$$\int_n^\infty e^{-x}\,dx = \lim_{t\to\infty}\int_n^t e^{-x}\,dx = \lim_{t\to\infty}\left[-e^{-x}\right]_n^t = \lim_{t\to\infty}\left(-e^{-t}+e^{-n}\right) = e^{-n}.$$

Since for $n=10$, $e^{-n}\approx 0.0000454$, we have $R_{10}\le 0.0000454$.

(d) Using the values of S_{10} and the bound on R_{10} from above,

$$S_{10}\approx 1.58190 \le \sum_{k=1}^\infty e^{-k} \le S_{10}+\int_{10}^\infty e^{-x}\,dx \approx 1.58190+0.0000454 \approx 1.58195.$$

(e) To get $R_n \le 10^{-6}$, we need $\int_n^\infty e^{-x} = e^{-n} \le 10^{-6}$. Solving numerically gives $n\approx 13.8155$, so that $n=14$ suffices.

47. (a) Let $f(x)=\frac{1}{x^2}$; then $f(x)$ is clearly continuous, positive, and decreasing on $[1,\infty)$, so the integral test applies. We have

$$\int_n^\infty \frac{1}{x^2}\,dx = \lim_{t\to\infty}\int_n^t \frac{1}{x^2}\,dx = \lim_{t\to\infty}\left[-x^{-1}\right]_n^t = \lim_{t\to\infty}\left(-t^{-1}+n^{-1}\right) = \frac{1}{n},$$

so that $\int_1^\infty \frac{1}{x^2}\,dx = 1$ converges and thus the series $\sum_{k=1}^\infty \frac{1}{k^2}$ does as well. (Note that for the purposes of this part of the exercise, we could just as well have integrated from 1 to ∞; however, this bound for the integral from n to ∞ will be useful in part (c)).

(b) The tenth term is the sequence of partial sums is

$$S_{10} = \sum_{k=1}^\infty \frac{1}{k^2} = \frac{1}{1}+\frac{1}{4}+\frac{1}{9}+\cdots+\frac{1}{100} \approx 1.54977.$$

(c) From part (a), $R_{10}\le \frac{1}{10}=0.1$ by Theorem 7.31.

(d) From Theorem 7.31, we have

$$S_{10}\approx 1.54977 \le \sum_{k=1}^\infty \frac{1}{k^2} \le S_{10}+\int_{10}^\infty \frac{1}{x^2}\,dx \approx 1.64977.$$

(e) To get a bound of 10^{-6} on R_n, we need $\frac{1}{n}\le 10^{-6}$, so that $n\ge 10^6$. Thus $n=10^6$ suffices.

49. Let $f(x)=\frac{1}{x(\ln x)^p}=x^{-1}(\ln x)^{-p}$. Then for $x\ge 2$, this function is continuous and positive. Additionally,

$$f'(x) = -x^{-2}(\ln x)^{-p} - px^{-1}(\ln x)^{-p-1}\cdot\frac{1}{x} = -\frac{(\ln x)^{-p-1}(p+\ln x)}{x^2}.$$

Thus for $\ln x > |p|$, we see that $f'(x) < 0$, so that $f(x)$ is eventually decreasing. So by the generalization of the integral test proved in Exercise 9, the series and $\int_1^\infty f(x)\,dx$ either both converge or both diverge. Using the substitution $u=\ln x$, we integrate as follows. First assume $p=1$. Then

$$\int_2^\infty \frac{1}{x\ln x}\,dx = \lim_{t\to\infty}\int_2^t \frac{1}{x\ln x}\,dx$$
$$= \lim_{t\to\infty}\int_{x=2}^{x=t} \frac{1}{u}\,du$$
$$= \lim_{t\to\infty}\left[\ln|u|\right]_{x=2}^{x=t}$$
$$= \lim_{t\to\infty}\left[\ln|\ln x|\right]_2^t$$
$$= \lim_{t\to\infty}\left(\ln\ln t - \ln\ln 2\right)$$
$$= \infty.$$

Since the integral diverges, the series diverges for $p = 1$. If instead $p \neq 1$, then

$$\int_2^\infty \frac{1}{x(\ln x)^p}\,dx = \lim_{t \to \infty} \int_2^t \frac{1}{x(\ln x)^p}\,dx$$

$$= \lim_{t \to \infty} \int_{x=2}^{x=t} \frac{1}{u^p}\,du$$

$$= \lim_{t \to \infty} \left[\frac{1}{1-p} u^{1-p} \right]_{x=2}^{x=t}$$

$$= \lim_{t \to \infty} \left[\frac{1}{1-p} (\ln x)^{1-p} \right]_2^t$$

$$= \frac{1}{1-p} \lim_{t \to \infty} \left((\ln t)^{1-p} - (\ln 2)^{1-p} \right).$$

For $p > 1$, the exponent on $\ln t$ is negative, so in the limit this is zero and the integral converges. For $p < 1$, the exponent is positive, so in the limit this is infinite and the integral diverges. Thus the series converges only for $p > 1$.

51. If $p \leq 0$, then the terms of the series are $k(1 + k^2)^{-p}$ where $-p$ is a nonnegative exponent. Thus the terms do not tend to zero, so by the divergence theorem the series does not converge. So assume $p > 0$, and let $f(x) = \frac{x}{(1+x^2)^p}$. Clearly $f(x)$ is continuous and positive for $x \geq 1$. Then

$$f'(x) = \frac{(1+x^2)^p(1) - x \cdot p(1+x^2)^{p-1} \cdot 2x}{(1+x^2)^{2p}} = \frac{1 + x^2 - 2px^2}{(1+x^2)^{p+1}} = -\frac{px^2 - 1}{(1+x^2)^{p+1}}.$$

Since $px^2 - 1 > 0$ for x large enough, $f'(x)$ is eventually negative, so that f is eventually decreasing and the integral test applies. Using the substitution $u = 1 + x^2$, assume first that $p = 1$. Then

$$\int_1^\infty \frac{x}{(1+x^2)^p}\,dx = \lim_{t \to \infty} \int_1^t \frac{x}{1+x^2}\,dx$$

$$= \lim_{t \to \infty} \frac{1}{2} \int_{x=1}^{x=t} \frac{1}{u}\,du$$

$$= \frac{1}{2} \lim_{t \to \infty} \left[\ln|u| \right]_{x=1}^{x=t}$$

$$= \frac{1}{2} \lim_{t \to \infty} \left[\ln(1+x^2) \right]_{x=1}^{x=t}$$

$$= \frac{1}{2} \lim_{t \to \infty} \left(\ln(t^2 + 1) - \ln 2 \right),$$

which diverges. Thus the series diverges for $p = 1$. For $p \neq 1$, we get

$$\int_1^\infty \frac{x}{(1+x^2)^p}\,dx = \lim_{t \to \infty} \int_1^t \frac{x}{(1+x^2)^p}\,dx$$

$$= \lim_{t \to \infty} \frac{1}{2} \int_{x=1}^{x=t} \frac{1}{u^p}\,du$$

$$= \frac{1}{2} \lim_{t \to \infty} \left[\frac{1}{1-p} u^{1-p} \right]_{x=1}^{x=t}$$

$$= \frac{1}{2(1-p)} \lim_{t \to \infty} \left[(1+x^2)^{1-p} \right]_1^t$$

$$= \frac{1}{2(1-p)} \lim_{t \to \infty} \left((1+t^2)^{1-p} - 2^{1-p} \right).$$

Now, for $p < 1$, the exponent on $1 + t^2$ is positive, so that the limit (and thus the integral) diverges. For $p > 1$, the exponent on $1 + t^2$ is negative, so that the limit (and thus the integral) diverges. Thus, by the integral test, the series converges only for $p > 1$.

Applications

53. (a) From the recurrence given, we see that if k is even then

$$
\begin{aligned}
q_{k+2} &= (0.14(-1)^{k+1} + 0.36)(q_{k+1} + h) \\
&= 0.22(q_{k+1} + h) \\
&= 0.22((0.14(-1)^k + 0.36)(q_k + h) + h) \\
&= 0.22(0.5(q_k + h) + h) \\
&= 0.11q_k + 0.33h.
\end{aligned}
$$

We prove by induction that $q_{2n} = 0.11^n q_0 + 3h \sum_{k=1}^n 0.11^k$. This holds for $n = 1$ since by the above formula, $q_2 = 0.11q_0 + 0.33h = 0.11^1 q_0 + 3h \sum_{k=1}^1 0.11$. Now, assume it holds for $n - 1$. Then

$$
\begin{aligned}
q_{2n} &= 0.11q_{2n-2} + 0.33h \\
&= 0.11\left(0.11^{n-1}q_0 + 3h \sum_{k=1}^{n-1} 0.11^k\right) + 0.33h \\
&= 0.11^n q_0 + 3h \sum_{k=1}^{n-1} 0.11^{k+1} + 3h \cdot 0.11 \\
&= 0.11^n q_0 + 3h \sum_{k=2}^{n} 0.11^k + 3h \cdot 0.11 \\
&= 0.11^n q_0 + 3h \sum_{k=1}^{n} 0.11^k.
\end{aligned}
$$

Then q_e, the limit of q_{2n}, is

$$
q_e = \lim_{n \to \infty} q_{2n} = \lim_{n \to \infty} 0.11^n q_0 + \lim_{n \to \infty}\left(3h \sum_{k=1}^{n} 0.11^k\right) = 3h \sum_{k=1}^{\infty} 0.11^k.
$$

(b) From the recurrence given, we see that if k is odd, then

$$
\begin{aligned}
q_{k+2} &= (0.14(-1)^{k+1} + 0.36)(q_{k+1} + h) \\
&= 0.5(q_{k+1} + h) \\
&= 0.5((0.14(-1)^k + 0.36)(q_k + h) + h) \\
&= 0.5(0.22q_k + 0.22h + h) \\
&= 0.11q_k + 0.61h.
\end{aligned}
$$

We prove by induction that $q_{2n+1} = 0.11^n q_1 + \frac{61}{11}h \sum_{k=1}^n 0.11^k$. This holds for $n = 1$, since

$q_3 = 0.11q_1 + 0.61h = 0.11^n q_1 + \frac{61}{11}h\sum_{k=1}^{n}0.11^k$. Now assume it holds for $n-1$. Then

$$q_{2n+1} = 0.11q_{2n-1} + 0.61h = 0.11q_{2(n-1)+1} + 0.61h$$

$$= 0.11\left(0.11^{n-1}q_1 + \frac{61}{11}h\sum_{k=1}^{n-1}0.11^k\right) + 0.61h$$

$$= 0.11^n q_1 + \frac{61}{11}h\sum_{k=1}^{n-1}0.11^{k+1} + \frac{61}{11}h\cdot 0.11$$

$$= 0.11^n q_1 + \frac{61}{11}h\sum_{k=2}^{n}0.11^k + \frac{61}{11}h\cdot 0.11$$

$$= 0.11^n q_1 + \frac{61}{11}h\sum_{k=1}^{n}0.11^k.$$

Then q_o, the limit of q_{2n+1}, is

$$q_o = \lim_{n\to\infty} q_{2n+1} = \lim_{n\to\infty} 0.11^n q_1 + \lim_{n\to\infty}\left(\frac{61}{11}h\sum_{k=1}^{n}0.11^k\right) = \frac{61}{11}h\sum_{k=1}^{\infty}0.11^k.$$

(c) Since $\sum_{k=1}^{\infty}0.11^k$ is a geometric series with initial term 0.11 and ratio 0.11, its sum is $\frac{0.11}{1-0.11} = \frac{11}{89}$. Thus

$$q_e = 3h\frac{11}{89} \approx 0.3708h, \qquad q_o = \frac{61}{11}h\cdot\frac{11}{89} \approx 0.6853h.$$

Clearly we cannot have both $0.3708h = P$ and $0.6853h = P$, but if we choose $h = 2P$, then $P = 0.5h$ will lie between the two in the long run.

Proofs

55. Suppose $a : [1,\infty) \to \mathbb{R}$ is continuous, positive, and decreasing, and that $\int_1^\infty a(x)\,dx$ converges. We wish to bound the nth remainder $R_n = \sum_{k=n+1}^{\infty} a_k$, where we write a_k for $a(k)$. Note that since a is positive and $\int_1^\infty a(x)\,dx$ exists that $\int_n^\infty a(x)\,dx$ also exists and is bounded above by $\int_1^\infty a(x)\,dx$. Choose $m > n$, and consider the right Riemann sum for $\int_n^m a(x)\,dx$ with $m-n+1$ equally spaced points $n, n+1, \ldots, m$. Then the area of the rectangle in that Riemann sum over $[k, k+1]$ is $a(k+1) = a_{k+1}$ for $k = n, n+1, \ldots, m-1$, since a is decreasing. Also since a is decreasing, the right Riemann sum is an underapproximation to $\int_n^m a(x)\,dx$. It follows that $\sum_{k=n}^{m-1} a_{k+1} = \sum_{n+1}^{m} a_k \leq \int_n^m a(x)\,dx$. Then, since $a(x)$ is positive, and since $\int_n^\infty a(x)\,dx$ converges, we have $\int_n^m a(x)\,dx \leq \int_n^\infty a(x)\,dx$. Thus for any m,

$$S_m = \sum_{k=n+1}^{m} a_k \leq \int_n^m a(x)\,dx \leq \int_n^\infty a(x)\,dx.$$

Then the sequence of partial sums S_m is bounded above by $\int_n^\infty a(x)\,dx$, and is monotonically increasing since $S_m - S_{m-1} = a_m > 0$. So by Theorem 7.19, the sequence of partial sums converges, and converges by definition to $\sum_{k=n+1}^{\infty} a_k$. Since each partial sum is bounded above by $\int_n^\infty a(x)\,dx$, so is $\sum_{k=n+1}^{\infty} a_k$ (this last statement follows from the proof of Theorem 7.19, which shows that a bounded monotonically increasing sequence converges to its least upper bound, which is no greater than any upper bound, in this case $\int_n^\infty a(x)\,dx$).

57. If $r \leq -1$, then $\lim_{k\to\infty} cr^k$ does not exist since the magnitudes increase but the signs of successive terms alternate. If $r = 1$, then $cr^k = c$, so that $\lim_{k\to\infty} cr^k = c$. Finally, if $r > 1$, then the magnitude of

cr^k increases without bound, so that $\lim\limits_{k\to\infty} cr^k = \infty$ if $c > 0$ and $-\infty$ if $c < 0$. Thus in each case the series diverges, by the divergence test.

Thinking Forward

A series of monomials

$\sum_{k=1}^{\infty}(4x)^k$ is a geometric series with ratio $4x$, so it converges if and only if $|4x| < 1$, so that $|x| < \frac{1}{4}$.

A series of monomials

This is a geometric series with initial term $\frac{x}{3}$ and ratio $\frac{x}{3}$. Thus it converges if and only if $\left|\frac{x}{3}\right| < 1$, so if and only if $|x| < 3$.

7.5   Comparison Tests

Thinking Back

The comparison test

▷ See Theorem 5.23. It says that if f and g are functions on some interval I, bounded or unbounded, if the integral of f converges on I, and $0 \le g(x) \le f(x)$ on I, then the integral of g also converges on I. Similarly, if the integral of f diverges on I and $0 \le f(x) \le g(x)$ on I, then the integral of g on I also diverges.

▷ For a complex integrand, it may be possible to bound it below by an integrand whose integral we know diverges, or to bound it above by an integrand whose integral we know converges, on the interval in question. Then Theorem 5.23 allows us to draw a conclusion about the convergence of the integral we are concerned with.

▷ Any type. The functions could be continuous on an unbounded interval, or we may be considering an integral over a bounded interval where one or both functions have infinite discontinuities.

The limit comparison test

▷ See the discussion preceding Exercise 65 in Section 5.6. The limit comparison test says that if $\lim\limits_{x\to\infty}\frac{f(x)}{g(x)} = L$ for some positive real number L, then $\int_a^\infty f(x)\,dx$ and $\int_a^\infty g(x)\,dx$ either both converge or both diverge.

▷ For a complex integrand, it may be possible to find a simpler integrand such that the ratios approach some limit. For example, if the integrand is a rational function, then an appropriate choice for the comparison function would be the ratio of the leading terms of the function; then the limit of the ratio of the two functions will be 1.

▷ Here f and g are functions that are integrable on $[a, \infty)$ for some real number a.

Concepts

1. (a) True. This is the statement of the first part of Theorem 5.23, the comparison test for improper integrals.

    (b) False. First, the condition $0 \le f(x) \le g(x)$ cannot be satisfied given the statement about the limit. But more importantly, the limit comparison test for improper integrals does not say that both integrals converge; it only says that the integrals either both converge or both diverge.

(c) False. The comparison test says that if $0 \leq a_k \leq b_k$ and $\sum_{k=1}^{\infty} b_k$ converges, then $\sum_{k=1}^{\infty} a_k$ also converges. It says nothing about the case where the sum of the b_k does not converge. In this case, $\sum_{k=1}^{\infty} \frac{1}{k}$ is divergent.

(d) False. The comparison test says that if $0 \leq a_k \leq b_k$ and $\sum_{k=1}^{\infty} a_k$ diverges, then $\sum_{k=1}^{\infty} b_k$ also diverges. It says nothing about the case where the sum of the a_k converges. In this case, $\sum_{k=1}^{\infty} \frac{1}{k^2}$ is convergent.

(e) False. The condition $0 \leq a_k$ for all k is also needed. Otherwise, for example, we could have $b_k = \frac{1}{k^2}$ and $a_k = -1$ for all k. Then $\sum_{k=1}^{\infty} b_k$ converges but clearly $\sum_{k=1}^{\infty} a_k$ does not.

(f) False. For example, let $a_k = \{0, 1, 0, 1, 0, 1, \dots\}$ and $b_k = \{1, 0, 1, 0, 1, 0, \dots\}$. Then both $\sum_{k=1}^{\infty} a_k$ and $\sum_{k=1}^{\infty} b_k$ diverge, but $\sum_{k=1}^{\infty} (a_k \cdot b_k) = \sum_{k=1}^{\infty} 0$ converges.

(g) False. Theorem 7.33 (a) says in this case that $\sum_{k=1}^{\infty} a_k$ and $\sum_{k=1}^{\infty} b_k$ either both converge or both diverge.

(h) False. For example, $\sum_{k=1}^{\infty} \frac{1}{k^2}$ and $\sum_{k=1}^{\infty} \frac{1}{k^3}$ both converge, but $\lim_{k \to \infty} \frac{1/k^2}{1/k^3} = \lim_{k \to \infty} k = \infty$.

3. If all the terms are negative, then the terms of $\sum_{k=1}^{\infty} (-a_k)$ are all positive. So if there is a series $\{-b_k\}$ of positive terms with $0 \leq -a_k \leq -b_k$ such that $\sum_{k=1}^{\infty} (-b_k)$ converges, then $\sum_{k=1}^{\infty} (-a_k)$ converges by the comparison test. But $\sum_{k=1}^{\infty} (-b_k) = -\sum_{k=1}^{\infty} b_k$ and $\sum_{k=1}^{\infty} (-a_k) = -\sum_{k=1}^{\infty} a_k$. Thus if $\sum_{k=1}^{\infty} b_k$ converges, so does $\sum_{k=1}^{\infty} a_k$, and also $0 \geq a_k \geq b_k$ for all k. So the corresponding statement is: if $\sum_{k=1}^{\infty} a_k$ and $\sum_{k=1}^{\infty} b_k$ are two series with nonpositive terms such that $0 \geq a_k \geq b_k$ for every positive integer k, and the series $\sum_{k=1}^{\infty} b_k$ converges, then so does the series $\sum_{k=1}^{\infty} a_k$.

5. The statement is: Suppose $\sum_{k=1}^{\infty} a_k$ and $\sum_{k=1}^{\infty} b_k$ are two series such that $0 \leq a_k \leq b_k$ for all $k > K$, where K is some positive integer. If $\sum_{k=1}^{\infty} b_k$ converges, then $\sum_{k=1}^{\infty} a_k$ converges.

To prove this, note that

$$\sum_{k=1}^{\infty} b_k = \sum_{k=1}^{K-1} b_k + \sum_{k=K}^{\infty} b_k, \qquad \sum_{k=1}^{\infty} a_k = \sum_{k=1}^{K-1} a_k + \sum_{k=K}^{\infty} a_k.$$

Now, reindexing gives

$$\sum_{k=1}^{\infty} b_k = \sum_{k=1}^{K-1} b_k + \sum_{k=1}^{\infty} b_{K+k-1}, \qquad \sum_{k=1}^{\infty} a_k = \sum_{k=1}^{K-1} a_k + \sum_{k=1}^{\infty} a_{K+k-1}.$$

Since $0 \leq a_{K+k-1} \leq b_{K+k-1}$ for all positive integers k, and $\sum_{k=1}^{\infty} b_{K+k-1}$ converges since $\sum_{k=1}^{\infty} b_k$ does, we have by the comparison test that $\sum_{k=1}^{\infty} a_{K+k-1}$ converges. But this sum is the same as $\sum_{k=K}^{\infty} a_k$, which thus converges, and hence $\sum_{k=1}^{\infty} a_k = \sum_{k=1}^{K} a_k + \sum_{k=K}^{\infty} a_k$ converges since the first sum on the right is finite and the second converges.

7. The statement remains exactly the same, except that the hypothesis is that the series have terms that are eventually positive. Since the conclusion of the theorem involves only the limit of the ratio of the terms of the two series, only the tail of the series matters in the determining the limits. But the tails of the series are also all that matters in determining convergence. Thus the revised statement also holds.

9. To show that a series $\sum_{k=1}^{\infty} a_k$ converges using the comparison test, you must show that it is smaller, term by term, than a convergent series $\sum_{k=1}^{\infty} b_k$. Thus if you suspect it converges, you must compare with a convergent series. Using the limit comparison test, with part (a), to show that $\sum_{k=1}^{\infty} a_k$ converges, you must have another series $\sum_{k=1}^{\infty} b_k$ which also converges if their ratio is to give the desired conclusion. Similarly, with part (b), $\sum_{k=1}^{\infty} b_k$ must converge. (Part (c) is not relevant since it applies only to proving divergence).

11. Let $f(x) = \frac{1}{x \ln x}$. Then for $x \geq 2$, we see that f is clearly continuous, positive, and decreasing, so the integral test applies, and (with the substitution $u = \ln x$)

$$\int_2^\infty \frac{1}{x \ln x}\, dx = \lim_{t \to \infty} \int_2^t \frac{1}{x \ln x}\, dx = \lim_{t \to \infty} \int_{x=2}^{x=t} \frac{1}{u}\, du = \lim_{t \to \infty} \left[\ln |u|\right]_{x=2}^{x=t} = \lim_{t \to \infty} \left[\ln |\ln x|\right]_2^t$$
$$= \lim_{t \to \infty} \left(\ln \ln t - \ln \ln 2\right),$$

which diverges since $\ln \ln t \to \infty$.

13. We have

$$\lim_{k \to \infty} \frac{\frac{1}{k \ln k}}{\frac{1}{k^p}} = \lim_{k \to \infty} \frac{k^p}{k \ln k} = \lim_{k \to \infty} \frac{1}{k^{1-p} \ln k}.$$

As $k \to \infty$, since $1 - p > 0$, both k^{1-p} and $\ln k$ go to infinity, so this limit is zero. But the limit comparison test where the limit is zero can only be used to show convergence, not divergence (see Theorem 7.33 (b)).

15. This is part (b) of Theorem 7.33. If $\lim\limits_{k \to \infty} \frac{a_k}{b_k} = 0$ and $\sum_{k=1}^\infty b_k$ converges, then $\sum_{k=1}^\infty a_k$ converges.

17. This is part (c) of Theorem 7.33. If $\lim\limits_{k \to \infty} \frac{a_k}{b_k} = \infty$ and $\sum_{k=1}^\infty b_k$ diverges, then $\sum_{k=1}^\infty a_k$ diverges.

19. Let $b_k = \frac{k^{3/2}}{k^3} = k^{-3/2}$. Then

$$\lim_{k \to \infty} \frac{a_k}{b_k} = \lim_{k \to \infty} \frac{\frac{k^{3/2} - k - 1}{5k^3 + 3}}{k^{-3/2}}$$
$$= \lim_{k \to \infty} \frac{k^{3/2}(k^{3/2} - k - 1)}{5k^3 + 3}$$
$$= \lim_{k \to \infty} \frac{k^3 - k^{5/2} - k^{3/2}}{5k^3 + 3}$$
$$= \lim_{k \to \infty} \frac{1 - k^{-1/2} - k^{-3/2}}{5 + 3/k^3}$$
$$= \frac{1}{5}.$$

Since the limit of the ratio of the terms is a positive rational number, and $\sum_{k=1}^\infty b_k$ converges since it is a p-series with $p > 1$, it follows that $\sum_{k=1}^\infty a_k$ converges as well.

Skills

21. Let $b_k = \frac{1}{k}$. Since both series have positive terms, the limit comparison test applies. Then

$$\lim_{k \to \infty} \frac{a_k}{b_k} = \lim_{k \to \infty} \frac{\frac{3k^2 + 1}{k^3 + k^2 + 5}}{\frac{1}{k}} = \lim_{k \to \infty} \frac{3k^3 + k}{k^3 + k^2 + 5} = \lim_{k \to \infty} \frac{3 + 1/k^2}{1 + 1/k + 5/k^3} = 3.$$

Since $\sum_{k=1}^\infty b_k$ diverges (it is the harmonic series), Theorem 7.33 (a) tells us that the original series diverges as well.

23. Let $b_k = \frac{1}{k^{3/2}}$. Since both series have positive terms for $k \geq 2$, the limit comparison test applies. Then

$$\lim_{k \to \infty} \frac{a_k}{b_k} = \lim_{k \to \infty} \frac{\frac{3k-5}{k\sqrt{k^3-4}}}{\frac{1}{k^{3/2}}}$$

$$= \lim_{k \to \infty} \frac{3k^{5/2} - 5k^{3/2}}{k\sqrt{k^3-4}}$$

$$= \lim_{k \to \infty} \frac{3 - 5/k}{k \cdot k^{-1} \cdot k^{-3/2}\sqrt{k^3-4}}$$

$$= \lim_{k \to \infty} \frac{3 - 5/k}{\sqrt{1 - 4/k^3}}$$

$$= 3.$$

Since $\sum_{k=1}^{\infty} b_k$ converges, being a p-series with $p > 1$, Theorem 7.33 (a) tells us that the original series converges as well.

25. Use the comparison test with $b_k = \frac{1}{k}$. Both series have positive terms, and since for $k \geq 1$ we have $1 + \ln k \geq 1$, also $b_k \leq a_k = \frac{1+\ln k}{k}$. Since $\sum_{k=1}^{\infty} b_k$ diverges since it is the harmonic series, the contrapositive form of the comparison test tells us that $\sum_{k=1}^{\infty} a_k$ diverges as well.

27. Let $b_k = \frac{\sqrt{k}}{k^2}$. Both series have positive terms, so the limit comparison test applies, and

$$\lim_{k \to \infty} \frac{a_k}{b_k} = \lim_{k \to \infty} \frac{\frac{1+\ln k}{k^2}}{\frac{\sqrt{k}}{k^2}} = \lim_{k \to \infty} \frac{1+\ln k}{\sqrt{k}} = \lim_{k \to \infty} \left(\frac{1}{\sqrt{k}} + \frac{\ln k}{\sqrt{k}} \right).$$

The first term clearly has limit zero, and since power functions dominate logs, the second term does as well. Thus the limit is zero, and $\sum_{k=1}^{\infty} b_k = \sum_{k=1}^{\infty} \frac{\sqrt{k}}{k^2} = \sum_{k=1}^{\infty} k^{-3/2}$ converges since it is a p-series with $p > 1$. Thus by Theorem 7.33 (b), the original series converges as well.

29. Let $b_k = \frac{\sqrt{k}}{k^3}$. Both series have positive terms, so the limit comparison test applies, and

$$\lim_{k \to \infty} \frac{a_k}{b_k} = \lim_{k \to \infty} \frac{\frac{1+\ln k}{k^3}}{\frac{\sqrt{k}}{k^3}} = \lim_{k \to \infty} \frac{1+\ln k}{\sqrt{k}} = \lim_{k \to \infty} \left(\frac{1}{\sqrt{k}} + \frac{\ln k}{\sqrt{k}} \right).$$

The first term clearly has limit zero, and since power functions dominate logs, the second term does as well. Thus the limit is zero, and $\sum_{k=1}^{\infty} b_k = \sum_{k=1}^{\infty} \frac{\sqrt{k}}{k^3} = \sum_{k=1}^{\infty} k^{-5/2}$ converges since it is a p-series with $p > 1$. Thus by Theorem 7.33 (b), the original series converges as well.

31. This is a p-series with $p = \frac{1}{2}$, which has magnitude less than 1, so that this series diverges.

33. The terms of the given series are $a_k = \frac{k^{-3/4}}{k+2} = \frac{1}{k^{7/4}+2k^{3/4}}$. Since $k^{7/4} + 2k^{3/4} > k^{7/4}$, we have $0 \leq a_k \leq \frac{1}{k^{7/4}} = b_k$. Now apply the comparison test. $\sum_{k=1}^{\infty} b_k$ converges since it is a p-series with $p = \frac{7}{4}$, so that $\sum_{k=1}^{\infty} a_k$ converges as well.

35. Let $b_k = \frac{1}{k}$. Then both $a_k = \sin\left(\frac{1}{k}\right)$ and b_k are positive for $k \geq 1$ (a_k is positive since for $k \geq 1$, $\frac{1}{k}$ is in the first quadrant), so the limit comparison test applies, and

$$\lim_{k \to \infty} \frac{a_k}{b_k} = \lim_{k \to \infty} \frac{\sin\left(\frac{1}{k}\right)}{\frac{1}{k}} \stackrel{L'H}{=} \lim_{k \to \infty} \frac{\cos\left(\frac{1}{k}\right)\left(-\frac{1}{k^2}\right)}{-\frac{1}{k^2}} = \lim_{k \to \infty} \cos\left(\frac{1}{k}\right) = 1.$$

But $\sum_{k=1}^{\infty} b_k$ diverges since it is the harmonic series, so by Theorem 7.33(a), the original series diverges as well.

37. For $k \geq 3$, we have $\ln k > 1$ so that $\frac{\ln k}{k} \geq \frac{1}{k} \geq 0$. Now, $\sum_{k=1}^{\infty} \frac{1}{k}$ diverges because it is the harmonic series. Since the given series is eventually bounded below by $\{\frac{1}{k}\}$, it diverges as well (see the variant of the comparison test in Exercise 5).

39. The terms of the series are

$$\frac{\sqrt{k+1}-\sqrt{k}}{k} = \frac{\sqrt{k+1}-\sqrt{k}}{k} \cdot \frac{\sqrt{k+1}+\sqrt{k}}{\sqrt{k+1}+\sqrt{k}} = \frac{1}{k(\sqrt{k+1}+\sqrt{k})} \leq \frac{1}{k \cdot 2\sqrt{k}} \leq \frac{1}{k^{3/2}}.$$

Thus the terms of the series are bounded by a convergent p-series with $p = \frac{3}{2}$, so that the given series also converges, by the comparison test.

41. Let $f(x) = x^2 e^{-x^3}$. Clearly f is positive and continuous for $x \geq 1$; since $f'(x) = 2xe^{-x^3} - 3x^2 \cdot x^2 e^{-x^3} = (2x - 3x^4)e^{-x^3}$, we also have $f'(x) < 0$ for $x \geq 1$, so that f is decreasing there and the integral test applies. Then using the substitution $u = x^3$, we have

$$\int_1^{\infty} x^2 e^{-x^3}\, dx = \lim_{t\to\infty} \int_1^t x^2 e^{-x^3}\, dx = \frac{1}{3}\lim_{t\to\infty}\int_{x=1}^{x=t} e^{-u}\, du = \frac{1}{3}\lim_{t\to\infty}\left[-e^{-u}\right]_{x=1}^{x=t}$$

$$= \frac{1}{3}\lim_{t\to\infty}\left[-e^{-x^3}\right]_1^t = \frac{1}{3}\lim_{t\to\infty}\left(e^{-1} - e^{-t^3}\right),$$

which converges. So by the integral test, the series converges as well.

43. Let $b_k = \frac{1}{k^2}$, so that $\sum_{k=1}^{\infty} b_k$ is a convergent p-series. Then since both $a_k = \frac{k^2}{e^k}$ and b_k are series of positive terms, the limit comparison test applies, and

$$\lim_{k\to\infty}\frac{a_k}{b_k} = \lim_{k\to\infty}\frac{\frac{k^2}{e^k}}{\frac{1}{k^2}} = \lim_{k\to\infty}\frac{k^4}{e^k} = 0$$

since exponential functions dominate power functions. Then by Theorem 7.33 (b), since $\sum_{k=1}^{\infty} b_k$ converges, $\sum_{k=1}^{\infty} a_k$ converges as well.

45. Let $f(x) = \frac{1}{\sqrt{x}e^{\sqrt{x}}}$. Then clearly f is continuous and positive on $[1, \infty)$; equally clearly, since the denominator is increasing, f is decreasing there. So the integral test applies, and then (using the substitution $u = \sqrt{x}$, so that $du = \frac{1}{2}x^{-1/2}$),

$$\int_1^{\infty} f(x)\, dx = \lim_{t\to\infty}\int_1^t \frac{1}{\sqrt{x}e^{\sqrt{x}}}\, dx = 2\lim_{t\to\infty}\int_{x=1}^{x=t}\frac{1}{e^u}\, du = 2\lim_{t\to\infty}\left[-e^{-u}\right]_{x=1}^{x=t}$$

$$= 2\lim_{t\to\infty}\left[-e^{-\sqrt{x}}\right]_1^t = 2\lim_{t\to\infty}\left(e^{-1} - e^{-\sqrt{t}}\right) = 2e^{-1}.$$

Since the integral converges, the series converges as well.

47. Use the limit comparison test with $b_k = \frac{1}{k}$. We have

$$\lim_{k\to\infty}\frac{\frac{1}{2k+7}}{\frac{1}{k}} = \lim_{k\to\infty}\frac{k}{2k+7} = \lim_{k\to\infty}\frac{1}{2+7/k} = \frac{1}{2}.$$

Since $\sum_{k=1}^{\infty} b_k$ is a p-series with $p = 1$, it is divergent, so by Theorem 7.33(a), $\sum_{k=1}^{\infty}\frac{1}{2k+7}$ is divergent as well.

49. This splits into several cases. For $p \leq 0$, we have

$$\frac{k^p}{e^k} = \frac{1}{k^{-p}e^k} \leq \frac{1}{e^k}$$

since $-p > 0$ so that $k^{-p} \geq 1$. But $\sum_{k=1}^{\infty} \frac{1}{e^k}$ converges by the integral test: let $f(x) = e^{-x}$; then f is continuous, positive, and decreasing for $x \geq 1$, and

$$\int_1^{\infty} f(x)\,dx = \lim_{t \to \infty} \int_1^t e^{-x}\,dx = \lim_{t \to \infty} \left[-e^{-x}\right]_1^t = \lim_{t \to \infty} \left(e^{-1} - e^{-t}\right) = e^{-1}.$$

Since the integral converges, so does $\sum_{k=1}^{\infty} \frac{1}{e^k}$. This shows that the original series converges for all $p \leq 0$.

Next assume $p > 1$, and use the limit comparison test with $b_k = k^{-p}$, so that $\sum_{k=1}^{\infty} b_k$ is a convergent p-series. Then since both $a_k = \frac{k^p}{e^k}$ and b_k are series of positive terms, the limit comparison test applies, and

$$\lim_{k \to \infty} \frac{a_k}{b_k} = \lim_{k \to \infty} \frac{\frac{k^p}{e^k}}{\frac{1}{k^p}} = \lim_{k \to \infty} \frac{k^{2p}}{e^k} = 0$$

since exponential functions dominate power functions. Then by Theorem 7.33 (b), since $\sum_{k=1}^{\infty} b_k$ converges, $\sum_{k=1}^{\infty} a_k$ converges as well.

If $p = 1$, we use the integral test again. Let $f(x) = \frac{x}{e^x}$; then ince $f'(x) = \frac{e^x - xe^x}{e^{2x}} = (1 - x)e^{-x}$, we see that $f(x)$ is decreasing for $x \geq 1$; clearly $f(x)$ is continuous and positive for $x \geq 1$. Thus the integral test applies, and (using integration by parts with $u = x$ and $dv = e^{-x}\,dx$),

$$\int_1^{\infty} \frac{x}{e^x}\,dx = \lim_{t \to \infty} \int_1^t xe^{-x}\,dx$$

$$= \lim_{t \to \infty} \left(\left[-xe^{-x}\right]_1^t + \int_1^t e^{-x}\,dx \right)$$

$$= \lim_{t \to \infty} \left(-te^{-t} + e^{-1} - \left[e^{-x}\right]_1^t \right)$$

$$= \lim_{t \to \infty} \left(-te^{-t} + e^{-1} - e^{-t} + e^{-1} \right)$$

$$= 2e^{-1}.$$

Since the integral converges, so does the series.

Finally, if $0 < p < 1$, note that $k^p \leq k$, so that $\frac{k^p}{e^k} \leq \frac{k}{e^k}$. Since $\sum_{k=1}^{\infty} \frac{k}{e^k}$ converges by the previous case, so does $\sum_{k=1}^{\infty} \frac{k^p}{e^k}$, by the comparison test.

In summary, $\sum_{k=1}^{\infty} \frac{k^p}{e^k}$ converges for all p.

Applications

51. Let $\alpha(k) = \sum_{j=1}^{n} a_k k^j$, and let $A = \max(r^{a_1}, r^{a_2}, \ldots r^{a_n})$. Then

$$Wr^{\alpha(k)} = Wr^{\sum_{j=1}^{n} a_k k^j} = W\prod_{j=1}^{n} \left(r^{a_k k^j}\right) = W\prod_{j=1}^{n} r^{a_k} r^{k^j} = Wr^{a_k} \prod_{j=1}^{n} (r^j)^k \leq WA \prod_{j=1}^{n} (r^j)^k.$$

In order to prove that $V = \sum_{k=0}^{\infty} w_k$ converges, and since

$$0 \leq w_k \leq Wr^{\alpha(k)} \leq WA \prod_{j=1}^{n} (r^j)^k,$$

it suffices to show that $\sum_{k=0}^{\infty} \left(WA \prod_{j=1}^{n} (r^j)^k \right) = WA \sum_{k=0}^{\infty} \left(\prod_{j=1}^{n} (r^j)^k \right)$ converges, by the comparison test. Let $R_j = \sum_{k=0}^{\infty} (r^j)^k$ be the geometric series with ratio r^j, for $j = 1, 2, \ldots, n$. Since $r < 1$, also $r^j < 1$, so that R_j is convergent. Now, Exercise 58 in Section 4 shows that if $\sum_{k=1}^{\infty} a_k$ and $\sum_{k=1}^{\infty} b_k$ are both convergent, then so is $\sum_{k=1}^{\infty} a_k b_k$. Since R_1 and R_2 are both convergent, so is $\sum_{k=0}^{\infty} (r^1)^k \cdot (r^2)^k$. Since R_3 is also convergent, so is $\sum_{k=0}^{\infty} (r^1)^k \cdot (r^2)^k \cdot (r^3)^k$. Continue to see that

$$\sum_{k=0}^{\infty} (r^1)^k \cdot (r^2)^k \cdots (r^n)^k = \sum_{k=0}^{\infty} \left(\prod_{j=1}^{n} (r^j)^k \right)$$

is convergent, and thus so is $WA \sum_{k=0}^{\infty} \left(\prod_{j=1}^{n} (r^j)^k \right)$. So by the comparison test $V = \sum_{k=0}^{\infty} w_k$ is also convergent.

Proofs

53. You can simply apply Exercise 58 in the preceding section. For a direct proof, use the limit comparison test:

$$\lim_{k \to \infty} \frac{a_k^2}{a_k} = \lim_{k \to \infty} a_k = 0,$$

where the limit is zero by the contrapositive of the divergence test since $\sum_{k=1}^{\infty} a_k$ converges. Since $\sum_{k=1}^{\infty} a_k$ converges, part (b) of Theorem 7.33 tells us that $\sum_{k=1}^{\infty} a_k^2$ converges as well.

55. Use the limit comparison test:

$$\lim_{k \to \infty} \frac{a_k b_k}{a_k} = \lim_{k \to \infty} b_k = 0,$$

where the limit is zero by the contrapositive of the divergence test since $\sum_{k=1}^{\infty} b_k$ converges. Since $\sum_{k=1}^{\infty} a_k$ converges, part (b) of Theorem 7.33 tells us that $\sum_{k=1}^{\infty} a_k b_k$ converges as well.

57. Suppose that $\lim_{k \to \infty} \frac{a_k}{b_k} = 0$ where a_k and b_k are nonnegative for all k, and assume that $\sum_{k=1}^{\infty} b_k$ converges. Choose $\epsilon > 0$, and then choose N such that for $k \geq N$, we have

$$\left| \frac{a_k}{b_k} - 0 \right| = \left| \frac{a_k}{b_k} \right| = \frac{a_k}{b_k} < \epsilon.$$

Thus $a_k < \epsilon b_k$ for $k \geq N$, and hence

$$\sum_{k=N}^{\infty} a_k < \sum_{k=N}^{\infty} (\epsilon b_k) = \epsilon \sum_{k=N}^{\infty} b_k.$$

Since the convergence of a series depends only on the tail of the series, and since $\sum_{k=1}^{\infty} b_k$ converges, we see that $\sum_{k=N}^{\infty} b_k$ converges. But then the partial sums of $\sum_{k=N}^{\infty} a_k$ are a monotonically increasing sequence that is bounded above by $\epsilon \sum_{k=N}^{\infty} b_k$, so the limit of the partial sums exists, and thus $\sum_{k=N}^{\infty} a_k$ converges. Again, since convergence depends only on the tail of the series, this implies that $\sum_{k=1}^{\infty} a_k$ converges.

Thinking Forward

A series of monomials

If $|x| > 1$, then $\sum_{k=1}^{\infty} \frac{x^{2k}}{k^2} = \sum_{k=1}^{\infty} \frac{(x^2)^k}{k^2} = \infty$ since exponential functions dominate power functions. If $|x| < 1$, use the limit comparison test with $b_k = \frac{1}{k^2}$:

$$\lim_{k \to \infty} \frac{\frac{x^{2k}}{k^2}}{\frac{1}{k^2}} = \lim_{k \to \infty} x^{2k} = 0$$

since $|x| < 1$. Since $\sum_{k=1}^{\infty} b_k$ converges, Theorem 7.33(b) tells us that $\sum_{k=1}^{\infty} \frac{x^{2k}}{k^2}$ converges as well. Finally, if $|x| = 1$, then $x^{2k} = 1$, so that the series is $\sum_{k=1}^{\infty} \frac{1}{k^2}$, which converges since it is a p-series with $p > 1$. Thus this series converges precisely when $|x| \le 1$.

A series of monomials

Note that $\sum_{k=0}^{\infty} \frac{|x|}{4^k} = |x| \sum_{k=0}^{\infty} \frac{1}{4^k}$. This is now a geometric series with ratio $\frac{1}{4}$, so it converges for all x.

7.6   The Ratio and Root Tests

Thinking Back

An indeterminate form of the type ∞^0

As $k \to \infty$, we see that $\frac{1}{k} \to 0$, so that $\lim_{k\to\infty} k^{1/k}$ has the indeterminate form ∞^0. To evaluate the limit, first take logs and then use L'Hôpital's rule to evaluate the result. $\ln\left(k^{1/k}\right) = \frac{1}{k}\ln k$, and then

$$\lim_{k\to\infty}\frac{\ln k}{k} \stackrel{\text{L'H}}{=} \lim_{k\to\infty}\frac{1/k}{1} = \lim_{k\to\infty}\frac{1}{k} = 0.$$

Since the limit of the log is zero, the original limit is $\lim_{k\to\infty} k^{1/k} = e^0 = 1$.

Basic functions

A *monomial* is an expression of the form ax^n where $a \in \mathbb{R}$, n is a nonnegative integer, and x is a variable. A *power function* is an expression of the form ax^n where $a \in \mathbb{R}$, n is a rational number, and x is a variable. A *polynomial* is a finite sum of monomials, so is an expression of the form $a_n x^n + a_{n-1}x^{n-1} + \cdots + a_1 x + a_0$ where all the a_i are real numbers and n is a nonnegative integer. Finally, a *rational function* is a quotient $\frac{p(x)}{q(x)}$ where $p(x)$ and $q(x)$ are polynomial functions and $q(x)$ is not identically zero.

Concepts

1. (a) True. Since a_k is a nonzero polynomial function, either it is a nonzero constant, in which case $\sum_{k=1}^{\infty} a_k$ clearly diverges by the divergence test, or it is a nonconstant, nonzero polynomial, in which case $\lim_{k\to\infty} a_k$ is determined by the leading term of a_k, which is of the form ax^n for some $n > 0$ and $a \ne 0$. But $\lim_{k\to\infty} ax^n = \pm\infty$ depending on the sign of a, so that the terms of the series do not converge to zero and thus the series diverges by the divergence test.

    (b) True. Recall that if $\frac{p(x)}{q(x)}$ is a rational function, then $\lim_{x\to\infty}\frac{p(x)}{q(x)}$ is the limit of the ratio of the leading terms of the two polynomials. If $a_k = b_n k^n + b_{n-1}k^{n-1} + \cdots + b_1 k + b_0$ where $b_n \ne 0$, then

$$\lim_{k\to\infty}\frac{b_n(k+1)^n + b_{n-1}(k+1)^{n-1} + \cdots + b_1(k+1) + b_0}{b_n k^n + b_{n-1}k^{n-1} + \cdots + b_1 k + b_0} \quad \lim_{k\to\infty}\frac{b_n k^n + \text{lower-degree terms}}{b_n k^n + \text{lower-degree terms}} = 1.$$

    since the leading terms are equal in numerator and denominator. Thus the ratio test produces 1, so it is inconclusive.

(c) True, for the same reason as in part (b): the limit of a rational function of k as $k \to \infty$ is determined by its leading terms. Let

$$a_k = \frac{b_n k^n + b_{n-1} k^{n-1} + \cdots + b_1 k + b_0}{c_m k^m + c_{m-1} k^{m-1} + \cdots + c_1 k + c_0},$$

where $b_n \neq 0$ and $c_m \neq 0$. Then

$$\lim_{k \to \infty} \frac{a_{k+1}}{a_k} = \lim_{k \to \infty} \frac{\frac{b_n(k+1)^n + b_{n-1}(k+1)^{n-1} + \cdots + b_1(k+1) + b_0}{c_m(k+1)^m + c_{m-1}(k+1)^{m-1} + \cdots + c_1(k+1) + c_0}}{\frac{b_n k^n + b_{n-1} k^{n-1} + \cdots + b_1 k + b_0}{c_m k^m + c_{m-1} k^{m-1} + \cdots + c_1 k + c_0}}$$

$$= \lim_{k \to \infty} \frac{(b_n(k+1)^n + b_{n-1}(k+1)^{n-1} + \cdots + b_1(k+1) + b_0)(c_m k^m + c_{m-1} k^{m-1} + \cdots + c_1 k + c_0)}{(b_n k^n + b_{n-1} k^{n-1} + \cdots + b_1 k + b_0)(c_m(k+1)^m + c_{m-1}(k+1)^{m-1} + \cdots + c_1(k+1) + c_0)}$$

$$= \lim_{k \to \infty} \frac{b_n c_m k^{m+n} + \text{lower-degree terms}}{b_n c_m k^{m+n} + \text{lower-degree terms}}$$

$$= 1.$$

Thus the ratio test produces 1, so it is inconclusive.

(d) False. While the ratio test is often effective in this situation, it may not always be effective. It depends on the characteristics of the rest of the function. For example, it would not be terribly useful when trying to determine if $\sum_{k=1}^{\infty} \frac{1}{1+\sin^2(k!)}$ converges.

(e) True. See the proof.

(f) False. The root test is useful primarily for series involving k^{th} powers. Using the root test on, for example, $\sum_{k=1}^{\infty} \frac{1}{k^2}$ leaves us with the problem of evaluating $\lim_{k \to \infty} k^{-2/k}$. The root test is generally inconclusive for testing a p-series.

(g) False. For instance, Example 4 uses the root test to show divergence of $\sum_{k=1}^{\infty} \frac{3^k}{k^5}$.

(h) False. $\frac{(n+2)!}{n!} = \frac{(n+2)(n+1)(n)(n-1)\ldots(2)(1)}{n(n-1)\ldots(2)(1)} = (n+2)(n+1)$.

3. Recall that if $\frac{r(x)}{s(x)}$ is a rational function, then $\lim_{x \to \infty} \frac{r(x)}{s(x)}$ is the limit of the ratio of the leading terms of the two polynomials. If $p(x) = b_n x^n + b_{n-1} x^{n-1} + \cdots + b_1 x + b_0$ where $b_n \neq 0$, then

$$\lim_{x \to \infty} \frac{p(x+1)}{p(x)} = \lim_{x \to \infty} \frac{b_n(x+1)^n + b_{n-1}(x+1)^{n-1} + \cdots + b_1(x+1) + b_0}{b_n x^n + b_{n-1} x^{n-1} + \cdots + b_1 x + b_0}$$

$$= \lim_{x \to \infty} \frac{b_n x^n + \text{lower-degree terms}}{b_n x^n + \text{lower-degree terms}} = 1$$

since the leading terms are equal in numerator and denominator.

5. Let

$$r(x) = \frac{b_n x^n + b_{n-1} x^{n-1} + \cdots + b_1 x + b_0}{c_m x^m + c_{m-1} x^{m-1} + \cdots + c_1 x + c_0},$$

where $b_n \neq 0$ and $c_m \neq 0$ be a rational function. Then the limit as $k \to \infty$ of a rational function of

k is determined by its leading terms,

$$
\begin{aligned}
\lim_{x\to\infty} \frac{r(x+1)}{r(x)} &= \lim_{x\to\infty} \frac{\frac{b_n(x+1)^n + b_{n-1}(x+1)^{n-1} + \cdots + b_1(x+1) + b_0}{c_m(x+1)^m + c_{m-1}(x+1)^{m-1} + \cdots + c_1(x+1) + c_0}}{\frac{b_n x^n + b_{n-1}x^{n-1} + \cdots + b_1 x + b_0}{c_m x^m + c_{m-1}x^{m-1} + \cdots + c_1 x + c_0}} \\
&= \lim_{x\to\infty} \frac{(b_n(x+1)^n + b_{n-1}(x+1)^{n-1} + \cdots + b_1(x+1) + b_0)(c_m x^m + c_{m-1}x^{m-1} + \cdots + c_1 x + c_0)}{(b_n x^n + b_{n-1}x^{n-1} + \cdots + b_1 x + b_0)(c_m(x+1)^m + c_{m-1}(x+1)^{m-1} + \cdots + c_1(x+1) + c_0)} \\
&= \lim_{x\to\infty} \frac{b_n c_m x^{m+n} + \text{lower-degree terms}}{b_n c_m x^{m+n} + \text{lower-degree terms}} \\
&= 1.
\end{aligned}
$$

7. Since $\lim\limits_{k\to\infty} \frac{a_{k+1}}{a_k} = \lim\limits_{k\to\infty} \frac{|a_{k+1}|}{|a_k|}$ because the ratio of two negative numbers is positive, the test does not change: If $\sum_{k=1}^{\infty} a_k$ is a series with negative terms, and $\rho = \lim\limits_{k\to\infty} \frac{a_{k+1}}{a_k}$, then the series converges if $\rho < 1$ and diverges if $\rho > 1$. If $\rho = 1$, this test gives no information.

9. The ratio test shows divergence since (in the language of Theorem 7.34) $\rho > 1$. The divergence test also shows divergence, since if $\lim\limits_{k\to\infty} \frac{a_{k+1}}{a_k} > 1$, then there is some N such that for all $k > N$, we have $a_k > a_N$. But this means that $\lim\limits_{k\to\infty} a_k \geq a_N > 0$, so that the terms of the series do not tend to zero.

11. Suppose $m(x) = Ax^p$ is a power function. Then

$$
\lim_{x\to\infty} (m(x))^{1/x} = \lim_{x\to\infty} \left(A^{1/x}(x^p)^{1/x} \right) = \lim_{x\to\infty} \left(A^{1/x}(x^{1/x})^p \right).
$$

But $\lim\limits_{x\to\infty} A^{1/x} = A^0 = 1$ and $\lim\limits_{x\to\infty} x^{1/x} = 1$ (see for example Exercise 65 in Section 7.2), so that

$$
\lim_{x\to\infty} \left(A^{1/x}(x^{1/x})^p \right) = \left(\lim_{x\to\infty} A^{1/x} \right) \left(\lim_{x\to\infty} x^{1/x} \right)^p = 1.
$$

13. Clearly $\sum_{k=1}^{\infty} a_k$ converges if and only if $\sum_{k=1}^{\infty} (-a_k)$ converges. If the a_k are all negative, then $-a_k$ are all positive, so that if $\rho = \lim\limits_{k\to\infty} (-a_k)^{1/k}$, then $\sum_{k=1}^{\infty} (-a_k)$ and thus $\sum_{k=1}^{\infty} a_k$ converge if $\rho < 1$ and diverge if $\rho > 1$. If $\rho = 1$, the test is inconclusive.

15. It would be difficult to use the root test because we have no way of evaluating the limit $\lim\limits_{k\to\infty} \frac{1}{(k!)^{1/k}}$.

17. The series converges since it is a geometric series. It also converges by the integral test with $f(x) = \frac{1}{3^x}$ since

$$
\int_1^{\infty} f(x)\,dx = \lim_{t\to\infty} \left[-\frac{1}{\ln 3} 3^{-x} \right]_1^t = \lim_{t\to\infty} \left(\frac{1}{3\ln 3} - \frac{1}{3^t \ln 3} \right) = \frac{1}{3\ln 3}.
$$

The root test also works, since $\lim\limits_{k\to\infty} \left(\frac{1}{3^k} \right)^{1/k} = \frac{1}{3} < 1$. The ratio test works as well, since

$$
\lim_{k\to\infty} \frac{a_{k+1}}{a_k} = \lim_{k\to\infty} \frac{\frac{1}{3^{k+1}}}{\frac{1}{3^k}} = \lim_{k\to\infty} \frac{3^k}{3^{k+1}} = \frac{1}{3} < 1.
$$

Finally, the comparison or limit comparison test against a p-series such as $\sum_{k=1}^{\infty} \frac{1}{k^2}$ can also be made to work.

19. Since $a_{2n-1} = 10^{-n}$ and $a_{2n} = \frac{1}{5}a_{2n-1}$, we have

$$\frac{a_{2n}}{a_{2n-1}} = \frac{1}{5}, \qquad \frac{a_{2n+1}}{a_{2n}} = \frac{10^{-(n+1)}}{\frac{1}{5}10^{-n}} = \frac{5}{10} = \frac{1}{2},$$

so that the sequence $\{\frac{a_{k+1}}{a_k}\}$ alternates between $\frac{1}{5}$ and $\frac{1}{2}$ and thus does not have a limit. However, this series is

$$1 + \frac{1}{5} + \frac{1}{10} + \frac{1}{50} + \frac{1}{100} + \frac{1}{500} + \cdots$$

$$= \left(1 + \frac{1}{5}\right) + \frac{1}{10}\left(1 + \frac{1}{5}\right) + \frac{1}{100}\left(1 + \frac{1}{5}\right) + \cdots$$

$$= \frac{6}{5}\sum_{k=0}^{\infty}\left(\frac{1}{10}\right)^k$$

$$= \frac{6}{5} \cdot \frac{1}{1 - \frac{1}{10}} = \frac{6}{5} \cdot \frac{10}{9} = \frac{4}{3}.$$

Skills

21. $\frac{8!}{10!} = \frac{1 \cdot 2 \cdot 3 \cdots 8}{1 \cdot 2 \cdot 3 \cdots 10} = \frac{1}{9 \cdot 10} = \frac{1}{90}$.

23. $\frac{k!}{(k+2)!} = \frac{1 \cdot 2 \cdot 3 \cdots k}{1 \cdot 2 \cdot 3 \cdots k \cdot (k+1) \cdot (k+2)} = \frac{1}{(k+1)(k+2)}$.

25. $\frac{4!}{(2!)!} = \frac{4!}{2!} = \frac{24}{2} = 12$.

27. $\frac{(n+1)!}{(n-2)!} = \frac{1 \cdot 2 \cdot 3 \cdots (n-2)(n-1)(n)(n+1)}{1 \cdot 2 \cdot 3 \cdots (n-2)} = n(n-1)(n+1)$.

29. Using the ratio test, we have

$$\rho = \lim_{k \to \infty} \frac{a_{k+1}}{a_k} = \lim_{k \to \infty} \frac{\frac{5^{k+1}}{(k+1)!}}{\frac{5^k}{k!}} = \lim_{k \to \infty} \frac{5}{k} = 0.$$

Since $\rho = 0 < 1$, the series converges by the ratio test.

31. Since $\frac{1}{k^3+1}$ is a rational function, the ratio test will be inconclusive, by Exercise 6. Instead, use the comparison test with $\frac{1}{k^3}$. Since $k^3 + 1 > k^3$ for $k \geq 1$, it follows that $0 < \frac{1}{k^3+1} < \frac{1}{k^3}$. Since $\sum_{k=1}^{\infty} \frac{1}{k^3}$ converges, being a p-series with $p > 1$, it follows that $\sum_{k=1}^{\infty} \frac{1}{1+k^3}$ also converges.

33. To compute $\lim_{k \to \infty} \frac{a_{k+1}}{a_k}$, note that a_{k+1} is just like a_k except that it has an extra factor of $(2k+2)$ in the numerator, and of $(2k+1)$ in the denominator. Thus

$$\rho = \lim_{k \to \infty} \frac{a_{k+1}}{a_k} = \lim_{k \to \infty} \frac{2k+2}{2k+1} = \lim_{k \to \infty} \frac{2 + 2/k}{2 + 1/k} = 1.$$

Since $\rho = 1$, the ratio test is inconclusive. However, the divergence test shows that the series diverges: the numerator and denominator each have k factors, and each factor in the numerator is greater than the corresponding factor in the denominator. Thus each term is at least 1, so the terms do not converge to zero and thus the series diverges by the divergence test.

35. Recalling that $\lim\limits_{k\to\infty} k^{1/k} = 1$, we have

$$\rho = \lim_{k\to\infty} \left(\frac{5^k}{k^5}\right)^{1/k} = \lim_{k\to\infty} \frac{(5^k)^{1/k}}{(k^5)^{1/k}} = \lim_{k\to\infty} \frac{5}{(k^{1/k})^5} = \frac{5}{(\lim\limits_{k\to\infty} k^{1/k})^5} = 5.$$

Since $\rho = 5 > 1$, the series diverges by the root test.

37. The root test gives

$$\rho = \lim_{k\to\infty} \left(\frac{1}{k^3+1}\right)^{1/k} = \lim_{k\to\infty} (k^3+1)^{-1/k}.$$

To evaluate this limit, evaluate the limit of the log, which is $-\frac{1}{k}\ln(k^3+1)$, using L'Hôpital's rule where needed:

$$-\lim_{k\to\infty} \frac{\ln(k^3+1)}{k} \overset{\text{L'H}}{=} -\lim_{k\to\infty} \frac{\frac{1}{k^3+1}\cdot 3k^2}{1} = -\lim_{k\to\infty} \frac{3k^2}{k^3+1} = -\lim_{k\to\infty} \frac{3}{k+1/k^2} = 0.$$

Since this limit is zero, $\rho = e^0 = 1$, so the root test is inconclusive. Instead use the comparison test with $\frac{1}{k^3}$. Since $k^3+1 > k^3$ for $k \geq 1$, it follows that $0 < \frac{1}{k^3+1} < \frac{1}{k^3}$. Since $\sum_{k=1}^{\infty} \frac{1}{k^3}$ converges, being a p-series with $p > 1$, it follows that $\sum_{k=1}^{\infty} \frac{1}{1+k^3}$ also converges.

39. Using the root test, we have

$$\rho = \lim_{k\to\infty} \left(\left(\frac{1+4k}{5k-1}\right)^k\right)^{1/k} = \lim_{k\to\infty} \frac{4k+1}{5k-1} = \lim_{k\to\infty} \frac{4+1/k}{5-1/k} = \frac{4}{5}.$$

Since $\rho = \frac{4}{5} < 1$, the series converges by the root test.

41. Use the root test:

$$\rho = \lim_{k\to\infty} \left(\left(\frac{k+1}{3k+1}\right)^k\right)^{1/k} = \lim_{k\to\infty} \frac{k+1}{3k+1} = \lim_{k\to\infty} \frac{1+1/k}{3+1/k} = \frac{1}{3}.$$

Since $\rho = \frac{1}{3} < 1$, the series converges by the root test.

43. Use the limit comparison test with $\frac{1}{k}$. We have

$$\lim_{k\to\infty} \frac{\frac{4}{k+3}}{\frac{1}{k}} = \lim_{k\to\infty} \frac{4k}{k+3} = \lim_{k\to\infty} \frac{4}{1+3/k} = 4.$$

The ratio is a nonzero real number, and $\sum_{k=1}^{\infty} \frac{1}{k}$ is the divergent harmonic series, so $\sum_{k=1}^{\infty} \frac{4}{k+3}$ diverges as well, by the limit comparison test.

45. Use the comparison test. We have

$$\frac{k!}{(k+3)!} = \frac{1}{(k+1)(k+2)(k+3)} < \frac{1}{k^3}.$$

Since $\sum_{k=1}^{\infty} \frac{1}{k^3}$ is a p-series with $p > 1$, it converges, so that the original series does as well.

47. Use the limit comparison test with $k^{-3/2}$:

$$\lim_{k\to\infty} \frac{\frac{\sqrt{k}}{k^2+3}}{k^{-3/2}} = \lim_{k\to\infty} \frac{k^2}{k^2+3} = \lim_{k\to\infty} \frac{1}{1+3/k^2} = 1.$$

Since the limit of the ratio is a nonzero real, and since $k^{-3/2}$ is a convergent p-series (with $p > 1$), the original series converges as well, by the limit comparison test.

49. Use the root test:

$$\rho = \lim_{k\to\infty} (a_k)^{1/k} = \lim_{k\to\infty} \left(\frac{3^k}{k^k}\right)^{1/k} = \lim_{k\to\infty} \frac{3}{k} = 0.$$

Since $\rho = 0 < 1$, the series converges by the root test.

51. Use the ratio test:

$$\rho = \lim_{k\to\infty} \frac{a_{k+1}}{a_k} = \lim_{k\to\infty} \frac{\frac{(2(k+1))!}{((k+1)!)!}}{\frac{(2k)!}{(k)!}}$$

$$= \lim_{k\to\infty} \frac{(2k+2)!(k!)!}{(2k)!((k+1)!)!}$$

$$= \lim_{k\to\infty} (2k+1)(2k+2) \cdot \frac{(k!)!}{((k+1)!)!}.$$

Now, look at the fraction. The numerator is the product of the integers from 1 up to $k!$, and the denominator is the product of the integers from 1 up to $(k+1)!$, which is $(k+1)k!$. Thus the fraction is $\frac{1}{p(k)}$, where $p(k)$ is a polynomial of degree $(k+1)$ in k. Thus for $k > 2$, the fraction is no larger than

$$\frac{(2k+1)(2k+2)}{\text{some polynomial of degree at least 3 in } k}.$$

But the limit of such a fraction as $k \to \infty$ is zero. So by the ratio test, $\rho = 0 < 1$, so the series converges.

53. Use the ratio test. a_{k+1} is the same as a_k except for an extra factor of $3k+2$ in the numerator and an extra factor of $2k+3$ in the denominator. Thus

$$\rho = \lim_{k\to\infty} \frac{a_{k+1}}{a_k} = \lim_{k\to\infty} \frac{3k+2}{2k+3} = \lim_{k\to\infty} \frac{3+2/k}{2+3/k} = \frac{3}{2}.$$

Since $\rho = \frac{3}{2} > 1$, the series diverges.

55. Use the root test.

$$\rho = \lim_{k\to\infty} \sqrt[k]{a_k} = \lim_{k\to\infty} \left(\left(\frac{k}{k+1}\right)^{k^2}\right)^{1/k} = \lim_{k\to\infty} \left(\frac{k}{k+1}\right)^{k} = \frac{1}{\lim_{k\to\infty}\left(\frac{k+1}{k}\right)^k} = \frac{1}{\lim_{k\to\infty}\left(1+\frac{1}{k}\right)^k} = \frac{1}{e}.$$

Thus $\rho = e^{-1} < 1$, so that the original series converges by the root test.

57. Use the ratio test. a_{k+1} is the same as a_k except for an additional factor of $k+1$ in the numerator and an additional factor of $2k+1$ in the denominator. Thus

$$\rho = \lim_{k\to\infty} \frac{a_{k+1}}{a_k} = \lim_{k\to\infty} \frac{k+1}{2k+1} = \frac{1}{2}.$$

Since $\rho = \frac{1}{2} < 1$, the series converges by the ratio test.

59. Write a_k as

$$a_k = \frac{1 \cdot 4 \cdot 7 \cdots (3k-2)}{3 \cdot 6 \cdot 9 \cdots (3k)} = \frac{1}{3k} \cdot \frac{4}{3} \cdot \frac{7}{6} \cdot \frac{10}{9} \cdots \frac{3k-2}{3k-3}.$$

Now, each of the fractions in that product after the first is greater than 1, so that $a_k \geq \frac{1}{3k}$ for $k \geq 1$. However, $\sum_{k=1}^{\infty} \frac{1}{3k} = \frac{1}{3} \sum_{k=1}^{\infty} \frac{1}{k}$ diverges since it is a p-series with $p = 1$. So by the comparison test, $\sum_{k=1}^{\infty} a_k$ diverges as well.

61. (a) Use the ratio test:

$$\rho = \lim_{k \to \infty} \frac{a_{k+1}}{a_k} = \lim_{k \to \infty} \frac{\frac{(k+1)^n}{r^{k+1}}}{\frac{k^n}{r^k}} = \lim_{k \to \infty} \left(\frac{1}{r} \left(\frac{k+1}{k} \right)^n \right) = \frac{1}{r} \left(\lim_{k \to \infty} \frac{k+1}{k} \right)^n = \frac{1}{r} \cdot 1^n = \frac{1}{r}.$$

Since $r > 1$, we have $\rho = \frac{1}{r} < 1$, so that the series converges by the ratio test.

(b) Since the series converges, the contrapositive of the divergence test tells us that the terms of the series converges to zero, which means that $\lim_{k \to \infty} \frac{k^n}{r^k} = 0$.

(c) Part (b) tells us that exponential growth dominates monomial growth. But a polynomial is simply a sum of monomials with coefficients, so that

$$\lim_{k \to \infty} \frac{a_n x^n + a_{n-1} x^{n-1} + \cdots + a_0}{r^k} = \sum_{j=0}^{n} \left(a_n \lim_{k \to \infty} \frac{x^n}{r^k} \right) = 0.$$

This says that exponential growth dominates polynomial growth.

63. (a) Use the ratio test:

$$\rho = \lim_{k \to \infty} \frac{a_{k+1}}{a_k} = \lim_{k \to \infty} \frac{\frac{(k+1)!}{(k+1)^{k+1}}}{\frac{k!}{k^k}} = \lim_{k \to \infty} \frac{(k+1)k^k}{(k+1)^{k+1}} = \lim_{k \to \infty} \left(\frac{k}{k+1} \right)^k = \frac{1}{\lim_{k \to \infty} \left(\frac{k+1}{k} \right)^k}$$

$$= \frac{1}{\lim_{k \to \infty} \left(1 + \frac{1}{k} \right)^k} = e^{-1}.$$

Since $\rho = e^{-1} < 1$, the series converges by the ratio test.

(b) Since the series converges, the contrapositive of the divergence test tells us that the terms of the series converges to zero, which means that $\lim_{k \to \infty} \frac{k!}{k^k} = 0$.

(c) The limit statement in part (b) is exactly what it means for k^k to dominate factorial growth.

Applications

65. (a) Since $r_j = 0.8$ for $1 \leq j \leq 10$, we have, using the formula for the sum of a finite geometric series,

$$f_{10} = s \left(1 + \sum_{i=1}^{10} (0.8)^i \right) = 420 \left(1 + \frac{0.8 - 0.8^{11}}{1 - 0.8} \right) \approx 420(1 + 3.5705) \approx 1919.61.$$

1920 is the estimate for f_∞ given these assumptions.

(b) Note that $r_k \leq 0.95$, so that the sum is bounded by $s\left(1 + \sum_{i=1}^{\infty}(0.95)^i\right)$. Theorem 7.31 says that the sum is bounded above by

$$
\begin{aligned}
f_{10} + \int_{10}^{\infty} 0.95^x \, dx &= f_{10} + \lim_{t \to \infty} \int_{10}^{t} 0.95^x \, dx \\
&= f_{10} + \lim_{t \to \infty} \left[\frac{1}{\ln 0.95} 0.95^x \right]_{10}^{t} \\
&= f_{10} + \lim_{t \to \infty} \left(\frac{1}{\ln 0.95} 0.95^t - \frac{1}{\ln 0.95} 0.95^{10} \right) \\
&= f_{10} - \frac{1}{\ln 0.95} 0.95^{10} \\
&\approx 1920 + 11.67.
\end{aligned}
$$

The maximum error in this approximation of f_{∞} is about 11.7.

Proofs

67. Let

$$
r(x) = \frac{b_n x^n + b_{n-1} x^{n-1} + \cdots + b_1 x + b_0}{c_m x^m + c_{m-1} x^{m-1} + \cdots + c_1 x + c_0},
$$

where $b_n \neq 0$ and $c_m \neq 0$, be a rational function. Then the limit of a rational function of k as $k \to \infty$ is determined by its leading terms,

$$
\begin{aligned}
\lim_{x \to \infty} \frac{r(x+1)}{r(x)} &= \lim_{x \to \infty} \frac{\frac{b_n(x+1)^n + b_{n-1}(x+1)^{n-1} + \cdots + b_1(x+1) + b_0}{c_m(x+1)^m + c_{m-1}(x+1)^{m-1} + \cdots + c_1(x+1) + c_0}}{\frac{b_n x^n + b_{n-1} x^{n-1} + \cdots + b_1 x + b_0}{c_m x^m + c_{m-1} x^{m-1} + \cdots + c_1 x + c_0}} \\
&= \lim_{x \to \infty} \frac{(b_n(x+1)^n + b_{n-1}(x+1)^{n-1} + \cdots + b_1(x+1) + b_0)(c_m x^m + c_{m-1} x^{m-1} + \cdots + c_1 x + c_0)}{(b_n x^n + b_{n-1} x^{n-1} + \cdots + b_1 x + b_0)(c_m(x+1)^m + c_{m-1}(x+1)^{m-1} + \cdots + c_1(x+1) + c_0)} \\
&= \lim_{x \to \infty} \frac{b_n c_m x^{m+n} + \text{lower-degree terms}}{b_n c_m x^{m+n} + \text{lower-degree terms}} \\
&= 1.
\end{aligned}
$$

Thus if $a_k = r(k)$ where r is a rational function, it follows that $\lim_{k \to \infty} \frac{a_{k+1}}{a_k} = 1$, so that the ratio test is inconclusive.

69. Recall that $\lim_{k \to \infty} k^{1/k} = 1$. Then

$$
\begin{aligned}
\lim_{k \to \infty} (c k^m)^{1/k} &= \lim_{k \to \infty} c^{1/k} (k^m)^{1/k} \\
&= \left(\lim_{k \to \infty} c^{1/k} \right) \left(\lim_{k \to \infty} (k^m)^{1/k} \right) \\
&= 1 \cdot \left(\lim_{k \to \infty} (k^{1/k})^m \right) \\
&= \left(\lim_{k \to \infty} k^{1/k} \right)^m = 1^m = 1.
\end{aligned}
$$

(Note that $\lim_{k \to \infty} c^{1/k} = 1$ since $c > 0$). Applying the root test to $\sum_{k=1}^{\infty} c k^m$ requires evaluating the limit $\rho = \lim_{k \to \infty} (c k^m)^{1/k}$, so by the above computation, $\rho = 1$ and the root test is inconclusive. Since our assumption was that $m \in \mathbb{Z}$, this shows that in particular the root test is inconclusive for every p-series.

71. (a) We prove by induction that

$$\left(\sum_{j=1}^{n} c_j\right) k^{m_1} \leq c_1 k^{m_1} + \cdots + c_n k^{m_n} \leq \left(\sum_{j=1}^{n} c_j\right) k^{m_n}.$$

Exercise 70 shows that the result holds for $n = 2$. Suppose it holds for $n = r$; then

$$\left(\sum_{j=1}^{r+1} c_j\right) k^{m_1} = \left(\sum_{j=1}^{r} c_j\right) k^{m_1} + c_{r+1} k^{m_1} \qquad \text{(algebra)}$$

$$\leq c_1 k^{m_1} + \cdots + c_r k^{m_r} + c_{r+1} k^{m_1} \qquad \text{(inductive hypothesis)}$$

$$\leq c_1 k^{m_1} + \cdots + c_r k^{m_r} + c_{r+1} k^{m_{r+1}} \qquad \text{(since } m_1 \leq m_{r+1}\text{)}$$

$$= (c_1 k^{m_1} + \cdots + c_r k^{m_r}) + c_{r+1} k^{m_{r+1}} \qquad \text{(regroup)}$$

$$\leq \left(\sum_{j=1}^{r} c_j\right) k^{m_r} + c_{r+1} k^{m_{r+1}} \qquad \text{(inductive hypothesis)}$$

$$\leq \left(\sum_{j=1}^{r+1} c_j\right) k^{m_{r+1}} \qquad \text{(inductive hypothesis)}.$$

Then raising the string of inequalities to the $\frac{1}{k}$ power as before, and taking limits, gives by Exercise 69

$$1 = \lim_{k \to \infty} \left(\left(\sum_{j=1}^{n} c_j\right) k^{m_1}\right)^{1/k} \leq \lim_{k \to \infty} (c_1 k^{m_1} + \cdots + c_n k^{m_n})^{1/k} \leq \lim_{k \to \infty} \left(\left(\sum_{j=1}^{n} c_j\right) k^{m_n}\right)^{1/k} = 1,$$

so that by the Squeeze Theorem, the middle limit is 1 as well.

(b) Using the root test to test convergence we get $\rho = \lim_{k \to \infty} (c_1 k^{m_1} + \cdots + c_n k^{m_n})^{1/k} = 1$. Since $\rho = 1$, the root test is inconclusive.

Thinking Forward

A series of monomials

By the ratio test, we have

$$\rho = \lim_{k \to \infty} \frac{a_{k+1}}{a_k} = \lim_{k \to \infty} \frac{\frac{x^{2(k+1)}}{(k+1)!}}{\frac{x^{2k}}{k!}} = \lim_{k \to \infty} \frac{x^2}{k+1} = x^2 \lim_{k \to \infty} \frac{1}{k+1}.$$

This limit is obviously zero for all x, so that $\rho = 0 < 1$ and the series converges for any x.

A series of monomials

By the root test, we have

$$\rho = \lim_{k \to \infty} \left(\left(\frac{x}{k}\right)^k\right)^{1/k} = \lim_{k \to \infty} \frac{x}{k} = x \lim_{k \to \infty} \frac{1}{k}.$$

This limit is obviously zero for all x, so that $\rho = 0 < 1$ and the series converges for any x.

7.7   Alternating Series

Thinking Back

A strictly increasing sequence of partial sums

Since $S_n = \sum_{k=1}^{n} a_k$, we have, for $n > 1$,

$$S_n - S_{n-1} = \sum_{k=1}^{n} a_k - \sum_{k=1}^{n-1} a_k = a_n.$$

Since $a_n > 0$ for all n, we see that $S_n - S_{n-1} > 0$ for all $n > 1$, so that the sequence of partial sums is increasing.

A bounded sequence of partial sums

Saying that $\sum_{k=1}^{\infty} a_k$ converges to L is the same as saying, by definition, that the sequence of partial sums $\{S_n\}$ converges to L. But if a sequence converges, then it is bounded. Thus the sequence of partial sums is bounded.

Concepts

1. (a) False. It is an alternating series only if $a_k > 0$ for all k.

   (b) True, since $\sum_{k=1}^{\infty} (-1)^{k+1} a_k = -\sum_{k=1}^{\infty} (-1)^k a_k$.

   (c) True. Since $a_k > 0$, we have $\sum_{k=1}^{\infty} a_k = \sum_{k=1}^{\infty} |a_k|$, so that convergence is the same as absolute convergence.

   (d) False. All the integral test says is that $\sum_{k=1}^{\infty} f(k)$ and $\int_1^{\infty} f(x)\, dx$ either both converge or both diverge. However, if you also assume that $\int_1^{\infty} f(x)\, dx$ converges, then $\sum_{k=1}^{\infty} f(k)$ converges, so its terms go to zero. Since all the terms are positive, $\sum_{k=1}^{\infty} (-1)^{k+1} f(k)$ satisfies the conditions of the alternating series test, so that in this case $\sum_{k=1}^{\infty} (-1)^{k+1} f(k)$ converges.

   (e) True. See Definition 7.40.

   (f) False. For example, take $a_k = 1$ for all k or take $a_k = (-1)^k$ for all k.

   (g) True. Absolute convergence means that $\sum_{k=1}^{\infty} |a_k|$ converges. Applying the root test gives $\rho = \lim_{k \to \infty} (|a_k|)^{1/k} < 1$, so by the root test, the series converges, so the original series converges absolutely.

   (h) True. See the discussion in the text.

3. See Definition 7.36. An alternating series is a series of the form $\sum_{k=1}^{\infty} (-1)^{k+1} a_k$ or $\sum_{k=1}^{\infty} (-1)^k a_k$ where $\{a_k\}$ is a sequence of positive numbers.

5. It need not satisfy condition (i), since for example $a_k = (-1)^{k+1} \frac{1}{k^2}$ fails condition (i), yet

$$\sum_{k=1}^{\infty} (-1)^{k+1} a_k = \sum_{k=1}^{\infty} (-1)^{k+1} (-1)^{k+1} \frac{1}{k^2} = \sum_{k=1}^{\infty} \frac{1}{k^2}$$

converges. It also need not satisfy condition (ii). To see this, take a_k to be any increasing sequence for $k \le 100$, say, and then $a_k = \frac{1}{k^2}$. Then all the a_k are positive, and $\lim_{k \to \infty} a_k = 0$, but the sequence is not decreasing. Since convergence of a series is determined by the tail of the series, we see that $\sum_{k=101}^{\infty} \frac{1}{k^2}$ converges so that $\sum_{k=1}^{\infty} \frac{1}{k^2}$ converges.

However, it must satisfy condition (iii); if it does not, then $\lim_{k \to \infty} (-1)^{k+1} a_k$ is nonzero as well, and then the divergence test tells us that $\sum_{k=1}^{\infty} (-1)^{k+1} a_k$ cannot converge.

7. The proof of the alternating series test showed that the odd-numbered partial sums are a decreasing sequence converging to the limit of the series. A similar argument (see Exercise 67) shows that the even-numbered partial sums are an increasing sequence converging to the limit of the series. Since consecutive partial sums are one odd and one even, one of them is below the series limit and the other is above the series limit.

9. In order to be absolutely convergent, the series $\sum_{k=1}^{\infty} |a_k|$ must be convergent. See Definition 7.39.

11. The definition of absolute convergence requires only that we test the convergence of $\sum_{k=1}^{\infty} |a_k|$. (Note also that if $\sum_{k=1}^{\infty} |a_k|$ converges, then $\sum_{k=1}^{\infty} a_k$ does as well, by Theorem 7.41, although we do not require that fact as part of the definition of absolute convergence). For the corresponding fact about conditional convergence, see the solution to the next exercise.

13. If $\sum_{k=1}^{\infty} a_k$ is convergent and $a_k > 0$ for all k, then $|a_k| = a_k$, so that $\sum_{k=1}^{\infty} |a_k| = \sum_{k=1}^{\infty} a_k$, which converges, so that $\sum_{k=1}^{\infty} a_k$ is absolutely convergent.

15. Let f be a function with domain $[1, \infty)$. If the function $|f|$ is continuous and decreasing, and if $\int_1^{\infty} |f(x)|\, dx$ converges, then the series $\sum_{k=1}^{\infty} f(k)$ converges absolutely. To see that this is true, apply the integral test to the series $\sum_{k=1}^{\infty} |f(k)|$. By hypothesis, $|f|$ is continuous and decreasing, and it is obviously positive, so that $\sum_{k=1}^{\infty} |f(k)|$ and $\int_1^{\infty} |f(x)|\, dx$ either both converge or both diverge. But we are given that the integral converges, so the series does as well. Convergence of $\sum_{k=1}^{\infty} |f(k)|$ means that $\sum_{k=1}^{\infty} f(k)$ is absolutely convergent.

17. Since the sequence is a sequence of positive numbers, we want to show that if $\sum_{k=1}^{\infty} a_{2k-1}$ and $\sum_{k=1}^{\infty} a_{2k}$ converge, then so does $\sum_{k=1}^{\infty} a_k$. Recall that if $\sum_{k=1}^{\infty} b_k$ and $\sum_{k=1}^{\infty} c_k$ both converge, then so does $\sum_{k=1}^{\infty} (b_k + c_k)$. Thus $\sum_{k=1}^{\infty} (a_{2k-1} + a_{2k})$ converges, but this sum is just $\sum_{k=1}^{\infty} a_k$.

19. For example, let $a_k = b_k = \frac{1}{k}$. Then $\sum_{k=1}^{\infty} a_k$ and $\sum_{k=1}^{\infty} b_k$ are both divergent p-series with $p = 1$, but $\sum_{k=1}^{\infty} a_k b_k = \sum_{k=1}^{\infty} \frac{1}{k^2}$ is a p-series with $p = 2$, so is convergent.

Skills

21. This series diverges, since it can be rewritten (without reordering the terms) as

$$1 + \left(-\frac{1}{2} - \frac{1}{2} \right) + \left(\frac{1}{3} + \frac{1}{3} + \frac{1}{3} \right) + \left(-\frac{1}{4} - \frac{1}{4} - \frac{1}{4} - \frac{1}{4} \right) + \cdots = 1 - 1 + 1 - 1 + \ldots,$$

which is obviously divergent.

23. Rewriting the series as

$$\left(1 - \frac{1}{2^3} \right) + \left(\frac{1}{3^2} - \frac{1}{4^3} \right) + \left(\frac{1}{5^2} - \frac{1}{6^3} \right) + \left(\frac{1}{7^2} - \frac{1}{8^3} \right) + \cdots,$$

we see that each term is positive and the kth term is less than $\frac{1}{(2k-1)^2}$. Inserting a zero between each pair of terms gives us a sequence in which the kth term is less than $\frac{1}{k^2}$, so that by the comparison test, since $\sum_{k=1}^{\infty} \frac{1}{k^2}$ converges, so does the given series.

25. This is an alternating series, and $\frac{1}{1+k^2} < \frac{1}{k^2}$, so that since $\sum_{k=1}^{\infty} \frac{1}{k^2}$ converges, so does $\sum_{k=1}^{\infty} \frac{1}{1+k^2} = \sum_{k=1}^{\infty} \left| \frac{(-1)^{k+1}}{1+k^2} \right|$. Hence the series converges absolutely.

27. We have

$$\sum_{k=0}^{\infty} \frac{(-2)^{k+1}}{1 + 3^k} = \sum_{k=0}^{\infty} (-1)^{k+1} \frac{2^{k+1}}{1 + 3^k} = 2 \sum_{k=0}^{\infty} (-1)^{k+1} \frac{2^k}{1 + 3^k}.$$

We show that the series is absolutely convergent, i.e., that $2\sum_{k=0}^{\infty}\frac{2^k}{1+3^k}$ converges. Using the ratio test,

$$\rho = \lim_{k\to\infty}\frac{a_{k+1}}{a_k} = \lim_{k\to\infty}\frac{\frac{2^{k+1}}{1+3^{k+1}}}{\frac{2^k}{1+3^k}} = \lim_{k\to\infty}\left(2\cdot\frac{1+3^k}{1+3^{k+1}}\right) = 2\lim_{k\to\infty}\frac{3^{-k}+1}{3^{-k}+3} = \frac{2}{3}.$$

Since $\rho = \frac{2}{3} < 1$, the series converges, so by definition, the original series converges absolutely.

29. Since $\lim_{k\to\infty}\frac{k^2}{1+k^2} = \lim_{k\to\infty}\frac{1}{k^{-2}+1} = 1$, the terms of the sequence do not converge to zero. Rather, the even-numbered terms converge to 1 while the odd-numbered terms converge to -1. By the divergence test, the series diverges.

31. Using the ratio test for absolute convergence, we have

$$\rho = \lim_{k\to\infty}\left|\frac{b_{k+1}}{b_k}\right| = \lim_{k\to\infty}\frac{\frac{3^{k+2}}{(2k+2)!}}{\frac{3^{k+1}}{(2k)!}} = \lim_{k\to\infty}\frac{3}{(2k+1)(2k+2)} = 0.$$

Since $\rho = 0 < 1$, the series converges absolutely.

33. Using the ratio test for absolute convergence, we have

$$\rho = \lim_{k\to\infty}\left|\frac{b_{k+1}}{b_k}\right| = \lim_{k\to\infty}\frac{\frac{7^{k+1}}{(2k+3)!}}{\frac{7^k}{(2k+1)!}} = \lim_{k\to\infty}\frac{7}{(2k+3)(2k+2)} = 0.$$

Since $\rho = 0 < 1$, the series converges absolutely.

35. The absolute value of the terms is $\frac{1\cdot3\cdot5\cdots(2k+1)}{1\cdot4\cdot7\cdots(3k+1)}$. Then the $(k+1)^{\text{st}}$ term differs from the kth term in that it has an extra factor of $2k+3$ in the numerator and an extra factor of $3k+4$ in the denominator. Thus

$$\rho = \lim_{k\to\infty}\left|\frac{b_{k+1}}{b_k}\right| = \lim_{k\to\infty}\frac{2k+3}{3k+4} = \lim_{k\to\infty}\frac{2k+3}{3k+4} = \frac{2}{3}.$$

Since $\rho = \frac{2}{3} < 1$, the series converges absolutely.

37. Note that $\frac{1}{k^2-1} < \frac{1}{k^{3/2}}$ for $k \geq 2$. Since $\sum_{k=2}^{\infty}\frac{1}{k^{3/2}}$ is a convergent p-series, it follows from the comparison test that the original series converges. Since the terms are all positive, it also converges absolutely.

39. This series diverges. The even terms are $\{\frac{2k}{2k+1}\}$, and $\lim_{k\to\infty}\frac{2k}{2k+1} = \lim_{k\to\infty}\frac{2}{2+1/k} = 1$. Thus the terms of the series do not converge to zero, so the series diverges by the divergence test.

41. Since $\frac{\ln k}{k+1} \leq \frac{\ln k}{k}$, and power functions dominate exponential, $\{\frac{\ln k}{k+1}\} \to 0$. Since the terms are decreasing and positive, $\sum_{k=1}^{\infty}(-1)^k\frac{\ln k}{k+1}$ converges by the alternating series test. However, the series converges conditionally, since using the limit comparison test for absolute convergence against $\frac{1}{k}$ gives

$$\lim_{k\to\infty}\frac{\frac{\ln k}{k+1}}{\frac{1}{k}} = \lim_{k\to\infty}\left(\ln k\cdot\frac{k}{k+1}\right) = \infty$$

since $\frac{k}{k+1} \to 1$ and $\ln k \to \infty$. Since $\sum_{k=1}^{\infty}\frac{1}{k}$ diverges, so does $\sum_{k=1}^{\infty}\frac{\ln k}{k+1}$, by the limit comparison test.

43. This series converges absolutely. To see this, use the limit comparison test with $\sum_{k=1}^{\infty} \frac{1}{k^{3/2}}$:

$$\lim_{k \to \infty} \frac{\frac{\ln k}{k^2-1}}{\frac{1}{k^{3/2}}} = \lim_{k \to \infty} \frac{\ln k}{k^{1/2} - k^{-3/2}} = 0,$$

since power functions dominate log functions. Since $\sum_{k=1}^{\infty} \frac{1}{k^{3/2}}$ is a convergent p-series, $\sum_{k=1}^{\infty} \frac{\ln k}{k^2-1}$ converges and thus the original series converges absolutely.

45. Since

$$\lim_{k \to \infty} \frac{\sqrt[3]{k^2 + 3k + 1}}{k+1} = \lim_{k \to \infty} \frac{\sqrt[3]{1/k + 3/k^2 + 1/k^3}}{1 + 1/k} = 0,$$

the terms of the underlying positive series converge to zero, and the sequence is (eventually) decreasing. Thus by the alternating series test, it converges. However, it converges only conditionally: use the limit comparison test with $\sum_{k=2}^{\infty} \frac{1}{k^{1/3}}$ to get

$$\lim_{k \to \infty} \frac{\frac{\sqrt[3]{k^2+3k+1}}{k+1}}{\frac{1}{k^{1/3}}} = \lim_{k \to \infty} \frac{\sqrt[3]{k} \cdot \sqrt[3]{k^2 + 3k + 1}}{k+1} = \lim_{k \to \infty} \frac{\sqrt[3]{k^3 + 3k^2 + k}}{k+1} = \lim_{k \to \infty} \frac{\sqrt[3]{1 + 3/k + 1/k^2}}{1 + 1/k} = 1,$$

and since $\sum_{k=2}^{\infty} \frac{1}{k^{1/3}}$ diverges since it is a p-series with $p < 1$, the original series fails to converge absolutely, so it converges only conditionally.

47. This series converges absolutely. Using the ratio test for absolute convergence gives

$$\rho = \lim_{k \to \infty} \left| \frac{(-1)^{k+1} e^{-(k+1)^2}}{(-1)^k e^{-k^2}} \right| = \lim_{k \to \infty} \frac{e^{-k^2 - 2k - 1}}{e^{-k^2}} = \lim_{k \to \infty} \frac{1}{e^{2k+1}} = 0.$$

Since $\rho = 0 < 1$, the series converges absolutely.

49. This series converges absolutely. To see this, note that absolute convergence means that $\sum_{k=1}^{\infty} \frac{|\cos k|}{k^2}$ converges. But $0 \le |\cos k| \le 1$, so that $\frac{|\cos k|}{k^2} \le \frac{1}{k^2}$. Since $\sum_{k=1}^{\infty} \frac{1}{k^2}$ is a p-series with $p > 1$, it is convergent, so that by the comparison test, the original series is absolutely convergent.

51. Use the limit comparison test with $\frac{1}{k^3}$. Since for k large, $\sin\left(\frac{1}{k}\right)$ is the sine of an angle that is near zero but positive, it is also positive, so that this series has eventually positive terms and the limit comparison test applies. Then

$$\lim_{k \to \infty} \frac{\sin^3\left(\frac{1}{k}\right)}{\frac{1}{k^3}} = \lim_{k \to \infty} \left(\frac{\sin\left(\frac{1}{k}\right)}{\frac{1}{k}} \right)^3 = \lim_{x \to 0} \left(\frac{\sin x}{x} \right)^3 = 1$$

since $\lim_{x \to 0} \frac{\sin x}{x} = 1$. Since $\sum_{k=1}^{\infty} \frac{1}{k^3}$ is a p-series with $p > 1$, it converges, so the original series converges (absolutely) as well.

53. (a) This series converges absolutely since $\sum_{k=1}^{\infty} \frac{1}{k^2}$ is a p-series with $p > 1$, so it converges.

  (b) $S_{10} = \sum_{k=1}^{10} \frac{(-1)^{k+1}}{k^2} = 1 - \frac{1}{4} + \frac{1}{9} - \cdots + \frac{1}{100} = \frac{5194387}{6350400} \approx 0.817962$. Since $a_{11} = \frac{1}{11^2} \approx 0.008264$, we know that $|L - S_{10}| < a_{11}$, so that $S_{10} - a_{11} < L < S_{10} + a_{11}$, or $0.809698 < L < 0.826226$.

  (c) To get S_n within 10^{-6} of L, we need $a_{n+1} = \frac{1}{(n+1)^2} < \frac{1}{10^6}$, so that $(n+1)^2 > 10^6$ and $n > 999$. Thus $n = 1000$ works.

55. (a) This series converges absolutely by the ratio test for absolute convergence since

$$\lim_{k\to\infty} \frac{\frac{1}{(2k+2)!}}{\frac{1}{(2k)!}} = \lim_{k\to\infty} \frac{1}{(2k+2)(2k+1)} = 0.$$

(b) $S_{10} = \sum_{k=0}^{10} \frac{(-1)^k}{(2k)!} = 1 - \frac{1}{2!} + \frac{1}{4!} - \cdots + \frac{1}{20!} \approx 0.540302$. Since $a_{11} = \frac{1}{22!} \approx 8.897 \times 10^{-22}$, we know that $|L - S_{10}| < a_{11}$, so that $S_{10} - a_{11} < L < S_{10} + a_{11}$, so that to the accuracy of our computation, $L = 0.540302$.

(c) To get S_n within 10^{-6} of L, we need $a_{n+1} = \frac{1}{(2n+2)!} < \frac{1}{10^6}$, so that $(2n+2)! > 10^6$. The first factorial greater than one million is $10!$, so that the first n satisfying this inequality is $n = 4$.

57. (a) This series converges absolutely by the ratio test for absolute convergence, since

$$\lim_{k\to\infty} \frac{\frac{k+2}{(k+1)!}}{\frac{k+1}{k!}} = \lim_{k\to\infty} \frac{k+2}{(k+1)(k+1)} = \lim_{k\to\infty} \frac{k+2}{k^2+2k+1} = \lim_{k\to\infty} \frac{1+2/k}{k+2+1/k} = 0.$$

(b) $S_{10} = \sum_{k=0}^{10} \frac{(-1)^k(k+1)}{k!} = 1 - \frac{2}{1!} + \frac{3}{2!} - \cdots - \frac{11}{10!} = \frac{1}{3628800} \approx 2.75573 \times 10^{-7}$. Since $a_{11} = \frac{12}{11!} \approx 3.00625 \times 10^{-7}$, we know that $|L - S_{10}| < a_{11}$, so that $S_{10} - a_{11} < L < S_{10} + a_{11}$, or $-2.50521 \times 10^{-8} < L < 5.76198 \times 10^{-7}$.

(c) To get S_n within 10^{-6} of L, we need $a_{n+1} = \frac{n+2}{(n+1)!} < \frac{1}{10^6}$, so that $(n+1)! > 10^6(n+2)$. The first n satisfying this inequality is $n = 10$. (Note by the way that if you decompose the series terms as

$$\frac{(-1)^k(k+1)}{k!} = \frac{(-1)^k k}{k!} + \frac{(-1)^k}{k!} = -\frac{(-1)^{k-1}}{(k-1)!} + \frac{(-1)^k}{k!}$$

for $k > 0$, then the series telescopes and it is easy to see that the limit is in fact zero).

59. For $p > 1$, the series converges absolutely. Choose some q with $1 < q < p$, and use the limit comparison test between $\sum_{k=1}^{\infty} \frac{\ln k}{k^p}$ and $\sum_{k=1}^{\infty} \frac{1}{k^q}$:

$$\lim_{k\to\infty} \frac{\frac{\ln k}{k^p}}{\frac{1}{k^q}} = \lim_{k\to\infty} \frac{\ln k}{k^{p-q}} = 0$$

since power functions for positive exponents, which $p - q$ is, dominate logs.

For $0 < p \leq 1$ the series converges conditionally. It converges since $\frac{\ln k}{k^p}$ is decreasing and has limit zero (again since power functions dominate logs). However, it fails to converge absolutely since $\frac{\ln k}{k^p} > \frac{1}{k^p}$, and $\sum_{k=1}^{\infty} \frac{1}{k^p}$ diverges since $0 < p \leq 1$. So by the comparison test, the original series fails to converge absolutely.

For $p = 0$, the series becomes $\sum_{k=1}^{\infty} (-1)^{k+1} \ln k$, which clearly diverges by the divergence test. Similarly, for $p < 0$, the series is $\sum_{k=1}^{\infty} (-1)^{k+1} k^{|p|} \ln k$, which also obviously diverges by the divergence test.

So in summary, the series converges absolutely for $p > 1$, converges conditionally for $0 < p \leq 1$, and diverges otherwise.

61. This series converges absolutely for all p. To see this, use the ratio test for absolute convergence:

$$\rho = \lim_{k\to\infty} \frac{(k+1)^p e^{-(k+1)}}{k^p e^{-k}} = \lim_{k\to\infty} \frac{(k+1)^p}{k^p e} = e^{-1} \left(\lim_{k\to\infty} \frac{k+1}{k} \right)^p = e^{-1} < 1$$

since $\lim_{k\to\infty} \frac{k+1}{k} = 1$. Since the ratio is less than one, the series converges absolutely.

Applications

63. (a) We have

$$\frac{A_k - A_{k-1}}{A_{k-1} - A_{k-2}} = \frac{\left(8069 - \frac{1}{7}A_{k-1}\right) - \left(8069 - \frac{1}{7}A_{k-2}\right)}{A_{k-1} - A_{k-2}} = \frac{\frac{1}{7}\left(A_{k-2} - A_{k-1}\right)}{A_{k-1} - A_{k-2}} = -\frac{1}{7}.$$

(b) Expand the sum on the right and note that it telescopes:

$$\sum_{k=1}^{N}(A_k - A_{k-1}) = (A_1 - A_0) + (A_2 - A_1) + \cdots + (A_{N-1} - A_{N-2}) + (A_N - A_{N-1})$$

$$= A_N - A_0 = A_N,$$

since $A_0 = 0$.

(c) Use the ratio test for absolute convergence on $\sum_{k=1}^{\infty}(A_k - A_{k-1})$ together with part (a):

$$\rho = \lim_{k \to \infty}\left|\frac{A_k - A_{k-1}}{A_{k-1} - A_{k-2}}\right| = \left|-\frac{1}{7}\right| = \frac{1}{7} < 1.$$

Since $\rho < 1$, the series converges absolutely, so it converges.

Proofs

65. By exercise 7.5.53, if $\sum_{k=1}^{\infty} b_k$ converges, with $b_k \geq 0$ for all k, then $\sum_{k=1}^{\infty} b_k^2$ converges. Now let $b_k = |a_k|$. Since $\sum_{k=1}^{\infty} a_k$ is absolutely convergent, then $\sum_{k=1}^{\infty} |a_k|$ converges, so $\sum_{k=1}^{\infty} |a_k|^2 = \sum_{k=1}^{\infty} a_k^2$ converges as well.

67. To show that S_{2n} is strictly increasing, note that S_{2n} is

$$S_{2n} = (a_1 - a_2) + (a_3 - a_4) + \cdots + (a_{2n-3} - a_{2n-2}) + (a_{2n-1} - a_{2n}).$$

Then $S_{2n+2} - S_{2n} = a_{2n+1} - a_{2n+2} > 0$ since $\{a_n\}$ is decreasing. Hence $\{S_{2n}\}$ is strictly increasing. It is bounded above by a_1 since (rewriting S_{2n})

$$S_{2n} = a_1 - (a_2 - a_3) - (a_4 - a_5) + \cdots - (a_{2n-2} - a_{2n-1}) - a_{2n}.$$

Since $a_{2n} > 0$ and since each of the other subtracted terms is positive since $\{a_k\}$ is decreasing, we see that $S_{2n} \leq a_1$, so that $\{S_{2n}\}$ is bounded above by a_1.

69. This is the contrapositive of Theorem 7.41. If $\sum_{k=1}^{\infty} a_k$ diverges, then it does not converge absolutely by Theorem 7.41, so that $\sum_{k=1}^{\infty} |a_k|$ diverges. Another, more direct, proof notes that since $a_k \leq |a_k|$, by the comparison test, if $\sum_{k=1}^{\infty} a_k$ diverges, then so does $\sum_{k=1}^{\infty} |a_k|$.

71. Since $2k - 1 \leq 2k$, we get $\frac{1}{2k-1} > \frac{1}{2k}$. Then since $\sum_{k=1}^{\infty} \frac{1}{2k} = \frac{1}{2}\sum_{k=1}^{\infty} \frac{1}{k}$ diverges since it is the harmonic series, we see that $\sum_{k=1}^{\infty} \frac{1}{2k-1}$ diverges by the comparison test.

73. Using the hint, take some partial sum of $\sum_{k=1}^{\infty} \frac{1}{2k-1}$ that is greater than α. This is possible since $\sum_{k=1}^{\infty} \frac{1}{2k-1}$ diverges. Call this S_1. Then take some partial sum of $\sum_{k=1}^{\infty}\left(-\frac{1}{2k}\right)$ to get the sum back to just below α. That is possible since the second series diverges as well. Call this S_2. We are left with tails of both series; those tails diverge since the original series do. So take just enough of a partial sum of the remaining tail of $\sum_{k=1}^{\infty} \frac{1}{2k-1}$ to get the sum back to above α (call this S_3), and then just enough of a partial sum of the remaining tail of $\sum_{k=1}^{\infty}\left(-\frac{1}{2k}\right)$ to get back below α (call this S_4). Clearly this process can be continued ad infinitum. Thus we get a sequence $\{S_k\}$, and because of the way we constructed the S_k, we know that $|S_k - \alpha|$ is less than $\frac{1}{n}$ for some n, and that n increases with k. But since $\frac{1}{n} \to 0$, so does $|S_k - \alpha|$, so that $\lim_{k \to \infty} S_k = \alpha$ — that is, this rearrangement converges to α.

Thinking Forward

A series of monomials

Using the ratio test for absolute convergence, we have

$$\rho = \lim_{k \to \infty} \left| \frac{\frac{x^{k+1}}{k+1}}{\frac{x^k}{k}} \right| = \lim_{k \to \infty} \frac{k\,|x|}{k+1} = |x| \lim_{k \to \infty} \frac{k}{k+1} = |x|.$$

So by the ratio test for absolute convergence, the series converges absolutely, so converges, for $|x| < 1$, i.e., for $-1 < x < 1$. For $|x| > 1$, since exponential functions with a base greater than 1 dominate power functions, the terms do not converge to zero, so that the series diverges.

A series of monomials

Using the ratio test for absolute convergence, we have

$$\rho = \lim_{k \to \infty} \left| \frac{\frac{x^{k+1}}{(k+1)!}}{\frac{x^k}{k!}} \right| = \lim_{k \to \infty} \frac{|x|}{k} = |x| \lim_{k \to \infty} \frac{1}{k} = 0.$$

So by the ratio test for absolute convergence, the series converges absolutely for all x.

Chapter Review and Self-Test

1. This sequence is monotonically increasing since $\frac{\frac{e^{k+1}}{k+1}}{\frac{e^k}{k}} = e \frac{k}{k+1}$. But $\frac{k}{k+1}$ is increasing, and for $k = 1$, the ratio is $\frac{1}{2}e > 1$, so that the ratio is always greater than 1 and the sequence is monotonically increasing. Clearly the sequence is bounded below by zero, but it is not bounded above since exponential functions dominate power functions; it diverges to ∞.

3. We have

$$\lim_{k \to \infty} \frac{2k^2 + 1}{3k^2 + 1} = \lim_{k \to \infty} \frac{2 + 1/k^2}{3 + 1/k^2} = \frac{2}{3},$$

so that the series converges to $\frac{2}{3}$ and thus is bounded. Since

$$a_{k+1} - a_k = \frac{2(k+1)^2 + 1}{3(k+1)^2 - 1} - \frac{2k^2 + 1}{3k^2 - 1} = \frac{(2k^2 + 4k + 3)(3k^2 - 1) - (2k^2 + 1)(3k^2 + 6k + 2)}{(3(k+1)^2 - 1)(3k^2 - 1)}$$

$$= \frac{-10k - 5}{(3(k+1)^2 - 1)(3k^2 - 1)},$$

and since the denominator is positive for $k \geq 1$, the difference is negative and thus the sequence is monotonically decreasing.

5. Simplifying this sequence gives

$$\frac{k^2}{k+3} - \frac{k^2}{k+4} = \frac{k^2(k+4) - k^2(k+3)}{(k+3)(k+4)} = \frac{k^2}{k^2 + 7k + 12}.$$

Then

$$\lim_{k \to \infty} \frac{k^2}{k^2 + 7k + 12} = \lim_{k \to \infty} \frac{1}{1 + 7/k + 12/k^2} = 1,$$

so that the sequence converges to 1 and thus is bounded. It is also monotonically increasing; to see this, look at $f(x) = \frac{x^2}{x^2+7x+12}$; then

$$f'(x) = \frac{(x^2+7x+12)(2x) - x^2(2x+7)}{(x^2+7x+12)^2} = \frac{7x^2+24x}{(x^2+7x+12)^2} > 0.$$

Since $f(k)$ is the kth term of the sequence, it follows that the sequence is monotonically increasing.

7. The kth term is

$$a_k = \frac{k!}{1 \cdot 3 \cdot 5 \cdots (2k-1)} = \frac{1 \cdot 2 \cdot 3 \cdots k}{1 \cdot 3 \cdot 5 \cdots (2k-1)} = \frac{2}{3} \cdot \frac{3}{5} \cdots \frac{k}{2k-1}.$$

Since each term multiplies the previous term by a factor less than 1, the sequence is monotonically decreasing. It is bounded below by 0, so it converges. Note that $\frac{n}{2n-1} > \frac{n+1}{2(n+1)-1}$ since

$$\frac{n}{2n-1} - \frac{n+1}{2(n+1)-1} = \frac{n(2(n+1)+1) - (n+1)(2n-1)}{(2n-1)(2(n+1)-1)} = \frac{1}{(2n-1)(2(n+1)-1)} > 0.$$

Thus each factor in a_k is bounded above by $\frac{2}{3}$, so that $0 \leq \lim_{k \to \infty} a_k \leq \lim_{k \to \infty} \left(\frac{2}{3}\right)^k = 0$. So the sequence converges to zero and thus is bounded.

9. (a) We have

$$S_0 = \frac{1}{0+3} - \frac{1}{0+4} = \frac{1}{3} - \frac{1}{4} = \frac{1}{12}$$

$$S_1 = \frac{1}{0+3} - \frac{1}{0+4} + \frac{1}{1+3} - \frac{1}{1+4} = \frac{1}{3} - \frac{1}{5} = \frac{2}{15}$$

$$S_2 = S_1 + \frac{1}{2+3} - \frac{1}{2+4} = \frac{2}{15} + \frac{1}{5} - \frac{1}{6} = \frac{1}{6}$$

$$S_3 = S_2 + \frac{1}{3+3} - \frac{1}{3+4} = \frac{1}{6} + \frac{1}{6} - \frac{1}{7} = \frac{4}{21}$$

$$S_4 = S_3 + \frac{1}{4+3} - \frac{1}{4+4} = \frac{4}{21} + \frac{1}{7} - \frac{1}{8} = \frac{5}{24}.$$

(b) This series telescopes:

$$S_n = \sum_{k=0}^{n} \left(\frac{1}{k+3} - \frac{1}{k+4} \right)$$
$$= \left(\frac{1}{3} - \frac{1}{4} \right) + \left(\frac{1}{4} - \frac{1}{5} \right) + \cdots + \left(\frac{1}{n+2} - \frac{1}{n+3} \right) + \left(\frac{1}{n+3} - \frac{1}{n+4} \right)$$
$$= \frac{1}{3} - \frac{1}{n+4}.$$

(c) Since $\lim_{n \to \infty} \frac{1}{n+4} = 0$, we get $\lim_{n \to \infty} S_n = \frac{1}{3}$.

11. (a) We can rewrite a_k as

$$a_k = \frac{\left(\frac{1}{2}\right)^k}{\ln(k+2)} - \frac{\left(\frac{1}{2}\right)^{k+1}}{\ln(k+3)} = \frac{1}{2^k \ln(k+2)} - \frac{1}{2^{k+1} \ln(k+3)}.$$

Then

$$S_0 = \frac{1}{\ln 2} - \frac{1}{2\ln 3}$$

$$S_1 = S_0 + \frac{1}{2\ln 3} - \frac{1}{4\ln 4} = \frac{1}{\ln 2} - \frac{1}{4\ln 4}$$

$$S_2 = S_1 + \frac{1}{4\ln 4} - \frac{1}{8\ln 5} = \frac{1}{\ln 2} - \frac{1}{8\ln 5}$$

$$S_3 = S_2 + \frac{1}{8\ln 5} - \frac{1}{16\ln 6} = \frac{1}{\ln 2} - \frac{1}{16\ln 6}$$

$$S_4 = S_3 + \frac{1}{16\ln 6} - \frac{1}{32\ln 7} = \frac{1}{\ln 2} - \frac{1}{32\ln 7}.$$

(b) Clearly the series telescopes:

$$S_n = \sum_{k=0}^{n} \left(\frac{1}{2^k \ln(k+2)} - \frac{1}{2^{k+1} \ln(k+3)} \right)$$

$$= \left(\frac{1}{\ln(2)} - \frac{1}{2\ln 3} \right) + \left(\frac{1}{2\ln 3} - \frac{1}{4\ln 4} \right) + \cdots$$

$$+ \left(\frac{1}{2^{n-1}\ln(n+1)} - \frac{1}{2^n \ln(n+2)} \right) + \left(\frac{1}{2^n \ln(n+2)} - \frac{1}{2^{n+1}\ln(n+3)} \right)$$

$$= \frac{1}{\ln 2} - \frac{1}{2^{n+1}\ln(n+3)}.$$

(c) Since $\lim_{n\to\infty} \frac{1}{2^{n+1}\ln(n+3)} = 0$, we see that $\lim_{n\to\infty} S_n = \frac{1}{\ln 2}$.

13. This is a geometric series with first term $\left(\frac{1}{3}\right)^{4+2}$ and ratio $\frac{1}{3}$, so its sum is

$$\frac{\frac{1}{729}}{1 - \frac{1}{3}} = \frac{\frac{1}{729}}{\frac{2}{3}} = \frac{1}{486}.$$

15. Since all terms of this series are $(1)^k = 1$, the terms do not converge to zero, so the series diverges by the divergence test.

17. Since $\lim_{k\to\infty} \ln k = \infty$, the terms of the series do not converge to zero, so the series diverges by the divergence test.

19. Since power functions dominate logs, $\lim_{k\to\infty} \frac{k}{\ln k} = \infty$. Since the terms of the series do not converge to zero, the series diverges by the divergence test.

21. Using the ratio test, we get

$$\rho = \lim_{k\to\infty} \frac{\frac{k+1}{3^{k+1}}}{\frac{k}{3^k}} = \lim_{k\to\infty} \frac{k+1}{3k} = \lim_{k\to\infty} \frac{1+1/k}{3} = \frac{1}{3}.$$

Since $\rho < 1$, the series converges by the ratio test.

23. Use the limit comparison test against $\sum_{k=1}^{\infty} \frac{1}{k}$:

$$\lim_{k\to\infty} \frac{\frac{k}{1+k^2}}{\frac{1}{k}} = \lim_{k\to\infty} \frac{k^2}{k^2+1} = \lim_{k\to\infty} \frac{1}{1+1/k} = 1.$$

Since the ratio is a nonzero real number, either both series converge or both diverge. But $\sum_{k=1}^{\infty} \frac{1}{k}$ diverges since it is the harmonic series, so that $\sum_{k=1}^{\infty} \frac{k}{1+k^2}$ diverges as well.

25. Since $\lim\limits_{k\to\infty} \sqrt[k]{e} = 1$, the terms of the series converge to $\frac{1}{1} = 1$. Since they do not converge to zero, the series diverges by the divergence test.

27. Use the limit comparison test with $\sum_{k=1}^{\infty} \frac{1}{k}$:

$$\lim_{k\to\infty} \frac{\frac{1}{k+\sqrt{k}}}{\frac{1}{k}} = \lim_{k\to\infty} \frac{k}{k+k^{1/2}} = \lim_{k\to\infty} \frac{1}{1+k^{-1/2}} = 1.$$

Since the ratio is a nonzero real number, either both series converge or both diverge. Since $\sum_{k=1}^{\infty} \frac{1}{k}$ is the harmonic series, it diverges, so the original series diverges as well.

29. Use the limit comparison test with $\sum_{k=1}^{\infty} \frac{1}{2^k}$:

$$\lim_{k\to\infty} \frac{\frac{1}{2^k+\sin k}}{\frac{1}{2^k}} = \lim_{k\to\infty} \frac{2^k}{2^k + \sin k} = \lim_{k\to\infty} \frac{1}{1 + \frac{\sin k}{2^k}} = 1$$

since $|\sin k| \le 1$ and $\lim\limits_{k\to\infty} 2^k = \infty$. Since the ratio is a nonzero real number, either both series converge or both diverge. But $\sum_{k=1}^{\infty} \frac{1}{2^k}$ is a geometric series with ratio $r = \frac{1}{2}$, so it converges and thus the original series converges as well.

31. Since the terms of the series diverge to ∞, they do not converge to zero, so that by the divergence test, this series diverges.

33. This series converges absolutely. To see this, use the ratio test on $\sum_{k=1}^{\infty} e^{-k}$:

$$\rho = \lim_{k\to\infty} \frac{e^{-(k+1)}}{e^{-k}} = \lim_{k\to\infty} \frac{1}{e} = \frac{1}{e}.$$

Since $\rho < 1$, the series converges, so the original series converges absolutely.

35. Since $\lim\limits_{k\to\infty} \frac{2^k}{1+2^k} = \lim\limits_{k\to\infty} \frac{1}{2^{-k}+1} = 1$, the terms do not converge to zero, so that by the divergence test, the series diverges.

37. This series converges absolutely. Use the ratio test for absolute convergence:

$$\rho = \lim_{k\to\infty} \left| \frac{\frac{(-1)^{k+1}}{3^{k+1}}}{\frac{(-1)^k}{3^k}} \right| = \frac{1}{3}.$$

Since $\rho < 1$, the series converges absolutely.

39. Since $\lim\limits_{k\to\infty} \frac{\sqrt{k}}{3k-1} = \lim\limits_{k\to\infty} \frac{1}{3\sqrt{k}-1/\sqrt{k}} = 0$, the terms converge to zero. To see that the terms are decreasing, let $f(x) = \frac{\sqrt{x}}{3x-1}$ and look at $f'(x) = \frac{-3x-1}{2\sqrt{x}(3x-1)^2}$. Since $f'(x) < 0$ for $x > 0$, the function, and thus the terms of the series, are decreasing. Thus the series converges by the alternating series test. However, it does not converge absolutely. To see this, use the limit comparison test with $\frac{1}{\sqrt{k}}$:

$$\lim_{k\to\infty} \frac{\frac{\sqrt{k}}{3k-1}}{\frac{1}{\sqrt{k}}} = \lim_{k\to\infty} \frac{k}{3k-1} = \lim_{k\to\infty} \frac{1}{3-1/k} = \frac{1}{3}.$$

Since $\sum_{k=1}^{\infty} \frac{1}{\sqrt{k}}$ is a p-series with $p < 1$, it diverges, so by the limit comparison test, so does $\sum_{k=1}^{\infty} \frac{\sqrt{k}}{3k-1}$, so that the original series converges conditionally.

Chapter 8

Power Series

8.1 Power Series

Thinking Back

Translating the graph of a function

The graph of $f(x - x_0)$ is the graph of f shifted right by x_0 units when x_0 is a positive constant, and shifted left by $|x_0| = -x_0$ units when x_0 is a negative constant. See the discussions of Transformations and Symmetry in Section 0.2.

Odd and Even Functions

An *odd* function is a function for which $f(-x) = -f(x)$ for all x. An *even* function is a function for which $f(-x) = f(x)$ for all x. The graph of every odd function is symmetric about the origin. The graph of every even function is symmetric about the y-axis. See the discussions of Transformations and Symmetry in Section 0.2.

Concepts

1. (a) False. A power series in x uses only nonnegative powers of x.

   (b) True. See Definition 8.4.

   (c) True. Even though it has only odd powers of x, it is a power series in x in which the coefficients of even powers are zero.

   (d) False. When $x = 3$, the series is $\sum_{k=0}^{\infty} 6^k k!$, which obviously diverges since the terms themselves diverge.

   (e) True. Since the series converges conditionally at $x = 3$, the radius of convergence ρ must be 3 (see Theorem 8.3 and the discussion following); since $-2 \in (-3, 3)$, the series converges absolutely at $x = -2$.

   (f) True. Since the series diverges at $x = 3$, the radius of convergence ρ must be less than 3 (see Theorem 8.3 and the discussion following); since $-4 < -\rho$, the series diverges at $x = -4$.

   (g) True. Since the power series converges at $x = 10$, the radius of convergence must be at least $10 - 4 = 6$, so the series must converge absolutely for all $x \in (4 - 6, 4 + 6) = (-2, 10)$. Since -1 is in this interval, the series converges absolutely at $x = -1$.

   (h) False. Since the power series converges conditionally at $x = 3$, the radius of convergence must be equal to $3 - (-5) = 8$, so that the series converges absolutely for $x \in (-5 - 8, -5 + 8) = (-13, 3)$. However, Theorem 8.3 says nothing about convergence at the two endpoints of the interval of convergence, and the series could either converge conditionally, converge absolutely, or diverge at -13. (See the discussion following Theorem 8.3).

3. A power series in x is a series of the form $\sum_{k=0}^{\infty} a_k x^k$ where the a_k are real numbers. See Definition 8.1.

5. In Definition 8.4, x_0 is required to be a (fixed) real number. In the case of $\sum_{k=0}^{\infty}(x-k)^k$, k is not fixed, so this is not a power series expansion at any fixed point x_0.

7. By Theorem 8.5, a power series converges absolutely in some open interval of radius ρ around x_0. It may also converge at one or both endpoints. The interval of convergence is the set of values of x for which the power series converges, so it is one of $(x_0 - \rho, x_0 + \rho)$, $[x_0 - \rho, x_0 + \rho)$, $(x_0 - \rho, x_0 + \rho]$, or $[x_0 - \rho, x_0 + \rho]$. The interval of convergence is determined by using a convergence test (usually the ratio test) on the terms of the series to get a value for the limit of the ratio of terms that usually depends on $x - x_0$. Those values of x which make the ratio less than 1 determine the points at which the series converges. That gives an open interval $(x_0 - \rho, x_0 + \rho)$; testing convergence at the endpoints then gives the full interval of convergence.

9. When $x = 2$, the series becomes

$$\sum_{k=0}^{\infty}(-1)^k \frac{k}{3k+1}(2 \cdot 2 - 3)^k = \sum_{k=0}^{\infty}(-1)^k \frac{k}{3k+1}.$$

Since

$$\lim_{k \to \infty}\left|(-1)^k \frac{k}{3k+1}\right| = \lim_{k \to \infty}\frac{k}{3k+1} = \lim_{k \to \infty}\frac{1}{3+1/k} = \frac{1}{3},$$

the terms of the series do not converge to zero, so that the series diverges by the divergence test.

11. When $x = 1$, the series is $\sum_{k=1}^{\infty}\frac{(-1)^k}{k}$, which is the alternating harmonic series. We know from the previous chapter that this series converges conditionally. When $x = -1$, the series evaluates to $\sum_{k=1}^{\infty}\frac{(-1)^k}{k}(-1)^k = \sum_{k=1}^{\infty}\frac{1}{k}$, which is the harmonic series, so is divergent. Thus the interval of convergence is $(-1, 1]$.

13. Since the interval of convergence is symmetric around x_0, it follows that x_0 is halfway between 2 and 10 so that $x_0 = 6$.

15. Since convergence is conditional at both $x = -4$ and $x = 8$, the interval of convergence must be $[-4, 8]$, since if either point were in the interior of the interval, convergence would be absolute, and if either point were outside the closed interval, the series would diverge. Since the interval of convergence is $[-4, 8]$, we must have $x_0 = \frac{-4+8}{2} = 2$.

17. The radius of convergence is normally computed by the ratio test for absolute convergence; the result of applying that test to the power series is an expression in $|x|$, and the radius of convergence ρ is the value of $|x|$ which makes the expression equal to 1. Thus applying the ratio test for absolute convergence when $x = \pm\rho$, i.e., when $|x| = \rho$, will give 1, which is inconclusive.

19. The radius of convergence is ρ^{-1}. To see this, note that because the radius of convergence of the original series is ρ, it follows that

$$\lim_{k \to \infty}\left|\frac{a_{k+1}x^{k+1}}{a_k x^k}\right| = |x| \lim_{k \to \infty}\left|\frac{a_{k+1}}{a_k}\right| = \frac{1}{\rho}|x|.$$

Thus $\lim_{k \to \infty}\left|\frac{a_{k+1}}{a_k}\right| = \frac{1}{\rho}$, so that $\lim_{k \to \infty}\left|\frac{a_k}{a_{k+1}}\right| = \rho$. Applying the ratio test for absolute convergence to the series of reciprocals gives

$$\lim_{k \to \infty}\left|\frac{\frac{1}{a_{k+1}}x^{k+1}}{\frac{1}{a_k}x_k}\right| = |x| \lim_{k \to \infty}\left|\frac{a_k}{a_{k+1}}\right| = \rho|x|.$$

This is less than 1 for $|x| < \frac{1}{\rho}$, so that the radius of convergence is $\frac{1}{\rho}$.

Skills

21. By the ratio test for absolute convergence, we have

$$\lim_{k\to\infty} \left| \frac{\frac{1}{(k+1)!}x^{k+1}}{\frac{1}{k!}x^k} \right| = \lim_{k\to\infty} \frac{|x|}{k} = |x| \lim_{k\to\infty} \frac{1}{k} = 0.$$

Since the ratio is always less than 1, the radius of convergence is infinite, so that the interval of convergence is all real numbers.

23. By the ratio test for absolute convergence, we have

$$\lim_{k\to\infty} \left| \frac{\frac{(-1)^{k+1}x^{2(k+1)}}{(2(k+1))!}}{\frac{(-1)^kx^{2k}}{(2k)!}} \right| = \lim_{k\to\infty} \frac{|x|^2}{(2k+1)(2k+2)} = |x|^2 \lim_{k\to\infty} \frac{1}{(2k+1)(2k+2)} = 0.$$

Since the ratio is always less than 1, the radius of convergence is infinite, so that the interval of convergence is all real numbers.

25. By the ratio test for absolute convergence, we have

$$\lim_{k\to\infty} \left| \frac{\frac{(-1)^{k+1}x^{k+1}}{k+2}}{\frac{(-1)^kx^k}{k+1}} \right| = \lim_{k\to\infty} \frac{|x|(k+1)}{k+2} = |x| \lim_{k\to\infty} \frac{k+1}{k+2} = |x|.$$

Thus the series converges absolutely for $|x| < 1$, so the radius of convergence is 1. Checking the endpoints, we have at $x = -1$ the series $\sum_{k=1}^{\infty} \frac{(-1)^k}{k+1}(-1)^k = \sum_{k=1}^{\infty} \frac{1}{k+1}$, which is divergent by the limit comparison test against the harmonic series, so the series diverges at $x = -1$. At $x = 1$, the series is $\sum_{k=1}^{\infty} \frac{(-1)^k}{k+1}$, which is an alternating series satisfying the conditions of the alternating series test, so it converges. However, it does not converge absolutely, since $\sum_{k=1}^{\infty} \frac{1}{k+1}$ diverges by the above. Hence the interval of convergence is $(-1, 1]$, with conditional convergence at $x = 1$.

27. By the ratio test for absolute convergence, we have

$$\lim_{k\to\infty} \left| \frac{\frac{1}{k+1}(x+2)^{k+1}}{\frac{1}{k}(x+2)^k} \right| = \lim_{k\to\infty} \frac{|x+2|k}{k+1} = |x+2| \lim_{k\to\infty} \frac{k}{k+1} = |x+2|.$$

Thus the series converges absolutely for $|x + 2| < 1$, i.e., for $-1 < x + 2 < 1$, so for $-3 < x < -1$. Checking the endpoints, at $x = -3$ we get the series $\sum_{k=1}^{\infty} \frac{1}{k}(-1)^k$, which is the convergent alternating harmonic series. Convergence is conditional, since the corresponding positive series is the divergent harmonic series. At $x = -1$ we get the divergent harmonic series $\sum_{k=1}^{\infty} \frac{1}{k}$. Thus the interval of convergence is $[-3, -1)$, with conditional convergence at -3.

29. By the ratio test for absolute convergence, we have

$$\lim_{k\to\infty} \left| \frac{\frac{(2(k+1)+1)(x-\pi)^{k+1}}{(k+1)^3}}{\frac{(2k+1)(x-\pi)^k}{k^3}} \right| = \lim_{k\to\infty} \frac{|x-\pi|(2k+3)k^3}{(2k+1)(k+1)^3} = |x-\pi| \lim_{k\to\infty} \frac{2k^4 + 3k^3}{2k^4 + 7k^3 + 9k^2 + 5k + 1}$$

$$= |x - \pi|.$$

Thus the series converges absolutely for $|x - \pi| < 1$, i.e., for $-1 < x - \pi < 1$, so for $\pi - 1 < x < \pi + 1$. Checking the endpoints, at $x = \pi - 1$ we get the series $\sum_{k=1}^{\infty} \frac{2k+1}{k^3}(-1)^k$. The corresponding positive series is $\sum_{k=1}^{\infty} \frac{2k+1}{k^3}$, which converges by the limit comparison test with $\sum_{k=1}^{\infty} \frac{1}{k^2}$. At $x = \pi + 1$ we again get the series $\sum_{k=1}^{\infty} \frac{2k+1}{k^3}$. Thus the given series is absolutely convergent at both endpoints, and its interval of convergence is $[\pi - 1, \pi + 1]$.

31. By the ratio test for absolute convergence, we have

$$\lim_{k \to \infty} \left| \frac{\frac{(-1)^{k+1}\sqrt{k+1}\left(x-\frac{3}{2}\right)^{k+1}}{(k+1)^3}}{\frac{(-1)^k \sqrt{k}\left(x-\frac{3}{2}\right)^k}{k^3}} \right| = \lim_{k \to \infty} \frac{\left|x - \frac{3}{2}\right| k^3 \sqrt{k+1}}{(k+1)^3 \sqrt{k}} = \left|x - \frac{3}{2}\right| \lim_{k \to \infty} \frac{\sqrt{1+1/k}}{(1+1/k)^3 \sqrt{1/k}} = \left|x - \frac{3}{2}\right|.$$

Thus the series converges absolutely for $\left|x - \frac{3}{2}\right| < 1$, so that $-1 < x - \frac{3}{2} < 1$, or $\frac{1}{2} < x < \frac{5}{2}$. Checking the endpoints, at $x = \frac{1}{2}$ we get $\sum_{k=1}^{\infty}(-1)^k \frac{\sqrt{k}}{k^3}(-1)^k = \sum_{k=1}^{\infty} \frac{\sqrt{k}}{k^3} = \sum_{k=1}^{\infty} \frac{1}{k^{5/2}}$, which converges since it is a p-series with $p > 1$. At $x = \frac{5}{2}$, we get $\sum_{k=1}^{\infty}(-1)^k \frac{\sqrt{k}}{k^3}$, which converges absolutely since we get the same series in absolute value as for $x = \frac{1}{2}$. Thus the interval of convergence is $\left[\frac{1}{2}, \frac{5}{2}\right]$, with absolute convergence at both endpoints.

33. Using the ratio test for absolute convergence, we have

$$\lim_{k \to \infty} \left| \frac{\frac{(-1)^{k+1}(k+1)^2\left(x-\frac{1}{2}\right)^{k+1}}{3^{k+1}}}{\frac{(-1)^k k^2 \left(x-\frac{1}{2}\right)^k}{3^k}} \right| = \lim_{k \to \infty} \frac{(k+1)^2 \left|x - \frac{1}{2}\right|}{3k^2} = \left|x - \frac{1}{2}\right| \lim_{k \to \infty} \frac{(k+1)^2}{3k^2} = \frac{1}{3}\left|x - \frac{1}{2}\right|.$$

Thus the series converges absolutely for $\frac{1}{3}\left|x - \frac{1}{2}\right| < 1$, which means $\left|x - \frac{1}{2}\right| < 3$, so that $-\frac{5}{2} < x < \frac{7}{2}$. Checking the endpoints, at $x = -\frac{5}{2}$ we get $\sum_{k=0}^{\infty}(-1)^k \frac{k^2}{3^k}(-3)^k = \sum_{k=0}^{\infty} k^2$, which clearly diverges. At $x = \frac{7}{2}$ we get $\sum_{k=0}^{\infty}(-1)^k k^2$, which also diverges since the terms do not converge to zero. Thus the interval of convergence is $\left(-\frac{5}{2}, \frac{7}{2}\right)$.

35. By the ratio test for absolute convergence, we have

$$\lim_{k \to \infty} \left| \frac{\frac{((k+1)!)^2(x+2)^{k+1}}{(2(k+1))!}}{\frac{(k!)^2(x+2)^k}{(2k)!}} \right| = \lim_{k \to \infty} \frac{((k+1)!)^2(2k)! \, |x+2|}{(2k+2)!(k!)^2} = |x+2| \lim_{k \to \infty} \frac{(k+1)^2}{(2k+1)(2k+2)} = \frac{1}{4}|x+2|.$$

Thus the series converges absolutely for $\frac{1}{4}|x+2| < 1$, so for $|x+2| < 4$, and thus for $-6 < x < 2$. Checking the endpoints, at $x = 2$ we get $\sum_{k=0}^{\infty} \frac{(k!)^2}{(2k)!} 4^k$. But

$$\frac{(k!)^2}{(2k)!}4^k = \frac{(k!)^2}{(1\cdot 3\cdot 5 \cdots (2k-1))(2\cdot 4\cdot 6 \cdots (2k))}4^k$$

$$= \frac{(k!)^2}{(1\cdot 3\cdot 5 \cdots (2k-1))\cdot 2^k \cdot k!}4^k$$

$$= \frac{2^k k!}{1\cdot 3\cdot 5 \cdots (2k-1)}$$

$$= \frac{2\cdot 4\cdot 6 \cdots (2k)}{1\cdot 3\cdot 5 \cdots (2k-1)}$$

$$= \frac{2}{1}\cdot\frac{4}{3}\cdot\frac{6}{5} \cdots \frac{2k}{2k-1},$$

so that the terms of the series obviously cannot converge to zero. Thus the series diverges at $x = 2$. At $x = -6$, we get the series $\sum_{k=0}^{\infty}(-1)^k \frac{(k!)^2}{(2k)!}4^k$, which diverges for the same reason. Thus the interval of convergence is $(-6, 2)$.

37. By the ratio test for absolute convergence, we have

$$\lim_{k\to\infty}\left|\frac{(k+1)!x^{k+1}}{k!x^k}\right| = \lim_{k\to\infty} k\,|x| = \infty$$

unless $x = 0$. Thus the interval of convergence consists of the single point $x = 0$. (Note that by combining Exercise 21 and a modification of Exercise 19 you could arrive at the same result without computation).

39. By the ratio test for absolute convergence, we have

$$\lim_{k\to\infty}\left|\frac{\frac{\ln(k+1)\cdot\left(x+\frac{5}{3}\right)^{k+1}}{(k+1)^2}}{\frac{\ln k\cdot\left(x+\frac{5}{3}\right)^{k}}{k^2}}\right| = \lim_{k\to\infty}\frac{\left|x+\frac{5}{3}\right|k^2\ln(k+1)}{(k+1)^2\ln k} = \left|x+\frac{5}{3}\right|\lim_{k\to\infty}\left(\frac{k^2}{(k+1)^2}\cdot\frac{\ln(k+1)}{\ln k}\right)$$

$$= \left|x+\frac{5}{3}\right|.$$

(Note that the limit of the second factor above is 1 by L'Hôpital's Rule). Thus the series converges absolutely for $\left|x+\frac{5}{3}\right| < 1$, so for $-\frac{8}{3} < x < -\frac{2}{3}$. Checking the endpoints, at $x = -\frac{2}{3}$ we get $\sum_{k=1}^{\infty}\frac{\ln k}{k^2}$, which converges for example by the integral test. At $x = -\frac{8}{3}$, we get $\sum_{k=1}^{\infty}(-1)^k\frac{\ln k}{k^2}$, which converges absolutely for the same reason. Thus the interval of convergence is $\left[-\frac{8}{3}, -\frac{2}{3}\right]$ with absolute convergence at both endpoints.

41. By the ratio test for absolute convergence, we have

$$\lim_{k\to\infty}\left|\frac{\frac{(k+2)(2x-5)^{k+1}}{3^{k+1}}}{\frac{(k+1)(2x-5)^{k}}{3^k}}\right| = \lim_{k\to\infty}\frac{|2x-5|\,(k+2)}{3k+3} = |2x-5|\lim_{k\to\infty}\frac{k+2}{3k+3} = \frac{1}{3}|2x-5|.$$

Thus the series converges absolutely for $\frac{1}{3}|2x-5| < 1$, or $|2x-5| < 3$. This means that $-3 < 2x-5 < 3$, so that $2 < 2x < 8$ or $1 < x < 4$. Checking the endpoints gives at $x = 4$ the series $\sum_{k=0}^{\infty}\frac{k+1}{3^k}\cdot 3^k = \sum_{k=0}^{\infty}(k+1)$, which diverges. At $x = 1$ we get $\sum_{k=0}^{\infty}\frac{k+1}{3^k}(-3)^k = \sum_{k=0}^{\infty}(-1)^k(k+1)$, which also diverges. Thus the interval of convergence is $(1, 4)$.

43. By the ratio test for absolute convergence, we have

$$\lim_{k\to\infty}\left|\frac{\frac{(4x+7)^{k+1}\sqrt{k+1}}{(k+1)!}}{\frac{(4x+7)^{k}\sqrt{k}}{k!}}\right| = \lim_{k\to\infty}\frac{|4x+7|\sqrt{k+1}}{k\sqrt{k}} = |4x+7|\lim_{k\to\infty}\frac{\sqrt{k+1}}{k\sqrt{k}} = |4x+7|\lim_{k\to\infty}\frac{\sqrt{1+1/k}}{k} = 0.$$

Since the ratio is always less than 1, the interval of convergence is $(-\infty, \infty)$.

45. The coefficient of x^k is $\frac{1}{2\cdot4\cdot6\cdots(2k)} = \frac{1}{2^k k!}$. Then using the ratio test for absolute convergence, we have

$$\lim_{k\to\infty}\left|\frac{\frac{x^{k+1}}{2^{k+1}(k+1)!}}{\frac{x^k}{2^k k!}}\right| = \lim_{k\to\infty}\frac{|x|}{2(k+1)} = 0.$$

Since the ratio is always less than 1, the interval of convergence is $(-\infty, \infty)$.

47. By the ratio test for absolute convergence, we have

$$\lim_{k \to \infty} \left| \frac{\frac{1}{(k+1)^{k+1}}(x-3)^{k+1}}{\frac{1}{k^k}(x-3)^k} \right| = \lim_{k \to \infty} \left(\frac{1}{k+1} \left(\frac{k}{k+1} \right)^k |x-3| \right)$$

$$= \lim_{k \to \infty} \frac{1}{k+1} \left(1 + \frac{1}{k} \right)^{-k} |x-3| = e^{-1} |x-3| \lim_{k \to \infty} \frac{1}{k+1} = 0.$$

Since the limit is zero, the radius of convergence is infinite and this series converges for any x.

49. Using the ratio test for absolute convergence, we get (since $m > 0$)

$$\lim_{k \to \infty} \left| \frac{\frac{(k+1)! x^{k+1}}{(k+m+1)!}}{\frac{k! x^k}{(k+m)!}} \right| = \lim_{k \to \infty} \frac{(k+1)|x|}{k+m} = |x| \lim_{k \to \infty} \frac{k+1}{k+m} = |x|.$$

Then the series converges absolutely for $|x| < 1$, so that the radius of convergence is 1.

51. Using the ratio test for absolute convergence, we have (since $m, n > 0$)

$$\lim_{k \to \infty} \left| \frac{\frac{(k+1)! x^{k+1}}{(k+m+1)!(k+n+1)!}}{\frac{k! x^k}{(k+m)!(k+n)!}} \right| = \lim_{k \to \infty} \frac{(k+1)|x|}{(k+m+1)(k+n+1)} = |x| \lim_{k \to \infty} \frac{k+1}{(k+m+1)(k+n+1)} = 0.$$

Since the ratio is always less than 1, this series converges everywhere; its radius of convergence is ∞.

53. This is not a power series in x since it is not a sum of multiples of positive powers of x. However, using the ratio test for absolute convergence, we get

$$\lim_{k \to \infty} \left| \frac{\frac{1}{(x+1)^{k+1}}}{\frac{1}{(x+1)^k}} \right| = \lim_{k \to \infty} \frac{1}{|x+1|} = \frac{1}{|x+1|}.$$

We know that the series converges when $\frac{1}{|x+1|} < 1$, so when $|x+1| > 1$; this happens for $x > 0$ and for $x < -2$. We know it diverges when $-2 < x < 0$ since then the ratio is greater than 1. It remains to look at the two boundary points. At $x = -2$, this becomes the divergent series $\sum_{k=0}^{\infty}(-1)^k$; at $x = 0$ we get the divergent series $\sum_{k=0}^{\infty} 1^k$. Thus the series converges on $(-\infty, -2) \cup (0, \infty)$.

55. This is not a power series in x since we must have terms including x^k, not $f(x)^k$ for some function f. However, using the ratio test for absolute convergence, we have

$$\lim_{k \to \infty} \left| \frac{\frac{(\sin x)^{k+1}}{(k+1)!}}{\frac{(\sin x)^k}{k}} \right| = \lim_{k \to \infty} \frac{|\sin x|}{k} = 0.$$

Thus this series converges for all x.

57. This is not a power series in x since it includes x in the denominator. However, using the ratio test for absolute convergence, we have

$$\lim_{k \to \infty} \left| \frac{\frac{(x+2)^{k+1}}{(k+1)^3(x-3)^{k+1}}}{\frac{(x+2)^k}{k^3(x-3)^k}} \right| = \lim_{k \to \infty} \frac{k^3 |x+2|}{(k+1)^3 |x-3|} = \left| \frac{x+2}{x-3} \right| \lim_{k \to \infty} \frac{k^3}{(k+1)^3} = \left| \frac{x+2}{x-3} \right|.$$

So the series converges absolutely when $\left|\frac{x+2}{x-3}\right| < 1$, so when $-1 < \frac{x+2}{x-3} < 1$. This quotient is undefined at $x = 3$. For $x > 3$, it is always greater than 1. For $x < 3$, we have $\frac{x+2}{x-3} < 1$ always, and $\frac{x+2}{x-3} > -1$ for $x < \frac{1}{2}$. Thus the series converges absolutely for $x < \frac{1}{2}$ and diverges for $x > \frac{1}{2}$ (except at $x = 3$ where it is undefined). When $x = \frac{1}{2}$, then $\frac{x+2}{x-3} = -1$, and the series becomes $\sum_{k=1}^{\infty} (-1)^k \frac{1}{k^3}$, which is absolutely convergent since it is a p-series with $p > 1$. So in summary, the series converges absolutely on $\left(-\infty, \frac{1}{2}\right]$ and diverges otherwise.

Applications

59. (a) Using the ratio test, we get

$$\lim_{n\to\infty} \frac{\frac{w^{n+1}}{19200^{n+1}(n+1)!}}{\frac{w^n}{19200^n n!}} = \lim_{n\to\infty} \frac{w}{19200(n+1)} = 0.$$

Since the ratio is zero, this series converges for all w.

(b) The denominator grows very quickly. For $w = 500$, $\frac{w}{19200}$ is about 0.02, so already when $n = 1$ we will have an approximation exact to within 0.02, or to within 1 decimal place. Since we are counting an integral number of beavers, this should be sufficient. A plot of $b_\sigma(w)$ is

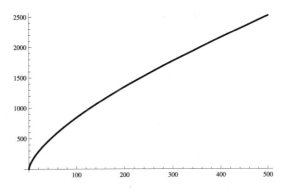

Proofs

61. Suppose that $|d| > |c|$ and that the series converges at d. Then by Theorem 8.2(a), the series converges at c. Thus the contrapositive of this implication holds, which is that if the series diverges at c, and $|d| > |c|$, then the series diverges at d.

63. Note that at $x_0 \pm \rho$, the series is $\sum_{k=0}^{\infty} a_k((x_0 \pm \rho) - x_0)^k = \sum_{k=0}^{\infty} a_k(\pm\rho)^k$, so that to determine absolute convergence we look at $\sum_{k=0}^{\infty} a_k \left|\pm\rho\right|^k = \sum_{k=0}^{\infty} a_k \left|\rho\right|^k$. Thus at both endpoints, the underlying positive series are the same, so that absolute convergence at either endpoint implies absolute convergence at the other.

65. Suppose that $\sum_{k=0}^{\infty} a_k(x - x_0)^k$ converged absolutely at $x_0 + \rho$. Then by Exercise 63, it would also converge absolutely at $x_0 - \rho$, so that the interval of convergence would be $[x_0 - \rho, x_0 + \rho]$. But the series diverges there, so that since $x_0 + \rho$ is in the interval of convergence, convergence must be conditional there.

67. If ρ is not infinite and is nonzero, then by Exercise 66, if the radius of convergence of $\sum_{k=0}^{\infty} a_k x^k$ is ρ, then the radius of convergence of $\sum_{k=0}^{\infty} a_k x^{2k}$ is $\sqrt{\rho}$. Since these two are equal, we have $\rho = \sqrt{\rho}$ and, since $\rho > 0$, it follows that $\rho = 1$.

69. This series is $\sum_{k=0}^{\infty} b^k x^k = \sum_{k=0}^{\infty} (bx)^k$, so it is a geometric series with ratio bk. Since $b \neq 0$, in order for this series to converge, we must have $|bx| < 1$, so that $|x| < \frac{1}{|b|}$. This says that the radius of convergence of the power series is $\frac{1}{|b|}$.

71. (a) Rewrite the series as follows:

$$\sum_{k=0}^{\infty} a_k x^k = (a_0 + a_1 x) + (a_0 + a_1 x)x^2 + (a_0 + a_1 x)x^4 + \cdots = \sum_{k=0}^{\infty} (a_0 + a_1 x)x^{2k}.$$

It is thus seen to be a geometric series with initial term $a_0 + a_1 x$ and ratio x^2, so that it converges for $x^2 < 1$ and thus for $|x| < 1$. The radius of convergence of the series must be 1.

(b) At $x = \pm 1$, the series becomes

$$\sum_{k=0}^{\infty} (a_0 \pm a_1)(-1)^{2k} = \sum_{k=0}^{\infty} (a_0 \pm a_1),$$

which is a sum of nonzero terms except in the case where $a_0 = a_1$ at $x = -1$.

(c) The sum of this geometric series at $x \in (-\rho, \rho)$ is

$$\frac{c}{1-r} = \frac{a_0 + a_1 x}{1 - x^2}.$$

Thinking Forward

Constructing a power series of the form $\sum_{k=0}^{\infty} \frac{f^{(k)}(0)}{k!} x^k$

We know that

$$\sin(x)^{(n)} = \begin{cases} \cos x, & n \equiv 1 \pmod 4, \\ -\sin x, & n \equiv 2 \pmod 4, \\ -\cos x, & n \equiv 3 \pmod 4, \\ \sin x, & n \equiv 0 \pmod 4 \end{cases} \quad \text{so} \quad \sin(x)^{(n)}\big|_{x=0} = \begin{cases} 1, & n \equiv 1 \pmod 4, \\ -1, & n \equiv 3 \pmod 4, \\ 0, & \text{otherwise.} \end{cases}$$

Thus the power series in question has zero coefficients for even k, and $f^{(2k+1)}(0)$ is 1 for k even and -1 for k odd. So the series is

$$\sum_{k=0}^{\infty} \frac{(-1)^k}{(2k+1)!} x^{2k+1}.$$

Graphing $\sin x$ and a polynomial approximation

The graphs are:

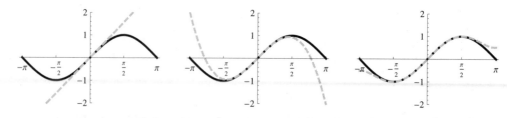

All three graphs track the curve to some extent, but the first two fall away from the curve noticeably at or before the first local extremum on either side of zero.

8.2   Maclaurin Series and Taylor Series

Thinking Back

Polynomials

A polynomial is a function of the form $a_n x^n + a_{n-1} x^{n-1} + \cdots + a_1 x + a_0$ where n is a nonnegative integer and the a_i are real numbers. The domain of any such function is $\mathbb{R}$.

Derivatives and linear approximations

The equation of the tangent to the graph of f at x_0 is $y = f'(x_0)(x - x_0) + f(x_0)$. This is a good approximation for f near x_0 since it has the same value as f at x_0, and since the slopes of f and the tangent line are the same at x_0, the two functions should change by roughly the same amount near x_0.

Concepts

1. (a) True. By Definition 8.6, the first-order Maclaurin polynomial for f is defined if f has a first derivative at $x = 0$, i.e. if f is differentiable there.

    (b) True, by Definition 8.7.

    (c) False. $f(x)$ is not differentiable at $x = 0$, so it does not have a Maclaurin polynomial there.

    (d) True. The kth Taylor polynomial at $x = 3$ is simply the terms of the nth Taylor polynomial up through the x^k term. It exists since if f is n times differentiable, then it is k times differentiable for $k \leq n$.

    (e) False. In the Taylor polynomial, every term except the term for $k = 0$ has a factor of $x - \pi$, so it vanishes at π. The term for $k = 0$ is $f(\pi)$. Thus $P_n(\pi) = f(\pi)$, not $f^{(n)}(\pi)$.

    (f) False. All that is required is that f have derivatives of all orders at $x = 0$. See Definition 8.7.

    (g) True. See Definition 8.7.

    (h) False. 0 might not be in the domain of the function.

3. When $x = 0$, the series is

$$\ln 2 + \sum_{k=1}^{\infty} \frac{(-1)^{k-1}}{k2^k} (-2)^k = \ln 2 + \sum_{k=1}^{\infty} \frac{(-1)^{k-1}(-1)^k}{k} = \ln 2 - \sum_{k=1}^{\infty} \frac{1}{k},$$

which diverges since it is the harmonic series. When $x = 4$, we get

$$\ln 2 + \sum_{k=1}^{\infty} \frac{(-1)^{k-1}}{k2^k} \cdot 2^k = \ln 2 + \sum_{k=1}^{\infty} \frac{(-1)^{k-1}}{k}.$$

This converges by the Alternating Series test, since the terms go to zero, but it converges only conditionally, since the underlying positive series is the divergent harmonic series.

5. Since $P_2(x) = f(x_0) + f'(x_0)(x - x_0) + \frac{f''(x_0)}{2}(x - x_0)^2$, we see that $P_2(x_0) = f(x_0)$, so that the values of the two functions are equal. Also, $P_2'(x) = f'(x_0) + f''(x_0)(x - x_0)$, so that $P_2'(x_0) = f'(x_0)$. Thus the slopes of P_2 and f are equal at x_0. Finally, $P_2''(x) = f''(x_0)$, so that the concavity of P_2 matches that of f at x_0. So f and P_2 are the same in terms of these three measures of what a graph looks like.

7. The Maclaurin polynomial of order n for f is simply the Taylor polynomial of order n for f at $x_0 = 0$. A Taylor polynomial may be around any x_0 where f is differentiable to the appropriate degree; Maclaurin polynomials are always around $x_0 = 0$.

9. The Taylor series for f at x_0 is an infinite series. It is the limit as $n \to \infty$ of the Taylor polynomials of order n.

11. Since the Maclaurin series is a power series in x, it can be convergent only at $x = 0$, or it can converge for all $\mathbb{R}$. Finally, it may have a radius of convergence $0 < R < \infty$; the series will converge absolutely for $|x| < R$, and it may converge absolutely or conditionally, or diverge, for $x = R$ and for $x = -R$.

13. We have $f(0) = 5$. Next, $f'(x) = 6x - 2$, so that $f'(0) = -2$. Then $f''(x) = 6$, so that $f''(0) = 6$. Finally, $f'''(x) = f'''(0) = 0$. So by Definition 8.6, the Maclaurin polynomials of orders one to three for f are

$$P_1(x) = f(x_0) + f'(x_0)x = 5 - 2x,$$
$$P_2(x) = f(x_0) + f'(x_0)x + \frac{f''(x_0)}{2}x^2 = 5 - 2x + 3x^2,$$
$$P_3(x) = f(x_0) + f'(x_0)x + \frac{f''(x_0)}{2}x^2 + \frac{f'''(x_0)}{6}x^3 = 5 - 2x + 3x^2.$$

$P_2(x) = P_3(x)$ since $f'''(0) = 0$. Further, since $f(x)$ is a quadratic, the best quadratic or higher-order approximation to $f(x)$ is $f(x)$ itself, so that $f(x) = P_2(x)$. Graphs of $f(x)$, $P_1(x)$, and $P_2(x)$ are below (of course, $P_2(x) = f(x)$, so its graph is invisible on top of the graph of $f(x)$):

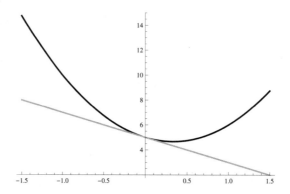

15. We have $f(1) = 6$. Next, $f'(x) = 6x - 2$, so that $f'(1) = 4$. Then $f''(x) = 6$, so that $f''(1) = 6$. Finally, $f'''(x) = f'''(1) = 0$. So by Definition 8.6, the Taylor polynomials of orders one to three for f are

$$P_1(x) = f(x_0) + f'(x_0)(x - 1) = 6 + 4(x - 1) = 4x + 2,$$
$$P_2(x) = f(x_0) + f'(x_0)(x - 1) + \frac{f''(x_0)}{2}(x - 1)^2 = 6 + 4(x - 1) + 3(x - 1)^2 = 3x^2 - 2x + 5$$
$$P_3(x) = f(x_0) + f'(x_0)x + \frac{f''(x_0)}{2}x^2 + \frac{f'''(x_0)}{6}x^3 = 6 + 4(x - 1) + 3(x - 1)^2 = 3x^2 - 2x + 5.$$

$P_2(x) = P_3(x)$ since $f'''(0) = 0$. Further, since $f(x)$ is a quadratic, the best quadratic or higher-order approximation to $f(x)$ is $f(x)$ itself, so that $f(x) = P_2(x)$. Graphs of $f(x)$, $P_1(x)$, and $P_2(x)$ are below (of course, $P_2(x) = f(x)$, so its graph is invisible on top of the graph of $f(x)$):

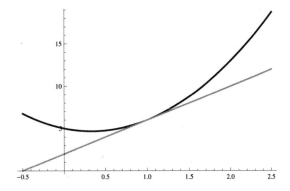

17. We have $f(x_0) = ax_0^3 + bx_0^2 + cx_0 + d$. Then

$$f'(x) = 3ax^2 + 2bx + c \quad \Rightarrow \quad f'(x_0) = 3ax_0^2 + 2bx_0 + c$$
$$f''(x) = 6ax + 2b \quad \Rightarrow \quad f''(x_0) = 6ax_0 + 2b$$
$$f'''(x) = 6a \quad \Rightarrow \quad f'''(x_0) = 6a$$
$$f^{(4)}(x) = 0 \quad \Rightarrow \quad f^{(4)}(x_0) = 0.$$

Then

$$P_1(x) = f(x_0) + f'(x_0)(x - x_0)$$
$$= ax_0^3 + bx_0^2 + cx_0 + d + (3ax_0^2 + 2bx_0 + c)(x - x_0),$$
$$P_2(x) = f(x_0) + f'(x_0)(x - x_0) + \frac{f''(x_0)}{2}(x - x_0)^2$$
$$= ax_0^3 + bx_0^2 + cx_0 + d + (3ax_0^2 + 2bx_0 + c)(x - x_0) + (3ax_0 + b)(x - x_0)^2,$$
$$P_3(x) = f(x_0) + f'(x_0)(x - x_0) + \frac{f''(x_0)}{2}(x - x_0)^2 + \frac{f'''(x_0)}{6}(x - x_0)^3$$
$$= ax_0^3 + bx_0^2 + cx_0 + d + (3ax_0^2 + 2bx_0 + c)(x - x_0) + (3ax_0 + b)(x - x_0)^2 + a(x - x_0)^3,$$
$$P_4(x) = f(x_0) + f'(x_0)(x - x_0) + \frac{f''(x_0)}{2}(x - x_0)^2 + \frac{f'''(x_0)}{6}(x - x_0)^3 + \frac{f^{(4)}(x_0)}{4!}(x - x_0)^4$$
$$= ax_0^3 + bx_0^2 + cx_0 + d + (3ax_0^2 + 2bx_0 + c)(x - x_0) + (3ax_0 + b)(x - x_0)^2 + a(x - x_0)^3.$$

Note that $P_3(x) = P_4(x)$ since $f^{(4)}(x)$ is identically zero, so it vanishes at x_0. Also, expanding $P_3(x)$ shows that in fact $P_3(x) = f(x)$. This is reasonable since f itself is a cubic, so that the best cubic or higher-order approximation to f is f itself.

19. (a) $f(0) = 0$, and $f'(x) = \frac{7}{3}x^{4/3}$, so that $f'(0) = 0$. Next, $f''(x) = \frac{28}{9}x^{1/3}$, so that $f''(0) = 0$. Thus

$$P_1(x) = f(0) + f'(0)x = 0, \qquad P_2(x) = f(0) + f'(0)x + \frac{f''(0)}{2}x^2 = 0.$$

A graph of $f(x)$, $P_1(x)$, and $P_2(x)$ is below (since $P_1(x) = P_2(x) = 0$, their graphs lie on the x axis):

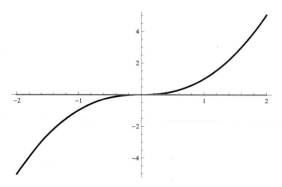

(b) Since $f'''(x) = \frac{28}{27}x^{-2/3}$, f''' is not defined at 0, so f does not have a third-order Maclaurin polynomial.

Skills

21. Since $(\cos x)' = -\sin x$, $(\cos x)'' = -\cos x$, $(\cos x)''' = \sin x$, and $(\cos x)^{(4)} = \cos x$, we have

$$P_4(x) = \cos 0 - (\sin 0)x - \frac{\cos 0}{2}x^2 + \frac{\sin 0}{6}x^3 + \frac{\cos 0}{24}x^4 = 1 - \frac{1}{2}x^2 + \frac{1}{24}x^4.$$

23. Since $(\sin x)' = \cos x$, $(\sin x)'' = -\sin x$, $(\sin x)''' = -\cos x$, and $(\sin x)^{(4)} = \sin x$, we have

$$P_4(x) = \sin 0 + (\cos 0)x - \frac{\sin 0}{2}x^2 - \frac{\cos 0}{6}x^3 + \frac{\sin 0}{24}x^4 = x - \frac{1}{6}x^3.$$

25. With $f(x) = \cos 2x$, we have $f(0) = 1$, and

$$
\begin{aligned}
f'(x) &= -2\sin 2x & f'(0) &= 0, \\
f''(x) &= -4\cos 2x & f''(0) &= -4, \\
f'''(x) &= 8\sin 2x & f'''(0) &= 0, \\
f^{(4)}(x) &= 16\cos 2x & f^{(4)}(0) &= 16.
\end{aligned}
$$

Then

$$P_4(x) = 1 + 0 \cdot x - \frac{4}{2}x^2 + \frac{0}{6}x^3 + \frac{16}{24}x^4 = 1 - 2x^2 + \frac{2}{3}x^4.$$

27. With $f(x) = (1-x)^{1/2}$, we have $f(0) = 1$, and

$$
\begin{aligned}
f'(x) &= -\frac{1}{2}(1-x)^{-1/2} & f'(0) &= -\frac{1}{2}, \\
f''(x) &= -\frac{1}{4}(1-x)^{-3/2} & f''(0) &= -\frac{1}{4}, \\
f'''(x) &= -\frac{3}{8}(1-x)^{-5/2} & f'''(0) &= -\frac{3}{8}, \\
f^{(4)}(x) &= -\frac{15}{16}(1-x)^{-7/2} & f^{(4)}(0) &= -\frac{15}{16}.
\end{aligned}
$$

Then

$$P_4(x) = 1 - \frac{1}{2}x - \frac{1}{4 \cdot 2}x^2 - \frac{3}{8 \cdot 6}x^3 - \frac{15}{16 \cdot 24}x^4 = 1 - \frac{1}{2}x - \frac{1}{8}x^2 - \frac{1}{16}x^3 - \frac{5}{128}x^4.$$

29. With $f(x) = x \sin x$, we have $f(0) = 0$, and

$$
\begin{aligned}
f'(x) &= x \cos x + \sin x & f'(0) &= 0, \\
f''(x) &= 2 \cos x - x \sin x & f''(0) &= 2, \\
f'''(x) &= -x \cos x - 3 \sin x & f'''(0) &= 0, \\
f^{(4)}(x) &= -4 \cos x + x \sin x & f^{(4)}(0) &= -4.
\end{aligned}
$$

Then

$$
P_4(x) = 0 + 0 \cdot x + \frac{2}{2}x^2 + \frac{0}{6}x^3 - \frac{4}{24}x^4 = x^2 - \frac{1}{6}x^4.
$$

31. From Exercise 21, we know that with $f(x) = \cos x$, we have

$$
f^{(n)}(0) = \begin{cases}
\cos 0 = 1, & n \equiv 0 \pmod 4, \\
-\sin 0 = 0, & n \equiv 1 \pmod 4, \\
-\cos 0 = -1, & n \equiv 2 \pmod 4, \\
\sin 0 = 0, & n \equiv 3 \pmod 4.
\end{cases}
$$

Thus only the even powers of x appear in the Maclaurin series, and the signs alternate. Thus the Maclaurin series is

$$
1 - \frac{1}{2!}x^2 + \frac{1}{4!}x^4 - \frac{1}{6!}x^6 + \cdots = \sum_{k=0}^{\infty} \frac{(-1)^k}{(2k)!}x^{2k}.
$$

33. From Exercise 23, we know that with $f(x) = \sin x$, we have

$$
f^{(n)}(0) = \begin{cases}
\sin 0 = 0, & n \equiv 0 \pmod 4, \\
\cos 0 = 1, & n \equiv 1 \pmod 4, \\
-\sin 0 = 0, & n \equiv 2 \pmod 4, \\
-\cos 0 = -1, & n \equiv 3 \pmod 4.
\end{cases}
$$

Thus only the odd powers of x appear in the Maclaurin series, and the signs alternate. Thus the Maclaurin series is

$$
x - \frac{1}{3!}x^3 + \frac{1}{5!}x^5 - \frac{1}{7!}x^7 + \cdots = \sum_{k=0}^{\infty} \frac{(-1)^k}{(2k+1)!}x^{2k+1}.
$$

35. With $f(x) = \cos 2x$, the first four derivatives are shown in Exercise 25. From the pattern, it is clear that

$$
f^{(n)}(0) = \begin{cases}
0, & n \text{ odd}, \\
-2^n, & n \equiv 2 \pmod 4, \\
2^n, & n \equiv 0 \pmod 4,
\end{cases}
$$

and that this formula is valid for $n = 0$ as well. Thus the Maclaurin series is

$$
\sum_{k=0}^{\infty} \frac{f^{(2k)}(0)}{(2k)!}x^{2k} = \sum_{k=0}^{\infty} (-1)^k \frac{2^{2k}}{(2k)!}x^{2k}.
$$

37. With $f(x) = (1-x)^{1/2}$, we have $f(0) = 1$ and $f'(x) = -\frac{1}{2}(1-x)^{-1/2}$ so that $f'(0) = -\frac{1}{2}$. We claim that

$$
f^{(n)}(x) = -\frac{1 \cdot 3 \cdot 5 \cdots (2n-3)}{2^n}(1-x)^{-(2n-1)/2}
$$

for $n \geq 2$. Clearly for $n = 2$ we have $f''(x) = -\frac{1}{4}(1-x)^{-3/2}$, so that the formula holds for $n = 2$. Now assume it holds for n; then

$$f^{(n+1)}(x) = \left(f^{(n)}(x)\right)' = \left(-\frac{1 \cdot 3 \cdot 5 \cdots (2n-3)}{2^n}(1-x)^{-(2n-1)/2}\right)'$$

$$= -\frac{1 \cdot 3 \cdot 5 \cdots (2n-3)}{2^n} \cdot \left(-\frac{2n-1}{2}\right)(1-x)^{-(2n-1)/2-1}(-1)$$

$$= -\frac{1 \cdot 3 \cdot 5 \cdots (2n-1)}{2^{n+1}}(1-x)^{-(2n+1)/2},$$

so that the formula holds for $n+1$. This proves the claim. Thus for $n \geq 2$,

$$f^{(n)}(0) = -\frac{1 \cdot 3 \cdot 5 \cdots (2n-3)}{2^n},$$

so that the Maclaurin series is

$$1 - \frac{1}{2}x - \sum_{k=2}^{\infty} \frac{1 \cdot 3 \cdot 5 \cdots (2k-3)}{2^k k!} x^k.$$

39. With $f(x) = x \sin x$, the first four derivatives are given in Exercise 29. We claim that for $n \geq 1$,

$$f^{(n)}(x) = \begin{cases} n \sin x + x \cos x, & n \equiv 1 \pmod 4, \\ -x \sin x + n \cos x, & n \equiv 2 \pmod 4, \\ -n \sin x - x \cos x, & n \equiv 3 \pmod 4, \\ x \sin x - n \cos x, & n \equiv 0 \pmod 4. \end{cases}$$

From Exercise 29, this formula holds for $n = 1, 2, 3, 4$. Since

$$(n \sin x + x \cos x)' = n \cos x + \cos x - x \sin x = -x \sin x + (n+1) \cos x,$$
$$(-x \sin x + n \cos x)' = -\sin x - x \cos x - n \sin x = -(n+1) \sin x - x \cos x,$$
$$(-n \sin x - x \cos x)' = -n \cos x - \cos x + x \sin x = x \sin x - (n+1) \cos x,$$
$$(x \sin x - n \cos x)' = \sin x + x \cos x + n \sin x = (n+1) \sin x + x \cos x,$$

the formula holds in general by induction. Thus

$$f^{(n)}(0) = \begin{cases} 0 & n \equiv 1 \pmod 4, \\ n, & n \equiv 2 \pmod 4, \\ 0 & n \equiv 3 \pmod 4, \\ -n, & n \equiv 0 \pmod 4. \end{cases}$$

It follows that only the even power terms in the Maclaurin series are nonzero and that the terms alternate in sign. The Maclaurin series is (since $f(0) = 0$)

$$\sum_{k=1}^{\infty} \frac{f^{(2k)}(0)}{(2k)!} x^{2k} = \sum_{k=1}^{\infty} \frac{(-1)^{k+1}(2k)}{(2k)!} x^{2k} = \sum_{k=1}^{\infty} \frac{(-1)^{k+1}}{(2k-1)!} x^{2k}.$$

41. Using the derivatives of $\cos x$ from Exercise 21, we have

$$\cos \frac{\pi}{2} = 0, \quad (\cos x)'\left(\frac{\pi}{2}\right) = -1, \quad (\cos x)''\left(\frac{\pi}{2}\right) = 0, \quad (\cos x)'''\left(\frac{\pi}{2}\right) = 1, \quad (\cos x)^{(4)}\left(\frac{\pi}{2}\right) = 0,$$

so that the Taylor polynomial is

$$P_4(x) = 0 - \frac{1}{1!}\left(x - \frac{\pi}{2}\right) + \frac{0}{2!}\left(x - \frac{\pi}{2}\right)^2 + \frac{1}{3!}\left(x - \frac{\pi}{2}\right)^3 + \frac{0}{4!}\left(x - \frac{\pi}{2}\right)^4$$
$$= -\left(x - \frac{\pi}{2}\right) + \frac{1}{3!}\left(x - \frac{\pi}{2}\right)^3.$$

43. Using the derivatives of $\sin x$ from Exercise 23, we have

$$\sin \pi = 0, \qquad (\sin x)'(\pi) = -1, \qquad (\sin x)''(\pi) = 0, \qquad (\sin x)'''(\pi) = 1, \qquad (\sin x)^{(4)}(\pi) = 0.$$

Thus the Taylor polynomial is

$$P_4(x) = 0 - \frac{1}{1!}(x - \pi) + \frac{0}{2!}(x - \pi)^2 + \frac{1}{3!}(x - \pi)^3 - \frac{0}{4!}(x - \pi)^4 = -(x - \pi) + \frac{1}{3!}(x - \pi)^3.$$

45. With $f(x) = \ln x$, we have $f(3) = \ln 3$, and

$$f'(x) = \frac{1}{x} \qquad f'(3) = \frac{1}{3},$$
$$f''(x) = -\frac{1}{x^2} \qquad f''(3) = -\frac{1}{9},$$
$$f'''(x) = \frac{2}{x^3} \qquad f'''(3) = \frac{2}{27},$$
$$f^{(4)}(x) = -\frac{6}{x^4} \qquad f^{(4)}(3) = -\frac{6}{81} = -\frac{2}{27}.$$

Then the Taylor polynomial is

$$P_4(x) = \ln 3 + \frac{1}{3 \cdot 1!}(x - 3) - \frac{1}{9 \cdot 2!}(x - 3)^2 + \frac{2}{27 \cdot 3!}(x - 3)^3 - \frac{2}{27 \cdot 4!}(x - 3)^4.$$

47. With $f(x) = \cos 2x$, we have $f\left(\frac{\pi}{4}\right) = \cos \frac{\pi}{2} = 0$, and

$$f'(x) = -2\sin 2x \qquad f'\left(\frac{\pi}{4}\right) = -2,$$
$$f''(x) = -4\cos 2x \qquad f''\left(\frac{\pi}{4}\right) = 0,$$
$$f'''(x) = 8\sin 2x \qquad f'''\left(\frac{\pi}{4}\right) = 8,$$
$$f^{(4)}(x) = 16\cos 2x \qquad f^{(4)}\left(\frac{\pi}{4}\right) = 0.$$

Then the Taylor polynomial is

$$P_4(x) = -\frac{2}{1!}\left(x - \frac{\pi}{4}\right) + \frac{8}{3!}\left(x - \frac{\pi}{4}\right)^3 = -2\left(x - \frac{\pi}{4}\right) + \frac{4}{3}\left(x - \frac{\pi}{4}\right)^3.$$

49. From Exercise 41 and the fact that $(\cos x)^{(4)} = \cos x$, we know that

$$(\cos x)^{(n)}\left(\frac{\pi}{2}\right) = \begin{cases} 0, & n \equiv 0 \pmod 4, \\ -1, & n \equiv 1 \pmod 4, \\ 0, & n \equiv 2 \pmod 4, \\ 1, & n \equiv 3 \pmod 4. \end{cases}$$

Thus only the odd power terms are nonzero in the Taylor series, and those alternate in sign. The Taylor series is (noting that $\cos \frac{\pi}{2} = 0$)

$$\sum_{k=0}^{\infty} \frac{(-1)^{k+1}}{(2k+1)!} \left(x - \frac{\pi}{2}\right)^{2k+1}.$$

51. From Exercise 43 and the fact that $(\sin x)^{(4)} = \sin x$, we know that

$$(\sin x)^{(n)}(\pi) = \begin{cases} 0, & n \equiv 0 \pmod 4, \\ -1, & n \equiv 1 \pmod 4, \\ 0, & n \equiv 2 \pmod 4, \\ 1, & n \equiv 3 \pmod 4. \end{cases}$$

Thus only the odd power terms are nonzero in the Taylor series, and those alternate in sign. The Taylor series is (noting that $\sin \pi = 0$)

$$\sum_{k=0}^{\infty} \frac{(-1)^{k+1}}{(2k+1)!}(x - \pi)^{2k+1}.$$

53. The first four derivatives of $f(x) = \ln x$ are given in Exercise 45. We claim that

$$f^{(n)}(x) = (-1)^{n-1}\frac{(n-1)!}{x^n} \text{ for } n \geq 1.$$

From the table in Exercise 45, this formula holds through $n = 4$. If it holds for n, then

$$f^{(n+1)}(x) = \left(f^{(n)}(x)\right)' = (-1)^{n-1}(n-1)!(x^{-n})' = (-1)^{n-1}(n-1)!(-n)(x^{-n-1}) = (-1)^n \frac{n!}{x^{n+1}},$$

so the formula holds for all $n \geq 1$. Thus $f(3) = \ln 3$ and for $n \geq 1$

$$f^{(n)}(3) = (-1)^{n-1}\frac{(n-1)!}{3^n},$$

so that the Taylor series is

$$\ln 3 + \sum_{k=1}^{\infty} \frac{f^{(k)}(3)}{k!}x^k = \ln 3 + \sum_{k=1}^{\infty}(-1)^{k-1}\frac{(k-1)!}{3^k k!}(x-3)^k = \ln 3 + \sum_{k=1}^{\infty}(-1)^{k-1}\frac{1}{k3^k}(x-3)^k.$$

55. With $f(x) = \cos 2x$, the first four derivatives at $x = \frac{\pi}{4}$ are shown in Exercise 47. From the pattern, it is clear that

$$f^{(n)}\left(\frac{\pi}{4}\right) = \begin{cases} 0, & n \text{ even}, \\ -2^n, & n \equiv 1 \pmod 4, \\ 2^n, & n \equiv 3 \pmod 4, \end{cases}$$

and that this formula holds for $n = 0$ as well. Thus the Taylor series is

$$\sum_{k=0}^{\infty} \frac{f^{(2k+1)}\left(\frac{\pi}{4}\right)}{(2k+1)!}\left(x - \frac{\pi}{4}\right)^{2k+1} = \sum_{k=0}^{\infty}(-1)^{k+1}\frac{2^{2k+1}}{(2k+1)!}\left(x - \frac{\pi}{4}\right)^{2k+1}.$$

57. If p is an integer, then for $r > p$, the coefficient

$$\binom{p}{r} = \frac{p(p-1)\ldots(p-p)(p-(p+1))\ldots(p-r)}{r!} = 0$$

since it includes a factor of $p - p$. Thus the sum $\sum_{k=0}^{\infty}\binom{p}{k}x^k$ is in fact equal to the finite sum $\sum_{k=0}^{p}\binom{p}{k}x^k$, so it is a polynomial since all powers of x are nonnegative integers.

59. $\sqrt[3]{1+x} = (1+x)^{1/3} = \sum_{k=0}^{\infty} \binom{1/3}{k} x^k$.

61. $\frac{1}{\sqrt{1+x}} = (1+x)^{-1/2} = \sum_{k=0}^{\infty} \binom{-1/2}{k} x^k$.

63. (a) We must first show that f is continuous at 0. But $\lim_{h\to 0} e^{-1/h^2} = e^{-\infty} = 0$. Thus f is continuous at 0. Now, $f(x)$ is differentiable at 0 if $\lim_{h\to 0} \frac{f(h)-f(0)}{h}$ exists. This limit is

$$\lim_{h\to 0} \frac{f(h)-f(0)}{h} = \lim_{h\to 0} \frac{e^{-1/h^2}}{h} = \lim_{h\to 0} \frac{1/h}{e^{1/h^2}} \overset{\text{L'H}}{=} \lim_{h\to 0} \frac{-h^{-2}}{-2h^{-3}e^{1/h^2}} = \lim_{h\to 0} \frac{1}{2} h e^{-1/h^2} = 0,$$

so that f is also differentiable at 0.

(b) Since $f'(x) = 2x^{-3}e^{-1/x^2} = 2x^{-3}f(x)$, it is clear from the product rule that taking successive derivatives will result in a formula for $f^{(n)}(x)$ that is a sum of terms each of which is of the form a constant times a power of x times $f(x)$. For example,

$$f''(x) = -6x^{-4}f(x) + 2x^{-3}f'(x) = -6x^{-4}f(x) + 2x^{-3}(+2x^{-3}f(x))$$
$$= -6x^{-4}f(x) + 4x^{-6}f(x).$$

Then $f^{(n)}(0)$ is the limit of $f^{(n)}(x)$ as $x \to 0$. But in any term, if the power of x is nonnegative, then both that factor and $f(x)$ go to zero, so the term does. If the power of x is negative, then the term is of the form $\frac{e^{-1/x^2}}{x^k}$, which also goes to zero by the limit computation in part (a). Thus $f^{(n)}(0) = 0$ for all $n \geq 0$.

(c) Since all derivatives of $f(x)$ vanish at 0, the Maclaurin series is 0:

$$\sum_{k=0}^{\infty} \frac{f^{(k)}(0)}{k!} x^k = \sum_{k=0}^{\infty} \frac{0}{k!} x^k = 0.$$

(d) Since the Maclaurin series is a constant, it clearly converges for any value of x. However, since $f(x) \neq 0$ for $x \neq 0$, $f(c)$ is not equal to the Maclaurin series at c for any $c \neq 0$.

Applications

65. The first four terms are

$$1, \quad 1 - \frac{1}{4}x^2, \quad 1 - \frac{1}{4}x^2 + \frac{1}{64}x^4, \quad 1 - \frac{1}{4}x^2 + \frac{1}{64}x^4 - \frac{1}{2304}x^6.$$

Plots of the first four partial sums of $J_0(x)$, shown with increasing line thickness, are

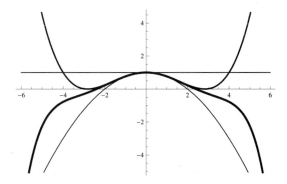

67. The first four terms of $J_1(x)$ are

$$\frac{1}{2}x, \quad \frac{1}{2}x - \frac{1}{16}x^3, \quad \frac{1}{2}x - \frac{1}{16}x^3 + \frac{1}{384}x^5, \quad \frac{1}{2}x - \frac{1}{16}x^3 + \frac{1}{384}x^5 - \frac{1}{18432}x^7.$$

Plots of the first four partial sums of $J_1(x)$, shown with increasing line thickness, are

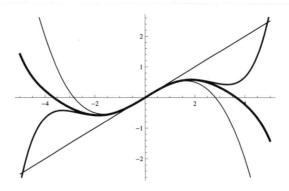

Proofs

69. Let $f(x) = \ln x$. Then

$$\frac{d}{dx}(f(x)) = \frac{1}{x} = (-1)^{1-1}\frac{(1-1)!}{x^1},$$

so that the formula holds for $k = 1$. If it holds for n, then

$$f^{(n+1)}(x) = \left(f^{(n)}(x)\right)' = (-1)^{n-1}(n-1)!(x^{-n})' = (-1)^{n-1}(n-1)!(-n)(x^{-n-1}) = (-1)^n\frac{n!}{x^{n+1}},$$

so it holds for $n + 1$. By induction, the formula holds for all $k \geq 1$.

Thinking Forward

Graphing Differences

Graphs of the differences are:

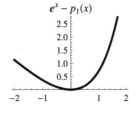

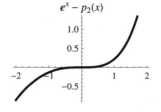

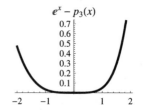

Note that the vertical scale is getting finer with each graph, indicating that the approximations are getting better (since the maximum difference is smaller). But it does appear that for each of the graphs, the differences start to increase quite a bit as you get further from 0. For the first graph, it is hard to tell for what range of x the difference is less than 0.1, but it is small, say about $|x| \leq 0.1$. For the second graph, the range of x values is larger, about $|x| \leq 0.6$. Finally, for the third graph, the difference is less than 0.1 for about $|x| \leq 1.4$.

Graphing More Differences

Graphs of the differences are:

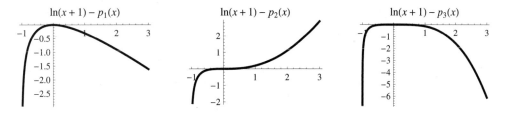

$\ln(x+1) - p_1(x)$ $\ln(x+1) - p_2(x)$ $\ln(x+1) - p_3(x)$

Each successive graph stays close to zero for a wider range of values of x. However, once you are outside about $|x| = 1$, the difference increases very quickly. For the first graph, it is hard to tell for what range of x the difference is less than 0.1, but it is small, say about $|x| \leq 0.1$. For the second graph, the range of x values is larger, about $|x| \leq 0.8$. Finally, for the third graph, the difference is less than 0.1 for about $-0.8 \leq x \leq 1$.

8.3   Convergence of Power Series

Thinking Back

Analyzing the quality of an approximation for a definite integral

Theorem 5.27 gives a bound for the error of a midpoint approximation of a definite integral on the interval $[a, b]$:

$$\left| E_{\text{MID}(n)} \right| \leq \frac{M(b-a)^3}{24n^2}$$

Here M is an upper bound on the magnitude of the second derivative of the function being integrated, and n is the number of intervals used in the approximation.

Analyzing the quality of an approximation for the sum of an alternating series

By Theorem 7.38, if $L = \sum_{k=1}^{\infty}(-1)^{k+1}a_k$ is the sum of the series and $S_n = \sum_{k=1}^{n}(-1)^{k+1}a_k$ is the nth partial sum, then $|L - S_n| < a_{n+1}$, and the sign of the difference $L - S_n$ is $(-1)^n$, which is the sign of the coefficient of a_{n+1}.

Concepts

1.   (a) False. For example, see the function in Exercise 63, Section 8.2.

   (b) True. Every Taylor series around x_0 is equal to the function at the point x_0.

   (c) False. This is true only if the Taylor series actually converges to f.

   (d) True. See Theorem 8.9.

   (e) True. See Theorem 8.10.

   (f) True. It is the difference between the true value of $f(x)$ and the value of the polynomial approximation $P_n(x)$, so it measures the error in the approximation.

   (g) False. If the Maclaurin series for $f(x)$ converges for $x \in (-\rho, \rho)$, then the Maclaurin series for $f(x^3)$ converges for $x^3 \in (-\rho, \rho)$, i.e., for $x \in \left(-\rho^{1/3}, \rho^{1/3}\right)$. See the discussion in the text.

   (h) True. When computing the limit of the ratio between successive terms, the x^3 will cancel out in numerator and denominator, so the ratio will be the same as that used to compute the radius of convergence of the Maclaurin series for $f(x)$.

3. $\sum_{k=0}^{\infty} a_k x^k$ is equal to the Maclaurin series for $f(x)$ on $(-2, 2)$, so that $a_k = \frac{f^{(k)}(0)}{k!}$.

5. $\sum_{k=0}^{\infty} c_k (x - x_0)^k$ is the Taylor series of $h(x)$ around x_0, and it converges everywhere. Thus $c_k = \frac{h^{(k)}(x_0)}{k!}$.

7. If $b \neq 0$, then

$$\frac{1}{b} f\left(-\frac{a}{b} x\right) = \frac{1}{b} \cdot \frac{1}{1 - \left(-\frac{a}{b} x\right)} = \frac{1}{b - (-ax)} = \frac{1}{ax + b} = g(x).$$

9. Since $f(x)$ is a cubic, its third and fourth degree Taylor polynomials are equal to $f(x)$, since f is clearly its own best cubic approximation. Thus $R_3(x) = R_4(x) = 0$.

11. Taylor's Theorem gives a formula for the remainder in a Taylor polynomial approximation; see Theorem 8.9.

13. Since series for sums, products, and quotients of functions can be built up out of the series for those functions, it may be helpful to know the Maclaurin series in order to determine Maclaurin series for more complicated functions. Also, being able to recognize Maclaurin series may help evaluate certain infinite sums.

15. If the Maclaurin series for $f(x)$ is $\sum_{k=0}^{\infty} a_k x^k$, then the Maclaurin series for $cx^m f(x)$ is the product of the two: $cx^m \sum_{k=0}^{\infty} a_k x^k = \sum_{k=0}^{\infty} c a_k x^{k+m}$. The interval of convergence of the new series is the same as that for the old series, since if you consider using the ratio test on the new series to determine the radius of convergence, the factor of cx^m appears both in the numerator and denominator, so it cancels out, and what is left is the ratio test for the original series.

17. The series looks like the Maclaurin series $\sum_{k=0}^{\infty} \frac{1}{k!} x^k = e^x$ with $x = -\pi$, so this sum is $e^{-\pi}$.

19. This series looks like the Maclaurin series $\sum_{k=0}^{\infty} \frac{(-1)^k}{2k+1} x^{2k+1} = \tan^{-1} x$ with $x = \frac{1}{\sqrt{3}}$, so the sum is $\tan^{-1} \frac{1}{\sqrt{3}} = \frac{\pi}{6}$.

21. This series looks like the Maclaurin series $\sum_{k=0}^{\infty} \frac{(-1)^k}{(2k+1)!} x^{2k+1} = \sin x$ for $x = \pi$, so the sum is $\sin \pi = 0$.

Skills

23. Lagrange's form says that

$$R_4(x) = \frac{(\cos x)^{(5)}(c)}{5!} x^5 = -\frac{\sin c}{5!} x^5$$

for some c between 0 and x.

25. Lagrange's form says that

$$R_4(x) = \frac{(\sin x)^{(5)}(c)}{5!} x^5 = \frac{\cos c}{5!} x^5$$

for some c between 0 and x.

27. Note that

$$(\tan^{-1} x)' = \frac{1}{1 + x^2}$$

$$(\tan^{-1} x)'' = -\frac{2x}{(1 + x^2)^2}$$

$$(\tan^{-1} x)''' = -\frac{(1 + x^2)^2(2) - 2x\left(2(1 + x^2)(2x)\right)}{(1 + x^2)^4} = -\frac{2x^2 + 2 - 8x^2}{(1 + x^2)^3} = \frac{6x^2 - 2}{(1 + x^2)^3}$$

$$(\tan^{-1} x)^{(4)} = \frac{(1 + x^2)^3(12x) - (6x^2 - 2)\left(3(1 + x^2)^2(2x)\right)}{(1 + x^2)^6} = \frac{(1 + x^2)(12x) - (6x^2 - 2)6x}{(1 + x^2)^4}$$

$$= \frac{24x - 24x^3}{(1 + x^2)^4}$$

$$(\tan^{-1} x)^{(5)} = \frac{(1 + x^2)^4(24 - 72x^2) - (24x - 24x^3)\left(4(1 + x^2)^3(2x)\right)}{(1 + x^2)^8}$$

$$= \frac{(1 + x^2)(24 - 72x^2) - (24x - 24x^3)(8x)}{(1 + x^2)^5}$$

$$= \frac{120x^4 - 240x^2 + 24}{(1 + x^2)^5} = \frac{24(1 - 10x^2 + 5x^4)}{(1 + x^2)^5}$$

Then Lagrange's form says that

$$R_4(x) = \frac{(\tan^{-1} x)^{(5)}(c)}{5!} x^5 = \frac{24(1 - 10c^2 + 5c^4)}{5!(1 + c^2)^5} x^5 = \frac{1 - 10c^2 + 5c^4}{5(1 + c^2)^5} x^5$$

for some c between 0 and x.

29. Note that

$$\left((1 - x)^{1/2}\right)' = \frac{1}{2}(1 - x)^{-1/2}(-1) = -\frac{1}{2}(1 - x)^{-1/2}$$

$$\left((1 - x)^{1/2}\right)'' = \frac{1}{4}(1 - x)^{-3/2}(-1) = -\frac{1}{4}(1 - x)^{-3/2}$$

$$\left((1 - x)^{1/2}\right)''' = \frac{3}{8}(1 - x)^{-5/2}(-1) = -\frac{3}{8}(1 - x)^{-5/2}$$

$$\left((1 - x)^{1/2}\right)^{(4)} = \frac{15}{16}(1 - x)^{-7/2}(-1) = -\frac{15}{16}(1 - x)^{-7/2}$$

$$\left((1 - x)^{1/2}\right)^{(5)} = \frac{105}{32}(1 - x)^{-9/2}(-1) = -\frac{105}{32}(1 - x)^{-9/2}.$$

Then Lagrange's form says that

$$R_4(x) = \frac{(\sqrt{1 - x})^{(5)}(c)}{5!} x^5 = -\frac{105}{5! \cdot 32(1 - c)^{9/2}} x^5 = -\frac{7}{256(1 - c)^{9/2}} x^5$$

for some c between 0 and x.

31. Note that

$$(x \sin x)' = \sin x + x \cos x$$

$$(x \sin x)'' = \cos x + \cos x - x \sin x = 2 \cos x - x \sin x$$

$$(x \sin x)''' = -2 \sin x - \sin x - x \cos x = -3 \sin x - x \cos x$$

$$(x \sin x)^{(4)} = -3 \cos x - \cos x + x \sin x = -4 \cos x + x \sin x$$

$$(x \sin x)^{(5)} = 4 \sin x + \sin x + x \cos x = 5 \sin x + x \cos x.$$

Then Lagrange's form says that

$$R_4(x) = \frac{(x \sin x)^{(5)}(c)}{5!} x^5 = \frac{c \cos c + 5 \sin c}{5!} x^5$$

for some c between 0 and x.

33. Note that $\cos x$ is infinitely differentiable everywhere on $\mathbb{R}$, so Lagrange's form says that for any $x \in \mathbb{R}$

$$R_n(x) = \frac{(\cos x)^{(n+1)}(c)}{(n+1)!} x^{n+1}$$

for some c between 0 and x. To compute the limit, note that derivatives of $\cos x$ are either $\pm \cos x$ or $\pm \sin x$, so that their magnitudes are all at most 1. Then

$$\lim_{n \to \infty} |R_n(x)| = \lim_{n \to \infty} \left| \frac{(\cos x)^{(n+1)}(c)}{(n+1)!} x^{n+1} \right| \leq \lim_{n \to \infty} \frac{|x|^{n+1}}{(n+1)!} = 0,$$

where the final equality comes from Exercise 6.

35. Note that $\sin x$ is infinitely differentiable everywhere on $\mathbb{R}$, so Lagrange's form says that for any $x \in \mathbb{R}$

$$R_n(x) = \frac{(\sin x)^{(n+1)}(c)}{(n+1)!} x^{n+1}$$

for some c between 0 and x. To compute the limit, note that derivatives of $\sin x$ are either $\pm \cos x$ or $\pm \sin x$, so that their magnitudes are all at most 1. Then

$$\lim_{n \to \infty} |R_n(x)| = \lim_{n \to \infty} \left| \frac{(\sin x)^{(n+1)}(c)}{(n+1)!} x^{n+1} \right| \leq \lim_{n \to \infty} \frac{|x|^{n+1}}{(n+1)!} = 0,$$

where the final equality comes from Exercise 6.

37. Lagrange's form says that

$$R_4(x) = \frac{(\cos x)^{(5)}(c)}{5!} \left(x - \frac{\pi}{2} \right)^5 = -\frac{\sin c}{5!} \left(x - \frac{\pi}{2} \right)^5$$

for some c between $\frac{\pi}{2}$ and x.

39. Lagrange's form says that

$$R_4(x) = \frac{(\sin x)^{(5)}(c)}{5!} (x - \pi)^5 = \frac{\cos c}{5!} (x - \pi)^5$$

for some c between π and x.

41. Lagrange's form says that

$$R_4(x) = \left| \frac{(\ln x)^{(5)}(c)}{5!} (x - 3)^5 \right| = \frac{24}{5! c^5} (x - 3)^5 = \frac{1}{5c^5} (x - 3)^5$$

for some c between 3 and x.

43. See Exercise 28 for derivatives of $\tan x$. Then Lagrange's form says that

$$R_4(x) = \frac{(\tan x)^{(5)}(c)}{5!} \left(x - \frac{\pi}{4} \right)^5 = \frac{16 \sec^6 c + 88 \sec^4 c \tan^2 c + 16 \sec^2 c \tan^4 c}{5!} \left(x - \frac{\pi}{4} \right)^5$$

for some c between $\frac{\pi}{4}$ and x.

45. Note that $\cos x$ is infinitely differentiable everywhere on $\mathbb{R}$, so Lagrange's form says that for any $x \in \mathbb{R}$

$$R_n(x) = \frac{(\cos x)^{(n+1)}(c)}{(n+1)!} \left(x - \frac{\pi}{2}\right)^{n+1}$$

for some c between $\frac{\pi}{2}$ and x. To compute the limit, note that derivatives of $\cos x$ are either $\pm \cos x$ or $\pm \sin x$, so that their magnitudes are all at most 1. Then

$$\lim_{n \to \infty} |R_n(x)| = \lim_{n \to \infty} \left| \frac{(\cos x)^{(n+1)}(c)}{(n+1)!} \left(x - \frac{\pi}{2}\right)^{n+1} \right| \leq \lim_{n \to \infty} \frac{\left|x - \frac{\pi}{2}\right|^{n+1}}{(n+1)!} = 0,$$

where the final equality comes from Exercise 6.

47. Note that $\sin x$ is infinitely differentiable everywhere on $\mathbb{R}$, so Lagrange's form says that for any $x \in \mathbb{R}$

$$R_n(x) = \frac{(\sin x)^{(n+1)}(c)}{(n+1)!}(x - \pi)^{n+1}$$

for some c between π and x. To compute the limit, note that derivatives of $\sin x$ are either $\pm \cos x$ or $\pm \sin x$, so that their magnitudes are all at most 1. Then

$$\lim_{n \to \infty} |R_n(x)| = \lim_{n \to \infty} \left| \frac{(\sin x)^{(n+1)}(c)}{(n+1)!}(x - \pi)^{n+1} \right| \leq \lim_{n \to \infty} \frac{|x - \pi|^{n+1}}{(n+1)!} = 0,$$

where the final equality comes from Exercise 6.

49. Note that $\ln x$ is infinitely differentiable on $(2,4)$, so Lagrange's form says that for any $x \in (2,4)$ there is some c between 3 and x such that

$$R_n(x) = \frac{(\ln x)^{(n+1)}(c)}{(n+1)!}(x - 3)^{n+1}$$

To compute the limit, note that

$$\left|(\ln x)^{(n)}\right| = \left|(-1)^{n+1}\frac{(n-1)!}{x^n}\right| = \frac{(n-1)!}{x^n}.$$

For any n, this expression is bounded above on $(2,4)$ by its value at $x = 2$, at which point its value is $\frac{(n-1)!}{2^n}$. Thus

$$\lim_{n \to \infty} |R_n(x)| = \lim_{n \to \infty} \left| \frac{(\ln x)^{(n+1)}(c)}{(n+1)!}(x - 3)^{n+1} \right| \leq \lim_{n \to \infty} \frac{n!}{2^{n+1}(n+1)!}|x - 3|^{n+1}$$

$$= \lim_{n \to \infty} \left(\frac{1}{n+1} \cdot \left(\frac{|x-3|}{2}\right)^{n+1} \right).$$

The first factor in the product clearly goes to zero as $n \to \infty$. The second does as well, since $x \in (2,4)$ and thus $|x - 3| < 1$ so that the fraction is less than 1. Thus $\lim_{n \to \infty} R_n(x) = 0$.

51. The Maclaurin series is

$$e^{x^3} = \sum_{k=0}^{\infty} \frac{1}{k!}(x^3)^k = \sum_{k=0}^{\infty} \frac{1}{k!}x^{3k}.$$

Since the series for e^x converges everywhere, so does the series for e^{x^3}.

53. The Maclaurin series is

$$\sin(-5x^2) = \sum_{k=0}^{\infty} \frac{(-1)^k}{(2k+1)!}(-5x^2)^{2k+1} = \sum_{k=0}^{\infty} \frac{(-1)^k(-1)^{2k+1}5^{2k+1}}{(2k+1)!}x^{4k+2} = \sum_{k=0}^{\infty} \frac{(-1)^{k+1}5^{2k+1}}{(2k+1)!}x^{4k+2}.$$

Since the series for $\sin x$ converges everywhere, so does the series for $\sin(-5x^2)$.

55. Note that

$$\frac{1}{8+x^3} = \frac{1}{8}\left(\frac{1}{1+\left(\frac{x}{2}\right)^3}\right) = \frac{1}{8}\left(\frac{1}{1-\left(-\frac{x}{2}\right)^3}\right).$$

Substitute into the Maclaurin series for $\frac{1}{1-x}$ to get

$$\frac{1}{8+x^3} = \frac{1}{8}\cdot\sum_{k=0}^{\infty}\left(\left(-\frac{x}{2}\right)^3\right)^k = \frac{1}{8}\cdot\sum_{k=0}^{\infty}(-1)^k\frac{1}{2^{3k}}x^{3k} = \sum_{k=0}^{\infty}(-1)^k\frac{1}{2^{3k+3}}x^{3k}.$$

The series for $\frac{1}{1-x}$ converges for $|x| < 1$. Since we substituted $\frac{1}{8}x^3$ for x, by the discussion in the text, the new radius of convergence is $\sqrt[3]{\frac{1}{1/8}} = 2$.

57. The series will be the sum of the series for e^x and e^{-x} divided by 2, or

$$\frac{1}{2}\left(\sum_{k=0}^{\infty}\frac{1}{k!}x^k + \sum_{k=0}^{\infty}\frac{(-1)^k}{k!}x^k\right) = \frac{1}{2}\sum_{k=0}^{\infty}\frac{1+(-1)^k}{k!}x^k = \sum_{k=0}^{\infty}\frac{1}{(2k)!}x^{2k}$$

Since e^x and e^{-x} converge everywhere, so does this series.

59. Since $\cos^2 x = \frac{1}{2}(1+\cos 2x)$, we can compute this series by substituting $2x$ for x in the series for $\cos x$:

$$\frac{1}{2}\left(1+\sum_{k=0}^{\infty}\frac{(-1)^k}{(2k)!}(2x)^{2k}\right) = \frac{1}{2}\left(1+\sum_{k=0}^{\infty}\frac{(-1)^k 2^{2k}}{(2k)!}x^{2k}\right) = \frac{1}{2}+\sum_{k=0}^{\infty}\frac{(-1)^k 2^{2k-1}}{(2k)!}x^{2k}$$

$$= 1+\sum_{k=1}^{\infty}\frac{(-1)^k 2^{2k-1}}{(2k)!}x^{2k}.$$

Since the series for $\cos x$ converges everywhere, this series does as well.

61. The Maclaurin series for $e^x \sin x$ is the product of the Maclaurin series for e^x and $\sin x$. Since we are interested in only the first four terms, we need only compute the product involving the first few nonzero terms of each, since after that the powers of x are too high:

$$\left(1+x+\frac{x^2}{2}+\frac{x^3}{6}+\frac{x^4}{24}+\dots\right)\left(x-\frac{x^3}{6}+\frac{x^5}{120}+\dots\right) = x+x^2+\frac{1}{3}x^3-\frac{1}{30}x^5+\dots$$

Since the series for e^x and $\sin x$ converge everywhere, so does the series for $e^x \sin x$.

63. The Maclaurin series for $e^x \ln(1+x^3)$ is the product of the Maclaurin series for e^x and $\ln(1+x^3)$. Since we are interested in only the first four terms, we need only compute the product involving the first few nonzero terms of each, since after that the powers of x get too high. The Maclaurin series for $\ln(1+x^3)$ is derived from the Maclaurin series for $\ln(1+x)$ by substituting x^3 for x:

$$\ln(1+x^3) = \sum_{k=1}^{\infty}\frac{(-1)^{k+1}}{k}(x^3)^k = \sum_{k=1}^{\infty}\frac{(-1)^{k+1}}{k}x^{3k} = x^3 - \frac{1}{2}x^6 + \frac{1}{3}x^9 + \dots$$

so we get for the product

$$\left(1 + x + \frac{1}{2}x^2 + \frac{1}{6}x^3 + \dots\right)\left(x^3 - \frac{1}{2}x^6 + \frac{1}{3}x^9 + \dots\right) = x^3 + x^4 + \frac{1}{2}x^5 - \frac{1}{3}x^6 + \dots$$

Since the series for $\ln(1+x)$ converges on $(-1, 1)$, we know that the series for $\ln(1+x^3)$ converges on $(\sqrt[3]{-1}, \sqrt[3]{1}) = (-1, 1)$. The series for e^x converges everywhere, so the product converges on the intersection of the convergence intervals, or $(-1, 1)$.

65. The Maclaurin series for $\frac{\sin x}{1-x^2}$ is the product of the Maclaurin series for $\sin x$ and for $\frac{1}{1-x^2}$. The latter series is found by substituting x^2 for x in the series for $\frac{1}{1-x}$:

$$\frac{1}{1-x^2} = \sum_{k=0}^{\infty}(x^2)^k = \sum_{k=0}^{\infty} x^{2k} = 1 + x^2 + x^4 + x^6 + \dots.$$

Thus for the product we get

$$\left(x - \frac{1}{6}x^3 + \frac{1}{120}x^5 - \frac{1}{5040}x^7 + \dots\right)\left(1 + x^2 + x^4 + x^6 + \dots\right) = x + \frac{5}{6}x^3 + \frac{101}{120}x^5 + \frac{4241}{5040}x^7.$$

Since $\sin x$ converges everywhere and $\frac{1}{1-x^2}$ converges on $(-1, 1)$, this series converges on $(-1, 1)$ as well.

67. Substitute -0.3 in the Maclaurin series for e^x to get

$$e^{-0.3} = \sum_{k=0}^{\infty} \frac{1}{k!}(-0.3)^k = \sum_{k=0}^{\infty}(-1)^k\frac{0.3^k}{k!}.$$

This is an alternating series with terms decreasing to zero, so by Theorem 7.38, to approximate $e^{-0.3}$ to within 0.001 we must find k such that $\frac{0.3^k}{k!} < 0.001$. Since $\frac{0.3^3}{6} = 0.0045$ and $\frac{0.3^4}{24} \approx 0.0003$, $k = 4$ is the remainder term, so that S_3 approximates the value to within 0.001:

$$S_3 = 1 - 0.3 + \frac{0.3^2}{2} - \frac{0.3^3}{6} = 0.7405.$$

To approximate to within 10^{-6}, we want k such that $\frac{0.3^k}{k!} < 10^{-6}$, which first happens for $k = 7$. Thus S_6 approximates $e^{-0.3}$ to within 10^{-6}.

69. Substitute 0.5 in the Maclaurin series for $\ln(1+x)$ to get

$$\ln 1.5 = \sum_{k=1}^{\infty} \frac{(-1)^{k+1}}{k}0.5^k.$$

This is an alternating series with terms decreasing to zero, so by Theorem 7.38, to approximate $\ln 1.5$ to within 0.001 we must find k such that $\frac{0.5^k}{k} < 0.001$. Since $\frac{0.5^7}{7} \approx 0.00112$ and $\frac{0.5^8}{8} \approx 0.00049$, $k = 8$ is the remainder term, so that S_7 approximates the value to within 0.001:

$$S_7 = \sum_{k=1}^{7} \frac{(-1)^{k+1}}{k}0.5^k \approx 0.4058.$$

To approximate to within 10^{-6}, we want k such that $\frac{0.5^k}{k} < 10^{-6}$, which first happens for $k = 16$. Thus S_{15} approximates $\ln 1.5$ to within 10^{-6}.

71. Substitute 0.4 in the Maclaurin series for $\tan^{-1} x$ to get

$$\tan^{-1} 0.4 = \sum_{k=0}^{\infty} \frac{(-1)^k}{2k+1} 0.4^{2k+1}.$$

This is an alternating series with terms decreasing to zero, so by Theorem 7.38, to approximate $\tan^{-1} 0.4$ to within 0.001 we must find k such that $\frac{0.4^{2k+1}}{2k+1} < 0.001$. Since $\frac{0.4^5}{5} \approx 0.0020$ and $\frac{0.4^7}{7} \approx 0.0002$, $k = 3$ is the remainder term, so that S_2 approximates the value to within 0.001:

$$S_2 = \sum_{k=0}^{2} \frac{(-1)^k}{2k+1} 0.4^{2k+1} \approx 0.3807.$$

To approximate to within 10^{-6}, we want k such that $\frac{0.4^{2k+1}}{2k+1} < 10^{-6}$, which first happens for $k = 6$. Thus S_5 approximates $\tan^{-1} 0.4$ to within 10^{-6}.

73. Convert $2°$ to radians to get $2 \cdot \frac{\pi}{180} \approx 0.0349$. Then substitute 0.0349 in the Maclaurin series for $\sin x$ to get

$$\sin 0.0349 = \sum_{k=0}^{\infty} \frac{(-1)^k}{(2k+1)!} (0.0349)^{2k+1}.$$

This is an alternating series with terms decreasing to zero, so by Theorem 7.38, to approximate $\sin 0.0349$ to within 0.001 we must find k such that $\frac{0.0349^{2k+1}}{(2k+1)!} < 0.001$. Since $\frac{0.0349^1}{1} = 0.0349$ and $\frac{0.0349^3}{3!} \approx 7.1 \times 10^{-6}$, $k = 1$ is the remainder term, so that S_0 approximates the value to within 0.001:

$$S_0 = \sum_{k=0}^{0} \frac{(-1)^k}{(2k+1)!} (0.0349)^{2k+1} = 0.0349.$$

To approximate to within 10^{-6}, we want k such that $\frac{0.0349^{2k+1}}{(2k+1)!} < 10^{-6}$, which first happens for $k = 2$. Thus S_1 approximates $\sin 0.0349$ to within 10^{-6}.

75. Convert $5°$ to radians to get $5 \cdot \frac{\pi}{180} \approx 0.0873$. Then substitute 0.0873 in the Maclaurin series for $\cos x$ to get

$$\cos 0.0873 = \sum_{k=0}^{\infty} \frac{(-1)^k}{(2k)!} (0.0873)^{2k}.$$

This is an alternating series with terms decreasing to zero, so by Theorem 7.38, to approximate $\cos 0.0873$ to within 0.001 we must find k such that $\frac{0.0873^{2k}}{(2k)!} < 0.001$. Since $\frac{0.0873^2}{2!} \approx 0.0038$ and $\frac{0.0873^4}{4!} \approx 2.4 \times 10^{-6}$, $k = 2$ is the remainder term, so that S_1 approximates the value to within 0.001:

$$S_1 = \sum_{k=0}^{1} \frac{(-1)^k}{(2k)!} (0.0873)^{2k} = 0.9962.$$

To approximate to within 10^{-6}, we want k such that $\frac{0.0873^{2k}}{(2k)!} < 10^{-6}$, which first happens for $k = 3$. Thus S_2 approximates $\cos 5°$ to within 10^{-6}.

77. (a) Substitute 0.3 into the series for e^x to get $e^{0.3} = \sum_{k=0}^{\infty} \frac{1}{k!} 0.3^k$.

    (b) Using Lagrange's form, the error in using the 5th degree Maclaurin polynomial is

$$R_5(0.3) = \frac{(e^x)^{(6)}(c)}{6!} 0.3^6 = \frac{e^c}{6!} 0.3^6$$

for some $c \in [0, 0.3]$. Since e^x is increasing, we could choose $c = 0.3$ as the maximum value, but that would not help as we would need to evaluate $e^{0.3}$ to compute the remainder! So choose $e^1 \approx 2.7$; then we get

$$R_5(0.3) \le \frac{2.7}{6!}0.3^6 \approx 2.73 \times 10^{-6}.$$

(c) Again using $c = 1$ in the approximation of the error, we get

$$R_n(0.3) \le \frac{2.7}{(n+1)!}0.3^{n+1},$$

and this is less than 10^{-6} first when $n + 1 = 7$, so $n = 6$.

79. (a) Substitute -0.5 into the series for $\ln(1 + x)$ to get

$$\ln 0.5 = \sum_{k=1}^{\infty} \frac{(-1)^{k+1}}{k}(-0.5)^k = \sum_{k=1}^{\infty} \frac{(-1)^{k+1}(-1)^k}{k}0.5^k = -\sum_{k=1}^{\infty} \frac{0.5^k}{k}.$$

(b) The error using the 5th degree Maclaurin polynomial is given, by Lagrange's form, by

$$|R_5(-0.5)| = \left| \frac{(\ln(1+x))^{(6)}(c)}{6!}0.5^6 \right| = \left| \frac{-120}{6!(1+c)^6}0.5^6 \right| = \frac{1}{6}\left(\frac{1}{2(1+c)} \right)^6$$

for some c between 0 and -0.5. This takes its maximum value when c is as small as possible, which is for $c = -0.5$, in which case $|R_5(-0.5)| \le \frac{1}{6}$.

(c) Since $\left| (\ln(x+1))^{(n)} \right| = \frac{(n-1)!}{(1+x)^n}$, we have

$$|R_n(-0.5)| = \left| \frac{(\ln(1+x))^{(n+1)}(c)}{(n+1)!}0.5^{n+1} \right| = \frac{n!}{(n+1)!(1+c)^{n+1}}0.5^{n+1}$$

$$= \frac{1}{n+1}\left(\frac{1}{2(1+c)} \right)^{n+1} \le \frac{1}{n+1}.$$

Then the smallest n for which $R_n(-0.5) \le 10^{-6}$ is $n = 999999$.

81. (a) Substitute 1 into the series for $\sin x$ to get

$$\sin 1 = \sum_{k=0}^{\infty} \frac{(-1)^k}{(2k+1)!}1^{2k+1} = \sum_{k=0}^{\infty} \frac{(-1)^k}{(2k+1)!}.$$

(b) The error using the 5th Maclaurin polynomial is given, by Lagrange's form, by

$$|R_5(1)| = \left| \frac{(\sin x)^6(c)}{(6)!}1^6 \right| = \left| \frac{-\sin c}{6!} \right| = \frac{|\sin c|}{6!} \le \frac{1}{6!}$$

for some c between 0 and 1, so that the maximum error is $\frac{1}{6!} \approx 0.0014$.

(c) Since all derivatives of $\sin x$ are $\pm \sin x$ or $\pm \cos x$, it follows that

$$|R_n(1)| = \left| \frac{(\sin x)^{n+1}(c)}{(n+1)!}1^{n+1} \right| \le \frac{1}{(n+1)!}.$$

Then the smallest n for which $R_n(1) < 10^{-6}$ is $n = 9$.

83. (a) We know that $\lim\limits_{x \to 0} \frac{\sin x}{x} = 1$ from previous work (or, apply L'Hôpital's rule to evaluate this limit). Thus f is continuous at 0. To show it is differentiable, compute

$$\lim_{x \to 0} \frac{f(x) - f(0)}{x - 0} = \lim_{x \to 0} \frac{\frac{\sin x}{x} - 1}{x - 0} = \lim_{x \to 0} \frac{\sin x - x}{x^2} \overset{\text{L'H}}{=} \lim_{x \to 0} \frac{\cos x - 1}{2x} \overset{\text{L'H}}{=} \lim_{x \to 0} \frac{-\sin x}{2} = 0,$$

so that the limit of the difference quotient as $x \to 0$ exists and thus f is differentiable at 0.

(b) The Maclaurin series is simply $\frac{1}{x}$ times the Maclaurin series for $\sin x$, so it is

$$\frac{1}{x} \sum_{k=0}^{\infty} \frac{(-1)^k}{(2k+1)!} x^{2k+1} = \sum_{k=0}^{\infty} \frac{(-1)^k}{(2k+1)!} x^{2k}.$$

Applications

85. (a) From the first two wells, the slope is $\frac{38-35}{300} = \frac{1}{100}$. Thus the equation of the line describing the depth of groundwater at a distance of x feet is $w(x) - 35 = \frac{1}{100}(x - 0)$, or $w(x) = \frac{1}{100}x + 35$.

(b) By Definition 8.8, the error in this approximation is

$$R_1(x) = w(x) - P_1(x) = w(x) - \frac{1}{100}x - 35.$$

Alternatively, using Taylor's Theorem and assuming that $w(x)$ is at least twice differentiable (a reasonable assumption), we have

$$R_1(x) = \int_0^x (x - t)w''(t)\,dt.$$

(c) Evaluating the quadratic at these three points gives

$$w_2(0) = \frac{1}{180000} \cdot 0^2 + \frac{1}{120} \cdot 0 + 35 = 35,$$

$$w_2(300) = \frac{1}{180000} \cdot 300^2 + \frac{1}{120} \cdot 300 + 35 = \frac{90000}{180000} + \frac{5}{2} + 35 = 38,$$

$$w_2(600) = \frac{1}{180000} \cdot 600^2 + \frac{1}{120} \cdot 600 + 35 = \frac{360000}{180000} + \frac{600}{120} = 42.$$

(d) With $f(x) = w_2(x)$, Lagrange's form for the remainder tells us that the error in the linear approximation, which is $R_1(x)$, is such that for any x, there is some c between 0 and x such that

$$R_1(x) = \frac{f''(c)}{2!} x^2 = \frac{1}{2 \cdot 90000} x^2 = \frac{1}{180000} x^2.$$

Proofs

87. The base case ($n = 1$) as well as the case $n = 2$ were proved as part of the proof in the text. Assume the formula holds for n, so that

$$f(x) = \sum_{k=0}^{n} \frac{f^{(k)}(x_0)}{k!} (x - x_0)^k + \int_{x_0}^{x} \frac{(x - t)^n}{n!} f^{(n+1)}(t)\,dt.$$

Evaluate the integral using integration by parts, with $u = f^{(n+1)}(t)$ and $dv = \frac{(x-t)^n}{n!}\,dt$. Then $du = f^{(n+2)}(t)\,dt$ and $v = -\frac{(x-t)^{n+1}}{(n+1)n!} = -\frac{(x-t)^{n+1}}{(n+1)!}$. Thus

$$\int_{x_0}^x \frac{(x-t)^n}{n!} f^{(n+1)}(t)\,dt = \left[-f^{(n+1)}(t)\frac{(x-t)^{n+1}}{(n+1)!}\right]_{t=x_0}^{t=x} - \int_{x_0}^x \left(-\frac{(x-t)^{n+1}}{(n+1)!}\right) f^{(n+2)}(t)\,dt$$

$$= \frac{f^{(n+1)}(t)}{(n+1)!}(x-x_0)^{n+1} + \int_{x_0}^x \frac{(x-t)^{n+1}}{(n+1)!} f^{(n+2)}(t)\,dt.$$

Substitute this into the equation for $f(x)$ to get

$$f(x) = \sum_{k=0}^n \frac{f^{(k)}(x_0)}{k!}(x-x_0)^k + \frac{f^{(n+1)}(t)}{(n+1)!}(x-x_0)^{n+1} + \int_{x_0}^x \frac{(x-t)^{n+1}}{(n+1)!} f^{(n+2)}(t)\,dt$$

$$= \sum_{k=0}^{n+1} \frac{f^{(k)}(x_0)}{k!}(x-x_0)^k + \int_{x_0}^x \frac{(x-t)^{n+1}}{(n+1)!} f^{(n+2)}(t)\,dt,$$

proving the inductive step.

89. The Maclaurin series for $\cos x$ converges for every real number if and only if $R_n(x) \to 0$ as $n \to \infty$ for any real number x. It suffices to show that $|R_n(x)| \to 0$. Note that $(\cos x)^{(n)}$ is $\pm \sin x$ or $\pm \cos x$, so that in absolute value, it is at most 1. It follows that for some c between 0 and x,

$$\lim_{n\to\infty} |R_n(x)| = \lim_{n\to\infty} \left|\frac{(\cos x)^{n+1}(c)}{(n+1)!}(x-0)^{n+1}\right| \le \lim_{n\to\infty} \frac{|x|^{n+1}}{(n+1)!} = 0$$

using Exercise 6. Since the limit is zero for every x, the Maclaurin series for $\cos x$ converges for all x.

91. By definition, $\sinh x = \frac{e^x - e^{-x}}{2}$. Thus the Maclaurin series for $\sinh x$ can be computed from those for e^x and e^{-x}:

$$\sinh x = \frac{1}{2}\left(\sum_{k=0}^\infty \frac{1}{k!} x^k - \sum_{k=0}^\infty \frac{(-1)^k}{k!} x^k\right) = \frac{1}{2}\sum_{k=0}^\infty \frac{1-(-1)^k}{k!} x^k = \sum_{k=0}^\infty \frac{1}{(2k+1)!} x^{2k+1}$$

Thinking Forward

The derivative of a series

$$\sum_{k=0}^\infty \frac{d}{dx}\left(\frac{x^k}{k!}\right) = \frac{d}{dx}(1) + \sum_{k=1}^\infty \frac{d}{dx}\left(\frac{x^k}{k!}\right) = \sum_{k=1}^\infty \frac{kx^{k-1}}{k!} = \sum_{k=1}^\infty \frac{x^{k-1}}{(k-1)!} = \sum_{k=0}^\infty \frac{x^k}{k!}.$$

The derivative of a series

$$\sum_{k=0}^\infty \frac{d}{dx}\left(\frac{(-1)^k}{(2k+1)!} x^{2k+1}\right) = \sum_{k=0}^\infty \left(\frac{(-1)^k}{(2k+1)!}\cdot(2k+1)x^{2k}\right) = \sum_{k=0}^\infty \frac{(-1)^k}{(2k)!} x^{2k}.$$

8.4   Differentiating and Integrating Power Series

Thinking Back

The derivative of a sum

We have

$$
\begin{aligned}
\frac{d}{dx}(f_1(x)) + \frac{d}{dx}(f_2(x)) &= \lim_{h\to 0} \frac{(f_1(x+h)+f_2(x+h)) - (f_1(x)+f_2(x))}{h}\\
&= \lim_{h\to 0} \frac{(f_1(x+h)-f_1(x)) + (f_2(x+h)-f_2(x))}{h}\\
&= \lim_{h\to 0} \left(\frac{f_1(x+h)-f_1(x)}{h} + \frac{f_2(x+h)-f_2(x)}{h} \right)\\
&= \lim_{h\to 0} \frac{f_1(x+h)-f_1(x)}{h} + \lim_{h\to 0} \frac{f_2(x+h)-f_2(x)}{h}\\
&= \frac{d}{dx}(f_1(x)) + \frac{d}{dx}(f_2(x)).
\end{aligned}
$$

The proof for a finite sum of functions $f_1, f_2, \ldots, f_n$ is the same, writing out the definition of the derivative and regrouping.

The integral of a sum

If F_1 is an antiderivative of f_1 and F_2 is an antiderivative of f_2, then $(F_1 + F_2)' = F_1' + F_2'$, so that $F_1' + F_2'$ is an antiderivative of $f_1 + f_2$. It follows from the definition of indefinite integrals that

$$
\int f_1(x)\, dx + \int f_2(x)\, dx = F_1 + F_2 = \int (f_1 + f_2)(x)\, dx = \int (f_1(x)+f_2(x))\, dx.
$$

The proof for a finite sum of functions $f_1, f_2, \ldots, f_n$ is the same.

Concepts

1. (a) True. By Theorem 8.11, the interval of convergence of the Taylor series for f' is the same as that for f (except possibly at the endpoints, but in this case f converges everywhere).

    (b) True. By Theorem 8.12, the interval of convergence of $G(x)$ is the same as that for f (except possibly at the endpoints, but in this case f converges everywhere), and $G'(x) = f(x)$ by Theorem 8.11.

    (c) True, since by Theorem 8.11,

    $$
    \frac{d}{dx}\left(\sum_{k=0}^{\infty} a_k x^k \right) = \frac{d}{dx}(a_0) + \frac{d}{dx}\sum_{k=1}^{\infty} a_k x^k = \sum_{k=1}^{\infty} \frac{d}{dx}(a_k x^k) = \sum_{k=1}^{\infty} \left(k a_k x^{k-1} \right) = \sum_{k=0}^{\infty} (k+1) a_{k+1} x^k.
    $$

    (d) True. See Theorem 8.12 (b) with $x_0 = 0$.

    (e) False. The formula is correct, by Theorem 8.12 with $x_0 = 0$, but there is a missing constant, $+C$ on the right-hand side.

    (f) False. The second series is the derivative of the first series, so its radius of convergence is also 3, but it may have different convergence properties at the endpoints -3 and 3 (see Theorem 8.11).

    (g) False. The second series is the integral of the first series, so its radius of convergence is also 3, but there is no guarantee that the series will converge at the endpoints (see Theorem 8.12).

(h) True. By Theorem 8.12, the interval of convergence of the second series is the same as that for the first series, except at the endpoints, so in particular the second series converges on $(-3, 3)$.

3. Since all terms except the term for $k = 0$ have a factor of $x - x_0$, they all vanish at $x = x_0$, leaving only $f(x_0) = a_0(x - x_0)^0 = a_0$. Differentiating gives $f'(x) = \sum_{k=1}^{\infty} k a_k (x - x_0)^{k-1}$ and again terms for $k > 1$ vanish at $x = x_0$, leaving $1 a_1 = a_1$. Differentiating again gives $f''(x) = \sum_{k=2}^{\infty} k(k-1) a_k (x - x_0)^{k-2}$; terms for $k > 2$ vanish at $x = x_0$, leaving $2(2-1) a_2 = 2 a_2$. Continuing this process gives $f^{(k)}(x_0) = k! a_k$.

5. Let $F(x) = \sum_{k=0}^{\infty} \frac{a_k}{k+1} (x - x_0)^{k+1}$; then F is an antiderivative for f, and $F(x_0) = 0$ by Exercise 3. So let $G(x) = F(x) + 7$; then G is also an antiderivative for f, and its Taylor series is

$$G(x) = 7 + \sum_{k=0}^{\infty} \frac{a_k}{k+1} (x - x_0)^{k+1}.$$

7. If $f(x) = \frac{1}{1-x}$, then $f^{(k-1)}(x) = \frac{(k-1)!}{(1-x)^k} = (k-1)! \cdot \frac{1}{(1-x)^k}$, so that

$$\frac{1}{(1-x)^k} = \frac{1}{(k-1)!} f^{(k-1)}(x)$$

So applying Theorem 8.10 $k - 1$ times, we can differentiate the Maclaurin series for $\frac{1}{1-x}$ $k - 1$ times and divide by $(k-1)!$ to get the Maclaurin series for $\frac{1}{(1-x)^k}$.

9. Since $f'(x) = -3f(x)$, it follows that $f''(x) = (f'(x))' = (-3f(x))' = 9f(x)$, and in general, $f^{(k)}(x) = (-3)^k f(x)$ for all x. Since $f(0) = 2$, we get $f^{(k)}(0) = 2(-3)^k$, so that the Maclaurin series is

$$f(x) = \sum_{k=0}^{\infty} \frac{f^{(k)}(0)}{k!} x^k = \sum_{k=0}^{\infty} \frac{2(-3)^k}{k!} x^k = \sum_{k=0}^{\infty} (-1)^k \frac{2 \cdot 3^k}{k!} x^k.$$

11. (a) Using the ratio test, we have

$$\rho = \lim_{k \to \infty} \left| \frac{2^{k+1} x^{k+1}}{2^k x^k} \right| = \lim_{k \to \infty} 2 |x| = 2 |x|.$$

Thus the radius of convergence is $\frac{1}{2}$. At $x = \frac{1}{2}$, the series is $\sum_{k=0}^{\infty} 1$, which diverges; at $x = -\frac{1}{2}$, the series is $\sum_{k=0}^{\infty} (-1)^k$, which also diverges. Thus the interval of convergence is $\left(-\frac{1}{2}, \frac{1}{2} \right)$.

(b) By Theorem 8.11, the power series for f' on I is the term-by-term derivative of the original power series, which is $\sum_{k=1}^{\infty} k 2^k x^{k-1}$.

(c) By Theorem 8.12, the power series for $\int_0^x f(t) \, dt$ is the term-by-term integral of the original power series, which is $\sum_{k=0}^{\infty} \frac{2^k}{k+1} x^{k+1}$.

13. (a) Using the Ratio test, we have

$$\rho = \lim_{k \to \infty} \left| \frac{\frac{(-1)^{k+1}}{(2(k+1))!} x^{k+1}}{\frac{(-1)^k}{(2k)!} x^k} \right| = \lim_{k \to \infty} \frac{1}{(2k+1)(2k+2)} |x| = 0.$$

Thus the series converges for all $x \in \mathbb{R}$.

(b) By Theorem 8.11, the power series for f' on $\mathbb{R}$ is the term-by-term derivative of the original power series, which is $\sum_{k=1}^{\infty} \frac{(-1)^k k}{(2k)!} x^{k-1}$.

(c) By Theorem 8.12, the power series for $\int_0^x f(t)\,dt$ is the term-by-term integral of the original power series, which is $\sum_{k=0}^{\infty} \frac{(-1)^k}{(k+1)(2k)!} x^{k+1}$.

15. (a) Using the Ratio test, we have

$$\rho = \lim_{k \to \infty} \left| \frac{\frac{(-1)^{k+1}}{k+1}(x+5)^{k+1}}{\frac{(-1)^k}{k}(x+5)^k} \right| = \lim_{k \to \infty} \frac{k}{k+1}|x+5| = |x+5|.$$

Thus the series converges on $(-6, -4)$. Checking the endpoints, at $x = -4$ we get $\sum_{k=1}^{\infty} \frac{(-1)^k}{k}$, which is a convergent alternating harmonic series. At $x = -6$ we get $\sum_{k=1}^{\infty} \frac{(-1)^k}{k}(-1)^k = \sum_{k=1}^{\infty} \frac{1}{k}$, which diverges. Thus the interval of convergence is $(-6, -4]$.

(b) By Theorem 8.11, the power series for f' on $\mathbb{R}$ is the term-by-term derivative of the original power series, which is $\sum_{k=1}^{\infty} (-1)^k (x+5)^{k-1}$.

(c) By Theorem 8.12, the power series for $\int_5^x f(t)\,dt$ is the term-by-term integral of the original power series, which is $\sum_{k=0}^{\infty} \frac{(-1)^k}{k(k+1)}(x+5)^{k+1}$.

17. This integral exists because $\frac{1}{1+x^3}$ is continuous and therefore bounded on $[0, 0.5]$, say by M. It is also positive, so a sequence of lower sums for the integral will be positive, increasing, and bounded above by $M(0.5 - 0) = 0.5M$, so it converges. Thus the integral exists.

(a) By Theorem 5.30, the error of a Simpson's method approximation to the integral on $[0, 0.5]$ with n parabolas will be bounded in magnitude by $\frac{M(0.5-0)^5}{180n^4} = \frac{M}{32 \cdot 180 n^4}$ where M is an upper bound for the fourth derivative of $\frac{1}{1+x^3}$. So to approximate the integral to within 0.001 using Simpson's method, first determine an upper bound for the fourth derivative of $\frac{1}{1+x^3}$, set the formula equal to 0.001, and solve for n.

(b) Since $\frac{1}{1+x^3} = \frac{1}{1-(-x^3)}$, we can substitute $-x^3$ for x in the Maclaurin series for $\frac{1}{1-x}$:

$$\frac{1}{1+x^3} = \sum_{k=0}^{\infty} (-x^3)^k = \sum_{k=0}^{\infty} (-1)^k x^{3k}.$$

Since the Maclaurin series for $\frac{1}{1-x}$ converges on $(-1, 1)$, so does the series for $\frac{1}{1+x^3}$, so in particular it converges on $[0, 0.5]$.

(c) By Theorem 8.12 (b),

$$\int_0^{0.5} \frac{dx}{1+x^3} = \sum_{k=0}^{\infty} \frac{(-1)^k}{k+1}(0.5^{k+1} - 0^{k+1}) = \sum_{k=0}^{\infty} (-1)^k \frac{1}{(k+1)2^{k+1}}.$$

This is an alternating series that satisfies the hypotheses of the Alternating Series test, so we could approximate the integral to within 0.001 by finding the first k such that $\frac{1}{(k+1)2^{k+1}} < 0.001$.

19. This integral exists because $\sin(x^2)$ is continuous and therefore bounded on $[0, 0.5]$, say by M. It is also positive, so a sequence of lower sums for the integral will be positive, increasing, and bounded above by $M(0.5 - 0) = 0.5M$, so it converges. Thus the integral exists.

(a) By Theorem 5.30, the error of a Simpson's method approximation to the integral on $[0, 0.5]$ with n parabolas will be bounded in magnitude by $\frac{M(0.5-0)^5}{180n^4} = \frac{M}{32 \cdot 180 n^4}$ where M is an upper bound for the fourth derivative of $\sin(x^2)$. So to approximate the integral to within 0.001 using Simpson's method, first determine an upper bound for the fourth derivative of $\sin(x^2)$, set the formula equal to 0.001, and solve for n.

(b) Substitute x^2 for x in the Maclaurin series for $\sin x$:

$$\sin(x^2) = \sum_{k=0}^{\infty} \frac{(-1)^k}{(2k+1)!}(x^2)^{2k+1} = \sum_{k=0}^{\infty} \frac{(-1)^k}{(2k+1)!}x^{4k+2}.$$

Since the Maclaurin series for $\sin x$ converges everywhere, so does the series for $\sin(x^2)$.

(c) By Theorem 8.12 (b),

$$\int_0^{0.5} \sin(x^2) = \sum_{k=0}^{\infty} \frac{(-1)^k}{(4k+2)(2k+1)!}(0.5^{4k+3} - 0^{4k+3}) = \sum_{k=0}^{\infty} \frac{(-1)^k}{(4k+2)(2k+1)!}0.5^{4k+3}.$$

This is an alternating series that satisfies the hypotheses of the Alternating Series test, so we could approximate the integral to within 0.001 by finding the first k such that we get $\frac{1}{(4k+2)(2k+1)!}0.5^{4k+3} < 0.001$.

21. $\sum_{k=0}^{\infty} \frac{d}{dx}\left(\frac{(-1)^k}{(2k+1)!}x^{2k+1}\right) = \sum_{k=0}^{\infty} \frac{(-1)^k(2k+1)}{(2k+1)!}x^{2k} = \sum_{k=0}^{\infty} \frac{(-1)^k}{(2k)!}x^{2k}$.

23. $\sum_{k=0}^{\infty} \frac{d}{dx}\left(\frac{1}{k!}x^k\right) = \frac{d}{dx}1 + \sum_{k=1}^{\infty} \frac{d}{dx}\left(\frac{1}{k!}x^k\right) = \sum_{k=1}^{\infty} \frac{k}{k!}x^{k-1} = \sum_{k=1}^{\infty} \frac{1}{(k-1)!}x^{k-1} = \sum_{k=0}^{\infty} \frac{1}{k!}x^k$. This result makes sense since by Theorem 8.11, the result of differentiating this series term by term gives the derivative of e^x, which we know to be e^x.

Skills

25. (a) We have

$$\frac{1}{1+x^2} = \frac{1}{1-(-x^2)} = \sum_{k=0}^{\infty}(-x^2)^k = \sum_{k=0}^{\infty}(-1)^k x^{2k}.$$

To determine the interval of convergence, use the Ratio test:

$$\rho = \lim_{k\to\infty}\left|\frac{(-1)^{k+1}x^{2k+2}}{(-1)^k x^{2k}}\right| = \lim_{k\to\infty}|x|^2 = |x|^2.$$

Thus the radius of convergence is 1. Checking the endpoints, at $x = -1$ we get the series $\sum_{k=0}^{\infty}(-1)^k(-1)^{2k} = \sum_{k=0}^{\infty}(-1)^k$, which diverges by the Divergence test. Similarly, at $x = 1$ we get the same series: $\sum_{k=0}^{\infty}(-1)^k 1^{2k} = \sum_{k=0}^{\infty}(-1)^k$ Thus the interval of convergence is $(-1,1)$.

(b) Since $\frac{d}{dx}\frac{1}{1+x^2} = -2\frac{x}{(1+x^2)^2}$, we have by Theorem 8.11

$$\frac{x}{(1+x^2)^2} = -\frac{1}{2}\frac{d}{dx}\frac{1}{1+x^2} = -\frac{1}{2}\sum_{k=0}^{\infty}\frac{d}{dx}\left((-1)^k x^{2k}\right) = -\frac{1}{2}\sum_{k=1}^{\infty}(-1)^k(2k)x^{2k-1}$$

$$= \sum_{k=1}^{\infty}(-1)^{k-1}kx^{2k-1} = \sum_{k=0}^{\infty}(-1)^k(k+1)x^{2k+1}.$$

By Theorem 8.11, this series also has radius of convergence 1, so it converges on $(-1,1)$. Checking the endpoints, at $x = -1$ we get the series

$$\sum_{k=0}^{\infty}(-1)^k(k+1)(-1)^{2k+1} = \sum_{k=0}^{\infty}(-1)^{k+1}(k+1),$$

which diverges by the Divergence test. Similarly, at $x = 1$ we get $\sum_{k=0}^{\infty}(-1)^k(k+1)1^{2k+1} = \sum_{k=0}^{\infty}(-1)^k(k+1)$, which diverges by the Divergence Test. Thus the interval of convergence is $(-1,1)$.

27. (a) We have

$$\frac{1}{1+8x^3} = \frac{1}{1-(-8x^3)} = \sum_{k=0}^{\infty}(-8x^3)^k = \sum_{k=0}^{\infty}(-1)^k 8^k x^{3k}.$$

Since the series for $f(x) = \frac{1}{1-x}$ has radius of convergence 1, the series for $f(-8x^3) = \frac{1}{1+8x^3}$ converges when $|-8x^3| < 1$, so for $|x|^3 < \frac{1}{8}$, so for $|x| < \frac{1}{2}$. So the radius of convergence is $\frac{1}{2}$. At the endpoints we get the divergent series $\sum_{k=0}^{\infty}(-1)^k$ and $\sum_{k=0}^{\infty}1^k$, so the interval of convergence is $\left(-\frac{1}{2}, \frac{1}{2}\right)$.

(b) Since $\frac{d}{dx}\left(\frac{1}{1+8x^3}\right) = -24\frac{x^2}{(1+8x^3)^2}$, we have by Theorem 8.11

$$\frac{x^2}{(1+8x^3)^2} = -\frac{1}{24}\frac{d}{dx}\frac{1}{1+8x^3} = -\frac{1}{24}\sum_{k=0}^{\infty}\frac{d}{dx}\left((-1)^k 8^k x^{3k}\right)$$

$$= -\frac{1}{24}\sum_{k=1}^{\infty}(-1)^k 3k \cdot 8^k x^{3k-1} = \sum_{k=1}^{\infty}(-1)^{k+1}8^{k-1}kx^{3k-1}.$$

By Theorem 8.11, this series also has radius of convergence $\frac{1}{2}$, so that it converges on $\left(-\frac{1}{2}, \frac{1}{2}\right)$. Checking the endpoints gives at $x = -\frac{1}{2}$ the series

$$\sum_{k=1}^{\infty}(-1)^{k+1}8^{k-1}k\left(-\frac{1}{2}\right)^{3k-1} = \sum_{k=1}^{\infty}(-1)^{k+1}(-1)^{3k-1}k2^{3k-3}\left(\frac{1}{2}\right)^{3k-1} = \sum_{k=1}^{\infty}(-1)^{4k}\frac{k}{4} = \sum_{k=1}^{\infty}\frac{k}{4},$$

which diverges by the divergence test. Similarly, at $x = \frac{1}{2}$, we get

$$\sum_{k=1}^{\infty}(-1)^{k+1}8^{k-1}k\left(\frac{1}{2}\right)^{3k-1} = \sum_{k=1}^{\infty}(-1)^{k+1}k2^{3k-3}\left(\frac{1}{2}\right)^{3k-1} = \sum_{k=1}^{\infty}(-1)^{k+1}\frac{k}{4},$$

which also diverges by the divergence test. Thus the interval of convergence is $\left(-\frac{1}{2}, \frac{1}{2}\right)$.

29. (a) We have

$$\frac{1}{4-x^2} = \frac{1}{4}\cdot\frac{1}{1-\left(\frac{x^2}{4}\right)} = \frac{1}{4}\sum_{k=0}^{\infty}\left(\frac{x^2}{4}\right)^k = \sum_{k=0}^{\infty}\frac{x^{2k}}{4^{k+1}}.$$

Since the series for $f(x) = \frac{1}{1-x}$ has radius of convergence 1, the series for $f\left(\frac{x^2}{4}\right)$ converges when $\left|\frac{x^2}{4}\right| < 1$, so when $|x| < 2$, so the radius of convergence is 2. Checking the endpoints gives both at $x = -2$ and at $x = 2$ the series $\sum_{k=0}^{\infty}\frac{1}{4}$, which diverges by the divergence test. Thus the interval of convergence is $(-2, 2)$.

(b) Since $\frac{d}{dx}\frac{1}{4-x^2} = 2\frac{x}{(4-x^2)^2}$, we have by Theorem 8.11

$$\frac{x}{(4-x^2)^2} = \frac{1}{2}\frac{d}{dx}\frac{1}{4-x^2} = \frac{1}{2}\sum_{k=0}^{\infty}\frac{d}{dx}\left(\frac{x^{2k}}{4^{k+1}}\right) = \frac{1}{2}\sum_{k=1}^{\infty}\frac{2k}{4^{k+1}}x^{2k-1} = \sum_{k=1}^{\infty}\frac{k}{4^{k+1}}x^{2k-1}.$$

By Theorem 8.11, this series also has radius of convergence 2. Checking the endpoints gives at $x = -2$ the series $\sum_{k=1}^{\infty}(-1)^{2k-1}\frac{k}{8}$ and at $x = 2$ the series $\sum_{k=1}^{\infty}\frac{k}{8}$, both of which diverge by the divergence test. Thus the interval of convergence is $(-2, 2)$.

31. (a) Substituting x^2 into the Maclaurin series for $\sin x$ gives

$$\sin(x^2) = \sum_{k=0}^{\infty} \frac{(-1)^k}{(2k+1)!}(x^2)^{2k+1} = \sum_{k=0}^{\infty} \frac{(-1)^k}{(2k+1)!}x^{4k+2}.$$

(b) Since $\frac{d}{dx}\left(\sin(x^2)\right) = 2x\cos(x^2)$, we have by Theorem 8.11

$$x\cos(x^2) = \frac{1}{2}\frac{d}{dx}\left(\sin(x^2)\right) = \frac{1}{2}\sum_{k=0}^{\infty}\frac{d}{dx}\left(\frac{(-1)^k}{(2k+1)!}x^{4k+2}\right) = \sum_{k=0}^{\infty}\left(\frac{(-1)^k(4k+2)}{2(2k+1)!}x^{4k+1}\right)$$

$$= \sum_{k=0}^{\infty}\frac{(-1)^k}{(2k)!}x^{4k+1}.$$

(c) Substitute x^2 into the series for $\cos x$ and multiply by x to get

$$x\cos(x^2) = x\sum_{k=0}^{\infty}\frac{(-1)^k}{(2k)!}(x^2)^{2k} = \sum_{k=0}^{\infty}\frac{(-1)^k}{(2k)!}x^{4k+1}.$$

The answers from parts (b) and (c) are equal.

33. (a) Substituting $-x^2$ into the Maclaurin series for e^x gives

$$e^{-x^2} = \sum_{k=0}^{\infty}\frac{1}{k!}(-x^2)^k = \sum_{k=0}^{\infty}\frac{(-1)^k}{k!}x^{2k}.$$

(b) Since $\frac{d}{dx}\left(e^{-x^2}\right) = -2xe^{-x^2}$, we have by Theorem 8.11

$$xe^{-x^2} = -\frac{1}{2}\frac{d}{dx}\left(e^{-x^2}\right) = -\frac{1}{2}\sum_{k=0}^{\infty}\frac{d}{dx}\left(\frac{(-1)^k}{k!}x^{2k}\right) = \sum_{k=1}^{\infty}\frac{2k(-1)^{k+1}}{2k!}x^{2k-1}$$

$$= \sum_{k=1}^{\infty}\frac{(-1)^{k+1}}{(k-1)!}x^{2k-1} = \sum_{k=0}^{\infty}\frac{(-1)^k}{k!}x^{2k+1}.$$

(c) Multiply x by the series in part (a) to get

$$xe^{-x^2} = x\sum_{k=0}^{\infty}\frac{(-1)^k}{k!}x^{2k} = \sum_{k=0}^{\infty}\frac{(-1)^k}{k!}x^{2k+1}.$$

The two series are equal.

35. (a) Since $\frac{d^n}{dx^n}\frac{1}{1-x} = \frac{n!}{(1-x)^{n+1}}$, we have $\frac{d^n}{dx^n}\frac{1}{1-x}\Big|_{x=3} = \frac{(-1)^{n+1}n!}{2^{n+1}}$. Thus the Taylor series at $x_0 = 3$ is (including the $k = 0$ term which is $-\frac{1}{2}$)

$$\sum_{k=0}^{\infty}\frac{(-1)^{k+1}k!}{k!2^{k+1}}(x-3)^k = \sum_{k=0}^{\infty}\frac{(-1)^{k+1}}{2^{k+1}}(x-3)^k.$$

To determine the radius of convergence, use the Ratio test:

$$\rho = \left|\frac{\frac{(-1)^{k+2}}{2^{k+2}}(x-3)^{k+1}}{\frac{(-1)^{k+1}}{2^{k+1}}(x-3)^k}\right| = \frac{|x-3|}{2},$$

so that the radius of convergence is 2 around 3. Checking the endpoints, at $x = 1$ we get $\sum_{k=0}^{\infty}\frac{-1}{2}$; at $x = 5$ we get $\sum_{k=0}^{\infty}\frac{(-1)^{k+1}}{2}$, both of which diverge by the divergence test. Thus the interval of convergence is $(1, 5)$.

(b) This function is the derivative of the function in part (a), so by Theorem 8.11,

$$\frac{1}{(1-x)^2} = \sum_{k=0}^{\infty} \frac{d}{dx}\left(\frac{(-1)^{k+1}}{2^{k+1}}(x-3)^k\right) = \sum_{k=1}^{\infty} \frac{(-1)^{k+1}k}{2^{k+1}}(x-3)^{k-1}$$

$$= \sum_{k=0}^{\infty} \frac{(-1)^k(k+1)}{2^{k+2}}(x-3)^k.$$

Since the radius of convergence of the original series is 2, the radius of convergence of this series is 2 as well. Checking the endpoints gives the series $\sum_{k=0}^{\infty} \frac{k+1}{4}$ and $\sum_{k=0}^{\infty} \frac{(-1)^k(k+1)}{4}$, both of which diverge by the divergence test. Thus the interval of convergence is $(1, 5)$.

37. (a) Since $\frac{d^n}{dx^n}\frac{1}{2-x} = \frac{n!}{(2-x)^{n+1}}$, we have $\frac{d^n}{dx^n}\frac{1}{2-x}\Big|_{x=3} = (-1)^{n+1}n!$. Thus the Taylor series at $x_0 = 3$ is (including the $k=0$ term which is -1)

$$\sum_{k=0}^{\infty} \frac{(-1)^{k+1}k!}{k!}(x-3)^k = \sum_{k=0}^{\infty}(-1)^{k+1}(x-3)^k.$$

To determine the radius of convergence, use the Ratio test:

$$\rho = \left|\frac{(-1)^{k+2}(x-3)^{k+1}}{(-1)^{k+1}(x-3)^k}\right| = |x-3|,$$

so that the radius of convergence is 1 around 3. Checking the endpoints, at $x = 2$ we get $\sum_{k=0}^{\infty}(-1)$; at $x = 4$ we get $\sum_{k=0}^{\infty}(-1)^{k+1}$, both of which diverge by the divergence test. Thus the interval of convergence is $(2, 4)$.

(b) This function is the derivative of the function in part (a), so by Theorem 8.11,

$$\frac{1}{(2-x)^2} = \sum_{k=0}^{\infty} \frac{d}{dx}\left((-1)^{k+1}(x-3)^k\right) = \sum_{k=1}^{\infty}(-1)^{k+1}k(x-3)^{k-1} = \sum_{k=0}^{\infty}(-1)^k(k+1)(x-3)^k.$$

Since the radius of convergence of the original series is 1, the radius of convergence of this series is 1 as well. Checking the endpoints gives the series $\sum_{k=0}^{\infty}(k+1)$ and $\sum_{k=0}^{\infty}(-1)^k(k+1)$, both of which diverge by the divergence test. Thus the interval of convergence is $(2, 4)$.

39. (a) A short computation shows that $\frac{d^n}{dx^n}\frac{1}{1-3x} = \frac{3^n n!}{(1-3x)^{n+1}}$, we have $\frac{d^n}{dx^n}\frac{1}{1-3x}\Big|_{x=3} = \frac{(-1)^{n+1}3^n n!}{8^{n+1}}$. Thus the Taylor series at $x_0 = 3$ is (including the $k=0$ term which is $-\frac{1}{8}$)

$$\sum_{k=0}^{\infty} \frac{(-1)^{k+1}3^k k!}{k!8^{k+1}}(x-3)^k = \sum_{k=0}^{\infty} \frac{(-1)^{k+1}3^k}{8^{k+1}}(x-3)^k.$$

To determine the radius of convergence, use the Ratio test:

$$\rho = \left|\frac{\frac{(-1)^{k+2}3^{k+1}}{8^{k+2}}(x-3)^{k+1}}{\frac{(-1)^{k+1}3^k}{8^{k+1}}(x-3)^k}\right| = \frac{3}{8}|x-3|,$$

so that the radius of convergence is $\frac{8}{3}$ around 3. Checking the endpoints, at $x = \frac{1}{3}$ we get $\sum_{k=0}^{\infty} \frac{-1}{8}$; at $x = \frac{17}{3}$ we get $\sum_{k=0}^{\infty} \frac{(-1)^{k+1}}{8}$, both of which diverge by the divergence test. Thus the interval of convergence is $\left(\frac{1}{3}, \frac{17}{3}\right)$.

(b) Since $\frac{d}{dx}\frac{1}{1-3x} = \frac{3}{(1-3x)^2}$, we have by Theorem 8.11

$$\frac{1}{(1-3x)^2} = \frac{1}{3}\sum_{k=0}^{\infty}\frac{d}{dx}\left(\frac{(-1)^{k+1}3^k}{8^{k+1}}(x-3)^k\right) = \frac{1}{3}\sum_{k=1}^{\infty}\frac{(-1)^{k+1}k3^k}{8^{k+1}}(x-3)^{k-1}$$

$$= \sum_{k=1}^{\infty}\frac{(-1)^{k+1}k3^{k-1}}{8^{k+1}}(x-3)^{k-1} = \sum_{k=0}^{\infty}\frac{(-1)^k(k+1)3^k}{8^{k+2}}(x-3)^k.$$

Since the radius of convergence of the original series is $\frac{8}{3}$, the radius of convergence of this series is $\frac{8}{3}$ as well. Checking the endpoints gives the series $\sum_{k=0}^{\infty}\frac{k+1}{8}$ and $\sum_{k=0}^{\infty}(-1)^k\frac{k+1}{8}$, both of which diverge by the divergence test. Thus the interval of convergence is $\left(\frac{1}{3},\frac{17}{3}\right)$.

41. (a) Substitute x^3 for x in the Maclaurin series for $\sin x$ to get

$$\sin(x^3) = \sum_{k=0}^{\infty}\frac{(-1)^k}{(2k+1)!}(x^3)^{2k+1} = \sum_{k=0}^{\infty}\frac{(-1)^k}{(2k+1)!}x^{6k+3}.$$

(b) By Theorem 8.12 (a),

$$F(x) = \int \sin(x^3)\,dx = \sum_{k=0}^{\infty}\int\left(\frac{(-1)^k}{(2k+1)!}x^{6k+3}\,dx\right) + C = \sum_{k=0}^{\infty}\frac{(-1)^k}{(6k+4)(2k+1)!}x^{6k+4} + C.$$

When $x = 0$, all terms in the sum vanish, since they all have positive powers of x, and we are left with $F(0) = C = 2$, so that $C = 2$ and the result is

$$F(x) = \sum_{k=0}^{\infty}\frac{(-1)^k}{(6k+4)(2k+1)!}x^{6k+4} + 2.$$

43. (a) Substitute $-\frac{x^2}{3}$ for x in the Maclaurin series for e^x to get

$$e^{-x^2/3} = \sum_{k=0}^{\infty}\frac{1}{k!}\left(-\frac{x^2}{3}\right)^k = \sum_{k=0}^{\infty}\frac{(-1)^k}{3^k k!}x^{2k}.$$

(b) By Theorem 8.12(a),

$$F(x) = \int\left(e^{-x^2/3}\right)dx = \sum_{k=0}^{\infty}\int\left(\frac{(-1)^k}{3^k k!}x^{2k}\right)dx + C = \sum_{k=0}^{\infty}\frac{(-1)^k}{(2k+1)3^k k!}x^{2k+1} + C.$$

When $x = 0$, all the terms in the sum vanish, leaving $C = 0$, so the result is

$$F(x) = \sum_{k=0}^{\infty}\frac{(-1)^k}{(2k+1)3^k k!}x^{2k+1}.$$

45. (a) Substitute $-x^3$ in the Maclaurin series for $\frac{1}{1-x}$ to get

$$\frac{1}{1+x^3} = \sum_{k=0}^{\infty}(-x^3)^k = \sum_{k=0}^{\infty}(-1)^k x^{3k}.$$

(b) By Theorem 8.12(a),

$$F(x) = \int \frac{1}{1+x^3}\,dx = \sum_{k=0}^{\infty} \int \left((-1)^k x^{3k} \right)\,dx + C = \sum_{k=0}^{\infty} \frac{(-1)^k}{3k+1} x^{3k+1} + C.$$

When $x = 0$, all the terms in the sum vanish, leaving $C = -5$, so the result is

$$F(x) = \sum_{k=0}^{\infty} \frac{(-1)^k}{3k+1} x^{3k+1} - 5.$$

47. (a) Substitute $5x^2$ for x in the Maclaurin series for $\sin x$ and multiply by x^2 to get

$$x^2 \sin(5x^2) = x^2 \sum_{k=0}^{\infty} \frac{(-1)^k}{(2k+1)!} (5x^2)^{2k+1} = \sum_{k=0}^{\infty} \frac{(-1)^k 5^{2k+1}}{(2k+1)!} x^{4k+4}.$$

(b) By Theorem 8.12(a),

$$F(x) = \int \left(x^2 \sin(5x^2) \right)\,dx = \sum_{k=0}^{\infty} \int \left(\frac{(-1)^k 5^{2k+1}}{(2k+1)!} x^{4k+4} \right)\,dx + C$$

$$= \sum_{k=0}^{\infty} \frac{(-1)^k 5^{2k+1}}{(4k+5)(2k+1)!} x^{4k+5} + C.$$

When $x = 0$, all terms except for the C vanish, leaving $C = 1$, so that the result is

$$F(x) = \sum_{k=0}^{\infty} \frac{(-1)^k 5^{2k+1}}{(4k+5)(2k+1)!} x^{4k+5} + 1.$$

49. (a) Substitute $-3x^2$ for x in the Maclaurin series for e^x and multiply by x^2 to get

$$x^2 e^{-3x^2} = x^2 \sum_{k=0}^{\infty} \frac{1}{k!} (-3x^2)^k = \sum_{k=0}^{\infty} \frac{(-1)^k 3^k}{k!} x^{2k+2}.$$

(b) By Theorem 8.12(a),

$$F(x) = \int \left(x^2 e^{-3x^2} \right)\,dx = \sum_{k=0}^{\infty} \int \left(\frac{(-1)^k 3^k}{k!} x^{2k+2} \right)\,dx + C = \sum_{k=0}^{\infty} \frac{(-1)^k 3^k}{(2k+3)k!} x^{2k+3} + C.$$

When $x = 0$, all terms except for the C vanish, leaving $C = 1$, so that the result is

$$F(x) = \sum_{k=0}^{\infty} \frac{(-1)^k 3^k}{(2k+3)k!} x^{2k+3} + 1.$$

51. Since the indefinite integral is

$$F(x) = \sum_{k=0}^{\infty} \frac{(-1)^k}{(6k+4)(2k+1)!} x^{6k+4} + 2,$$

the definite integral is, by Theorem 8.12(b),

$$\int_0^2 \sin\left(x^3 \right)\,dx = F(2) - F(0) = \sum_{k=0}^{\infty} \frac{(-1)^k}{(6k+4)(2k+1)!} (2^{6k+4} - 0^{6k+4})$$

$$= \sum_{k=0}^{\infty} \frac{(-1)^k}{(3k+2)(2k+1)!} 2^{6k+3}.$$

53. Since the indefinite integral is

$$F(x) = \sum_{k=0}^{\infty} \frac{(-1)^k}{(2k+1)3^k k!} x^{2k+1},$$

the definite integral is, by Theorem 8.12(b),

$$\int_{-1}^{1} e^{-x^2/3} \, dx = \sum_{k=0}^{\infty} \frac{(-1)^k}{(2k+1)3^k k!} (1^{2k+1} - (-1)^{2k+1}) = \sum_{k=0}^{\infty} \frac{2(-1)^k}{(2k+1)3^k k!}.$$

55. Since the indefinite integral is

$$F(x) = \sum_{k=0}^{\infty} \frac{(-1)^k}{3k+1} x^{3k+1} - 5,$$

the definite integral is, by Theorem 8.12(b),

$$\int_{0}^{0.3} \frac{dx}{1+x^3} = \sum_{k=0}^{\infty} \frac{(-1)^k}{3k+1} \left(0.3^{3k+1} - 0^{3k+1}\right) = \sum_{k=0}^{\infty} \frac{(-1)^k}{3k+1} 0.3^{3k+1}.$$

57. Since the indefinite integral is

$$F(x) = \sum_{k=0}^{\infty} \frac{(-1)^k 5^{2k+1}}{(4k+5)(2k+1)!} x^{4k+5} + 1,$$

the definite integral is, by Theorem 8.12(b),

$$\int_{1}^{2} x^2 \sin(5x^2) \, dx = \sum_{k=0}^{\infty} \frac{(-1)^k 5^{2k+1}}{(4k+5)(2k+1)!} (2^{4k+5} - 1^{4k+5}) = \sum_{k=0}^{\infty} \frac{(-1)^k 5^{2k+1}}{(4k+5)(2k+1)!} (2^{4k+5} - 1).$$

59. Since the indefinite integral is

$$F(x) = \sum_{k=0}^{\infty} \frac{(-1)^k 3^k}{(2k+3)k!} x^{2k+3} + 1.$$

the definite integral is, by Theorem 8.12(b),

$$\int_{0.5}^{1} x^2 e^{-3x^2} \, dx = \sum_{k=0}^{\infty} \frac{(-1)^k 3^k}{(2k+3)k!} (1^{2k+3} - 0.5^{2k+3}) = \sum_{k=0}^{\infty} \frac{(-1)^k 3^k}{(2k+3)k!} (1 - 0.5^{2k+3}).$$

61. By Theorem 7.38, we must find the first k such that the $(k+1)^{\text{st}}$ term of the underlying positive series is less than 0.001, i.e. such that

$$\frac{2^{6(k+1)+3}}{(3(k+1)+2)(2(k+1)+1)!} = \frac{2^{6k+9}}{(3k+5)(2k+3)!} < 0.001.$$

For $k = 9$, the value is ≈ 0.0056, while for $k = 10$ it is ≈ 0.0007. So $k = 10$ is the smallest k, and the integral is approximately

$$\sum_{k=0}^{10} \frac{(-1)^k 2^{6k+3}}{(3k+2)(2k+1)!} \approx 0.45254.$$

63. By Theorem 7.38, we must find the first k such that the $(k+1)^{\text{st}}$ term of the underlying positive series is less than 0.001, i.e. such that

$$\frac{2}{(2(k+1)+1)3^{k+1}(k+1)!} = \frac{2}{(2k+3)3^{k+1}(k+1)!} < 0.001.$$

For $k = 2$, the value is ≈ 0.0018, while for $k = 3$ it is ≈ 0.0001. So $k = 3$ is the smallest k, and the integral is approximately

$$\sum_{k=0}^{3} \frac{2(-1)^k}{(2k+1)3^k(k)!} \approx 1.79824.$$

65. By Theorem 7.38, we must find the first k such that the $(k+1)^{\text{st}}$ term of the underlying positive series is less than 0.001, i.e. such that

$$\frac{1}{3(k+1)+1}0.3^{3(k+1)+1} = \frac{1}{3k+4}0.3^{3k+4} < 0.001.$$

For $k = 0$, the value is ≈ 0.0020, while for $k = 1$ it is ≈ 0.00003. So $k = 1$ is the smallest k, and the integral is approximately

$$\sum_{k=0}^{1} \frac{(-1)^k}{3k+1}0.3^{3k+1} \approx 0.29798.$$

67. By Theorem 7.38, we must find the first k such that the $(k+1)^{\text{st}}$ term of the underlying positive series is less than 0.001, i.e. such that

$$\frac{5^{2(k+1)+1}}{(4(k+1)+5)(2(k+1)+1)!}(2^{4(k+1)+5}-1) = \frac{5^{2k+3}}{(4k+9)(2k+3)!}(2^{4k+9}-1) < 0.001.$$

For $k = 26$, the value is ≈ 0.002, while for $k = 27$ it is ≈ 0.0002. So $k = 27$ is the smallest k, and the integral is approximately

$$\sum_{k=0}^{27} \frac{(-1)^k 5^{2k+1}}{(4k+5)(2k+1)!}(2^{4k+5}-1) \approx -0.039348.$$

69. By Theorem 7.38, we must find the first k such that the $(k+1)^{\text{st}}$ term of the underlying positive series is less than 0.001, i.e. such that

$$\frac{3^{k+1}}{(2(k+1)+3)(k+1)!}(1-0.5^{2(k+1)+3}) = \frac{3^{k+1}}{(2k+5)(k+1)!}(1-0.5^{2k+5}) < 0.001.$$

For $k = 8$, the value is ≈ 0.002, while for $k = 9$ it is ≈ 0.0007. So $k = 9$ is the smallest k, and the integral is approximately

$$\sum_{k=0}^{9} \frac{(-1)^k 3^k}{(2k+3)k!}(1-0.5^{2k+3}) \approx 0.48101.$$

Applications

71. The right-hand side converges by definition to $y_0(x)$. The radius of convergence of the power series is computed by noting that the $(k+1)^{\text{st}}$ term differs from the kth term by a factor of x^3 in the numerator and a factor of $(3k+2)(3k+3)$ in the denominator. Thus

$$\lim_{k \to \infty} \left| \frac{a_{k+1}}{a_k} \right| = \left| x^3 \right| \lim_{k \to \infty} \frac{1}{(3k+2)(3k+3)} = 0.$$

Since the ratio is less than zero everywhere, the radius of convergence is infinite. Now, Theorem 8.11 says that we can compute $y_0'(x)$ by differentiating the power series. Then

$$y_0(x) = 1 + \sum_{k=1}^{\infty} \frac{x^{3k}}{(2 \cdot 3)(5 \cdot 6) \ldots ((3k-1) \cdot 3k)} = 1 + \frac{x^3}{2 \cdot 3} + \sum_{k=2}^{\infty} \frac{x^{3k}}{(2 \cdot 3)(5 \cdot 6) \ldots ((3k-1) \cdot 3k)}$$

$$y_0'(x) = \frac{x^2}{2} + \sum_{k=2}^{\infty} \frac{3kx^{3k-1}}{(2 \cdot 3)(5 \cdot 6) \ldots ((3k-1) \cdot 3k)} = \frac{x^2}{2} + \sum_{k=2}^{\infty} \frac{x^{3k-1}}{(2 \cdot 3)(5 \cdot 6) \ldots ((3k-1))}$$

$$y_0''(x) = x + \sum_{k=2}^{\infty} \frac{(3k-1)x^{3k-2}}{(2 \cdot 3)(5 \cdot 6) \ldots ((3k-1))} = x + \sum_{k=2}^{\infty} \frac{x^{3k-2}}{(2 \cdot 3)(5 \cdot 6) \ldots ((3k-4) \cdot (3k-3))}.$$

Then

$$y_0''(x) = x + \sum_{k=2}^{\infty} \frac{x^{3k-2}}{(2 \cdot 3)(5 \cdot 6) \ldots ((3k-4) \cdot (3k-3))}$$

$$= x \left(1 + \sum_{k=2}^{\infty} \frac{x^{3k-3}}{(2 \cdot 3)(5 \cdot 6) \ldots ((3k-4) \cdot (3k-3))} \right)$$

$$= x \left(1 + \sum_{k=1}^{\infty} \frac{x^{3k}}{(2 \cdot 3)(5 \cdot 6) \ldots ((3k-1) \cdot (3k))} \right)$$

$$= xy.$$

Proofs

73. Let $f(x) = \sum_{k=0}^{\infty} a_k x^k$ and $g(x) = \sum_{k=0}^{\infty} b_k x^k$. Then $f(x) = g(x)$ by hypothesis, so that in particular $a_0 = f(0) = g(0) = b_0$. Further, by Exercise 3, $f^{(k)}(x_0) = k!a_k$ and $g^{(k)}(x_0) = k!b_k$. But since $f^{(k)}(x_0) = g^{(k)}(x_0)$, it follows that $a_k = b_k$. So the coefficients of the two power series are equal.

75. Since f is odd, we have $f(-x) = -f(x)$, or $-f(-x) = f(x)$. The power series for $-f(-x)$ is $-\sum_{k=0}^{\infty} a_k(-x)^k = \sum_{k=0}^{\infty}(-1)^{k+1}a_k x^k$. By Exercise 73, since $f(x) = -f(-x)$, we know that $a_k = (-1)^{k+1}a_k$ for all k. If k is even, this becomes $a_k = -a_k$, so that $a_k = 0$ for k even.

Thinking Forward

▷ $i^2 = -1$ by definition, since $i = \sqrt{-1}$. Then $i^3 = i \cdot i^2 = i \cdot (-1) = -i$ and $i^4 = (i^2)^2 = (-1)^2 = 1$.

Euler's Formula

▷ The Maclaurin series for e^{ix} is obtained by substituting ix for x in the series for e^x:

$$e^{ix} = \sum_{k=0}^{\infty} \frac{1}{k!}(ix)^k = \sum_{k=0}^{\infty} \frac{i^k}{k!} x^k.$$

The real part of this series (the terms whose coefficients are real) are the terms where k is even:

$$\sum_{\substack{k=0 \\ k \text{ even}}}^{\infty} \frac{i^k}{k!} x^k = \sum_{\substack{k=0 \\ k \text{ even}}}^{\infty} \frac{(-1)^{k/2}}{k!} x^k = \sum_{k=0}^{\infty} \frac{(-1)^k}{(2k)!} x^{2k} = \cos x.$$

The imaginary part of this series (the terms whose coefficients are imaginary) are the terms where k is odd:

$$\sum_{\substack{k=1 \\ k \text{ odd}}}^{\infty} \frac{i^k}{k!} x^k = i \sum_{\substack{k=1 \\ k \text{ odd}}}^{\infty} \frac{i^{k-1}}{k!} x^k = i \sum_{\substack{k=1 \\ k \text{ odd}}}^{\infty} \frac{(-1)^{(k-1)/2}}{k!} x^k = i \sum_{k=1}^{\infty} \frac{(-1)^{k-1}}{(2k-1)!} x^{2k-1} = i \sum_{k=0}^{\infty} \frac{(-1)^k}{(2k+1)!} x^{2k+1} = i \sin x.$$

Thus $e^{ix} = \cos x + i \sin x$.

Chapter Review and Self-Test

1. To find the radius of convergence, use the Ratio test:

$$\rho = \lim_{k \to \infty} \left| \frac{\frac{1}{\ln(k+1)} x^{k+1}}{\frac{1}{\ln k} x^k} \right| = \lim_{k \to \infty} \frac{\ln k}{\ln(k+1)} |x| \overset{L'H}{=} |x| \lim_{k \to \infty} \frac{1/k}{1/(k+1)} = |x| \lim_{k \to \infty} \frac{k+1}{k} = |x|.$$

Thus the series converges absolutely when $\rho = |x| < 1$, so the radius of convergence is 1. At -1 the series becomes $\sum_{k=2}^{\infty} \frac{(-1)^k}{\ln k}$, which is an alternating series satisfying the conditions of the Alternating Series test, so it converges. At $x = 1$ the series becomes $\sum_{k=2}^{\infty} \frac{1}{\ln k}$. Since $\ln k < k$ for $k \geq 2$, we have $\frac{1}{\ln k} > \frac{1}{k}$; since $\sum_{k=2}^{\infty} \frac{1}{k}$ diverges, so does $\sum_{k=2}^{\infty} \frac{1}{\ln k}$. So the interval of convergence is $[-1, 1)$.

3. To find the radius of convergence, use the Ratio test:

$$\rho = \lim_{k \to \infty} \left| \frac{\ln(k+1)(x-3)^{k+1}}{\ln k (x-3)^k} \right| = \lim_{k \to \infty} \frac{\ln(k+1)}{\ln k} |x-3|$$

$$\overset{L'H}{=} \lim_{k \to \infty} \frac{1/(k+1)}{1/k} |x-3| = \lim_{k \to \infty} \frac{k}{k+1} |x-3| = |x-3|.$$

Thus the series converges absolutely when $\rho = |x-3| < 1$, so the radius of convergence is 1 around $x_0 = 3$. At 2 the series becomes $\sum_{k=1}^{\infty} (-1)^k \ln k$ and at 4 it becomes $\sum_{k=1}^{\infty} \ln k$; both of these diverge by the divergence test. So the interval of convergence is $(2, 4)$.

5. To find the radius of convergence, use the Ratio test:

$$\rho = \lim_{k \to \infty} \left| \frac{\frac{5^{k+1}}{(2k+2)!} (x+2)^{k+1}}{\frac{5^k}{(2k)!} (x+2)^k} \right| = \lim_{k \to \infty} \frac{5}{(2k+1)(2k+2)} |x+2| = 0.$$

This series converges for all real numbers.

7. To find the radius of convergence, use the Ratio test:

$$\rho = \lim_{k \to \infty} \left| \frac{\frac{(-1)^{k+1}}{(k+1)2^{k+1}} (x-2)^{k+1}}{\frac{(-1)^k}{k 2^k} (x-2)^k} \right| = \lim_{k \to \infty} \frac{k}{2(k+1)} |x-2| = \lim_{k \to \infty} \frac{k}{2k+2} |x-2| = \frac{1}{2} |x-2|.$$

The series converges when $\frac{1}{2}|x-2| < 1$, so when $|x-2| < 2$, so the radius of convergence is 2 and the series converges absolutely on $(0, 4)$. At $x = 0$ we get $\sum_{k=1}^{\infty} \frac{1}{k}$, which diverges. At $x = 4$ we get $\sum_{k=1}^{\infty} \frac{(-1)^k}{k}$, which converges by the Alternating Series test. Thus the interval of convergence is $(0, 4]$.

9. With $f(x) = \tan^{-1} x$, we have $f(0) = 0$, and

$$f'(x) = \frac{1}{1+x^2} \qquad f'(0) = 1,$$

$$f''(x) = -\frac{2x}{(1+x^2)^2} \qquad f''(0) = 0,$$

$$f'''(x) = \frac{2(3x^2 - 1)}{(1+x^2)^3} \qquad f'''(0) = -2.$$

Then

$$P_3(x) = 0 + 1 \cdot x + \frac{0}{2}x^2 - \frac{2}{6}x^3 = x - \frac{1}{3}x^3.$$

11. With $f(x) = x \sin x$, we have $f(0) = 0$, and

$$\begin{aligned}
f'(x) &= x \cos x + \sin x & f'(0) &= 0, \\
f''(x) &= 2 \cos x - x \sin x & f''(0) &= 2, \\
f'''(x) &= -x \cos x - 3 \sin x & f'''(0) &= 0.
\end{aligned}$$

Then

$$P_3(x) = 0 + 0 \cdot x + \frac{2}{2}x^2 + \frac{0}{6}x^3 = x^2.$$

13. With $f(x) = (x^2 + 1)^{3/2}$, we have $f(0) = 1$, and

$$\begin{aligned}
f'(x) &= 3x(x^2 + 1)^{1/2} & f'(0) &= 0, \\
f''(x) &= 3(2x^2 + 1)(x^2 + 1)^{-1/2} & f''(0) &= 3, \\
f'''(x) &= 3(2x^3 + 3x)(x^2 + 1)^{-3/2} & f'''(0) &= 0.
\end{aligned}$$

Then

$$P_3(x) = 1 + 0 \cdot x + \frac{3}{2}x^2 + \frac{0}{6}x^3 = 1 + \frac{3}{2}x^2.$$

15. From Exercise 12 together with the fact that $(\sin x)^{(4)} = \sin x$, we know that with $f(x) = \sin x$, we have

$$f^{(n)}(0) = \begin{cases}
\sin 0 = 0, & n \equiv 0 \pmod 4, \\
\cos 0 = 1, & n \equiv 1 \pmod 4, \\
-\sin 0 = 0, & n \equiv 2 \pmod 4, \\
-\cos 0 = -1, & n \equiv 3 \pmod 4.
\end{cases}$$

Thus only the odd powers of x appear in the Maclaurin series, and the signs alternate. Thus the Maclaurin series is

$$x - \frac{1}{3!}x^3 + \frac{1}{5!}x^5 - \frac{1}{7!}x^7 + \cdots = \sum_{k=0}^{\infty} \frac{(-1)^k}{(2k+1)!}x^{2k+1}.$$

To determine convergence, use the Ratio test:

$$\rho = \lim_{k \to \infty} \left| \frac{\frac{(-1)^{k+1}}{(2k+3)!}x^{2k+3}}{\frac{(-1)^k}{(2k+1)!}x^{2k+1}} \right| = \lim_{k \to \infty} \frac{1}{(2k+2)(2k+3)} |x|^2 = 0.$$

The radius of convergence is infinite.

17. Since all derivatives of e^x are equal to e^x, each derivative evaluated at 0 is 1, so the Maclaurin series is

$$1 + \frac{1}{1!}x + \frac{1}{2!}x^2 + \frac{1}{3!}x^3 + \frac{1}{4!}x^4 + \cdots = \sum_{k=0}^{\infty} \frac{1}{k!}x^k.$$

To determine convergence, use the Ratio test:

$$\rho = \lim_{k \to \infty} \left| \frac{\frac{1}{(k+1)!}x^{k+1}}{\frac{1}{k!}x^k} \right| = \lim_{k \to \infty} \frac{1}{k+1} |x| = 0.$$

The radius of convergence is infinite.

19. With $f(x) = \frac{1}{1-x} = (1-x)^{-1}$, we have $f'(x) = (1-x)^{-2}$, $f''(x) = 2(1-x)^{-3}$, and in general $f^{(n)}(x) = n!(1-x)^{-n-1}$. Thus $f^{(n)}(0) = n!$ for $n \geq 0$, so that the Maclaurin series is

$$0! + \frac{1!}{1!}x + \frac{2!}{2!}x^2 + \frac{3!}{3!}x^3 + \cdots = \sum_{k=0}^{\infty} x^k.$$

Since this is a geometric series with ratio x, it converges for $|x| < 1$, so the radius of convergence is 1.

21. We have $f^{(n)}(x) = (-1)^n n! x^{-n-1}$, so that $f^{(n)}(1) = (-1)^n n!$. Thus the Taylor series is

$$0! - \frac{1!}{1!}(x-1) + \frac{2!}{2!}(x-1)^2 - \frac{3!}{3!}(x-1)^3 + \cdots = \sum_{k=0}^{\infty} (-1)^k (x-1)^k.$$

Since this is a geometric series with ratio $-(x-1)$, it has radius of convergence 1.

23. With $f(x) = \sqrt{x} = x^{1/2}$, we have $f'(x) = \frac{1}{2}x^{-1/2}$ and $f''(x) = -\frac{1}{4}x^{-3/2}$. We claim that

$$f^{(n)}(x) = (-1)^{n-1} \frac{1 \cdot 3 \cdot 5 \cdots (2n-3)}{2^n} x^{-(2n-1)/2} \text{ for } n \geq 2.$$

From the above computation, this holds for $n = 2$. If it holds for n, then

$$f^{(n+1)}(x) = \left(f^{(n)}(x) \right)' = (-1)^{n-1} \frac{1 \cdot 3 \cdot 5 \cdots (2n-3)}{2^n} \left(x^{-(2n-1)/2} \right)'$$

$$= (-1)^{n-1} \frac{1 \cdot 3 \cdot 5 \cdots (2n-3)}{2^n} \left(-\frac{2n-1}{2} x^{-(2n-1)/2-1} \right)$$

$$= (-1)^n \frac{1 \cdot 3 \cdot 5 \cdots (2n-1)}{2^{n+1}} x^{-(2n+1)/2},$$

so the result holds for all $n \geq 2$ by induction. Thus

$$f^{(n)}(1) = (-1)^{n-1} \frac{1 \cdot 3 \cdot 5 \cdots (2n-3)}{2^n} \text{ for } n \geq 2,$$

so together with $f(1) = 1$ and $f'(1) = \frac{1}{2}$, we get for the Taylor series

$$1 + \frac{1}{2}(x-1) + \sum_{k=2}^{\infty} (-1)^{k-1} \frac{1 \cdot 3 \cdot 5 \cdots (2k-3)}{2^k k!} (x-1)^k.$$

To determine the radius of convergence, use the Ratio test. The term for $k+1$ differs from that for k by having an extra $2k-1$ in the numerator and an extra $2(k+1)$ in the denominator, along with an extra factor of $x-1$. Thus

$$\rho = \lim_{k \to \infty} \left| \frac{2k-1}{2(k+1)}(x-1) \right| = \lim_{k \to \infty} \frac{2k-1}{2k+2} |x-1| = |x-1|.$$

Thus the series converges for $|x-1| < 1$, so that the radius of convergence is 1.

25. This is the Maclaurin series for $\sin x$, which is $\sum_{k=0}^{\infty} \frac{(-1)^k}{(2k+1)!} x^{2k+1}$, with $x = \pi$, so the result is $\sin \pi = 0$.

27. This is the Maclaurin series for $\cos x$, which is $\sum_{k=0}^{\infty} \frac{(-1)^k}{(2k)!} x^{2k}$, with $x = \frac{\pi}{2}$, so the result is $\cos \frac{\pi}{2} = 0$.

29. The Maclaurin series for $x^2 e^x$ is found by multiplying the series for e^x by x^2, so that

$$g(x) = x^2 e^x = x^2 \sum_{k=0}^{\infty} \frac{1}{k!} x^k = \sum_{k=0}^{\infty} \frac{1}{k!} x^{k+2}.$$

By Theorem 8.11,

$$g'(x) = \sum_{k=0}^{\infty} \frac{d}{dx}\left(\frac{1}{k!} x^{k+2}\right) = \sum_{k=0}^{\infty} \frac{k+2}{k!} x^{k+1}.$$

By Theorem 8.12(a),

$$\int g(x)\, dx = \sum_{k=0}^{\infty} \int \left(\frac{1}{k!} x^{k+2}\right) dx = \sum_{k=0}^{\infty} \frac{1}{k!(k+3)} x^{k+3} + C.$$

31. The Maclaurin series for $\frac{1}{1+x}$ is found by substituting $-x$ into the Maclaurin series for $\frac{1}{1-x}$, so

$$\frac{1}{1+x} = \sum_{k=0}^{\infty} (-x)^k = \sum_{k=0}^{\infty} (-1)^k x^k.$$

By Theorem 8.12(a),

$$\ln(1+x) = \int \frac{1}{1+x}\, dx = \sum_{k=0}^{\infty} \int \left((-1)^k x^k\right) dx = \sum_{k=0}^{\infty} \frac{(-1)^k}{k+1} x^{k+1} + C.$$

At $x = 0$, the left-hand side is $\ln(1+0) = \ln 1 = 0$, and the right-hand side is $0 + C$, so that $C = 0$. Then applying Theorem 8.12(a) again gives

$$\int \ln(1+x)\, dx = \sum_{k=0}^{\infty} \int \left(\frac{(-1)^k}{k+1} x^{k+1}\right) dx = \sum_{k=0}^{\infty} \frac{(-1)^k}{(k+1)(k+2)} x^{k+2} + C.$$

Finally, by Theorem 8.12(b), we have

$$\int_0^{1/2} \ln(1+x)\, dx = \sum_{k=0}^{\infty} \frac{(-1)^k}{(k+1)(k+2)} \left(\left(\frac{1}{2}\right)^{k+2} - 0^{k+2}\right) = \sum_{k=0}^{\infty} \frac{(-1)^k}{2^{k+2}(k+1)(k+2)}.$$

33. Since the Maclaurin series for e^x is $\sum_{k=0}^{\infty} \frac{1}{k!} x^k$, the Maclaurin series for e^{-x} is $\sum_{k=0}^{\infty} \frac{(-1)^k}{k!} x^k$. Setting $x = \frac{1}{2}$ gives

$$\frac{1}{\sqrt{e}} = e^{-1/2} = \sum_{k=0}^{\infty} (-1)^k \frac{1}{2^k k!}.$$

By the alternating series test, to approximate $e^{-1/2}$ to within 0.001, we must find the first k such that the $(k+1)^{\text{st}}$ term of the underlying positive series is less than 0.001, i.e., such that

$$\frac{1}{2^{k+1}(k+1)!} < 0.001.$$

For $k = 3$ this gives ≈ 0.0026, while for $k = 4$ it is ≈ 0.00026, so that $k = 4$ is the smallest value of k, and then the approximate value of $e^{-1/2}$ is

$$e^{-1/2} \approx \sum_{k=0}^{4} (-1)^k \frac{1}{2^k k!} \approx 0.60677.$$

To approximate within 10^{-6} requires that the $(k+1)^{\text{st}}$ term be less than 10^{-6}, which first happens for $k = 7$.

35. Substitute x^2 for x in the Maclaurin series for $\sin x$ to get

$$\sin(x^2) = \sum_{k=0}^{\infty} \frac{(-1)^k}{(2k+1)!}(x^2)^{2k+1} = \sum_{k=0}^{\infty} \frac{(-1)^k}{(2k+1)!}x^{4k+2}.$$

Then by Theorem 8.12(b),

$$\int_0^3 \sin(x^3)\,dx = \sum_{k=0}^{\infty} \frac{(-1)^k}{(4k+3)(2k+1)!}\left(3^{4k+3} - 0^{4k+3}\right) = \sum_{k=0}^{\infty}(-1)^k \frac{3^{4k+3}}{(4k+3)(2k+1)!}.$$

By the alternating series test, to approximate the integral to within 0.001, we must find the first k such that the $(k+1)^{\text{st}}$ term of the underlying positive series is less than 0.001, i.e., such that

$$\frac{3^{4k+7}}{(4k+7)(2k+3)!} < 0.001.$$

For $k = 11$ this gives ≈ 0.0027, while for $k = 12$ it is ≈ 0.00029, so that $k = 12$ is the smallest value of k, and then the approximate value of the integral is

$$\int_0^3 \sin(x^3)\,dx = \sum_{k=0}^{12}(-1)^k \frac{3^{4k+3}}{(4k+3)(2k+1)!} \approx 0.773829.$$

To approximate within 10^{-6} requires that the $(k+1)^{\text{st}}$ term be less than 10^{-6}, which first happens for $k = 15$.

Part IV

Vector Calculus

Chapter 9

Parametric Equations and Polar Functions

9.1 Parametric Equations

Thinking Back

Eliminating a variable

▷ Since $x = t^2$, we get $y = t^6 = (t^2)^3 = x^3$.

▷ $x = \sin t = y$, so the curve is $x = y$.

▷ Since $\sin^2 t + \cos^2 t = 1$, the relationship between x and y is $x^2 + y^2 = 1$.

▷ Since $\cosh^2 t - \sinh^2 t = 1$, the relationship between x and y is $y^2 - x^2 = 1$.

Arc length

Using the arc length formula from Theorem 6.7, we have

▷ The arc length is $\int_0^5 \sqrt{1^2 + ((3x - 4)')^2}\, dx = \int_0^5 \sqrt{1^2 + 3^2}\, dx = 5\sqrt{10}$.

▷ With $f(x) = \sqrt{4 - x^2}$, we get $f'(x) = \frac{1}{2}(4 - x^2)^{-1/2} \cdot (-2x) = -x(4 - x^2)^{-1/2}$. Then the arc length is

$$
\int_0^2 \sqrt{1 + (f'(x))^2}\, dx = \int_0^2 \sqrt{1 + \left(-x(4 - x^2)^{-1/2}\right)^2}\, dx
$$

$$
= \int_0^2 \sqrt{1 + \frac{x^2}{4 - x^2}}\, dx
$$

$$
= \int_0^2 \sqrt{\frac{4}{4 - x^2}}\, dx
$$

$$
= \int_0^2 \frac{2}{\sqrt{4 - x^2}}\, dx
$$

$$
= \left[2 \sin^{-1} \frac{x}{2}\right]_0^2
$$

$$
= 2 \sin^{-1} 1 - 2 \sin^{-1} 0 = \pi.
$$

▷ With $f(x) = x^{3/2}$ we get $f'(x) = \frac{3}{2}\sqrt{x}$, so that the arc length is

$$\int_0^4 \sqrt{1 + \frac{9}{4}x}\, dx = \left[\frac{8}{27}\left(1 + \frac{9}{4}x\right)^{3/2} \right]_0^4 = \frac{8}{27}\left(10\sqrt{10} - 1\right).$$

▷ With $f(x) = x^2$ we get $f'(x) = 2x$, so that the arc length is

$$\int_0^1 \sqrt{1 + 4x^2}\, dx.$$

To integrate, use the trigonometric substitution $x = \frac{1}{2}\tan u$, so that $dx = \frac{1}{2}\sec^2 u$. Then (using the integral of $\sec^3 u$ from previous chapters)

$$
\begin{aligned}
\int_0^1 \sqrt{1 + 4x^2}\, dx &= \frac{1}{2}\int_{x=0}^{x=1} \sqrt{1 + \tan^2 u}\,\sec^2 u\, du \\
&= \frac{1}{2}\int_{x=0}^{x=1} \sec^3 u\, du \\
&= \frac{1}{4}\left[(\sec u \tan u + \ln|\sec u + \tan u|) \right]_{x=0}^{x=1} \\
&= \frac{1}{4}\left[\left(2x\sqrt{1 + 4x^2} + \ln\left|\sqrt{1 + 4x^2} + 2x\right| \right) \right]_0^1 \\
&= \frac{1}{4}(2\sqrt{5} + \ln(2 + \sqrt{5})).
\end{aligned}
$$

Concepts

1. (a) True. One way of writing it is $x = t$, $y = f(t)$.

   (b) False. It may not be possible to solve either $x(t)$ or $y(t)$ for t.

   (c) False. For example, $x = \cos t$ and $y = \sin t$ for $t \in [0, 2\pi)$ is the unit circle, which does not pass the vertical line test.

   (d) False. Almost any curve has an infinite number of parameterizations. For example, the unit circle can be parametrized as $x = \cos(at)$, $y = \sin(at)$ for $t \in \left[0, \frac{2\pi}{a}\right]$ for any positive real number a.

   (e) False. If $f'(t) = 0$ for some value of t, then the slope of the parametric curve is infinite, so the curve is not differentiable there.

   (f) False. It could be that $x'(t) = 0$ as well, in which case the slope of the tangent line is undefined. For example, let $x(t) = t^2$ and $y(t) = t^3$; at $t = 0$, we have $x'(t) = y'(t) = 0$, and in fact the curve, which is $y = x^{3/2}$, is not differentiable at $x = 0$.

   (g) True. By Definition 9.2, the slope of the tangent line is $m = \frac{y'(t_0)}{x'(t_0)}$, so in the case given, $m = 0$.

   (h) False. If each loop were a semicircle, then clearly the center of the first circle would be at $(\pi r, 0)$. But the points $(0, 0)$ and $(\pi r, 2r)$, which both lie on the curve, are not equidistant from $(\pi r, 0)$. Thus this is not a semicircle.

3. (a) Since $x'(t) > 0$, the direction of motion is to the right. Since $y'(t) > 0$, the direction of motion is also up. Thus the direction of motion is up and to the right.

   (b) Since $x'(t) > 0$, the direction of motion is to the right. Since $y'(t) < 0$, the direction of motion is also down. Thus the direction of motion is down and to the right.

   (c) Since $x'(t) < 0$, the direction of motion is to the left. Since $y'(t) > 0$, the direction of motion is also up. Thus the direction of motion is up and to the left.

(d) Since $x'(t) < 0$, the direction of motion is to the left. Since $y'(t) < 0$, the direction of motion is also down. Thus the direction of motion is down and to the left.

5. Since $x'(t) = 2t$ and $y'(t) = 3t^2$, we see that for $t < 0$, $x'(t) < 0$ while $y'(t) > 0$, so that the particle is moving up and to the left. For $t > 0$, both derivatives are positive, so the particle is moving up and to the right. Finally, at $t = 0$, both derivatives are zero, so the particle is stopped.

7. Since $x'(t) = e^t$ and $y'(t) = \frac{1}{t}$, both derivatives are positive everywhere, so that the particle is always moving up and to the right.

9. (i) For (a), substitute x for t in $y = t^2 - 1$ to get $y = x^2 - 1$. For (b), substitute $-x$ for t in $y = t^2 - 1$ to again get $y = x^2 - 1$.

   (ii) In part (a), since $t \geq 0$, we only get the portion of $y = x^2 - 1$ corresponding to $x = t \geq 0$, so the right half of the parabola. For part (b), since $t \geq 0$, we get the portion of $y = x^2 - 1$ corresponding to $x = -t \leq 0$, so the left half of the parabola.

   (iii) For both parts, since $y'(t) = 2t$ and $t \geq 0$, the particle is always rising. In part (a), since $x'(t) = 1$, the particle is moving to the right, and in part (b), since $x'(t) = -1$, the particle is moving to the left.

11. (i) For part (a), substitute x for t in $y = \sin t$ to get $y = \sin x$ for $x \geq 0$. For part (b), substitute $t = x + 1$ for t in $y = \sin(t - 1)$ to get $y = \sin x$. In this substitution, since $t \geq 1$ and $x = t - 1$, we again have $x \geq 0$.

   (ii) Since the parametrizations and ranges of the independent variables are the same, these two systems parametrize the exact same curve over the same domain.

   (iii) For both parts, $x'(t) = 1$, so that both curves are always moving to the right. For part (a), $y'(t) = \cos t$, so that the particle oscillates between moving up and moving down as $\cos t$ changes sign. For part (b), $y'(t) = \cos(t - 1)$, so that again the particle oscillates between moving up and moving down (as it must, since the two parametrizations represent the same function over the same domain).

13. Horizontal tangent lines occur when $\frac{y'(t_0)}{x'(t_0)} = 0$, so when $y'(t_0) = 0$ but $x'(t_0) \neq 0$. Vertical tangent lines occur when $\frac{y'(t_0)}{x'(t_0)} = \pm\infty$, so when $y'(t_0) \neq 0$ but $x'(t_0) = 0$. Note that if both derivatives vanish at some point, then the tangent to the parametric curve at that point is not defined.

15. Since

$$x(t)^2 - 2x(t) - 3 = (\sin t + 1)^2 - 2(\sin t + 1) - 3 = \sin^2 t + 2\sin t + 1 - 2\sin t - 2 - 3$$
$$= \sin^2 t - 4 = y(t),$$

the given parametrization indeed traces out points on the graph of the function. But since no matter what t is, $\sin t$ lies between -1 and 1, it will trace out only portions of the graph for which $x = \sin t + 1$ is between 0 and 2. Since $\sin$ is periodic, it will trace out the same region over and over.

Skills

17. We have

t	-2	-1	0	1	2	3	4	5
$x = 3t + 1$	-5	-2	1	4	7	10	13	16
$y = t$	-2	-1	0	1	2	3	4	5

so a graph connecting these points is

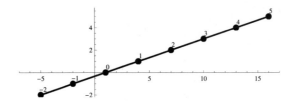

19. We have

t	-2	-1	0	1	2	3
$x = 1 - 2t$	5	3	1	-1	-3	-5
$y = 5 - 3t$	11	8	5	2	-1	-4

so a graph connecting these points is

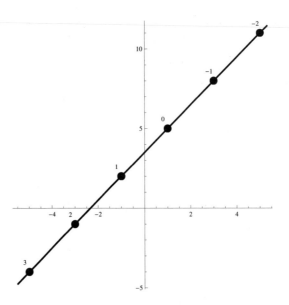

21. We have

t	-2.5	-2	-1	0	1	2	3
$x = t^3 - t$	-13.125	-6	0	0	0	6	24
$y = t^3 + t$	-18.125	-10	-2	0	2	10	30

so a graph connecting these points is

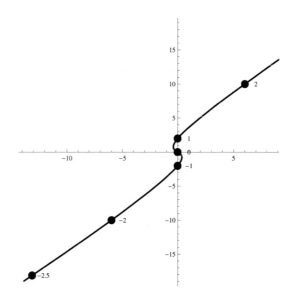

23. We have

t	0	$\frac{\pi}{6}$	$\frac{\pi}{2}$	$\frac{2\pi}{3}$	π	$\frac{5\pi}{4}$	$\frac{3\pi}{2}$	$\frac{7\pi}{4}$
$x = \cos^5 t$	1	$\frac{9\sqrt{3}}{32}$	0	$-\frac{1}{32}$	-1	$-\frac{\sqrt{2}}{8}$	0	$\frac{\sqrt{2}}{8}$
$y = \sin^5 t$	0	$\frac{1}{32}$	1	$\frac{9\sqrt{3}}{32}$	0	$-\frac{\sqrt{2}}{8}$	-1	$-\frac{\sqrt{2}}{8}$

so a graph connecting these points is

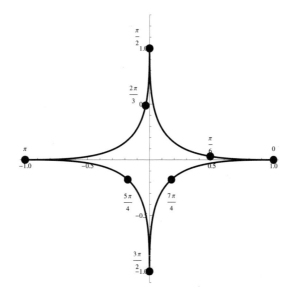

25. Solve $x = 2t - 1$ for t to get $t = \frac{x+1}{2}$; substitute into

$$y = 3t^2 + 5 = 3\left(\frac{x+1}{2}\right)^2 + 5 = 3\left(\frac{1}{4}x^2 + \frac{1}{2}x + \frac{1}{4}\right) + 5 = \frac{3}{4}x^2 + \frac{3}{2}x + \frac{23}{4}.$$

A plot is

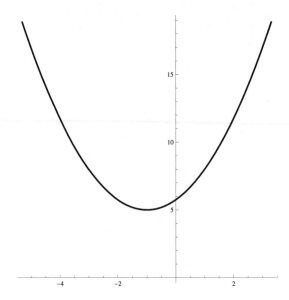

27. Note that $x = y$. The range of $\tan t$ for $t \in \left(-\frac{\pi}{2}, \frac{\pi}{2}\right)$ is $(-\infty, \infty)$, so this is $y = x$ on all of $\mathbb{R}$. A plot is

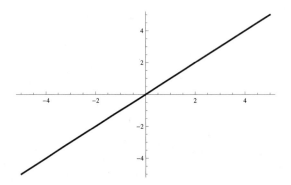

29. Use the identity $\cos^2 t + \sin^2 t = 1$. We have $\frac{x}{3} = \cos t$ and $\frac{y}{4} = \sin t$, so that the equation is

$$\frac{x^2}{9} + \frac{y^2}{16} = 1.$$

This is an ellipse with semimajor axis 4 and semiminor axis 3, oriented vertically:

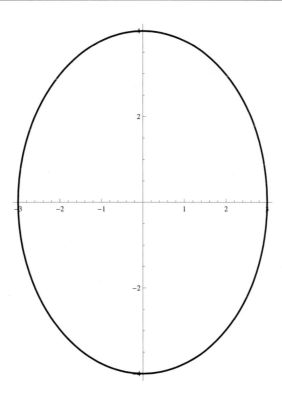

31. Since $\cosh^2 t - \sinh^2 t = 1$, we get for the equation $x^2 - y^2 = 1$. This is a hyperbola whose graph is

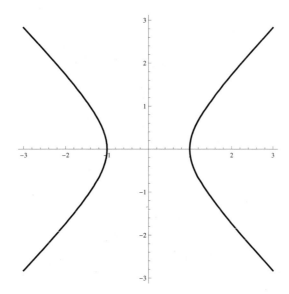

33. Since $1 + \cot^2 t = \csc^2 t$, we get $1 + y^2 = x^2$, so that $x^2 - y^2 = 1$. As $t \in (0, \pi)$, $x = \csc t$ ranges over $[1, \infty)$. This is the same function as in the previous exercise, the right branch of the hyperbola from Exercise 31:

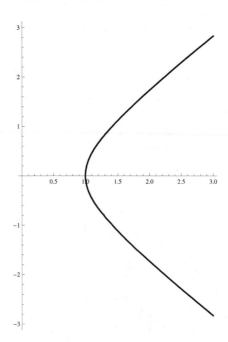

35. In Exercise 1(a), eliminating t gave the equation $x^2 + y^2 = 1$. Now, for $t \in [\pi, 2\pi]$, we see that $y = \sin t \leq 0$, while $x = \cos t$ ranges from -1 to 1. Thus this must be the lower half-circle. Since $x = \cos t$ increases monotonically from -1 to 1 as t goes from π to 2π, the semicircle is traversed counterclockwise.

37. From Exercise 36, we see that $x = 3\sin t$, $y = 3\cos t$ is a circle of radius 3 starting at $(0,3)$ and traversed *clockwise* for $t \in [0, 2\pi]$. To turn this into a counterclockwise parametrization, we must replace x by its negative everywhere (reflecting across the y axis). We must also use $2\pi t$ instead of t to change the range of the parameter to $[0, 1]$. Thus the required parametrization is $x = -3\sin 2\pi t$, $y = 3\cos 2\pi t$.

39. A line through $(0, 1)$ with slope t has equation $y = tx + 1$. Each such line, for $t \in \mathbb{R}$, intersects the unit circle in exactly one point, but $(0, -1)$ is not covered by these lines since the line through $(0, 1)$ and $(0, -1)$ has infinite slope. We can use this information to derive an explicit parametrization. To find the point of intersection of $y = tx + 1$ and the unit circle $x^2 + y^2 = 1$, substitute $tx + 1$ for y to get

$$x^2 + (tx + 1)^2 = (1 + t^2)x^2 + 2tx + 1 = 1.$$

Thus $x = -\frac{2t}{1+t^2}$. Since $y = tx + 1$, we get

$$y = tx + 1 = t \cdot \frac{-2t}{t^2 + 1} + 1 = \frac{1 - t^2}{1 + t^2}.$$

41. The slope of the tangent line is

$$m = \frac{y'(-1)}{x'(-1)} = \frac{3}{2},$$

so that the equation of the tangent line at $t = -1$ is $y - y(-1) = \frac{3}{2}(x - x(-1))$, or $y - 2 = \frac{3}{2}(x - (-3))$. This simplifies to $y = \frac{3}{2}x + \frac{13}{2}$.

43. The slope of the tangent line is

$$m = \frac{y'\left(\frac{1}{2}\right)}{x'\left(\frac{1}{2}\right)} = \frac{-2(2 - t)|_{t=1/2}}{2t|_{t=1/2}} = -3,$$

so that the equation of the tangent line at $t = \frac{1}{2}$ is $y - y\left(\frac{1}{2}\right) = -3\left(x - x\left(\frac{1}{2}\right)\right)$, or

$$y - \left(2 - \frac{1}{2}\right)^2 = -3\left(x - \left(\frac{1}{2}\right)^2\right),$$ which simplifies to $y = -3x + 3$.

45. Since $\frac{d^2y}{dt^2} = \frac{d^2x}{dt^2} = 0$ and $\frac{dx}{dt} = 2 \neq 0$, we have

$$\frac{d^2y}{dx^2} = \frac{\frac{dx}{dt}\frac{d^2y}{dt^2} - \frac{d^2x}{dt^2}\frac{dy}{dt}}{\left(\frac{dx}{dt}\right)^3} = 0.$$

47. Since $\frac{d^2x}{dt^2} = 2 = \frac{d^2y}{dt^2}$, we have

$$\frac{d^2y}{dx^2} = \frac{\frac{dx}{dt}\frac{d^2y}{dt^2} - \frac{d^2x}{dt^2}\frac{dy}{dt}}{\left(\frac{dx}{dt}\right)^3} = \frac{(2t)(2) - 2(-2(2-t))}{8t^3} = \frac{1}{t^3}.$$

Thus at $t = \frac{1}{2}$ we get $\frac{d^2y}{dx^2} = \frac{1}{(1/2)^3} = 8$.

49. On $t \in [0,1]$, the curve is

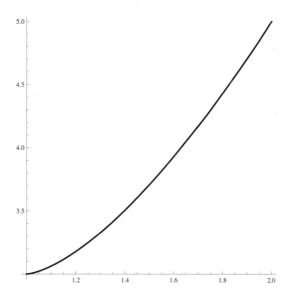

By Theorem 9.4, the arc length is given by

$$\int_0^1 \sqrt{x'(t)^2 + y'(t)^2}\, dt = \int_0^1 \sqrt{(2t)^2 + (6t^2)^2}\, dt$$
$$= \int_0^1 \sqrt{4t^2 + 36t^4}\, dt$$
$$= 2\int_0^1 t\sqrt{1 + 9t^2}\, dt$$
$$= \frac{2}{27}\left[(1 + 9t^2)^{3/2}\right]_0^1$$
$$= \frac{2}{27}(10^{3/2} - 1) = \frac{2}{27}(10\sqrt{10} - 1).$$

51. For $\theta \in [0, 2\pi]$, the curve is

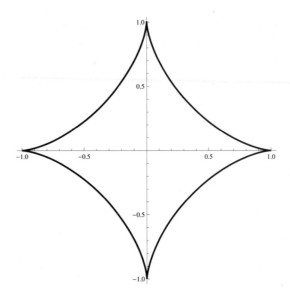

By Theorem 9.4, the arc length is given by

$$\int_0^{2\pi} \sqrt{x'(t)^2 + y'(t)^2} \, dt = \int_0^{2\pi} \sqrt{(3\sin^2 t \cos t)^2 + (-3\cos^2 t \sin t)^2} \, dt$$

$$= \int_0^{2\pi} \sqrt{9\sin^4 t \cos^2 t + 9\cos^4 t \sin^2 t} \, dt$$

$$= \int_0^{2\pi} \sqrt{9\sin^2 t \cos^2 t (\sin^2 t + \cos^2 t)} \, dt$$

$$= 3\int_0^{2\pi} \sqrt{\sin^2 t \cos^2 t} \, dt.$$

Now we must be careful, since $\sin t$ and $\cos t$ are both positive and negative on this range, so compute the integral from 0 to $\frac{\pi}{2}$ using the positive square roots and multiply by 4:

$$3\int_0^{2\pi} \sqrt{\sin^2 t \cos^2 t} \, dt = 12\int_0^{\pi/2} \sin t \cos t \, dt = 6\int_0^{\pi/2} \sin 2t \, dt = 6\left[-\frac{1}{2}\cos 2t\right]_0^{\pi/2} = 6.$$

53. For $t \in [0, 1]$, the curve is

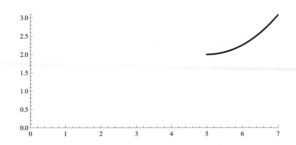

By Theorem 9.4, the arc length is given by

$$\int_0^1 \sqrt{x'(t)^2 + y'(t)^2}\, dt = \int_0^1 \sqrt{2^2 + (e^t - e^{-t})^2}\, dt$$

$$= \int_0^1 \sqrt{e^{2t} + 2 + e^{-t}}\, dt$$

$$= \int_0^1 \sqrt{(e^t + e^{-t})^2}\, dt$$

$$= \int_0^1 (e^t + e^{-t})\, dt$$

$$= \left[e^t - e^{-t}\right]_0^1 = e - \frac{1}{e}.$$

55. Using Exercise 54, parametric equations are

$$x = 1 + (6 - 1)t,\ y = -3 + (7 - (-3))t, \quad \text{or } x = 5t + 1, y = 10t - 3,\ t \in [0,1].$$

57. Using Exercise 54, parametric equations are

$$x = 1 + (-3 - 1)t,\ y = 4 + (5 - 4)t, \quad \text{or } x = -4t + 1, y = t + 4,\ t \in [0,1].$$

59. Using Exercise 54, parametric equations are

$$x = \pi + (\pi - \pi)t,\ y = 3 + (8 - 3)t, \quad \text{or } x = \pi, y = 5t + 3,\ t \in [0,1].$$

61. Using the parametrization from Exercise 55, we get for the arc length

$$\int_0^1 \sqrt{x'(t)^2 + y'(t)^2}\, dt = \int_0^1 \sqrt{5^2 + 10^2}\, dt = \int_0^1 \sqrt{125}\, dt = 5\sqrt{5}\, [t]_0^1 = 5\sqrt{5}.$$

Using the distance formula we get

$$\sqrt{(7 - (-3))^2 + (6 - 1)^2} = \sqrt{10^2 + 5^2} = \sqrt{125} = 5\sqrt{5}.$$

The two answers are the same.

63. Consider the following diagram of the situation. At the start, the rolling circle lies along the x axis (the lower of the two small circles), and it rolls counterclockwise.

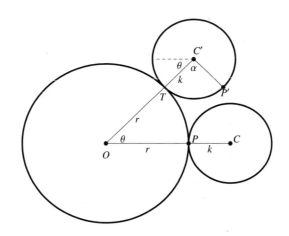

The coordinates of P' relative to C' are $(-k\cos(\theta + \alpha), -k\sin(\theta + \alpha))$, so that the coordinates of P' relative to O are

$$P' = ((r + k)\cos\theta, (r + k)\sin\theta) + (-k\cos(\theta + \alpha), -k\sin(\theta + \alpha))$$
$$= ((r + k)\cos\theta - k\cos(\theta + \alpha), (r + k)\sin\theta - k\sin(\theta + \alpha)).$$

We now need to eliminate α, in order to get a formula involving only r, k, and θ. However, note that when the point of tangency is T, the arcs TP' and TP are equal. Since the radius of the larger circle is r, we know that $TP = r\theta$, so that $TP' = r\theta$. But also $TP' = k\alpha$, giving $\alpha = \frac{r}{k}\theta$. Substituting gives

$$P' = \left((r + k)\cos\theta - k\cos\left(\theta + \frac{r}{k}\theta\right), (r + k)\sin\theta - k\sin\left(\theta + \frac{r}{k}\theta\right)\right)$$
$$= \left((r + k)\cos\theta - k\cos\left(\frac{r + k}{k}\theta\right), (r + k)\sin\theta - k\sin\left(\frac{r + k}{k}\theta\right)\right).$$

Applications

65. (a) Using Theorem 9.4, we have $f(\theta) = \sin\theta$ and $g(\theta) = 3\cos\theta$, so that $f'(\theta)^2 = \cos^2\theta$ and $g'(\theta)^2 = 9\sin^2\theta$. Thus the length of the track is

$$\int_0^{2\pi} \sqrt{f'(\theta)^2 + g'(\theta)^2}\, d\theta = \int_0^{2\pi} \sqrt{\cos^2\theta + 9\sin^2\theta}\, d\theta = \int_0^{2\pi} \sqrt{1 + 8\sin^2\theta}\, d\theta.$$

(b) Using 20 subintervals, we have $\Delta\theta = \frac{2\pi}{20} = \frac{\pi}{10}$ and $\theta_k = \frac{k\pi}{10}$ for $k = 0, 1, 2, \ldots, 20$. Then the midpoints are $\theta_k^* = \frac{(2k-1)\pi}{20}$ for $k = 1, 2, \ldots, 20$, and the midpoint sum is

$$\sum_{k=1}^{20} \sqrt{1 + 8\sin^2\frac{(2k - 1)\pi}{20}}\, \frac{\pi}{10} = \frac{\pi}{10}\sum_{k=1}^{20} \sqrt{1 + 8\sin^2\frac{(2k - 1)\pi}{20}}.$$

Evaluating numerically gives an approximate value of 13.3651.

67. (a) Integrating gives $x(t) = 2t\cos\frac{\pi}{12} + C$; applying the initial condition $x(0) = 0$ gives $C = 0$ so that $x(t) = 2t\cos\frac{\pi}{12}$. Then

$$y'(t) = 0.444x(t)(x(t) - 3) + 2\sin\frac{\pi}{12}$$
$$= 0.444 \cdot 2t\cos\frac{\pi}{12}\left(2t\cos\frac{\pi}{12} - 3\right) + 2\sin\frac{\pi}{12}$$
$$= 1.776t^2\cos^2\frac{\pi}{12} - 2.664t\cos\frac{\pi}{12} + 2\sin\frac{\pi}{12}.$$

Integrating gives (again with a zero constant of integration due to the initial condition $y(0) = 0$)

$$y(t) = 0.592t^3\cos^2\frac{\pi}{12} - 1.332t^2\cos\frac{\pi}{12} + 2t\sin\frac{\pi}{12}.$$

(b) Since the channel is 3 miles wide, she makes landfall when $2t\cos\frac{\pi}{12} = 3$, or $t = \frac{3}{2\cos\frac{\pi}{12}}$. At that time, her north-south position is

$$y\left(\frac{3}{2\cos\frac{\pi}{12}}\right) = 0.592\left(\frac{3}{2\cos\frac{\pi}{12}}\right)^3\cos^2\frac{\pi}{12} - 1.332\left(\frac{3}{2\cos\frac{\pi}{12}}\right)^2\cos\frac{\pi}{12} + 2\left(\frac{3}{2\cos\frac{\pi}{12}}\right)\sin\frac{\pi}{12}$$
$$\approx -0.2304 \text{ miles},$$

so about 0.23 miles south of her starting point.

(c) She paddles for $\frac{3}{2\cos\frac{\pi}{12}} \approx 1.553$ hours. The distance she travels is approximately

$$\int_0^{1.553} \sqrt{x'(t)^2 + y'(t)^2}\, dt \approx 3.044 \text{ miles}.$$

Proofs

69. Using the parametrization $x = t$, $y = f(t)$, the arc length from $x = a$ to $x = b$ is the arc length from $t = a$ to $t = b$, which is

$$\int_a^b \sqrt{x'(t)^2 + y'(t)^2}\, dt = \int_a^b \sqrt{(t')^2 + (f'(t))^2}\, dt = \int_a^b \sqrt{1 + f'(t)^2}\, dt = \int_a^b \sqrt{1 + f'(x)^2}\, dx.$$

71. The arc length of the curve parametrized by $x = kf(t)$ and $y = kg(t)$ from $t = a$ to $t = b$ is

$$\int_a^b \sqrt{((kf(t))')^2 + ((kg(t))')^2}\, dt = \int_a^b \sqrt{k^2 f'(t) + k^2 g'(t)}\, dt = k \int_a^b \sqrt{f'(t)^2 + g'(t)^2}\, dt,$$

since $k > 0$, and this is just k times the arc length of the curve parametrized by $x = f(t)$ and $y = g(t)$ from $t = a$ to $t = b$.

The arc length of the curve parametrized by $x = f(kt)$ and $y = g(kt)$ from $t = \frac{a}{k}$ to $t = \frac{b}{k}$ is

$$\int_{a/k}^{b/k} \sqrt{((f(kt))')^2 + ((g(kt))')^2}\, dt = \int_{a/k}^{b/k} \sqrt{k^2 f'(kt) + k^2 g'(kt)}\, dt = \int_{a/k}^{b/k} k\sqrt{f'(kt) + g'(kt)}\, dt.$$

Now use the change of variables $u = kt$, so that $du = k\, dt$. The new bounds of integration are from $t = a/k$, or $u = a$, to $t = b/k$ or $u = b$. Thus the arc length is

$$\int_a^b \sqrt{f'(u)^2 + g'(u)^2}\, du,$$

which is the arc length of the curve parametrized by $x = f(t)$ and $y = g(t)$ from $t = a$ to $t = b$.

73. We have

$$\frac{dy}{dx} = \frac{\frac{dy}{d\theta}}{\frac{dx}{d\theta}} = \frac{r \sin \theta}{-r \cos \theta} = -\cot \theta.$$

Then the slope of the tangent at any even multiple of π is $-\lim_{\theta \to 2k\pi} \cot \theta = -\infty$, so this limit does not exist.

Thinking Forward

Parametric equations in three dimensions

▷ Since for every value of t we have $z = 2x$ and $y = -3x$, this is a line through the origin pointing in the direction $(1, -3, 2)$.

▷ As t increases, x and y go around the unit circle, but the curve is rising in the z coordinate as well, so this looks like a coiled spring with radius 1.

▷ As t increases, x and y go around the origin, but form an expanding helix. In addition, the curve is rising. So this curve looks something like what you would get if you wound a piece of string around an inverted cone.

9.2   Polar Coordinates

Thinking Back

Plotting in a rectangular coordinate system

▷ The plot looks like

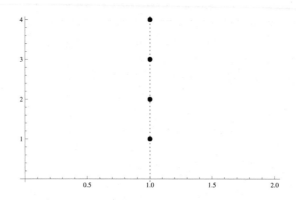

These points all lie on the same vertical line.

▷ The plot looks like

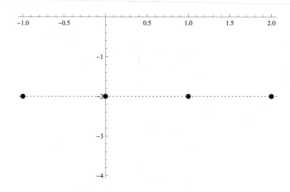

These points all lie on the same horizontal line.

▷ The plot looks like

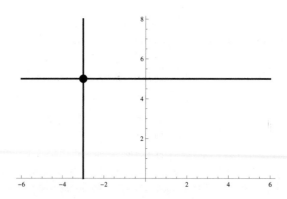

The two lines intersect at the point $(-3, 5)$.

Completing the square

▷ Subtract $4x$ from both sides, and add 4 to both sides, to get $x^2 - 4x + 4 + y^2 = 4$. Now $x^2 - 4x + 4 = (x-2)^2$, so the equation becomes $(x-2)^2 + y^2 = 4$.

▷ Regroup to get $(x^2 + 6x) + (y^2 - 8y) = 44$. Then add 9 to the part involving x and 16 to the part involving y. To keep the equation balanced, add $9 + 16 = 25$ to the right-hand side. This gives $(x^2 + 6x + 9) + (y^2 - 8y + 16) = 44 + 25 = 69$, so $(x+3)^2 + (y-4)^2 = 69$. This is a circle with center $(-3, 4)$ and radius $\sqrt{69}$.

Concepts

1. (a) True. The x and y coordinates are determined by its position in the plane.

   (b) False. For example, $(r, \theta) = (r, 2\pi + \theta)$.

   (c) False. If $b = -c$, they are the same.

   (d) False. For example, if $\beta = \alpha + 2\pi$, the two graphs are identical — they are both lines through the origin at angle α.

   (e) True. Since $r = \csc \theta$, we have $r \sin \theta = 1$. But $r \sin \theta = y$, so this is just the graph of $y = 1$, and since $\theta \in \left(-\frac{\pi}{2}, \frac{\pi}{2}\right)$ means that $\csc \theta \in (-\infty, \infty)$, we get the entire line $y = 1$.

   (f) True. By Theorem 9.5(b), (r, θ) and $(-r, \theta + \pi)$ represent the same point, so if this is also the point $(r, \theta + \pi)$, we must have $r = -r$ so that $r = 0$.

   (g) True. Multiply through by r, giving $Ar \sin \theta + Br \cos \theta = r^2$, so that $Ay + Bx = x^2 + y^2$. Collect terms and complete the square to get

   $$\left(x - \frac{B}{2}\right)^2 + \left(y - \frac{A}{2}\right)^2 = \frac{A^2 + B^2}{4},$$

   which is the equation of a circle.

   (h) True. Using Theorems 9.6 and 9.7, we can always translate polar to rectangular coordinates. However, the form of the expression in rectangular coordinates may not be particularly simple.

3. A point in the polar plane has a radius and an angle. However, the angle is not unique, since you can add any multiple of 2π to it and get the same angle with respect to the polar axis. Similarly, the radius is not unique, since $(r, \theta) = (-r, \theta + \pi)$.

5. Consider the diagram

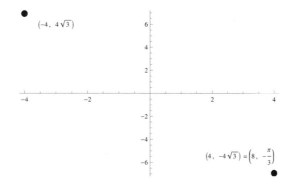

Even though with $(r, \theta) = \left(8, -\frac{\pi}{3}\right)$ and $(x, y) = (-4, 4\sqrt{3})$ we have

$$x^2 + y^2 = 16 + 48 = 64 = r^2 \qquad \text{and} \qquad \tan\left(-\frac{\pi}{3}\right) = -\sqrt{3} = \frac{y}{x},$$

the two points are not the same because they are reflections of each other in the origin. The reason for this is that $\frac{y}{x} = \tan\theta$ has two solutions for θ, which differ by π. In order to decide which one to choose, you must look at the quadrant of the point to determine the proper signs for x and y. In this case, since the point is in the fourth quadrant, we want $\frac{y}{x} = -\sqrt{3}$ with $x > 0$ and $y < 0$, so the correct point is $(4, -4\sqrt{3})$.

7. These two graphs are the same for any value of c. If the point (c, θ) is on the graph of $r = c$, then since $(c, \theta) = (-c, \pi + \theta)$, we see that (c, θ) is also on the graph of $r = -c$. Similarly, if the point $(-c, \theta)$ is on the graph of $r = -c$, then since $(-c, \theta) = (c, \pi + \theta)$, we see that $(-c, \theta)$ is also on the graph of $r = c$. Thus the two graphs are identical. This argument holds equally well if $c = 0$.

9. We know from the previous exercise that $\theta = \alpha$ and $\theta = -\alpha$ are the same for $\alpha = \frac{\pi}{2}$. So assume that $\alpha \neq \frac{\pi}{2}$, and assume $\alpha \in [0, 2\pi]$. Then $\tan\alpha = \frac{y}{x}$, and $\tan(-\alpha) = -\tan\alpha = \frac{y}{x}$. Thus $\tan\alpha = 0$, so that $\alpha = 0$ or $\alpha = \pi$. So the only possible values for α are $0 + 2n\pi$, $\pi + 2n\pi$, and $\frac{\pi}{2} + 2n\pi$ (or, more simply, $\alpha = n\frac{\pi}{2}$), where n is an integer.

11. A point (r, θ) with $r > 0$ and $0 < \theta < \frac{\pi}{2}$ is a positive distance from the origin along a first-quadrant angle, so it lies in the first quadrant. One possible description of third-quadrant points is $r > 0$ and $\pi < \theta < \frac{3\pi}{2}$; another is $r < 0$ and $0 < \theta < \frac{\pi}{2}$.

13. We must have $r = 1$ or $r = -1$. If $r = 1$, then the angle is 0 plus a multiple of 2π; if $r = -1$, then the angle is π plus a multiple of 2π. Thus the possible representations are $(1, 2n\pi)$ and $(-1, \pi + 2n\pi)$ where n is an integer.

15. The point $(-1, 0)$ in polar coordinates is the same as $(1, \pi)$ in polar coordinates. Other representations for this point add multiples of 2π to either of these, so we get $(-1, 2n\pi)$ and $(1, \pi + 2n\pi)$.

Skills

17. We have

$$\left(3, \frac{\pi}{6}\right): \quad \left(3\cos\frac{\pi}{6}, 3\sin\frac{\pi}{6}\right) = \left(\frac{3\sqrt{3}}{2}, \frac{3}{2}\right)$$

$$\left(-3, \frac{\pi}{6}\right): \quad \left(-3\cos\frac{\pi}{6}, -3\sin\frac{\pi}{6}\right) = \left(-\frac{3\sqrt{3}}{2}, -\frac{3}{2}\right)$$

$$\left(3, -\frac{\pi}{6}\right): \quad \left(3\cos\left(-\frac{\pi}{6}\right), 3\sin\left(-\frac{\pi}{6}\right)\right) = \left(\frac{3\sqrt{3}}{2}, -\frac{3}{2}\right)$$

$$\left(-3, -\frac{\pi}{6}\right): \quad \left(-3\cos\left(-\frac{\pi}{6}\right), -3\sin\left(-\frac{\pi}{6}\right)\right) = \left(-\frac{3\sqrt{3}}{2}, \frac{3}{2}\right).$$

A plot of all four points is

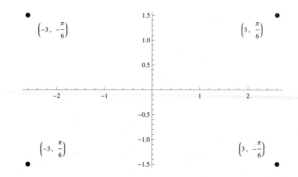

19. We have

$$\left(5, \frac{\pi}{2}\right): \qquad \left(5\cos\frac{\pi}{2}, 5\sin\frac{\pi}{2}\right) = (0, 5)$$

$$\left(-5, \frac{-3\pi}{2}\right): \qquad \left(-5\cos\frac{-3\pi}{2}, -5\sin\frac{-3\pi}{2}\right) = \left(-5\cos\frac{\pi}{2}, -5\sin\frac{\pi}{2}\right) = (0, -5)$$

$$\left(5, \frac{5\pi}{2}\right): \qquad \left(5\cos\frac{5\pi}{2}, 5\sin\frac{5\pi}{2}\right) = \left(5\cos\frac{\pi}{2}, 5\sin\frac{\pi}{2}\right) = (0, 5)$$

$$\left(-5, \frac{-\pi}{2}\right): \qquad \left(-5\cos\frac{-\pi}{2}, -5\sin\frac{-\pi}{2}\right) = (0, 5).$$

A plot of all four points is

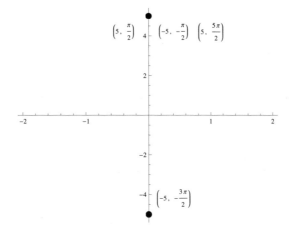

21. The point $\left(r, \frac{\pi}{4}\right)$ has rectangular coordinates $\left(r\cos\frac{\pi}{4}, r\sin\frac{\pi}{4}\right) = \left(r\frac{\sqrt{2}}{2}, r\frac{\sqrt{2}}{2}\right)$. Thus the rectangular coordinates of the four given points are

$$\left(\frac{\sqrt{2}}{2}, \frac{\sqrt{2}}{2}\right), \quad \left(\sqrt{2}, \sqrt{2}\right), \quad \left(\frac{3\sqrt{2}}{2}, \frac{3\sqrt{2}}{2}\right), \quad \left(2\sqrt{2}, 2\sqrt{2}\right).$$

A plot of all four points is

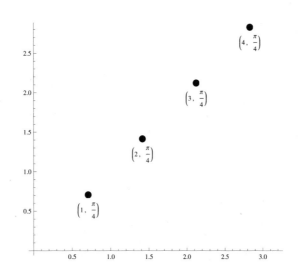

23. The point $\left(r, -\frac{\pi}{2}\right)$ has rectangular coordinates $\left(r\cos\left(-\frac{\pi}{2}\right), r\sin\left(-\frac{\pi}{2}\right)\right) = (0, -r)$, so the rectangular coordinates of the four given points are $(0, -1)$, $(0, -2)$, $(0, -3)$, and $(0, -4)$. A plot of all four points is

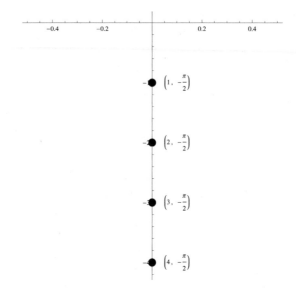

25. The angle representing this point must be either 0 plus a multiple of 2π or (by Theorem 9.5(b)) π plus a multiple of 2π; in the latter case, r will be -1. Thus the representations are $(1, 2n\pi)$ and $(-1, \pi + 2n\pi)$ where n is an integer.

27. The angle representing this point must be either $\frac{\pi}{2}$ plus a multiple of 2π or (by Theorem 9.5(b)) $-\frac{\pi}{2}$ plus a multiple of 2π; in the latter case, r will be -2. Thus the representations are $\left(2, \frac{\pi}{2} + 2n\pi\right)$ and $\left(-2, -\frac{\pi}{2} + 2n\pi\right)$ where n is an integer.

29. The angle θ representing this point is such that $\tan\theta = \frac{2\sqrt{3}}{6} = \frac{1}{\sqrt{3}}$; since x and y are both positive, we know that θ is a first quadrant angle and thus $\theta = \tan^{-1}\frac{1}{\sqrt{3}} = \frac{\pi}{6}$. We also have $r = \sqrt{6^2 + (2\sqrt{3})^2} = \sqrt{48} = 4\sqrt{3}$, so that one set of representations is $\left(4\sqrt{3}, \frac{\pi}{6} + 2n\pi\right)$ where n is an integer. By Theorem 9.5(b), the other set of representations is then $\left(-4\sqrt{3}, \frac{\pi}{6} + 2n\pi + \pi\right) = \left(-4\sqrt{3}, \frac{7\pi}{6} + 2n\pi\right)$ where n is an integer.

31. The angle θ representing this point is such that $\tan\theta = \frac{-3}{3} = -1$; since $x > 0$ and $y < 0$, we know that θ is a fourth quadrant angle, so that $\theta = \tan^{-1}(-1) = -\frac{\pi}{4}$. We also have $r = \sqrt{9+9} = 3\sqrt{2}$, so that one set of representations is $\left(3\sqrt{2}, -\frac{\pi}{4} + 2n\pi\right)$ where n is an integer. By Theorem 9.5(b), the other set of representations is then $\left(-3\sqrt{2}, -\frac{\pi}{4} + 2n\pi + \pi\right) = \left(-3\sqrt{2}, \frac{3\pi}{4} + 2n\pi\right)$ where n is an integer.

33. $\theta = \frac{\pi}{4}$ is the line through the origin at an angle of $\frac{\pi}{4}$, so for any point (x, y) on that line we have $\frac{y}{x} = \tan\frac{\pi}{4} = 1$ so that $y = x$. This is the line $y = x$.

35. $\theta = \frac{7\pi}{6}$ is the line through the origin at an angle of $\frac{7\pi}{6}$, so for any point (x, y) on that line we have $\frac{y}{x} = \tan\left(\frac{7\pi}{6}\right) = \frac{1}{\sqrt{3}}$. Thus the equation of the line is $y = \frac{1}{\sqrt{3}}x$.

37. By Theorem 9.8 with $a = \frac{5}{2}$, this is the circle $x^2 + \left(y - \frac{5}{2}\right)^2 = \frac{25}{4}$.

39. $r = 6 \csc \theta$ means that $r \sin \theta = 6$, so that $y = 6$ since $y = r \sin \theta$ in polar coordinates.

41. Since $\sin 2\theta = 2 \sin \theta \cos \theta = 2 \frac{y}{r} \frac{x}{r}$, we get $r^3 = 2xy$. But $r = \pm\sqrt{x^2 + y^2}$, so substituting gives $\pm(x^2 + y^2)^{3/2} = 2xy$.

43. Multiply both sides by r to get $r^3 = r \cos \theta$. Now use the fact that $r = \pm\sqrt{x^2 + y^2}$ and $x = r \cos \theta$ to get $\pm(x^2 + y^2)^{3/2} = x$.

45. There are several ways to do this. For example, since $r = \theta$, take the sine of both sides and multiply through by r to get $r \sin r = r \sin \theta = y$. Then $y = \sqrt{x^2 + y^2} \sin \sqrt{x^2 + y^2}$. There is no satisfying formula for this spiral in cartesian coordinates, but there is a reasonable parametrized solution given by $x = \pm\theta \cos \theta$ and $y = \pm\theta \sin \theta$.

47. Multiply through by r^3 to get $r^4 = r^3 \sin^3 \theta = (r \sin \theta)^3$, so that $(x^2 + y^2)^2 = y^3$.

49. $y = 0$ is $r \sin \theta = 0$, so we get $\sin \theta = 0$, so this is the line $\theta = 0$ (or $\theta = \pi$). (Note that we divided through by r, so we assumed $r \neq 0$. However, the origin is on the graph of the resulting equation, so that case is included in the answer).

51. $y = x$ means $r \sin \theta = r \cos \theta$, so that $\sin \theta = \cos \theta$, or $\tan \theta = 1$. Thus the equation is $\theta = \frac{\pi}{4}$ or $\theta = -\frac{\pi}{4}$. (Note that we divided through by r, so we assumed $r \neq 0$. However, the origin is on the graph of the resulting equation, so that case is included in the answer).

53. $y = -3$ means $r \sin \theta = -3$, so that $r = -3 \csc \theta$.

55. $y = mx$ gives upon substitution $r \sin \theta = rm \cos \theta$, so that $\tan \theta = m$ and then $\theta = \tan^{-1} m$.

57. In each case, substitute $r = \pm\sqrt{x^2 + y^2}$ and square the result:

$$r = 1: \quad \pm\sqrt{x^2 + y^2} = 1 \quad \Rightarrow \quad x^2 + y^2 = 1,$$

$$r = 2: \quad \pm\sqrt{x^2 + y^2} = 2 \quad \Rightarrow \quad x^2 + y^2 = 4,$$

$$r = 3: \quad \pm\sqrt{x^2 + y^2} = 3 \quad \Rightarrow \quad x^2 + y^2 = 9,$$

$$r = 4: \quad \pm\sqrt{x^2 + y^2} = 4 \quad \Rightarrow \quad x^2 + y^2 = 16,$$

$$r = 5: \quad \pm\sqrt{x^2 + y^2} = 5 \quad \Rightarrow \quad x^2 + y^2 = 25,$$

$$r = 6: \quad \pm\sqrt{x^2 + y^2} = 6 \quad \Rightarrow \quad x^2 + y^2 = 36,$$

$$r = 7: \quad \pm\sqrt{x^2 + y^2} = 7 \quad \Rightarrow \quad x^2 + y^2 = 49,$$

$$r = 8: \quad \pm\sqrt{x^2 + y^2} = 8 \quad \Rightarrow \quad x^2 + y^2 = 64,$$

$$r = 9: \quad \pm\sqrt{x^2 + y^2} = 9 \quad \Rightarrow \quad x^2 + y^2 = 81.$$

59. For most values of k, take the tangent of both sides of the equation and simplify:

$$\theta = 0: \quad \tan\theta = 0 = \frac{y}{x} \quad \Rightarrow \quad y = 0,$$

$$\theta = \frac{\pi}{6}, \frac{7\pi}{6}: \quad \tan\theta = \frac{1}{\sqrt{3}} = \frac{y}{x} \quad \Rightarrow \quad y = \frac{1}{\sqrt{3}}x,$$

$$\theta = \frac{2\pi}{6} = \frac{\pi}{3}, \frac{8\pi}{6} = \frac{4\pi}{3}: \quad \tan\theta = \sqrt{3} = \frac{y}{x} \quad \Rightarrow \quad y = x\sqrt{3},$$

$$\theta = \frac{3\pi}{6} = \frac{\pi}{2}, \frac{9\pi}{6} = \frac{3\pi}{2}: \quad \tan\theta = \infty, \text{ so this is the vertical line } x = 0,$$

$$\theta = \frac{4\pi}{6} = \frac{2\pi}{3}, \frac{10\pi}{6} = \frac{5\pi}{3}: \quad \tan\theta = -\sqrt{3} = \frac{y}{x} \quad \Rightarrow \quad y = -x\sqrt{3},$$

$$\theta = \frac{5\pi}{6}, \frac{11\pi}{6}: \quad \tan\theta = -\frac{1}{\sqrt{3}} = \frac{y}{x} \quad \Rightarrow \quad y = -\frac{1}{\sqrt{3}}x.$$

Applications

61. (a) From the diagram, assuming the arms of the cam remain symmetrically placed, each of the $\frac{3}{4}$ length arms will cover half the crack. So for each arm, we have a right triangle whose hypotenuse is $\frac{3}{4}$ and one of whose angles is θ. The leg opposite that angle is half the width of the crack, so is $\frac{1}{2}$. Then $\theta = \arcsin\frac{1/2}{3/4} = \arcsin\frac{2}{3} \approx 41.8°$.

(b) As in the previous problem, the downward force on the rope results in a rightward force inside the crack, which again causes the cams to wedge more tightly into the crack.

Proofs

63. Multiply both sides by $\sin\theta$ to get $r\sin\theta = k$. Since $r\cos\theta = y$, this graph has equation $y = k$ in rectangular coordinates, so it is a horizontal line.

65. If $a \neq 0$ and $b > 0$, multiply through by $1 - b\cos\theta$ to get $r - br\cos\theta = a$, so that $r = a + br\cos\theta$. Substituting gives $\pm\sqrt{x^2 + y^2} = a + bx$. Now square both sides to get $x^2 + y^2 = a^2 + 2abx + b^2x^2$. Rearrange to get

$$(1 - b^2)x^2 - 2abx + y^2 = a^2, \quad \text{so that} \quad x^2 - \frac{2ab}{1 - b^2}x + \frac{y^2}{1 - b^2} = \frac{a^2}{1 - b^2}.$$

Complete the square for x on the left to get

$$\left(x - \frac{ab}{1 - b^2}\right)^2 + \frac{y^2}{1 - b^2} = \frac{a^2}{1 - b^2} + \frac{a^2b^2}{(1 - b^2)^2} = \frac{a^2}{(1 - b^2)^2}.$$

The right-hand side of this equation is always positive, since $a \neq 0$ and $b \neq 1$. On the left, if $0 < b < 1$ then $1 - b^2 > 0$, so we have an equation of the form $x^2 + \frac{y^2}{c^2} = d^2$, which is an ellipse. If $b > 1$ then $1 - b^2 < 0$, so we have an equation of the form $x^2 - \frac{y^2}{c^2} = d^2$, which is a hyperbola.

67. Multiply this equation through by r to get $r^2 = kr\cos\theta$. Substituting gives $x^2 + y^2 = kx$, so that $x^2 - kx + y^2 = 0$. Complete the square to get $\left(x - \frac{k}{2}\right)^2 + y^2 = \frac{k^2}{4}$, so this is a circle with radius $\frac{k}{2}$ and center $\left(\frac{k}{2}, 0\right)$. Comparing the radius and the center, we see that the circle is tangent to the y axis at the origin.

69. Multiply $r = k \sin \theta + l \cos \theta$ through by r to get $r^2 = kr \sin \theta + lr \cos \theta$. Now substitute to get $x^2 + y^2 = ky + lx$, so that $x^2 - lx + y^2 - ky = 0$. Complete both squares to get

$$\left(x - \frac{l}{2}\right)^2 + \left(y - \frac{k}{2}\right)^2 = \frac{k^2 + l^2}{4}.$$

This is a circle of radius $\frac{\sqrt{k^2 + l^2}}{2}$ and center $\left(\frac{l}{2}, \frac{k}{2}\right)$.

Thinking Forward

Understanding symmetry

$r = f(\theta) = f(-\theta)$ means that the distance of the graph from the origin is the same for θ as it is for $-\theta$. Since $-\theta$ is θ reflected through the x axis, this equality means that the graph is symmetric around the x axis.

Understanding symmetry

If $f(\theta) = r$, then $f(-\theta) = -f(\theta) = -r$. Thus the points (r, θ) and $(-r, -\theta)$ are both on the graph. But $(-r, -\theta) = (r, \pi - \theta)$ by Theorem 9.5, so that (r, θ) and $(r, \pi - \theta)$ are both on the graph. But these two points are mirror images across the y axis, since in rectangular coordinates $(r, \pi - \theta)$ is $(r \cos(\pi - \theta), r \sin(\pi - \theta)) = (-r \cos \theta, r \sin \theta)$. Thus the graph is symmetric with respect to the y axis.

9.3   Graphing Polar Equations

Thinking Back

Finding points of intersection

Graphs of the two functions ($1 + \sin x$ is the solid graph) are:

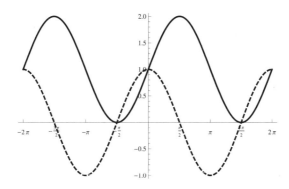

It seems that the two intersect at $2n\pi$ and at $-\frac{\pi}{2} + 2n\pi$, and in fact, equating the two gives

$$1 + \sin x = \cos x = \sqrt{1 - \sin^2 x}, \quad \text{so} \quad (1 + \sin x)^2 = 1 - \sin^2 x.$$

Expanding the left-hand side gives $1 + 2 \sin x + \sin^2 x = 1 - \sin^2 x$. Now simplify to get $\sin^2 x + \sin x = 0$, so that $\sin x = 0$ (which happens for $x = \pm 2n\pi$ and $x = \pi \pm 2n\pi$) or $\sin x = -1$ (which happens for $x = -\frac{\pi}{2} + 2n\pi$). The solution $x = \pi \pm 2n\pi$ is an extraneous solution that was introduced when we squared the equation, so we reject it.

Finding points of intersection

Graphs of the two functions ($\frac{1}{2} + \cos x$ is the solid graph) are:

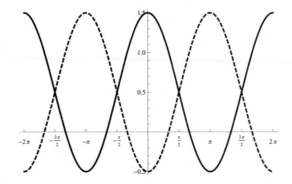

It seems that the two intersect at $\frac{\pi}{2} + n\pi$, and in fact, equating the two gives

$$\frac{1}{2} + \cos x = \frac{1}{2} - \cos x, \quad \text{so} \quad 2\cos x = 0.$$

So the solutions correspond to $\cos x = 0$, which is for $x = \frac{\pi}{2} + n\pi$.

Translations of graphs

The graph of $y = f(x - k)$ is the graph of $y = f(x)$ shifted right by k units, since when (say) $x = k$, the first graph contains the point $(k, f(0))$ while the second contains the point $(0, f(0))$.

Translations of graphs

The graph of $y = f(x) + k$ is the graph of $y = f(x)$ shifted up by k units, since for any value of x, the point $(x, f(x))$ is on the graph of $y = f(x)$, while the point $(x, f(x) + k)$ is on the graph of $y = f(x) + k$.

Concepts

1. (a) False. It is true if $r > 0$, but if $r < 0$, then this point is in the fourth quadrant.

    (b) True. Its graph is

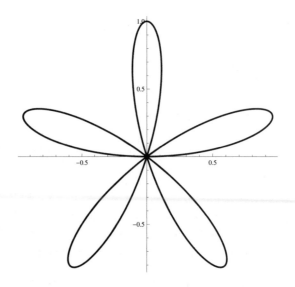

(c) False; it has twelve petals. Its graph is

(d) True. If a graph is symmetric with respect to the origin, then the points (r, θ) is on the graph if and only if the point $(-r, \theta)$ is on the graph, by Theorem 9.11(c). But $(-r, \theta) = (-r, \theta + 2\pi)$ by Theorem 9.5(a).

(e) False. By Theorem 9.11, this condition expresses symmetry around the x axis. Symmetry around the y axis is given by the condition that (r, θ) is on the graph if and only if $(-r, -\theta)$ is on the graph.

(f) True. Let G be the graph of $r = \sin k\theta$. This graph is always symmetric around the y axis, since if $(r, \theta) \in G$, then $\sin(-k\theta) = -\sin k\theta = -r$, so that $(-r, -\theta) \in G$. For x axis symmetry, note that

$$(r, -\theta) \in G \iff (-r, \pi - \theta) \in G \iff -r = \sin(k(\pi - \theta)) \iff r = -\sin(k\pi - k\theta)$$
$$\iff r = -\sin k\pi \cos k\theta + \cos k\pi \sin k\theta = \cos k\pi \sin k\theta.$$

Now, if $(r, \theta) \in G$, then $r = \sin \theta$, so that $(r, -\theta) \in G$ if and only if $\cos k\pi = 1$, i.e., if k is even. The graph G of $r = \cos k\theta$ is always symmetric around the x axis, since if $(r, \theta) \in G$, then $\cos(-k\theta) = \cos k\theta = r$, so that $(r, -\theta) \in G$. For y axis symmetry, note that

$$(-r, -\theta) \in G \iff (r, \pi - \theta) \in G \iff r = \cos(k(\pi - \theta))$$
$$\iff r = \cos k\pi \cos k\theta + \sin k\pi \sin k\theta = \cos k\pi \cos k\theta.$$

Now, if $(r, \theta) \in G$, then $r = \cos \theta$, so that $(-r, -\theta) \in G$ if and only if $\cos k\pi = 1$, i.e., if k is even.

(g) True. Converting to polar coordinates gives $r^4 = k(r^2 \cos^2 \theta - r^2 \sin^2 \theta)$, or $r^2 = k(\cos^2 \theta - \sin^2 \theta) = k \cos 2\theta$. This is a lemniscate.

(h) True. If $y = f(x)$ is symmetric with respect to the y axis, then $f(x) = f(-x)$ for all x. If it is symmetric with respect to the origin, then $f(x) = -f(-x)$ for all x, so that $f(-x) = -f(x)$. It follows that for all x, $f(x) = f(-x)$, so that $f(x) = 0$ for all x and $y = 0$.

3. See the discussion in the text for an extended explanation. Basically, the areas above and below the θ axis for each multiple of $\frac{\pi}{2}$ correspond to quadrants in which the graph will lie for those values of θ.

5. (a) The pole in the polar plane corresponds to $r = 0$, and θ is arbitrary. This is the θ axis in the $r\theta$ plane.

(b) The horizontal axis in the polar plane is the line where θ is a multiple of π. This corresponds in the $r\theta$ plane to the vertical lines $\theta = k\pi$ for k an integer.

(c) The vertical axis in the polar plane is the line where θ is an odd multiple of $\frac{\pi}{2}$, so this corresponds in the $r\theta$ plane to the vertical lines $\theta = \frac{k\pi}{2}$ for k an odd integer.

7. (a) By Theorem 9.11(a), the point $\left(2, -\frac{\pi}{3}\right)$ is also on the graph.

   (b) By Theorem 9.11(b), the point $\left(-2, -\frac{\pi}{3}\right)$ — or $\left(2, \frac{2\pi}{3}\right)$ — is also on the graph.

   (c) Symmetry with respect to the y axis means that the function is even, so that $(-3, 5)$ is also on the graph.

   (d) By Theorem 9.11(c), the point $\left(-2, \frac{\pi}{3}\right)$ — or $\left(2, \frac{4\pi}{3}\right)$ — is also on the graph.

   (e) Symmetry with respect to the origin means that the function is odd, so that $(-3, -5)$ is also on the graph.

9. (a) For example, $r = \cos\theta$, since $\cos(-\theta) = \cos\theta$, so that (r, θ) is on the graph if and only if $(r, -\theta)$ is on the graph.

   (b) For example, $r = \sin\theta$, since $\sin(-\theta) = -\sin\theta$, so that (r, θ) is on the graph if and only if $(-r, -\theta)$ is on the graph.

   (c) For example, $r = 1$, since for any point (r, θ), the point $(r, \theta + \pi) = (-r, \theta)$ is also on the graph, since the graph contains all points whose radial coordinate is r.

   (d) For example, the lemniscate $r^2 = \sin 2\theta$. Since if (r, θ) is on the curve, then $r^2 = \sin 2\theta$, so that also $(-r)^2 = \sin 2\theta$ and thus $(-r, \theta)$ is on the graph. Thus the graph is symmetric around the origin. However, since $\sin(-2\theta) = -2\sin 2\theta$, the curve is not symmetric around the x axis.

   (e) For example, $r = 2$; this is a circle centered at the origin, so it has symmetry around the x axis, the y axis, and the origin.

   (f) Not possible. By Theorem 9.10, and Exercise 56, any curve that has two of these three symmetries also has the third.

11. The cardioid is named for the heart; the limaçon for a snail, and the lemniscate for a ribbon.

13. The graph G of $r = \cos 5\theta$ is always symmetric around the x axis, since if $(r, \theta) \in G$, then $\cos(-5\theta) = \cos 5\theta = r$, so that $(r, -\theta) \in G$. However, let $\theta = 0$. Then $(\cos 5 \cdot 0, 0) = (1, 0) \in G$, but $(-1, 0) \notin G$. Thus G is not symmetric with respect to the y axis. By the table in the text, this rose has 5 petals. (It is also rotationally symmetric through an angle of $\frac{2\pi}{5}$).

   Next, let G be the graph of $r = \sin 8\theta$. This graph is always symmetric around the y axis, since if $(r, \theta) \in G$, then $\sin(-8\theta) = -\sin 8\theta = -r$, so that $(-r, -\theta) \in G$. For x axis symmetry, note that

   $$(r, -\theta) \in G \iff (-r, \pi - \theta) \in G \iff -r = \sin(8(\pi - \theta)) \iff r = -\sin(8\pi - 8\theta)$$
   $$\iff r = -\sin 8\pi \cos 8\theta + \cos 8\pi \sin 8\theta = \cos 8\pi \sin\theta = \sin 8\theta.$$

   But the last equation simply expresses the fact that $(r, \theta) \in G$. Thus $(r, -\theta) \in G \iff (r, \theta) \in G$, so that the graph is symmetric with respect to the x axis as well. Since it is symmetric around both axes, it is symmetric around the origin as well. By the table in the text, this rose has 16 petals. (It is also rotationally symmetric through an angle of $\frac{2\pi}{16} = \frac{\pi}{8}$.)

15. If (r, θ) is on the graph of $r = \cos 2\theta$, then since $\cos(-2\theta) = \cos 2\theta = r$, we see that $(r, -\theta)$ is also on the graph, so that the graph is symmetric around the x (horizontal) axis.

Skills

17. A graph of the curve is:

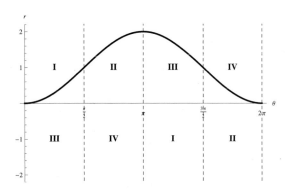

19. A graph of the curve is:

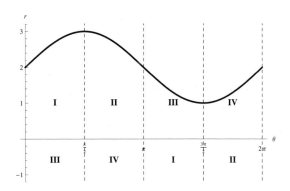

21. A graph of the curve is:

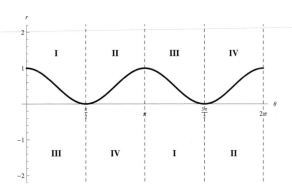

23. A graph of the curve is:

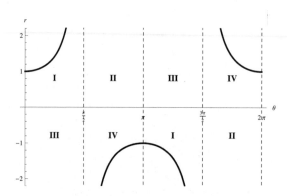

25. A graph of the curve is:

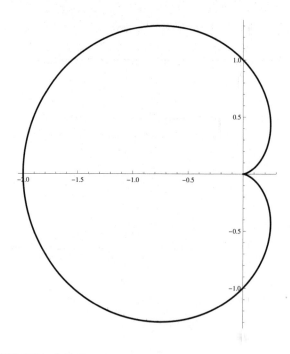

The curve moves through the quadrants in increasing order, as the $r\theta$ plot implies, and the radii are larger in the second and third quadrants, again in agreement with the $r\theta$ plot.

27. A graph of the curve is:

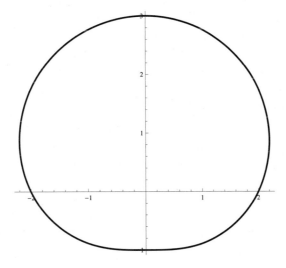

As implied by the $r\theta$ plot, the curve moves through all four quadrants in order.

29. A graph of the curve is:

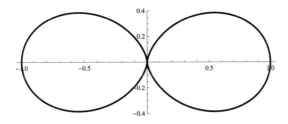

The quadrant I loop corresponds to the first arc of the $r\theta$ plot; then for the next arc, the radii are still positive, so we move into the second quadrant, and so forth. This again corresponds to the $r\theta$ plot.

31. A graph of the curve is:

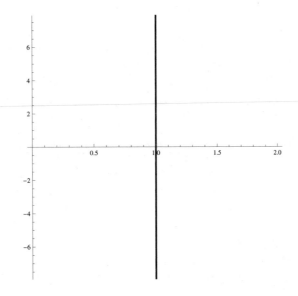

The polar graph is wholly in quadrants I and IV, which is what the $r\theta$ plot tells us. The upper half of the line is traced for $0 \leq \theta < \frac{\pi}{2}$ and the lower half is traced, starting at $-\infty$ for $\frac{\pi}{2} < \theta \leq \pi$.

33. The function first repeats for θ a multiple of 2π when $\theta = 10\pi$, so this is the period. The polar graph is

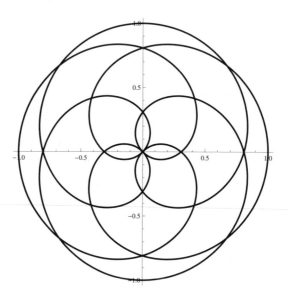

35. From the $r\theta$ plot

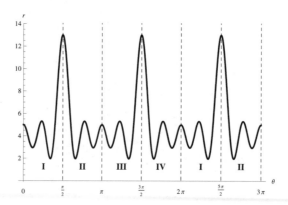

we see that the graph first repeats in quadrant I after 2π, so this is the period. A polar graph is

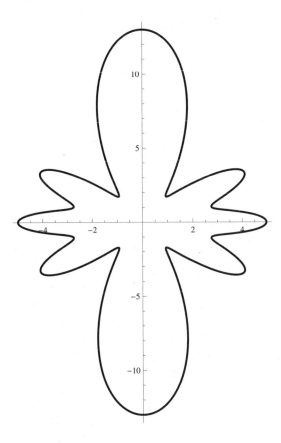

37. All three functions repeat first for θ a multiple of 2π when $\theta = 24\pi$, so this is the period. The polar graph is

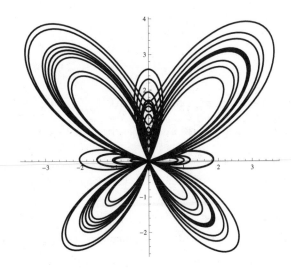

39. The period is 2π. The graph is symmetric around the y axis, if $r = \sin\theta$, then $-r = -\sin\theta = \sin(-\theta)$, so that $(-r, -\theta)$ is also on the graph. Plotting, for example, $-\frac{\pi}{2} < \theta < \frac{\pi}{2}$ and reflecting it gives

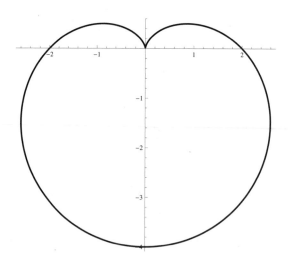

41. (a) $f_1(\theta) = f_2(\theta)$ if and only if $\sin\theta = \cos\theta$, i.e., when $\theta = \frac{\pi}{4} + n\pi$.

(b) Plots of the two curves, with $\sin\theta$ the solid line, are

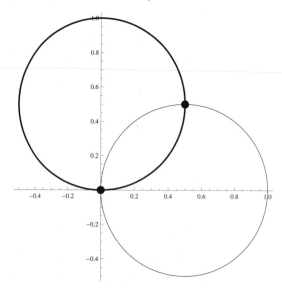

(c) The equations in rectangular coordinates are, by Theorem 9.8, $\left(x - \frac{1}{2}\right)^2 + y^2 = \frac{1}{4}$ and $x^2 + \left(y - \frac{1}{2}\right)^2 = \frac{1}{4}$. These two curves intersect when the two are equal, so that

$$\left(x - \frac{1}{2}\right)^2 + y^2 = x^2 + \left(y - \frac{1}{2}\right)^2 \Rightarrow -x = -y \Rightarrow x = y;$$

then setting y to x in the first equation and simplifying gives $2x^2 - x = 0$, so that $x = 0$ and $x = \frac{1}{2}$. Since $y = x$, the two points of intersection are $(0,0)$ and $\left(\frac{1}{2}, \frac{1}{2}\right)$, marked on the graph. In polar coordinates, these are the points $(0,0)$ and $\left(\frac{\sqrt{2}}{2}, \frac{\pi}{4}\right)$.

43. (a) $f_1(\theta) = f_2(\theta)$ if and only if $1 + \sin\theta = 1 - \cos\theta$, so if and only if $\sin\theta = -\cos\theta$. this happens for $\theta = -\frac{\pi}{4} + n\pi$ for n an integer.

(b) Plots of the two curves, with $1 + \sin\theta$ the solid line, are

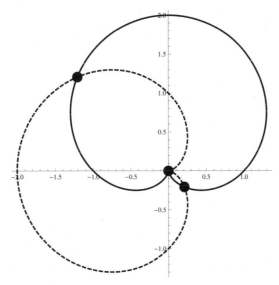

(c) Multiply the equations through by r to get $r^2 = r + r\sin\theta$ and $r^2 = r - r\cos\theta$; converting to rectangular coordinates gives

$$x^2 + y^2 = \sqrt{x^2 + y^2} + y, \quad x^2 + y^2 = \sqrt{x^2 + y^2} - x,$$

so that the two are equal for $y = -x$. Substitute $-x$ for y in the first equation to get $2y^2 = \sqrt{2y^2} + y$. Move the y to the left, square, and simplify to get $4y^4 - 4y^3 - y^2 = y^2(4y^2 - 4y - 1) = 0$. This has solutions $y = 0$ and $y = \frac{1}{2}(1 \pm \sqrt{2})$. Thus the points of intersection are $(0,0)$ and $\left(-\frac{1}{2}(1 \pm \sqrt{2}), \frac{1}{2}(1 \pm \sqrt{2})\right)$, which are marked on the graph. In polar coordinates, these are the points $(0,0)$, $\left(1 + \frac{\sqrt{2}}{2}, \frac{3\pi}{4}\right)$, and $\left(1 - \frac{\sqrt{2}}{2}, \frac{7\pi}{4}\right)$.

45. (a) $f_1(\theta) = f_2(\theta)$ if and only if $\sin^2\theta = \cos^2\theta$, so if and only if $\sin\theta = \pm\cos\theta$. This happens exactly when $\theta = \frac{\pi}{4}$ plus a multiple of $\frac{\pi}{2}$, i.e. when θ is an odd multiple of $\frac{\pi}{4}$.

(b) Plots of the two curves, with $\sin\theta$ the solid line, are

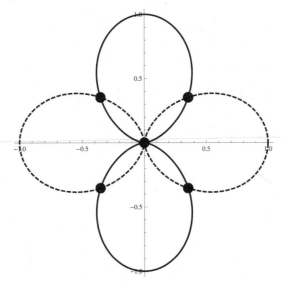

(c) To convert these equations to rectangular coordinates, multiply each through by r^2 to get $r^3 = (r\sin\theta)^2$ and $r^3 = (r\cos\theta)^2$, so that $\pm(x^2 + y^2)^{3/2} = y^2$ and $\pm(x^2 + y^2)^{3/2} = x^2$.

The points of intersection satisfy $x^2 = y^2$ (so that $y = \pm x$) Substituting x^2 for y^2 in the first equation gives $\pm(2x^2)^{3/2} = x^2$, or $\pm 2\sqrt{2}\,x^3 = x^2$. Thus $x = 0$ or $x = \pm\frac{1}{2\sqrt{2}}$. Hence the intersection points are $(0,0)$ together with the four points $\left(\pm\frac{1}{2\sqrt{2}}, \pm\frac{1}{2\sqrt{2}}\right)$, marked on the graph. In polar coordinates, these are the points $(0,0)$ and $\left(\frac{1}{2}, \frac{\pi}{4} + k\frac{\pi}{2}\right)$ for $k = 0, 1, 2, 3$.

47. The graphs are below, with $r = 1 + \sin\theta$ in black and $r = 1 - \sin\theta$ in gray.

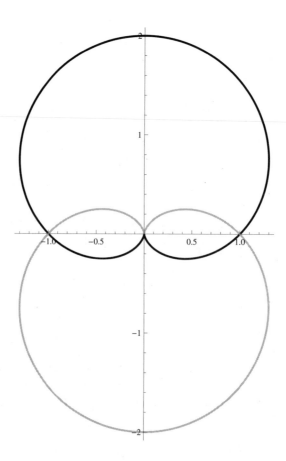

$1 - \sin\theta$ appears to be the reflection of $1 + \sin\theta$ through the x axis. To see this, note that $r = 1 + \sin\theta$ if and only if $r = 1 - \sin(-\theta)$, so that (r, θ) is on the graph of $1 + \sin\theta$ if and only if $(r, -\theta)$ is on the graph of $1 - \sin\theta$ — that is, the two are reflections of each other through the x axis (which is $\theta \leftrightarrow -\theta$).

49. Graphs of $r = 3 + b\sin\theta$ for various values of b are below:

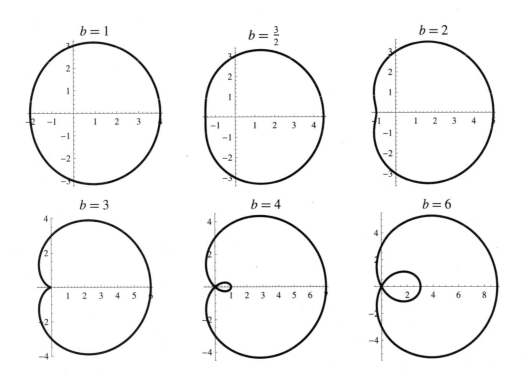

From these, it appears that for $b > 3$, the limaçon has an inner loop, for $\frac{3}{2} < b < 3$ it has a dimple, for $b = \frac{3}{2}$ the left side is flat, and for $b < \frac{3}{2}$ it is convex. For $b = 3$, of course, we get a cardioid, not a limaçon.

51. (a) Parametrize the circle by $(\cos\theta, \sin\theta)$. Then the midpoint of the chord extending from $(1,0)$ to $(\cos\theta, \sin\theta)$ is

$$\left(\frac{1 + \cos\theta}{2}, \frac{\sin\theta}{2} \right).$$

(b) In rectangular coordinates, if the coordinates of a point on the unit circle are (x, y), the midpoint of the chord between such a point and $(1,0)$ is

$$\left(\frac{x+1}{2}, \frac{y}{2} \right).$$

(c) Consider a point $B = (r, \theta)$ on the unit circle:

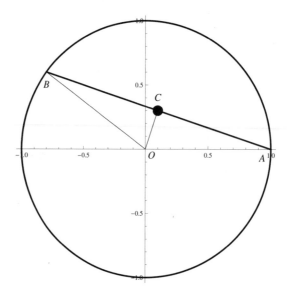

With the parametrization in part (a), the coordinates of the midpoint C are $\left(\frac{1+\cos\theta}{2}, \frac{\sin\theta}{2}\right)$, where $\theta = \angle AOB$. Then the length of OC, which is r in polar coordinates, is

$$r = \sqrt{\left(\frac{1+\cos\theta}{2}\right)^2 + \left(\frac{\sin\theta}{2}\right)^2} = \frac{1}{2}\sqrt{1 + 2\cos\theta + \cos^2\theta + \sin^2\theta} = \frac{1}{2}\sqrt{2 + 2\cos\theta}.$$

Also, since $\triangle AOB$ is an isosceles triangle and $AC = CB$ since C is the midpoint, we know that $\triangle OCB$ and $\triangle OCA$ are congruent, so that $\angle BOC = \angle AOC = \frac{1}{2}\angle AOB = \frac{\theta}{2}$. Thus the polar coordinates of point C are

$$\left(\frac{1}{2}\sqrt{2 + 2\cos\theta}, \frac{\theta}{2}\right).$$

Applications

53. When $A = 1$, this is the plot of $r = 1 - 0.5\sin\theta$:

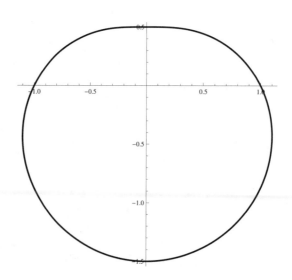

Proofs

55. Consider $r = \cos n\theta$. Note that $\cos(n(\theta + \pi)) = \cos(n\pi + n\theta)$; since n is odd, this is the same as $\cos(\pi + n\theta) = -\cos n\theta$. Thus the point on the graph associated with $\theta + \pi$ is $(-r, \theta + \pi) = (r, \theta)$. Hence every point is traced at least twice on $[0, 2\pi]$. But if $\alpha, \beta \in [0, \pi)$, with corresponding points on the graph (r_1, α) and (r_2, β), these points can be equal only if $\alpha = \beta$ plus a multiple of 2π or of π. But since α and β differ by less than π, they are equal. Thus $(r_1, \alpha) = (r_2, \beta)$ if and only if $\alpha = \beta$, so the curve is traced exactly once on $[0, \pi)$. Finally, to see that the curve has n petals, note that the tip of any petal corresponds to a point where $\cos n\theta = \pm 1$, so that θ must be a multiple of π. For $\theta \in [0, \pi)$, this happens at the points $\frac{k}{n}\pi$ for $k = 0, 1, \ldots, n - 1$. Since there are n of these points, and no two are the same since the curve is traced only once, there are exactly n petals.

If instead $r = \sin n\theta$, note that $r = \sin n\left(\theta + \frac{\pi}{2}\right)$ is the graph of $\sin n\theta$ rotated clockwise by $\frac{\pi}{2}$ (see Exercise 6), so it has the same number of petals as $\sin n\theta$. But since n is odd, we have

$$\sin n\left(\theta + \frac{\pi}{2}\right) = \sin n\theta \cos \frac{n\pi}{2} + \cos n\theta \sin \frac{n\pi}{2} = \pm \cos n\theta.$$

Thus $r = \sin n\theta$ has the same number of petals as either $r = \cos n\theta$ or $r = -\cos n\theta$, both of which have n petals by the first part of this exercise.

57. Let G be the graph of $r = \sin n\theta$. Then since n is even, we have

$$(r, -\theta) \in G \iff (-r, \pi - \theta) \in G \iff -r = \sin(n(\pi - \theta)) \iff r = -\sin(n\pi - n\theta)$$
$$\iff r = -\sin(-n\theta) = \sin n\theta \iff (r, \theta) \in G.$$

So by Theorem 9.11(a), the graph is symmetric about the x axis.

59. Let G be the graph of $r = \cos n\theta$. Then since n is even, we have

$$(-r, -\theta) \in G \iff (r, \pi - \theta) \in G \iff r = \cos(n(\pi - \theta)) = \cos(n\pi - n\theta)$$
$$\iff r = \cos n\pi \cos n\theta + \sin n\pi \sin n\theta = \cos n\theta \iff (r, \theta) \in G.$$

So by Theorem 9.11(b), G is symmetric about the y axis.

Thinking Forward

Volume of a cylinder

By Theorem 9.8, the cross-sectional circle is a circle of radius $r = 2$. The volume of a right circular cylinder is $\pi r^2 h$, so the volume of this cylinder is $\pi \cdot 2^2 \cdot 2 = 8\pi$.

Volume of a sphere

By Theorem 9.8, the cross-sectional circle is a circle of radius $r = 2$. The volume of a sphere is $\frac{4}{3}\pi r^3$, so the volume of this sphere is $\frac{4}{3}\pi \cdot 2^3 = \frac{32}{3}\pi$.

9.4   Arc Length and Area Using Polar Functions

Thinking Back

▷ Use the fact that $\cos^2 2\theta = \frac{1}{2}(1 + \cos 4\theta)$; then

$$\frac{1}{2} \int_0^{\pi/4} \cos^2 2\theta \, d\theta = \frac{1}{4} \int_0^{\pi/4} (1 + \cos 4\theta) \, d\theta = \frac{1}{4}\left[\theta + \frac{1}{4}\sin 4\theta\right]_0^{\pi/4} = \frac{\pi}{16}.$$

▷ Use the fact that $\cos^2 3\theta = \frac{1}{2}(1 + \cos 6\theta)$; then

$$\frac{1}{2}\int_0^\pi \cos^2 3\theta \, d\theta = \frac{1}{4}\int_0^\pi (1 + \cos 6\theta) \, d\theta = \frac{1}{4}\left[\theta + \frac{1}{6}\sin 6\theta\right]_0^\pi = \frac{\pi}{4}.$$

▷ Use the fact that $\cos^2 \theta = \frac{1}{2}(1 + \cos 2\theta)$; then

$$\begin{aligned}
\frac{1}{2}\int_0^{2\pi/3}\left(\frac{1}{2} + \cos\theta\right)^2 d\theta &= \frac{1}{2}\int_0^{2\pi/3}\left(\frac{1}{4} + \cos\theta + \cos^2\theta\right) d\theta \\
&= \frac{1}{2}\int_0^{2\pi/3}\left(\frac{1}{4} + \cos\theta + \frac{1}{2}(1 + \cos 2\theta)\right) d\theta \\
&= \frac{1}{2}\int_0^{2\pi/3}\left(\frac{3}{4} + \cos\theta + \frac{1}{2}\cos 2\theta\right) d\theta \\
&= \frac{1}{2}\left[\frac{3}{4}\theta + \sin\theta + \frac{1}{4}\sin 2\theta\right]_0^{2\pi/3} \\
&= \frac{1}{2}\left(\frac{3}{4}\cdot\frac{2\pi}{3} + \sin\frac{2\pi}{3} + \frac{1}{4}\sin\frac{4\pi}{3}\right) \\
&= \frac{1}{2}\left(\frac{\pi}{2} + \frac{\sqrt{3}}{2} - \frac{\sqrt{3}}{8}\right) \\
&= \frac{\pi}{4} + \frac{3\sqrt{3}}{16}.
\end{aligned}$$

▷ From the previous problem, the indefinite integral of the integrand is $\frac{3}{4}\theta + \sin\theta + \frac{1}{4}\sin 2\theta$, so we get

$$\begin{aligned}
\frac{1}{2}\int_\pi^{4\pi/3}\left(\frac{1}{2} + \cos\theta\right)^2 d\theta &= \frac{1}{2}\left[\frac{3}{4}\theta + \sin\theta + \frac{1}{4}\sin 2\theta\right]_\pi^{4\pi/3} \\
&= \frac{1}{2}\left(\frac{3}{4}\cdot\frac{4\pi}{3} + \sin\frac{4\pi}{3} + \frac{1}{4}\sin\frac{8\pi}{3} - \frac{3}{4}\pi\right) \\
&= \frac{1}{2}\left(\frac{1}{4}\pi - \frac{\sqrt{3}}{2} + \frac{\sqrt{3}}{8}\right) \\
&= \frac{\pi}{8} - \frac{3\sqrt{3}}{16}.
\end{aligned}$$

▷ Use the fact that $\cos^2\theta = \frac{1}{2}(1 + \cos 2\theta)$; then

$$\begin{aligned}
\frac{1}{2}\int_0^{\pi/3}\left((3\cos\theta)^2 - (1 + \cos\theta)^2\right) d\theta &= \frac{1}{2}\int_0^{\pi/3}\left(8\cos^2\theta - 2\cos\theta - 1\right) d\theta \\
&= \frac{1}{2}\int_0^{\pi/3}(4 + 4\cos 2\theta - 2\cos\theta - 1) \, d\theta \\
&= \frac{1}{2}[3\theta + 2\sin 2\theta - 2\sin\theta]_0^{\pi/3} \\
&= \frac{1}{2}\left(\pi + 2\sin\frac{2\pi}{3} - 2\sin\frac{\pi}{3}\right) = \frac{\pi}{2}.
\end{aligned}$$

Concepts

1. (a) True. See the discussion following Theorem 9.12.

   (b) True. $\Delta\theta = \frac{\frac{\pi}{2} - \frac{\pi}{4}}{4} = \frac{\frac{\pi}{4}}{4} = \frac{\pi}{16}$.

   (c) False. By Theorem 9.13, it is given by $\frac{1}{2} \int_a^b \left(f(\theta) \right)^2 d\theta$.

   (d) True. Here we are graphing $r = f(\theta)$ in a rectangular coordinate plane, so formulas for that situation apply, and the area between the curve and the θ axis is $\int_a^b |f(\theta)|\, d\theta$.

   (e) False. $r = 2\cos\theta$ traces this circle twice from 0 to 2π, since $2\cos(\pi + \theta) = -2\cos\theta$, so that the point $(-r, \pi + \theta) = (r, \theta)$ is traced for θ and for $\theta + \pi$. Thus the integral is twice the area of the circle, so its value is 2π.

   (f) False. Since $\cos(4(\theta + \pi)) = \cos(4\theta + 4\pi) = \cos(4\theta)$, the points (r, θ) and $(r, \pi + \theta)$ are both on the curve, and are different. So the curve is only traced once.

   (g) True. If $r = f(\theta)$, then $x = r\cos\theta = f(\theta)\cos\theta$ and $y = r\sin\theta = f(\theta)\sin\theta$.

   (h) False. This is only true if $f(\theta)$ is one-to-one on $[0, 2\pi]$; otherwise parts of the curve are traced more than once.

3. Since we are approximating area, which is summing up areas from the origin to the curve, the area between a small segment of the curve and the origin is a small triangle, which approximates a small wedge. This is the analog of the situation in rectangular coordinates where the area between a small segment of the curve and the x axis is a small rectangle.

5. A picture is:

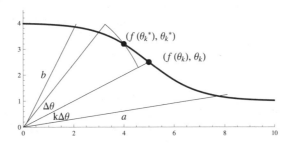

   The circular sector is bounded by the rays at angles $a + k\Delta\theta = \theta_k$ and $a + (k+1)\Delta\theta = \theta_{k+1}$, and we have chosen an angle θ_k^* between the two rays to use as the value of the function throughout that sector.

7. If $\beta - \alpha > 2\pi$, then the region bounded by the rays $\theta = \alpha$ and $\theta = \beta$ is the same region as that bounded by $\theta = \alpha$ and $\theta = \beta - 2\pi$. If we use α and β to compute the integral, we will be tracing the region multiple times.

9. If a graph is symmetric about the x axis, then the area need only be calculated for the region above the x axis, and then doubled. Similarly, if a graph is symmetric about the y axis, then we need only double the calculation of the area of the region to the right of the y axis. Finally, if a graph is symmetric about both axes, it suffices to calculate the area in the first quadrant and multiply by 4. Depending on the specifics of the graph, there may be further simplifications (for example, Example 1 discussed a situation in which we only needed to integrate from 0 to $\frac{\pi}{4}$ and then multiply by 8).

11. Using t as the parameter, let $x(t) = t$; then $y(t) = f(t)$.

13. First, this approach fails since $r = \sin\theta$ is not one-to-one on $[0, 2\pi]$ — the circle is traced twice on this interval. So his integral should have gone from 0 to π. So he gets a result that is twice the correct result. His second mistake, though, is that $r = \sin\theta$ is not a unit circle — it is a circle of radius $\frac{1}{2}$. Thus its true circumference is π. So he got an answer that was twice the correct circumference of a smaller circle. The two errors cancelled to give apparently a correct result for a unit circle.

15. A plot of the region is shown below:

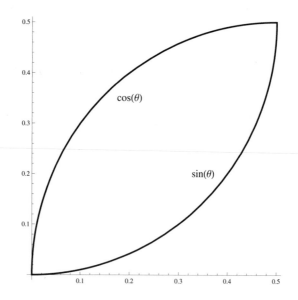

This region is clearly symmetric around the line $\theta = \frac{\pi}{4}$, since $\cos\left(\frac{\pi}{2} - \theta\right) = \sin\theta$. So it suffices to find the area under the polar curve $r = \sin\theta$ from 0 to $\frac{\pi}{4}$ and double it. Thus the required area is

$$2 \cdot \frac{1}{2} \int_0^{\pi/4} \sin^2\theta \, d\theta = \int_0^{\pi/4} \frac{1}{2}(1 - \cos 2\theta) \, d\theta = \frac{1}{2}\left[\theta - \frac{1}{2}\sin 2\theta\right]_0^{\pi/4} = \frac{\pi}{8} - \frac{1}{4}.$$

Skills

17. For $0 \leq \theta \leq \pi$, the graph of $r = \theta$ lies above the x axis, so the area between the curve and the x axis is the area of the polar region bounded by $\theta = 0$ and $r = \theta$, which is

$$\frac{1}{2}\int_0^{\pi} \theta^2 \, d\theta = \frac{1}{2}\left[\frac{1}{3}\theta^3\right]_0^{\pi} = \frac{\pi^3}{6}.$$

19. A graph of the limaçon is:

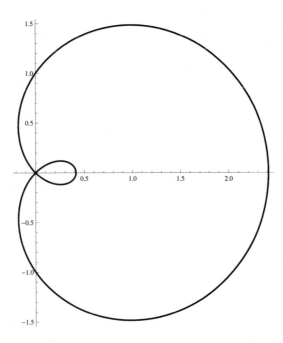

We need only determine the region between the two loops above the x axis and then double it. The rightmost point on the curve corresponds to $\theta = 0$; the curve then loops up and returns to the origin where $\cos \theta = -\frac{1}{\sqrt{2}}$, which is at $\theta = \frac{3\pi}{4}$. So the area under the curve between those two points is the area of the upper half of the outer loop. The upper half of the inner loop corresponds to $\pi \leq \theta \leq \frac{5\pi}{4}$. So we want to compute

$$2\left(\frac{1}{2}\int_0^{3\pi/4} (1+\sqrt{2}\cos\theta)^2\, d\theta - \frac{1}{2}\int_\pi^{5\pi/4} (1+\sqrt{2}\cos\theta)^2\, d\theta\right)$$

$$= \int_0^{3\pi/4} (1+\sqrt{2}\cos\theta)^2\, d\theta - \int_\pi^{5\pi/4} (1+\sqrt{2}\cos\theta)^2\, d\theta$$

$$= \int_0^{3\pi/4} (1+2\sqrt{2}\cos\theta + 2\cos^2\theta)\, d\theta - \int_\pi^{5\pi/4} (1+2\sqrt{2}\cos\theta + 2\cos^2\theta)\, d\theta$$

$$= \left[\theta + 2\sqrt{2}\sin\theta + \left(\theta + \frac{1}{2}\sin 2\theta\right)\right]_0^{3\pi/4} - \left[\theta + 2\sqrt{2}\sin\theta + \left(\theta + \frac{1}{2}\sin 2\theta\right)\right]_\pi^{5\pi/4}$$

$$= \left[2\theta + 2\sqrt{2}\sin\theta + \frac{1}{2}\sin 2\theta\right]_0^{3\pi/4} - \left[2\theta + 2\sqrt{2}\sin\theta + \frac{1}{2}\sin 2\theta\right]_\pi^{5\pi/4}$$

$$= \left(\frac{3\pi}{2} + 2 - \frac{1}{2}\right) - \left(\frac{5\pi}{2} - 2 + \frac{1}{2} - (2\pi + 0 + 0)\right)$$

$$= 3 + \pi.$$

21. Graphs of the two cardioids (with $3 - 3\sin\theta$ in black) are

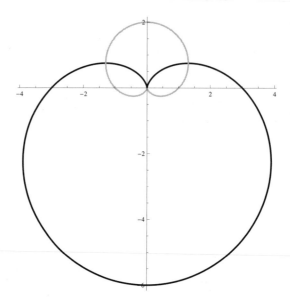

The curves intersect at the origin. The other two intersection points may be found by solving $3 - 3\sin\theta = 1 + \sin\theta$, which gives $4\sin\theta = 2$, so that $\theta = \frac{\pi}{6}$ and $\frac{5\pi}{6}$. It suffices to compute the area to the right of the vertical axis that is inside the larger cardioid and outside the smaller, and then double it. This area is found by integrating from $-\frac{\pi}{2}$ to $\frac{\pi}{6}$, and doubling it.

$$
\begin{aligned}
2 \cdot \frac{1}{2} \int_{-\pi/2}^{\pi/6} \left((3 - 3\sin\theta)^2 - (1 + \sin\theta)^2 \right) d\theta &= \int_{-\pi/2}^{\pi/6} (8 - 20\sin\theta + 8\sin^2\theta)\, d\theta \\
&= \left[8\theta + 20\cos\theta + (4\theta - 2\sin 2\theta) \right]_{-\pi/2}^{\pi/6} \\
&= \left[12\theta + 20\cos\theta - 2\sin 2\theta \right]_{-\pi/2}^{\pi/6} \\
&= \left(2\pi + 10\sqrt{3} - \sqrt{3} \right) - (-6\pi + 0 - 0) \\
&= 8\pi + 9\sqrt{3}.
\end{aligned}
$$

23. The region is shown below:

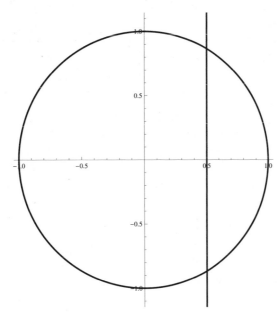

In polar coordinates, the two functions are $r = 1$ and $r = \frac{1}{2}\sec\theta$, so the two intersect where $\sec\theta = 2$, which is at $\theta = \pm\frac{\pi}{3}$. By symmetry, it suffices to find the area of the region between the two above the horizontal axis and double it, so for the range $0 \le \theta \le \frac{\pi}{3}$. Thus the area we want is

$$2 \cdot \frac{1}{2}\int_0^{\pi/3}\left(1^2 - \frac{1}{4}\sec^2\theta\right)d\theta = \int_0^{\pi/3}\left(1 - \frac{1}{4}\sec^2\theta\right)d\theta = \left[\theta - \frac{1}{4}\tan\theta\right]_0^{\pi/3}$$

$$= \frac{\pi}{3} - \frac{1}{4}\tan\frac{\pi}{3} = \frac{\pi}{3} - \frac{\sqrt{3}}{4}.$$

25. The curve looks like

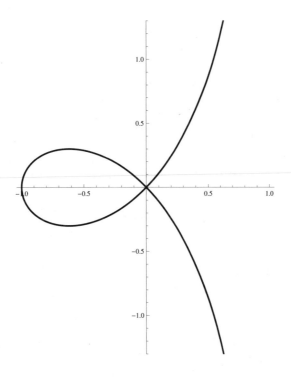

The vertical axis range has been restricted in order to show clearly the loop. With a larger vertical range, it would be clear that the equation has a vertical asymptote at $x = 1$ (this is not relevant to the problem, but is an interesting fact). The curve passes through the origin when $\sec\theta = 2\cos\theta$. Multiplying through by $\sec\theta$ gives $1 = 2\cos^2\theta$, so that in the range $-\frac{\pi}{2} < \theta < \frac{\pi}{2}$ we get $\theta = \pm\frac{\pi}{4}$. So the area enclosed by the loop is

$$
\begin{aligned}
\frac{1}{2}\int_{-\pi/4}^{\pi/4}(\sec\theta - 2\cos\theta)^2\,d\theta &= \frac{1}{2}\int_{-\pi/4}^{\pi/4}(\sec^2\theta + 4\cos^2\theta - 4)\,d\theta \\
&= \frac{1}{2}\left[\tan\theta + (2\theta + \sin 2\theta) - 4\theta\right]_{-\pi/4}^{\pi/4} \\
&= \frac{1}{2}\left[\tan\theta + \sin 2\theta - 2\theta\right]_{-\pi/4}^{\pi/4} \\
&= \frac{1}{2}\left(\left(1 + 1 - \frac{\pi}{2}\right) - \left(-1 - 1 + \frac{\pi}{2}\right)\right) \\
&= 2 - \frac{\pi}{2}.
\end{aligned}
$$

27. Since the function is one-to-one on its domain, the arc length is

$$
\int_0^{2\pi}\sqrt{((e^\theta)')^2 + (e^\theta)^2}\,d\theta = \int_0^{2\pi}\sqrt{2e^{2\theta}}\,d\theta = \sqrt{2}\int_0^{2\pi}e^\theta\,d\theta = \sqrt{2}\left[e^\theta\right]_0^{2\pi} = \sqrt{2}(e^{2\pi} - 1).
$$

29. Since the function is one-to-one on its domain, the arc length is

$$
\begin{aligned}
\int_0^{2\pi}\sqrt{((e^{\alpha\theta})')^2 + (e^{\alpha\theta})^2}\,d\theta &= \int_0^{2\pi}\sqrt{(\alpha^2 + 1)e^{2\alpha\theta}}\,d\theta = \sqrt{\alpha^2 + 1}\int_0^{2\pi}e^{\alpha\theta}\,d\theta \\
&= \frac{\sqrt{\alpha^2 + 1}}{\alpha}\left[e^{\alpha\theta}\right]_0^{2\pi} = \frac{\sqrt{\alpha^2 + 1}}{\alpha}(e^{2\alpha\pi} - 1).
\end{aligned}
$$

31. One loop of the rose is traced for $-\frac{\pi}{4} \leq \theta \leq \frac{\pi}{4}$, since $r = 0$ at each of these points. With $r = \cos 2\theta$ we have $r' = -2\sin 2\theta$, so that the arc length is

$$
\int_{-\pi/4}^{\pi/4}\sqrt{(-2\sin 2\theta)^2 + \cos^2 2\theta}\,d\theta = \int_{-\pi/4}^{\pi/4}\sqrt{4\sin^2 2\theta + \cos^2 2\theta}\,d\theta \approx 2.42211.
$$

33. One loop of the rose is traced for $-\frac{\pi}{8} \leq \theta \leq \frac{\pi}{8}$, since $r = 0$ at each of these points. With $r = \cos 4\theta$ we have $r' = -4\sin 4\theta$, so that the arc length is

$$
\int_{-\pi/8}^{\pi/8}\sqrt{(-4\sin 4\theta)^2 + \cos^2 4\theta}\,d\theta = \int_{-\pi/8}^{\pi/8}\sqrt{16\sin^2 4\theta + \cos^2 4\theta}\,d\theta \approx 2.14461.
$$

35. The limaçon passes through the origin when $r = 1 + 2\sin\theta = 0$, which is when $\sin\theta = -\frac{1}{2}$, so for $\theta = \frac{7\pi}{6}$ and $\frac{11\pi}{6}$. Thus the arc length of the inner loop is

$$
\int_{7\pi/6}^{11\pi/6}\sqrt{(2\cos\theta)^2 + (1 + 2\sin\theta)^2}\,d\theta = \int_{7\pi/6}^{11\pi/6}\sqrt{5 + 4\sin\theta}\,d\theta \approx 2.68245.
$$

37. We have

$$\int_0^{2\pi/3} \left(\frac{1}{2} + \cos\theta\right)^2 d\theta - \int_{\pi}^{4\pi/3} \left(\frac{1}{2} + \cos\theta\right)^2 d\theta$$

$$= \int_0^{2\pi/3} \left(\frac{1}{4} + \cos\theta + \cos^2\theta\right) d\theta - \int_{\pi}^{4\pi/3} \left(\frac{1}{4} + \cos\theta + \cos^2\theta\right) d\theta$$

$$= \left[\frac{\theta}{4} + \sin\theta + \frac{\theta}{2} + \frac{1}{4}\sin 2\theta\right]_0^{2\pi/3} - \left[\frac{\theta}{4} + \sin\theta + \frac{\theta}{2} + \frac{1}{4}\sin 2\theta\right]_{\pi}^{4\pi/3}$$

$$= \left[\frac{3\theta}{4} + \sin\theta + \frac{1}{4}\sin 2\theta\right]_0^{2\pi/3} - \left[\frac{3\theta}{4} + \sin\theta + \frac{1}{4}\sin 2\theta\right]_{\pi}^{4\pi/3}$$

$$= \left(\frac{3\sqrt{3}}{8} + \frac{\pi}{2}\right) - 0 - \left(\left(-\frac{3\sqrt{3}}{8} + \pi\right) - \frac{3\pi}{4}\right)$$

$$= \frac{\pi}{4} + \frac{3\sqrt{3}}{4}.$$

39. Here the function $f(\theta)$ is $1 + \cos\theta$, so this is twice the area enclosed by $r = 1 + \cos\theta$ for $0 \leq \theta \leq \pi$, which is the area enclosed by $r = 1 + \cos\theta$ for $0 \leq \theta \leq 2\pi$ (this is a cardioid):

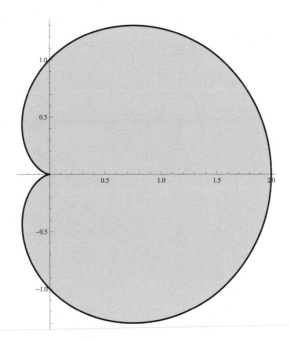

The area of the cardioid is the value of the integral:

$$\int_0^{\pi} (1 + \cos\theta)^2 \, d\theta = \int_0^{\pi} \left(1 + 2\cos\theta + \cos^2\theta\right) d\theta = \left[\frac{3\theta}{2} + 2\sin\theta + \frac{1}{4}\sin 2\theta\right]_0^{\pi} = \frac{3\pi}{2}.$$

41. The function $f(\theta)$ is $\sqrt{\sin 2\theta}$, so that this is twice the area enclosed by $r = \sqrt{\sin 2\theta}$ for $0 \leq \theta \leq \frac{\pi}{2}$, which is the area of the curve enclosed by $r = \sqrt{\sin 2\theta}$ for $0 \leq \theta \leq 2\pi$ (note that actually the valid range for θ is $0 \leq \theta \leq \frac{\pi}{2}$ and $\pi \leq \theta \leq \frac{3\pi}{2}$):

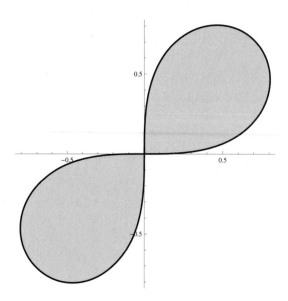

The area is the value of the integral:

$$\int_0^{\pi/2} \sin 2\theta \, d\theta = \left[-\frac{1}{2}\cos 2\theta\right]_0^{\pi/2} = 1.$$

43. The function $f(\theta)$ is $2 + \sin 4\theta$. This curve is traced once for $0 \le \theta \le 2\pi$, so that this integral represents the area enclosed by the curve:

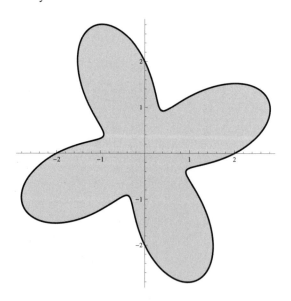

The area is

$$\frac{1}{2}\int_0^{2\pi}(2 + \sin 4\theta)^2 \, d\theta = \frac{1}{2}\int_0^{2\pi}(4 + 4\sin 4\theta + \sin^2 4\theta) \, d\theta$$

$$= \frac{1}{2}\left[4\theta - \cos 4\theta + \frac{\theta}{2} - \frac{1}{16}\sin 8\theta\right]_0^{2\pi}$$

$$= \frac{9\pi}{2}.$$

45. Since $r = f(\theta) = a$, the area of the circle is

$$\frac{1}{2} \int_0^{2\pi} a^2 \, d\theta = \frac{1}{2} \left[a^2 \theta \right]_0^{2\pi} = \pi a^2.$$

47. Since $r = f(\theta) = a$, $r'(\theta) = 0$, so that the circumference of the circle is just the arc length from 0 to 2π, which is

$$\int_0^{2\pi} \sqrt{r'(\theta)^2 + r(\theta)^2} \, d\theta = \int_0^{2\pi} a \, d\theta = [a\theta]_0^{2\pi} = 2\pi a.$$

49. The circle $r = a$ is the circle with radius a and center at the origin; the circle $r = 2a \cos \theta$ is the circle with radius a and center at $(a, 0)$, so that each circle passes through the center of the other. A diagram of this situation (with $a = 1$) is:

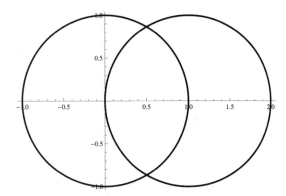

The upper point of intersection of the circles is at $\theta = \frac{\pi}{3}$, since $2a \cos \frac{\pi}{3} = a$, which is on the circle $r = a$ as well. We would like to find the area by integrating the squared difference between the two from 0 to $\frac{\pi}{3}$, but that does not work, as it fails to include the portion of the right-hand circle from $\frac{\pi}{3}$ to $\frac{\pi}{2}$. Instead, use the following approach: the area that is inside both circles is the entire area of the right-hand circle minus the area that is inside $r = 2a \cos \theta$ but outside $r = a$. This we *can* compute by finding the area between the curves from 0 to $\frac{\pi}{3}$ (and doubling it to get the region below the x axis as well), so the area is

$$\pi a^2 - 2 \cdot \frac{1}{2} \int_0^{\pi/3} \left((2a \cos \theta)^2 - a^2 \right) d\theta = \pi a^2 - a^2 \int_0^{\pi/3} (4 \cos^2 \theta - 1) \, d\theta$$

$$= \pi a^2 - a^2 \left[\theta + \sin 2\theta \right]_0^{\pi/3} = a^2 \left(\frac{2\pi}{3} - \frac{\sqrt{3}}{2} \right).$$

Applications

51. As in the previous exercise, we must compute the arc length of this curve from 0 to 2π. With $f(\theta) = 0.83(1 - 0.5 \sin \theta)(1 + 0.2 \cos^2 \theta)$ we get

$$f'(\theta) = \cos \theta (0.166 \sin^2 \theta - 0.332 \sin \theta - 0.083 \cos \theta^2 - 0.0415),$$

so that

$$\int_0^{2\pi} \sqrt{f(\theta)^2 + f'(\theta)^2} \, d\theta \approx 6.149 \text{ feet}.$$

Since

$$2 \text{ meters} \approx 2 \cdot \frac{39.37 \text{ in/m}}{12 \text{ in/ft}} \approx 6.562 \text{ feet},$$

she now can cover the kayak with two widths of the fabric.

Proofs

53. Let the circle be $r = a$. Then the area of the sector from $\theta = 0$ to $\theta = \phi$ is given, by Theorem 9.13, by

$$\frac{1}{2}\int_0^\phi a^2\,d\theta = \frac{1}{2}\left[\theta a^2\right]_0^\phi = \frac{1}{2}\phi a^2.$$

Unfortunately, we used the area of a sector in deriving that integral formula, so that proof may not be too satisfying. A more geometric proof is the following: Clearly the fraction of the area of the entire circle that is in a sector with central angle ϕ is $\frac{\phi}{2\pi}$. Since the area of the entire circle is πr^2, the area of a sector with central angle ϕ is $\frac{\phi}{2\pi}\cdot\pi r^2 = \frac{1}{2}\phi r^2$.

55. Note that $r = \cos 2n\theta$ has $4n$ petals whose tips are separated by an angle of $\frac{2\pi}{4n}$, and that

$$\cos\left(2n\left(\theta - \frac{2\pi}{4n}\right)\right) = \cos\left(2n\theta - \pi\right) = -\cos 2n\theta.$$

By Exercise 6 in Section 9.3, the graph of $r = \cos\left(2n\left(\theta - \frac{2\pi}{4n}\right)\right)$ is the graph of $r = -\cos 2n\theta$ rotated counterclockwise through $\frac{2\pi}{4n}$. But by Exercises 59 and 60 in Section 9.3, $\cos 2n\theta$ is symmetric about both the x and y axes, so it is symmetric around the origin. Thus the graph of $r = -\cos 2n\theta$ is the same as the graph of $r = \cos 2n\theta$, so that the rose $r = \cos 2n\theta$ is moved on top of itself by a rotation through $\frac{2\pi}{4n}$. But since the graph has $4n$ petals, this rotation moves each petal on top of another petal. Since rotations do not change area, the areas of all the petals must be the same.

57. With $f(\theta) = \frac{1}{\theta}$ we get $f'(\theta) = -\frac{1}{\theta^2}$. The portion of the curve inside $r = 1$ is the portion for $\theta \geq 1$, so that the length is

$$\int_1^\infty \sqrt{f(\theta)^2 + f'(\theta)^2}\,d\theta = \lim_{t\to\infty}\int_1^t \sqrt{\frac{1}{\theta^2} + \frac{1}{\theta^4}}\,d\theta = \lim_{t\to\infty}\int_1^t \frac{1}{\theta^2}\sqrt{\theta^2 + 1}\,d\theta.$$

To integrate, use integration by parts with $u = \sqrt{\theta^2 + 1}$ and $dv = \frac{1}{\theta^2}\,d\theta$; then $du = \frac{\theta}{\sqrt{\theta^2+1}}$ and $v = -\frac{1}{\theta}$, so that we get

$$\lim_{t\to\infty}\int_1^t \frac{1}{\theta^2}\sqrt{\theta^2 + 1}\,d\theta = \lim_{t\to\infty}\left(\left[-\frac{\sqrt{\theta^2+1}}{\theta}\right]_1^t + \int_1^t \frac{1}{\sqrt{\theta^2+1}}\,d\theta\right)$$

$$= \lim_{t\to\infty}\left(\sqrt{2} - \frac{\sqrt{t^2+1}}{t} + \left[\sinh^{-1}\theta\right]_1^t\right)$$

$$= \lim_{t\to\infty}\left(\sqrt{2} - \frac{\sqrt{t^2+1}}{t} + \sinh^{-1}t - \sinh^{-1}1\right)$$

$$= \infty$$

since the first and fourth terms are constants, the second term goes to 1, and the third term grows without bound. Since the integral diverges to ∞, the curve has infinite length.

Thinking Forward

▷ The area of a right cylinder is the area of its base times its height. Here the base is the cardioid $r = 1 + \cos\theta$, which has area

$$\frac{1}{2}\int_0^{2\pi}(1+\cos\theta)^2\,d\theta = \frac{1}{2}\int_0^{2\pi}(1 + 2\cos\theta + \cos^2\theta)\,d\theta$$

$$= \frac{1}{2}\left[\frac{3}{2}\theta + 2\sin\theta + \frac{1}{4}\sin 2\theta\right]_0^{2\pi} = \frac{3\pi}{2}.$$

Since the height is 1 unit, its volume is $\frac{3\pi}{2}$.

▷ The surface area is twice the area of the base (for the two ends) plus the height times the circumference of the base. To compute the circumference of the base, we have $r(\theta) = 1 + \cos\theta$, so that $r'(\theta) = -\sin\theta$, and the circumference is

$$\int_0^{2\pi} \sqrt{(1 + \cos\theta)^2 + (-\sin\theta)^2}\, d\theta = \int_0^{2\pi} \sqrt{2 + 2\cos\theta}\, d\theta.$$

This integral arose in Example 4, and was evaluated in Exercise 16 to equal 8. So the circumference of the base is 8; since the height is 1, the lateral surface area is 8, so that the total surface area is

$$2 \cdot \frac{3\pi}{2} + 8 = 8 + 3\pi.$$

9.5   Conic Sections

Thinking Back

Finding the center and radius for a circle by completing the square

▷ We have $x^2 + y^2 - y - \frac{3}{4} = x^2 + \left(y^2 - y + \frac{1}{4}\right) - \frac{1}{4} - \frac{3}{4} = 0$, so that $x^2 + \left(y - \frac{1}{2}\right)^2 = 1$. This is a circle of radius 1 centered at $\left(0, \frac{1}{2}\right)$.

▷ We have $x^2 + 4x + y^2 - 21 = (x^2 + 4x + 4) + y^2 - 4 - 21 = 0$, so that $(x + 2)^2 + y^2 = 25$. This is a circle of radius 5 centered at $(-2, 0)$.

▷ We have $x^2 + 3x + y^2 + y + \frac{3}{2} = (x^2 + 3x + \frac{9}{4}) + \left(y^2 + y + \frac{1}{4}\right) + \frac{3}{2} - \frac{9}{4} - \frac{1}{4} = 0$, so that $\left(x + \frac{3}{2}\right)^2 + \left(y + \frac{1}{2}\right)^2 = 1$. This is a circle of radius 1 centered at $\left(-\frac{3}{2}, -\frac{1}{2}\right)$.

Completing the square

▷ Complete the square to get $(x - 3)^2 + (y + 1)^2 = -10 + 9 + 1 = 0$. The only point satisfying the equation is the point $(3, -1)$.

▷ Complete the square to get $(x - 4)^2 + (y - 5)^2 = -40 + 25 + 16 = 1$. This is the circle of radius 1 centered at $(4, 5)$, so those are the points that satisfy the equation.

▷ Rearrange terms to get $y = -x^2 + 6x - 40$. This is an upward-opening parabola; the points on that parabola satisfy the equation.

▷ Rearrange terms to get $x = -y^2 + 4y + 5$. This is a leftward-opening parabola; the points on that parabola satisfy the equation.

Concepts

1. (a) True. See the discussion at the beginning of the chapter.

   (b) True. A point is a degenerate conic section.

   (c) False. An ellipse is the set of points in the plane the *sum* of whose distances from two distinct points is a constant. See Definition 9.15.

   (d) True. See Definition 9.17.

(e) False. Parabolas and hyperbolas have different definitions and different shapes; see Definitions 9.17 and 9.19.

(f) True. The set of all points equidistant from a line $\mathcal{L}$ and a point P on that line is the set of points on a line perpendicular to $\mathcal{L}$ through P.

(g) False. A hyperbola is the set of points in the plane the *difference* of whose distances from two distinct points is a constant. See Definition 9.19.

(h) True. The eccentricity of $\frac{1}{1+\sin\theta} = \frac{eu}{1+e\sin\theta}$ is clearly 1, so this is a parabola.

3. Given two points in the plane, called foci, an ellipse is the set of points for which the sum of the distances from the foci is a constant.

5. Given two points in a plane, called foci, a hyperbola is the set of points in the plane for which the difference between the distances to the foci is a constant.

7. Since a hyperbola is the set of points the difference of whose distances from two foci is a constant, by varying the constant we can get different hyperbolas. For example, with the foci $(1,0)$ and $(-1,0)$, if the constant difference is 1, then $\left(\frac{1}{2},0\right)$ is on the hyperbola, while if the constant difference is $\frac{2}{3}$, then $\left(\frac{1}{3},0\right)$ is on the hyperbola.

9. (a) By Definition 9.21 (b), the eccentricity is $\frac{\sqrt{A^2-B^2}}{A}$.

(b) By Definition 9.21 (b), the eccentricity is $\frac{\sqrt{B^2-A^2}}{B}$.

(c) In either of the above cases, the numerator is the square root of some number that is smaller than the denominator. Thus the fraction is always less than 1. If $A \neq B$, then the numerator is nonzero, so the eccentricity is greater than 0. Finally, both numerator and denominator are positive. Thus $0 < e < 1$.

(d) $\lim\limits_{A \to B} \frac{\sqrt{A^2-B^2}}{B} = \frac{\sqrt{B^2-B^2}}{B} = 0$. The limit of the eccentricity is zero. As A approaches B, the length of the two axes approach each other, so the conic approaches a circle.

(e) $\lim\limits_{A \to \infty} \frac{\sqrt{A^2-B^2}}{B} = \infty$. As A gets larger, the ellipse gets longer and longer; it gets arbitrarily stretched out, and when viewed in a large enough scale, becomes indistinguishable from a line.

11. Rewrite the equation as $\frac{y^2}{B^2} + \frac{x^2}{A^2} = 1$ with $B > A$; by Definition 9.22 (with x and y reversed) tells us that the directrices of the ellipse are $y = \pm\frac{B}{e}$.

13. Graphs of the two equations are

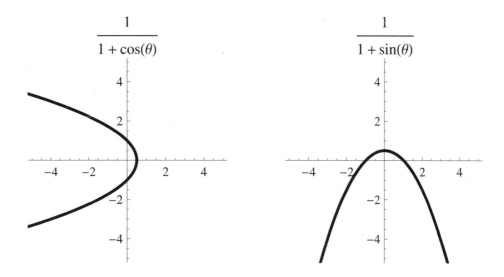

Since the eccentricity (the coefficient of $\sin\theta$ or $\cos\theta$) is 1 in both cases, these are both parabolas, with focus at the origin. From the graphs, one is vertically oriented and the other is horizontally oriented. They are rotated versions of one another.

15. Graphs of the two equations are

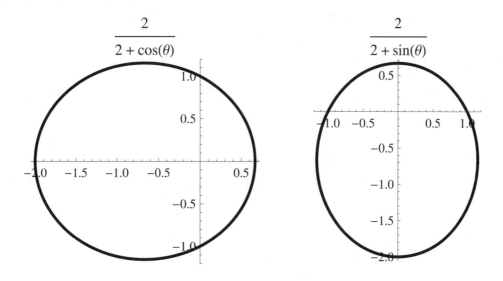

Rewriting these equations in standard form gives $r = \frac{1}{1+\frac{1}{2}\cos\theta}$ and $r = \frac{1}{1+\frac{1}{2}\sin\theta}$. Since the eccentricity (the coefficient of $\sin\theta$ or $\cos\theta$) is $\frac{1}{2}$ in both cases, these are both ellipses. From the graphs, one is vertically oriented and the other is horizontally oriented. They are rotated versions of one another.

17. Divide top and bottom of the right hand side by β to get $r = \frac{\frac{\alpha}{\beta}}{1+\frac{\gamma}{\beta}\sin\theta}$. Thus the eccentricity is $\left|\frac{\gamma}{\beta}\right|$.

The directrix (or directrices) are $y = \frac{\alpha}{\beta} \cdot \frac{\beta}{\gamma} = \frac{\alpha}{\gamma}$.

Skills

19. We get

$$2x^2 + 4x - y^2 - 6y - 23 = (2x^2 + 4x + 2) - (y^2 + 6y + 9) - 23 - 2 + 9$$
$$= 2(x+1)^2 - (y+3)^2 - 16 = 0,$$

so that

$$2(x+1)^2 - (y+3)^2 = 16, \quad \text{or} \quad \frac{(x+1)^2}{8} - \frac{(y+3)^2}{16} = 1.$$

This is a hyperbola with foci $\left(-1 \pm \sqrt{8+16}, -3\right) = \left(-1 \pm 2\sqrt{6}, -3\right)$ and asymptotes $y + 3 = \pm\sqrt{2}(x+1)$, or $y = \pm\sqrt{2}(x+1) - 3$.

21. We get

$$y^2 - 8y - 4x^2 - 8x - 13 = (y^2 - 8y + 16) - 4(x^2 + 2x + 1) - 13 - 16 + 4 = 0,$$

so that

$$(y-4)^2 - 4(x+1)^2 = 25, \quad \text{or} \quad \frac{(y-4)^2}{25} - \frac{(x+1)^2}{25/4} = 1.$$

This is a hyperbola with foci $\left(-1, 4 \pm \sqrt{25 + \frac{25}{4}}\right) = \left(-1, 4 \pm \frac{5}{2}\sqrt{5}\right)$ and asymptotes $y - 4 = \pm 2(x+1)$, or $y = 2x + 6$ and $y = -2x + 2$.

23. Using Theorem 9.18, since the directrix is $y = 0 = y_0 - \frac{1}{4a}$ and the focus is $(0,1) = \left(x_0, y_0 + \frac{1}{4a}\right)$, we have $x_0 = 0$ and $y_0 = \frac{1}{4a}$, so that $2y_0 = 1$ and $y_0 = \frac{1}{4a} = \frac{1}{2}$. Thus $a = \frac{1}{2}$, and the equation is $y = \frac{1}{2}x^2 + \frac{1}{2}$.

25. Use Theorem 9.18. Since the directrix is $y = -6 = y_0 - \frac{1}{4a}$ and the focus is $(2, -8) = \left(x_0, y_0 + \frac{1}{4a}\right)$, we have $x_0 = 2$, $y_0 - \frac{1}{4a} = -6$, and $y_0 + \frac{1}{4a} = -8$. Thus $y_0 = -7$ and $a = -\frac{1}{4}$, so the equation is $y = -\frac{1}{4}(x-2)^2 - 7$.

27. Using Theorem 9.18, we construct an equation $y = a(x - x')^2 + y'$. Since the directrix is $y = y_0 = y' - \frac{1}{4a}$ and the focus is $(x_1, y_1) = \left(x', y' + \frac{1}{4a}\right)$, we get $x' = x_1$, $y' - \frac{1}{4a} = y_0$, and $y' + \frac{1}{4a} = y_1$, so that $y' = \frac{y_0 + y_1}{2}$ and $a = \frac{1}{2(y_1 - y_0)}$. Thus the equation is

$$y = \frac{1}{2(y_1 - y_0)}(x - x_1)^2 + \frac{y_0 + y_1}{2}.$$

29. By Theorem 9.24, the focus is at the pole; since the numerator is $4 = eu$ and the denominator uses a sin rather than a cos, the directrix is $y = 4$. Then by Exercise 27 we get for the equation

$$y = \frac{1}{2(0-4)}(x-0)^2 + 2 = -\frac{1}{8}x^2 + 2.$$

31. By Theorem 9.24, the focus is at the pole; since the numerator is $\alpha = eu$ and the denominator uses sine rather than cosine, the directrix is $y = \alpha$. Then by Exercise 27 we get for the equation

$$y = \frac{1}{2(0-\alpha)}(x-0)^2 + \frac{\alpha}{2} = -\frac{1}{2\alpha}x^2 + \frac{\alpha}{2}.$$

33. Since the foci are on the line $y = 1$, the ellipse is horizontally oriented, and is the ellipse with foci $\left(\pm\sqrt{5}, 0\right)$ and major axis 6 shifted up by 1 unit. To determine the equation of such an ellipse, Theorem 9.16 tells us that $\sqrt{A^2 - B^2} = \sqrt{5}$ and the major axis is $2A = 6$. Thus $A^2 = 9$ and thus $B^2 = 4$, so the equation of the ellipse is (shifting it up by one unit in the y direction)

$$\frac{x^2}{A^2} + \frac{(y-1)^2}{B^2} = \frac{x^2}{9} + \frac{(y-1)^2}{4} = 1.$$

35. Since the foci are on the y axis, the ellipse is vertically oriented. By Theorem 9.16, we have $\sqrt{B^2 - A^2} = \frac{3\sqrt{3}}{2}$, so that $B^2 - A^2 = \frac{27}{4}$; also, the minor axis is $2A = 3$, so that $A^2 = \frac{9}{4}$. Thus $B^2 = \frac{9}{4} + \frac{27}{4} = \frac{36}{4} = 9$. The equation of the ellipse is

$$\frac{x^2}{A^2} + \frac{y^2}{B^2} = \frac{4x^2}{9} + \frac{y^2}{9} = 1.$$

37. Since the foci are on the y axis, the ellipse is vertically oriented. By Theorem 9.16, we have $\sqrt{B^2 - A^2} = \alpha$, so that $B^2 - A^2 = \alpha^2$; also, the minor axis is $2A = 2\alpha$, so that $A = \alpha$. Then $B^2 = A^2 + \alpha^2 = 2\alpha^2$, so that the equation of the ellipse is

$$\frac{x^2}{A^2} + \frac{y^2}{B^2} = \frac{x^2}{\alpha^2} + \frac{y^2}{2\alpha^2} = 1.$$

39. The directrices are

$$x = \pm\frac{A}{e} = \pm\frac{A}{\frac{\sqrt{A^2+B^2}}{A}} = \pm\frac{A^2}{\sqrt{A^2+B^2}}.$$

For this exercise, $\sqrt{A^2 + B^2} = 6$, and the directrices are $x = \pm 1$, so that $1 = \frac{A^2}{6}$. Thus $A^2 = 6$ and $B^2 = 30$, so the equation of the hyperbola is

$$\frac{x^2}{A^2} - \frac{y^2}{B^2} = \frac{x^2}{6} - \frac{y^2}{30} = 1.$$

41. The directrices are

$$y = \pm\frac{B}{e} = \pm\frac{B}{\frac{\sqrt{A^2+B^2}}{B}} = \pm\frac{B^2}{\sqrt{A^2+B^2}}.$$

For this exercise, $\sqrt{A^2 + B^2} = 4$, and the directrices are $y = \pm 2$, so that $2 = \frac{B^2}{4}$. Thus $B^2 = 8$ and $A^2 = 8$, so the equation of the hyperbola is

$$\frac{y^2}{B^2} - \frac{x^2}{A^2} = \frac{y^2}{8} - \frac{x^2}{8} = 1.$$

43. The foci are at $(3, 5 \pm 4)$, so this is the hyperbola with foci $(0, \pm 4)$ and directrix $y = 4 - 5 = -1$ shifted right by 3 units and up by 5 units. The directrices are

$$y = \pm\frac{B}{e} = \pm\frac{B}{\frac{\sqrt{A^2+B^2}}{B}} = \pm\frac{B^2}{\sqrt{A^2+B^2}}.$$

For this exercise, $\sqrt{A^2 + B^2} = 4$, and the directrices are $y = \pm 1$, so that $1 = \frac{B^2}{4}$. Thus $B^2 = 4$ and $A^2 = 12$, so the equation of the hyperbola is (remembering to shift back)

$$\frac{(y-5)^2}{B^2} - \frac{(x-3)^2}{A^2} = \frac{(y-5)^2}{4} - \frac{(x-3)^2}{12} = 1.$$

45. Since $eu = 2$ and $e = 3$, this is a hyperbola with one focus at the origin and directrix $x = -\frac{2}{3}$ (due to the minus sign in the denominator). Its graph is

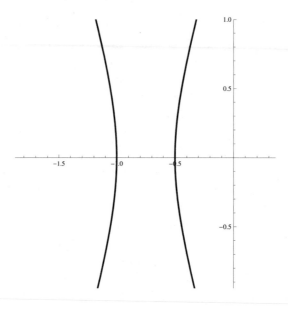

47. Since $e = 1$, this is a parabola; the numerator is $eu = 5$, so it is a parabola with directrix $x = 5$ and focus at the origin. Thus it is a leftward-opening parabola; its graph is

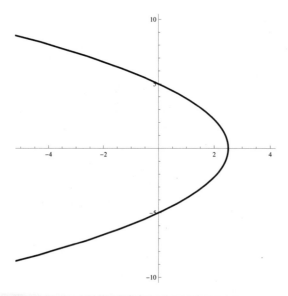

49. Dividing numerator and denominator by 4 gives $r = \frac{1/2}{1+\frac{1}{4}\sin\theta}$. Thus $e = \frac{1}{4}$ and this is an ellipse. Since the denominator includes $\sin\theta$, it is vertically oriented. Its graph is

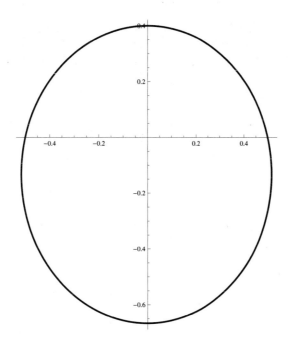

51. Dividing numerator and denominator by 2 gives $r = \dfrac{1}{1+\frac{1}{2}\sin\theta}$. Thus $e = \frac{1}{2}$ and this is an ellipse. Since the denominator includes $\sin\theta$, it is vertically oriented. Its graph is

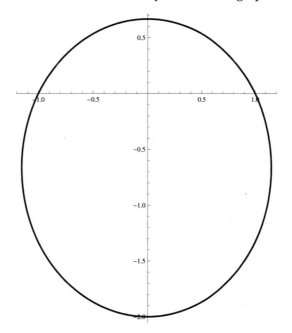

Applications

53. (a) The semiminor axis is given, by Exercise 52, as

$$B^2 = A^2(1 - e^2) = 1.00000011^2(1 - 0.0167^2) \approx 0.999721.$$

Thus the equation of the earth's orbit is approximately

$$\frac{x^2}{A^2} + \frac{y^2}{B^2} = \frac{x^2}{1.00000022} + \frac{y^2}{0.999721} = 1.$$

(b) Placing the center of the ellipse at the origin and the sun at the focus where $\theta = 0$ (i.e., in Cartesian coordinates, where $x > 0$), the associated directrix is (by Definition 9.22) the line $x = u = \frac{A}{e}$. Then by Theorem 9.24, the equation of the earth's orbit in polar coordinates is

$$r = \frac{eu}{1 + e\cos\theta} = \frac{A}{1 + e\cos\theta} = \frac{1.00000011}{1 + 0.0167\cos\theta}.$$

Proofs

55. Let $C = \sqrt{A^2 - B^2}$ as in the text. If (x,y) lies on the curve, then solving for y^2 gives $y^2 = B^2\left(1 - \frac{x^2}{A^2}\right) = B^2 - \frac{B^2 x^2}{A^2}$. Then the square of the distance D_2 from (x,y) to $(-C,0)$ is

$$
\begin{aligned}
D_2^2 &= (x - (-C))^2 + (y - 0)^2 \\
&= (x + C)^2 + y^2 \\
&= x^2 + 2Cx + C^2 + B^2 - \frac{B^2 x^2}{A^2} \\
&= x^2 + 2Cx + A^2 - \frac{B^2 x^2}{A^2} \\
&= \frac{A^4 + 2A^2 Cx + (A^2 - B^2)x^2}{A^2} \\
&= \frac{A^4 + 2A^2 Cx + C^2 x^2}{A^2} \\
&= \frac{(A^2 + Cx)^2}{A^2}.
\end{aligned}
$$

Thus $D_2 = \frac{A^2 + Cx}{A}$.

57. Let G be the graph of the curve with equation $\frac{x^2}{A^2} - \frac{y^2}{B^2} = 1$ with $A, B > 0$. We show that if $(x,y) \in G$, then

$$d((x,y), (-\sqrt{A^2 + B^2}, 0)) - d((x,y), (\sqrt{A^2 + B^2}, 0)),$$

where d is the Euclidean distance function, is a constant independent of x and y. By Definition 9.19, this will show that G is the graph of a hyperbola with foci $(\pm\sqrt{A^2 + B^2}, 0)$.

Let $C = \sqrt{A^2 + B^2}$, so that $C^2 = A^2 + B^2$, and let (x,y) be a point on G. Solving the equation of the curve for y^2 gives

$$y^2 = \frac{B^2 x^2}{A^2} - B^2.$$

Let D_1 represent the distance from (x,y) to $(C,0)$ and D_2 the distance from (x,y) to $(-C,0)$. We want to show that $D_2 - D_1$ is constant. Then

$$D_1^2 = (x - C)^2 + (y - 0)^2 \qquad\qquad D_2^2 = (x - (-C))^2 + (y - 0)^2$$
$$= (x - C)^2 + y^2 \qquad\qquad\qquad = (x + C)^2 + y^2$$
$$= x^2 - 2Cx + C^2 + \frac{B^2 x^2}{A^2} - B^2 \qquad = x^2 + 2Cx + C^2 + \frac{B^2 x^2}{A^2} - B^2$$
$$= x^2 - 2Cx + A^2 + \frac{B^2 x^2}{A^2} \qquad\quad = x^2 + 2Cx + A^2 + \frac{B^2 x^2}{A^2}$$
$$= \frac{(A^2 + B^2)x^2 - 2A^2 Cx + A^4}{A^2} \qquad = \frac{(A^2 + B^2)x^2 + 2A^2 Cx + A^4}{A^2}$$
$$= \frac{C^2 x^2 - 2A^2 Cx + A^4}{A^2} \qquad\quad = \frac{C^2 x^2 + 2A^2 Cx + A^4}{A^2}$$
$$= \left(\frac{Cx - A^2}{A} \right)^2 \qquad\qquad\quad = \left(\frac{Cx + A^2}{A} \right)^2 .$$

Note that $C > A$ and that $|x| \geq A$ (since if $|x| < A$, the equation has no solutions). Thus $Cx \pm A^2$ are both positive, so taking square roots gives

$$D_1 = \frac{Cx - A^2}{A}, \qquad D_2 \frac{Cx + A^2}{A},$$

so that $D_2 - D_1 = 2A$, which is a constant independent of x and y. So G is a hyperbola with $(\pm C, 0)$ as the foci.

59. For an ellipse of the form $\frac{x^2}{A^2} + \frac{y^2}{B^2} = 1$ where $A > B > 0$, the foci are $(\pm\sqrt{A^2 - B^2}, 0)$ (by Definition 9.15) and the vertices are the places where the curve intersects the x axis, which are $x = \pm A$. Thus

$$\frac{\text{distance between the foci}}{\text{distance between the vertices}} = \frac{2\sqrt{A^2 - B^2}}{2A} = \frac{\sqrt{A^2 - B^2}}{A} = e$$

by Definition 9.21.

For an ellipse of the form $\frac{x^2}{A^2} + \frac{y^2}{B^2} = 1$ where $B > A > 0$, the foci are $(\pm\sqrt{B^2 - A^2}, 0)$ (by Definition 9.15) and the vertices are the places where the curve intersects the y axis, which are $y = \pm B$. Thus

$$\frac{\text{distance between the foci}}{\text{distance between the vertices}} = \frac{2\sqrt{B^2 - A^2}}{2B} = \frac{\sqrt{B^2 - A^2}}{B} = e$$

by Definition 9.21.

For a hyperbola of the form $\frac{x^2}{A^2} - \frac{y^2}{B^2} = 1$, the foci are $(\pm\sqrt{A^2 + B^2}, 0)$ (by Definition 9.19) and the vertices are the places where the curve intersects the x axis, which are $x = \pm A$. Thus

$$\frac{\text{distance between the foci}}{\text{distance between the vertices}} = \frac{2\sqrt{A^2 + B^2}}{2A} = \frac{\sqrt{A^2 + B^2}}{A} = e$$

by Definition 9.21.

For a hyperbola of the form $\frac{y^2}{B^2} - \frac{x^2}{A^2} = 1$, the foci are $(0, \pm\sqrt{A^2 + B^2})$ (by Definition 9.19) and the vertices are the places where the curve intersects the y axis, which are $y = \pm B$. Thus

$$\frac{\text{distance between the foci}}{\text{distance between the vertices}} = \frac{2\sqrt{A^2 + B^2}}{2B} = \frac{\sqrt{A^2 + B^2}}{B} = e$$

by Definition 9.21.

61. Suppose $\frac{x^2}{A^2} - \frac{y^2}{B^2} = 1$ is a hyperbola, and let F, e, and l be as in the problem statement. If $P = (x, y)$ is a point on the curve, then the point D on l closest to (x, y) to l is clearly $\left(\frac{A}{e}, y\right)$, so that $DP = \frac{A}{e} - x$, so that $DP \cdot e = A - ex = A - x\frac{\sqrt{A^2+B^2}}{A}$. Now, we have

$$FP = \sqrt{(\sqrt{A^2 + B^2} - x)^2 + y^2}$$

$$= \sqrt{A^2 + B^2 - 2x\sqrt{A^2 + B^2} + x^2 + \left(\frac{B^2 x^2}{A^2} - B^2\right)}$$

$$= \sqrt{A^2 - 2x\sqrt{A^2 + B^2} + \frac{(A^2 + B^2)x^2}{A^2}}$$

$$= \sqrt{\frac{A^4 - 2xA^2\sqrt{A^2 + B^2} + (A^2 + B^2)x^2}{A^2}}$$

$$= \sqrt{\frac{(A^2 - x\sqrt{A^2 + B^2})^2}{A^2}} = \frac{A^2 - x\sqrt{A^2 + B^2}}{A} = A - x\frac{\sqrt{A^2 + B^2}}{A}.$$

So $FP = e \cdot DP$ and thus $\frac{FP}{DP} = e$.

Thinking Forward

Analogs of conic sections in three dimensions

▷ The figure below is a plot of the ellipsoid consisting of points the sum of whose distances from $(-5, 0, 0)$ and $(5, 0, 0)$ is 20.

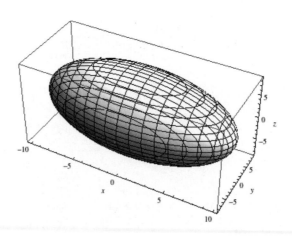

▷ The figure below is a plot of the paraboloid consisting of points whose distances from $(1, 0, 0)$ and the plane $z = -1$ are equal.

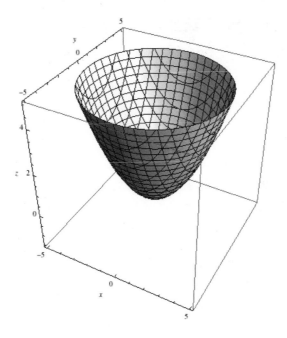

▷ The figure below is a plot of the hyperboloid consisting of points the difference of whose distances from $(5, 0, 0)$ and $(-5, 0, 0)$ is 4.

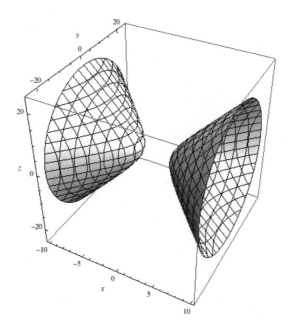

Chapter Review and Self-Test

1.

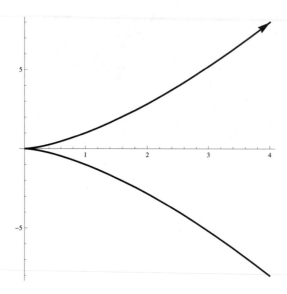

3. Note that the curve is traced twice as t ranges from 0 to 4π.

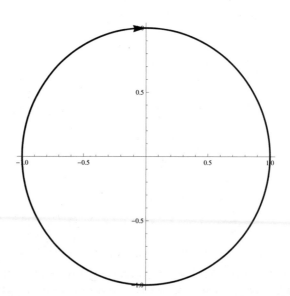

5.

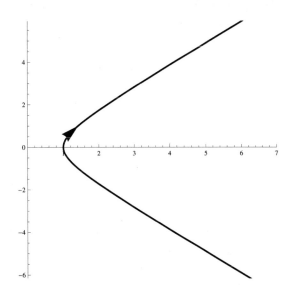

7.

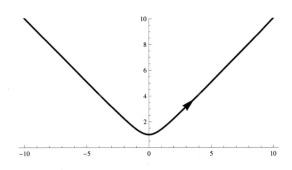

9. From the graph (see Exercise 1), or by symmetry, it is clear that the arc length on $[-2, 2]$ is twice the arc length on $[0, 2]$, so we get (with the substitution $u = 4 + 9t^2$)

$$2 \int_0^2 \sqrt{x'(t)^2 + y'(t)^2}\, dt = 2 \int_0^2 \sqrt{(2t)^2 + (3t^2)^2}\, dt$$

$$= 2 \int_0^2 \sqrt{4t^2 + 9t^4}\, dt$$

$$= 2 \int_0^2 t\sqrt{4 + 9t^2}\, dt$$

$$= \frac{1}{9} \int_{t=0}^{t=2} \sqrt{u}\, du$$

$$= \frac{1}{9} \left[\frac{2}{3} u^{3/2} \right]_{t=0}^{t=2}$$

$$= \frac{2}{27} \left[(4 + 9t^2)^{3/2} \right]_0^2 = \frac{2}{27} \left(80\sqrt{10} - 8 \right).$$

11. This is a circle of radius k centered at the origin, so its arc length is its circumference, which is $2\pi k$. To evaluate using an integral, note that we can compute the arc length (using symmetry) as 8 times

the arc length from 0 to $\frac{\pi}{4}$, which is

$$8 \int_0^{\pi/4} \sqrt{x'(t)^2 + y'(t)^2}\, dt = 8 \int_0^{\pi/4} \sqrt{(k\cos kt)^2 + (-k\sin kt)^2}\, dt$$

$$= 8 \int_0^{\pi/4} \sqrt{k^2(\cos^2 kt + \sin^2 kt)}\, dt = 8 \int_0^{\pi/4} k\, dt = 2k\pi.$$

13. This is a circle of radius $\frac{1}{2}$ with center at $\left(0, \frac{1}{2}\right)$. It is traced once for $\theta \in [0, \pi)$. Also, it is symmetric around the y axis, since $\sin(-\theta) = -\sin\theta$, so if (r, θ) is on the curve, so is $(-r, -\theta)$. Thus it suffices to plot the curve from 0 to $\frac{\pi}{2}$ and reflect it.

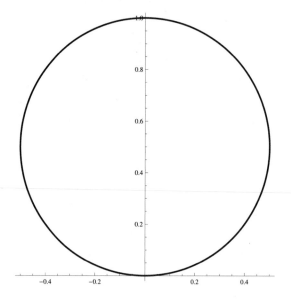

15. This is a three-petaled rose. It is traced once for $\theta \in [0, \pi)$. Also, it is symmetric around the y axis, since $\sin(-3\theta) = -\sin 3\theta$, so if (r, θ) is on the curve, so is $(-r, -\theta)$. Thus it suffices to plot the curve from 0 to $\frac{\pi}{2}$ and reflect it.

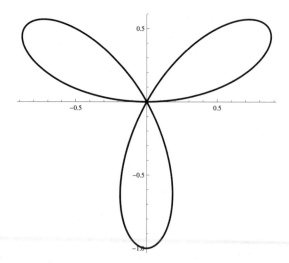

17. This is a cardioid. It is traced once for $\theta \in [0, 2\pi)$. It is symmetric around the y axis, since if (r, θ) is on the curve, then $2 - 2\sin(\pi - \theta) = 2 - 2\sin\theta$, so that $(r, \pi - \theta) = (-r, -\theta)$ is on the curve. So we need only plot the curve from 0 to π and reflect it across the y axis.

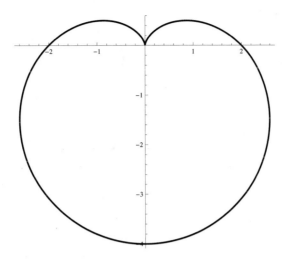

19. This is a limaçon with an inner loop. It is traced once for $\theta \in [0, 2\pi)$. It is symmetric around the y axis, since if (r, θ) is on the curve, then $2 - \sqrt{8} \sin(\pi - \theta) = 2 - \sqrt{8} \sin \theta$, so that $(r, \pi - \theta) = (-r, -\theta)$ is on the curve. So we need only plot the curve from 0 to π and reflect it across the y axis.

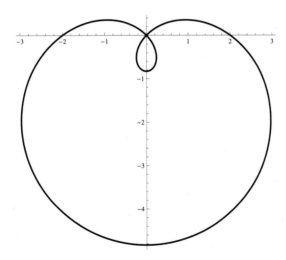

21. This is a lemniscate. It is traced once for $\theta \in \left[\frac{\pi}{2}, \pi\right) \cup \left[\frac{3\pi}{2}, 2\pi\right)$ since $\sin 2\theta$ is positive elsewhere. The curve is symmetric around the origin, since if (r, θ) is on the curve, then $r^2 = -\sin 2\theta$, so that also $(-r)^2 = -\sin 2\theta$ and $(-r, \theta)$ is on the curve. So we need only plot the curve from $\frac{\pi}{2}$ to π and reflect it through the origin.

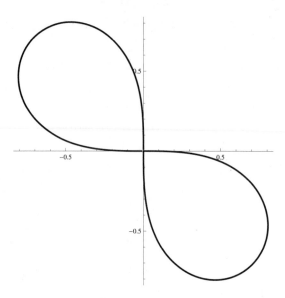

23. By Theorem 9.14, since $f(\theta) = a$ so that $f'(\theta) = 0$, we get for the arc length

$$\int_0^{2\pi} \sqrt{0^2 + a^2}\, d\theta = \int_0^{2\pi} a\, d\theta = 2\pi a.$$

25. This is a circle of radius $\frac{a}{2}$ that is traced once for $0 \le \theta < \pi$. With $f(\theta) = a \sin \theta$ we have $f'(\theta) = a \cos \theta$, so that the arc length is

$$\int_0^{\pi} \sqrt{(a \cos \theta)^2 + (a \sin \theta)^2}\, d\theta = \int_0^{\pi} \sqrt{a^2(\cos^2 \theta + \sin^2 \theta)}\, d\theta = \int_0^{\pi} a\, d\theta = a\pi.$$

27. This cardioid is symmetric around the y axis. The right half is traced from $-\frac{\pi}{2}$ to $\frac{\pi}{2}$, so we compute that arc length and double it. With $f(\theta) = 1 - \sin \theta$, we have $f'(\theta) = -\cos \theta$, so that the arc length is

$$2 \int_{-\pi/2}^{\pi/2} \sqrt{(1 - \sin \theta)^2 + (-\cos \theta)^2}\, d\theta = 2 \int_{-\pi/2}^{\pi/2} \sqrt{1 - 2 \sin \theta + \sin^2 \theta + \cos^2 \theta}\, d\theta$$

$$= 2 \int_{-\pi/2}^{\pi/2} \sqrt{2 - 2 \sin \theta}\, d\theta.$$

To evaluate this integral, make the substitution $u = 1 - \sin \theta$; then $du = -\cos \theta\, d\theta$. But

$$\cos \theta = \sqrt{1 - \sin^2 \theta} = \sqrt{1 - (1 - u)^2} = \sqrt{2u - u^2}$$

(note that $\cos \theta$ is nonnegative on $\left[-\frac{\pi}{2}, \frac{\pi}{2}\right]$, so we can use the positive square root), so that we get $-\frac{du}{\sqrt{2u-u^2}} = d\theta$. Then

$$2 \int_{-\pi/2}^{\pi/2} \sqrt{2 - 2 \sin \theta}\, d\theta = -2 \int_{\theta=-\pi/2}^{\theta=\pi/2} \frac{\sqrt{2u}}{\sqrt{2u - u^2}}\, du$$

$$= -2\sqrt{2} \int_{\theta=-\pi/2}^{\theta=\pi/2} \frac{1}{\sqrt{2 - u}}\, du$$

$$= -2\sqrt{2} \int_{\theta=-\pi/2}^{\theta=\pi/2} (2 - u)^{-1/2}\, du$$

$$= 4\sqrt{2} \left[(2 - u)^{1/2} \right]_{\theta=-\pi/2}^{\theta=\pi/2}$$

$$= 4\sqrt{2} \left[(1 + \sin \theta)^{1/2} \right]_{-\pi/2}^{\pi/2} = 8.$$

29. This is a segment of a vertical line. With $f(\theta) = \sec\theta$, we have $f'(\theta) = \sec\theta\tan\theta$, so that the arc length is

$$\int_{-\pi/6}^{\pi/6} \sqrt{\sec^2\theta\tan^2\theta + \sec^2\theta}\, d\theta = \int_{-\pi/6}^{\pi/6} \sqrt{\sec^2\theta(\tan^2\theta + 1)}\, d\theta$$

$$= \int_{-\pi/6}^{\pi/6} \sqrt{\sec^4\theta}\, d\theta = \int_{-\pi/6}^{\pi/6} \sec^2\theta\, d\theta = [\tan\theta]_{-\pi/6}^{\pi/6} = \frac{2\sqrt{3}}{3}.$$

31. Since $r = f(\theta) = a$, the area of the circle is

$$\frac{1}{2}\int_0^{2\pi} a^2\, d\theta = \frac{1}{2}\left[a^2\theta\right]_0^{2\pi} = \pi a^2.$$

33. $r = \cos 2\theta$ is a four-petaled rose; one rose is traced for $-\frac{\pi}{4} \le \theta \le \frac{\pi}{4}$, so we compute the area inside the function from 0 to $\frac{\pi}{4}$ and double it. This is

$$2 \cdot \frac{1}{2}\int_0^{\pi/4} (\cos 2\theta)^2\, d\theta = \int_0^{\pi/4} \cos^2 2\theta\, d\theta = \left[\frac{\theta}{2} + \frac{1}{8}\sin 4\theta\right]_0^{\pi/4} = \frac{\pi}{8}.$$

35. This cardioid is traced once for $0 \le \theta \le 2\pi$, so the area inside the cardioid is

$$\frac{1}{2}\int_0^{2\pi} (1 - \sin\theta)^2\, d\theta = \frac{1}{2}\int_0^{2\pi} (1 - 2\sin\theta + \sin^2\theta)\, d\theta = \frac{1}{2}\left[\theta + 2\cos\theta + \frac{\theta}{2} - \frac{1}{4}\sin 2\theta\right]_0^{2\pi} = \frac{3\pi}{2}.$$

37. The lemniscate is traced completely for $\frac{\pi}{2} \le \theta \le \pi$ if we take both square roots of r^2. So taking just the positive square root will plot the loop of the lemniscate in the second quadrant. The area is then

$$\frac{1}{2}\int_{\pi/2}^{\pi} (\sqrt{-\sin 2\theta})^2\, d\theta = \frac{1}{2}\int_{\pi/2}^{\pi} (-\sin 2\theta)\, d\theta = \frac{1}{2}\left[\frac{1}{2}\cos 2\theta\right]_{\pi/2}^{\pi} = \frac{1}{2}.$$

39. Plots of both curves, with $r^2 = \cos 2\theta$ the dashed graph, are:

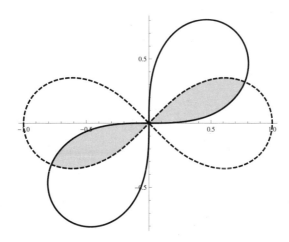

By symmetry, we can compute the area between the curves in the first quadrant and double it. The two curves intersect where $\cos 2\theta = \sin 2\theta$, or where $\tan 2\theta = 1$. This is at $\theta = \frac{\pi}{8}$. So we can compute the area between the curves by computing the area enclosed by $r^2 = \sin 2\theta$ for $\theta \in \left[0, \frac{\pi}{8}\right]$ and $r^2 = \cos^2 2\theta$ for $\theta \in \left[\frac{\pi}{8}, \frac{\pi}{4}\right]$ and doubling it:

$$2\left(\frac{1}{2}\int_0^{\pi/8}\left(\sqrt{\sin 2\theta}\right)^2 d\theta + \frac{1}{2}\int_{\pi/8}^{\pi/4}\left(\sqrt{\cos 2\theta}\right)^2 d\theta\right) = \int_0^{\pi/8}\sin 2\theta\, d\theta + \int_{\pi/8}^{\pi/4}\cos 2\theta\, d\theta$$

$$= \left[-\frac{1}{2}\cos 2\theta\right]_0^{\pi/8} + \left[\frac{1}{2}\sin 2\theta\right]_{\pi/8}^{\pi/4}$$

$$= -\frac{\sqrt{2}}{4} + \frac{1}{2} + \left(\frac{1}{2} - \frac{\sqrt{2}}{4}\right)$$

$$= 1 - \frac{\sqrt{2}}{2}.$$